Geography of the World's Major Regions

Geography of the World's Major Regions discusses today's most pressing issues through a comprehensive examination of human activity in twelve regions of the world. Environmental degradation, natural catastrophes, population pressures and human conflict all impact in different ways to different degrees on the society and environment of each of the world's regions.

The emphasis is placed on the varying attributes of different parts of the world and the significance of relative locations. Within each region the author describes economic and political restructuring, the quantity of natural resources and production from agriculture, industry and services, the role of the military, as well as the impact of global economic change. The author not only charts the course of each region's development to its present state, but also discusses the implications of current issues and problems, examining a possible new political and economic environment for the twenty-first century.

Geography of the World's Major Regions is an exciting introduction to students studying the changing geography of the world. With over 250 photographs, maps and figures, boxed case studies and key points and further reading by chapter, this book provides an invaluable overview to students researching specific regions, seeking comparative analysis of regions, or following general courses on the economic and political geography of both the developed and the developing worlds.

John Cole is Emeritus Professor of Human and Regional Geography at the University of Nottingham. He has worked in several of the major regions of the world used in this book and has published widely in various aspects of geography.

John Cole

Geography of the World's Major Regions

MOUNT ROYAL COLLEGE
LIBRARY

LONDON AND NEW YORK

To my grandchildren Melanie, Josephine, Paul, Simon, Sebastian, Maximilian and James for their contribution to the population explosion. Hopefully the future will be brighter for them than some of my projections indicate.

First published 1996
by Routledge
11 New Fetter Lane, London EC4P 4EE

Simultaneously published in the USA and Canada
by Routledge
29 West 35th Street, New York, NY 10001

© 1996 John Cole

Typeset in Sabon by Solidus (Bristol) Limited

Printed and bound in Great Britain by
Butler & Tanner Ltd, Frome and London

All rights reserved. No part of this book may be reprinted or reproduced or utilized in any form or by any electronic, mechanical, or other means, now known or hereafter invented, including photocopying and recording, or in any information storage or retrieval system, without permission in writing from the publishers.

British Library Cataloguing in Publication Data

A catalogue record for this book is available from the British Library

Library of Congress Cataloguing in Publication Data

A catalogue record for this book has been requested

ISBN 0-415-11742-9
0-415-11743-7 (pbk)

Contents

List of boxes

Preface

Great changes have been taking place in the mid-1990s in many parts of the world while this book was being written. Delays in the publication of sets of data for the countries of the world mean that there is a considerable gap in time between some of the material used and the publication of the book. The discerning reader will be able to distinguish the more durable physical, demographic and economic features of the regions from the sometimes abrupt political changes. Various publications cited and used in the present book appear annually and can be referred to by the reader to update situations.

The support and help of a number of people during the preparation of this book is greatly appreciated. My wife, Isabel, showed her customary patience as well as encouragement. My son, Francis, made useful suggestions about the content of the first version. Mrs E. O. Wigginton typed the first draft, while Mrs R. Hoole produced various later versions on a word processor. About half of the maps and diagrams were drawn by Mr C. Lewis and Mrs E. Watts. Dr J. C. Doornkamp kindly checked the material on the environment in Chapter 4 and Dr I. V. Filatochev provided the material on Yekaterinburg in Chapter 7. Several years of undergraduates in the Department of Geography, Nottingham University, who took my courses on the developing world, the USSR and the European Union, also made a major contribution, mainly unknowingly. The encouragement of Tristan Palmer of Routledge and of an anonymous reviewer was a vital part of the support, as also was the patience of the editors, crucial in shaping the final product.

List of acronyms

ACP	African, Caribbean and Pacific (Countries)
AIDS	Acquired immune deficiency syndrome
Asean	Association of South-East Asian Nations
ASSR	Autonomous Soviet Socialist Republic (in former USSR)
Benelux	Economic Union of Belgium, the Netherlands and Luxembourg
CFC	Chlorofluorocarbons
CIS	Commonwealth of Independent States (Russia and its satellites)
CMEA	Council for Mutual Economic Assistance (principally USSR and Eastern Europe)
COMECON	Earlier contraction of Council for Mutual Economic Assistance, CMEA
CPSU	Communist Party of the Soviet Union
EAEC	European Atomic Energy Commission
EBRD	European Bank for Reconstruction and Development
EC	European Community or Communities
ECSC	European Coal and Steel Community
ECU	European Currency Unit
EEC	European Economic Community
EFTA	European Free Trade Association
EMS	European Monetary System
ERDF	European Regional Development Fund
EU	European Union (formerly European Community)
FAOPY	*Food and Agriculture Organisation Production Yearbook*
FAOTY	*Food and Agriculture Organisation Trade Yearbook*
FRG	Federal Republic of Germany (former West Germany)
GATT	General Agreement on Tariffs and Trade
GDP	Gross Domestic Product
GDR	German Democratic Republic (former East Germany)
GNP	Gross National Product
HDI	Human Development Index (of the UNDP)

HDR	*Human Development Report*
HIV	Human immunodeficiency virus
IMF	International Monetary Fund
ITSY	*International Trade Statistics Yearbook*
LAIA	Latin American Integration Association
LDC	Least developed countries
LNG	Liquefied Natural Gas
NAFTA	North American Free Trade Agreement
NATO	North Atlantic Treaty Organisation
NGO	Non-Government Organisation
NIC	Newly Industrialised Countries
OECD	Organisation for Economic Cooperation and Development
OPEC	Organisation of Petroleum Exporting Countries
ppp	Purchasing power parity
pps	Purchasing power standard
PRB	Population Reference Bureau
PRC	People's Republic of China
RSFSR	Russian Soviet Federal Socialist Republic (now Russia)
UK	United Kingdom of Great Britain and Northern Ireland
UN	United Nations
UNDP	United Nations Development Programme
UNDYB	*United Nations Demographic Yearbook*
UNSYB	*United Nations Statistical Yearbook*
UNYITS	*United Nations Yearbook of International Trade Statistics*
USA	United States of America
USSR	Union of Soviet Socialist Republics
WDR	*World Development Report*
WPDS	*World Population Data Sheet* of the Population Reference Bureau, annual publication
WTO	World Trade Organisation (succeeds GATT)

Conversions

See *Statistical Yearbook* of the United Nations for a detailed coverage of conversion coefficients and factors (Annex II in the thirty-eighth issue, UN, New York, 1993)

Length (1 kilometre = 1,000 metres)

1 centimetre	= 0.394 inch
1 inch	= 2.540 cm
1 metre (m)	= 3.281 feet
1 foot	= 30.48 cm = 0.305 m
1 kilometre (km)	= 0.621 mile
1 mile	= 1609 m = 1.609 km

Area (1 square kilometre = 100 hectares, 1 hectare = 10,000 square metres)

1 hectare (ha)	= 2.471 acres
1 acre	= 0.405 hectare
1 square kilometre	= 0.386 square mile
1 square mile	= 2.590 square kilometres

Weight (tonne is the usual spelling in English for metric ton, 1 tonne = 1,000 kilograms)

1 kilogram (kg)	= 2.205 pounds
1 tonne	= 1.102 short tons
	= 0.984 long ton

List of changed place-names

Abyssinia	Ethiopia
Aden	Yemen
Bechuanaland	Botswana
Belorussian SSR	Belarus
British Honduras	Belize
Burma	Myanmar
Cameroon	United Republic of Cameroon
Ceylon	Sri Lanka
Congo (Belgian)	Zaire
Czechoslovakia	Czech Republic
	Slovakia
Dahomey	Benin
Eastern Europe	Central Europe
Ethiopia	Ethiopia and Eritrea
Formosa	Taiwan
German Democratic Republic (GDR)	East Germany, New Länder (of all Germany)
German Federal Republic	West Germany
Guiana, British	Guyana
Guiana, Dutch	Suriname
Ivory Coast	Côte d'Ivoire
Kampuchea	Cambodia
Korea	Democratic People's Republic of North Korea
	Republic of South Korea
Leningrad	St Petersburg
Malaya	Malaysia (part of)
Moldavian SSR	Moldova
Pakistan, East	Bangladesh
Pakistan, West	Pakistan
Persia	Iran
Rhodesia, Northern	Zambia

Rhodesia, Southern	Zimbabwe
Siam	Thailand
Southwest Africa	Namibia
Spanish Sahara	Western Sahara
Tanganyika	Tanzania
Upper Volta	Burkina Faso

Chapter one

Introduction

If the earth were uniform – well polished, like a billiard ball – there probably would not be any such science as geography.

J. Gottmann (1951)

1.1 The geographical approach

The content of geography is very large and is not clearly defined at its margins. Geographers study the distribution of various phenomena and the relationships between. They work on scales ranging in size from a farm, a village or part of a city, to the whole of the earth's surface. This book has been compiled and presented according to a tradition that was particularly strong in the latter part of the nineteenth century and early in the present century in Germany and France as well as Britain and the United States. The theme is based on the idea of comparing regions in the context of a universal or global geography. Some geographers chose to identify regions of the world distinguished by climatic features and to compare differences in the human response to regions with similar environmental conditions in different parts of the world. Others focused on cultural regions, contrasting features of the human geography in various parts of the world. The emphasis in the present book is on regional differences in political, economic and social conditions and on the influence of demographic, cultural, natural resource and technological factors on these differences. For this purpose the world has been divided into a number of 'major regions'.

To produce a geography of the major regions of the world is a daunting task. Depending on the context and requirements of any particular issue or problem to be studied, the surface of the earth can be subdivided on many different scales and according to a variety of methods and criteria. The smaller the subdivisions, the more detailed can be the differentiation between them. When the whole of the world is being considered within the size constraints of a single volume, the initial regional subdivisions cannot be too numerous and small. The first-level regional breakdown which forms the basis of the present book consists of twelve major regions.

For some decades now, geographers have for many purposes largely abandoned the traditional division of the world into five continents. Initially three continents were recognised, Europe, Asia and Africa. The 'discovery' of the Americas brought in a fourth, that of Australia a fifth. For convenience the Americas are often divided into North and South, while the land mass of Antarctica, although almost entirely covered by ice, is arguably a continent too. The twelve regions used in the present book, shown in Figure 1.1, are based on human activities and cultural features rather than on physical and environmental conditions. The choice of these twelve regions will be discussed in detail in Chapter 3.

The remainder of this section is intended primarily for the reader with little or no background in the subject of geography and may be left by those with such a background. In the 1960s, W. D. Pattison (1964), one of many geographers attempting to define his or her subject, expressed the view that, as perceived and practised over the centuries, geographical research and applications could be broadly classified under four traditions: earth sciences, humans and the environment (people/land relationships), spatial and locational aspects, and area studies. Although these fields overlap

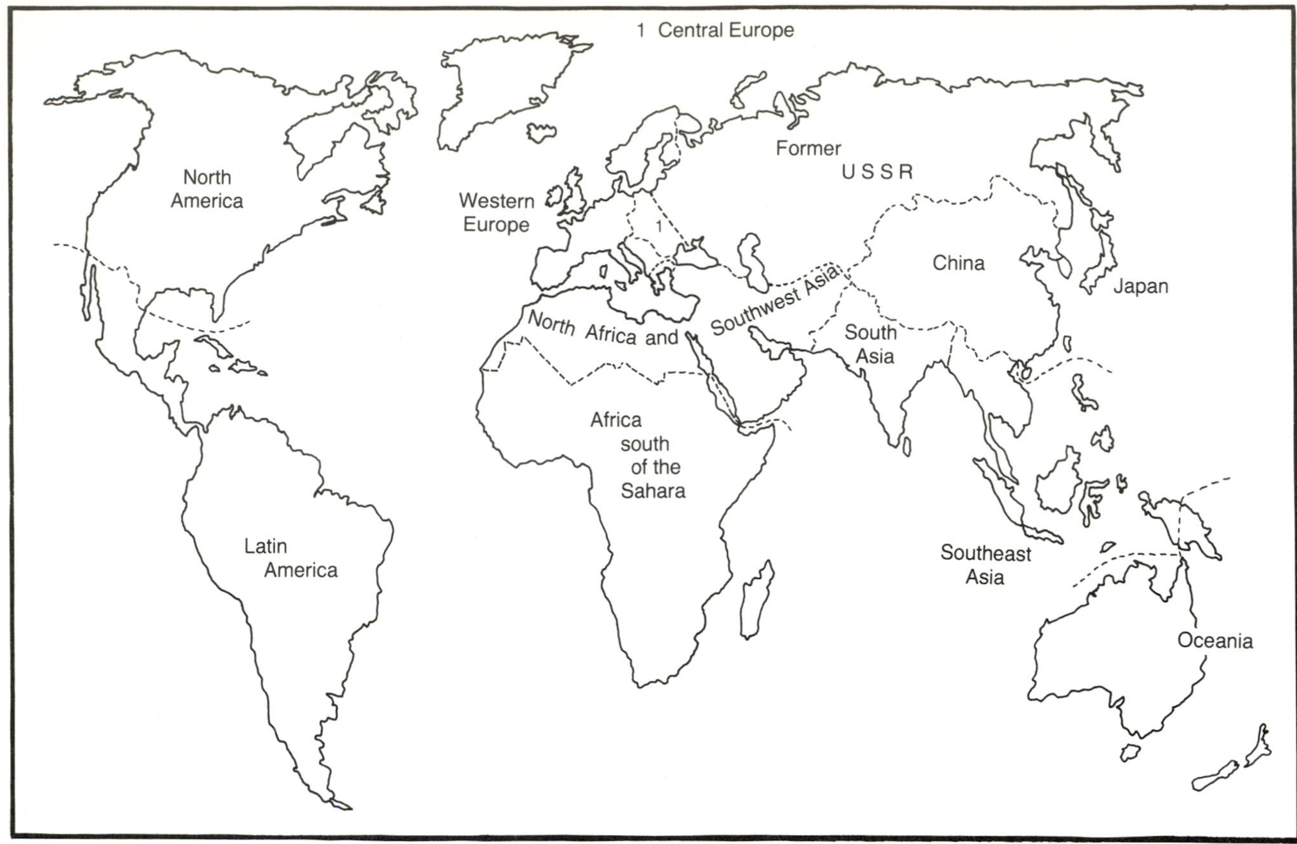

FIGURE 1.1 The twelve major regions used in this book as a division of the world on mainly cultural grounds

and also miss out some issues covered by geographers, they should give the non-geographical reader a broad idea of the scope of the subject. The area studies or regional approach forms the basis of the present book.

In contrast to Pattison's version of the scope of geography, in everyday language the word 'geography' (or 'geographical') is usually used for one of two situations: the location of places, or their attributes and characteristics. The locational aspects of places can be illustrated by the following statement: on account of its 'geographical' position, Singapore is a focus of sea routes in peacetime and a strategic locality in wartime. In similar fashion, the reports of transnational companies usually contain a description of the 'geographical' distribution of their activities and sales. Concern with what places are like can be illustrated by the following statement: the 'geographical' conditions of the Amazon rainforest make it difficult to develop. Similarly, one might link the influence of surrounding mountains and the presence of some 20 million people in a small area to account for serious pollution in Mexico City. In the present book, these two 'popular' views of geography will serve as starters for the study of the major world regions: where, for example, is Australia in relation to other regions of the world, and what physical conditions and human activities does it have? Similar questions can be asked about places at different scales, whether administrative units within a state, such as Florida or California, or cities such as Bombay or Bangkok.

Ever since the sixteenth century, one self-appointed task of European geographers has been to locate places on the world map as they were 'discovered' by explorers (see Glossary for a note on the term 'discovery'). In the nineteenth century, geography developed into a more extensive and sophisticated subject, with geographical societies and research centres in such countries as the USA, Britain, France, Russia and Brazil, associated closely in many countries with the military establishment. The results of exploration and travel were analysed, the new findings of geology, biology and climatic data were taken into account, and the workings of human activities on the earth's surface led to the formulation of various theories. Some US and British geographers have stressed and criticised the role of practising cartographers and geographers in supporting the military and economic conquest of colonies and the establishment and administration of empires.

Geographical information has proliferated in the twentieth century, with new sources of data like large-

PLATE 1.1
Art and crafts in the Caribbean

a Naive painting of a school in Haiti. Many aspects of life in Haiti are portrayed in such work. Virtually all are sold to tourists or abroad

b Figurines for sale at a roadside stall in the Dominican Republic, also with the tourist in mind. Some people lament the commercialisation of such local work in developing countries

scale mapping, the regular holding of censuses in almost every country of the world, aerial photography, and satellite imagery. With computer-based Geographical Information Systems maps can now be updated rapidly and used to experiment with alternative future distributions, comparing the results to be expected on the ground of the application of alternative policies and decisions.

This book is not intended to provide detailed coverage of the twelve major regions but to highlight their present outstanding features and, with reference where appropriate to historical trends, to consider their future prospects. Chapter 2 provides a brief summary of global trends in human activities since the end of the fifteenth century. In Chapters 3 and 4, demographic, economic and social aspects of the twelve regions are outlined with the help of appropriate data sets. In Chapters 5–16 each region is described in turn. In the two final chapters attention is focused on the prospects for each region and for the world as a whole in the next few decades.

The outstanding aspects of the geography of each region are noted, but emphasis is given to recent and current issues, and to the region's problems and characteristics. Thus, for example, in the region of North Africa and Southwest Asia, Islam, water supply and the oil industry are given prominence, in China the effects of central planning and the problem of regional disparities, and in Western and Central Europe the enlargement of the European Union. The author is aware that some of the regions are very diverse, and that none is homogeneous. While, for example, Japan with South Korea, or Australia with New Zealand, are fairly uniform regions of the world, at least in their human geographical conditions, enormous variations are found, for example, in Latin America and in Southeast Asia. In a single book on a global scale, considerable generalisations and simplifications therefore have to be made.

It should be appreciated that the approach used here is only one of many that could be chosen to compile a geography of the world. Some geographers use regions to illustrate models, others take a systematic approach, looking globally at given topics. In this book a comprehensive region-by-region analysis is used to describe the current issues and differences in the world as a whole.

1.2 The physical world

For simplicity, the earth's crust can be subdivided into four main elements: the solid base (the lithosphere), water (the hydrosphere), air (the atmosphere) and all forms of organic life (the biosphere). Given its crucial importance to the biosphere, soil might be considered a fifth sphere, consisting, however, of particles of the lithosphere, hydrosphere and atmosphere as well as live elements and dead remains of the biosphere. Throughout the history of the planet, internal and external forces (energy) have produced changes in the first three of these 'spheres' noted above and in the fourth, once there was life on the planet. Living organisms have also modified the planet in ways that benefited themselves. In the last few millennia, however, humans have modified the natural environment far more intensively, systematically and consciously than any other known species.

Before a brief description is given of some outstanding aspects of the impact of humans on the four 'spheres' of the physical world referred to above, it is appropriate to consider here the position of the human species itself. Human beings are in the ambivalent position of forming part of the biosphere, yet of behaving differently from the rest of it. According to the anthropocentric view of several influential religions, humans are indeed special. The Christian view, as expressed in the Books of Genesis and Esdras (see Box 1.1), gives the impression that humans have been given a mandate by God to make use of the rest of the biosphere at will. *Homo sapiens* is, however, just one among millions of plant and animal species. Our increasing command of various means of utilising, exploiting and destroying other species, and our rapidly increasing numbers, could eventually lead us to destroy ourselves and to reduce the number of other species in the process.

In increasing order of their vulnerability to the

BOX 1.1

Biblical and Apocryphal references to the creation

In view of the great influence on the Christian world of the Bible it may be argued that the passages quoted below could be used to generate complacency about the availability of natural resources in the world and about the size of the world's population. While it is increasingly appreciated by Christians that the account of the creation should not be taken literally, widely known statements such as those quoted below could still be taken as a mandate to exploit the resources at will.

- Genesis 1

27 So God created man in his own image, in the image of God created he him; male and female created he them.

28 And God blessed them, and God said unto them, Be fruitful, and multiply, and replenish the earth, and subdue it: and have dominion over the fish of the sea, and over the fowl of the air, and over every living thing that moveth upon the earth.

- 2 Esdras (Apocrypha) 6[1]

42 On the third day thou didst command the waters to be gathered together in the seventh part of the earth; six parts thou didst dry up and keep so that some of them might be planted and cultivated and be of service before thee. For thy word went forth and at once the work was done.

47 On the fifth day thou didst command the seventh part, where the water had been gathered together, to bring forth living creatures, birds and fishes, as it was commanded, that therefore the nations might declare thy wondrous works.

Note: [1]See B. M. Metzger (1965).

various endeavours of humans, the following elements of the natural environment may be noted:

1. **The lithosphere** Human activities such as quarrying and mining, underground nuclear blasting, and the construction of roads, railways and dock basins modify the relief of the earth's surface, although big changes are usually confined to comparatively small areas. According to Miller (1992), 'miners dig some 24 billion tonnes of minerals from the ground each year, including metals, sand, gravel and phosphates', almost 5 tonnes per inhabitant on average and about eighty times the weight of a human body. Natural events such as earthquakes also cause changes, e.g. landslides, capable of damaging human settlements and other works of construction. Water and wind erosion affect much of the earth's land surface to varying degrees. The extent and shape of the land and sea areas alter only gradually, however, with movements of tectonic 'plates' of lighter material moving distances varying between 2 and 18 centimetres (1 and 7 inches) a year in some areas. At present only 27 per cent of the earth's surface is land, some of which is covered by thick ice sheets, mainly in Antarctica and Greenland, and some by inland water surfaces, such as the Great Lakes in North America.

2. **The hydrosphere** The total amount of water on the earth's surface and in the ground below hardly changes over long periods of time, but the water itself takes different forms, according to where it is. At any one time, roughly 97.3 per cent of the water is in the oceans and adjoining seas, and is saline. Almost all of the rest is fresh water in lakes and rivers or in rocks beneath the surface of the land. At any given time a minute proportion, about one hundred-thousandth, or 0.001 per cent, is held in the earth's atmosphere, having evaporated off water surfaces, and being destined in due course to form precipitation (rain, snow, hail), ending up once again on the land or in the oceans. Forming a vital natural resource, the natural hydrological system of the world has been modified considerably by human activities, both in the control and direction of flows of water and by the deposition of pollutants, in inland waters, seas and oceans. The general configuration of the floor of the oceans is now mapped, the circulation of ocean currents is understood, and the presence of economic minerals such as manganese on the sea bed is known. Even so, much is still to be discovered in and about the oceans.

3. **The atmosphere** consists roughly of four-fifths nitrogen, one-fifth oxygen and much smaller quantities of carbon dioxide and other gases, as well as water and dust. Atmospheric temperature, moisture content and other elements affect human activities in various ways, but most profoundly by providing the thermal and moisture resources for plant growth. It has been hoped that changes in the weather could be forecast over long periods, and weather conditions actually modified. Cloud 'seeding', achieved by dispersing substances from aircraft into an appropriate place in the atmosphere to induce condensation, can artificially produce rain where required. It has also been assumed, perhaps optimistically, that planting trees in drier areas would produce a more humid climate. Neither practice has produced spectacular results. There is, however, growing concern over both the local and the global weather prospects in the long term, because of trends that could lead to greater pollution and to a comparatively rapid rise in the temperature of the atmosphere, the result of the so-called 'greenhouse' effect (see Glossary), with consequences that are still a matter of speculation.

4. **The biosphere** Early humans depended for their subsistence on hunting, fishing, and gathering plant products. Fire and simple tools gave them limited opportunities to modify the natural environment. For several millennia now, cultivators have cleared the natural vegetation for crops, herders have grazed their livestock on natural pastures, forests have been cut for fuelwood and construction timber, and river flows diverted for irrigation purposes. By the sixteenth century, much of the natural vegetation of the world, and consequently also the soil, had already been modified in some way by humans.

The scale of the human activities that were modifying the natural environment even in the eighteenth century has been dwarfed by the combined effects of demographic and economic growth in the world in the last 200 years. Between about 1750 and 1870 the population of the world doubled from about 720 million to 1,400 million (McEvedy and Jones 1985) while the consumption of fuel and raw materials increased about tenfold during that period. Between 1870 and 1990 the population of the world increased almost fourfold, and consumption again about tenfold. There will be a discussion in Chapter 17 of the implications of these trends for the future of the human species and in Chapter 4 reference is made to the various concerns voiced about the environment, especially since the 1970s.

One immediate effect of the growing pressure on the biosphere has been the disappearance of many species of plant and animal. May (1992) estimates that there are between 3 and 30 million plant, animal and other species, of which less than 2 million have been catalogued over the last 250 years. While it is known that some species have already disappeared, it is impossible to calculate the precise number lost, and difficult to estimate how many will be threatened in the near future. Holloway (1993, p. 78) refers to an estimate of

TABLE 1.1 The area of the land, the oceans, and the types of land use of the world

	Area in thousands of sq km/(miles)		*% of land only*	*% of all surface*
Land areas				
Asia without the former USSR	27,580	(10,618)	18.5	5.4
Former USSR	22,403	(8,650)	15.0	4.4
Europe without the former USSR	4,877	(1,883)	3.3	1.0
Africa	30,293	(11,696)	20.3	5.9
North and Central America	22,407	(8,651)	15.0	4.4
South America	17,819	(6,880)	11.9	3.5
Oceania	8,537	(3,296)	5.7	1.7
Antarctica and Greenland	15,353	(5,928)	10.3	3.0
Land total	149,269	(57,633)	100.0	29.3
Ocean areas				
Pacific Ocean	179,650	(69,363)		35.2
Atlantic Ocean	106,100	(40,965)		20.8
Indian Ocean	74,900	(28,919)		14.7
Sea total	360,650	(139,247)		70.7
World total	509,919	(196,880)		100.0
Land uses (excludes Antarctica and Greenland)				
All 'land'	133,916	(51,705)	100.0	
Inland water	3,124	(1,206)	2.3	1.0
Arable and permanent crops	14,442	(5,576)	10.8	2.8
Permanent pasture	34,021	(13,136)	25.4	6.7
Forest and woodland	40,276	(15,551)	30.1	7.9
Other	42,053	(16,237)	31.4	8.2

Sources of data: Land areas and uses, *FAOPY* (1991), vol. 45, Table 1; ocean areas: *Calendario Atlante de Agostini* (1994), p. 19

TABLE 1.2 Percentage cover of main vegetation types in the continents and in the four countries with the largest populations

	(1) Desert and semi-desert	*(2) Polar and alpine*	*(3) Grass and shrub*	*(4) Crop and settlements*	*(5) Interrupted woods*	*(6) Major forests*	*(7) Major wetlands*	*(8) Other, coastal, aquatic*
Asia[1]	16	9	24	17	10	18	1	4
Former USSR	5	26	10	8	21	26	2	3
Europe[1]	0	9	4	35	22	23	0	6
Africa	30	0	28	7	14	17	2	2
N & C America	3	33	9	10	17	21	2	5
South America	5	2	32	8	14	33	3	2
Oceania	18	0	18	5	38	16	1	4
World	13	12	20	11	17	22	2	4
China	14	22	21	17	5	18	1	1
India	2	2	12	44	23	14	0	3
USA	5	15	16	18	24	18	1	5

Note: [1] Excluding USSR
Source: Adapted from Groombridge (1992), pp. 251–2

27,000 species being consigned to extinction every year in the tropical rainforests of the world alone. This environment contains about half of all the species in the world. From statements by politicians and economists criticising those showing concern about the loss of species, it would seem that they do not realise that new species evolve only on very long timescales.

To prepare the way for more detailed studies of

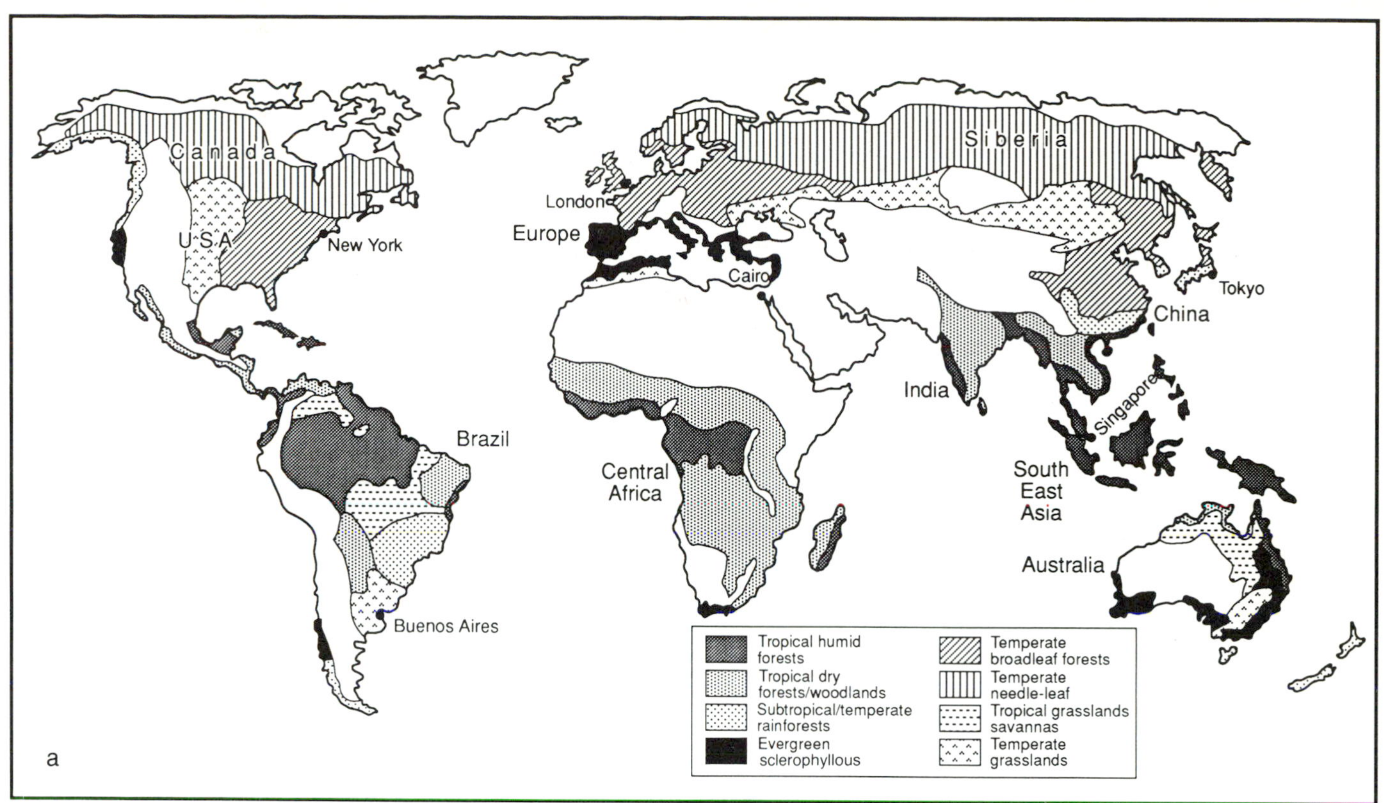

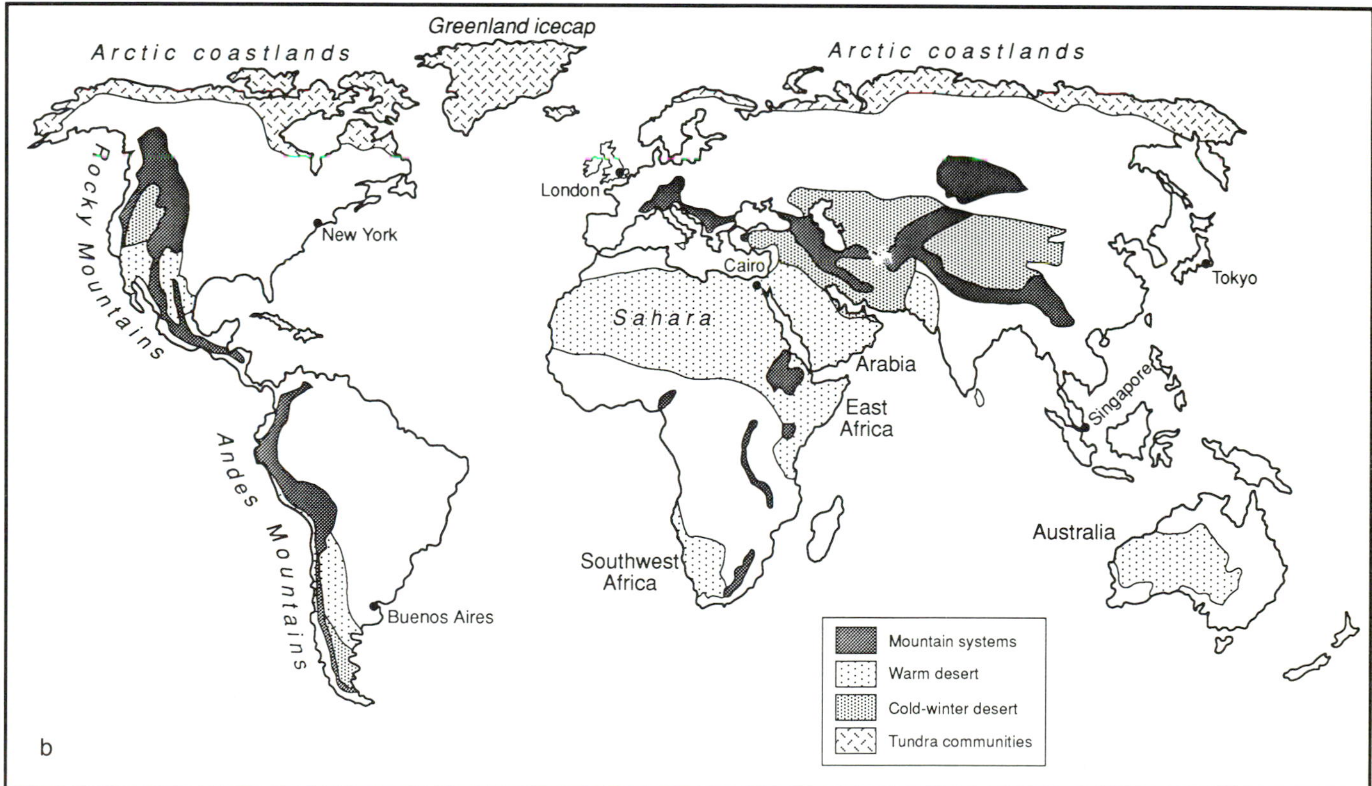

FIGURE 1.2 Biogeographical biomes of the world

a Distribution of broadly 'friendly' environments from the viewpoint of bioclimatic resources

b Environments in which extremely dry, cold and/or rugged conditions restrict cultivation, natural pasture and forest. Antarctica is not shown

Source: Groombridge (1992), Plate 1

parts of the world's land surface later in the book, Table 1.1 shows the areas of the conventional continents and oceans. The main types of land use are also shown. Antarctica and Greenland, although land masses, are in effect almost entirely water surfaces, being covered by thick ice. In addition, over 2 per cent of the rest of the 'land' area of the world is covered by inland waters. Although medieval European maps of the world showed six-sevenths of the surface of the earth as land, following the Book of Esdras in the Apocrypha (see Box 1.1), only about one-quarter actually is. Even now only about 11 per cent of the land surface of the world is cultivated, under arable or under permanent crops, that is, less than 3 per cent of the total surface of the globe. Estimates differ as to how much more could be brought under cultivation, from little more at all to perhaps an increase of 150 per cent. Table 1.2 shows the proportion of the land area of each continent and of the four largest countries of the world in population (before the break-up of the USSR) under each main type of natural vegetation, including the area transformed by human activities. The main types of vegetation are shown in Figure 1.2a.

1.3 The organisation of human activities

As this chapter has already stressed, humans have 'taken over' the planet. Virtually all of the earth's land surface is already recognised as 'belonging' to some sovereign state, while some is disputed. Much of the sea and the ocean adjacent to the land areas is also claimed as economically under the control of given sovereign states. Throughout the rest of the book the word 'country' is generally used as a convenient umbrella term to include both sovereign states and colonial units (which are not sovereign). The word country also avoids confusion with the use of 'state' for the internal divisions in many countries (e.g. the USA, Brazil). Subdivisions or regions have been mapped out on land and sea to serve as areas of administration or for economic activities. The management of territory or space, a commonly used concept in French regional planning for several decades, is now recognised as a concern of many governments.

In their attempt to describe and account for spatial variations on the earth's surface, geographers have commonly used either natural regions such as river basins and climatic regions, or man-made regions such as politico-administrative regions. A famous example in the inter-war period was the creation of the Tennessee Valley Authority area in the USA. In 1933 a region delimited by the watershed of the Tennessee river drainage system, cutting across state boundaries, was created to help solve some of the problems of the poorer Appalachian states. Hydroelectric, flood control, irrigation and other schemes were implemented. Features of the natural environment still greatly influence human activities, although, since the nineteenth century, the power of humans to tame and exploit nature has been stressed, and environmental influences have been played down. On the other hand, much of the economic activity in the world is organised and managed within the boundaries of each country, notwithstanding the increasing exchanges of goods and, to a lesser extent, of services, across international boundaries.

Recently, some have argued that there is now a single global economy, in which world bodies such as the World Bank and transnational companies operate. Indeed the General Agreement on Tariffs and Trade (GATT) has progressively reduced barriers to trade between the countries of the world. Information flows rapidly, producing frequent changes in the system. National boundaries still serve, however, to control the migration of people, to ensure that customs duties are levied on selected goods and that import quotas are respected and, with diminishing success, to prevent the diffusion of information regarded as harmful.

The countries of the world, the bases for the management of space, vary greatly in size, whether this is measured by population, area or Gross Domestic Product (hereafter usually referred to as GDP). It is appropriate next to examine some of the outstanding features of the countries of the world and the implications of the differences in size.

The exact number of countries in the world is difficult to calculate. Freedman (1993) counted 187 sovereign states early in 1993, 179 of which were United Nations (UN) Members. The *1994 World Population Data Sheet* of the Population Reference Bureau includes 197 countries. In the first edition of the *Statistical Yearbook of the United Nations* to include the fifteen newly independent states that emerged from the former USSR, 234 separate countries are listed (*UNSYB*, vol. 38, Table 11). Hong Kong and various smaller colonial territories are included, as well as the small 'pseudo-sovereign' states (e.g. Andorra). According to Binyon (1993), these numbered twenty-nine at the end of 1993. The UN total of 234 does not include a divided Czechoslovakia or Yugoslavia, or Taiwan, excluded from UN publications despite being a very successful independent economy. The total, then, is *about* 240, but as of February 1993, only 180 were Member States of the UN, any colonial countries being automatically excluded, as also are some very small and/or special countries such as Andorra and the British Channel Islands.

The amalgamation of countries since the Second World War has been rare. Examples include Syria and Egypt, which formed the United Arab Republic for a time, North Yemen and South Yemen (formerly Aden),

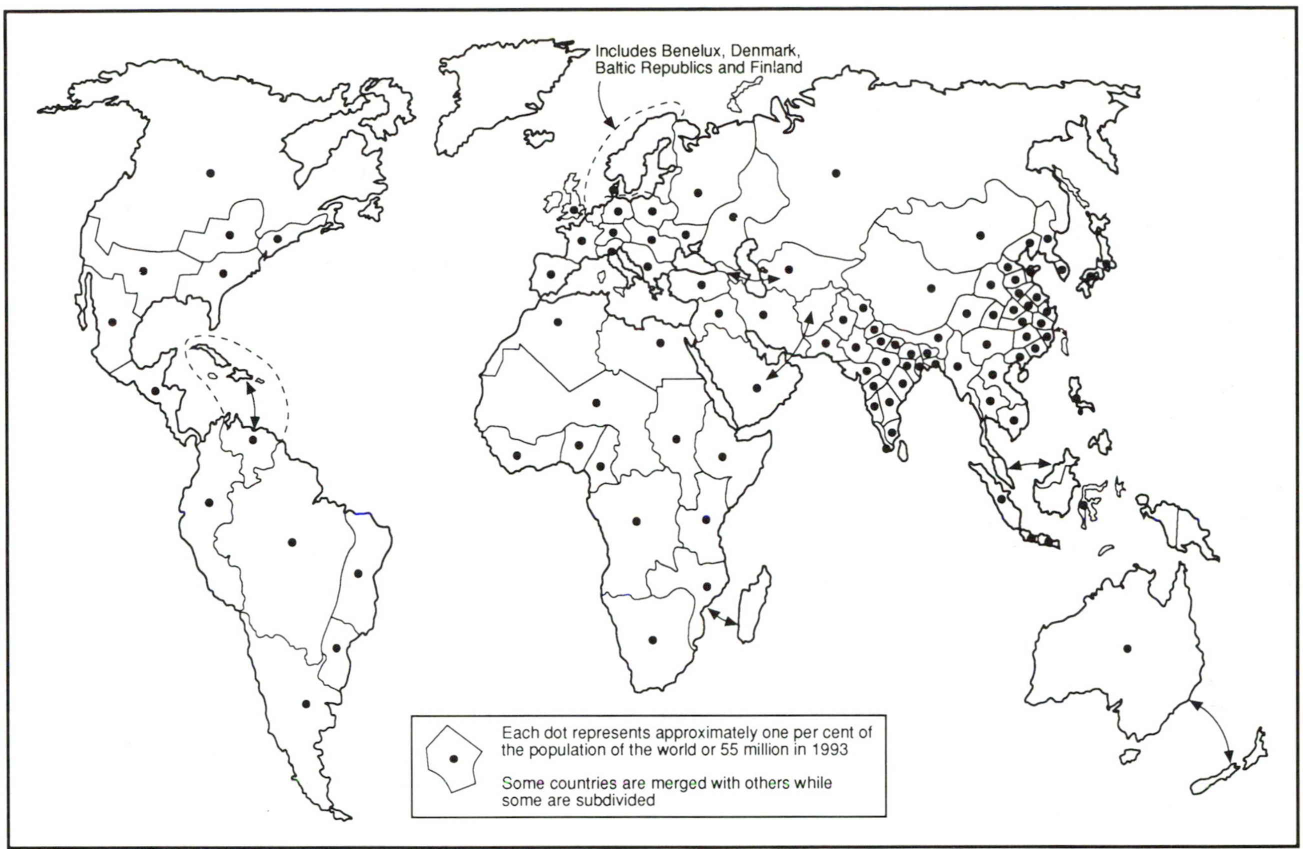

FIGURE 1.3 The distribution of population in the world in 100 'equal' population 'cells'. In practice it was not appropriate to fit exactly 55 million people into each cell. Oceania, for example, was kept separate from Indonesia and therefore has fewer than 30 million

Source: calculations by the author

which were unified in 1990, as also was Germany. In contrast, the membership of the United Nations has increased more than threefold since its foundation in 1944–5. In 1992 the UN Secretary-General (Boutros Boutros-Ghali) anticipated 'that the world could splinter into 400 economically crippled mini-states unless the rights of minorities move to the top of the international agenda' (*China Daily*, 21 September 1992, p. 8).

The imbalance in population size among the countries of the world is illustrated by the fact that the six largest countries in population, China, India, the USA, Indonesia, Brazil and the Russian Federation together have more than half of the total. The smallest in the UN list have only a few hundred inhabitants (e.g. the Holy See in Italy has 890, the Cocos Islands and Pitcairn in the Pacific Ocean have 555 and 51 respectively). A comparable disparity is found in the territorial size of countries, with the largest six having about half of the land area of the world: the Russian Federation, Canada, China, the USA, Brazil and Australia. With an area of almost 17.1 million sq km (6.56 million sq miles), Russia is 17,000 times as large in area as Hong Kong, although the latter manages to contain on its 1,000 sq km (386 sq miles) 5.4 million people, compared with Russia's population of 147 million.

For the geographer, the disparities in *size* among units of the same *status* cause problems in the interpretation of data in comparisons between different parts of the world. In an administratively ideal world all the countries would be roughly equal in population size. Figure 1.3 shows the result of a series of calculations made to produce 100 'equal' population 'cells', each with approximately 1 per cent of the total population of the world, or about 55.5 million inhabitants. A similar network of 100 cells based on the population of the world in 1975 was used for the collection of economic and demographic data by Cole (1981). At that time the average population of each cell was only about 40 million.

An example of the side-effects of having countries of very unequal size is the uneven disbursement of development assistance in the world. Small countries generally attract much greater attention per inhabitant than very large ones. In India and China, for example, there are states/provinces which, if they were separate

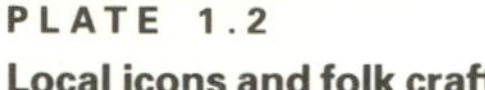

PLATE 1.2
Local icons and folk craft

a Roadside shrine in northern Thailand, Buddhist religion

b Since *glasnost* in the former USSR in the late 1980s it has been possible for local wood-carvers to branch out from the traditional peasant women nested dolls. The choice of individuals and topics reveals much about the new way of thinking in Russia. Front row left to right are the members of one set: the last tsar, Lenin, Stalin, Khrushchev, Brezhnev, Yeltsin, Gorbachev. Middle pair: OMON, the security forces, and Yazov, one of the leaders of the failed 1991 coup against Gorbachev. Back row: The Orthodox church is back, the traditional peasant woman doll, a nest of dictators, with Saddam Hussein the outer doll, plus Hitler, Stalin and Mao inside, and Yeltsin, destroyer of the Communist Party (see Chapter 7)

countries, would be included among the least developed countries (see Figure 4.2, p. 79). Because they are embedded in such very large countries they are overlooked. More seriously for world affairs in general, the concentration of economic and military strength among a few large 'world powers' gives them great potential influence over the many small ones.

It is possible to obtain accurate data for the area of countries and reasonably accurate data for the size of population and for GDP. It is more difficult to calculate the relationship between public and private sectors in their contribution to the total economy. The matter is of particular relevance now, however, as most of the former 'socialist' countries (as defined by the USSR), including the former USSR itself, are changing. They are moving from a situation in which the public sector was dominant, and a command economy, with central planning, formed the basis of management, to a market economy and to one with at least a substantial private sector. In the case of some industrial countries also, including for example the UK, France and Sweden, there have been moves to privatise some activities in the 1980s and early 1990s. In the UK in about ten years, approximately 10 per cent of the total capitalisation value of companies was transferred from public to private ownership, including utilities such as telecommunications and electricity, and industries such as steel. According to Riley (1993), between 1985 and

TABLE 1.3 Central government expenditure as a percentage of total GNP in 1991

OECD[1] selected countries		*Other selected countries*	
Japan	16	India	18
Canada	24	Mexico	18
USA	25	Indonesia	21
Australia	28	Sri Lanka	29
Germany	33	Brazil	35
UK	38	Egypt	40
New Zealand	44	Bulgaria	77
France	44		
Sweden	44		
Italy	50		
Netherlands	53		

Note: [1] Organisation for Economic Cooperation and Development
Source: World Bank, *World Development Report 1993*, Table 11

1993, state-owned companies worth about 300 billion US dollars were sold off in some fifty countries, reversing a trend that began in the 1930s when capitalism 'failed' in the depression. One way of showing the influence of the state in the economic life of countries is to express central government expenditure as a proportion (percentage) of total GDP. Unfortunately, data are not readily available for many countries of the world. The data in Table 1.3 do not therefore give a comprehensive overview of the subject but they are sufficient to give an idea of the contrasts between countries.

1.4 Cultural features of human communities

If everyone lived in a single world state, spoke the same language, had the same skin colour and received the same income, the subjects of economics, sociology and politics would be far less complicated and controversial than they are. As it is, human beings vary in so many different respects that in order to generalise about so many individuals it is convenient to put them into classes. In the context of people, class has come to refer above all to groups resulting from socio-economic stratification. More generally, classification in the social sciences is related to the mathematical theory of sets. In mathematics it is possible to assume that sets are exact, or extensionally definite. For example, the members of the set of positive integers between and including 6 and 14 (in base 10) are 6, 8, 10, 12, 14. For most practical purposes it is unrealistic to assume that sets are always exact. In this section attention focuses on the problems of defining the kinds of sets or classes recognised and used in socio-economic and cultural spheres of human activity.

For simplicity, three types of class are identified and illustrated below with examples. The ideas are then used to examine the class stratification in a selection of large and influential countries. Cultural matters are further discussed in various chapters of the book.

1. Some sets are reasonably clearly defined, with their membership unambiguous. The members are usually discrete elements with (or without) a given attribute or characteristic. For example, traditionally males and females are distinguishable and sex/gender classification is central in demographic data. Similarly, speakers of a given language are for practical purposes identifiable, since most people have one language as their mother tongue, rather than none or more than one. Thus, for example, in Canada, French and English speakers are two distinct sets or classes, although many Canadians use both languages and truly bilingual people fall in the intersection of the two sets. Similarly, in Belgium the Dutch-speaking Flemish and French-speaking Walloons are separate sets. With religion, again, it is expected that each person professes and practises only one religion or none at all. With some simplification, types of employment can for convenience also be placed in distinct classes.

2. Some sets appear to be clearly defined, but in effect the apparency of the existence of discrete elements is misleading. The proliferation of shareholders in western industrial countries has drawn attention to the existence of a shareholding class. It may be noted, however, that in reality a shareholder with shares worth 1,000 dollars is much more similar in terms of wealth to a person without any shares at all than to one with shares worth 1 million dollars. Similarly, the practice in the Americas and South Africa of dividing the population into Black and White on the basis of skin colour, and thus implicitly by continent of origin, is convenient, but does not necessarily reflect the actual situation. In the Americas, African and European contributions to the make-up of the individuals concerned occur along a scale from all of one, through various mixtures, to all of the other (see Section 9.2).

3. Many classes are defined by arbitrary cut-off points on a continuum. Examples are classes based on income and age. For example, the 'poor' may be those with an income below the poverty line (see Section 4.1), so that their standard of living as defined in a given context is inadequate. Conventionally, the young and the elderly are separated from adults of working age (see Box 3.3) but cut-off ages vary from country to country. Some examples of classes now follow, with specific reference to the situation in selected countries.

Although the influence of the caste system on economic and social life is diminishing in India, and officially it should not be used to discriminate, for example, among candidates for employment, in practice it still affects a very large number of people. The caste of an individual is determined by a specific social rank, which depends on descent, marriage and occupation. Four basic castes, or varnas, are recognised: priests (Brahmans), barons or warriors (Kshatriyas), commoners or merchants (Vaisyas) and artisans and labourers (Sudras). Differentiation depends on the degree to which individuals are 'ritually polluted' according to defilements resulting from their occupations, dietary habits and customs. A large residue of 'outcastes', described as 'untouchables', but tactfully referred to now as children of God (Harijans), perform work that is so defiling that they do not graduate even to the lowest caste.

Mobility between castes is possible, but it depends on a change in dietary habits or on having the means to hire someone else to carry out one's polluting work. In spite of caste reform, it would be surprising if a cultural feature that has been in place for three thousand years, and is closely tied to the Hindu religion (see Box 14.1), could disappear in a few decades, at most two generations. Mitchell (1968, p. 182) gives a concise definition: 'A pure caste system is rooted in the religious order and may be thought of as a hierarchy of hereditary, endogamous, occupational groups with positions fixed and mobility barred by ritual distances between each caste.' Although primarily a feature of India, castes are found elsewhere in Southeast and East Asia. For example, the class of workers carrying out menial tasks in Japan, the Bukharamin, is likened by some sociologists to a caste. The slaves brought into the Americas from the sixteenth to the nineteenth century (see Box 2.1) were not strictly a caste, although, like the serfs in tsarist Russia, in effect they were tied to certain localities and types of employment until, coincidentally, 'freed' in the 1860s in both Russia and the USA.

TABLE 1.4 The class composition of the population of Russia and the USSR (percentages)

	1913	*1939*	*1959*	*1970*	*1986*
Office workers (*sluzhashchiye*)	2.4	16.7	18.8	22.7	26.2
Industrial workers (*rabochiye*)	14.6	33.5	49.5	56.8	61.7
Collective farm peasants	–	47.2	31.4	20.5	12.1
Individual peasants	66.7	2.6	0.3	–	–
Bourgeois, traders, (*Kulaki*)	16.3	–	–	–	–
Total	100.0	100.0	100.0	100.0	100.0

Source: Narodnoye khozyaystvo SSSR v 1978 godu, p. 9 and *1985 godu*, p. 7, Moscow 'Statistika', 1979, and 'Finansy i Statistika', 1986

Social stratification in Europe has been based on systems of estates, defined by laws made by humans, in contrast to the caste system of India, based on the laws of religious ritual. According to Mitchell (1968, p. 82), the universal characteristic of these laws was 'that they defined not merely the rights but the duties of members of estates and thus provided a clear system of social order based on responsibilities enforceable either in the courts or by military strength'. In both the caste and the estate systems, social position is normally ascribed. Since the eighteenth century, the industrial revolution and the impact of 'industrial capitalism' have led to a shift from a rigid traditional system to one in which individual merit is valued and (theoretically) rewarded. Karl Marx argued a century and a half ago that new social classes were emerging and that these classes could be defined in their relationship to the means of production and the (uneven) distribution of wealth. His expectation that class-consciousness would lead to class-conflict in Western Europe has to some extent been justified by events, but with economic growth, opportunities for social mobility and a welfare net to provide for the basic needs of most, if not all, of the economically deprived, social revolutions did not occur.

In the former Soviet Union, the Communist Party appropriated the concept of the class system as expounded by Marx for its own use. It pre-empted the class struggle that it expected or professed to expect by creating classes that would take the Soviet Union smoothly through socialism to a final stage of communism. For example, in *Narodnoye khozyaystvo SSSR v 1978 godu* (p. 9), the class composition of the population is given (see Table 1.4). The data show that the agricultural propulation was collectivised, mainly during the early 1930s, 'capitalist' elements disappeared, and the working class, consisting of production workers in industry and on state farms, as well as office workers, grew relative to the less socially advanced collective farm population. The intellectuals or 'intelligentsia' formed a small but influential class not tabulated in the yearbooks consulted.

During the seven decades in which the Communist Party ruled the USSR, type of employment was, therefore, the main determinant of socio-economic class. The Marxist-Leninist class stratification system was adopted to some extent in Eastern Europe and China as well as in developing countries influenced by the USSR such as Cuba and Viet Nam. In Western Europe, classes were and still are less rigidly defined, with terms such as lower, middle and upper classes, as well as the working class, a set frequently referred to, but not clearly defined. Social classes in British statistical sources are based on type of employment, which correlates fairly closely with income.

In contrast, in the United States class is not a

commonly used term, presumably because the USA is theoretically a classless society, one in which anyone could run for the presidency provided he or she could amass enough money to conduct an election campaign. Possibly also the term class, as used in a socio-economic sense, is off-putting to Americans because of its connection this century with communism. Nevertheless, for statistical purposes the population of the USA is subdivided into various sets. In *Statistical Abstract of the United States, 1994* (State rankings, pp. xii–xxi) brief but illuminating insights are given into the way 'classes' are distinguished in the USA. The following are the main criteria used: sex, age, educational attainment, farm/non-farm, disposable income and retail sales, immigrants, metropolitan and non-metropolitan, and race. As used in the source referred to above, race is largely based on colour of skin, associated broadly with different parts of the world. The races are White, Black, American Indian/Eskimo/Aleut ('Red') and Asian/Pacific islander ('Yellow'). The sizeable Hispanic population in the USA is given no separate category, but is included under 'White'. The concern with highlighting the presence of Blacks among the total population, and also in each state and city of the USA, may be justified by the policy of implementing affirmative action against ethnic discrimination in obtaining employment. However, the fact that statistically Blacks do not behave as other Americans do is also publicised. For example Yanagashita and MacKellar (1995) quote FBI crime data showing that the 10 per cent of the US population defined as Black accounted in 1993 for over 50 per cent of all homicide victims *and* offenders.

The existence of classes tends to lead to confrontation, often with one class in a stronger position than another or others. In Western Europe and the USA, confrontation between the owners of means of production, together with management on the one hand, and the workers on the other, has been common at least during the last two centuries, even when the state and therefore the workers themselves are the owners of public sector enterprises. There follows a brief account of management–labour relations in Japan. Japanese industry itself is described in some detail in Section 8.4. It is a matter of speculation as to what extent emphasis on collaboration rather than confrontation in working relations and practices between labour and management has contributed to the success of the Japanese industrial sector in the period since the Second World War.

According to *Nippon* (1993, p. 50):

> Japanese management is markedly different from that of Europe and the United States in terms of labour management. There exist in Japan frameworks such as lifetime employment, seniority-based ranking and wage systems, company-based labour unions, corporate housing, and interdependence among employees for support during personal ceremonial occasions. These systems have allowed companies to secure loyal workers who are willing to work long hours.

The system with the characteristics described above was introduced by some corporations before the Second World War to secure adequate numbers of experienced workers and to make investment in their training worth while. The system was widespread in Japanese industry, especially in larger companies, until the 1980s, but in changed circumstances workers are less resistant to changing jobs and companies can no longer assign higher positions according to age.

Unemployment in Japan remains low in the 1990s. The wealthy keep a low profile, while the management mix with the workers. Some features of Japanese labour management have been transferred to Japanese transplants in Western Europe and the USA. It remains to be seen whether the approach is relevant, satisfactory or acceptable in the many new enterprises being set up by Japan in Newly Industrialised Countries in Southeast Asia, China and elsewhere. In *Nippon* (1993, pp. 50–1) it is noted that

> labour unions organized within companies have not adequately protected the rights of their members .... The most prominent issue currently being discussed is the ending of long working hours and to decrease the time worked per year to 1,800 hours, a level equal to that of Europeans and Americans. These goals are of such importance that they are included in the government's economic plans.

In some respects, unions, labour management relations, and levels of unemployment in Japan have resembled those found in the USSR, at least until the late 1980s. Are both the Japanese and the Russian systems to change to the confrontational, to some extent class-based, situation common if not universal in western industrial countries?

The hope of achieving equality between males and females has been a major issue in many countries since the nineteenth century but the subject has been neglected by geographers until the 1980s. Given the recent proliferation of papers and books on gender it is appropriate to note here some aspects of the subject and to draw attention to the extent of the female–male gap in different countries and major world regions.

In its *Human Development Report 1993*, the UNDP includes much evidence of the way in which females lose out in the provision of education and healthcare, in the competition for the better jobs, and in political systems, especially in developing countries. In *HDR*

1993 (p. 25) women are described as the non-participating majority:

> women who are not in paid employment are, of course, far from idle. Indeed, they tend to work much longer hours than men. The problem is that the work they do, in domestic chores and caring for the elderly, does not get the recognition it deserves in national account statistics .... Women are often invisible in statistics. If women's unpaid housework were counted as productive output in national income statistics, global output would increase by 20–30 per cent.

This section concludes with a review of some of the main sex and gender differences between the two classes. For convenience, a number of broad features are listed and discussed. Attention is drawn to three contrasts: between males and females in the developed countries, between males and females in the developing countries, and between females in the developed and developing countries. Some variables, such as female mortality, are not directly attributable to any sex or gender conflict or gap but are comparable only among females.

1. **Infanticide** is one of the most difficult practices to assess because data are not readily available. When all family planning measures fail and abortion is not available, infanticide is the ultimate lethal measure for influencing population growth. It is widely accepted that infanticide is applied more extensively to female babies than to male ones, and for at least two reasons. First, male children are more valued than female children in some societies because they stay with the family, form part of the workforce, and support their parents in their old age. Girls usually marry outside the family, often in another settlement. Second, if the aim is to restrict population growth, it is more effective to dispose of females rather than males. Infanticide was practised in Europe in Roman times, was common in Western Europe in the nineteenth century (see e.g. Langer 1972) and is reported in India, China and other developing countries today. Infanticide is rare in the developed countries since various forms of birth control are now extensively practised.

2. **Life expectancy** On average, women live considerably longer than men, in spite of the general consensus (see, e.g., Holloway 1994) that healthcare facilities are orientated more towards the needs of men than of women. According to *WPDS 1994*, the average life expectancy at birth in years in developed countries is 75 years, but it is 71 years for males and 78 years for females, a female–male gap of 7 years. In less developed countries, the respective figures are 63, 61 and 64, leaving a gap of only three years. Among the developed countries Russia has the greatest female–male gap, with a life expectancy of only 62 years for males (blame accidents, alcoholism?) compared with 74 years for women. At the other extreme, Japan's figures are the highest in the world, 76 and 82, and the gap is much narrower.

In some developing countries the statistics for life expectancy need interpreting carefully. According to *HDR 1993* (p. 25): 'Women tend on average to live longer than men. But in some Asian and North African countries the discrimination against women – through neglect of their health or nutrition – is such that they have a shorter life expectancy [albeit still higher than that of men]. Indeed, comparing the populations who should be alive, based on global mortality patterns, it seems that 100 million Asian women are "missing".' Among developing countries there are marked differences in female life expectancy, ranging from 72 years for China through 59 years for India to under 50 years in about twenty countries, all of them but Afghanistan situated in Africa.

3. **Maternal mortality** is characterised by marked differences between the countries of the world, and reflects the lack of healthcare for women giving birth. Maternal mortality rates are measured per 100,000 live births. They are extremely rare in most developed countries, being below 10 in Scandinavia and several other European Union countries, 13 in the USA and 16 in Japan. In Bulgaria, Albania and Romania conditions are the worst in Europe. In all but a few developing countries, however, the level is above 100 per 100,000 births (e.g. China 130, Mexico 150, India 550, Nigeria 750, some parts of West Africa 1,000).

4. **Education** The female–male gap in educational enrolment is considerable in many parts of the world, being most marked in some Muslim countries. In most of the developed countries the female–male gap in mean years of schooling is small, with females spending slightly longer than males in, for example, the Netherlands and the USA (104 and 102 to 100) but less time, for example, in Spain and Israel (92 and 82 to 100). In upper secondary and tertiary education, female enrolment exceeds male enrolment in many developed countries (e.g. in tertiary Sweden 130, USA 116 to 100) but in others male enrolment exceeds female enrolment (e.g. Switzerland 48, Ireland 84 to 100). The most conspicuous female–male gap is in the arts/science divide – female enrolment in science in most developed countries is between a quarter and a half of male enrolment.

In developing countries the female–male gap in mean years of schooling is much more in favour of males in the countries with low human development scores, with girls receiving only a quarter as much schooling as boys and in some countries even less (e.g. Afghanistan 12 per cent, Yemen 18 per cent). Between 1960 and 1990, the female–male gap narrowed in almost every developing country in primary enrolment, but it still remains very marked in secondary and especially tertiary enrolment. Between 1970 and 1990, overall literacy rates have improved in developing countries, but the female–male gap has changed little overall, being highest in Africa and lowest in Latin America.

5. **Labour force** While it is accepted that virtually all women perform some kind of work during their lifetime, those working in their own homes are not usually counted as employed. Even those officially recorded as employed tend to have lower average incomes than men. One reason is that they are paid less for doing similar work to men, in spite of legislation against this situation. Also, through lack of skills, appropriate education or the interference of child-bearing, they tend to predominate in poorly paid types of work.

An index of females as a percentage of males in the labour force, although very approximate in developing countries, is available for almost every country of the world. Compared with the score of 100 for males, the score for females is 77 for all developed countries, compared with 52 for all developing countries. The Nordic countries of Europe score 88, making them the most balanced in the world in terms of male–female employment. The US figure is 83, those of the European Union and Japan about 70. The employment of women is also at a high level in Central Europe, and the former USSR (mostly 75–87), but considerably lower in southern Europe.

For those concerned with the growth of population in the developing world, the availability of formal employment and of a money income for women is regarded, like education, as a means of making it easier for them to practise some kind of family planning, thereby reducing fertility. The female–male gap in the labour force in developing countries does not correlate with the level of human development, or relate to major regions. There are few variables on which the great variations among developing countries stand out so clearly. For this reason, the scores of a selection of countries are shown in Table 1.5. Very small countries have not been considered for inclusion. Some tentative generalisations are possible.

Following the Soviet model, the policy of maintaining a high level of female participation in the labour force has been applied in most of the countries that have been dominated or influenced by the Soviet Union since 1945. At the other extreme, the influence of Islam, with its emphasis on the home as the place for women to spend most of their time, is illustrated by the virtual absence of women in the labour force in most

TABLE 1.5 The ratio of women to men in the labour force (men = 100)

Communist/socialist influence		*Relatively large*		*Muslim influence*	
Tanzania	93	Thailand	88	Turkey	49
Mozambique	92	Colombia	69	Morocco	26
Czechoslovakia	89	Indonesia	66	Iran	21
Viet Nam	88	Zaire	56	Pakistan	13
Romania	87	Brazil	54	Egypt	12
North Korea	85	South Korea	51	Afghanistan	9
Poland	83	Mexico	46	Saudi Arabia	8
China	76	India	34	Bangladesh	7
Ethiopia	71	Argentina	27	Iraq	6
Cuba	46	Nigeria	25	Algeria	5

Source: UNDP, *Human Development Report 1993*, Tables 8, 9, 34

countries, with the exception of Turkey which is in transition to a more western lifestyle. Ideology, whether religious (Islam) or secular (Marxism), seems to be a strong influence in many countries, even, indeed, at work in Western Europe, where predominantly Protestant countries have higher participation of women in the labour force than Catholic countries.

In addition to the above indicators of female–male differences, many others should be taken into account if a reasonably complete picture of the state of women in the world is to be assessed objectively. Average age at first marriage, related to the possibilities available for young women to receive education and to gain experience, varies greatly. This variable is closely connected with fertility. In the western industrial countries, the average age is 25 years, ranging between 22 (e.g. Belgium) and 27 (e.g. Sweden), whereas in the developing countries with high and medium human development the average age at first marriage is 22 years and for those with low human development, 19 years (e.g. Bangladesh 17, India 19).

In the developed world, the political influence of women may be roughly gauged by the percentage of seats they occupy in the executive body of parliament, assuming such an institution exists and is functioning. Finland (39 per cent), Sweden (38 per cent), Norway (36 per cent) and Denmark (33 per cent) head the ranking, while Spain (15 per cent) and Italy (13 per cent) have a larger proportion than the more *macho* USA, UK, France (all 6 per cent) and Japan (2 per cent). Among developing countries, those with communist governments, with single-party systems, have the highest participation of women among members of parliament: Cuba (34 per cent), China (21 per cent), North Korea (20 per cent), Viet Nam (18 per cent). There is some representation of women in Latin American parliaments but a minimal presence in some African and Asian countries: Kenya and Pakistan a token 1 per cent, Algeria 2 per cent.

From the considerable variety of evidence about the status of women in developed and developing countries, the main generalisation that can be made is that everywhere they are in some way underprivileged and underpaid, but the reasons vary from country to country. In the last resort it is a matter of speculation if, where and when perfect equality can be achieved.

1.5 Global issues

An introduction has been given in this chapter to the scope of geography. Problems related to the spatial organisation of human activities in over 200 countries of greatly different size have been discussed. In this section some of the issues that recur frequently in this book are outlined. They are numbered for convenience, but are not placed in any order of priority. They all relate with varying degrees of significance to the twelve major regions.

1. The great disparity in income per capita between the rich and poor countries of the world is sometimes referred to as the 'development gap' (see Box 1.2 for some guidance on the use of the term 'development'). The economies of many former European colonies, particularly those in tropical regions, had been managed in such a way that the export of primary products (see Glossary) was a priority, while manufacturing was prevented or discouraged. After the Second World War, the view was widely held that the 'developing' countries, many of them former or existing colonies of Europe, would sooner or later catch up economically with the developed ones. A contribution could be made to narrowing the gap through transfers of financial resources and technology from rich to poor countries. The evidence over the last five decades shows that the gap has tended to widen. For example, in the UNDP *Human Development Report 1992* (p. 1) it is pointed out that 'In 1960, the richest 20% of the world's population had incomes 30 times greater than the poorest 20%. By 1990 the richest 20% were getting 60 times more.' International bodies such as the United Nations now seem resigned to damage limitation with regard to the 'development gap'. Box 1.3 shows how basic the aims are.

2. Sporadic concern had been expressed about the availability of non-renewable natural resources for many decades. In the early 1970s the idea that there were 'limits to growth' suddenly gained prominence and publicity through the huge rise in oil prices in 1973 and the publication of several influential documents and books on the subject. Growing concern since the

BOX 1.2

Political correctness with special reference to development

In the 1980s it became customary to question and even ban the use of terms that could be offensive to certain subsets of the human race. Beard and Cerf (1992) provide numerous examples. Sensitive issues that affect large sections of the population include gender, physical build and skin colour. In the USA variations on a theme include the terms 'black', 'non-white', 'Afro-American' and 'negro' referring to citizens all or some of whose ancestors came from Africa. 'Eurocentrism' is a concept to be avoided, e.g. describing the Near East, Middle East and Far East in terms of their distance from Europe.

In the context of this book the author feels compelled to justify his choice of terms related to development, a concept generally referring to the material levels of different regions. Terms include 'more' and 'less' developed, underdeveloped, misdeveloped (of the USSR), developing and even undeveloped. The term 'develop' is usually economic in flavour, assuming 'progress' through a number of stages. It can be taken as offensive by members of 'less developed' countries. Developed is sometimes replaced by industrialised or modernised. Although the term Third World (and Fourth World) avoids the word 'development', its origins were ideological and political rather than economic, since it consisted of the 'neutrals' (including e.g. Egypt and Yugoslavia (*sic*)), left over after the allocation of countries to the First World (western, capitalist, market economy) and the Second World (eastern, communist, command economy).

North and South, which have more of a geographical connotation, are preferred by some people because they do not have economic or political connotations. These compass points are not used in this book because they are geographically incorrect, and this is a geography book. When North and South are changed to northern and southern hemispheres for rich and poor it is time for all good geographers to point out that about three-quarters of the population of the developing/Third/South world actually live in the northern hemisphere (consult a globe!).

Presumably with the general acquiescence of its member states, the UN uses the terms developing and developed regions/countries/economies (e.g. *UNYSB* vol. 38, p. 1078 for definition) and reference is made, for example, to 'development assistance'. The UN terminology will be used in this book.

TABLE 1.6 Selected man-made disasters

	Location	Origin	Product(s)	Deaths	Population injured	Evacuated
1974	UK, Flixborough	Explosion	Cyclohexane	28	104	3,000
1976	Italy, Seveso	Air release	TCCD (Dioxine)	–	>200	730
1979	USA, Three Mile Island	Reactor failure	Nuclear	–	–	200,000
1984	Mexico, San Juan Ixhuatepec	Explosion	Gas (LPD)	>500	2,500	>200,000
1984	India, Bhopal	Leakage	Methyl isocyanate	2,800	50,000	200,000
1986	USSR, Chernobyl	Reactor explosion	Nuclear	31	299	135,000
1988	UK, North Sea	Explosion on oil rig	Oil, gas	167	–	–
1989	USSR, Acha Ufa	Pipeline explosion	Gas	575	623	–
1989	USA, Alaska	Exxon Valdez tanker	Oil spillage		Environmental	

Source: OECD (1991), pp. 200–3

1960s has been expressed about the impact of human activities on the natural environment, the degradation of vegetation and soil, and the emission both locally and globally of various harmful substances into the atmosphere and the waters of the world. Table 1.6 contains some examples of man-made catastrophes that have caused loss of life, as well as damage to human works and activities and to the natural environment. It has been argued by scientists for two decades that a substantial rise in the temperature of the atmosphere could cause ice to melt from the ice sheets of Antarctica and Greenland, and sea level to rise, threatening coastal populations. Another view (Ryan 1994) is that warmer conditions would result in an increase in the moisture-carrying capacity of the atmosphere and even lead to a drop in sea level as snowfall increases and accumulates on the existing ice sheets.

BOX 1.3

The human dimension of development

In the first *Human Development Report 1990* (*HDR 1990*) it is stated that the United Nations Development Programme (UNDP) will produce an annual report on the human dimension of development.

> The central message ... is that while growth in national production (GDP) is absolutely necessary to meet all essential human objectives, it is important also to study how this growth translates – or fails to translate – into human development in various societies. Some societies have achieved high levels of human development at modest levels of per capita income.
>
> (*HDR 1990*, p. iii)

Health and education, as well as basic material needs such as adequate food and water supply, are the focus of attention.

The setting of global targets for socio-economic progress has advantages and disadvantages. According to *HDR 1990* (p. 67) global target setting should be realistic and operational. The following global targets for the year 2000 do exist, although, in the view of the author, present global trends make their achievement highly unlikely.

- Complete immunisation of all children
- Reduction of the under-five child mortality rate by half or to 70 per 1,000 live births, whichever is less
- Elimination of severe malnutrition, and a 50 per cent reduction in moderate malnutrition
- Universal primary enrolment of all children of primary-school age
- Reduction of the adult illiteracy rate in 1990 by half, with the female illiteracy rate to be no higher than the male illiteracy rate
- Universal access to safe water

Modest though the above targets are if compared with standards in the richer countries, progress in the poorest countries is at least hoped for. That is not so for the physical environment. In *HDR 1990* (p. 7) it is stated: 'But the concept of sustainable development is much broader than the protection of natural resources and the physical environment. After all, it is people, not trees, whose future choices have to be protected.' This anthropocentric view of the world is repeated later, presumably in case the reader missed the first statement (*HDR 1990*, pp. 61–2): 'After all, it is people, not trees, whose future options need to be protected.'

TABLE 1.7 The estimated and expected population (in millions) of the world and the continents in selected years from 1500 to 2025

	1500	*1750*	*1900*	*1950*	*2000*	*2025*
The world[1]	425	720	1,625	2,500	6,178	8,425
Europe[2]	81	140	390	515	739	756
Asia[2]	280	495	970	1,450	3,689	5,026
Africa	46	65	110	205	886	1,552
The Americas	14	16	145	325	833	1,053
Oceania	0.5	0.7	5	11	30	39

Notes:
[1] World totals do not tally precisely with the sum of the populations of the continents for reasons not explained in the sources
[2] Former USSR divided in 2000 and 2025 into 75 per cent Europe, 25 per cent Asia
Sources: McEvedy and Jones (1985) for 1500, 1750, 1900, 1950; *WPDS 1988* for 2000; *WPDS 1993* for 2025

3. Population growth is blamed for many of the problems of the world outlined in the previous two paragraphs. Population is growing fastest in many of the poorest countries of the world, those at first sight least able to support it. The increase in population requires an equivalent increase in the production and consumption of sources of energy and of raw materials even to keep present levels per inhabitant the same. Economic growth is the aim of many governments. Similarly, there are more people to 'produce' waste and to generate pollution indirectly through industries and other activities and directly from their homes. Table 1.7 shows the estimated and expected population of the continents of the world at intervals from 1500 to 2025.

4. The prospect of a third world war has influenced world affairs profoundly since the late 1940s. The prospect of a nuclear conflict causing loss of human life, damage to structures, and a long 'nuclear winter' caused by the blacking out of the sun by material in the atmosphere preoccupied the world for more than three decades. One positive recent change has been the 'official' ending of the 'Cold War' by Russia in the late 1980s. Nuclear weapons are still available in a considerable number of countries, and it would be premature to exclude the possibility of their use in the future. Meanwhile, the peace dividend may produce limited savings in expenditure on arms, but since numerous regional and local conflicts continue, no major power is in a hurry to disarm completely. Switching output quickly from arms to products for civil use is not easy.

Given the various experiences of the last fifty years,

BOX 1.4

The hazards of forecasting: six examples

1. Spokesman for Daimler-Benz, *c.* 1900: 'There will probably be a mass market for no more than a thousand cars in Europe. There is, after all, a limit to the number of chauffeurs who could be found to drive them.'

2. J. P. Cole, Lima, Peru, 1955: 'The population of Greater Lima (then 1.2 million) could exceed 5 million by the year 2000.' This forecast was dismissed with scorn in the Peruvian press. (Five million was passed in about 1985.)

3. Sir Harold Spencer Jones, Astronomer Royal, 1957: 'Man will never set his foot on the moon.' (Proved incorrect in 1969)

4. Education Minister, Margaret Thatcher on BBC-TV's 'Blue Peter', 1973: 'I do not think there will be a woman prime minister in my life-time.' (Proved incorrect in 1979.)

5. P. W. Richards, 'The tropical rain forest', *Scientific American*, December 1973, p. 58: 'One of the oldest ecosystems and a reservoir of genetic diversity, the wet evergreen tropical forest is threatened by the activities of man and may virtually disappear by the end of the 20th century.' (By 1994 only about 10 per cent of the vast tropical rainforest of Brazilian Amazonia had been cleared.)

6. A. R. Flower, 'World oil production', *Scientific American*, March 1978, p. 42: 'The supply of oil will fail to meet increasing demand before the year 2000.' (In 1980 world oil reserves were 90 billion tonnes and annual production 3 billion, 'life' expectancy without new reserves therefore being thirty years, but in 1993 there were 137 billion tonnes of reserves, production was 2.8 billion, life expectancy forty-three years.)

Postscript: The danger of 'crying wolf' too often can dull the message of ardent forecasters, who often foresee doom unless their recommendations are taken on board at once. M. Mesarovic and E. Pestel (1975), for example, stated in *Mankind at the Turning Point*: 'Starting early enough on a new path of development can save mankind from traumatic experiences if not from catastrophes.' In spite of the urgency to change the ways of the world, nothing much has happened in two decades to change the situation, yet disaster has not struck globally.

is it possible to speculate constructively about the future? The present is a moving 'line' cutting across time, separating what has happened from what will happen. For convenience the present is often itself taken to be a short period of time. It can be argued that it is of more practical value to know what will or could happen in the future than to study the past. Alternatively, the study of the past can be justified as of practical value if it throws light on prospects for the future. There is little that is certain about the future, although it is reasonable to make such assumptions as, for example, that the solar system will continue to function unchanged for the foreseeable future and that humans and other species are mortal. The problems of speculating about the future are discussed in more detail in Chapters 17 and 18. Box 1.4 contains some serious and less serious forecasts and predictions, the aim of the examples being to show how easily people can get things wrong.

1.6 Global solutions

In the previous section, some of the issues and problems that concern all or large parts of the world were briefly outlined. Attitudes to the issues, and solutions proposed to confront the problems, will now be discussed. Two fundamentally different attitudes to the world and to human existence stand in contrast. The first, represented by several widely practised religions, assumes the presence of a super-being (or beings), with humans central in the universe. The other assumes that humans are just one of many species and that they are not central in the universe. Neither can be proved or disproved, but they *are* mutually exclusive.

According to many religions, life on earth is only a part of a grand scheme, or a stage in a progression towards a higher form of life. Guidelines (such as the Ten Commandments in the Old Testament, the five Obligations of Muslims) are provided for human conduct. Reference is made at appropriate places in this book to some of the assumptions made by different religions that affect the varied geography of the world's regions. In most religions there is a tendency for material possessions to be regarded as of limited importance and for a fatalistic attitude to prevail towards natural disasters or conflicts.

Science is concerned primarily with increasing the understanding of how the natural world works. Scientific discoveries can lead, through the development of technology and its application in inventions, to both beneficial and harmful changes in people's life styles and material well-being. Here again, humans are in practice regarded as the special species, for which other species, as well as natural resources, can be used or exploited, depending how one sees it, by the human species. Given the enormous size of the universe and the minute fraction of the time since the earth came into being that humans have existed, it hardly seems reasonable to assume that they are any more important, except in their own eyes, than other species. They just happen to hold the cards in the game of exploiting the earth's natural resources. While religions provide rules or guidelines for human conduct, science focuses on the acquisition of knowledge and technology to improve material conditions. The two approaches are closely related, and together cover many of the economic, social, political and cultural aspects of human activity.

Since the latter part of the nineteenth century, attention has focused on the contrasts between two fundamentally different approaches to the organisation of human activity. Capitalism is the term widely applied now to a market economy, in which the state plays a limited role and most of the means of production are in private hands, usually very unevenly (or unequably) shared out. The interests of the individual are stressed. Socialism (or communism) refers to a command economy, in which the state, or the public sector, owns and controls most if not all of the means of production and, through central planning, organises production. Strictly, as pointed out by Wallerstein (1983, p. 13), when describing capitalism as a historical social system, 'all historical systems back to those of Neanderthal man could be said to have been capitalist, since they all had accumulated wealth that incarnated past labour'.

From the various views of the world briefly introduced above, a large number of different explanations have been offered as to how human society (as opposed to the natural world) works, and how human activities should be managed. In considering the validity of the examples given below, the reader should be careful to distinguish between fact, speculation and opinion. Care should also be taken to bear in mind the temptation to produce and to adhere to a reductionist or procrustean viewpoint or framework, simple and perhaps oversimplified, in which one or a few influences and variables account for all or much of human activity. Here it is proposed that two issues have a large role in several of the '-isms' to be described: economic growth is vital (the 'cake' should be bigger), and material production should be shared out in society reasonably equably (the problem of how to 'slice the cake').

1. **The biological approach** Like other species, humans are tied to the availability of natural resources. These may run out (e.g. minerals) or may be required to support too many people (e.g. land). Access to and the acquisition of territory and its natural resources are major influences in human activities.

2. **Environmental concern** Some now argue that the prime concern of human organisation and activity should be protection of the natural environment. Development should be sustainable. For example, forests should be cut only at the rate at which they can be replaced; fishing should likewise have regard for fish stocks. What, then, should be done about exhaustible natural resources, particularly minerals? It is difficult to equate the drive to achieve economic growth (a bigger 'cake') with conservation of the environment.

3. **The economic approach** Human activities and relations are determined by economic forces, in particular the ownership of means of production, the central principle of marxism. Society is evolving, it is argued, through stages towards the ultimate form, communism.

4. **The role of science and technology** Changes in the organisation of human societies are influenced by changes in technology. Advances in medicine lead to improvements in healthcare. Animals and inanimate sources of energy (wind, water power, fossil fuels) used to drive machines replace and greatly exceed the strength of humans themselves. The human senses have been extended, vision by the invention of microscopes and telescopes, hearing by the telephone. The capacity of the human brain in the sorting and storage of information and in making calculations has been enormously enhanced in some areas by the development of electronic computers. Computers have, for example, been used since the 1950s for complex global models of physical processes, have been blamed for catastrophic events in the world stock market, and were crucial to the abortive 'Star Wars' (Ballistic-Missile Defense) programmes. Breakthroughs in astronomy, physics and chemistry from the sixteenth to the eighteenth century led to a broad appreciation of the working of the solar system and the nature of matter. Through discoveries in geology and biology long timescales were properly appreciated and the statements in the Bible about the timing of the creation were finally attributed to the realm of allegory, subject to a different set of criteria.

5. In *The Causes of Progress*, Todd (1987) stresses a different set of major influences on demographic trends and on progress in general in different regions of the world. In his view, family structure, especially age at marriage or childbearing, and the level of education of women compared with that of men, are crucial features of progress.

6. **Cultural features** Common language, religion and tribal allegiances should not be ignored. The collapse of the ethnically diverse Soviet Union, the existence of which it was assumed would continue because *Homo sovieticus* behaved according to economic influences, a marxist view, resulted in the resurfacing of strong nationalist feelings focused on language groups. In Africa, the underlying tribal systems have frequently reasserted themselves, as for example in South Africa and Rwanda in 1994. Long past battles are re-fought. The Serbs of Yugoslavia recall their defeat by the Turks at the Battle of Kosovo in 1389, with great bitterness, and the Poles have clung to their national identity against great odds. In the European Union, the UK and Denmark – both of which have tended to be on the margin of West European affairs through the centuries – are among the least willing Member States to accept changes expected to lead to greater integration and loss of sovereignty and individuality.

7. Do individuals play an important role in history, or are the political and military leaders just puppets or robots following paths predetermined by underlying economic and social forces? Are successful scientists and inventors just the fortunate ones, alive in the right place when the next discoveries and inventions are ready to be made? If Hitler had not come to power in Germany, would German expansionist policies and preparations have been built up in the 1930s as they were by the Nazis, and would the Second World War have take place?

Grand theories and models of human society should be accepted with reservations. In his satirical book *Sartor Resartus*, published in 1836, Thomas Carlyle did not find it difficult to argue the case that the wearing of clothes was the outstanding feature of the human species and that human behaviour was governed by what was worn. Astrologers seriously claim to account for much of human behaviour through the positions of the planets and stars. There is an ongoing dispute about the relative importance of genetic characteristics and the external influence of the environment on human behaviour, the 'nature versus nurture' debate.

By the 1990s, many matters of earlier speculation and controversy have been resolved, but many others remain. Indeed, much that is taken for granted in one century may be overturned in the next, a situation applicable both to the way the natural world 'works' and to the way human society should be organised. It is therefore appropriate to consider whether past situations and standards are being judged by late twentieth-century standards.

1.7 Technical aspects of the study of world regions

Living on a spherical surface

Gottmann's analogy of the globe and a billiard ball cited at the beginning of this chapter draws attention,

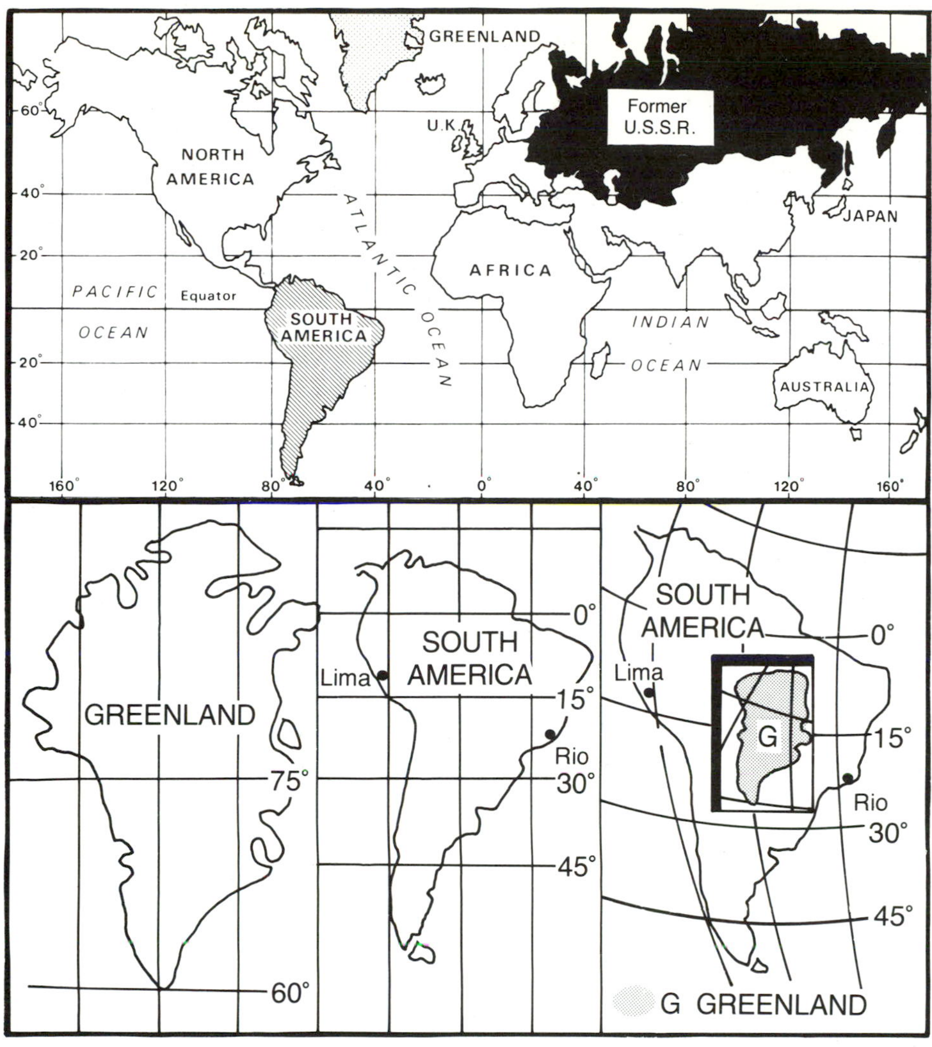

FIGURE 1.4 Effects of different projections:

a The world on Mercator's projection. On this projection, shape is correct but scale becomes increasingly exaggerated towards the North and South Poles

b *Left, centre*: Greenland and South America are compared as they appear on Mercator's projection. *Right*: as they are with the scale correct

perhaps unintentionally, to the fact that the earth's surface is finite in area, but unbounded. For both cartographical and geopolitical viewpoints there are implications and complications from this situation.

Cartographically it is a challenge to find how best to reproduce part or all of a spherical surface on the flat surface of a map. A useful analogy is to consider the problem of flattening a piece of orange peel. Either bearings and therefore shape, or distance and therefore scale, or both, will be distorted on a flat map, but the distortions are slight for the map of a country of moderate size and tolerable for an area the size of the USA or Australia. In Figure 1.4 selected parts of the world are compared on different projections. Most of the maps of the world used in this book give priority to correctness of area rather than of shape.

One feature of a spherical surface is the fact that there is no centre on the whole of the earth's surface, as there is for a bounded portion of it, such as the USA. On a flat unbounded surface, expansion could go on indefinitely outwards in all directions from a given centre, every given distance added taking in a larger area than the previous one. On a spherical surface, once expansion from a point has reached out far enough to occupy half of the total surface (i.e. a hemisphere), the length of the 'frontier' would shorten with added distance.

Just as mapping the spherical surface poses cartographical problems, so distance can have several different meanings to the geographer. The shortest distance over a flat surface between two points is a straight line, but in the real world, people, goods and information rarely move between two points exactly along such a line. According to the rules of the spherical geometry of the surface of the globe, the 'shortest' distance between two points without going through the globe itself is a 'great circle', part of the circumference of any circle on the earth's surface with its centre at the centre of the earth. The path can be found roughly by placing a piece of string on the surface of the globe between any two places and adjusting it until it is at its shortest between them. Meridians (lines of longitude) are great circles,

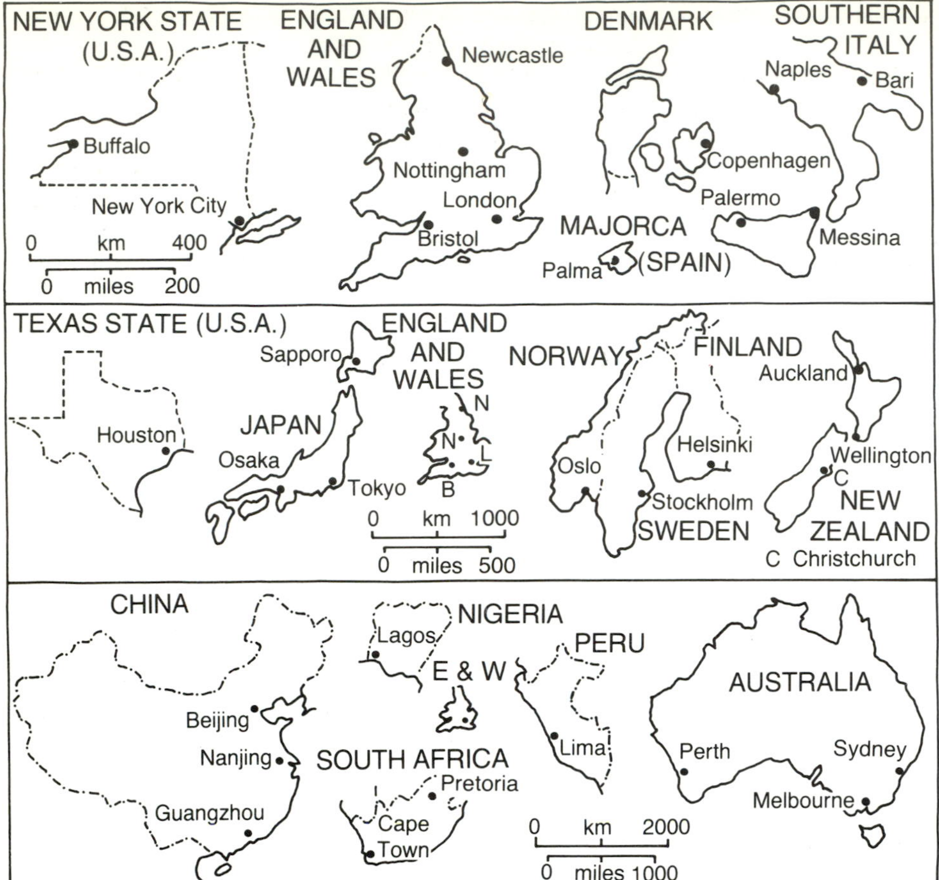

FIGURE 1.5 Selected countries of the world and states of the USA compared on three different scales

BOX 1.5

A Comedy of Errors

'Who suppes with the Devylle sholde have a long spoone' ('A Lay of St Nicholas', in *The Ingoldsby Legends*) is a useful thought to keep in mind when dealing with any kind of technical and numerical information. Below are a few examples of errors found by the author.

1. In the *1979/80 Statistical Yearbook* of the UN (New York 1981) the author spotted two gross numerical errors. In Table 125, 'Motor vehicle assembly' (p. 365), Zaire is credited with assembling astronomical numbers of motor vehicles in the 1970s, for example some 450,000 in 1970 and over 2 million in 1975. Zaire did not actually assemble any motor vehicles! The rogue line of data may have 'escaped' from some other country. In another table in the same Yearbook, it is shown that in the mid-1970s there were fewer than 200,000 motor vehicles in use in Zaire. Where did all the assembled vehicles go?

2. In Table 193 of the same *Yearbook*, 'Hospital establishments and health personnel' (p. 844), Senegal had only ten people per physician in 1977 (the USA figure was 569). Since Senegal actually had only 334 physicians and a total population of over 5 million, the population:physician ratio should have been about 1:15,000, not 1:10.

3. More recently, the *1990/91 Statistical Yearbook* of the UN (New York 1993) contains at least one error, the population of Afghanistan (Table 11, 'Population and human settlements', p. 62). The population for 1985 is given as 18,136,000 while the estimate for 1990 has dropped to 1,612,000, a striking reduction not even the conflict there in the 1980s could have produced. A slippage of a zero from the end of the number? How can one trust any number? Considering the *Statistical Yearbooks* of the UN contain around a quarter of a million pieces of numerical data and perhaps a million actual digits, it is not surprising that some printing errors occur. The user would do well to attempt to work out roughly what the size of the number is likely to be before relying on any particular bit of data.

but, apart from the Equator, parallels (lines of latitude) are not.

In practice, journeys are rarely able to follow either kind of shortest distance described above. Fixed links such as railways, roads, pipelines and electricity transmission lines usually have bends and may be forced to make detours round obstacles such as mountains, inlets of the sea, or cities, with the result that travel distance may be considerably greater than direct distance. Even shipping and air routes, which do not follow fixed links, may make detours. This may be because of traffic control requirements to use given lanes, or to avoid physical obstacles, especially by shipping, or political obstacles, particularly for air routes, which may be forced to avoid hostile countries.

Three other kinds of distance may influence the choice of mode of transport used and route followed on given journeys: time taken (speed of travel), cost of travel, and a more abstract set of considerations related to comfort, stress (e.g. not having to change) and perceived distance. Over longer distances, for those passengers who can afford it, air travel is preferable to road and rail travel. For the long-distance movement of bulky goods, sea transport is comparatively cheap and therefore preferable to rail (if available) or air, assuming that time is not a constraint. For the movement of many kinds of information, 'journey' and time distance are no longer an issue, so cost is the dominant consideration. Yet another kind of distance of a more abstract nature is economic distance. The gap in income per inhabitant between rich and poor countries is an issue of global significance, one that will be referred to frequently in this book.

In some atlases it is customary to give the user an idea of the comparative area of a given country or region that he or she is considering by showing a familiar country (e.g. his or her own) on the same scale at the side of the main map. There is a story that during the First World War German leaders were complacent about going to war with the USA in 1917 because it appeared in German atlases to be the same size as Germany, each occupying one page. Figure 1.5 shows the comparative sizes of a selection of countries and regions. On some of the maps of different parts of the world in this book, familiar countries or US states are included for comparison.

Handling and interpreting numerical data

Many questions of an abstract and general nature, especially involving opinion and speculation rather than fact, are not measurable numerically. On the other hand, the economy of a country cannot be run without the use of quantitative data. The reader is reminded that the term 'statistics', first used in 1787, originates from 'state', that is data collected for the state to use.

4. When incorrect numbers and instructions are given, results can be devastating. Rounding is often useful to give a clearer idea of a number and avoid spurious precision, but to round 999 to 1,000 when making an emergency telephone call would be disastrous. When aircraft were refuelled it used to be customary to check the level manually. Planes have run out of fuel in mid-ocean because gauges have failed mechanically, erroneously reading full before take-off. H. Lin (1985) gives two examples of errors to illustrate the problems facing the development of a foolproof Ballistic-Missile Defense system ('Star Wars'), especially the scope for errors in the required 10-million-line software program. During the Falklands War (1982), the British warship *Sheffield* was sunk by a French-manufactured Exocet missile launched by the Argentinians. The ship's radar warning systems were programmed to recognise the Exocet missile as 'friendly' because this type of missile was used by Britain itself. An example of a mission failure due to a sloppy numerical instruction had less lethal consequences. 'In 1985 the crew of a US space shuttle was to position the shuttle so that a mirror mounted on its side could reflect a laser beamed from the top of a mountain 10,023 feet above sea level.' The elevation was misinterpreted in a computer program as being 10,023 nautical miles.

The recommendation is: be watchful. Otherwise misunderstanding instructions can lead to bizarre situations, as with the presence of two rows of elegant early nineteenth-century town houses built in the Lincolnshire village of Wainfleet All Saints in the heart of the English Fenlands from architects' plans intended for Stepney in the heart of London.

When using numerical data it is useful to keep in mind a checklist of points to query. What unit of measurement is being used – for example, miles or kilometres, pounds or kilograms, Fahrenheit or Centigrade? At least the above measures remain constant, whereas, as noted below, currencies inflate or deflate, requiring particular care in comparisons through time. Is the number an absolute amount or is it a ratio such as a quantity per inhabitant or a proportion of a notional quantity such as a percentage? How accurately was the information collected? Does the number have the right number of digits? Is the number above your 'numeracy ceiling' whereby anything over a thousand, a million, is just big? If so, work on it.

On international and world scales production, consumption and even the quality of life are quantified and compared for the deliberations of such international bodies as the World Bank and the UN. In this subsection attention is drawn in particular to some of the pitfalls often encountered in the handling and interpretation of numerical data. The reader may prefer to return to it after proceeding further in the book.

In practice, many data sets are estimates, others even falsified or fictitious. For example, in the United States census of 1960 (taken on 1 April) 179,323,175 people were enumerated and that figure was published officially, but quite spuriously, as the total population of the country on that day. Post-census checks in enumeration districts showed that the enumerators had actually missed out about 3.1 per cent of the total population of the USA or about 4.5 million people (Hauser 1971). Much greater levels of error have recently been found in population estimates and census results from some African countries. No less disturbingly, misprints of numbers are not uncommon even in the more responsible newspapers, and they occasionally occur in such prestigious publications as the *United Nations Statistical Yearbook*. Two precautions can be taken to help in the use and interpretation of numerical data sets. First, it is useful to have a rough idea of what a number in question should be. Second, it is an advantage to round numbers, removing superfluous digits, which confuse the eye as it passes along a row or down a column of numbers. Box 1.5 contains some examples of numerical errors in published sources.

Care should also be taken in the interpretation of numerical data that refer to aggregations of many units. For example, a fairly common demographic statistic is life expectancy, the average age at which the individuals of a given population die. In practice, a considerable number of deaths take place shortly after birth, while most deaths are spread over a span of age groups above the age of 60. When national income is expressed in terms of *per capita* (see Glossary) or *per inhabitant*, usually the majority of the population have less than the average, while some have many times more than the average. There is a story of a south Italian farmer whose comment when told that EEC statisticians calculated that on average each Italian consumed 1.3 rabbits a year was: there are no rabbits around here.

A serious problem in the measurement in money terms of production, consumption, transactions and other aspects of the economy of countries or regions is the propensity of exchange rates between different currencies to alter frequently and sometimes drastically. This question is illustrated in Table 1.8 by GDP, the total value of goods and services produced in an economy during a given period, usually a year. Each year during the period 1982–90 the GDP of the UK in billions of pounds was as indicated in column (2) of Table 1.8. Fluctuations about the average were even greater over shorter periods within each year than from year to year. Thus, because the pound weakened against the dollar between 1982 and 1984, it appeared that the GDP of the UK actually fell, if measured in dollars. Later, from 1988 to 1990, it appeared to grow very fast. International comparisons of production, consumption, trade and other economic indicators are for convenience often made in US dollars, but in the European Union, the ECU (European Currency Unit) is

TABLE 1.8 US dollars and UK pounds compared, 1982–90

	(1)	*(2)*	*(3)*
		UK GDP in billions of	
	Dollars per pound	*pounds*	*dollars*
1982	1.75	278	487
1983	1.52	303	460
1984	1.34	324	433
1985	1.30	355	460
1986	1.47	382	560
1987	1.64	420	688
1988	1.53	466	711
1989	1.64	510	836
1990	1.79	549	980

Source: UNSYB 90/91 Tables 29, 30, 126

TABLE 1.9 Comparison of exchange rate and purchasing power parity (ppp) dollars in selected countries

	1987 US dollars per capita		*1989 US dollars per capita*	
	GNP	*GNP in ppp dollars*	*GNP*	*GNP in ppp dollars*
USA	18,530	17,620	21,100	21,000
Japan	15,760	13,140	23,730	14,310
Switzerland	21,330	15,400	30,270	16,840
Germany	14,460	14,370	16,200	14,510
UK	10,420	12,270	14,750	13,730
Spain	6,010	8,990	9,150	8,720
Brazil	2,020	4,310	2,550	4,950
Nigeria	370	670	250	1,160
India	300	1,050	350	910
Indonesia	450	1,660	490	2,030
China	290	2,120	360	2,660

Source: UNDP, *Human Development Reports 1990* and *1992*, Table 1

now used for some purposes in place of the national currencies of the Member States, yet another source of confusion.

Exchange rates between the US dollar and many other currencies fluctuate even more widely than those between the USA and the UK. On the other hand, some are fixed against the dollar at an artificially low rate. To make the GDP of different countries more comparable with regard to what can be bought *in a given country* with the national currency, a measure referred to as ppp (purchasing power parity) has been devised. 'Real' GDP, as the device is also termed, removes the effect of exchange rates when conversion is made to the dollar for comparative purposes. Table 1.9 shows the effect of the application of the above calculation to the GDP of selected countries. GDP is expressed in per inhabitant terms here to eliminate the effect of the different size of countries.

Given the importance of national accounts statistics in the study of regional differences and developments, attention is drawn now to some aspects that deserve thought.

- Total population is changing markedly in almost every country of the world, and is growing fast in many of them. Comparability is made easier if GDP and other items of production are expressed in per inhabitant terms. The direct effect of the population size of countries is thereby removed.

- Inflation of the currency is common and should be taken into account when comparisons are made through time. Thus, for example, in Table 1.8, it appears that the total GDP of the UK almost doubled between 1982 and 1990 from £278 billion to £549 billion, but almost all of the apparent increase was due to inflation.

- The contrast between GDP according to exchange rates with the US dollar and real GDP as calculated to achieve purchasing power parity is evident in Table 1.9. For example, in 1987 and in 1989 the GDP of many of the developing countries was greatly understated by the use of the exchange rates of the time. Thus the US figure of 21,100 dollars per inhabitant in 1989 was almost sixty times as high as China's 360 whereas, in real terms, the US 21,000 against China's 2,660 was only about eight times as high. The effect of reconsidering the gap between rich and poor countries in terms of real GDP has been in many cases to reduce it appreciably, a feature noted in China with some dismay. At the higher level China's case for receiving aid seems less convincing than at the lower level, although alternatively the higher level could show the achievements of the Communist Party in a more favourable light.

Fact, opinion and speculation

At the end of each chapter all or some of three sets of items are provided for possible further work, thought and discussion. They are key facts, matters of opinion and subjects for speculation.

- Facts are not always accurate or even correct but are assumed to be so either because they are self-evident or for convenience. For example, at one time the surface of the earth was considered (for a fact) to be flat, while many pieces of 'official' numerical data on demographic matters are estimates.

- 'Opinion' is the term widely used for statements that more appropriately could be termed 'speculation'. Here opinions are regarded as individual or collective views on moral or ideological issues, as for example whether (or not) economic production should be shared out reasonably equably in a given society.

- 'Speculation' refers in this book to statements that could become facts if more information were available. Speculation may be about the past (were the Americas populated exclusively by people from Asia?), the present (are there remains of human settlement under the Antarctic ice cap?) and the future (will the population of the world exceed 8 billion in the year 2020?). To answer the first two questions is difficult, while the answer to the third should be known in due course.

It is not claimed that the three categories of fact, opinion and speculation cover everything, but it is considered that they are useful working categories.

KEY FACTS

1. Only about a quarter of the earth's surface is not covered by water or ice sheets.

2. Of the land area, little more than 10 per cent is regularly cultivated.

3. At the level of sovereign states the world is divided into about 240 political units of greatly varying area and population size. As a result, international comparisons are difficult. A few large countries tend to dominate world affairs.

4. The population of the world has increased about threefold in the last hundred years. The average consumption per capita of materials and goods has increased several fold.

5. Many natural resources have already been used and the rate of pollution of the environment has been increasing sharply both locally and globally.

6. The consumption of goods and services per capita in the world varies greatly from country to country and within countries.

MATTERS OF OPINION

1. What justification is there for arguing that the wealthy countries of the world should help the poor ones? This discussion can be resumed in Chapters 4 and 17 of the book, in which evidence is given of the huge disparities in the production and consumption of goods and services among the regions and countries of the world.

2. Should there be a global 'taxation' system to ensure transfers take place from rich to poor countries?

3. What are the relative merits and drawbacks of democratic and authoritarian systems of government?

4. What are the advantages and disadvantages of market economies and centrally planned/command economies?

5. Is family planning through the use of contraceptives justified?

SUBJECTS FOR SPECULATION

(Return to 1 after finishing Chapter 17)

1. How feasible is it to make the necessary transfers from rich to poor countries (a) to prevent the 'development' gap from widening further? (b) to narrow the existing gap?

2. Can a contribution to the material well-being of the citizens of this planet be expected from exploitation of (a) the moon (about 400,000 km or 250,000 miles away)? (b) the planets of our solar system (varying distances, several hundred times the distance of the moon)? (c) planets in other solar systems? The nearest star to our sun, not expected to have any planets, is 4 light years away, approximately 2.35×10^{13} or 23,500,000,000,000 miles, one billion times the distance round the Equator.

Chapter two

A brief history of the world since 1500

From the present time forth, in the post-Columbian age, we shall again have to deal with a closed political system, and none the less that it will be one of world-wide scope.

H. J. Mackinder (1904)

2.1 The world around 1500

The Europeanisation of most of the rest of the world will be described in this chapter at some length. In the view of the author, the present number, distribution and composition of the countries of the world cannot be fully appreciated without some knowledge of their origins. Each country bears traces of its history, some at least through several centuries. Sometimes the present attitude or 'behaviour' in a given country may be accounted for by some past experience.

In the fifteenth century several civilisations flourished in the eastern hemisphere or 'Old World', in Asia, Africa and Europe. These were known to one another but were linked only tenuously through trade, and transactions were slow and limited. In the western hemisphere or 'New World' there were two civilisations at this time, the Aztec in what is now Mexico and the Inca, centred on Peru in South America. The total population of the world at that time was about 425 million, roughly a tenth of its population in 1980. The population of the Americas was only about 15 million, while about 280 million, or almost two-thirds of the world's population, lived in Asia.

Figure 2.1 shows the distribution of economies of different kinds in the world around 1500. All depended heavily on the cultivation of field and tree crops, on natural pastures and on the forests and woodlands. For the most part, the food gatherers and pastoralists were very few in number, but they needed extensive territories. Fertile soils could support quite high densities of cultivators. The main civilisations of Europe, Africa and Asia in the eastern hemisphere were based on plough cultivation, with livestock making a major contribution to the economy in providing work animals, food and raw materials. Many of the more densely populated areas of cultivation depended on irrigation, as did parts of the Inca Empire in South America.

The two main civilisations of the New World of the Americas were largely based on hand cultivation with the hoe, and domesticated livestock had only a limited economic function. They lagged behind the civilisations of the 'Old World' in the application of technology and techniques; for example, the wheel had not been invented, sophisticated writing was not used, and iron was not smelted. In contrast, late in the fifteenth century, the Ming Empire of China was technologically ahead of Europe in many respects.

The Ottoman Empire, centred on what is now Turkey, and with its capital at Constantinople, had been expanding for some time before 1500, by which date it held extensive territories in northeast Africa and southwest Asia, as well as roughly the southeastern quarter of Europe, reaching close to Prague and Vienna. In the Middle Ages there had been frequent if tenuous trade links between Europe and China, but by late in the fifteenth century the Ottoman Empire and, further north, the Khanate of Kazan and other states hostile to Christian Europe lay across trade routes. The rivalry between Christian Europe and Islam meant that the trade routes were at risk and could be blocked.

The period of European exploration that began towards the end of the fifteenth century was to some

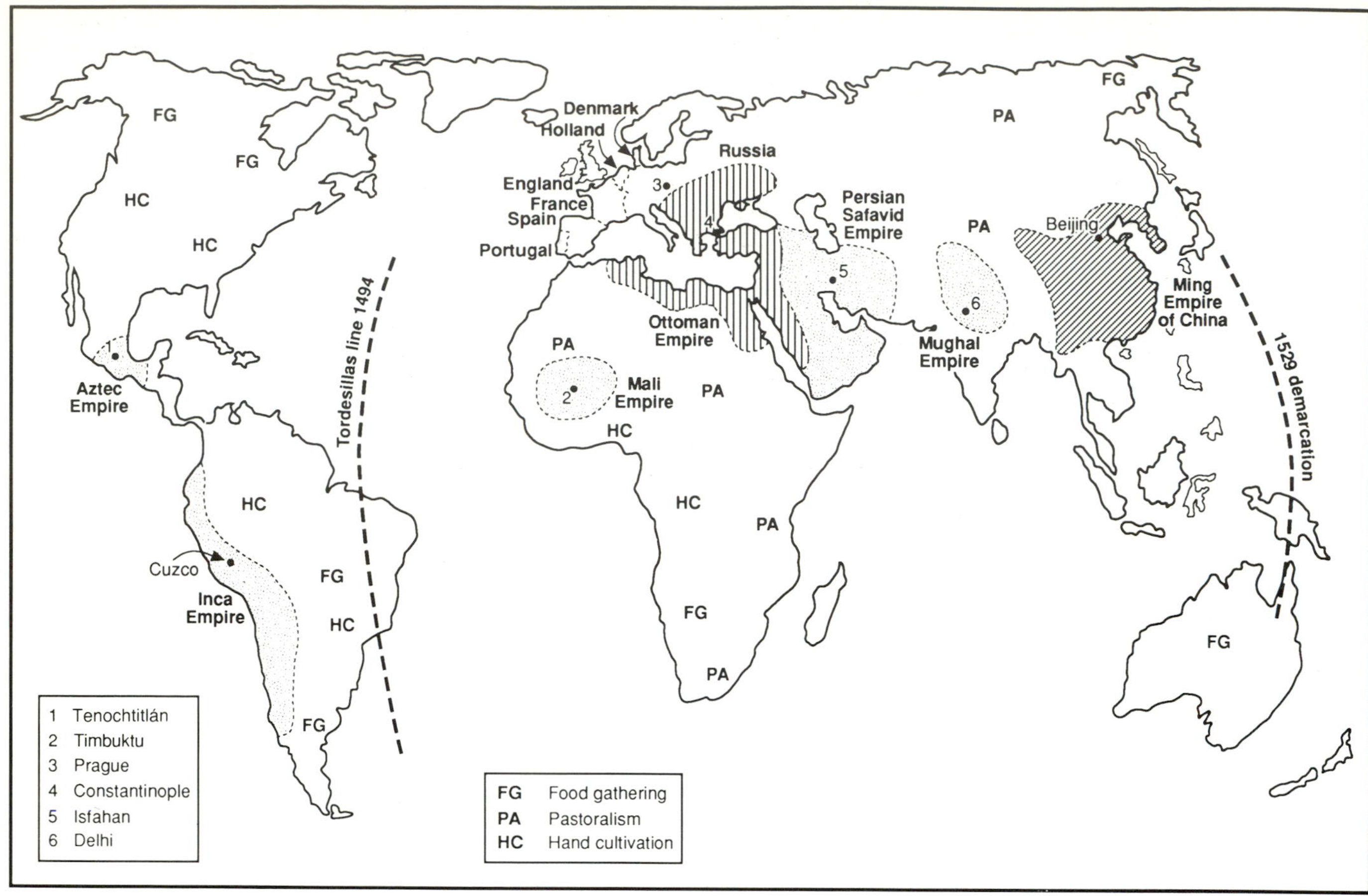

FIGURE 2.1 The state of the world around 1500. The predominant form of subsistence is indicated for areas of the world not included in the principal empires and European states of the time

extent a response to this situation, a result of the attempt to find trade routes across the oceans between Europe and southern and eastern Asia. For several centuries during the later Middle Ages the Christian culture of Europe had been confined to a small part of the world, but in the later part of the fifteenth century Europeans started to extend their influence across the oceans and across northern Asia. The present form and distribution of countries, religions, languages and economic systems throughout the world can only be fully appreciated through a study of the global spread of European influence during what is sometimes referred to as the Columbian era after the explorer Christopher Columbus.

Before the sixteenth century extensive empires covering large portions of the various continents had come and gone. Between 1405 and 1433 the Chinese admiral Cheng-ho 'explored' and carried the Chinese presence briefly into the Indian Ocean, as far as the shores of Africa and Arabia, but the possibility that China might have become the region of the world to build up a global influence ended abruptly as support for this endeavour suddenly stopped. It was the Europeans who made the breakthrough.

In 1522 the crew of the *Victoria* completed the first (known) circumnavigation of the globe, executed in a westerly direction. The ship set sail from Seville (Spain) in 1519 under the command of the Portuguese navigator and explorer Ferdinand Magellan (1480–1521), working then for Spain. Magellan himself was killed in the Philippines in 1521, but it is his name that is traditionally connected with the achievement. The return of the *Victoria* to Seville in 1522 symbolically 'completed' the work of various European explorers, most starting their voyages from Portugal or Spain (see Table 2.1 for selected dates of events). Arguably the two outstanding contributions were those of Christopher Columbus (1451–1506), the Italian (Genoese) explorer, sponsored by Spain, who reached the 'New World' from Spain in 1492, and Vasco da Gama (1460–1524) who, sailing from Portugal, rounded southern Africa to reach India in 1499.

Why did Portugal, Spain and then several other West European powers set out to conquer much of the rest

FIGURE 2.2 The political geography of Europe around 1500

of the world during the four centuries following the discoveries of the early explorers? Various reasons have been proposed to account for the acquisition of territory: the establishment of trading posts to exchange products, the acquisition of land to control agricultural and mineral resources, and in some areas the establishment of colonies for settlement by emigrants from Europe. Conquests of territory outside Europe by European powers continued from the sixteenth to the twentieth century, the Italian occupation of Abyssinia in the mid-1930s arguably being the last case of note. During this long period, European powers were frequently at war with one another both within Europe and elsewhere in the world, and colonies of one country were won or purchased by other countries, or exchanges were made.

Figure 2.2 shows the situation in Europe itself towards the end of the fifteenth century. The continent could be subdivided into three parts. The west-facing periphery had direct access to the Atlantic Ocean, while the east-facing periphery, controlled by Russia, bordered the nearly empty lands of Siberia. In contrast, the central part of Europe, bounded by France in the west, Russia in the east and the Ottoman Empire in the southeast, was less well placed to send explorers and settlers out of Europe. From the central area, indicated by fine dotted shading in Figure 2.2, many countries had access only to the Baltic, North Sea or Mediterranean, not directly to the Atlantic Ocean. Even Holland, when it emerged as an influential oceanic power in the seventeenth century, had access to the Atlantic only via the English Channel or North Sea, both routes being adjacent to England and easily obstructed.

The location of countries in Figure 2.2 shows that it would have been very difficult for a country such as Hungary or Poland to undertake the exploration and conquest of lands outside Europe, because other European countries could block the way. Indeed, when Sweden and, more recently, Germany attempted to conquer new territories, they tended to do so in Europe itself rather than elsewhere. Some areas like Iceland, Scotland, Italy and Germany were either too small in population or were politically too fragmented to take much part in the process of European expansion that was starting around 1500, but Germany and Italy were able to join in the carve-up of Africa late in the nineteeth century.

TABLE 2.1 Colonial calendar of selected key events, 1493 to 1775

1493	Columbus discovers islands in the Americas
1494	Treaty of Tordesillas establishes demarcation line dividing tropical world into Spanish and Portuguese domains
1499	Vasco da Gama reaches India
1519–20	Spanish invasion of Mexico
1519–22	Magellan-Elcano complete first circumnavigation of the world
1531–3	Spanish invasion of Peru, opening the way to the silver reserves of Peru and Bolivia
1530s	Portuguese settlements in Brazil
1552	Russians capture Kazan on the Volga, opening the way to eastward expansion into Siberia
1557	Portuguese settlement in Macao, establishing base for trade in the Far East
1571	Spanish found Luzon in the Philippines
Early seventeenth century	First English settlements in Virginia and New England
1608	French settlements in New France, now part of Quebec
1649	Russians reach the Pacific coast at Okhotsk
1697	Haiti ceded by Spain to France
1715	English East India Company secures concessions in India
1754	French renounce aspirations in India
1760	French lose all of Canada to the British

2.2 Europeanisation in the sixteenth to eighteenth centuries (see Table 2.1)

Figure 2.1 shows the meridian agreed by Pope Alexander IV to subdivide the world into Spanish and Portuguese spheres of influence. A treaty was signed between the two countries in 1494 in Tordesillas, a small town in Spain close to the Portuguese border. This was the first division of the world, although in effect it covered only the tropical world. Lands in the tropics west of a meridian passing 370 leagues west of the Cape Verde Islands (about 48° West of Greenwich) were allocated to Spain to colonise, while those to the east went to Portugal, superseding one chosen in 1493, only 100 leagues west of the islands. As the meridian actually passes close to the mouth of the Amazon, eastern South America was situated in the Portuguese sphere. The Tordesillas line in effect allocated a western hemisphere to Spain and an eastern one to Portugal. The division was completed in 1529 by the Treaty of Saragossa, whereby another meridian in the Pacific Ocean, off East Asia, gave the Philippines to Spain, but left the rest of Asia and the whole of Africa as a Portuguese preserve. The treaty was never accepted by the other West European powers.

As a result of the division of the tropics between Spain and Portugal the latter initially concentrated on the establishment of a route round southern Africa to gain access to the spices and manufactures of the East. Spanish explorers sought a route to Asia in a westerly direction from Europe, but actually reached the Americas, where Spain's energies were duly concentrated, although later a trans-Pacific link was established between Spain and what became its only Asian colony, the Philippines. The precious metals of Mexico and Peru remained its main economic interest outside Europe. In the sixteenth and seventeenth centuries both Portugal and Spain began to establish plantations to grow such crops as sugar, cotton and coffee in the Americas, especially in Northeast Brazil and the Caribbean, and slaves were brought across the Atlantic from Africa to supplement the supply of indigenous American Indian (Amerind) workers. Between 1580 and 1640 Spain and Portugal were united, and their colonial endeavours were conducted jointly. There has been widespread use of the term and concept 'globalisation' since the 1970s, but already around 1600 Spain and Portugal had a global colonial and trading system encircling the world completely, even if transactions and communication were far more limited and slower than today.

Following the lead of the two Iberian powers, England, France and Holland sent out explorers from Europe, but for a long time they failed to gain control of large territories, of extensive natural resources, and of subject populations whose labour could be harnessed. Not until the nineteenth century did the British and French Empires reach any very great territorial extent, by which time nearly every Spanish colony had already become independent.

Britain and France competed extensively in the eighteenth century for control of North America and India. Figure 2.3 shows the comparatively limited extent of the English colonies of North America before the War of American Independence (1775–83). They were in danger of being outflanked by French conquests in Canada, Louisiana and the Mississippi lowlands. For both England and France, their Caribbean colonies were of special importance. Haiti, in particular, was intensively developed by France in the eighteenth century with the use of slaves from Africa to grow tropical crops. The British likewise developed Jamaica and other islands. Great importance was given to tropical colonies, especially to forested areas, the dense vegetation of the tropical rainforest being attributed to the great fertility of the soil.

On the other side of Europe, another colonising

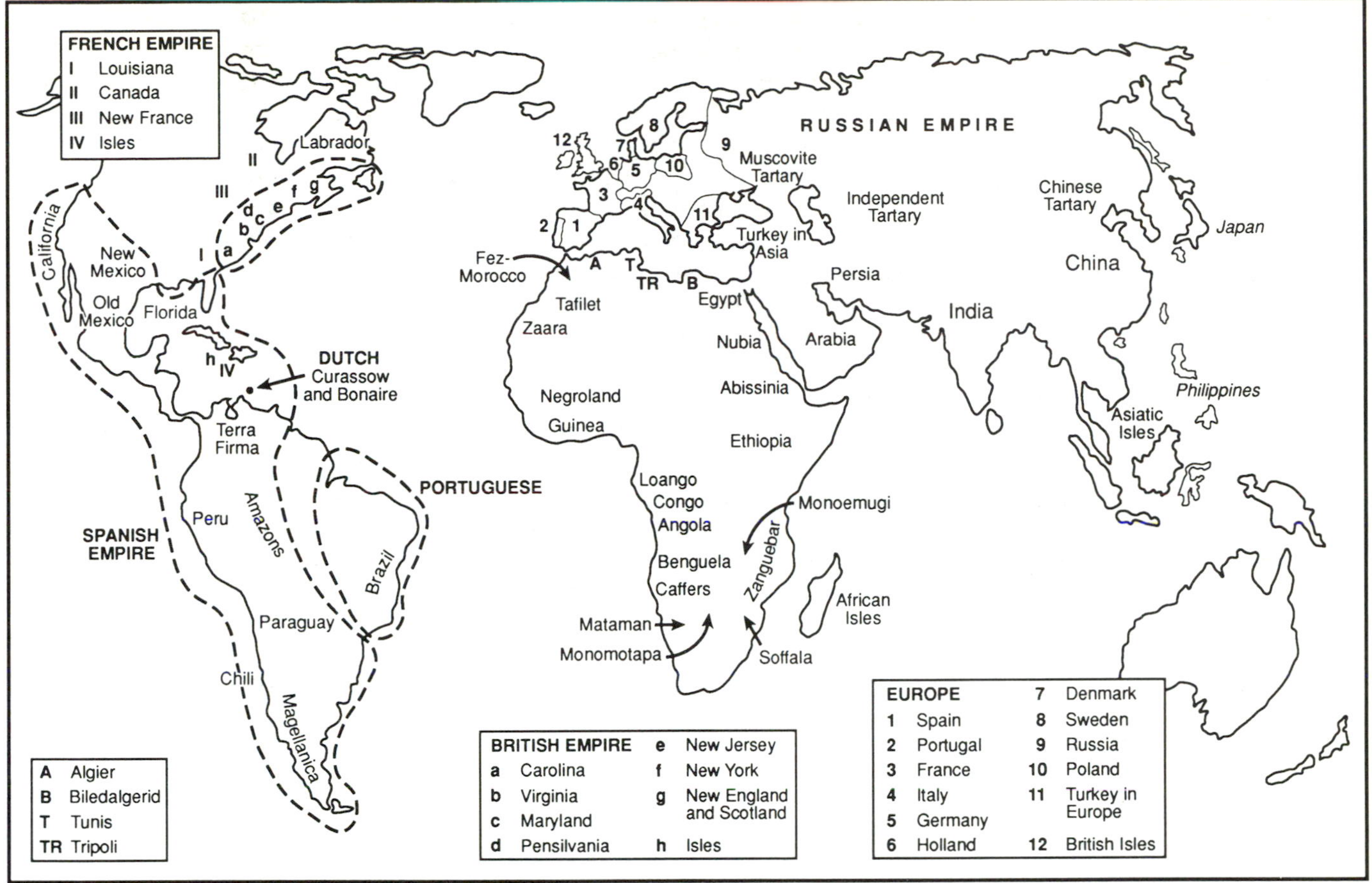

FIGURE 2.3 The world in the 1760s on the eve of the War of American Independence. Based on maps and a list of countries in *Encyclopaedia Britannica*, vol. 2 (Edinburgh, 1771, pp. 682–4). (Only the two italicised names have been added to the original)

power was at work. The influence of the Pope in Rome did not extend into the lands of Russia and its Orthodox Church. As Russia grew in strength in the fifteenth century, it developed its own policy of territorial expansion, conquering lands first to the north and then to the east. It had access to the Atlantic Ocean via Archangel and the White Sea, but this sea freezes for several months of the year. Russia did not use this difficult outlet to send expeditions to the New World or to tropical Africa and Asia. The conquest of Siberia was carried out almost entirely by land, with penetration facilitated by many stretches of river. By the middle of the seventeenth century Russia had outlets on the Pacific coast and in the eighteenth century was establishing settlements in what is now Alaska. A vast territory, at the time very thinly populated, was thus acquired, providing natural resources of enormous potential in the twentieth century. The lands conquered by Russia differed from those acquired by the West European powers in that they extended over a compact area, cut off from the rest of the world on most sides by mountains, deserts and seas frozen for much of the year. They could not immediately be made into profitable ventures, unlike the cotton and sugar plantations in the Caribbean or the gold and silver mines of Central and South America.

The map in Figure 2.3 is copied from one published shortly before the British colonies in North America became the new independent USA and only a few decades before Spanish control of its empire in the Americas was broken. Over two centuries ago many of the place-names that are well known in world affairs today already existed. Only the names of Japan and the Philippines have been added by the author to the 1771 map, but they also existed at that time. Africa is the area that changed most in the nineteenth century, being carved up by European powers into some fifty colonial units, giving an entirely different political structure from that based on the loosely delimited tribes and kingdoms shown on the 1771 map. The 1771 map, therefore, is evidence, except in Africa, of the durability of countries, given that so many can be identified at such an early date.

Between the period 1775–83 and the 1820s the British colonies in what became the USA, the key French colony of Haiti in the Caribbean, and almost all

PLATE 2.1
Colonial legacy, Latin America

a Haiti in the Caribbean remains traditional. Only the rubber tyres and metalled road surface show this scene is twentieth century, not eighteenth. Oxen haul sugar-cane to the refinery. In the eighteenth century Haiti was France's main colony in the tropics and, with the help of slaves, exported large quantities of products from tropical crops (see also Chapter 11)

b The cathedral of Mexico City was hardly affected by the 1985 earthquake. It remains a symbol of Spanish domination over the Aztec Empire, representing the religious wing of the sixteenth-century conquerors (see also Chapter 11)

of the vast Spanish and Portuguese empires in what is now Latin America, achieved independence. The history of the War of American Independence is well known in the USA and UK. At its conclusion in 1783 the Great Lakes in the north and the Mississippi River in the west were recognised in the Treaty of Versailles as its landward limits. The original thirteen new states thus formed the nucleus of what became acknowledged as one of the two superstates of the late twentieth century. After a slow start, the USA expanded westwards to the Pacific in the nineteenth century, purchasing Louisiana from France in 1803 and later Alaska from Russia in 1867. Canada remained a British colony.

The details of the history of the emergence of new countries in Latin America is not so well known outside the Iberian culture area, although the British helped in the process. By 1770 the Spanish Empire had become subdivided for administrative purposes. Much of the area had originally been administered under the Viceroyalties of New Spain (Mexico) and Peru (Lima). Already in the eighteenth century, however, smaller colonial subdivisions had been established, such as Chile, Colombia and Guatemala. When the Spanish American Empire finally broke up in the period following the Napoleonic Wars in Europe, ten separate countries emerged. These later split further to give eighteen countries of Spanish origin early in the twen-

c Cannon balls stacked outside La Citadelle in north Haiti. The fortress was built by ex-slaves to repel any French invasion attempting to retake the country after its independence in 1810. No French comeback took place after the Napoleonic Wars, but thousands of Haitians died in the construction work

tieth century. Spain continued to hold Cuba, Puerto Rico and the Philippines as colonies until the end of the nineteenth century.

Portugal's most important colony was Brazil, the territory of which was gradually pushed west, at the expense of Spain, across the Tordesillas line (see Figure 2.1) into Amazonia, the interior of South America. The various Portuguese settlements in Brazil stayed together as a single political unit to form a new independent state early in the nineteenth century. Portugal also controlled stretches of coast in Africa, territories behind which the colonies of Angola and Portuguese East Africa (now Mozambique) expanded inland late in the nineteenth century.

2.3 Europeanisation in the nineteenth and twentieth centuries (see Table 2.2)

The period from the 1770s to the 1830s saw for the first time the emergence of new sovereign states outside Europe. In addition to the USA itself, by 1830 the following countries had formed from the European colonies of the Americas: Mexico, Peru, Bolivia, Chile, Argentina, Brazil, Central America (which later split into five separate countries), Santo Domingo (and Haiti later), Colombia (and Ecuador, Venezuela and Panama later). Almost all the countries named consisted of a mix of the original indigenous population (American Indians), Europeans, whose origin depended on the colonial power, and Africans, brought in after about 1550 as slaves especially to serve in areas of tropical and subtropical agriculture. Box 2.1 contains an outline of the main features of the slave trade.

As European colonial influence in the Americas was challenged and thrown off in the way described above, European powers were becoming more involved in Asia. In the eighteenth century the power of Spain declined and Portuguese activities in the tropics were increasingly carried out in association with Holland. France and Britain were the major maritime colonial powers of the late eighteenth and nineteenth centuries. Russia continued to expand its empire, penetrating southwards in the nineteenth century into the Caucasus, Middle Asia and the Far East. South Asia (especially India) and parts of Southeast Asia (especially the East Indies, now Indonesia) were the areas most extensively colonised by the Western European maritime powers in Asia in the nineteenth century.

Several European powers had gained control over stretches of the coast of the continent of Africa during the sixteenth to eighteenth centuries. Such territories were used for the supply of provisions to ships passing between Europe and Asia, for collecting slaves for shipment to the Americas, and for trading in a limited way in such products as ivory. In the late nineteenth century, especially in the 1880s, European powers pushed into the interior of Africa, each one aspiring to acquire as much territory as possible behind its coastal footholds. By the end of the nineteenth century most of Africa had been partitioned, in places by mutual agreement among the seven European colonial powers.

During the nineteenth century two main regions of the world largely escaped direct colonisation by European powers – Southwest Asia and the Far East – but they did come under pressure from Europe. Persia (now Iran) and Afghanistan were next in line for conquest by Russia or Britain, but remained independent, serving as buffer zones. In the 1850s Japan was

TABLE 2.2 Colonial calendar 1775 to 1997

1775–83	War of American Independence
1788	First British settlement in Australia
1803	Louisiana purchased by the USA from the French
1804	Republic of Haiti independent from France
1808–26	Amost all of Latin America gains independence from Spain and Portugal
1840	British sovereignty declared over New Zealand
1846–76	Russia occupies oasis cultures of Middle Asia
1858	Britain takes over India (including present-day Pakistan, Bangladesh and Sri Lanka)
1867	Russia sells Alaska to the USA
By 1870	French had occupied much of Algeria
By 1880	Whites had occupied much of the habitable land of South Africa
1880–1900	During these two decades European powers, including Belgium, Britain, France, Germany, Italy, Portugal and Spain, occupy most of Africa
1895	Japan occupies Taiwan (Formosa)
1898	USA ceded Cuba, Puerto Rico and the Philippines by Spain
1910	Japan occupies Korea
1918	Germany loses colonies in Africa Turkish (Ottoman) Empire lost
1935	Italian occupation of Abyssinia
1945	Japan's possessions lost
1947	India gains independence and breaks up into India and Pakistan
1949	Netherlands recognises independence of Indonesia
1958	French give independence to Algeria
c. 1960	Most British, French and Belgian colonies in Africa become independent
1991	Break-up of the USSR. Russian Federation and fourteen other independent republics emerge
1997	Hong Kong reverts to China

forced by the USA and Europe to trade with the outside world. China was under great pressure in the later part of the nineteenth century from Russia, various West European powers and the USA to do likewise and was forced to allow the establishment of a large number of treaty ports over several decades. Late in the eighteenth century and early in the nineteenth, Australia and New Zealand were unobtrusively added to the British Empire, the small numbers of the indigenous population of Australia being unable to resist British conquest. To complete the picture, the building of a substantial empire by Japan should be noted. Although never a European colony, Japan began to adopt Western technology in the 1860s and was later able to conquer Formosa (now Taiwan), Korea, and in the 1930s Manchuria and other parts of China.

During the nineteenth and early twentieth century virtual independence was given to those parts of the British Empire in which European settlers were in the majority or, as in South Africa, were politically dominant, although the new countries subsequently stayed within the British Commonwealth: Canada, Australia, New Zealand and South Africa. Thus a second phase of decolonisation started with the granting of dominion status to Canada in 1867 through the British North America Act.

As a result of the First World War, Germany lost its colonies, most of which were in Africa. In 1935 Italy attacked Abyssinia (now Ethiopia) and gained control of the last sizeable part of Africa to escape colonisation. By 1900, however, the acquisition of extensive territories by European powers had largely come to an end as there were not many regions left that could be easily annexed, and the dominance of Europe was being challenged by emerging powers such as the USA, Japan and China. It is a matter of opinion whether or not the USA itself should actually be considered an extension of Europe since the great majority of its population is of European origin. By that argument, however, Canada, Australia, New Zealand and southern South America could likewise be regarded as European, if not geographically part of Europe.

Almost all the colonies of Europe were established on the initiative of the colonising powers, whose superior military strength and organisation was the key to their conquests. As the centuries went by, various reasons were given to justify the establishment and possession of colonies. One pretext was the assumed need and duty to christianise people in other parts of the world. A pretext for colonising Africa in the late nineteenth century was to stop the slave trade. Colonial powers generally brought peace, stability and sound, if highly authoritarian, administration to their colonies, but the way of life, customs and beliefs of the indigenous peoples were usually modified. The ex-colonial countries of today do therefore have access to technology originating mainly in Europe and the USA, but such technology is not necessarily appropriate for their needs. The legacy of improvements in healthcare and hygiene has been a population explosion.

In practice, colonies were sources of raw materials and of tropical foods and beverages that could not be produced in Europe. They also gave opportunities for some pressure of population to be relieved in Europe, but, in practice, comparatively few people left Europe until the second half of the nineteenth century, when the development of the steamship revolutionised ocean transport. Altogether about 60 million people left Europe for other continents between 1850 and 1930, most emigrating to the countries of the Americas *after* their independence.

It was common for colonial powers to discourage or prevent the development of manufacturing in their colonies, except for branches needed to satisfy local requirements in less sophisticated goods or to process materials for export. The colonies were captive markets for the manufactures of Europe. In the case of India, Britain even prevented the manufacture of cotton goods in its colony so that cotton grown in India could be taken to Britain for manufacture there and some of the products sent back.

Whatever is said for and against the process of colonisation by European powers, there is no doubt that at the height of European influence in the world Europeans rarely doubted their superiority over other 'races' and their self-confidence, if not arrogance, was impressive. The following extract from the entry, 'Colony', taken from *An Anthology of Pieces from Early Editions of Encyclopaedia Britannica* (London: Encyclopaedia Britannica, 1963, p. 42), illustrates the attitude:

> There is no doubt that increasing and multiplying the British stock increases the producing and trading populations over the earth. Hence, if we can conveniently spare a part of our people to inhabit distant regions, it may be politic to aid them by a government staff and other provisions in making a settlement. It happens that in general the country has not only been able to spare a portion of her citizens, but has found a relief in seeing them depart from her shores. Wherever we can plant a numerous and prosperous body of the British people, we create a trading population which will increase our own commerce; and if it be said that they take capital out of the country to employ it elsewhere, after the secondary answer that colonies do not generally remove much capital, the main answer is, that for a nation like the British, in the full employment of free trade, the best part of the world for their capital to be placed in, is that where it is most productive.

By the end of the nineteenth century the period of European conquests was coming to an end and the process by which colonies became independent was well under way. At the same time, the impact of improved communications in the nineteenth century and the further impact expected in the twentieth

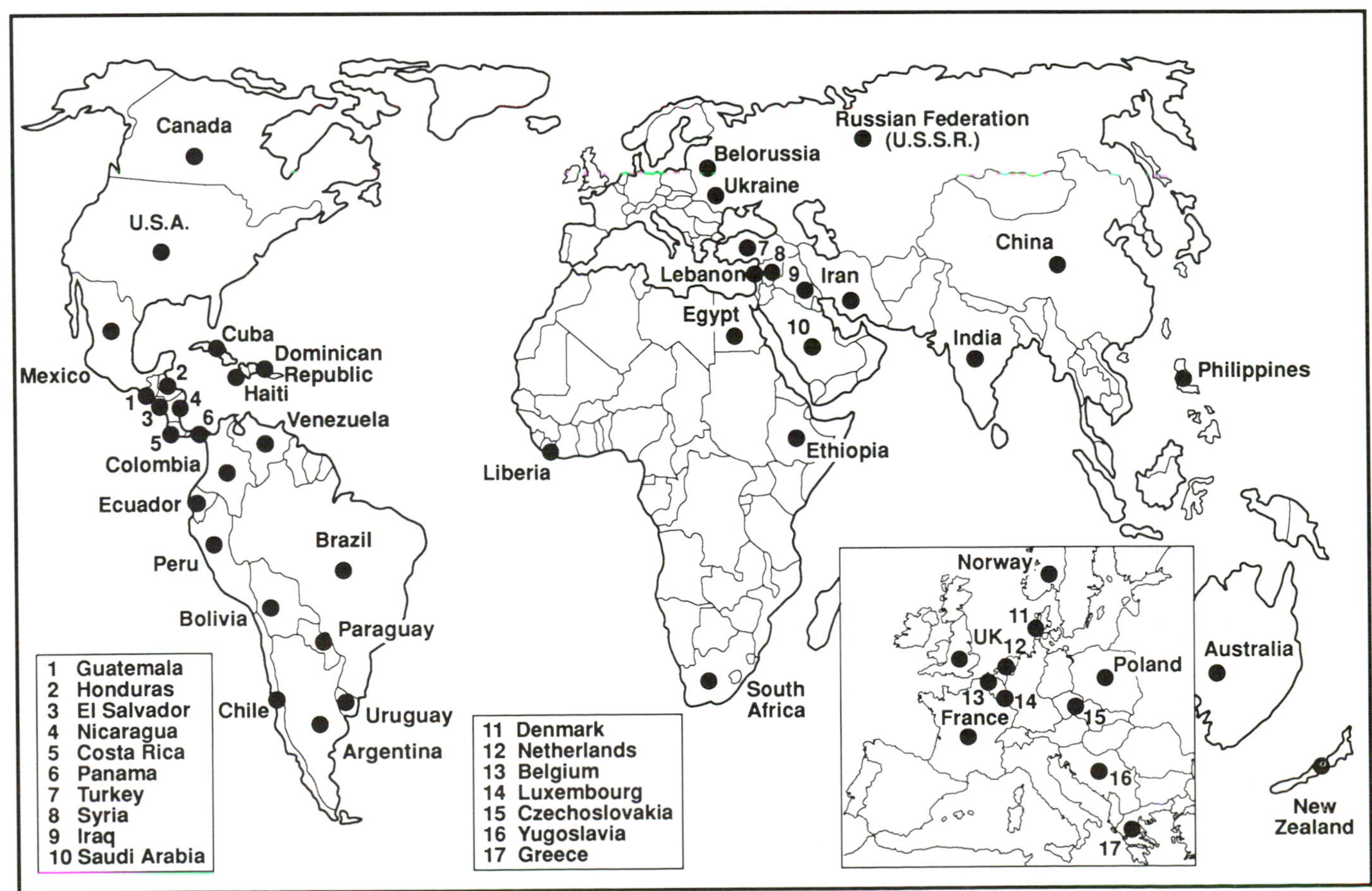

FIGURE 2.4 The fifty-one founder-members of the United Nations at the end of the Second World War. The current boundaries are shown of the countries of the world before the break-up of the USSR in 1991

century were changing the nature of world affairs, a prospect noted in a paper given in 1904 and published later in that year by H. J. (later Sir Halford) Mackinder, one of the first modern British geographers:

> Broadly speaking, we may contrast the Columbian epoch with the age which preceded it, by describing its essential characteristic as the expansion of Europe against almost negligible resistances, whereas medieval Christendom was pent into a narrow region and threatened by external barbarism. From the present time forth, in the post-Columbian age we shall again have to deal with a closed political system, and none the less that it will be one of world-wide scope. Every explosion of social forces, instead of being dissipated in a surrounding circuit of unknown space and barbaric chaos, will be sharply re-echoed from the far side of the globe, and weak elements in the political and economic organism of the world will be shattered in consequence. There is a vast difference of effect in the fall of a shell into an earthwork and its fall amid the closed spaces and rigid structure of a great building or ship. Probably some half-consciousness of this fact is at last diverting much of the attention of statesmen in all parts of the world from territorial expansion to the struggle for relative efficiency.
>
> (Mackinder, 1904, pp. 421–2)

In the BBC Reith Lectures given in 1953 the British historian Arnold Toynbee spoke on the theme of the World and the West. One point strongly argued in the series was that the day of reckoning had come for Europe. Those peoples of the world that it had for so long dominated and at times oppressed would retaliate. During the decades since the Second World War the response of the colonies to their ex-masters has generally been restrained and rather muted: Mexican anger is directed more towards the USA than to Spain. Italy has been one of Ethiopia's main trading partners since 1945 in spite of quite recent unpleasant memories. The UK is respected in India in some ways and the English language is widely spoken. Until the late 1980s the non-Russian Soviet Republics of Middle Asia accepted the Soviet Communist Party, and Uzbeks and Tadjiks there apparently accepted that the Russians should get many of the best jobs in their part of the world. That situation has, however, been changing since 1991.

The theme of this chapter so far has been the spread of European economic, political, cultural and technological influence over much of the rest of the world during the four centuries between 1500 and 1900. A counter-process started during the period 1770–1830, when most of the European colonies of the Americas became new, independent countries. It continued with the granting of dominion status to British possessions largely settled (Canada, Australia, New Zealand) or dominated (South Africa) by Europeans. The concluding stage has been the achievement of independence by virtually all the larger European colonies of Asia and Africa between 1945 and 1975. Many islands and groups of islands with very small populations in the Caribbean, Atlantic and South Pacific still retain colo-

BOX 2.1

The slave trade

A slave is defined as 'a person who is the property of, and entirely subject to, another person, whether by capture, purchase or birth; a servant completely divested of freedom and personal rights' (*Shorter Oxford English Dictionary*). Slavery has existed in various parts of the world for at least 2,500 years and is still found in the 1990s, especially in Southwest Asia. The term was even used in a loose sense by the press in the USA to describe the 'slave labour' used in Soviet labour camps in the 1950s and 1960s. The purpose of including some facts about the slave trade here is to emphasise how from 1500 to about 1880 African slaves were transported across the Atlantic Ocean to provide cheap labour in European colonies in the Americas and even in newly independent countries, including the USA and Brazil. Modern Canada, Argentina, Chile and Uruguay are the only areas in the Americas to which virtually no slaves were taken during the period in question. In time the African slaves mixed with the indigenous population of the region and with their European masters. Today being entirely or partly of African origin remains a drawback to varying degrees throughout the region. In the last fifty years many people of African origin have migrated from the Caribbean to Western Europe and, together with migrants from Africa itself, form considerable minority groups in many areas.

During medieval times, several slaving routes were used to take slaves from the lands immediately to the south of the Sahara Desert and from the eastern coastlands to North Africa and Southwest Asia. By 1500 the Portuguese established trading posts and forts along the western side of Africa, and the shipment of slaves to the Americas started soon after that. In the seventeenth century the Dutch had a major role in the slave trade and by 1700 were handling about 30,000 slaves per year. In the eighteenth century the British took the lead and some 75,000 per year were being nial status. The next section of this chapter covers events leading up to the present 'line-up' of countries of the world.

2.4 The world since the Second World War

Several types of country could be distinguished at the end of the Second World War, based on the study of the origins of modern countries given in this chapter. The following types may be proposed:

- European countries such as France, Hungary, Sweden, Russia
- Former European colonies, independent in 1945, including virtually all the Americas and also Australia. These could be broadly subdivided into:

(a) Those with a large proportion of people of European origin, including Canada, the USA, Argentina, Uruguay, Australia, New Zealand
(b) Those with a large proportion of non-European people, including the rest of Latin America (with regional exceptions such as Southeast Brazil), and South Africa

- Countries that have never been held directly as European colonies for any length of time. Such countries as Japan and China, Thailand, Afghanistan, most of those in Southwest Asia, Turkey and Ethiopia qualify, although virtually none has entirely avoided some conflict with European powers, and most have come to terms with the need in some way to change their culture and technology to exist in a Europeanised world
- European colonies, including almost all of Africa and much of South and Southeast Asia, as well as many small units in the Caribbean, and the Indian and Pacific Oceans. The latest former colonies to achieve independence, all the non-Russian parts of the former USSR, apart from those embedded in the Russian Federation, also belong to this group. In the 1920s various non-Russian peoples in the former Russian Empire were given the status of Soviet Socialist Republics, each with a boundary on the edge of the USSR, and with an option to secede. To the surprise of most 'experts' on the USSR (but not to some, e.g. Amalrik (1970), Todd (1976)) the hitherto monolithic Soviet state broke up in 1991, with each of the fifteen Soviet Socialist Republics becoming a sovereign state. Economic transactions between Russia and most of its former colonies remain considerable. In 1993 the former USSR outside the Russian Republic was being described by Russians as the Near Abroad (*blizhneye zarubezhye*). Interest in the other fourteen Soviet Socialist Republics is still intense, partly because some 25 million now expatriate Russians live there.

The world that emerged from the Second World War was one that was clearly divided into independent and dependent countries. The location of the fifty-one countries that in 1945 originally formed the new United Nations is given in Figure 2.4. They included

moved. It is estimated that between 1500 and 1810 almost 10 million Africans were taken across the Atlantic, 8 million of them in the eighteenth century alone. In the nineteenth century, the slave trade was suppressed, but the flow, estimated at over 2 million in total, continued to late in the century until Brazil and Cuba, among other Latin American countries, finally banned imports. Most of the Africans were shipped to Northeast Brazil, the Caribbean islands and coastlands, and the English colonies in North America.

The history of slavery in North America is well known to US and British readers but the attitude to the people of African origin in Latin America is less familiar. The following passage from Wagley (1952) was written by M. Harris in the early 1950s ('Race relations in Minas Velhas, a community in the mountain region of central Brazil'):

> In Minas Velhas, the superiority of the white man over the Negro is considered to be a scientific fact as well as the incontrovertible lesson of daily experience. Literacy only serves to reinforce the folk-opinion with the usual pseudo scientific re-working into more grammatical and hence more authoritative forms. A school textbook used in Minas Velhas plainly states the case:
>
>> Of all races the white race is the most intelligent, persevering, and the most enterprising .... The Negro race is much more retarded than the others ....
>
> None of the six urban teachers (who are all, incidentally, white females) could find ground to take exception with this view. They all contended that in their experience the intelligent Negro student was a great rarity. When asked to explain why this should be so, the invariable answer was: *E uma característica da raça negra* (it is a characteristic of the Negro race). Only one of the teachers thought that some other factor might be involved, such as the amount of interest which a child's parents took in his schoolwork. But of this she was very uncertain, since the textbook said nothing about it.

TABLE 2.3 Vital statistics for the 100 largest countries of the world in population

Population rank	*Colonising or occupying power(s)*[3]	*Chapter in book*	*Population in millions* 1993	(1994)	2025	*Area in thous. sq km/(miles)*		*Real GDP in billions of US dollars*
1 China	–	16	1,179	(1,192)	1,546	9,597	(3,705)	3,130
2 India	France, GB	14	897	(912)	1,380	3,288	(1,269)	817
3 USA	GB	9	258	(261)	335	9,373	(3,619)	5,424
4 Indonesia	Holland	15	188	(200)	278	1,905	(736)	382
5 Brazil	Portugal	11	152	(155)	205	8,512	(3,286)	753
6 Russian Federation	–	7	147	(148)	152	17,075	(6,593)	983[1]
7 Japan	–	8	125	(125)	126	378	(146)	1,786
8 Pakistan	GB	14	122	(126)	275	796	(307)	219
9 Bangladesh	GB	14	114	(117)	211	144	(56)	93
10 Nigeria	GB	12	95	(98)	246	924	(357)	110
11 Mexico	Spain	11	90	(92)	138	1,958	(756)	512
12 Germany	–	5	81	(81)	73	357	(138)	1,177
13 Viet Nam	France	15	72	(73)	107	332	(130)	72
14 Philippines	Spain, USA	15	65	(69)	101	300	(116)	147
15 Iran	–	13	63	(61)	162	1,648	(636)	196
16 Turkey	–	13	61	(62)	99	780	(301)	243
17 Egypt	–	13	58	(59)	105	1,002	(387)	113
18 UK	–	5	58	(58)	61	245	(95)	796
19 Italy	–	5	58	(57)	52	301	(116)	787
20 France	–	5	58	(58)	59	552	(213)	817
21 Thailand	–	15	57	(59)	76	513	(198)	204
22 Ethiopia	Italy	12	57	(55)	141	1,222	(472)	22
23 Ukraine	Russia	7	52	(52)	52	604	(233)	357[1]
24 South Korea	Japan	8	45	(45)	55	99	(38)	273
25 Myanmar (Burma)	GB	15	44	(45)	70	677	(261)	26
26 Zaire	Belgium	12	41	(43)	105	2,345	(905)	16
27 Spain	–	5	39	(39)	36	505	(195)	341
28 South Africa	Holland, GB	12	39	(41)	70	1,221	(471)	193
29 Poland	–	6	39	(39)	43	313	(121)	184
30 Colombia	Spain	11	35	(36)	51	1,139	(440)	142
31 Argentina	Spain	11	34	(34)	45	2,767	(1,068)	144
32 Canada	France, GB	10	28	(29)	36	9,976	(3,852)	524
33 Morocco	France	12	28	(29)	46	447	(173)	64
34 Tanzania	Germany, GB	12	28	(30)	73	945	(365)	15
35 Kenya	GB	12	28	(27)	62	580	(224)	28
36 Sudan	GB	12	27	(28)	61	2,506	(968)	29
37 Algeria	France	13	27	(28)	47	2,382	(920)	84
38 Romania	–	6	23	(23)	24	238	(92)	70
39 Peru	Spain	11	23	(23)	36	1,285	(496)	63
40 North Korea	Japan	16	23	(23)	32	121	(47)	49
41 Uzbekistan	Russia	7	21	(22)	37	447	(173)	89[1]
42 Taiwan	Japan	16	21	(21)	25	36	(14)	209
43 Venezuela	Spain	11	21	(21)	33	912	(352)	122
44 Nepal	–	14	20	(22)	41	141	(54)	18
45 Iraq	–	13	19	(20)	53	438	(169)	67
46 Malaysia	GB	15	18	(20)	34	330	(127	104
47 Uganda	GB	12	18	(20)	50	236	(91)	9
48 Sri Lanka	GB	14	18	(18)	24	66	(25)	40
49 Australia	GB	10	18	(18)	23	7,713	(2,978)	272
50 Saudi Arabia	–	13	18	(18)	49	2,150	(830)	181
51 Afghanistan	–	13	17	(18)	49	652	(252)	12
52 Kazakhstan	Russia	7	17	(17)	20	2,717	(1,049)	107[1]
53 Ghana	GB	12	16	(17)	38	239	(92)	16
54 Mozambique	Portugal	12	15	(16)	25	802	(310)	16
55 Netherlands	–	5	15	(15)	17	37	(14)	203
56 Chile	Spain	11	14	(14)	20	757	(292)	67
57 Syria	–	13	14	(14)	37	185	(71)	59
58 Côte d'Ivoire	France	12	13	(14)	40	323	(125)	19

TABLE 2.3 continued

Population rank	Colonising or occupying power(s)[3]	Chapter in book	Population in millions 1993	(1994)	2025	Area in thous. sq km/(miles)		Real GDP in billions of US dollars
59 Madagascar	France	12	13	(14)	34	587	(227)	9
60 Cameroon	Germany, GB	12	12	(13)	33	475	(183)	22
61 Yemen	GB (South Yemen)	13	11	(13)	32	528	(204)	18
62 Cuba	Spain	11	11	(11)	13	111	(43)	28
63 Zimbabwe	GB	12	11	(11)	23	391	(151)	16
64 Greece	–	5	11	(10)	11	132	(51)	71
65 Ecuador	Spain	11	10	(11)	17	284	(110)	31
66 Belarus	Russia	7	10	(10)	11	208	(80)	50[1]
67 Czech Republic	–	6	10	(10)	11	79	(31)	81[2]
68 Hungary	–	6	10	(10)	10	93	(36)	64
69 Belgium	–	5	10	(10)	10	33	(12)	134
70 Malawi	GB	12	10	(10)	26	119	(46)	6
71 Guatemala	Spain	11	10	(10)	22	109	(42)	25
72 Burkina Faso	France	12	10	(10)	26	274	(106)	6
73 Yugoslavia	–	6	10	(10)	11	88	(34)	40[2]
74 Portugal	–	5	9.8	(9.9)	9.8	92	(36)	61
75 Somalia	Italy, GB	12	9.5	(9.8)	23	638	(246)	8
76 Angola	Portugal	12	9.5	(11.2)	22	1,247	(481)	12
77 Cambodia	France	15	9.0	(10.3)	17	181	(70)	9
78 Bulgaria	–	6	9.0	(8.4)	8.7	111	(43)	46
79 Mali	France	12	8.9	(9.1)	24	1,240	(479)	5
80 Sweden	–	5	8.7	(8.8)	9.5	450	(174)	129
81 Zambia	GB	12	8.6	(9.1)	21	753	(291)	7
82 Tunisia	France	13	8.6	(8.7)	13	164	(63)	29
83 Niger	France	12	8.5	(8.8)	21	1,267	(489)	5
84 Bolivia	Spain	11	8.0	(8.2)	14	1,099	(424)	12
85 Austria	–	5	7.9	(8.0)	8.2	84	(32)	103
86 Senegal	France	12	7.9	(8.2)	17	197	(76)	10
87 Dominican Republic	Spain	11	7.6	(7.8)	11	49	(19)	19
88 Rwanda	Belgium	12	7.4	(7.7)	17	26	(10)	5
89 Azerbaijan	Russia	7	7.2	(7.4)	9.4	87	(34)	36[1]
90 Switzerland	–	5	7.0	(7.0)	7.8	41	(16)	130
91 Haiti	France	11	6.5	(7.0)	12	28	(11)	6
92 Guinea	France	12	6.2	(6.4)	13	246	(95)	4
93 Hong Kong	GB	16	5.8	(5.8)	6.2	1	(0.39)	88
94 Burundi	Belgium	12	5.8	(6.0)	15	28	(11)	4
95 Tadjikistan	Russia	7	5.7	(5.9)	11	143	(55)	27[1]
96 Honduras	Spain	11	5.6	(5.3)	12	112	(43)	8
97 Georgia	Russia	7	5.5	(5.5)	6.4	70	(27)	27
98 Chad	France	12	5.4	(6.5)	10	1,284	(496)	3
99 Slovakia	–	6	5.3	(5.3)	6.1	49	(19)	35[2]
100 Israel	–	13	5.3	(5.4)	8.0	21	(8)	55

Notes:
[1] Former USSR, weighted according to different levels among the republics in late 1980s
[2] Author's estimates of recently subdivided Yugoslavia and Czechoslovakia. Serbia and Montenegro rejoined in 1992 to form a 'new' reduced Yugoslavia
[3] Excludes the occupation of one European power by another European power or by Turkey; includes countries in the Russian Empire
Sources: Compiled from data in *WPDS 1993* and *1994* and *HDR 1992*

countries that had been victors in the Second World War and countries, such as those of Latin America, Southwest Asia and Africa, or of German-occupied Europe, that had been anti-German, or at least neutral. The USSR claimed fifteen seats, one for each of its Soviet Socialist Republics, but was given three, one each for Russia itself, Ukraine and Belorussia (now Belarus).

Between 1945 and the late 1980s the number of members of the UN increased more than threefold. Some countries that were already independent in 1945, such as Sweden, Finland and Japan, subsequently joined. Nearly all the additional members were, however, ex-colonies of varying shapes and sizes, almost half from Africa alone, while many of the others were small islands in the Pacific and Caribbean. The latest

TABLE 2.4 The countries ranked 101 to 128 in population size in 1993 (and 1994) (population in millions)

101	El Salvador	5.2 (5.2)	111	Norway	4.3 (4.3)	121	Lebanon	3.6 (3.6)
102	Denmark	5.2 (5.2)	112	Paraguay	4.2 (4.8)	122	Armenia	3.6 (3.7)
103	Benin	5.1 (5.3)	113	Togo	4.1 (4.3)	123	Ireland	3.6 (3.6)
104	Finland	5.1 (5.1)	114	Nicaragua	4.1 (4.3)	124	New Zealand	3.4 (3.5)
105	Libya	4.9 (5.1)	115	Turkmenistan	4.0 (4.1)	125	Costa Rica	3.3 (3.2)
106	Laos	4.6 (4.7)	116	Bosnia-Herzegovina	4.0 (4.6)	126	Albania	3.3 (3.4)
107	Kyrgyzstan	4.6 (4.5)	117	Papua-New Guinea	3.9 (4.0)	127	Uruguay	3.2 (3.2)
108	Sierra Leone	4.5 (4.6)	118	Jordan	3.8 (4.2)	128	C. African Republic	3.1 (3.1)
109	Croatia	4.4 (4.8)	119	Lithuania	3.8 (3.7)			
110	Moldova	4.4 (4.4)	120	Puerto Rico	3.6 (3.6)			

Source: Data from *WPDS 1993* and *1994*

additions have come mainly from Central Europe and the former USSR. The latest available list of members of the UN can be found in *Statesman's Yearbook*.

It is difficult to state exactly how many countries there are in the world in any given year, as has been explained in Chapter 1. The set of sovereign states in existence in the 1990s forms a mosaic of pieces of enormously different shapes and sizes. To alter the analogy, they are like the pieces in a kaleidoscope, shaken about to form new combinations following each of the two world wars and, lately, from the break-up of the USSR. The countries of the world have reached their present state in an unplanned way and do not form a balanced and effective system for the organisation of affairs at global level. Table 2.3 has vital statistics about the size and colonial origin of the countries: their population size in 1993 and expected population in 2025, territorial extent and total real GDP. The great range in the size of countries according to all three variables is evident even among the 100 with the largest populations, and is stretched even more when another 100 or more smaller countries are also taken into account. Table 2.4 shows the remaining countries with more than 3 million inhabitants.

Since the disappearance of most West European colonies by about 1970 and of most Russian colonies by the early 1990s, almost all of the land of the world and adjoining sea areas has been in the territory of some sovereign state. Much of the political power in the world is concentrated in the governments of these countries. Two levels of organisation above sovereign state level must, however, be taken into account in the geography of the world's major regions. First there is the supranational level, exemplified by the European Union, in which some of the powers of individual Member States are relinquished. Second, at global level, is the United Nations, with its various institutions within which all or many countries of the world consult and make recommendations, although strictly the UN was not originally intended to interfere in the internal affairs of countries.

Below sovereign state level, all but the smallest countries are divided for administrative purposes into one or more levels of areal units, while in some federal countries considerable political power is also vested in regional and local governments below the level of sovereign state. In larger countries there is usually a hierarchy of two or three levels, such as the states and counties in the USA, the states and *municipios* in Brazil and the provinces and counties in Canada.

The supranational and subnational units of today may be future sovereign states. These politico-administrative units are as much part of the global landscape as the mountains, waters, fields and cities. Their creation, survival, break-up and amalgamation relate partly at least to the fact that in the last five centuries Europe, rather than some other region of the world, has dominated the economic aspects of human activities and has imposed its own cultural features on those of many other cultures scattered round the rest of the world.

2.5 Eurocentrism and post-colonialism

The outstanding features of the process by which Europeans came to dominate and colonise most of the world have been outlined in this chapter. During the last five centuries, empires based outside Europe have also been created and have disappeared, in parallel with European empire-building. Throughout history, particular groups of people, united by race, language, religion or territorial proximity, have developed an urge to conquer, and so long as they have had the means to do so they have succeeded in annexing the territory of others.

Two sets of empire-builders can be distinguished:

- Those located in Europe: Spain, Portugal, England (later Britain), France, Holland, Russia, Sweden, Germany and Italy. The first five conquered territory outside Europe, while Sweden did so exclusively within Europe. Russia, Germany and Italy conquered territory both within and outside Europe, Russia expanding across northern Asia and into central Asia. The degree to which these various powers influenced the rest of the world has already been noted in Section 2.4.

- Outside Europe, the Ottoman Empire centred on Turkey expanded, as did Japan and China later. Arguably the English colonies of what is now the USA continued the process of empire-building after their independence.

For convenience, time and distance/space are often divided into quantities that look tidy in the base 10 number system used almost universally in the world today. Thus, although 10, 100, 1,000 are round numbers only in base 10 (100 in base 8, for example, is 64), decades, centuries and millennia affect the way past events are classified into periods of time and even the setting of targets for the future, the year 2000 in particular being the goal for objectives to be completed and for new goals to be established.

In spite of the reservations made above about round numbers, in order to summarise the impact of Europe on the rest of the world it is convenient to distinguish the three centuries roughly between 1500 and 1800 from the two centuries from 1800 to 2000. The explorers who set out westwards from Europe to sail to China and India across an unknown ocean might have expected to pass a few islands *en route* but they were not anticipating a whole 'new' continent, which actually blocked their way to eastern Asia. In a paper with the intriguing title 'No Incas in Hastings' Jared Diamond (1995) asks why Europeans, not the people of Africa, Australia or America, came to dominate the world. After 1500 most of the military strength of Spain not expended within Europe itself was devoted to the conquest of certain parts of the Americas, not as might have been expected to the establishment of trading posts in Asia, which came in the Portuguese sphere of influence. Thanks to their superiority in weaponry (steel swords, guns, crossbows), mobility (horses, tracker dogs), and organisation (helped by having a written language), and to the innocence or gullibility of the leaders of the Aztec and Inca Empires, in the sixteenth century Spain gained control of part of the newly discovered continent.

From late in the eighteenth century to 1815, Europe was engaged in the Napoleonic Wars and the colonies in the Americas almost all took advantage to become independent by 1820. About the middle of the nineteenth century, European powers had developed military means and mobility (steamships, railways) greatly superior to those of the Spaniards in the sixteenth century and of the Russians who loosely occupied Siberia between 1550 and 1650. From the middle of the nineteenth century to the start of the First World War (1914), large areas of Asia and Africa were conquered and several tens of millions of Europeans emigrated, mainly to the Americas, Australia and Siberia.

From 1914 to 1945 European powers were again in conflict within their own continent. During the first half of the twentieth century, other parts of the world, particularly the USA and Japan, rivalled and challenged the dominant position of Europe.

Finally, a second process of decolonisation took place in the second half of the twentieth century. By the mid-1990s, the colonies of the West European powers were reduced to a considerable number of small territories (e.g. Martinique to France, the Netherlands Antilles, Hong Kong to the UK until 1997). Some of the colonies of Russia were given independence in 1991, but in the Russian Federation itself, large numbers of non-Russians still have 'homelands' (see Chapter 7).

During the second half of the twentieth century, during which world affairs were dominated by the Cold War, an ideological West–East confrontation, political control over their remaining colonies was lost by Europe. Economic influence remains, however, and the wide difference in living standards between the countries of Western Europe and many of their colonies, accepted as inevitable before the Second World War, has been highlighted since the 1940s as the great 'development gap'. The terms of international trade, it is argued, still favour the industrial colonising powers over the producers of primary products. Giant transnational companies based in Western Europe, the USA and Japan (see Chapter 4) invest in former colonial countries explicitly to make a profit from their activities, while claiming also to bring benefits to the host countries. In practice, a small, relatively affluent elite, in some former colonial countries including many of the descendants of the European settlers, keep a disproportionate amount of the wealth and have life styles resembling those of the current 'middle classes' of the original colonising powers.

The 'globalisation' of the world economy, it is argued, perpetuates the situation described above and indeed may be increasing the gap between the rich and the poor in the world. To what extent should the former colonising powers accept responsibility for the actions of the colonial powers, whether based in Europe or elsewhere, during the last five centuries? To judge past events by present-day standards is neither appropriate nor helpful. Slavery was accepted widely, and implicitly accepted by Christianity, until the nineteenth century. Although explicit slavery of the indigenous population of the Spanish colonies was not permitted, the Spanish and Portuguese landowners and later the English colonists in the Americas had no qualms about acquiring millions of African slaves. The Africans were less susceptible to Old World diseases and therefore a better investment as a source of cheap or free labour than the American Indians.

Views on colonialism have changed dramatically in the last half century, although there are still political leaders (e.g. Iraq's Saddam Hussein) and politicians of influence (e.g. the Russian Vladimir Zhirinovsky – see Chapter 7) who practise or advocate the use of force to gain or regain territory. According to Beeston (1995), like Dante Alighieri (1265–1321), who condemned his enemies to hell, Zhirinovsky has been anticipating the dispatch of his political rivals to labour camps in the Siberian *gulag*.

While it is well to be aware of the 'bad' things Europeans did to the populations of other continents (and to each other), it is of more practical value to look ahead to the *next* half century and beyond. Throughout this book reference will be made to the economic conditions, contrasting living standards and demographic problems of the major regions of the world. It would be simplistic to blame *all* the negative features (however these are identified and defined) or, equally, *all* the positive features on the activities of the Europeans outside their own continent during the last five centuries. In the view of the author, not much can be changed in the next fifty years, but the relatively affluent West Europeans, Americans and Japanese would do well to appreciate global inequalities and to contemplate the altruistic transfer of greater quantities of development assistance (see Chapter 4), having assurances that it will be used particularly to help the 20 per cent of the world's population living in absolute poverty.

In March 1995 the United Nations staged the Summit for Social Development in Copenhagen. The summit communiqué contained the customary broad commitments, including

- The eradication of poverty
- The need to cut the debt burden of developing countries

BOX 2.2

Hart's 100 most influential persons in history

In 1992 the second edition was published of M. Hart's book *The 100, a Ranking of the Most Influential Persons in History*. The list is intended to be impartial and as far as possible objective. When each individual lived and was active, where they were born, and what they achieved are of interest to the present study. The strong influence of Europeans throughout the history of the last three millennia and particularly since the fifteenth century illustrates the dominance of Europe.

Seven broad types of person are represented (a few belong to more than one category). The numbers included by Hart in his 100 are: scientists and inventors (36), political and/or military leaders (31), secular philosophers (14), religious leaders (11), the arts (5), explorers (of territory, not science) (2), industrialists (1).

All the individuals who lived before AD 700 were in China, India, Southwest Asia, North Africa, ancient Greece or ancient Rome. Between 700 and 1750, all but one were associated with Europe. Since 1750 most have been Europeans, but the USA, the USSR and China are also represented. Italy is the only country represented during all three periods.

Broken down into separate countries or groups of countries, a remarkable bias towards the core of Western Europe emerges. Out of the 100, 50 were: English (12), Scottish (6), German or Austrian (15), French (9) or Italian (8). When other members of the present European Union (notably Portugal, Spain, the Netherlands and Greece) are added, over 60 out of 100 were from this small part of the world. To achieve great influence or distinction in the political/military category is virtually impossible for citizens of small European countries such as Denmark or Finland. To flourish in science and technology in the twentieth century it has been advantageous to be in a developed country with adequate resources for research and in one in which cultural and ideological conditions do not militate against innovation. Of the forty-two persons broadly falling into the class of scientists, mathematicians, inventors and explorers, two (Aristotle and Euclid) lived in ancient Greece, one (Ts'ai Lun, inventor of paper around AD 100) was Chinese, four were US citizens, one was born in New Zealand (Rutherford) and thirty-four were European (almost all German, English, French, Italian or Scottish), although several of the Europeans ended by working in the USA.

In the last fifty years European languages, religion, scientific discoveries and technological developments have dominated the world. It is of interest to speculate what the world would be like now if some other major region had built up a global influence and presence. Modern history might have taken a very different shape if the Chinese had found more extensive uses for their discoveries and inventions and if the Chinese fleet that set out early in the fifteenth century to conquer other lands had not been recalled. How fully other parts of the world have taken up European ways and whether or not the dwindling populations of remote parts of the world least influenced by Europe should now be protected before they disappear is a question not easily answered.

- Programmes designed to restructure the economies of developing countries should take into account social factors to avoid disruption
- Development assistance should be increased
- The so-called '20/20' proposal should be adopted. Twenty per cent of development aid should be combined with 20 per cent of national budgets to go to basic social programmes
- The following should be promoted: full employment, equality between men and women, universal access to education, decent healthcare and the protection of workers' rights

Correspondents at the conference were quick to draw attention to the fact that the participants seemed detached from the reality on the ground of the world's most poor. Krushelnycky (1995) reported:

> The 10,000 pampered politicians, bureaucrats and UN officials seem too isolated in their privileged lifestyles to be able to comprehend the hardships of the 1.2 billion people around the world living below the poverty line, or of the 120 million [only 120 million?] unemployed, or of those experiencing 'social exclusion' – marginalisation due to gender, race, disability and a myriad of other reasons.

The luxury hotels and lavish banquets of Copenhagen are indeed far removed from the Third World.

Even at the Copenhagen summit the development gap was noted. Prentice (1995) reported:

> Leaders of the world's most impoverished countries have, in effect, been sidelined at the United Nations summit staged to help the poor. Unable even to rent telephones, they could not take part in the crucial lobbying central to a successful summit . . . . Charity, press and interest groups found it almost impossible to contact people from the poorest countries and hear the views of the very people the gathering was supposed to help. . . . Countless mobile phones helped the richer nations to stay in touch.

Such is the world after five centuries of Europeanisation and Eurocentrism. In conclusion to this chapter it is thought-provoking to ask some questions.

What would the world be like now if the Chinese had conquered it in the fifteenth century? Suppose that Germany and Japan had won the Second World War and divided the world between them. What would the world be like now? Soviet leaders claimed that they had initiated a more advanced stage in the process of economic 'evolution'. What would a world run from Moscow be like? Those who condemn Europe's impact on the world must work out what should and could realistically be done to change the world to a new system. Rarely in history have the rich voluntarily given away much of their wealth. Can they be forced to do so? If not, then in the post-colonial world of the twenty-first century the gap between rich and poor could be wider, or at best there could be more of the same.

Postscript: Gulliver on colonies

> But I had another Reason which made me less forward to enlarge his Majesty's Dominions by my Discovery. To say the truth, I had conceived a few Scruples with relation to the Distributive Justice of Princes upon those Occasions. For instance, A Crew of Pirates are driven by a Storm they know not whither, at length a Boy discovers Land from the Topmast, they go on Shore to Rob and Plunder; they see an harmless People, are entertained with Kindness, they give the Country a new Name, they take formal Possession of it for their King, they set up a rotten Plank or a Stone for a Memorial, they murder two or three Dozen of the Natives, bring away a Couple more by Force for a Sample, return home, and get their Pardon. Here commence a new Dominion acquired with a Title by *Divine Right*. Ships are sent with the first Opportunity, the Natives driven out or destroyed, their Princes tortured to discover their Gold; a free License given to all Acts of Inhumanity and Lust, the Earth reeking with the Blood of its inhabitants: And this execrable Crew of Butchers employed in so pious an Expedition, is a *modern Colony* sent to convert and civilise an idolatrous and barbarous People.

This extract from *Gulliver's Travels* by Jonathan Swift (1667–1745) was published in 1726. Gulliver goes on to stress (with tongue in cheek?) that such was not the way the British conducted their colonial activities, but he also explains why he would not take possession in the name of his sovereign of any of the lands he visited.

KEY FACTS

1. An appreciation that the surface of the earth is spherical, not flat, encouraged navigators to sail westwards as well as eastwards to reach eastern Asia.

2. Portugal and Spain were the first two European countries with leaders ready to support 'exploration' of the world. France, England and Holland soon followed the Iberian example.

3. At the same time, the Russians explored eastwards across northern Asia.

4. The above and other European powers conquered and claimed territories over most of the land of the world, mainly during the period 1500 to 1900.

5. European culture (language, Christianity) extended widely round the world and the more advanced technology of Europe usually ensured the successful military and economic influence and control of other continents.

6. The independence of the USA marked the beginning of a period during which the political independence of virtually all the colonies of European powers has been achieved. Cultural and economic influence remains with varying degrees of intensity in many ex-colonies.

MATTERS OF OPINION

1. **Slavery** (see Box 2.1) Given the brutal treatment by Europeans of African slaves during their transfer across the Atlantic and subsequent life style (or lack of it) as slaves, should not contemporary Americans and Europeans show greater sympathy than they do towards the Africans and Caribbeans in their midst?

2. **Empires** Some western geographers avoid studying the developing countries because they would feel compromised by delving into the way their ancestors mistreated the rest of the world. Is this view justified?

3. Four of the six official languages of the United Nations are European: English, French, Russian and Spanish, the other two being Arabic and Chinese. English, French, Russian and Chinese were the languages of the victors at the end of the Second World War, when the United Nations was founded. Discuss the implications of the following comment by Coulmas (1991, pp. 20–1):

> Today, the colonial powers are dissolved, the Europeans have withdrawn to their old continent. But they continue to defend their linguistic heritage abroad, still clinging to the idea that these languages are at the same time universal and unique. European languages account for the bulk of the foreign languages taught anywhere in the world.

4. The changes to indigenous populations imposed directly by or resulting indirectly from European colonisation have been the subject of much discussion and some regret. The members of 'simple' societies in such regions as the Caribbean, the Amazon basin, Australia and Siberia have suffered losses of life and modifications to their cultures and technologies directly through intentional policies of extermination, as well as indirectly through diseases and through the introduction of more sophisticated technologies. Elimination, assimilation, removal to new (usually less favourable) homelands or relegation to reservations have been the fate of many tribes. Should the life styles of the remaining 'simple' societies of the world be preserved as far as possible, and if so, how can this aim be achieved? Refer to Ferguson (1992).

SUBJECTS FOR SPECULATION

1. Was there something distinctive about fifteenth/sixteenth-century Europe that gave it the conditions needed to dominate most of the rest of the world?

2. In the fifteenth century China briefly sent out expeditions to explore and establish trade with places on the coasts of the Indian Ocean. What would the world be like now if the Chinese, not the Europeans, had conquered and dominated much of the rest of the world after 1500?

3. Reference is made in Box 2.2 to a compilation of the 100 most influential people in the history of the world. Almost 70 of the 100 were Europeans, including Russians. Were so many of them Europeans because Europe has something special, or did the emergence of Europe as the dominant continent after 1500 provide the environment in which individuals could distinguish themselves in various ways?

Chapter three

Population and resources

However staggering the figures for total population may be, it is the present rapid net increase which has turned attention almost everywhere to the need for a stocktaking of resources, especially of land and its capacity for the production of food, raw materials and supplies of energy.

L. Dudley Stamp (1960, p. 19)

(The mid-1960 population of the world estimate of 3,000 million compares with the mid-1995 estimate of 5,700 million.)

3.1 The twelve major regions and seven elements

Reference was made at the beginning of Chapter 1 to the problem of dividing the world's land area into an appropriate set of regions for studying the geography of the world below the global level. Two questions have to be addressed: roughly how many regions are required? and how should their membership and limits be determined? Given that there is no objective method of working out the regions to be used, some subjective criteria have to be chosen.

For several centuries the land area of the world was divided conventionally into continents. Most atlases published in Europe still group their maps into these continents, usually starting with Europe (some atlases published in the USA start with the Americas). For many decades now, however, geographers have largely abandoned the traditional division of the world into continents. Before the sixteenth century three continents were identified by Europeans: Europe itself, Asia and Africa. The 'discovery' of the Americas by the Europeans introduced a fourth continent, the discovery of Australia a fifth. For convenience the Americas are generally regarded as two continents, North and South, while the large land mass of Antarctica, although mostly entirely covered by ice, is arguably a continent too.

For the purposes of this book it was decided that between ten and twelve regions would be appropriate. When several different criteria are used for purposes of regionalisation, it is not possible to satisfy all the criteria simultaneously. For example, Australia is large enough in area to merit consideration as a separate region but it is very small in population. Japan and Central Europe are large enough in population but very small in area. As far as possible, then, the following attributes of the regions were taken into account:

- They should be continuous rather than fragmented. For example Australia and South Africa should not be in the same region.
- They should be large enough either in area or in population to be broadly comparable with one another.

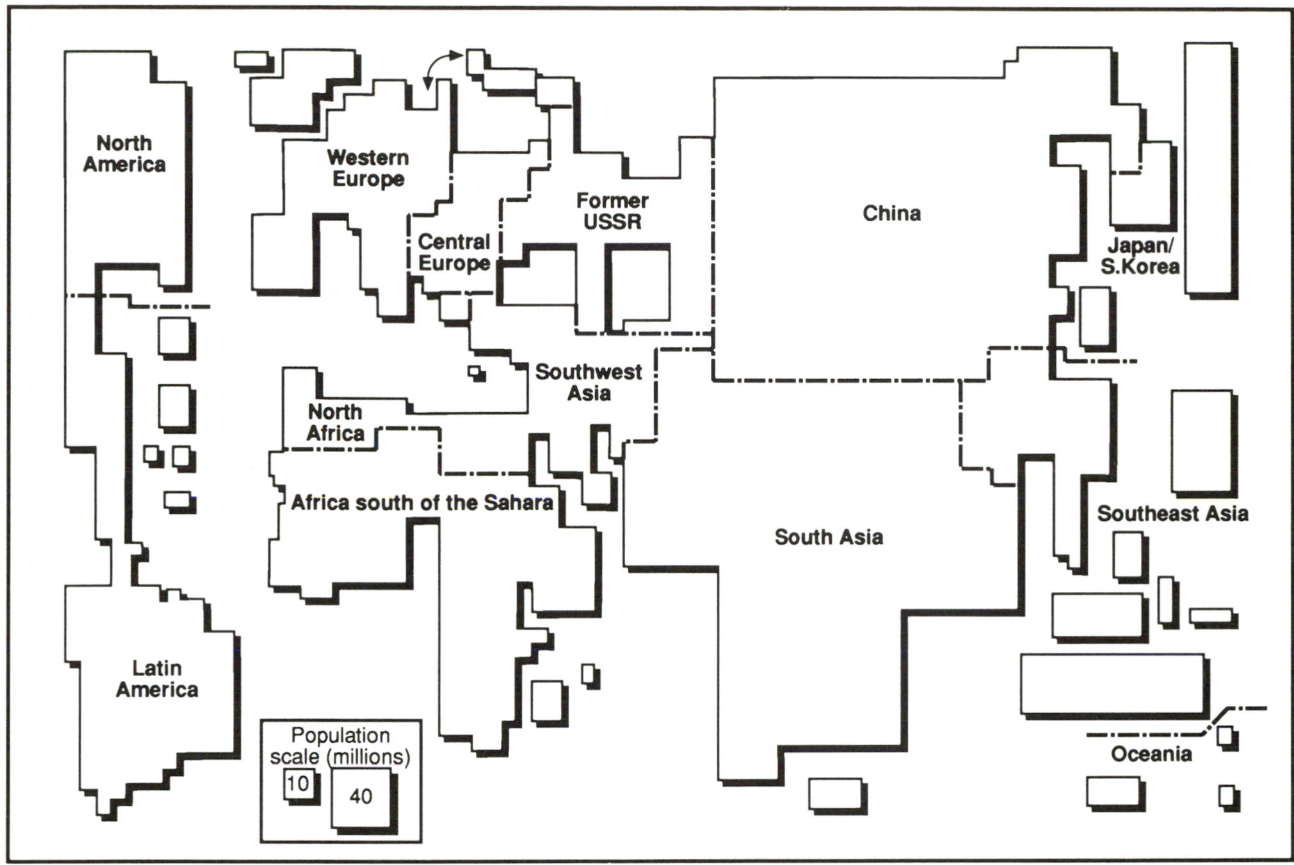

FIGURE 3.1 The twelve major regions of the world used in this book represented in the cartogram according to their population size. See Table 3.1 for the data used

- No sovereign state should be split between two regions.
- Where possible the sea, a thinly populated land area (e.g. the Sahara Desert) and/or a cultural break (e.g. between Anglo-America and Latin America) should be taken into account.
- The countries of each region should as far as possible share common cultural and/or economic features.

The set of regions to be used in the book are listed and described below. Many sets of 'major' regions covering the whole world have been devised and used by institutions such as the United Nations, as also by geographers. The process of regionalisation has sometimes been explicitly justified. There is general agreement about the choice of some major regions (e.g. Oceania, North or Anglo-America), but differences in detail in the choice of others. For the author or authors of a geography of the world's major regions it is crucial to decide on a choice of regions before embarking on writing the book and then to stick to the regions, making the most of the inevitable imperfections that occur in the system. For anyone undertaking this task in the early to mid-1990s, Europe presents the greatest headache, which is why in this book Western and Central Europe overlap, sharing Chapters 5 and 6. Again, balance and consistency in the length of chapters are expected by publishers. Canada has therefore been separated from the USA (Chapter 9) and placed alongside Australia (Chapter 10), since there is much more to say about the whole of North America than about Oceania.

1. **Western Europe** includes the fifteen Member States of the European Union (EU) (1995), as well as the three Member States of the European Free Trade Association (EFTA). These countries all have market economies, albeit with varying degrees of state influence. Apart from tiny Liechtenstein (the 'seventh' in EFTA) the relevant countries are listed in Table 5.2 (p. 106).

2. **Central Europe** was formerly commonly referred to as Eastern Europe, those countries *between* the

western bloc and the former USSR. Until recently all the countries had socialist, centrally planned economies. The three Baltic republics, free at the time of writing from any formal ties with the former Soviet Union, have for convenience been added to Central Europe, although Estonia has close affinities with Finland. In due course some or all of the countries of Central Europe may follow the former German Democratic Republic into the EU. With the break-up of Czechoslovakia into two separate states and of Yugoslavia into at least five, and with Malta included for convenience, in 1994 there were actually sixteen countries in Central Europe. The relevant countries apart from Malta are listed in Table 6.1 (p. 137).

3. **The former Soviet Union** was still loosely held together as the Commonwealth of Independent States in 1994 and, in the event of no further break-up of any of the former Soviet Socialist Republics, now consists of twelve separate countries, the three Baltic republics having been transferred to Central Europe.

4. **Japan and the Republic of Korea** have been placed together pending the possible reunification of Korea, to form a major region. The Republic of Korea, hereafter referred to as South Korea, is strikingly similar to Japan about three decades ago, and fits in more appropriately with its former colonial ruler than with the rest of East Asia.

5. **North America** consists of the USA and Canada. In a physical sense North America conventionally includes Mexico and Central America, but for the purposes of this book cultural considerations have overridden physical ones. In time the North American Free Trade Agreement (NAFTA) may draw Mexico economically even closer to North America than it is now. Although data are given here for the region of North America as a whole, Canada is placed in Chapter 10 with Oceania in order to draw attention to many largely unnoticed similarities between Canada and Australia. Both have very low densities of population, with generous natural resource endowments on large areas. British influence on both has been crucial in their emergence.

6. **Australia, New Zealand, Papua-New Guinea** and a very large number of smaller island countries in the Pacific form the region of Oceania. Although it contains only about 0.5 per cent of the total population of the world, this region is distinguished by the sharp contrast between Australia and neighbouring Indonesia, located in Southeast Asia

Regions 1–6 form the so-called developed world (see Box 1.2), although they include some very poor and backward countries, for example Albania in Central Europe, Tadjikistan in the former USSR and Papua-New Guinea in Oceania. The six remaining regions make up the developing world, although they in their turn include some relatively highly developed countries or regions, for example, Argentina, Uruguay and southern Brazil in Latin America, Israel in Southwest Asia and Singapore in Southeast Asia. They are as follows:

7. **Latin America** consists of all the Americas south of the USA. Spain and Portugal were the dominant cultural influences from the sixteenth to the eighteenth century (see Chapter 2).

8. **Africa south of the Sahara** excludes the five predominantly Arab/Muslim countries with their 'frontages' on the Mediterranean. South Africa is distinguished from the remaining countries by its influential white minority and relatively high level of industrialisation.

9. **North Africa and Southwest Asia** share the Islamic heritage (apart from Israel) but not all are Arab. The northern tier of countries, Turkey, Iran and Afghanistan, is included in the region, although in time Turkey may become more closely linked to the EU, or perhaps, together with Iran and the southern new independent states of the former USSR, may form a distinct major world region.

10. **South Asia** includes predominantly Hindu India as well as predominantly Muslim Pakistan (formerly West Pakistan) and Bangladesh (formerly East Pakistan). It is convenient to consider the '*Indian subcontinent*' as a major world region because of its spatial cohesion, its presence (apart from Nepal) in the British Empire for at least a century up to 1947, and its large population.

11. **Southeast Asia** is widely used as a major world region, although it contains countries with different cultural traditions and economic levels.

12. **China's** great size makes it a major region of the world. Whatever their ultimate fates, Hong Kong, Taiwan, Mongolia and the Democratic People's Republic of Korea (hereafter referred to as North Korea) have been included in this region.

In order to provide an introduction to the attributes and characteristics of each region of the world, initially at the level of the twelve regions and, later in the book, at a more detailed level, data have been collected to illustrate outstanding features. The following aspects provide a checklist of reasonably distinct elements of any major region, country, or subdivision of a country.

1. **Population** is divisible into a large number of clearly defined subsets, such as by sex (male/female), age (e.g. elderly subset of over-64s) and language, and

also into less precisely defined subsets related to income, location, religion and many other features.

2. **Natural resources** are considered to be any raw materials that are independent of human beings, such as air, water, soil, vegetation, fossil fuels, non-fuel minerals. With changes in technology and in people's needs and wants, different natural resources have been prominent at different times. For example, natural gas and water power to generate hydroelectricity have only been extensively used in the twentieth century, as also has bauxite, the raw material for the production of aluminium.

3. **Means of production of goods and services** range from the simplest tools used in hand cultivation to machines of many different kinds, that enhance and magnify the capabilities of the human body: cranes and cars the muscles, telescopes and telephones the senses, computers the brain. Most conspicuous on the ground are such constructions as farms, factories, hospitals, schools and fixed transportation links.

4. **Production** comes from the agricultural, industrial and service sectors, and gives rise to a very broad range of items, some much more tangible than others. These can conveniently be subdivided into products that are used for the replacement, maintenance or increase of the capacity of the means of production, and consumer items, which directly support population.

5. **Transport and communication links** are an essential element of any economy, even one in which each village is virtually self-sufficient, because even here tracks and paths are needed to reach fields and to link neighbouring settlements.

6. The **organisation** of economies.

7. At all but global level, **international transactions** between the members of any set of regions, such as international trade and flows of development assistance, must be taken into account.

In the rest of this chapter the first four elements will be considered, while chapter 4 covers the others.

3.2 General features of population

Many countries of the world have not had censuses of population at regular intervals even in the last few decades. As has been noted in Chapter 1, even in the

TABLE 3.1 Population trends in the world regions, 1931–93

	(1)	*(2)*	*(3)*	*(4)*	*(5)*	*(6)*	*(7)*	*(8)*	*(9)*
	Total population in millions and as percentage						*Population change (starting year = 100)*		
	1931		*1958*		*1993*				
	total	*%*	*total*	*%*	*total*	*%*	*1931–58*	*1958–93*	*1931–93*
Western Europe	280	13.8	330	11.4	381	6.9	118	115	136
Central Europe	101	5.0	104	3.6	131	2.4	103	126	130
Former USSR[1]	163	8.1	201	7.0	285	5.2	123	142	175
Japan/South Korea	80	6.5	116	4.0	169	3.1	145	146	211
North America	135	3.9	192	6.7	287	5.2	142	149	213
Oceania	10	0.5	15	0.5	28	0.5	150	187	280
Latin America	120	5.9	200	6.9	460	8.4	167	230	383
Africa S of Sahara	113	5.6	191	6.6	550	10.0	169	288	487
N Africa/SW Asia	77	3.8	140	4.8	353	6.4	182	252	458
South Asia	355	17.6	519	17.9	1,173	21.2	146	226	330
Southeast Asia	127	6.3	205	7.1	460	8.4	161	224	362
China plus	465	23.0	681	23.5	1,230	22.3	146	181	265
Total	2,026	100.0	2,893	100.0	5,506	100.0	143	190	272
Developed[2]	769	38	958	33	1,281	23	125	134	167
Developing	1,257	62	1,935	67	4,225	77	154	218	336

Notes:
[1] Excludes Baltic republics
[2] Includes South Korea
Sources of data: For 1931, League of Nations (1933); for 1958, *UNDYB* (1960); for 1993, *WPDS* (1993)

TABLE 3.2 Population features of the world regions in the early 1990s

	(1)	(2)	(3)	(4)	(5)	(6)	(7)	(8)	(9)
						Expected population in 2025			
	Total 1993	*Natural increase*	*TFR*	*Years life expectancy*	*% urban popn*	*millions*	*%*	*change 1993–2025 (1993 = 100)*	*millions gained*
Western Europe	381	0.2	1.6	76	78	384	4.6	101	3
Central Europe	131	0.1	1.8	71	60	132	1.6	101	1
Former USSR	285	0.6	2.2	70	66	320	3.8	112	35
Japan/South Korea	169	0.4	1.5	77	76	181	2.1	107	12
North America	287	0.8	2.0	76	75	371	4.4	129	84
Oceania	28	1.2	2.2	70	70	39	0.5	139	11
Latin America	460	1.9	3.2	68	71	682	8.1	148	222
Africa S of Sahara	550	3.0	6.5	52	26	1,326	15.7	241	776
N Africa/SW Asia	353	2.7	5.0	63	52	748	8.9	212	395
South Asia	1,173	2.2	4.1	58	25	1,933	22.8	165	760
Southeast Asia	460	1.9	3.4	62	29	696	8.3	151	236
China plus	1,230	1.2	1.9	70	28	1,614	19.2	131	384
Total	5,506	1.6	3.3	65	42	8,425	100.0	153	2,919
Developed	1,281	0.4	1.8	74	72	1,427	16.9	111	146
Developing	4,225	2.0	3.7	63	34	6,998	83.1	166	2,773

Description of variables:
(1) Population in millions
(2) Annual percentage rate of natural increase of population
(3) Total fertility rate: average number of children born to a woman
(4) Life expectancy in years at birth
(5) Note that there are considerable differences in the definition of 'urban' population from country to country
(6)–(9) The estimated population for 2025 in column (6) is the 'best' made by the Population Reference Bureau but is subject to upward or downward revision according to trends and new information. Even between 1993 and 1994, for example, the PRB estimate for the population of the world in 2025 was lowered from 8,425 million to 8,378 million
Source: WPDS 1993

most highly organised countries, people escape the count. At any given time, therefore, estimates of the total population of the world and of the rate of population change are very approximate. Table 3.1 shows that around 1930 the population of the world was about 2 billion. It had taken little more than 100 years to double from 1 billion early in the nineteenth century. It is expected that the population of the world will pass 6 billion before the year 2000. While the rate of increase is gradually slowing down, the absolute number added, 80–90 million a year, is not changing much in the 1990s.

A comparison of columns (1), (3) and (5) in Table 3.1 for each of the major regions and for the world as a whole shows how total population has changed. The rates of change during three periods are shown in columns (7)–(9). In the last six decades the fastest rates of growth have been in Africa and Southwest Asia, with Latin America, Southeast Asia and South Asia some way behind. Population increase in most of Western and Central Europe has been comparatively modest since the 1960s and in the early 1990s population size hardly changed at all. During the last six decades, the total population of the developed countries increased by less than 70 per cent, while that of the developing countries increased more than threefold. The developing countries are therefore gaining in relative importance in terms of population size and arguably, therefore, also in potential influence in the world as a whole.

Columns (2), (4) and (6) in Table 3.1 compare the share of world population in each region in 1931, 1958 and 1993. South Asia, Southeast Asia and China together have well over half of the population of the world. The share of Western Europe has halved over the last sixty years, decreasing from about 14 per cent in 1930 to about 7 per cent in 1990. In contrast, the shares of Africa and Southwest Asia have increased very markedly.

The study of past demographic trends and their projection into the future has led to some forecasts that turned out to be wildly wrong, making demographers more cautious than they were a few decades ago. In general, it is safer to extrapolate past demographic trends into the future for large populations than for small ones. Thus the forecasts for the total population of the world in Table 3.2 column (6), indicating a probable increase of nearly 3 billion between 1993 and 2025, assume fairly gradual changes in the overall demographic behaviour in each of the major regions of the world, and little interregional migration. Even China's drastic policy of limiting family size to one child (or at most two) has not been successfully implemented throughout the country. Non-Han peoples in particular have been allowed larger families, so the 1.6 billion estimate for China and its immediate neighbours in 2025 seems reasonable. Starvation and/or diseases such as AIDS could cut down population growth in Africa south of the Sahara. On the other hand, fertility rates could actually increase in some parts of the world. Indeed, according to Soviet data, in the 1970s and 1980s birthrates did rise slightly in the former republics of Soviet Central Asia (e.g. Tadjikistan), thereby not conforming to the demographic transition model, the essential features of which are outlined in Box 3.1.

In Table 3.2, columns (2)–(5) show some of the variables influencing population change. Column (2) shows that in Europe and Japan there is very little natural increase in population, the difference between the number of births and deaths per 1,000 population in a year. Column (3) shows that the total fertility rate has dropped below the theoretical replacement rate of an average of about 2.1 children per female in Europe, including the Russian Federation, Japan and North America. The decline of mortality rates in various age groups raises life expectancy (see column (4)), which increases as people live longer on average, thereby producing a growing proportion of 'elderly', unless, of course, the definition of elderly is changed. From the experience of the countries of Europe, at least several decades must elapse before birthrates and deathrates converge in the present developing countries, always assuming that they pass smoothly through the demographic transition process.

Various factors have been shown to influence the fertility rate of women in different parts of the world and at different times in the past. The influence of improvements in healthcare facilities has been widespread, reducing mortality rates in all age groups, but especially among the very young. It therefore becomes less necessary for parents to have as many children as possible to ensure that enough survive to support them in their old age. As infectious and parasitic diseases are eradicated (e.g. smallpox world-wide) or reduced (e.g. malaria in many areas), degenerative diseases are increasing as causes of death in an ageing population. Education in general, and especially the education of women, also affects family size, since, depending on the constraints imposed by some religions, family planning becomes easier to implement, and more widespread. In general, also, the type of settlement affects family size, with smaller families more common in urban centres than in rural areas in the same country or region. Column (5) in Table 3.2 shows the wide differences in the level of urbanisation in different parts of the world.

Material standards and the level of 'development' are conventionally measured in a given country or region by GDP per inhabitant. The expected correlation, a negative one, between total fertility rate and per capita GDP is not, however, high: for example, the two poorest countries in the EU, Portugal and Greece, each have a total fertility rate (TFR) of 1.5, with GDP per capita of 4,260 and 5,340 US dollars respectively. These low levels contrast with the high level of Saudi Arabia, which has a TFR of 7.2, with 6,320 dollars per capita, and Israel a TFR of 3.0 with 9,750 dollars per capita.

In this section the main demographic features of the twelve major regions in the early 1990s have been compared. In the following two sections, a broader view is taken of factors that influence population change. Given the central importance of the human population in the world as a whole and in each of the twelve regions, population behaviour and trends need to be appreciated in some depth.

3.3 Demographic structures affected by migration, war, genocide and ethnic cleansing

It is reasonable to assume that natural and human-made disasters have a large influence on population

TABLE 3.3 Examples of numbers of deaths from natural causes

- *1347–51 Pandemic:* the Black Death in Eurasia (bubonic, pneumonic, septicaemic plague) killed about 75 million. This was a substantial part of the total population of about 300 million in Eurasia at that time.
- *1959–61 Famine:* about 30 million deaths in northern China, about 5 per cent of the total population of the country (see Chapter 16).
- *1918–19 Influenza:* 21–22 million world wide, about 1 per cent of the total population of the world.
- *1970 Circular storm or hurricane:* about 1 million lives lost in the Ganges delta in Bangladesh, less than 2 per cent of the total population of the country at that time.
- *1994 Pneumonic plague:* several hundred in India, spreading from Surat in the state of Gujarat to several other places in the country.

BOX 3.1

The demographic transition model

The central feature of this model is the relationship between the birthrate and deathrate in given populations. When a population is stable and there is no net migration in or out, the number of births and deaths are the same in a given region. With no migration, if births exceed deaths, population grows, whereas if deaths exceed births, population declines. The diagram in Figure 3.2a shows a version of the model. The vertical scale measures the number of births and of deaths per thousand population. The horizontal scale represents time, the exact duration of which is not usually specified in the general model.

The same model, seen from the medical viewpoint of the causes of death, is referred to as the epidemiological transition model. In the last two centuries the effects of contagious and parasitic diseases have been drastically reduced in many parts of the world. As a result, a large proportion of the population lives long enough to be afflicted by and in due course die from degenerative illnesses.

Although population change in given countries and regions does not follow an exact path through time, many in the last two centuries or less have had the same broad experience, divided over time into a number of 'stages'.

Stage 1 is characteristic of populations throughout history. Birthrate and deathrate are both high, and the average age of the population is low, as is life expectancy. At times population is reduced when a famine or an epidemic occurs, but in due course the population is restored.

Stage 2 Improvements in healthcare, hygiene, food supply and other influences reduce the deathrate by prolonging life expectancy. Birthrate remains high.

Stage 3 Changes in economies, whereby realisation that a large number of children is a liability rather than an asset leads to a reduction in the number of births per female and then birthrate declines.

Stage 4 Eventually, birthrate and deathrate converge, and a situation with a stable population is restored.

Stage 5 It is unlikely that the population of a country or region would remain unchanged over more than a short period. Since the Second World War, near stability has been reached in much of Western and Central Europe but in the last two decades birthrate has actually dropped below deathrate in some countries, with, for example, zero growth in Italy in the early 1990s and a declining population in Hungary. Unless demographic behaviour changes markedly and/or the excess of deaths over births is compensated for by immigration, most of the countries of Europe and also Japan face a declining population in the future because total fertility rates have actually already fallen far below replacement rate, regarded as an average of about 2.1 children per female in developed countries.

A possible follow-up exercise to this account of the demographic transition model is to plot the position of selected countries not included in Figure 3.2 now and in the past. Relevant data can be found in *The Demographic Yearbook of the United Nations* and in appropriate volumes of Macmillan historical statistics edited by Mitchell (e.g. 1981).

FIGURE 3.2 (opposite) The demographic transition model

a Author's modification of the basic model adapted to allow greater flexibility in the presentation of the data on birthrates and deathrates with Sweden and Japan at different times in their history and all other countries as in 1994

b The relationship of birthrate and deathrate to annual rate of change of population for selected countries in 1994 and for Japan also in 1935, 1955 and 1975

a

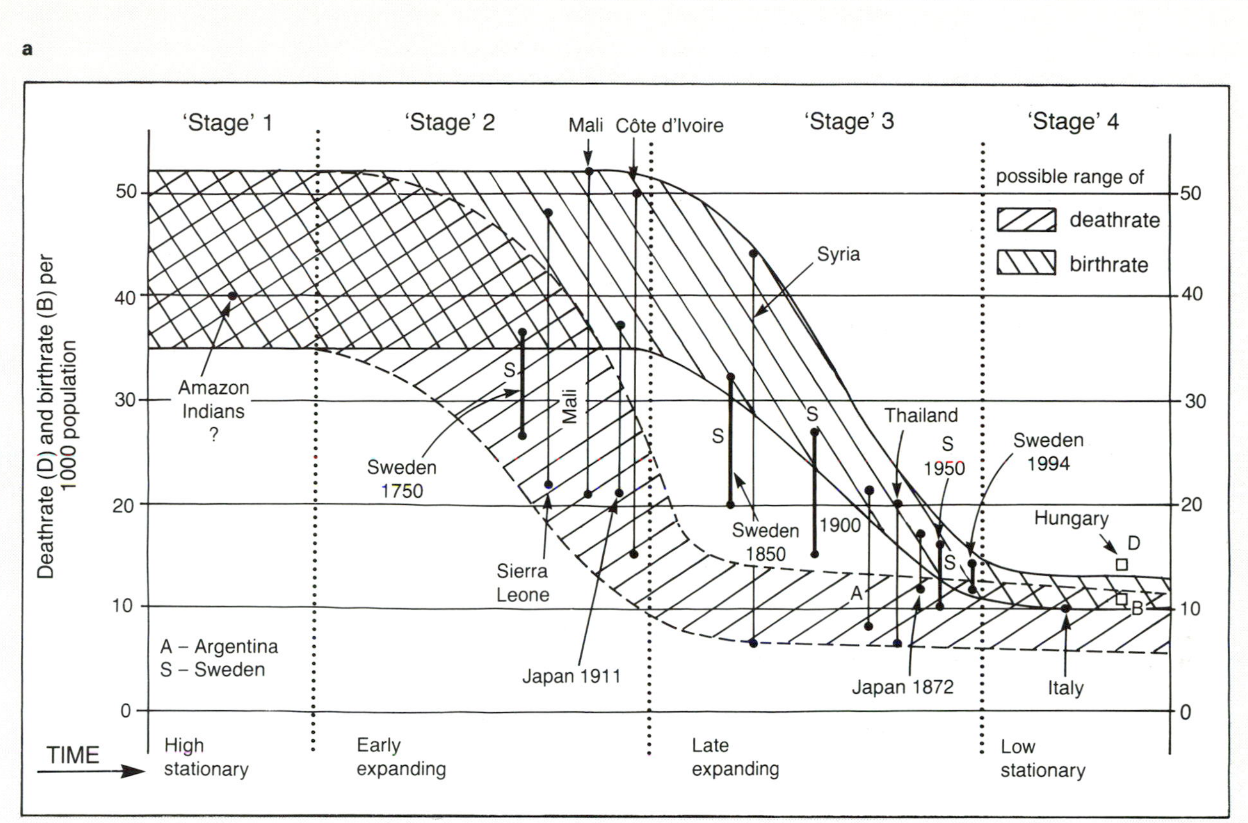

'Stage' 1
'Stage' 2
Mali
Côte d'Ivoire
'Stage' 3
'Stage' 4
possible range of
deathrate
birthrate
Syria
Amazon Indians ?
Sweden 1750
Mali
Sierra Leone
Japan 1911
Sweden 1850
1900
Thailand
S 1950
Sweden 1994
Hungary
Japan 1872
Italy
A – Argentina
S – Sweden
Deathrate (D) and birthrate (B) per 1000 population
TIME
High stationary
Early expanding
Late expanding
Low stationary

b

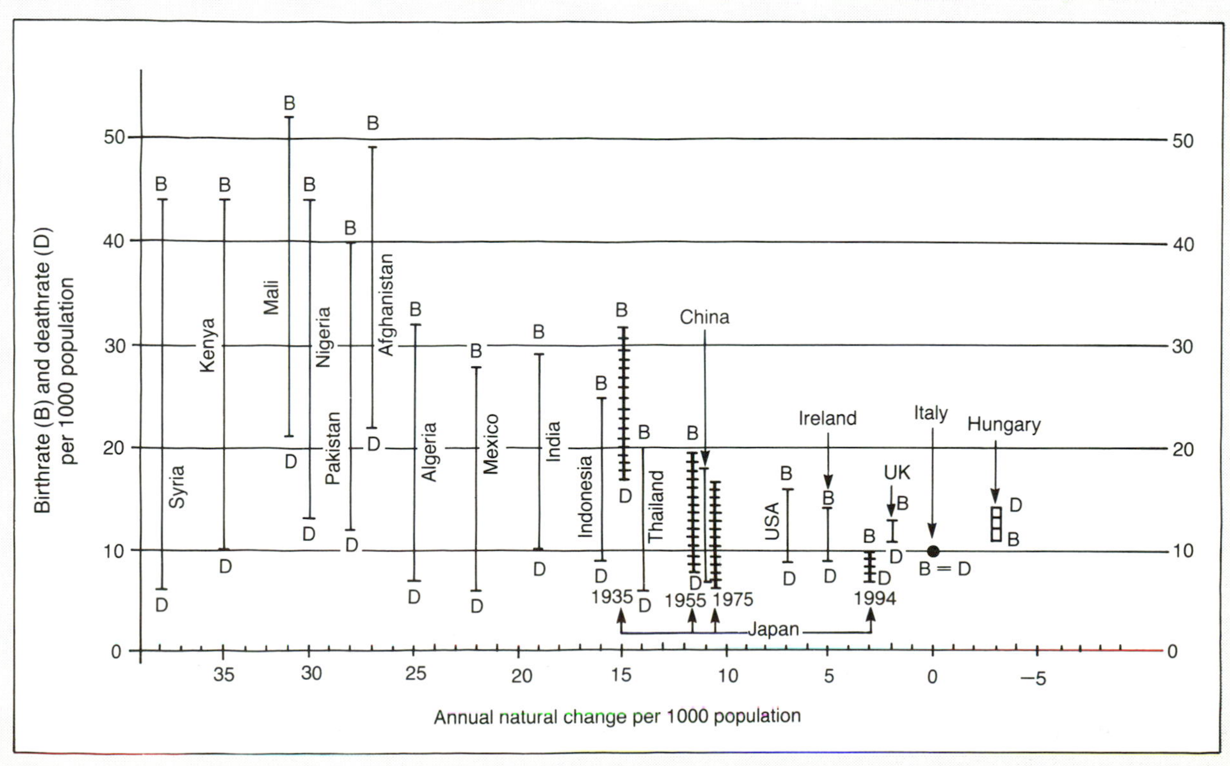

Birthrate (B) and deathrate (D) per 1000 population
Syria
Kenya
Mali
Nigeria
Pakistan
Afghanistan
Algeria
Mexico
India
Indonesia
Thailand
China
USA
Ireland
UK
Italy
Hungary
B = D
1935
1955
1975
1994
Japan
Annual natural change per 1000 population

change and size. In practice, loss of life resulting from such physical events as earthquakes, floods and droughts usually occurs at a local or regional level rather than at a global level. Similarly, with a few notable exceptions, in the last two centuries contagious diseases have usually had only a limited impact, the case so far with AIDS. Likewise, even the two world wars of the twentieth century hardly affected demographic processes in much of the world. Internal conflicts (civil wars) tend to have a devastating effect nationally or regionally, as exemplified in Spain during the Civil War of 1936–9, as recently as the early 1990s in the former Yugoslavia and in 1994 in Rwanda.

Some examples of the results of natural disasters and of human error are given in Tables 1.4 and 3.3, and in Box 11.1. The rest of this section deals with some striking cases of human conflicts. For simplicity such conflicts can be subdivided into two types. First, where two (or occasionally three) sides are fighting a war with the assistance of a professional army, the military personnel usually account for a large proportion of all casualties, a characteristic of the First World War. Second, in the case of genocide and ethnic cleansing, one side has the upper hand, thanks to its military superiority, and a large measure of control over the lives of some or all of the population of the other side.

- In 146 BC, after winning the Third Punic War, Rome exterminated the troublesome population of the city of Carthage, the expression associated with this act of genocide being remembered in history by the phrase 'Carthage has been destroyed' (*Cartago delenda est*). The city was demolished and the fields sown with salt. 'Success': 100 per cent.

- Between 1311 and 1340, following the Mongol conquest of the Sung Empire of China in 1279, the Mongols exterminated 35 million Chinese peasants, approximately a quarter of China's population, allegedly to make space for hunting reserves for the new rulers of the empire. The Mongol herders had a dislike of cultivators and cities. Result: 25 per cent reduction in population.

- Bailey and Nasatir (1960, p. 426) give a vivid summary of the effects on Paraguay of its all-out war against powerful neighbours. One can only wonder at the precision with which the remaining population was counted, given the disastrous situation in the war-torn country:

> Socially, the Paraguayan War, the War of Francisco Solano Lopez, was a disaster. It is estimated that there were 525,000 people in Paraguay when Francisco Lopez came to power. In 1871 the victorious allies counted 221,079, of whom 106,254 were women, 86,079 children and 28,746 adult men. Such a shortage of men led to promiscuity, to large numbers of illegitimate children, and to a race of lazy males in a land where the women did all the work. Also, the land which the women now had to farm was devastated, the orange orchards were neglected, the horses and cattle dead. Schools were closed and newspapers unprinted for a generation. There were no writers, for people had forgotten how to read.

Result: elimination of about 90 per cent of the men.

- During the Second World War (1939–45) some 6 million Jews were exterminated by the Nazis, including virtually all those who had been living in Germany, as well as many living in other countries occupied by the Germans, particularly Poland and Ukraine. Result: more than half killed.

- Evans (1994) examines new evidence regarding the loss of life in the Soviet Union as a result of the German invasion of 1941. Comparisons of population trends in the 1930s and 1950s were made and the gap between population size expected had there been no war, and that observed, indicated about 20–25 million losses, a figure revealed by Khrushchev in the 1950s. Erickson and Dilks (1994) have calculated the total loss of military and civilian population. This included those who died from hunger and disease as well as the loss resulting from a drop in birthrate after the war. They estimate a 'loss' of 49 million, not including vast numbers of wounded, especially among the military. 'Success': about 25 per cent of the population of the USSR eliminated. One result was a huge imbalance in the sexes. In theory it had been the intention of the Nazis to use Soviet citizens as virtual slaves in a new German-controlled East Europe.

- Between 1975 and 1978 some 2 million people were killed or died in Kampuchea (Cambodia) as a result of the civil war, out of a total population of 7 million. The term 'killing fields' seems to have originated from this disaster. Genocide is a polite way of referring to the events.

- The Iran–Iraq War 1979–89. In this conflict most of the losses were of men of extended military age. During the decade of the war, about 1 million people died in the conflict. During the same time, the population of Iran grew from 36 to 52 million, and that of Iraq grew from 13 million to 17.6 million, a net gain altogether of over 20 million. Thus it could be argued that the war only set the population growth back by about six months.

These examples show that large numbers of people in individual regions, and in some cases whole continents, can die over short periods, leaving irregularities and 'scars' on the population structure. Nevertheless, in comparison with the normal processes of demographic change in the world, conflicts and catastrophes usually have only a limited impact. This is partly because there is a tendency for a vacuum to be filled by increased birthrate. A small-scale example was the death of virtually all the junior school children in a narrow age group in the mining settlement of Aberfan, Wales, in 1966. A landslide from a colliery tip destroyed their school, while they were in it. Many of the parents quickly produced more children, in effect replacing those lost in the disaster. On a larger scale, the population of Viet Nam, a country devastated by war for a twenty-year period, is growing very rapidly.

With regard specifically to the effects of conflicts on human populations, some changes have been taking place in recent decades. The large-scale bombing of cities from the air in the Second World War led to an increase in the proportion of civilian deaths in comparison to military deaths. During the two world wars, most deaths resulting from the conflict occurred in the developed countries. In contrast, since 1945, very few people have died in conflicts in the developed countries, while in some parts of the developing world large numbers of both the military and civilian population have died, particularly in Korea, Viet Nam and Cambodia, and more recently in many parts of Africa. In the 1990s, serious conflicts have occurred in Europe for the first time since 1945, affecting the three former Soviet republics of Transcaucasia and also Yugoslavia.

Box 3.2 shows graphically some examples of irregularities in demographic structures from a number of situations while Box 3.3 highlights the position in the age–sex structure of important subsets of the total population (and see accompanying Figure 3.4).

It has been estimated that about 20 million people have died as a result of internal or international conflicts since the Second World War, almost all of them residents of developing countries. In 1945 the population of the world was about 2,400 million, in 1995 about 5,700, an increase of 3,300 million. Examples of the number of deaths caused by conflict and genocide have been discussed at some length in this section to show the kinds of situation in which these events affect regional and global populations; 20 million deaths against a gain of 3,300 million is a minute number.

With the risk of a nuclear war greatly reduced, and the increasing influence of the United Nations and of world opinion on world affairs, it is difficult to envisage situations occurring in the next five decades very different from those in the last few. One possibility seriously considered by some has been a 'North–South' confrontation, replacing the East–West confrontation. Two decades ago it would have seemed ludicrous to envisage millions of Mexicans and other Latin Americans swamping out Americans in Florida, Texas and California. Similarly, a decade ago few would have anticipated a large influx of migrants from northwest Africa or Central Europe into the European Union.

3.4 How many people?

One of the most widely debated demographic issues is the question of how many people there should be in the world. At one extreme are those who argue that the number is unlimited, that science and technology can provide for the increased material needs of the growing population indefinitely, and that when necessary people will colonise other planets. At the other extreme are those who argue that there are already too many people, at least in some regions of the world. Some views are quoted below. An appropriate starter is Thomas Malthus.

- Malthus (1766–1834) was an English economist and cleric. In his *Essay on the Principle of Population* (1798, revised 1803), he argued that population control was needed for the reasons stated below (pp. 71 and 79). His recommendation for family planning was abstention from intercourse, a prospect less acceptable 200 years later than in his time:

> Population, when unchecked increases in a geometrical ratio. Subsistence only increases in an arithmetical ratio. A slight acquaintance with numbers will shew the immensity of the first power in comparison with the second.
>
> By that law of our nature which makes food necessary to the life of man, the effects of these two unequal powers must be kept equal.

> But though the rich by unfair combinations contribute frequently to prolong a season of distress among the poor, yet no possible form of society could prevent the almost constant action of misery upon a great part of mankind, if in a state of inequality, and upon all, if all were equal.
>
> The theory on which the truth of this position depends appears to me so extremely clear that I feel at a loss to conjecture what part of it can be denied.
>
> That population cannot increase without the means of subsistence is a proposition so evident that it needs no illustration.
>
> That population does invariably increase where there are the means of subsistence, the history of every people that have ever existed will abundantly prove.
>
> And that the superior power of population cannot be checked without producing misery or vice,

BOX 3.2

Demographic irregularities

There is no such thing as an ideal or average population structure against which irregularities can be measured. The population structure of each region, large or small, has irregularities, due both to changes in demographic behaviour and to external influences. The diagrams in Figure 3.3 illustrate quirks of structures to be found in regions of various sizes.

1. The 1982 population structure of China is shown by yearly cohorts. Greater detail is thus revealed. The large increase in population in the period of Soviet influence shows the effect of a peaceful decade following the end of internal conflict in 1949 and of improved dietary and healthcare provision. The Great Leap Forward, with hindsight inappropriately so called, roughly 1958–62, can clearly be seen, with both high infant mortality and a drop in actual births. During the rest of the 1960s, partly at least the result of a response to the deaths in the preceding years, partly through the pro-natal policy of the Communist Party during the Cultural Revolution, large families were common. The effect of subsequent efforts to limit the number of children in the 1970s can be seen in the youngest cohorts.

2. Japan's population structure shows the effect of the last two years of the Second World War on population, when both military and civilian losses were escalating rapidly, and the ensuing fairly brief baby boom. The year-by-year detail also shows a feature that would not stand out in a diagram with five-yearly cohorts, the effect of the 1966 Year of the Bad Horse, the year in which it was regarded as undesirable to bear a child because if she was female, once adult and married, she would kill her husband. This particular unlucky year comes every sixty years, but in the diagram there is no trace of the Bad Horse effect on the number of births in 1906.

3. The Paraguayan War is referred to on p. 54. The author has made a reconstruction of the population of Paraguay at the end of the war in 1871. The excess of females over males except in the youngest age groups is overwhelming.

4. In contrast to Paraguay, the population structure of Kuwait shows a notable excess of males over females, the result of immigration of men to work in the oil and construction industries of the country.

5. The greatly distorted form of the structure of the population of the central part of Shanghai reflects both inward migration of young adults, mainly over a brief period (the 1970s), and the effect of intensive family planning, resulting in many one-child families and presumably even some with no children.

6. The flow of population from other parts of Peru into the capital Lima gathered momentum in the 1940s. Since the 1950s, about half of the population of Greater Lima has been born in other departments of Peru. The Department of Apurimac, situated in the Andes to the east of Lima, has been one of the contributors to the growth of Lima, and remains 'deprived' of young adults. Note the difference in the horizontal scales of the two diagrams.

FIGURE 3.3 (opposite) **Demographic irregularities**

a, b Population structures of China (1982) and Japan (1978) showing yearly cohorts and details of irregularities

c Author's reconstruction of the population of Paraguay in 1871 at the end of the Paraguayan War. Thousands: total 221, women 106, men 29, children 86

d Excess of males over females in groups of working age, reflecting large-scale immigration

e Central part of Shanghai, China, in 1980

f Comparison of the population structures of Apurímac and Lima Departments, Peru

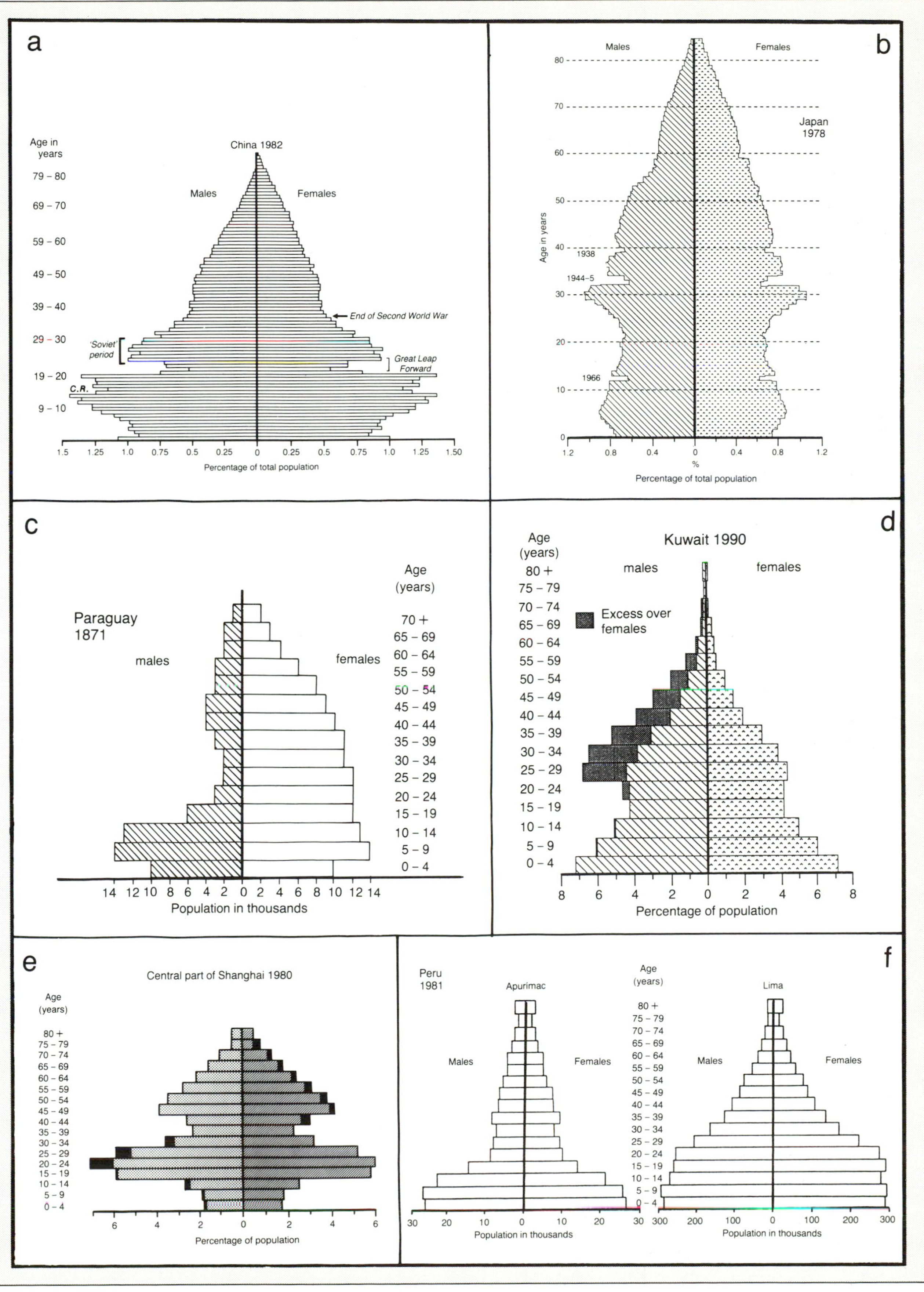

a
China 1982
Age in years
Males
Females
End of Second World War
'Soviet' period
Great Leap Forward
C.R.
Percentage of total population
b
Japan 1978
Males
Females
Age in years
1938
1944–5
1966
Percentage of total population
c
Paraguay 1871
males
females
Age (years)
Population in thousands
d
Kuwait 1990
Age (years)
males
females
Excess over females
Percentage of population
e
Central part of Shanghai 1980
Age (years)
Percentage of population
f
Peru 1981
Apurimac
Lima
Males
Females
Age (years)
Population in thousands

BOX 3.3

Demographic subsets

When referring to subsets of the total world population it is important to be clear on definitions. For geographers it is relevant to distinguish between subsets that are defined spatially and subsets that are not. For example, residence in a given area can be a definition: the citizens of the USA, of Texas, of Spain. Definition by age or by sex, on the other hand, is not explicitly spatial.

Various demographic issues and problems are related to particular subsets of total population. The diagrams in Figure 3.4 highlight some of the subsets.

A Women of **childbearing** age, actual and potential mothers, influence the potential for population change. In addition to its inherent relationship to reproduction, discrimination against females (mainly, presumably, by males) has become a very topical issue among geographers.

B Men eligible or liable to **military service** through conscription.

C The relationship of **dependent** to **working population** and the prospect of full employment or limited unemployment.

D The **elderly** and the expectation of support for them by the state in rich countries, given also that they take a disproportionate share of healthcare resources. **Children**, a declining proportion of total population in the rich countries, and therefore requiring fewer schools, but comprising a very high proportion in most poor countries, and putting strain on educational resources.

None of the above definitions applies consistently throughout the world. For example, retirement age in diagram D varies from country to country and there may not be an explicit age threshold in some. The subsets are also fuzzy, rather than sharply defined. For example, in A, women over the age of 44 give birth, especially in developing countries, as also do girls under the age of 15. In D young children under the age of 10 work, as they did, for example, in factories in Britain in the nineteenth century and as they do in some Asian countries now or in coal-mines in Colombia, while in some professions it is common for people to continue in employment well after the official age of retirement. Further points relating to three subsets follow: migrants, refugees, and children at risk of abuse.

• **Migrants** are found in all age groups, but generally with younger adults prominent. The predominant flow of international migrants in the latter part of the nineteenth century and into the 1920s was from Europe to the Americas and Australia. According to Kalish (1994) there were about 100 million international migrants in the mid-1980s. These were distributed as follows:

	% of world total
Asia and North Africa	36
Sub-Saharan Africa	10
Eastern and Western Europe	23
USA and Canada	20
Oceania	4
Latin America	6

In relation to their population sizes, the developed regions have far more migrants than the developing regions. Although much migration is either between pairs of developed regions or between pairs of developing ones, the richer regions remain attractive for migrants from poorer regions. Should such flows be encouraged, controlled, or as far as possible prevented altogether?

• **Refugees** According to Helgadottir (1994), the United National High Commission for Refugees (UNHCR) had in 1994 to deal with 19 million refugees world-wide, over 4 million of whom have emerged from three years of conflict in Yugoslavia. In mid-1994, from its small population of about 8 million, Rwanda produced 2–3 million refugees, fleeing mainly to Tanzania and Zaire. The official figure for refugees in 1970 was only 2.5 million. There are in addition some 25 million 'displaced persons'. Is the problem likely to continue to grow?

• **Children** Various forms of child abuse have recently been receiving much more publicity than in the past, at least in North America and Western Europe. Child labour is common in many countries. Mayes *et al.* (1992, p. 6) discuss various theories about the recent 'emergence' of child abuse. In the last resort, children belong to their parents, even if the state or local government can intervene in some countries:

> Those who maintain that child abuse is 'nothing new' would argue that there has never been a time when children have not been the subject of violence. Looked at in the long term historical perspective, it is plausible to argue that abuse is actually declining rather than increasing. Since ancient times children have suffered abuse in every conceivable form – physically, emotionally, through neglect and child labour and, not least, through sexual exploitation.

Child abuse is not the prerogative of the 20th century, parents have always beaten, whipped, burned, starved, neglected, over-worked and raped their children. The difference is that today this is recognised as a social problem worthy of our attention, and whilst it is recognised that children today still suffer very badly, great effort is expended to prevent this.

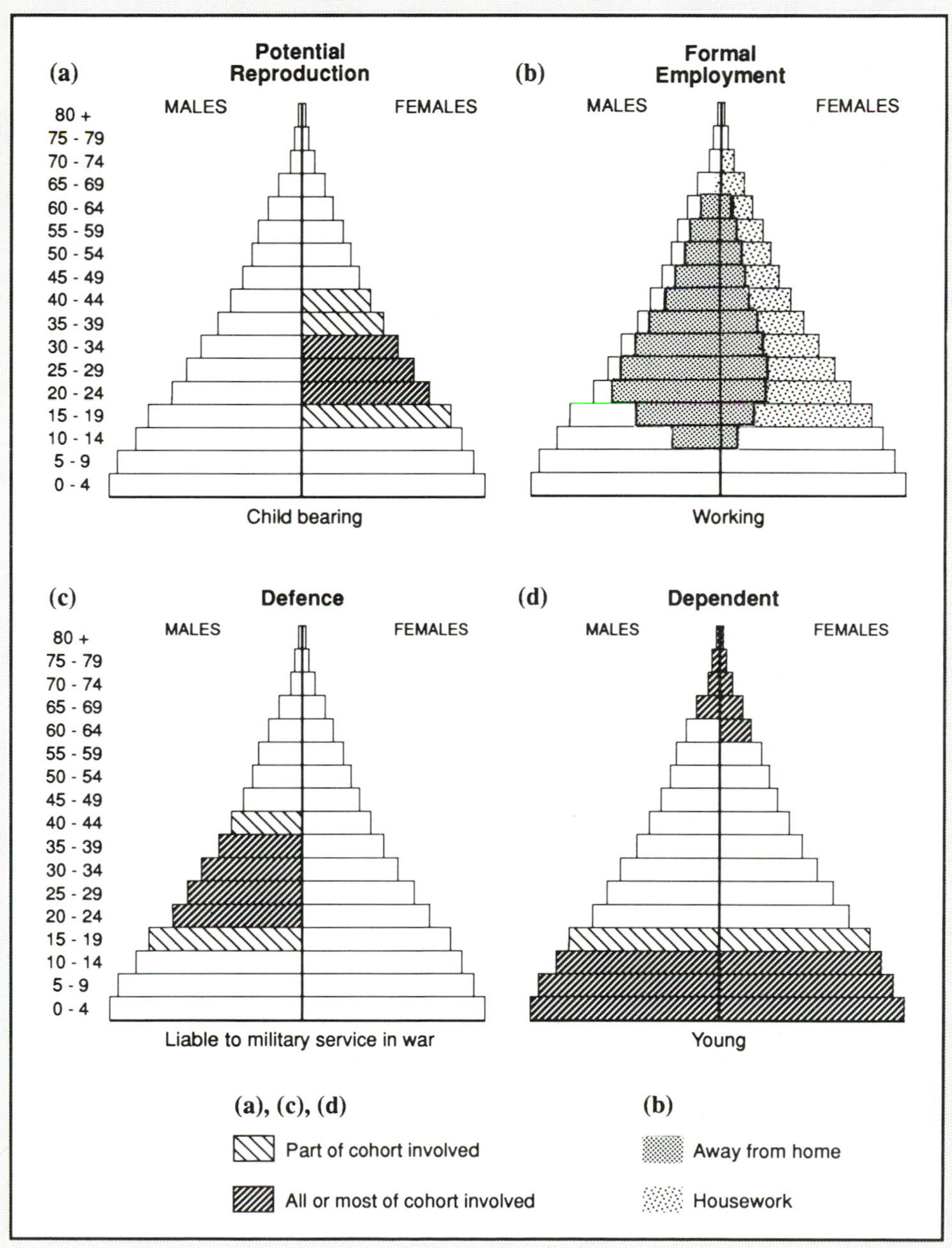

FIGURE 3.4 Key subsets of the total population of a hypothetical country

PLATE 3.1 Two thinkers who have made an enormous impact on affairs in the twentieth century and on geography

a Charles Robert Darwin (1809–82), English scientist who developed the modern theory of evolution and proposed the principle of natural selection. Darwin's findings disagreed with the literal interpretation of the Book of Genesis in the Bible. Rightly or wrongly, his ideas were transferred from biology to human society. (Photo by the author in Downe House, near Orpington, Kent)

b Karl Heinrich Marx (1818–83), German philosopher, economist and social theorist. Marx stressed the economic factors in politics. The way in which he perceived human society as changing (if not evolving) through time is not unlike Darwin's ideas of evolution. (Photo by author in Highgate Cemetery, London (fee paid £1!))

the ample portion of these too bitter ingredients in the cup of human life and the continuance of the physical causes that seem to have produced them bear too convincing a testimony.

- L. Dudley Stamp (1960, p. 19):

> However staggering the figures for total world population may be, it is the present rapid net increase which has turned attention almost everywhere to the need for a stocktaking of resources, especially of land and its capacity for the production of food, raw materials and supplies of energy.
>
> In the past, even the recent past, the rate of increase, evidenced usually by such readily visible signs as large numbers of young children, only called attention to itself in certain parts of the world. It became difficult to refer to China without mention of China's teeming millions – for some reason millions did not seem to teem elsewhere – or to the 'yellow peril' which took in the population expansion of both China and Japan. Other areas were India, with the seemingly hopeless poverty of its masses; in Europe one thought almost instinctively of the innumerable *bambinos* of Italy. These old ideas, in part misconceptions, die hard, but the world pattern of today is very different.

(Indeed! Italy now has the lowest birthrate in the world.)

- W. Zelinsky in Zelinsky *et al.* (1970); first published 1966, pp. 22–3):

> If anything is safely predictable for the final third of this most complex, dynamic, and crucial of mankind's many centuries, it is that the management and consequences of rapid, massive population increase will engross the attention of more and more of our best minds....
>
> The basic argument can be stated quite simply: We are about to confront practical decisions of the utmost gravity in our social, economic, ethical, political, and ecological affairs brought about, in large part, by very great population increments in recent years in most inhabited areas and by the even greater growth forecast for the immediate future....
>
> Attention is confined to the so-called 'underdeveloped' world, or 'developing lands', more for the sake of convenience than through conviction. It is quite possible that the arithmetically less alarming rates of population growth in the advanced nations, combined as they are with virtually unlimited expansion of economic production and consumption and by more mischievous manipulation of the

environment, may ultimately engender crises more pernicious and insoluble than those so visibly looming over less affluent countries.

- J Bongaarts (1994) raises the perennial question of how to feed the growing population of the world. He distinguishes environmentalists and ecologists, who see a catastrophe in the making from many economists and some agricultural scientists, who assert that technological innovation can produce enough food for well over the 10 billion people expected by about 2050. His conclusion is so vague that it seems to be designed to keep all readers of *Scientific American* happy (p. 24):

> Feeding a growing world population a diet that improves over time in quality and quantity is technologically feasible. But the economic and environmental costs incurred through bolstering food production may well prove too great for many poor countries. The course of events will depend crucially on their governments' ability to design and enforce effective policies that address the challenges posed by mounting human numbers, rising poverty and environmental degradation. Whatever the outcome, the task ahead will be made more difficult if population growth rates cannot be reduced.

- In 'What is overpopulation?' (Population Reference Bureau, *Interchange*, December 1988, vol. 17, no. 4), the difficulty of arriving at a simple, working definition is stressed (p. 2):

> In 1945, Frank Lorimer suggested three definitions of 'overpopulation': an area is overpopulated if fewer people would lead to a higher level of living, if population increase is more rapid than possible production increase, or if continuance of present growth trends would check economic advances. Thomlinson responds in 1965: 'using the first definition, about 95 percent of the world is overpopulated; using the second, there is little overpopulation, and using the third, about one-half of the world is overpopulated.'
>
> Obviously the term 'overpopulation' is not as simple to delineate as it may seem at first glance. Each community is different and cannot be judged by a universal standard. When debating the status of a community, consideration must be given to population growth rates, standard of living, lifestyle, culture, technology available, resources, type of economy, as well as a host of other variables.

3.5 Natural resources

Many data are available on the production and consumption of goods and services in various regions of the world and in the world as a whole. Data on the amount of natural resources are more limited and, since many mineral resources remain as yet undiscovered, inevitably more tentative. The annual report of British Petroleum keeps an up-to-date record of reserves of commercial sources of energy. At intervals the United States Bureau of Mines publishes estimates of reserves of a large number of non-fuel minerals. Publications of the Food and Agriculture Organization (FAO) show the area under crops, natural pasture and forest in the countries of the world. A rough guide to the availability of fresh water in different countries is given in Barney (1982). These sources have been used to compile an assessment of the overall natural resource endowment of the countries of the world, all of which have been appropriately allocated to the twelve major world regions. The way the resources were calculated, weighted and grouped is outlined in Box 3.4.

The weight and/or value of the production of goods and services can for practical purposes be measured reasonably accurately, but to put a value on, for example, the arable land of Canada or the tin reserves of Malaysia, involves subjective judgements. I have calculated the size of the world's natural resources in the late 1970s for 100 'cells' of the world, all roughly equal in population (Cole 1981), and also for the larger countries of the world around 1980 (Cole 1983). Instead of giving a monetary value to various natural resources, the share of the world total in each region or country was calculated, a method that will also be used here. Before the estimates used in this book are discussed, some general points will be made about natural resources.

Since the early 1970s, concern has grown in many quarters about the use of natural resources, the degradation of land resources and the depletion of minerals. The total quantity of water in the world changes only imperceptibly over time, but the availability of fresh water varies enormously from region to region. As with fresh water, changes in soil are very slow (it forms at a rate of about 2–3 cm or 0.8–1.12 inches a century). In practice, the inherent fertility of soil under given types of natural vegetation is modified by cultivation, irrigation and the use of fertilisers, and increasing areas of soil have been losing fertility or in extreme cases have been removed altogether by wind and water erosion.

Unlike water and soil, fossil fuel and non-fuel mineral resources are by their nature exhaustible, a term preferable to non-renewable because some minerals, including oil, are being formed, albeit on a very slow timescale. Some materials can, of course, be recycled, provided there are facilities for collecting scrap and waste. Another problem connected with the assessment of natural resources is how to take into account those that may be in the ground or under the sea but which have not yet been discovered. Some allowance should therefore be made for new discoveries, particularly in parts of the world where the search for minerals has been limited.

In the following assessment of all natural resources in the world, four familiar types of existing natural resource – water supply, bioclimatic resources, fossil fuels and non-fuel minerals – have each been given a weight of one-fifth. The total area of a region or country is also given a weight of one-fifth of the total natural resource score on the assumption that, all other things being equal, the larger the area of a region the more economic minerals it should have. Table 3.4 shows the scores for the 'population/natural resource' balance in the twelve regions of the world used in this book, while Table 3.5 shows scores for the twenty-three largest countries of the world in population. The methods used to calculate the quantity of each natural resource in each country, and the technical problems involved, are given in Box 3.4. Figure 3.5 shows the proportion of different 'uses' of the earth's surface.

Column (1) of Table 3.4 shows the percentage of the world's population in each major region. Columns (2)–(6) show the percentage of each of the five natural resources described above in each major region. Columns (2)–(6) can therefore each be compared with column (1) to find whether the share of each resource in a given region comes above or below its share of population. Each row therefore shows the resource 'profile' for the region. Thus, for example, in relation to their populations, Oceania, the former USSR and North America all do well on all five resources, Japan and South Asia very poorly. On the other hand, the region of North Africa and Southwest Asia has an uneven profile, with a very low score on water resources, but a very high one on fossil fuel reserves.

To arrive at an overall score combining all the five natural resources, the scores of each region in each of the five columns (2) to (6) have been summed (unweighted) in column (7) and divided by five to give in column (8) the percentage of all natural resources located in each region. The scores in column (8) are then divided by the scores for population in column (1) to remove the effect of differences in population size. Against the world average of 100, with its score of only 16 the region of Japan/South Korea is the most poorly endowed region with regard to natural resources,

BOX 3.4

Notes on the assessment of the natural resource endowment of the world

1. The *Food and Agriculture Production Yearbook* (1991) has been used as the source for **total land area.** In some countries, inland water surfaces make up a considerable proportion of the area, but these surfaces have been included. Greenland and Antarctica are omitted from the calculation since, even without a moratorium on the use of Antarctica's resources, the exploitation of the surface through mainly very thick ice is not a commercial proposition.

2. The calculation of the amount of **productive land** is a matter of greater subjectivity than the calculation of the other four resources. Only about 11 per cent of the total land surface of the world is under arable and permanent crops, compared with 25 per cent defined as natural pasture, 30 per cent as forest and woodland, and 34 per cent classed as 'other land' (waste, cities) or fresh water (2.5 per cent). The value of production from the 11 per cent under crops is, however, far greater per unit of area than the value of production from forest or from natural pasture, much of the latter being semi-desert, optimistically classed as pasture. In time, some of the natural pasture and forest will probably be improved or cleared for cultivation, but it would be naive to assume that the potential is very great or that changes could take place quickly, since human settlement has already spread to the farthest corners, taking in the best land, over millennia in some regions and centuries in others. A further complication in assessing the value of cultivated land is the fact that in South, Southeast and East Asia, about half of the land is double-cropped, with the result that the area harvested exceeds the area of cropland by as much as 50 per cent in some countries.

In order to take into account the differences in productivity of the different types of land use, in the calculations made for this chapter the area of arable land has not been changed, whereas forest has been reduced to a fifth of its area, and natural pasture to a tenth. Not surprisingly, the proportion of the national area of different countries under each of the four categories of land use noted above varies enormously. Thus, for example, 70 per cent of the area of Norway is in the 'other' category, as is 95 per cent of the area of Oman, almost all being waste land. In Uruguay, 76 per cent of the land is classed as natural pasture, in the Falkland Islands 97.5 per cent. In Finland 10 per cent of the total area is occupied by lakes and rivers. The overwhelming importance of herbaceous plants in the world food supply must be stressed. About 90 per cent of the direct human food comes from such plants, as does 90 per cent of animal food. Shrubs and small trees provide fruit, oils and beverages, while larger trees are almost exclusively used for fuelwood and construction.

having less than one-sixth of its 'share' of world natural resources, while Oceania, with a score of 1,040, has more than ten times its 'share'. A feature that does not receive the publicity it merits in many of the publications on development and on the world regions is clearly shown when the scores for natural resources per inhabitant for each region (column (9)) are compared with the scores for GDP per inhabitant (column (10)) (see Figure 3.6). GDP per inhabitant is a rough indicator of the rate at which primary products obtained from natural resources are being used, although some regions depend on imports from other regions for some of the fuel and materials they use. On the basis of this study, four broad types of region can be distinguished:

1. **Rich/developed regions with good natural resource scores** For convenience Oceania and North America have been amalgamated in Figure 3.6; with the former USSR, they are the members of this group. These regions use up exhaustible resources at a comparatively fast rate, but they also have large reserves of them, as well as good bioclimatic resources.

2. **Rich/developed regions with poor natural resource scores** Western Europe, Central Europe, and Japan with South Korea. These regions are large importers of primary commodities, mainly from group 1 or group 3 regions. Their high levels of material production and consumption are therefore maintained at the expense of the natural resources of other regions.

3. **Relatively poor/developing regions with a reasonable to good natural resource base** Latin America, Africa and Southwest Asia. These include most of the countries with the fastest growing populations in the world and, unless more natural resources are discovered (minerals) or 'created' (reclaiming land for cultivation), their scores in column (9) will gradually diminish.

4. **Relatively poor/developing regions with a poor natural resource base** South Asia, Southeast Asia and China. As population grows and economic growth takes place, increasing pressure will be put on their natural resources. Even the limited quantities of primary products they manage at present to export will

---

3. The availability of **fresh water** in a given area can be calculated very approximately by multiplying the area (in, for example, sq km) by the average 'depth' of precipitation. Over any large area there are considerable variations in precipitation, and these must be averaged out. Barney (1982) provides an estimate of the water supply of almost all the countries of the world.

4. Different types of **fossil fuel**, and also other primary sources of energy, can be compared by conversion to a common standard such as coal or oil equivalent. Reserves of fossil fuel can then be compared. For the estimation of this natural resource, only coal (various types), oil and natural gas have been used. Proved commercial reserves of coal are several times as large as proved reserves of oil and of natural gas, which are roughly similar. In the 1980s and early 1990s, more oil was consumed than coal or natural gas, but the consumption of natural gas has been increasing most rapidly. For simplicity, equal weight has been given to the reserves of each of the three fossil fuels, and they have been summed to give a final total. Nuclear fuels have not been considered.

5. **Non-fuel minerals** of commercial importance are so numerous and varied that the estimate of combined reserves of all non-fuel minerals is very approximate. For the calculation used here, 16 non-fuel minerals (12 metallic, 4 non-metallic) have been selected from over 70 covered in US Bureau of Mines (1985). They are weighted approximately according to their economic importance in the world economy. The weightings are as follows: a weight of 5 for iron ore, 4 each for bauxite and copper, 2 each for lead, zinc, phosphates, potash and industrial diamonds, and 1 each for tin, silver, gold, chrome, manganese, nickel, tungsten and sulphur. The relative importance of different non-fuel minerals can change over quite short time periods, for example when harmful effects on producers, consumers and the environment are recognised (e.g. asbestos, lead, uranium). The weightings chosen by the author therefore give only a broad picture of the present state of reserves of non-fuel minerals. Whatever minerals are included in the calculation and whatever reasonable weighting is given to them, the disparities in their availability among the countries of the world shown in Table 3.4 would remain. For example, the former USSR would always have 40–60 times the quantity of non-fuel minerals per inhabitant as Japan.

Air could be thought of as a natural resource, although it is usually taken for granted and, at least in its natural form, it is not 'produced', sold or transported and is available globally, even if it differs in quality from place to place.

TABLE 3.4 Population/natural resource balance in the world regions, early 1990s

	(1) Total popn	(2) Total area	(3) Productive land	(4) Water	(5) Fossil fuels	(6) Non-fuel minerals	(7) Total of all 5	(8) Total as %	(9) Nat res total ÷ popn	(10) Real GDP total ÷ popn
Western Europe	6.9	2.8	4.3	4.6	6.1	5.0	22.8	4.6	67	284
Central Europe	2.4	1.0	2.5	0.7	1.3	2.1	7.6	1.5	63	108
Former USSR	5.2	16.6	16.3	10.6	23.0	16.9	83.4	16.6	319	135
Japan/South Korea	3.1	0.4	0.5	1.1	0.0	0.3	2.3	0.5	16	261
North America	5.2	14.5	14.0	11.7	11.4	17.5	69.1	13.8	265	448
Oceania	0.5	6.4	4.5	2.0	3.1	10.1	26.1	5.2	1,040	260
Latin America	8.4	15.4	14.0	27.6	6.2	14.9	78.1	15.6	186	96
Africa S of Sahara	10.0	18.3	13.3	14.9	3.8	18.2	68.5	13.7	137	25
N Africa/SW Asia	6.4	9.4	4.8	1.5	34.8	5.0	55.5	11.1	173	81
South Asia	21.2	3.4	11.6	5.4	2.8	2.1	25.3	5.1	24	22
Southeast Asia	8.4	3.4	5.9	10.2	2.8	4.2	26.5	5.3	63	46
China plus	22.3	8.4	8.3	9.7	4.7	3.7	34.8	7.0	31	62
Total	100.0	100.0	100.0	100.0	100.0	100.0	500.0	100.0	100.0	100.0
N America + Oceania	5.7	20.9	18.5	13.7	14.5	27.6	95.2	19.0	334	437

Description of variables:
The data in columns (1)–(6) and (8) are percentages of the world total in each region
(7)–(8) The percentage shares located in each region of the fine natural resource categories are summed to give the overall natural resource share out of 500. These values are then divided by 5 to give the percentage of the world total in each region in column (8)
(9) The values in this column are the result of dividing the shares of natural resources in each region by the share of population (e.g. for North America 13.8 ÷ 5.2 gives 2.65) and then multiplying by 100 to remove the decimal point. The score of the region per inhabitant can then easily be compared with the world average of 100
(10) The share of GDP located in each region is similarly divided by the share of population in the region and multiplied by 100
Sources of data: See Box 3.4: *FAOPY* (1992) for (2) and (3); Barney (1982) for (4); BP (1993) for (5); US Bureau of Mines (1985) for (6)

diminish unless new natural resources are found. These last three regions have over half of the total population of the world but less than a fifth of the natural resources.

The diagram in Figure 3.6 shows the world situation around 1990. Regions have changed positions in this graph through time and will continue to do so in the future. If population growth occurs, regions will move towards the lower left of the space on the graph. If discoveries of natural resources increase more quickly than population, they will move to the right. If economic growth exceeds population growth, the regions will move upwards.

It is not difficult to tell when a local natural resource has run out, for example when a coal-mine closes because the remaining seams of coal are no longer commercial, or when erosion removes the soil from a field. Even at regional level some natural resources may run out. It seems to be a matter of time, for example, before fishing ceases completely in the Aral Sea in the former USSR because there will be no water. To visualise the running out of natural resources in the world as a whole is difficult although, given time, so long as humans are using up exhaustible resources, eventually they must run out. Box 3.5 shows a flaw in the view that a stable world population will solve the problem of natural resource exhaustion and environmental degradation. It will only delay the eventual exhaustion or the arrival at an unacceptable level of exhaustion or degradation.

One of the most spectacular examples in recent decades of the catastrophic exhaustion of a natural resource, the anchovy crisis of Peru, has been described by Idyll (1973). Nuttall (1994) cites a Worldwatch Institute estimate that, world-wide, nine out of ten fishing jobs, or the livelihoods of up to 19 million people, will have disappeared within twenty years unless governments act to curb the destruction of fish stocks. The peak-year catches of a number of commercially important and well-known species of fish were in the decade 1965–75. Since then, pollution, habitat destruction and over-exploitation have depleted stocks. To be sure, new sea areas and other species have been exploited since then, and between 1970 and 1990 (*UNSYB*) the catch has risen from 68 million to 97 million tonnes, keeping pace roughly with population growth.

TABLE 3.5 Natural resources of the countries with the largest populations

	(1)	(2)	(3)	(4)	(5)	(6)	(7)	(8)	(9)	(10)	(11)	(12)	(13)	(14)	(15)	(16)
				Bioclimatic					Fossil fuels							
	% share of popn	% of total area	% share of water	arable	forest	pasture	(4) + (5) + (6)	% of (4) + (5) + (6)	oil	gas	coal	(9) + (10) + (11) ÷ 3	Non-fuel minerals %	(2) + (3) + (8) + (12) + (13)	(14) ÷ 5	Share against world av.
China	21.41	7.17	9.35	966	253	400	1,619	6.23	2.4	0.8	11.0	4.73	3.70	31.18	6.24	30
India	16.30	2.46	4.27	1,691	133	12	1,836	7.06	0.6	0.6	6.0	2.40	2.00	18.19	3.64	20
Indonesia	3.41	1.42	4.22	220	227	12	459	1.77	0.7	1.5	3.1	1.77	1.40	10.58	2.12	60
Brazil	2.76	6.36	13.67	600	986	184	1,770	6.81	0.3	–	0.2	0.17	5.00	32.01	6.40	230
Pakistan	2.22	0.60	0.19	208	7	5	220	0.85	–	0.5	–	0.17	0.01	1.92	0.38	20
Bangladesh	2.07	0.11	0.36	91	4	1	96	0.37	–	0.6	–	0.20	0.10	1.14	0.23	10
Nigeria	1.73	0.69	0.75	323	24	40	387	1.49	1.8	2.4	–	1.40	0.20	4.53	0.91	50
Mexico	1.63	1.46	0.67	247	85	75	407	1.56	5.1	1.6	0.2	2.30	2.00	7.99	1.60	100
Viet Nam	1.30	0.25	1.24	66	20	1	87	0.33	–	–	–	0	–	1.82	0.36	30
Philippines	1.17	0.22	0.85	80	21	3	104	0.40	–	–	–	0	0.20	1.67	0.33	30
Iran	1.14	1.23	0.50	151	36	44	231	0.89	9.3	13.7	–	7.67	0.10	10.39	2.08	180
Turkey	1.10	0.58	0.48	279	40	9	328	1.26	–	–	0.7	0.23	0.10	2.65	0.53	50
Egypt	1.06	0.75	0.01	26	0	0	26	0.10	0.4	0.3	–	0.23	0.10	1.19	0.24	30
Thailand	1.04	0.38	0.38	221	35	2	258	0.99	–	–	–	–	0.10	1.85	0.37	40
Ethiopia	1.03	0.91	0.38	139	54	45	238	0.92	–	–	–	–	–	2.21	0.44	40
USA	4.69	7.00	4.78	1,899	587	242	2,728	10.49	3.4	3.9	23.1	10.13	8.00	40.40	8.08	170
Russia	2.71	12.75	10.60	1,320	1,600	80	3,000	11.54	5.0	35.0	18.0	19.33	14.00	68.22	13.64	500
Japan	2.27	0.36	1.11	46	50	1	97	0.37	–	–	0.1	0.03	0.20	2.07	0.41	20
Germany	1.47	0.27	0.28	124	21	6	151	0.58	–	0.2	7.7	2.63	0.13	3.89	0.78	50
UK	1.05	0.18	0.34	67	5	11	83	0.32	0.4	0.4	0.4	0.40	0.10	1.34	0.27	30
Italy	1.05	0.23	0.43	121	13	5	139	0.53	–	–	–	–	0.10	1.29	0.26	20
France	1.05	0.41	0.61	193	30	11	234	0.90	–	–	–	–	0.20	2.12	0.42	40
Ukraine	0.94	0.45	0.45	334	20	5	359	1.38	0.1	0.2	3.0	1.10	0.50	3.88	0.78	80

Note: – negligible quantity or none

Description of variables:

(1)–(3) The share of the world total in each country. Note that the sums of the values in these columns do not add up to 100 because only the largest countries are included in the table

(4)–(8) Bioclimatic points, summed in column (7) and expressed as a percentage of the world total (8)

(9)–(12) Fossil fuel percentages summed and divided by 3 to give overall fossil fuel reserves share of the world total

(13) Percentage share of non-fuel mineral reserves

(14), (15) The five categories of natural resources in columns (2), (3), (8), (12) and (13) summed up to give share of world grand total of 500, then divided by 5 in column (15) to give share of world total as a percentage

(16) Percentage share of natural resources in column (15) divided by percentage share of world population (column (1)), then multiplied by 100 (e.g. Pakistan 0.38 ÷ 2.22 gives 0.17, multiplied by 100 to remove the decimal point, then rounded, gives 20 compared with a world average of 100)

Sources of data: as for Table 3.4

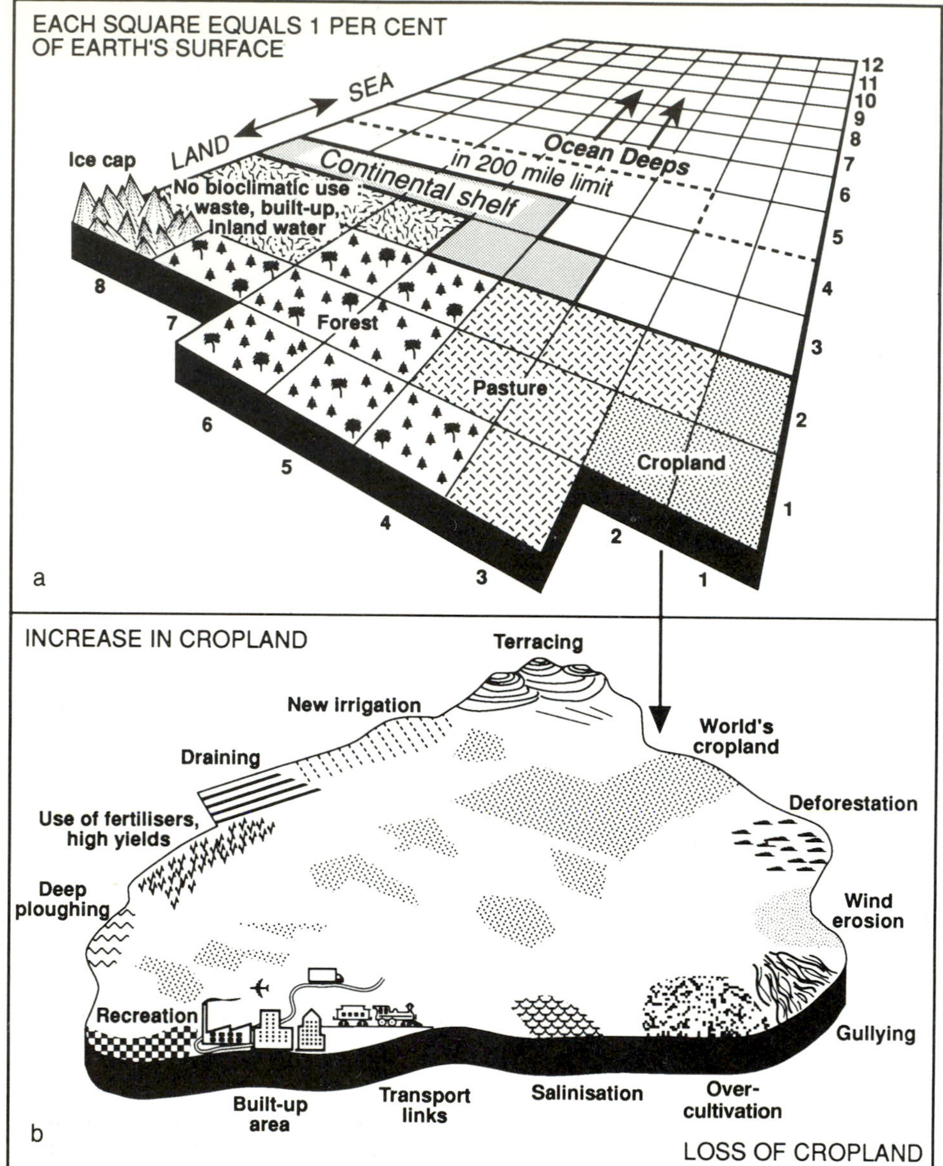

FIGURE 3.5

a The use of the earth's surface

b Potential changes in the use of the 3 per cent of the total surface currently used for cultivation

3.6 Production and consumption

In order to introduce the subject of production and consumption, four measures have been selected. *Gross Domestic Product* (GDP) per inhabitant is the overall indicator of the production of goods and services in an economy. *Energy* consumption, which differs greatly from production in some regions, reflects the general level of mechanisation of an economy, while including also the need for domestic heating in some regions. *Steel* production is central in heavy industry and is one of the main ingredients of the engineering industry and of some types of construction. The number of *passenger cars* in circulation is a guide to the level of affluence in different regions, and since cars have been widely used even in North America and Europe only from the inter-war period (1920s and 1930s), car ownership is a guide to differences in living standards between regions. In Chapter 17, past and possible future car ownership will also be studied. The above variables (see Table 3.6) will be discussed in turn.

- Columns (2)–(4). *Real GDP* compares the actual domestic purchasing power of national currencies rather than their value when converted at a particular exchange rate to US dollars or some other common currency (see Chapter 1). As shown in Table 1.7, the real measure of GDP narrows the gap between some rich and some poor regions considerably compared with the conventional measure using the exchange rate to dollars. A comparison of the share of population (column (1)) and share of real GDP (column (3)) shows that the average level of the six richer regions is only about six times as high as that of the six poorer regions,

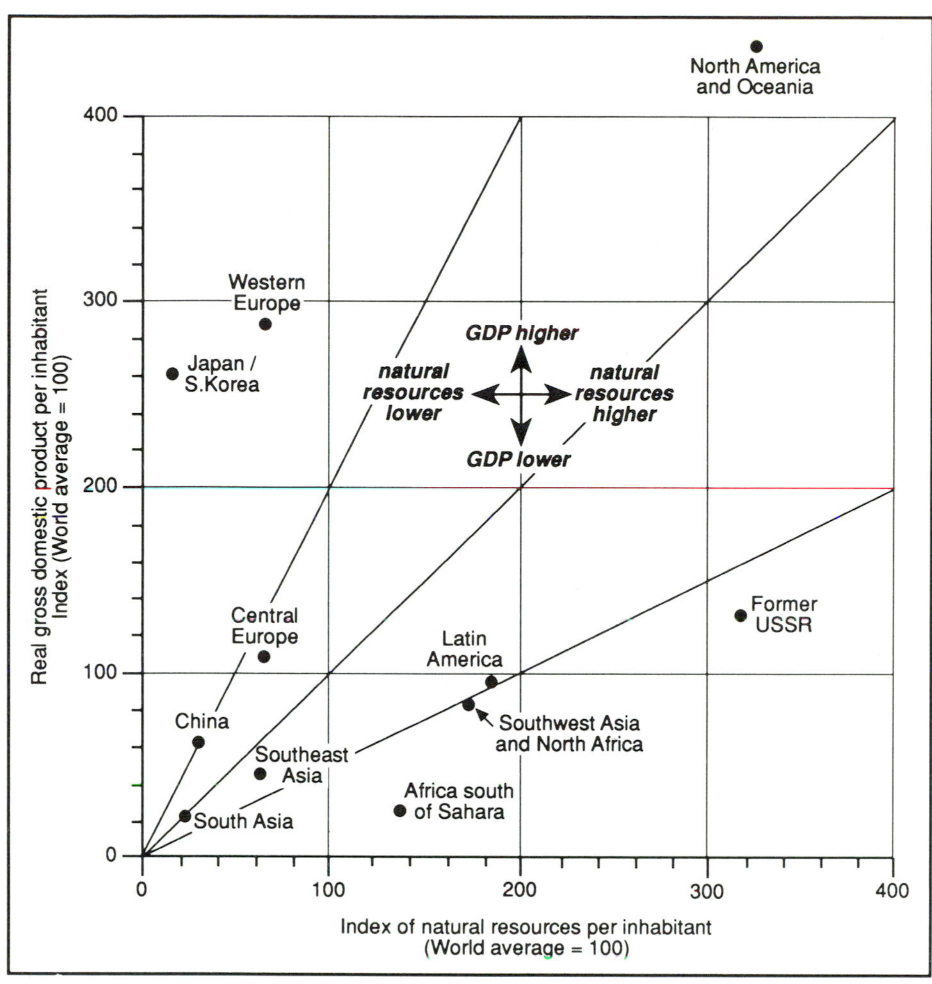

FIGURE 3.6 The relationship of natural resource availability to GDP per inhabitant in major regions of the world. Data are in Table 3.4

compared with a gap of 17 to 1 (*WPDS 1993*) for Gross National Product (GNP) at current exchange rates. This disparity illustrates how the *same situation* can be measured by two different indices to give widely different results; the situation does not change if it is described differently, but the perception of it may. Even a 6:1 gap is large, however, and it would indeed be wider still but for the fact that the group of developed regions is pulled down by the recently acknowledged poor performance of Central Europe and the former USSR. Comparisons between regions and between countries are less controversial and subjective if they are made on the basis of more concrete products or services rather than GDP, especially when changes over time are also to be compared.

• The consumption of commercial sources of *energy* given in Table 3.6 covers only the primary sources, which consist of fossil fuels, nuclear power and hydro-electric power. The electricity generated from fossil fuels is not counted as a primary source. Fuelwood is also widely used, especially in developing countries, and, if this is included, it boosts the level of energy consumption considerably in some of the poorer regions. With less than one-quarter of the total population of the world, the six richer regions use almost three-quarters of the energy of the world, almost ten times as much per inhabitant as is used in the six poorer regions. In relation to GDP per inhabitant, Central Europe and the former USSR are very wasteful users of energy, a consequence partly of climatic conditions, but mainly of the emphasis on the growth of heavy industrial capacity and of the lack of incentives in the centrally planned economy to economise with fuel. Indeed, high production targets were often set more because they were regarded as desirable features of economic growth than because the extra products were immediately needed.

• *Steel* production can be measured with considerable accuracy, but it should be noted that there are great variations in the quality of steel produced for different purposes, the developed countries accounting for most of the output of the special steels. The gap between richer and poorer countries is greater in steel output than for energy or total GDP, but it was even more marked several decades ago (see Chapter 17). At the level of individual regions, there is a striking contrast between Japan, which produces more than five times the world average level, and the region of

BOX 3.5

The fallacy of zero growth

A widely expressed view among those who are concerned about the increasing pressure of population on natural resources is that, if and when the population of the world stabilises, there will no longer be a problem. In other words, it is the growth of population that is the problem, not the fact that there are people. A stable population could be sustained indefinitely. With the help of the diagram in Figure 3.7 it can be shown that reaching zero growth of population in the world only delays the use of natural resources and the creation of pollution.

The message in Figure 3.7 is simple. Each unit of megapop is equivalent to 1 billion people in the world for 5 years. For simplicity it is assumed that, during the period shown, per capita consumption of natural resources and of goods and services remains constant. In reality the level actually rose considerably between 1975 and 1995, while economic growth is the goal of almost every government in the world.

In Figure 3.7 line *a* shows how a continuation of growth of population of 80–90 million per year or 2 billion every 25 years, gives a population of 10 billion in 2050. Line *b* shows what happens in the unlikely event of a complete moratorium on population growth in the year 2000 and an equally unlikely subsequent no-change situation for 50 years. The shaded area between lines *a* and *b* shows the megapops saved through zero growth of population. In situation *a*, between 1975 and 2050, 105 megapops (15 × 5 years by average 7 billion people) have been used, in situation *b* only 85. Assuming that population continues at 6 billion after 2050, then the 20 megapops saved by zero growth are used up in the next 17–18 years.

At least two strategies could be used to modify prospects for the future. First, population could actually be reduced, an unlikely prospect for some decades at least. Second, consumption levels per inhabitant could be reduced, an equally unlikely prospect, since the big consumers in the rich countries may prosper and be satisfied with present levels but the small consumers in the poor countries want more. Even if both the above changes were to occur, eventually the 105 megapops in projection *a* would be reached and passed, *so long as people are there.*

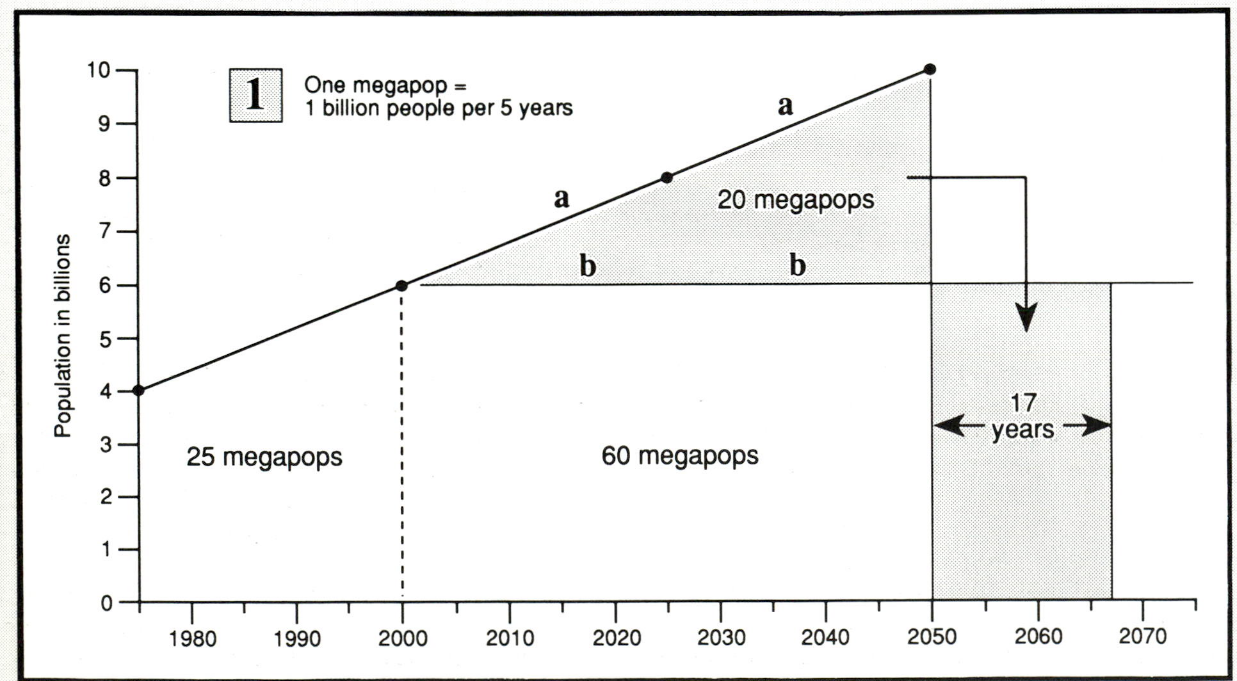

FIGURE 3.7 Illustration of the fallacy of assuming that zero growth of population will solve the problem of population pressure on non-renewable natural resources and the quality of the environment

TABLE 3.6 Aspects of the production and consumption of goods

	(1)	(2)	(3)	(4)	(5)	(6)	(7)	(8)	(9)	(10)	(11)	(12)	(13)
		GDP			*Energy consumption*			*Steel production*			*Cars in use*		
	% of popn	*real GDP billions of US$*	*% of real GDP*	*GDP % ÷ popn %*	*millions of tce* [1]	*% of energy*	*energy % ÷ popn %*	*millions of tonnes*	*% of steel*	*Steel % ÷ popn %*	*millions of cars*	*% of cars*	*cars % ÷ popn %*
Western Europe	6.9	5,005	19.6	284	1,682	16.4	238	161.9	21.1	306	143.5	33.6	487
Central Europe	2.4	659	2.6	108	543	5.3	221	52.7	6.9	288	15.5	3.6	150
Former USSR	5.2	1,787	7.0	135	1,921	18.7	360	160.1	20.8	400	16.3	3.8	73
Japan/South Korea	3.1	2,059	8.1	261	584	5.7	184	131.6	17.1	552	34.2	8.0	258
North America	5.2	5,948	23.3	448	2,792	27.2	523	107.0	13.9	267	155.8	36.5	702
Oceania	0.5	325	1.3	260	141	1.4	280	6.6	0.9	180	8.8	2.1	420
Latin America	8.4	2,055	8.1	96	505	4.9	58	41.4	5.4	64	27.9	6.5	77
Africa S of Sahara	10.0	645	2.5	25	163	1.6	16	9.2	1.2	12	6.0	1.4	14
N Africa/SW Asia	6.4	1,317	5.2	81	457	4.5	70	11.9	1.5	29	10.4	2.4	38
South Asia	21.2	1,188	4.7	22	298	2.9	14	12.9	1.7	8	3.4	0.8	4
Southeast Asia	8.4	995	3.9	46	160	1.6	19	0.9	0.1	1	4.8	1.1	13
China plus	22.3	3,485	13.7	61	999	9.8	44	72.2	9.4	42	0.8	0.2	1
World	100.0	25,464	100	100	10,245	100.0	100	768.4	100.0	100	427.6	100.0	100
Developed	23.3		61.9			74.7			80.7			87.6	
Less developed	76.7		38.1			25.3			19.3			12.4	

Note: [1] tce = tonnes of coal equivalent

Description of variables:

(1)) Percentage share of total world population in each region

(2)–(4), (5)–(7), (8)–(10), (11)–(13) The first variable in each set of three is the absolute production or consumption in the appropriate units of measurement. The second variable in each set of three is the percentage of the world total located in each region. The third variable in each set is the percentage in the second column of each set divided by the percentage of population in the region given in column (1) and then multiplied by 100 (e.g. for Latin American energy consumption, 4.9 per cent of energy divided by 8.4 per cent of population gives 0.58, which becomes 58 when multiplied by 100)

Southeast Asia, where steel production in modern plants began only in the 1980s and is still very limited. Almost all of the steel produced in the region of Africa south of the Sahara is accounted for by South Africa.

• The gap in *car ownership* levels between different regions of the world is even more marked than it is for GDP, energy or steel. With about 55 cars per 100 inhabitants, North America has about seven times the world average level of passenger car ownership, which is less than 8 per 100. Until the 1980s individuals were not permitted to own cars in China and even now the 800,000 passenger cars registered are mostly taxis and official vehicles. If every country in the world had a level of car ownership comparable with that in the United States there would be 3 billion cars in circulation instead of the 430 million at present, with enormous consequences for world oil supplies, urban environments and increases in carbon dioxide and other emissions into the atmosphere.

The production and consumption of other goods and the provision of services such as healthcare and education also vary greatly among the regions and countries of the world. For obvious reasons the intake of food does not vary as widely as the ownership of cars, although the composition and quality of diet, as opposed to quantity, do vary considerably.

PLATE 3.2
One of the traditional themes in geography has been the relationship of humans to the natural environment, not always a happy one

a Total clearance of the tropical rainforest, Amapá, North region of Brazil. The clearance of forest for cultivation and pastures has long been regarded as a mark of material progress and development. Many people are now concerned at the rapid cutting of the world's tropical rainforests (see also Chapter 11)

b Effective destruction of a hotel in the centre of Mexico City following the earthquake in September 1985 (see also Box 11.1, pp. 260–1)

For climatic, geological or technical reasons, many products are very limited in their distribution and places of production, whether they be certain species of tropical hardwood, comparatively rare minerals like chrome ore, or large jet aircraft.

In Chapters (5)–(16) of this book further data will be provided on the production and consumption of various goods and services for specific regions, this section being intended as an introduction to some of the outstanding examples of global disparities. Chapter 4 contains an overview of the 'development gap', summarising the variability in demographic features, natural resource endowment, and production and consumption levels discussed in this chapter with reference to the larger countries of the world. Transactions between the regions of the world are then considered with special reference to trade, transnational companies and development assistance. Finally the impact on the natural and human environment of the rapid increase in population and production in the world is considered.

PLATE 3.3
The environment at risk. One of the threats to conveniently located forest and scrub is the gathering of fuelwood for cooking and heating in developing countries

a Branches of cactus plants are stacked for collection at the roadside, Haiti

b Two Ethiopian women carry fuelwood to a nearby town. For two or three days' work they will get about 2 dollars (US). They would have to work for about thirty years to earn what the two onlookers, American tourists, paid to visit Ethiopia (see also Chapter 12)

KEY FACTS

1. The population of the world has grown about threefold between 1925 and 1995, roughly during the average lifetime of a citizen of the developed world.

2. In the 1980s and early 1990s the relative increase of population in the world as a whole was declining, but the absolute gain per year has remained at around 80–90 million.

3. Only a major nuclear disaster, the impact of an asteroid (meteorite) or some other global catastrophe could produce a sudden marked change in population numbers in the world.

4. The consumption of most primary products by the developed countries of the world has not increased greatly in the last two decades, partly because population is growing only slowly, partly because the efficiency with which materials and energy are used has improved. On the other hand, some developing countries have increased their consumption of primary products considerably in recent decades.

5. The developing countries of today are able to adopt and adapt technologies and means of production that have been developed over some two centuries in the industrialised countries of today. Such countries as Brazil, India, China and South Korea are therefore in a position to raise material standards over shorter periods than was possible in the eighteenth and nineteenth centuries in the present industrialised countries, requiring increasing use of reserves of natural resources.

MATTERS OF OPINION

1. Should governments enforce family planning on the population and, as in China, impose sanctions on families exceeding their 'ration' of babies?

2. Should the consumption of fuel and raw materials be rationed in some way?

3. Should the growth of very large cities be controlled? According to Cornish (1986), of the twenty largest cities in the world, eight were in the developed countries (Tokyo 17 million, New York/Northeast NJ 15 million), twelve in the developing countries (Mexico City 18 million, São Paulo 16 million). The eight in the developed world had a combined population of 85 million and are expected to grow to less than 90 million in the year 2000. The eight largest in the developing countries had 98 million in 1985, a total expected to grow to 137 million in the year 2000 (Mexico City with 26 million and São Paulo with 24 million heading the ranking).

4. Do we rely too heavily on the application of new scientific discoveries and the invention of new technologies to raise material standards and to solve environmental and other problems?

SUBJECTS FOR SPECULATION

1. When and with how many people do you expect the population of the world to peak?

2. Can economic growth in the world be sustained indefinitely?

3. Is it just a matter of time before various non-renewable natural resources run out, or can the development of renewable sources of energy, the conservation of bioclimatic resources, and the careful recycling of materials prolong the age of mass consumption indefinitely?

Chapter four

Global contrasts

The widening gap between the developed and developing countries has become a central issue of our time.
The effort to reduce it has inspired nations left behind by the technological revolution to mobilise their resources for economic growth. It has also produced a transfer of resources on an unprecedented scale from richer to poorer countries . . . .
However, international support for development is now flagging.

Lester Pearson (1969)

4.1 The development gap

Lester Pearson of Canada, like Willy Brandt of West Germany more recently, and a number of other eminent political figures, have lent their names to movements to publicise the great gap between rich and poor countries. More often the politicians who advocate a greater transfer of resources from rich to poor countries are those of lower rank, those no longer in high office, or those making election promises. Those in power are more concerned with winning the next election or, like President Reagan during his presidency in 1980–8, have little time for foreign aid anyway. Nevertheless, the United Nations solemnly continues to produce reports on global inequalities, putting countries into various categories according to their level of development or per capita income, and assessing their eligibility for development assistance.

Until the break-up of the USSR in 1991, UN publications recognised a fourfold division of the countries of the world into market economies, which were either developed (e.g. USA) or developing (e.g. India), and centrally planned economies, which were also either developed (e.g. USSR) or developing (e.g. China). There were still some centrally planned economies in the world in the mid-1990s (North Korea, Cuba, arguably China), but the four categories are less appropriate now. The precise grouping of countries into developed and developing varies according to who is using the terms and what their particular needs are. One example of a definition of the developing countries is the UN grouping for the tabulation of development assistance: all Africa except South Africa; Latin America and the Caribbean; Asia except for Japan and the Asian part of the USSR; and Oceania apart from Australia and New Zealand (i.e. Papua-New Guinea and numerous small Pacific islands).

The twenty-three countries of the world with over 50 million inhabitants in 1993 are shown in Table 4.1. They have been chosen to exemplify the gap between developed and developing countries. The range between extremes on most of the variables is even

TABLE 4.1 The development gap

	(1) Population (millions)	(2) Area in thous. sq km/(miles)		(3) Density per sq km/(mile)		(4) TFR	(5) Infant mortality	(6) Natural resources	(7) 1989 energy cons.	(8) Steel prodn	(9) Life expectancy	(10) Educational attainment	(11) Real GDP	(12) HDI
Developing														
1 China	1,179	9,597	(3,705)	123	(319)	1.9	53	30	1,210	54	70	175	2,660	612
2 India	897	3,288	(1,269)	273	(707)	3.9	91	20	310	14	59	93	910	297
3 Indonesia	188	1,905	(736)	99	(256)	3.0	68	60	310	0	62	177	2,030	491
4 Brazil	152	8,512	(3,286)	18	(47)	2.6	63	230	790	165	66	187	4,950	739
5 Pakistan	122	796	(307)	153	(396)	6.7	109	20	270	0	58	55	1,790	305
6 Bangladesh	114	144	(56)	792	(2,051)	4.9	116	10	70	0	52	58	820	185
7 Nigeria	95	924	(357)	103	(267)	6.6	84	50	210	0	52	89	1,160	241
8 Mexico	90	1,958	(756)	46	(119)	3.4	38	100	1,740	81	70	209	5,690	804
9 Viet Nam	72	333	(129)	216	(559)	4.0	45	30	130	0	63	209	1,000	464
10 Philippines	65	300	(116)	217	(562)	3.9	72	30	300	5	64	237	2,270	600
11 Iran	63	1,648	(636)	38	(98)	6.6	76	180	1,650	14	66	119	3,120	547
12 Turkey	61	780	(301)	78	(202)	3.6	59	50	1,030	130	65	182	4,000	671
13 Egypt	58	1,002	(387)	58	(150)	4.6	56	30	740	29	60	97	1,930	385
14 Thailand	57	513	(198)	111	(287)	2.4	40	40	640	10	66	216	3,570	685
15 Ethiopia	57	1,222	(472)	47	(122)	7.5	127	40	20	0	46	126	390	173
Developed														
16 USA	258	9,373	(3,619)	28	(73)	2.0	9	170	10,120	356	76	300	21,000	976
17 Russia	149	17,075	(6,593)	9	(23)	1.7	20	500	7,500[1]	622	71	262	6,270	873
18 Japan	125	477	(184)	262	(679)	1.5	4	20	4,000	865	79	287	14,310	981
19 Germany	81	357	(138)	227	(588)	1.4	7	50	5,850	603	75	290	14,510	955
20 UK	58	245	(95)	237	(614)	1.8	7	30	5,040	328	76	294	13,730	962
21 Italy	58	301	(116)	193	(500)	1.3	8	20	3,810	443	76	254	13,610	922
22 France	58	552	(213)	105	(272)	1.8	7	40	3,920	334	76	294	14,160	969
23 Ukraine	52	604	(233)	86	(223)	1.8	18	80	6,800[1]	1,056	71	262	6,270	873
World	5,506	133,916	(51,705)	41	(106)	3.3	70	100	1,860	141	65	150	4,620	526

Note: [1] Author's estimate

Description of variables:

(4) TFR total fertility rate
(5) Per thousand live births
(6) Per inhabitant compared with world average = 100
(7) Kilograms per inhabitant
(8) Kilograms per inhabitant
(9) In years
(10) Maximum score 300
(11) Dollars per inhabitant
(12) Human Development Index (maximum 1,000)

Sources of data: *WPDS 1993* for (1), (4), (5); (6) see Box 3.4; (7), (8) *UNSYB 90/91*; (9)–(12) UN, *Human Development Report 1992*, Table 1 greater for smaller countries than for the larger ones in the table. The data will now be discussed column by column.

Columns (1)–(3) show that in area, population size and population density the two groups overlap.

(4) Total fertility rate is a key demographic variable, giving an indication of population change expected over the next few decades. In the world as a whole in 1994, the averages for the developing and developed countries were 3.6 and 1.7 respectively. China and the USA have similar rates, but otherwise there is no marked overlap between developed and developing.

(5) Infant mortality rates vary greatly among the developing countries in the table, ranging from Ethiopia, officially with 127 per thousand (but probably more), to Mexico with 38 per thousand, but they are higher than for any of the developed countries. The inferior healthcare services of Russia and Ukraine distinguish them in the developed group.

(6) Natural resource scores for the twelve regions of the world were compared in Chapter 3. A similar distinction between 'resource-rich' and 'resource-poor' is clear in this table, with developed and developing countries included in both groups. Figure 4.1a shows that even among the twenty-three largest countries there is no correlation between natural resources per inhabitant and total production per inhabitant. On the other hand, in Figure 4.1b the broad correlation between energy consumption and total production is positive, although not strong, the main reason for the low correlation being that some countries use energy more efficiently than others.

(7) Although there are large disparities in levels of energy consumption among the developing countries, there is a clear gap between the highest, Mexico, with 1,740 kilograms (1.74 tonnes) of coal equivalent per inhabitant (kce), and the lowest level among the developed countries, that of Italy, with 3,810 (3.81 tonnes). (Conversion: 1 metric ton (also spelled tonne) is equal to 1.1023 short tons and 0.984 long tons.)

(8) Steel production has increased rapidly in many developing countries since the Second World War, with Brazil and Turkey near to the world average output per inhabitant. The gap between developed and developing is, however, very marked in this key sector of heavy industry.

A recently published measure of 'human development' has been produced by the UNDP and indices from *Human Development Report 1992* are used in columns (9)–(12) of Table 4.1. The method by which the index is calculated is explained below.

(9) Life expectancy has risen considerably in almost every part of the world since the end of the Second World War, with local relapses due to conflicts and food shortages. The gap between Japan (79 years) and Ethiopia (46 years) is very substantial, but on this variable there is a continuum of values from the highest developed to the lowest developing, with China and Mexico close to Russia and Ukraine around the 'interface' between the groups.

(10) Knowledge or educational attainment is measured by combining two widely used variables, adult literacy and mean years of schooling. In educational attainment there is no overlap of developed and developing, but a wide range among the developing countries.

(11) The adverse economic and financial state of Russia and Ukraine in the early 1990s leaves them far below the GDP level of the other developed countries listed here. Again, there are marked differences between the developing countries.

(12) The Human Development Index (HDI) is the consensus of the positions of each country on each of the three variables: (9) reflecting health (physical well-being), (10) education (representing mental attainment) and (11) purchasing power adequate to satisfy basic material requirements. Any dollars in excess of a purchasing power of about 5,000 per capita, the average poverty line in a selection of developed countries, are ignored. The maximum possible HDI score is 1,000, the position of a hypothetical country at the top of the range on all three variables. In the 1990 assessment, Canada came top, dropping only 18 points out of 1,000, while Japan was second, dropping 19. At the other end of the range, Guinea and Sierra Leone, both in Africa, scored 52/1,000 and 62/1,000, dropping 948 and 938 out of 1,000 respectively, while Afghanistan, the lowest scoring country in Asia, scored 65/1,000.

The UN HDI has been criticised as naive, unrepresentative and incomplete. Indeed, it was accepted in the 1990 and 1992 versions that it was not possible (or expedient) to add more abstract aspects such as individual freedoms of speech or mobility, or happiness, as opposed to material satisfaction. The reader may draw his or her own conclusions about the state of the development gap, illustrated only very broadly by the data in Table 4.1, but some features worth consideration in the view of the author are outlined in the next section.

4.2 Features of the development gap

With most variables in Table 4.1 there is a clear break between the developed and the developing countries, although if certain smaller countries were added to the list, spaces would be filled at the interface between the two groups. Often, however, the range between the highest and lowest scorers within each group is much greater than the extent of the gap itself, as with life expectancy and infant mortality. One inescapable conclusion is that it is not realistic in the 1990s to consider

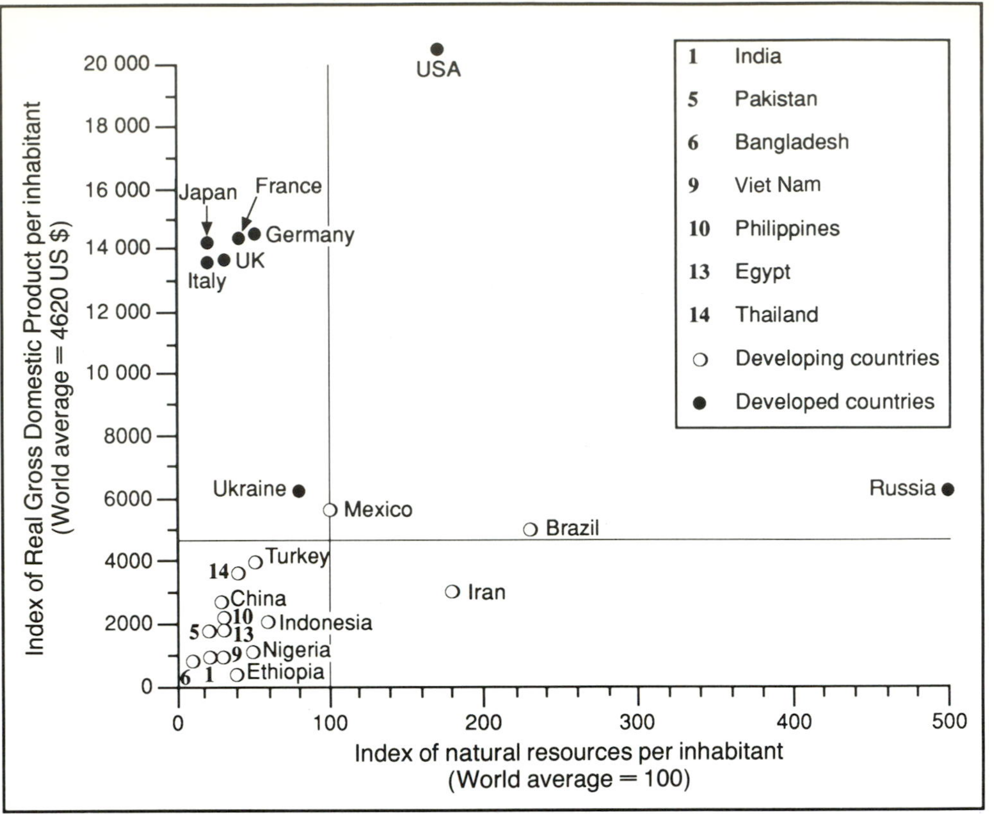

FIGURE 4.1 The relationship of natural resource availability (*a*) and of energy consumption (*b*) to GDP per inhabitant in the twenty-three countries of the world with the largest populations. Data are in Table 4.1

(a)

developed and developing countries as members of two discrete, homogeneous sets and, indeed, it never has been, although for simplicity, and for practical purposes, some dividing line has often been needed.

If the developed countries were isolated from the developing ones, a development gap would still be recognised in the remaining developing world itself. Top performers such as Mexico, Southeast Brazil, coastal China, the Middle East oil states and Singapore would stand at or near the top. There is a strong case, therefore, for directing development assistance towards the poorer developing countries, and indeed a set of Least Developed Countries, mostly in Africa and southern Asia, has been recognised by the World Bank (see Figure 4.2).

According to the UN Human Development Report (1992, see p. 1), since the Second World War the development gap has widened. 'In 1960, the richest 20% of the world's population had incomes 30 times greater than the poorest 20%. By 1990, the richest 20% were getting 60 times more.' The deterioration has taken place because in some of the least developed parts of the world there has been little change and, in some cases (e.g. possibly Bangladesh, Ethiopia, Haiti), conditions are actually worse than they were in the 1940s, whereas material standards have risen dramatically in most parts of the developed world, at least into the 1970s. The Report proposes two main reasons (p. 1):

First, where world trade is completely free and open – as in financial markets – it generally works to the benefit of the strongest. Developing countries enter the market as unequal partners – and leave with unequal rewards.

Second, in precisely those areas where developing countries may have a competitive edge – as in labour-intensive manufacturing and the export of unskilled labour – the market rules are often changed to prevent free and open competition.

It is difficult to identify any one recent trend or change in attitude and policy that would drastically alter previous trends. Even in the difficult times of the 1970s and 1980s, economic growth has been achieved in the developed countries, where an average increase of 2–3 per cent per year, accompanied by little or no population growth, is considerable in absolute terms. It has also been even faster in some developing countries, especially in some in Latin America and eastern Asia, collectively referred to as Newly Industrialising Countries (NICs), separating these, in their turn, from the sinks of the world, where economic growth has at times been slower than population growth, and at times even negative.

Although there are relatively poor people in the developed countries, the material standards of most of the population have risen greatly in the last few decades. What were once luxuries of the few (car,

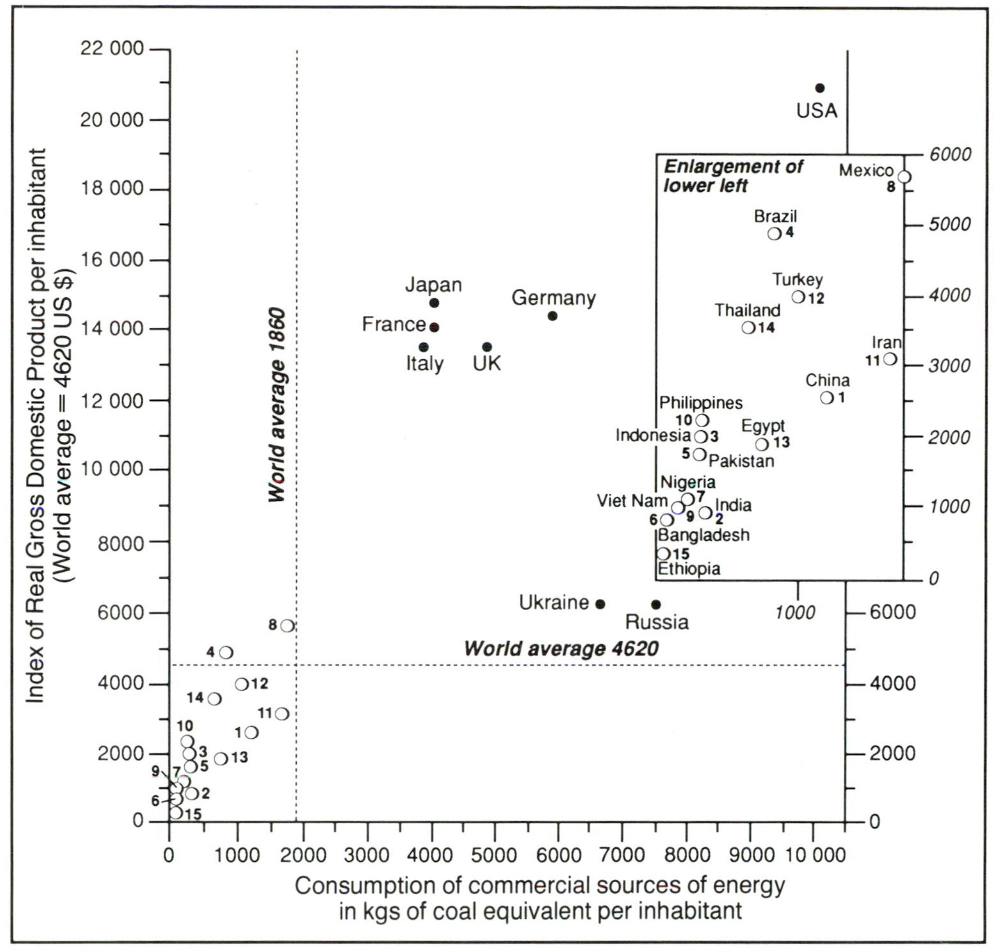

(b)

television, holiday abroad) have become widely available and are regarded as necessities. To basic 'needs' have been added what were previously 'wants'. In North America and Western Europe, for example, housing areas have increasingly been developed far enough from places of work for a car to become a need, a situation accentuated by cuts made by governments opposed to a strong public-sector transportation system. Any attempt in the affluent industrial societies of the market economy countries to curb material consumption (e.g. heavy fuel or vehicle taxes) would meet resistance and would also affect crucial sectors of the economy, reducing employment in industries such as motor-vehicle manufacturing.

In the developing countries, the average GDP per capita, and level of educational attainment, are highly misleading. Except in China and other less developed 'socialist' countries, there is an elite of 5–10 per cent of the population, much of it living in the capital city or other large cities, enjoying living standards near to or even above the average in the developed countries. Families in this elite minority commonly own a car or two, usually imported, have good housing and household appliances, and enjoy the services of low-paid domestic servants, who usually come from poor rural areas of their own countries, to carry out less palatable household chores such as cleaning, washing and nappy-changing. To be sure, many of these domestic employers are highly trained people such as doctors, engineers and accountants, whose work is indispensable to their country.

The affluent and influential people in developing world countries form an interface between the developed world and the poorer, mainly rural, population in their own countries. Arguably it is not in their interest that changes should take place in the distribution of wealth and income in their own countries. Some of them travel frequently to developed countries and have strong links with people living there. Even in China (see Chapter 16) big differences in living conditions exist between provinces, and between communities at more local levels. With the help of some fictitious countries Box 4.1 shows how in the interpretation of regional inequalities the effects of aggregation must be taken into account.

Figure 4.4 (pp. 82–3) shows two versions of the uneven world 'shareout' of goods and services. The first diagram (Cole 1959) shows the situation in the mid-1950s. The second, which is for the early 1990s, can be compared with the first. There is, however, a

TABLE 4.2 Importance of foreign trade to selected countries, 1989

	(1) Population (millions)	*(2) GDP ($ bn)*	*(3) Imports ($ bn)*	*(4) Imports ÷ popn*	*(5) Imports ÷ GDP*
Developing					
China	1,120	370	59.1	53	16
India	853	281	20.3	24	7
Indonesia	189	81	16.4	87	20
Brazil	150	342	18.3	122	5
Nigeria	119	34	3.4	29	10
Mexico	89	161	23.6	265	15
Egypt	55	36	7.4	135	21
Turkey	57	73	15.8	277	22
Kenya	25	9	2.1	84	23
Chile	13	20	6.7	515	34
Gabon	1.2	3.5	0.8	667	23
United Arab Emirates	1.6	25	16.0	10,000	64
Hong Kong	5.8	54	72.2	12,450	133
Singapore	2.7	25	52.2	19,330	209
Developed					
USA	251	4,965	493	1,964	10
USSR	291	1,746	115	395	7
Japan	124	2,601	211	1,701	8
West Germany	63	1,171	270	4,286	23
UK	57	735	198	3,474	27
Netherlands	15	216	104	6,933	48
Switzerland	7	183	58	8,286	32
Canada	27	446	114	4,222	26
Australia	17	212	40	2,300	19
New Zealand	3	32	9	3,000	28

Description of variables:
(1) Population in millions
(2) GDP in billions of US dollars
(3) Value of imports in billions of US dollars
(4) Per capita value of imports in US dollars (column (3) divided by column (1))
(5) Relationship of imports to total GDP (column (3) expressed as a percentage of column (2))
Source of data: UNSYB 1990/91, no. 38, Table 117

different ordering of the world's major regions and a different set of criteria for measuring global inequalities. The main change has been the relative decline of Anglo-America/North America and the rise of Western Europe and Japan. On the other hand, the contrast between developed and developing countries has not changed much.

Relationships and transactions between different regions of the world and between different countries will be discussed in the next four sections of this chapter with regard particularly to their influence on the development gap. Environmental issues and problems, and military expenditure and conflict are discussed in the last two sections.

4.3 International trade

International trade is a very complex aspect of the transactions between the countries of the world. One complicating factor is the fact that the countries of the world vary greatly in size, with the result that some of the internal trade within the larger countries of the world may in practice be over much greater distances (e.g. between Alaska and New York, St Petersburg and Vladivostok) than international trade (e.g. between Switzerland and France, or between adjoining parts of the USA and Canada).

The larger the country in population and area the more self-sufficient it is likely to be, so that the trade (imports, exports, or both) of a large country will tend

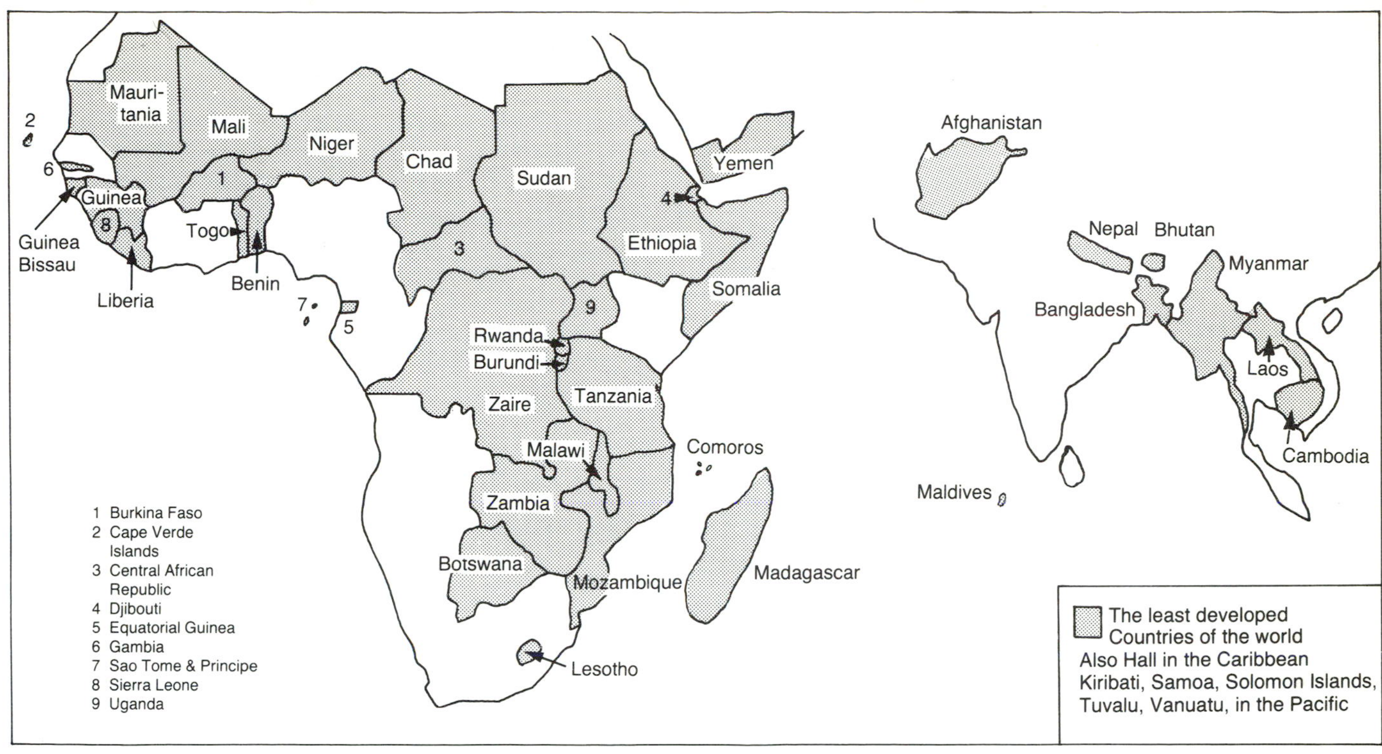

FIGURE 4.2 The *least* developed countries (LDC) as determined by the General Assembly of the United Nations (resolution 46/206)

Source: *UNSYB 90/91*, pp. 1089–90

to be a smaller proportion of its total GDP than will be the trade of a smaller country. The larger the country is, the more it can provide of the food, fuel and raw materials it needs for its economy. With some exceptions, the larger the home market, an advantage particularly for the production of some manufactured goods (e.g. aircraft, motor vehicles), the more the country can benefit from economies of scale. The development gap also influences the scale of foreign trade, with that of the richer countries generally being much larger per inhabitant and therefore accounting for a disproportionate share of all international trade.

Almost a third of the value of all international trade in the world is between the countries of Western Europe, yet in effect that trade is now virtually internal trade because the countries are in the same economic union. In complete contrast, the trade between the fifteen Soviet Socialist Republics of the USSR was internal trade until 1991 when, in spite of the creation of the Commonwealth of Independent States (CIS) and the continuation of heavy trade between Russia and its 'near abroad', such trade should now be recorded as foreign or international.

In the world of the market economy every country should ideally be able to specialise in what it is best at producing, using its comparative advantage in given areas of production. In theory it should exchange products freely with other countries when it would be cheaper and more advantageous to do so, even after the cost of transporting products between trading partners. In practice, national policies of self-sufficiency and security have led to the protection of various economic activities from foreign competition. In the EU, for example, agriculture and some industries are protected, as is industry in many developing countries, and as was the whole economic structure in the former USSR, thus 'distorting' or discouraging flows of trade. Some reductions in barriers to international trade were achieved in December 1993 with the conclusion of negotiations regarding GATT. What is often overlooked, however, is that in the real rather than the ideal world, many regions do not excel in the production of anything, and are therefore forced to concentrate on protecting high-cost production, for example of industrial goods in some developing countries, or to specialise in what they are least bad at producing.

Two outstanding aspects of geographical studies of foreign trade are its direction (which partners?) and its composition (what traded?). In this section, attention focuses on the direction of trade. In the chapters on the major regions, reference will also be made to its composition. For the purposes of this book, three aspects of global foreign trade will be illustrated in the rest of this section by Tables 4.2 and 4.3.

BOX 4.1

The pitfalls of spatial aggregation

In George Orwell's book *Nineteen Eighty-four* (1949), the main character, Winston Smith, reads *The Theory and Practice of Oligarchical Collectivism* (by Emmanuel Goldstein):

> Throughout recorded time ... there have been three kinds of people in the world, the High, the Middle and the Low.
>
> The aims of these three groups are entirely irreconcilable. The aim of the High is to remain where they are. The aim of the Middle is to change places with the High. The aim of the Low, when they have an aim – for it is an abiding characteristic of the Low that they are too much crushed by drudgery to be more than intermittently conscious of anything outside their daily lives – is to abolish all distinctions and create a society in which all men shall be equal.

Orwell's three groups, H, M and L will be used to illustrate how aggregation may or may not be a geographical problem (see Figure 4.3).

The country Equalia is subdivided into four provinces. By an ingenious piece of partitioning into districts, or by a quirk of good fortune, each province has exactly 10,000 inhabitants. Each province also has 1,000 Hs, 2,000 Ms and 7,000 Ls. While wealth and income are very unevenly divided among the population in each province, all provinces have the same average numbers of Income Units (IUs) per head. Since no regional inequality is evident when the populations are aggregated into the four provinces in Equalia, no regional problem is recognised and there is no national policy for one province to provide assistance to any other(s).

The country Irregula is also subdivided into four provinces, each with 10,000 inhabitants, but there are plenty of H and M people in Bend province, giving it an average of 5.8 IUs per inhabitant compared with only 1.0 in provinces Sedge and Wole, which are exclusively populated by L people. Whether for altruistic reasons, or to prevent people from Sedge and Wole migrating to Bend, the national government of Irregula is under great pressure to provide assistance to Sedge and Wole. It is a matter of speculation whether mainly the H people in Bend are the main providers of the assistance and also whether the assistance is distributed equally among all the L people in Sedge and Wole or whether some of the Ls are transformed into M or even H people thanks to the increased income.

Reduced to their simplest form, local, regional, national and global inequalities all have the features of Irregula, but without the convenience of districts of equal population size (or equal area) and exactly three distinct income groups.

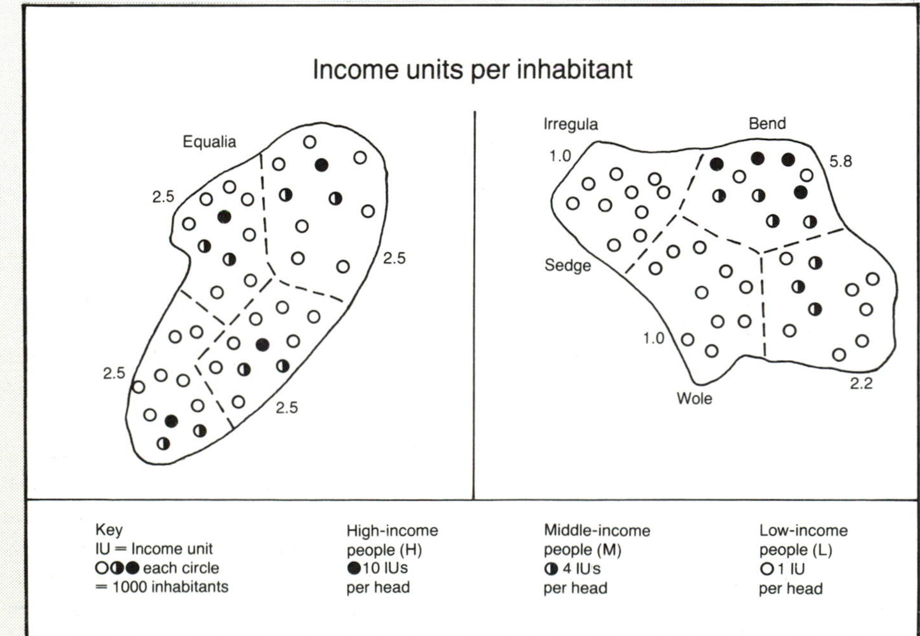

FIGURE 4.3 Two fictitious countries, Equalia and Irregula, with different distributions of income groups, to illustrate the subject of spatial aggregation

PLATE 4.1 West African schools

a and b The roll in class 6 is 41, of whom 38 are boys, evidence of Muslim influence on the bias of gender in providing education

c Children in Senegal, West Africa, take a break from working on the land. Do they go to school at all?

Table 4.2 includes the countries of the world with the largest populations, and other selected countries. The data illustrate in two ways the dependence of the countries listed on foreign trade; the data are for 1989.

Columns (1)–(3) show population, total GDP and imports in 1989. Since no country has an exact balance between the value of imports and exports, the picture would be somewhat different for exports, but in most cases not markedly so.

Column (4) shows the value of imports per inhabitant. With some exceptions, the value of imports is substantially greater for the developed countries than for the developing ones, the main reason being that noted earlier, namely that their GDPs per inhabitant are much larger.

Column (5) shows the trading coefficient, the relationship of trade (represented here by imports) to total production, in effect the level of self-sufficiency. In general, the smaller the country, the greater its dependence on foreign trade. The four countries of the world with the largest populations (in 1989) – China, India, the USSR and the USA – were among the most self-sufficient in the world. Indeed, before China and the USSR changed from a policy of maintaining as near self-sufficiency as possible to one of expanding foreign trade, their earlier coefficients were considerably lower

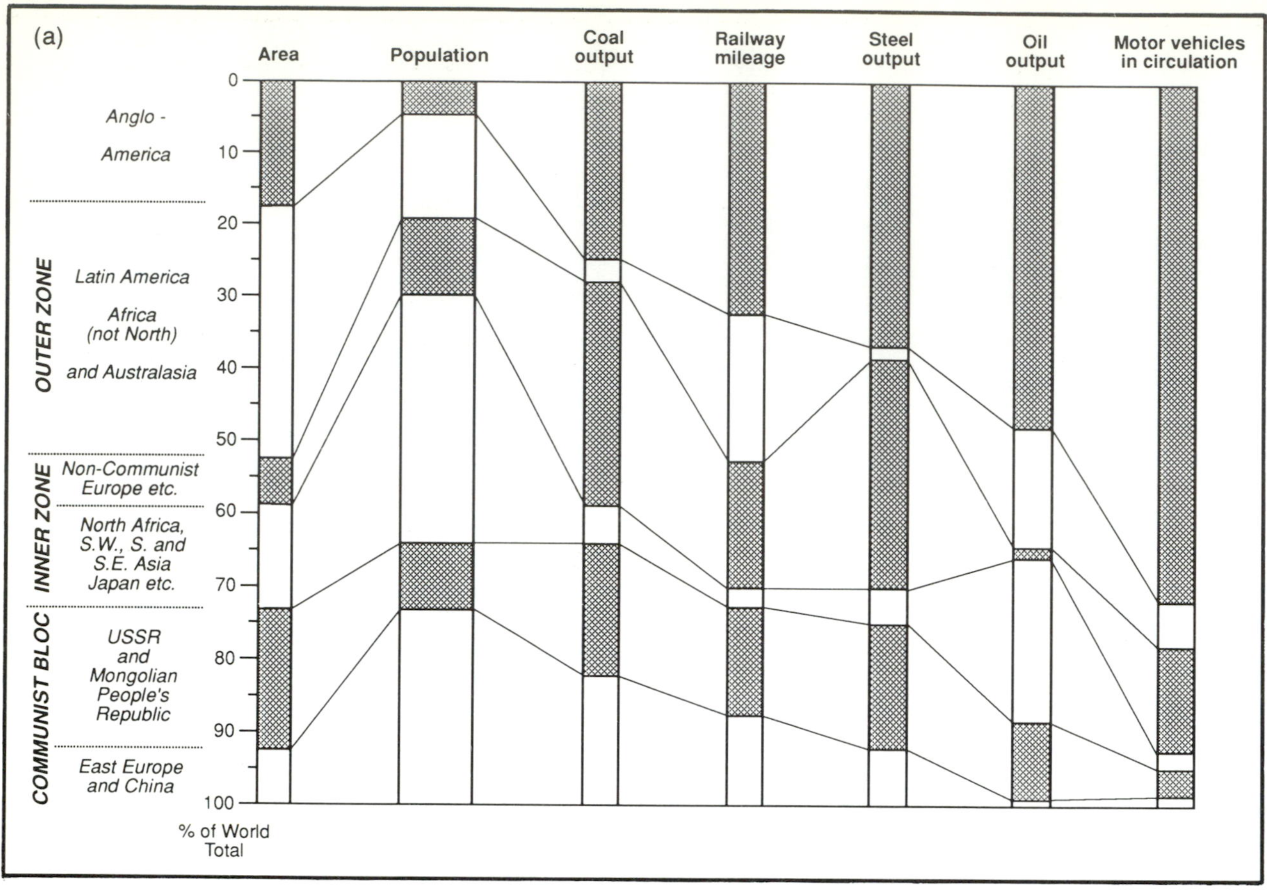

FIGURE 4.4 Comparison of the uneven shareout by major world regions of production and consumption in (*a*) the mid-1950s and (*b*) early 1990s. Note that the regions and the categories of production and consumption differ
Source of (a): Cole (1959)

than those of 1989. Hong Kong and Singapore are at the other extreme, as is Luxembourg, all three cases of small, highly urbanised and industrialised economies, entirely or heavily dependent on the outside world for their primary products. The broad features of international trade shown by the countries selected for inclusion in Table 4.2 were similar a decade and even two decades ago (see e.g. Cole 1983).

The *United Nations Statistical Yearbooks* include data on the trade between major regions of the world. The twelve regions used in these sources differ considerably from the twelve used here but, since the data are complex and difficult to interpret even as they stand, it was not considered desirable to modify the UN regions. The main differences are that Canada and the USA are shown separately (not as North America), Africa is shown in total, leaving Southwest Asia on its own, and South Asia (mainly India) and Southeast Asia are combined. Table 4.3 shows transactions in 1990 between and, where relevant, within each of the twelve regions. On account of some small omissions and rounding effects, the sums of the row and column totals do not coincide and therefore somewhat different grand totals result. The inconsistencies have been left, in preference to massaging the data to produce a perfect match.

The outstanding feature of Table 4.3 is the dominant position of the developed regions of the world, which, with about a quarter of the population of the world, account for over three-quarters of all international trade. That gives overall roughly ten times as much trade per inhabitant for the developed as for the developing countries. Western Europe is dominant among the developed regions. In spite of Japan's almost complete dependence on foreign sources for its fuel and raw material needs, it forms a very large home market, not exchanging goods with neighbouring industrial countries to the extent that West European countries do, largely due to its protectionist import barriers but also thanks to the strong tendency to 'buy Japanese'.

Among the developing regions and countries, some countries in Latin America, Southwest Asia and Southeast Asia, together with Hong Kong, South Korea and Taiwan (not included because it is not recognised by the UN), make considerable contributions to the grand

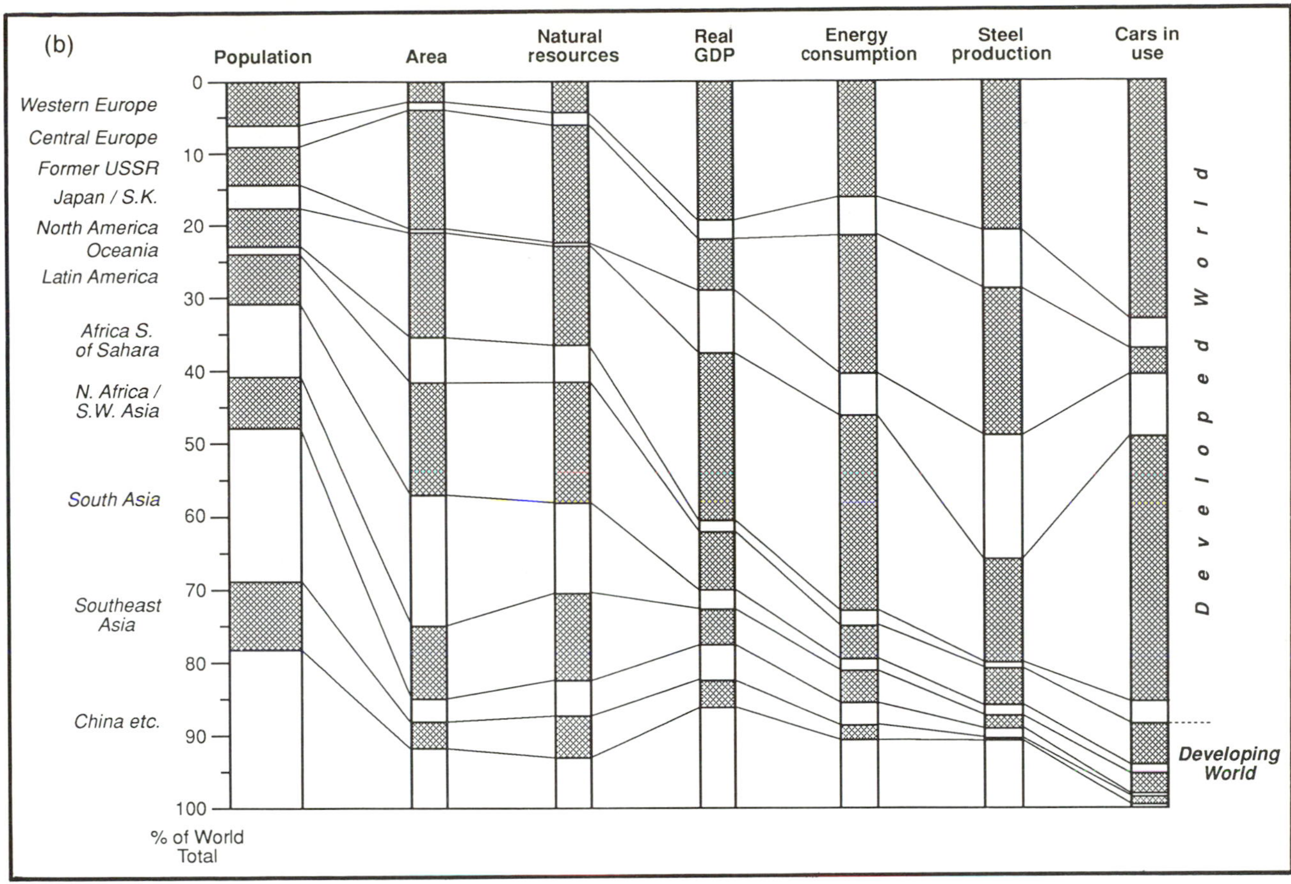

total of international trade in proportion to their population sizes. In contrast, China, India and Africa, having together almost half of the total population of the world, account for less than 5 per cent of all international trade, a fact that is thought-provoking but not particularly surprising or controversial.

The third aspect of foreign trade dealt with here is one of special interest to most readers of this book. As will be shown in later chapters, the composition of world trade has changed greatly since the Second World War, with manufactured goods accounting for an increasing share of the growing total of world trade. Many developing, formerly colonial, countries now export a greater value of manufactured products than of primary products. In parallel, traditional directions of trade flows have changed, with, for example, a more inbred trading picture in Western Europe in the 1990s than in the 1930s, Japan trading much more heavily with other developed countries than in the 1930s, and Australia's trade with southern and eastern Asia increasing enormously (see Chapter 10).

4.4 Development assistance

In theory, total international trade should roughly balance, with exports equal in value to imports, although the trade between any given pair of countries will not necessarily balance exactly. On the other hand, development assistance is generally directed from richer to poorer countries. In Table 4.4, columns (1)–(3) show data for Official Development Assistance (ODA), the first set of countries being net recipients, the second net donors. Some of the assistance is in the form of grants, some in the form of loans.

In 1991 (*WDR 1993*, Table 20) the net disbursement of overseas development totalled 49.6 billion US dollars. It was received as follows: by low-income economies 31.9, middle-income economies 15.5, high-income economies 2.2. In spite of their great size, having between them about half the total combined population of all the poorer countries, India and China received only 4.7 billion dollars.

The assistance to the receiving countries can be measured in two main ways, among other possible ones: according to the amount received per inhabitant and according to its size in relation to their GDP. From the data in columns (3) and (4) of Table 4.4 and from material on other countries some marked anomalies can be seen.

- On the whole, the larger the country, the smaller the amount of assistance it receives per inhabitant. Some very small countries receive very large amounts

TABLE 4.3 Total world exports in billions of US dollars, 1990, by regions of origin and destination

FROM \ *TO*	*World*	*(1) WEU*	*(2) EEU*	*(3) USSR*	*(4) Japan*	*(5) USA*	*(6) Canada*	*(7) Australia – NZ*	*(8) Latin America*	*(9) Africa*	*(10) Middle East*	*(11) and (12) S, SE Asia, China*
1 Western Europe	1,578	1,125	21	19	34	111	15	12	30	54	45	70
2 Eastern Europe[1]	68	26	9	20	1	1	0	0	1	2	2	2
3 USSR	104	34	36	–	2	1	0	0	7	2	3	11
4 Japan	287	62	1	3	–	91	7	8	10	7	9	89
5 USA	374	104	1	3	46	–	78	9	52	8	10	58
6 Canada	127	12	0	1	7	95	–	1	2	1	1	6
7 Australia, NZ	48	8	0	1	12	6	1	3	1	1	2	12
8 Latin America	134	30	2	4	7	44	2	1	18	2	2	6
9 Africa	67	40	1	1	1	12	1	0	1	4	2	2
10 Middle East	106	30	2	2	21	10	1	1	2	6	13	17
11 South, SE Asia	389	72	1	7	55	94	7	7	6	7	11	116
12 China[2]	62	6	1	1	9	5	0	1	1	1	1	34
World	3,392	1,573	78	65	200	476	111	44	132	95	102	426

Notes:
– not relevant, applies to one country
0 = less than 0.5 billion US dollars. All other numbers are rounded to the nearest billion
Due to rounding and the omission of some countries in this simplified table, row and column figures do not necessarily add up to world totals
[1] now referred to as Central Europe
[2] China can be distinguished from South and Southeast Asia as exporter but not as importer

Ten largest flows of exports, as percentage of world total:

	per cent
1 Within Western Europe	33.2
2 Within South, SE Asia	3.4
3 Western Europe to USA	3.3
4 USA to Western Europe	3.1
5 Canada to USA	2.8
6 South, SE Asia to USA	2.8
7 Japan to USA	2.7
8 Japan to South, SE Asia, China	2.6
9 USA to Canada	2.3
10 South, SE Asia to Western Europe	2.1

Source: Simplified from *UNSYB 1990/91*, no. 38, Table 118

per inhabitant, often for strategic rather than economic reasons. In 1991, Gabon received 121 dollars per inhabitant, Nicaragua 218 and Jordan 247. For comparable amounts to be given per inhabitant to much larger countries such as India or Indonesia would be far beyond the scope of current levels of disbursements.

• The income level of the receiving country is not closely correlated with the amount received. The following summarises the situation, the values being dollars per capita:

India and China	2.3
Other low-income economies	24.5
Lower-middle-income economies	24.3
Upper-middle-income economies	4.7
High-income economies	55.3

Since the mid-1980s, most upper-middle-income countries have, understandably, received very little assistance in relation to either population size or GDP. For example, in 1991 Venezuela received 1.7 dollars per inhabitant (equivalent to 0.1 per cent of its GDP), Brazil 1.2 dollars and Mexico 2.2 dollars. The high-income economies receive virtually nothing, apart from Israel, assisted for strategic and political reasons rather than because it needs economic or humanitarian assistance to the extent that poorer developing countries do. Israel receives 354 dollars per inhabitant (2.8 per cent of its GDP).

• A given amount of assistance in dollars per inhabitant makes more impact on the economy of a very poor country than on that of a middle-income country. Thus, for example, the 101 dollars per inhabitant in 1991 to Guinea-Bissau (West Africa) was equivalent to 48 per cent of its GDP per inhabitant, whereas the 100 dollars per inhabitant to the richer Papua-New Guinea was equivalent to 10.6 per cent of its GDP. Although in most transactions the amount of assistance is very small, there is a limit to the assistance that can be usefully absorbed by very poor countries over a short period.

TABLE 4.4 Selected recipients of development assistance in 1987 and 1989 (amount received in millions of dollars)

	(1) Population in millions	(2) Amount received	(3) As % of GNP	(4) Dollars per capita 1987	(5) Dollars per capita 1989
Somalia	8	580	57.0	73	58
Mozambique	15	649	40.9	43	59
Lesotho	2	108	29.4	54	78
Tanzania	24	882	25.2	37	42
Malawi	7	280	22.8	40	51
Zambia	7	429	21.1	61	51
Chad	5	198	20.3	40	55
Mauritania	2	178	19.0	89	102
Mali	8	364	18.6	46	50
Bhutan	2	17	16.7	9	28
Burkina Faso	7	283	16.2	40	34
Niger	7	348	16.1	50	46
Central African Republic	3	173	16.1	58	75
Madagascar	11	327	15.8	30	31
Burundi	5	192	15.3	38	47
Senegal	7	642	13.6	92	99
Nepal	18	345	12.7	19	20
Jordan	4	595	12.0	149	221
Ethiopia	46	635	11.8	14	18
Rwanda	7	243	11.6	35	39
Zaire	32	621	10.7	19	23
Papua-New Guinea	4	322	10.6	81	97
Sudan	24	902	10.5	38	30
Botswana	1	154	10.1	154	116
Togo	3	123	10.0	41	58
Bangladesh	107	1,637	9.3	15	18
Pakistan	105	858	2.4	8	9
Indonesia	175	1,245	1.8	7	9
Thailand	54	506	1.1	9	14
India	800	1,852	0.7	2	2
Peru	21	292	0.6	14	18
Turkey	51	417	0.6	8	23
China	1,062	1,449	0.5	1	2
Nigeria	109	69	0.3	1	2
Colombia	30	78	0.2	3	3
Brazil	142	288	0.1	2	1
Mexico	82	156	0.1	2	1
Venezuela	18	19	0.1	1	4

Sources: UN Human Development Report 1990, Table 19 for 1987; *1992*, Table 19 for 1989

With regard to the countries giving development assistance (see Table 4.5), the size of the donor country appears to have some influence on their contribution. Of the total of 55.5 billion US dollars of flow from the OECD countries, 11.4 billion, or 20 per cent, came from the USA, 11.0 billion from Japan, but 25.8 billion (46 per cent) from the eight richest EU countries. When the contribution of the rich countries is set against their total GDPs, it is seen that they are not all equally 'generous'. In 1991 only Norway set aside more than 1 per cent of its GDP for foreign assistance. Norway's 1.14 per cent was followed by 0.96 from Denmark, 0.92 from Sweden, and 0.88 from the Netherlands. Among the larger OECD countries, France's 0.62 per cent was the most generous, with the USA, at 0.2 per cent of its GDP, the lowest.

TABLE 4.5 Donors of foreign Official Development Assistance (ODA) (industrial to developing countries) (amount given in millions of dollars)

	Population in millions (1987)	Amount given	As % of GNP	Dollars per capita 1987	Dollars per capita 1990
Saudi Arabia	15	2,888	3.40	193	na
Kuwait	2	316	1.23	158	na
Norway	4	890	1.09	223	287
Netherlands	15	2,094	0.98	140	173
Sweden	8	1,337	0.88	167	238
Denmark	5	859	0.88	172	228
France	56	6,525	0.74	117	112
Finland	5	433	0.50	87	170
Belgium	10	689	0.49	69	91
Canada	26	1,885	0.47	73	93
German FR	61	4,391	0.39	72	81
Italy	57	2,615	0.35	46	59
Australia	16	627	0.33	39	57
Japan	122	7,454	0.31	61	73
Switzerland	7	547	0.31	78	113
Libya	4	76	0.30	19	na
Ireland	4	51	0.28	13	15
UK	57	1,865	0.28	33	46
New Zealand	3	87	0.26	29	27
USSR	284	4,321	0.25	15	na
USA	244	8,945	0.20	37	41
Austria	8	196	0.17	25	51
Spain	39	176	0.17	5	13

Note: na = not available
Sources: UN Human Development Report 1990, Table 19; later data from *UN Human Development Report 1992*, Tables 18, 39

The nature of official development assistance has changed since the 1950s and 1960s, when many colonies emerged as the poor countries of the developing world. The population of the rich countries has not grown much in the last four decades, while that of many developing countries has at least doubled. Thus if the membership of donor and recipient countries stays the same there will be less to go round as the share of total world population in the rich countries diminishes. In practice, the countries of the USSR and its CMEA (Council for Mutual Economic Assistance) partners in Europe, still net donors of assistance in the

PLATE 4.2 The development gap

a Expensive addition to the natural beauty of Sydney harbour. Acoustically the superstructure is of no importance. Controversial, an expensive eyesore or the icing on the cake in Sydney's beautiful harbour?

1980s, have now become recipients in the 1990s, bad news for the 'genuine' developing world. Since the 1960s some members of OPEC (Organisation of Petroleum Exporting Countries), particularly Saudi Arabia, Kuwait and the United Arab Emirates, have been generous if erratic suppliers of assistance, but much has been directed to fellow Arab countries, while the many non-Arab poor countries of the world that import oil from them, albeit in small quantities, have paid the normal world market prices – at very high levels in the 1970s and early 1980s. The OPEC countries have been less generous since the Gulf War in 1991.

Official development assistance is the main form of net transfer of resources from rich to poor countries and it should have a positive effect on the recipients. It amounts, however, to only between 0.3 and 0.4 per cent of the total GDP of the OECD countries. An even more modest sum is collected from gifts to private charities in rich countries for use in poor countries, as opposed to use among the poor in the rich countries themselves. Such sources are referred to as Non-Governmental Organisations (NGOs).

Development assistance is equivalent to only 2–3 per cent of the value of all exports in world trade from the OECD countries, and this includes some to other OECD countries. When some of the development assistance received in poor countries reaches and stays in the hands of the better-off, while some has a negative effect on the environment, it can be seen that the impact is not likely to change the relationship of rich to poor regions of the world substantially, except in some local cases. Military 'assistance' to developing countries, whether overt or covert, ostensibly for defence purposes, is often used in conflicts that do enormous local damage to people and property.

b Calcutta, India, temporary drain-dwellers

TABLE 4.6 Long-term external debt in six selected countries: rough estimates, 1991

	(1)	(2)	(3)	(4)	(5)	(6)
Peru	21	1,000	3.3	7 times	25	1 year
Brazil	117	800	32	4 "	450	¼ year
Ethiopia	3.3	< 100	0.3	10 "	6	½ year
Nigeria	34	300	12	3 "	30	1 year
Thailand	36	800	28	1.5 "	90	< ½ year
Indonesia	74	500	29	2.5 "	100	¾ year

Description of variables:
(1) Total external debt in 1991 in billions of US dollars
(2) Debt in dollars per head
(3) Exports in 1991 in billions of US dollars
(4) Debt *total* in relation to 1 year of exports (i.e. (1) ÷ (3))
(5) Total GDP in billions of US dollars in 1991
(6) Debt in comparison with one year's GDP (i.e. (1) ÷ (5))
Source of data: World Development Report 1993, Tables 21 (Total external debt) and 14 (Exports)

Another negative aspect of the transactions between rich and poor countries has been the great growth of foreign debt in many developing countries since the 1970s. Table 4.6 shows some of the features of the long-term external debt in six developing countries, representing the three developing world continents. Column (4) shows the extent of the variations in the ratio of debt to one year of exports. Column (6) also shows that the debt varies in relation to total GDP.

4.5 Transnational companies

In addition to international trade and development assistance, discussed in the previous two sections, investments form a major part of international transactions. Arguably the first modern international investments were those made by Spain and Portugal in their Latin American colonies to extract minerals such as silver and to process agricultural products such as sugar. Other European colonial powers followed suit, and by the nineteenth century, when Spain had declined as an economic power, Britain, France and other countries were making large investments both inside and outside Europe, most of the financial resources coming from private companies. For example, British capital formed the basis for rail building in many parts of the world while British, French and Belgian capital was used in the establishment of the modern iron and steel industry in Russia.

Before the Second World War, and indeed for a decade and a half after it, much of the investment from the main sources of capital of the time was still directed to the colonies of those countries with empires, particularly Britain, France and Japan. With no colonies of its own except the Philippines and Puerto Rico, the USA had become the largest source of foreign investment in the world, concentrating particularly on Canada, Latin America and a few European countries. Many transnational companies were already active, for example Esso, extracting oil in Venezuela, and Ford, with a motor-vehicle plant in the UK and for a time what proved to be unsuccessful rubber tree plantations in Brazil.

'Multinational' and 'transnational' are both terms widely used to describe companies that have investments and activities in more than one country. The terms are generally interchangeable, but it might be appropriate to use *multinational* to refer to companies based in more than one country, as is the oil company Shell, and to use *transnational* to refer to companies based in one country but with activities both inside and outside their own country, for example the Ford Motor Company.

The role of transnational companies in the world economy has increased since the 1950s for various reasons. With the break-up of empires, access to newly independent countries in Africa and Asia has made it easier to set up subsidiaries in some parts of the world. The break-up of the USSR and the emergence of market economies in the newly independent republics of the USSR since the early 1990s has already led to an increase in transactions between the western industrial countries and the former Soviet bloc. At the same time, improvements in communications and the proliferation of international financial services have facilitated the internationalisation of capital. The fact remains, however, that national governments can in the last resort control or influence the extent to which foreign investment is able to penetrate their economies. For example, neither North Korea nor Cuba has joined the international economic community yet, while China is still cautious about foreign investment.

According to Thrift (1986), capital has been exported for three distinct reasons:

- To gain access to primary products, whether raw

TABLE 4.7 Headquarters and sectors of the world's top fifty industrial companies

	Total	*Japan*	*USA*	*Western Europe*[1]
Sogo sosha (conglomerate)	9	9	0	0
Motor vehicles	10	3	4	3
Fuels	9	0	5	4
Communications	3	1	2	0
Tobacco	2	0	1	1
Stores (retailing)	3	0	3	0
Electronics, electricals	7	3	1	3
Contractors	1	0	0	1
Electricity	3	1	0	2
Food	2	0	0	2
Chemicals	1	0	0	1
Total	50	17	16	17

Note: [1] Germany 5, UK 3, Italy 3, Netherlands 2, Switzerland 2, France 2. For Shell and Unilever, jointly owned, the Netherlands and UK are each allocated half a point for each company
Source of data: Extel (1994)

materials or fuel, including food products, tobacco, timber, non-fuel minerals and oil. The large-scale extraction of minerals in particular has usually required more capital than could be provided internally in most developing countries.

- To gain entry into markets for manufactured products. Examples have been numerous in the history of the manufacture of motor vehicles. For example, the German firm Volkswagen set up plants in Mexico and Brazil. More recently, in order to gain access to the large market of the European Union, which has had quotas limiting the import of Japanese motor vehicles, Nissan, Toyota and other Japanese companies have established factories in the UK, producing vehicles defined as made in a Member State and therefore within the EU customs union.

- A more recent trend has been the export of capital to make use of cheap labour, particularly in developing countries. In addition to their sale in the host country, products can be re-exported to the country providing the capital and also to third countries. This type of transaction almost inevitably means the loss of jobs in the developed country. In the past, the term 'sweated labour' was sometimes used to refer to the exploitation of cheap labour in poor countries. Even now, attention is drawn to the use of low-paid labour, even of child labour. It should not be forgotten, however, that often people working in industry in developing countries, whether internally or foreign owned, earn much more than they would in agriculture or among the underemployed in services.

As noted above, transnational investments are not a new feature of the world economic scene, but there has been increasing criticism of the activities of transnational companies since the 1980s, at least some of it originating in the USSR and other socialist countries, whose own endeavours in developing countries, it was claimed, were entirely altruistic. The apparently unending precarious state of the world economy has been blamed by many throughout this century on the capitalist system and, in recent decades, increasingly in particular on transnational companies. The cartoon in Plate 4.4 drawn by Herbert Cole in 1901 (p. 92) shows that criticism of the profit-motivated entrepreneur and company is not new. The transnational companies are now large and powerful. They form a vital and intimately related part of the economy of almost every country in the world. Having weathered a century of criticism and having outlived, at least for the foreseeable future, the socialistically inspired state-run command economies, they seem set to play a major role in the world economy in the increasingly globalised capitalist system of the twenty-first century.

The term 'transnational' is applicable in the broadest sense to any company that employs capital in more than one country. However, transnational companies vary greatly in a number of respects. They differ in size, whether measured by capitalisation, turnover or size of labour force. Some operate in a small number of countries, some in virtually every country of the world. Some specialise in the production of a particular

TABLE 4.8 Large companies and countries compared: billions of US dollars (population in millions)

Selected US companies' turnover		*Selected countries GDP*		*Selected EU companies' sales*	
General Motors	131.3	Denmark	134.8 (5.2)		
Exxon	114.6	Norway	110.9 (4.3)		
Ford	99.2	South Africa	110.0 (41.2)	RD/Shell	82.5
IBM	63.9	Venezuela	61.8 (21.3)	BP	65.0
General Electric	37.5	New Zealand	42.2 (3.5)	Volkswagen	52.2
Boeing	29.9	Nigeria	31.4 (98.1)	Philips (Elec)	31.9

Sources: Extel (1993); *PRB/WPDS* (1994)

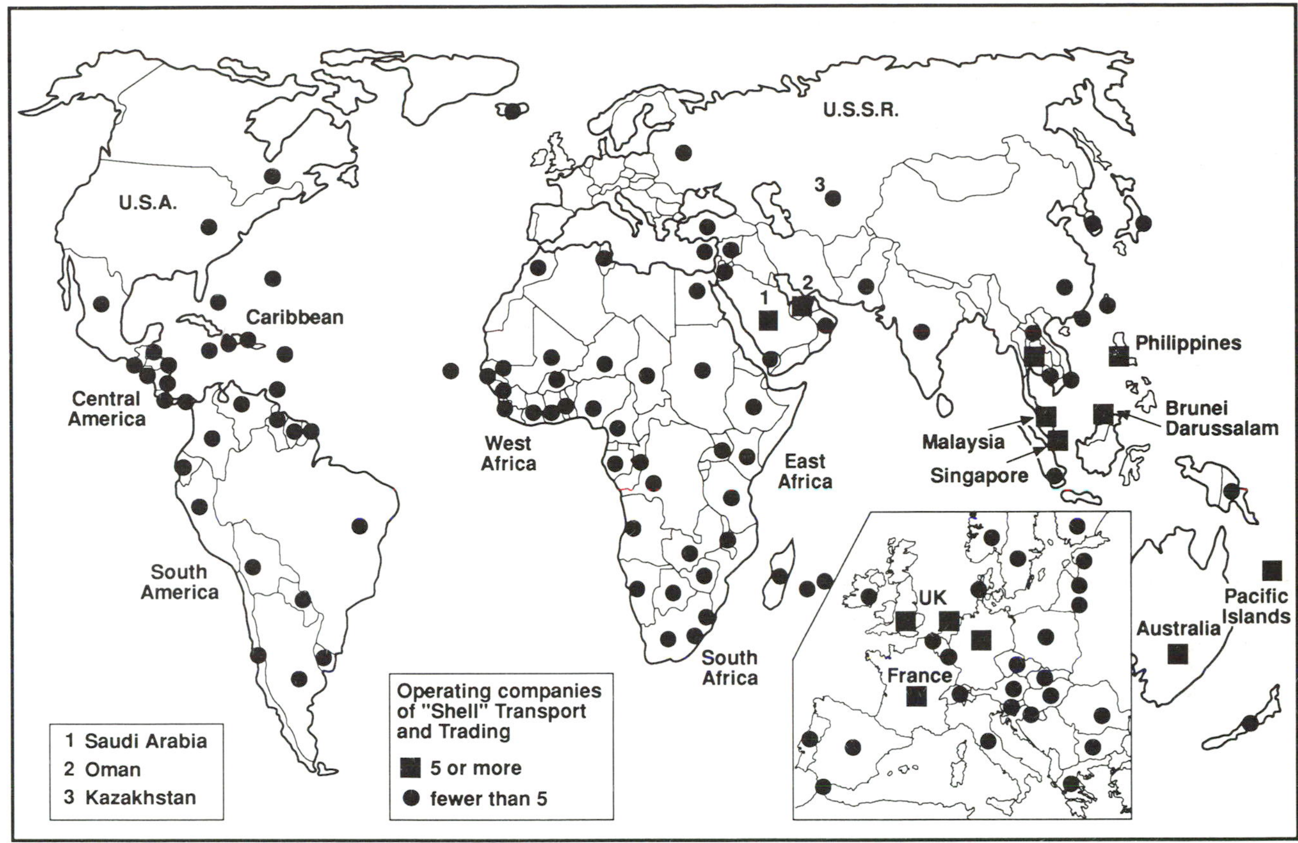

FIGURE 4.5 The global distribution of the operations of Shell Transport and Trading plc, the largest transnational company based in Western Europe

Source: *Annual Report 1993* of Shell (1994)

product while others, referred to as conglomerates or diversified industrials, have investments in a wide range of production, whether of goods or of services.

Most of today's largest transnational companies are based in the USA, Western Europe or Japan, but some are based in developing countries, including for example India, Brazil and Mexico. Others may emerge in Russia from former large state enterprises. Table 4.7 shows that of the 50 top industrial companies in the world according to value of sales, 17 have their headquarters in Japan, 17 in Western Europe and 16 in the USA. The huge size of the large companies is illustrated in Table 4.8 by a comparison of the turnover or sales of selected large companies in the USA and Western Europe with countries whose total GDP is roughly similar.

When the largest companies in each major industrial country are ranked according to size of turnover or sales they conform fairly closely to the rank–size rule (see Figure 4.6). The 100 largest companies in the USA and UK and the 10 largest in Italy are ranked according to size, which determines their position on the horizontal scale. They are placed on the vertical scale according to the size of turnover. Both scales are base-10 logarithmic. In all three countries the companies are situated close to a straight line. The three lines are separated on the vertical axis by the greater average size of the US companies than the size of those of the UK and Italy.

When individual companies are compared, striking contrasts emerge not only in their size, discussed above, but also in their organisation. Some examples follow:

- In some respects the Boeing Company cannot be regarded as transnational, in spite of its size. Of its 160,000 employees, about 90 per cent work in four locations in the USA: Puget Sound, Seattle (103,000), Wichita, Kansas (22,000), Philadelphia (6,000) and Huntsville, Alabama (3,700). Given the nature of Boeing's products, commercial jet transports, and military and space exploration equipment, it is to be expected that its activities are concentrated at home. Nevertheless, Boeing's influence is highly 'transnational' since in the early 1990s it has had the distinction of being the largest exporter of any US company, and its aircraft are used in virtually every major airline in the world. In conjunction with its sale of aircraft at home and abroad, it is closely involved with the training of aircrew and the provision of maintenance facilities world-wide.

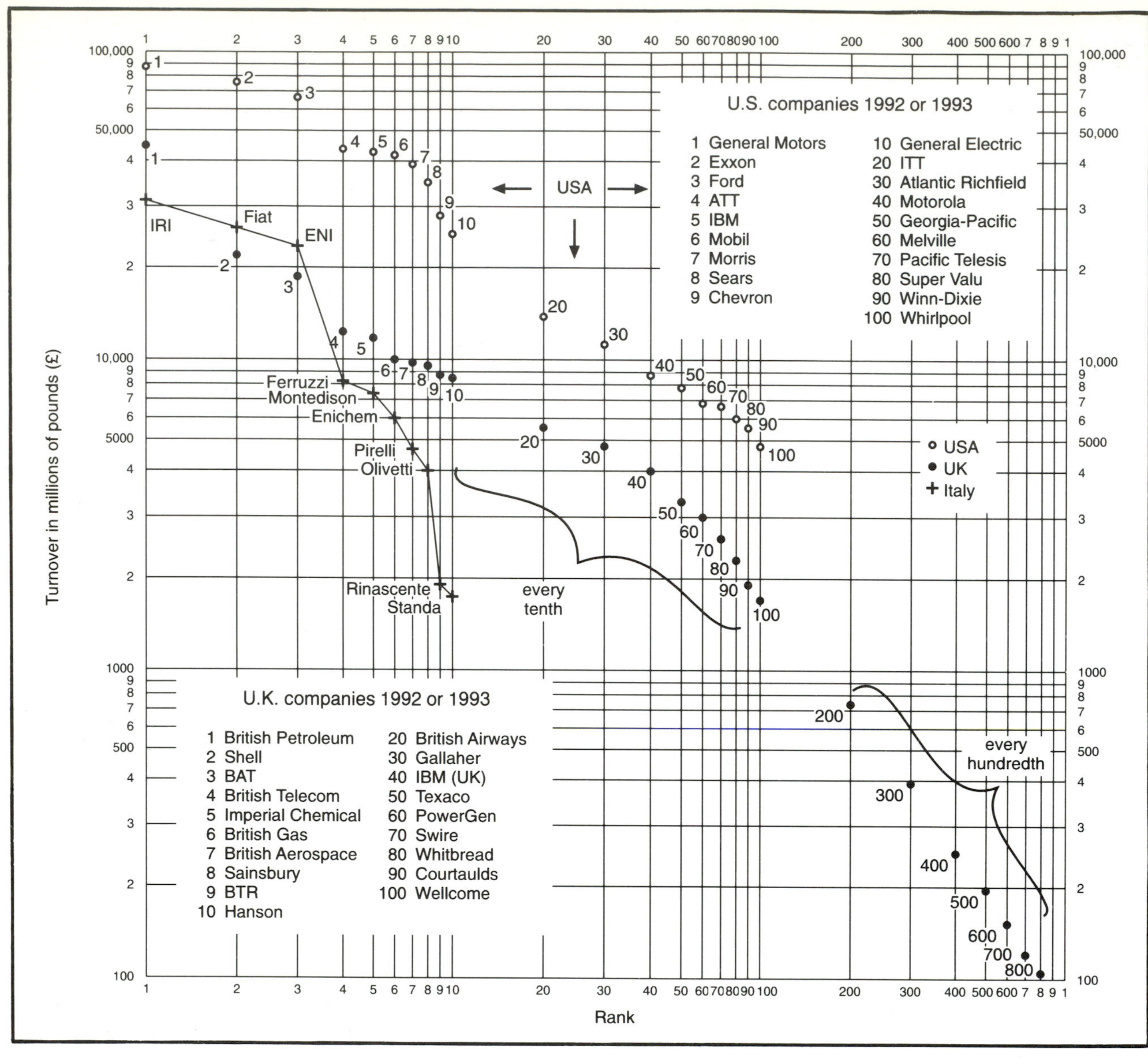

FIGURE 4.6 Comparative sizes of the largest companies in the USA, UK and Italy

- Shell Transport and Trading is based in London, while Royal Dutch Shell, concerned with oil exploration and extraction, is based in the Netherlands. While the organisation of this company is based in two countries, Shell has activities in well over 100 countries, the distribution of which is shown in Figure 4.5. Just as Boeing concentrates on the manufacture of aircraft, so Shell has specialised in the extraction, refining and distribution of oil and oil products, its early success partly due to its major role in developing the oil reserves in Venezuela in the 1920s. Unlike Boeing, directly, or indirectly through subsidiaries, Shell Transport and Trading gives employment to a large labour force, particularly in petrol service stations, throughout the world.

- Hanson PLC has its headquarters in London, but a large proportion of its 74,000 employees work in subsidiaries in the USA. Unlike Boeing and Shell, which specialise in a particular area of production, Hanson has major investments in a wide range of basic industries, including coal, chemicals, propane, building materials, forest products, tobacco and material handling. In this respect it resembles the large Japanese conglomerates. Two of the largest profit-making products of Hanson, coal and tobacco, have become potential risks, the former for environmental reasons, the latter for health reasons. As will be shown below, the company is conscious of these problems.

- Pioneer International Limited is an Australian

TABLE 4.9 The top fifteen industrial companies in North America and Western Europe, 1992

Company	Headquarters	Sales[1]	Sector
1 General Motors Corpn	USA	87.5	Transport equipment
2 Exxon Corpn	USA	76.4	Oil, gas and nuclear
3 Ford Motor Co.	USA	66.1	Transport equipment
4 Royal Dutch/Shell Group	Neth/UK	55.0	Oil, gas and nuclear
5 The British Petroleum Co. PLC	UK	43.4	Oil, gas and nuclear
6 American Telephone/graph Co.	USA	42.9	Communications
7 IBM[2]	USA	42.6	Communications
8 Mobil Corpn	USA	42.3	Oil, gas and nuclear
9 Philip Morris Cos Inc	USA	39.1	Tobacco
10 Volkswagen AG	Germany	34.8	Transport equipment
11 Sears Roebuck and Co.	USA	34.6	Retailing
12 Daimler-Benz AG	Germany	33.5	Transport equipment
13 IRI[3]	Italy	31.6	Contracting, construction
14 Siemens AG	Germany	31.1	Electricals
15 Chevron Corpn	USA	28.3	Oil, gas and nuclear

Notes: [1] Sales in billions of £s
[2] International Business Machines Corporation
[3] Istituto per la Ricostruzione Industriale
Source: Extel (1994)

company, based in Sydney. It started life only in 1949 but in 1993 was the eleventh company in Australia in turnover. It produces building materials and refines petroleum, having therefore two largely distinct areas of activity. The location of Pioneer's operations, employing about 10,000 altogether, qualifies it as transnational, but the fourteen countries in which it operates are restricted to three parts of the world. Singapore, Malaysia and Thailand, China, Hong Kong and Macau are developing countries comparatively near to Australia. The UK, Spain, Germany and the Netherlands are developed countries in the EU. Hungary, the Czech Republic and Israel are new ventures. Finally, Pioneer has operations in the USA, mainly in Texas. Like many other smaller developed economies and also newly industrialising countries, Australia is showing that it can export capital.

Some economic activities are mainly concentrated in large companies, while others continue to be successful in small enterprises. Although food processing and retailing is increasingly becoming part of the first group, agriculture itself is largely run on the basis of individual farms. Small companies get less publicity than large ones, but developments in North Italy in the post-war decades have shown that small enterprises can be successful in textiles, clothing and footwear. In contrast, Table 4.9 shows that fuel, transport and communications predominate among the largest industrial companies in North America and Western Europe.

Another reason why transnational companies vary greatly is the relationship between capital, labour and turnover. Table 4.10 shows marked variations among a selection of five large West European companies with varying activities. For example, France's electricity industry has a far higher ratio of capital to turnover than Switzerland's largest food company, Nestlé. Shell (oil) has a turnover per employee greatly in excess of that of Siemens, which produces electrical goods.

TABLE 4.10 Features of selected large West European companies, 1992

	Turnover (£ bn)	*Capital employed (£ bn)*	*Turnover/ capital ratio*	*No. of employees (thous.)*	*Turnover per employee (£ thous.)*	*Profit (£ bn)*	*Profit per employee (£ thous.)*
RD/Shell, UK	55.0	50.8	108	127.0	433	5.9	46
Volkswagen, Germany	34.8	26.4	132	281.6	124	1.3	5
Siemens, Germany	31.1	24.1	129	417.8	74	1.6	4
Nestlé, Switzerland	24.6	11.0	224	218.0	113	2.6	12
Electricité de France	21.1	66.4	32	118.6	178	2.9	24

Source: Extel (1994)

PLATE 4.3 'The Money Spider'
The unacceptable face of capitalism at the turn of the century by Herbert Cole (1901) in the magazine *Fun*. Almost a century later virtually every government in the world is looking to the market economy and private ownership of the means of production as the only way to achieve 'balanced' economic growth. (Herbert Cole (1867–1931) was the paternal grandfather of the author)

Shell's profit per employee is almost ten times as high as that of Volkswagen (motor vehicles).

In their annual reports and public relations brochures, large companies mix accounts of their activities with complimentary accounts about their concern for their workers, their customers and the environment. For example, the function of *Essoview*, a journal distributed freely, appearing four times a year, is defined as 'An insight into the company – its activities – its views'. Some examples follow of the terminology used in such publications.

- **Employees** Boeing (1994, p. 30) states: 'Boeing takes pride in the contributions of its workforce and is committed to providing employees with a safe and healthy workplace, opportunities for training and career advancement, and a wage and benefit package that is highly competitive.' With regard to employment, when a need is seen for greater efficiency, a private company, unlike a state-owned company, has no compunction in reducing its labour force. This procedure has been widely experienced in Western Europe in companies that have been privatised in the 1980s and in Central Europe and the former USSR in the 1990s, following the collapse of communism.

- **Customers** One of the most profitable branches of Hanson is the manufacture in Nottingham of cigarettes and handrolling tobaccos. Conscious of the controversy over smoking, *Hanson 1994* (1995, p. 15) states:

> The tobacco industry has concluded a new 5 year agreement with the Department of Health which has proved to be a more effective mechanism for dealing with the Government's concerns on tobacco advertising than the total ban practised in certain European countries. In return for very strict controls, monitored independently, the industry is permitted to compete strongly within the guidelines.

- **The environment** Concern over the environment is also widely expressed in company literature. One of the most successful sectors of Hanson is its forest products. The public is reassured by the following caption accompanying a picture of an area of forest in *Hanson 1994* (1995, p. 22): 'The importance of our continuing sensitivity to the environment is shown in the majesty of a Cavenham (the name of its subsidiary) second growth forest in the Pacific Northwest, preserved for the future through modern forestry replacement planting techniques.' Ryan (1993, p. 1) is frank about the priorities of Esso: 'Esso shares the public concern about the environment and is committed to continued efforts to improve performance.... The costs and practical aspects of all the alternatives need to be fully investigated and weighed against measurable environmental benefits.'

Throughout the twentieth century there have been critics of the 'capitalist system' (or lack of system). The news of the imminent death of capitalism has been greatly exaggerated. The role of large companies in the

PLATE 4.4 The unequal shareout of wealth, 1901
'Monopoly' anticipated by Herbert Cole. From the magazine *Fun*

geography of the world's major regions will be discussed in later chapters.

Progress in privatisation in Central Europe and the former USSR will be referred to in Chapter 6. The bastions of state socialism are falling as new transnational companies are formed in countries formerly run by state planners. It is difficult to see how in a few years Russia could create a sophisticated structure of large companies similar to that in the western world, whatever path it follows towards a market economy and private ownership. Nevertheless, according to Naudin (1994), in mid-1994 the private sector already accounted for half of Russia's GDP.

Large Japanese companies are discussed in Chapter 8 and are compared with companies in the newly industrialising countries of Southeast and East Asia. Canada, Australia, Brazil and Mexico are among other countries with companies of considerable size, some of them in the public sector. In Chapter 9, reference will be made to the Ford Motor Company of the USA.

4.6 The global distribution of capital

In this section attention focuses on the great concentration of capital resources in certain countries and cities of the world. Each of the companies in the world has a particular capitalisation value at any given time. That value changes with time as the assets of a company grow, shrink, are merged with those of another, and do well or badly in terms of sales.

It is possible to calculate roughly the total capitalisation of companies for each country. At the end of 1992, that value for the world was about 13,000 billion US dollars, equivalent to about half of one year's GDP for the world. Of the world total the USA had 34–5 per cent (with less than 5 per cent of the population of the world), Japan about 22 per cent (with 2.3 per cent of the world's population) and the UK about 9 per cent (with about 1 per cent of the world's population), these three countries having almost two-thirds of the world total. Much of the rest is in other western industrial countries and a little over 10 per cent in the 'emerging' countries of the developing world. Almost all of the capital in the former USSR and China is still not separated from the state sector and therefore is not included.

- Table 4.11(a) shows the overwhelming position of the developed world in the concentration of capital. The first six countries listed in the table have 75 per cent of the capital (9,700 out of 13,000 billion) with only 12 per cent of the population of the world.

An example of the influence of foreign investments not on single countries or industrial sectors, but globally, is given by Ham (1995). US investments in emerging markets (see also Chapter 18) have risen from a few billion US dollars annually in the early 1980s (see also Chapter 9) to over 50 billion in 1993. American fund managers, pursuing foreign equities for mutual and pension funds influenced world stock markets, investing almost 40 billion dollars in European stocks and about 10 billion each in Japan, other East Asian areas and Latin America. In 1994–5 much of the US money was withdrawn, leaving the economies of whole countries in financial chaos. According to Ham:

> The huge flows of American money can create havoc in already highly volatile, illiquid 'emerging' markets. Witness the unwarranted fall of the Brazilian stock market last month in the wake of the collapse in Mexico – the result of American funds withdrawing en masse from what was suddenly perceived as a highly risky market.

Such transactions are not new, but the increased speed with which they can be made, as a result of the improvement in communications and the flow of information, is producing a new situation, the complexity of which has been increased by the entry of developing countries and of former socialist countries into the market economy stakes.

- Table 4.11(b) shows, however, that there are considerable secondary concentrations in developing countries, with Hong Kong and Singapore outstanding, in view of the small size of their populations, and also Taiwan, with a mere fiftieth of the population of China but nearly ten times the capital.

- Table 4.11(c) compares the major financial cities of the world on a more limited measure, fund-management capital. Tokyo, London and Paris dominate their countries whereas several cities are prominent in the USA, and the international importance of Switzerland's Zurich and Geneva is evident. No German city appears in the top ten.

- Table 4.11(d) shows a remarkable similarity between the top ten financial cities of the world (c) and the top ten science cities. The sheer size of these cities raises their chances of being pre-eminent in many aspects of life. Both sets of ten together have less than 2 per cent of the total population of the world. One may speculate how many decades must pass before cities in Latin America, India and China come anywhere near.

- Table 4.6 is a reminder that many developing countries not only lack capital resources but have accumulated enormous long-term external debts, particularly since the 1970s. Some examples are given from various parts of the developing world.

TABLE 4.11 Market capitalisation and capitalisation per inhabitant in selected countries and cities

(a) Developed countries and regions, December 1993

	Total billion dollars	*Dollars per inhabitant*		*Total billion dollars*	*Dollars per inhabitant*
USA	4,467	17,290	Germany	441	5,440
Japan	2,883	23,160	Switzerland	247	35,290
UK	1,195	20,600	Other Western	1,961	–
France	454	7,870	Emerging	1,337	–
			World total:	12,985	

Note: – not calculable
Source: Wright (1994, p. 5.3)

(b) Developing countries/emerging markets

	Total billion dollars	*Dollars per inhabitant*		*Total billion dollars*	*Dollars per inhabitant*
Hong Kong	312	54,000	Thailand	111	1,900
Singapore	132	46,000	Argentina	42	1,200
Taiwan	191	9,100	Brazil	101	700
Malaysia	161	8,200	Philippines	34	500
South Africa	172	4,200	Indonesia	41	200
Chile	50	3,500	India	134	150
South Korea	153	3,400	China	22	20
Mexico	172	1,900			

Source: Miller (1994, p. 35)

(c) Top ten cities for fund-management capital value of equities in billions of US dollars

	1992	*1993*		*1992*	*1993*
1 Tokyo	1,367	1,163	6 Geneva	211	244
2 London	533	682	7 Paris	189	226
3 New York	657	584	8 San Francisco	171	218
4 Zurich	333	375	9 Los Angeles	141	184
5 Boston	269	369	10 Chicago	102	126

Source: Bennett (1994)

(d) Top ten science cities, scientific papers published, 1991

1 Moscow	14,541	6 Paris	7,964
2 London	14,051	7 Los Angeles	6,601
3 Boston[1]	12,480	8 Bethesda, Maryland	6,233
4 Tokyo	11,582	9 Philadelphia	6,183
5 New York	8,551	10 Osaka	5,403

Note: [1] Includes Cambridge, MA
Source: Hawkes (1993b)

4.7 The environment

In the world of nature, a general principle of self-interest is that parasites do not exceed the capacity of their hosts to sustain them, and predators do not destroy the species on which they depend. Human beings may be seen as superpredators, at the top of a hierarchy of species, and living off other animals, or directly from plants. Over the millennia, the natural world that has sustained the human species has been exploited, manipulated and in places destroyed. Concern about the natural environment, both from self-interest and (less often) for its own sake, has been mixed with complacency about the capacity of the environment to survive the onslaughts of settlers and developers, intent on achieving economic growth, and confidence about the capacity of scientific discoveries and applications to sort out problems.

Demographic and economic growth in the last two centuries have, however, greatly increased human exploitation of the natural world. The impact of European colonisation on small areas, especially island settlements in the tropics, led to concern about the clearance of forests and the extinction of animal species there as early as the seventeenth and eighteenth centuries. Grove (1992) gives many examples of environmental concern in the nineteenth century and earlier. Since the 1940s, the almost universal drive to achieve fast economic growth in developed and developing countries alike has generally overridden concern over

TABLE 4.12 Selected data for development assistance, the environment and military expenditure

	(1)	(2)	(3)	(4)	(5)	(6)	(7)	(8)	(9)
	Development assistance			*Environment*			*Military*		
	1991 total $US (millions)	*% US per capita*	*% of GNP 1991*	*CO_2 emissions per capita*	*CFCs/halons % of world*	*deforestation, annual rate*	*1989 % GDP on military*	*military vs. edn & health*	*arms imports/ exports*
Developing countries (received)									
China	1,954	1.7	0.5	0.61	1.16	na	3.7	97	0.4
India	2,747	3.2	1.1	0.22	0.34	2.3	3.3	80	16.7
Indonesia	1,854	10.2	1.6	0.21	0.16	0.8	2.0	143	1.7
Brazil	182	1.2	0.0	0.36	0.71	0.7	1.2	22	1.5
Pakistan	1,226	10.6	2.7	0.14	0.65	0.4	6.7	239	2.6
Bangladesh	1,636	14.8	7.0	0.04	–	0.9	1.6	57	0.4
Nigeria	262	2.6	0.8	0.21	–	2.7	1.1	65	0.4
Mexico	185	2.2	0.1	1.01	0.58	1.3	0.5	8	0.3
Viet Nam	na	na	na	0.10	–	1.7	na	na	0.4
Philippines	1,051	16.7	2.3	0.19	0.28	1.5	1.7	47	0.1
Iran	194	3.4	0.2	0.22	0.28	0.5	na	na	2.9
Turkey	1,675	29.2	1.6	0.69	0.18	–	3.9	118	4.3
Egypt	4,988	93.1	15.2	0.42	0.18	–	4.5	57	4.6
Thailand	722	12.6	0.8	0.46	0.15	2.5	3.2	74	2.3
Ethiopia	1,091	20.7	16.5	0.02	–	0.3	13.6	239	0.6
Developed countries (given)									
USA	11,362	45	0.20	5.26	23.47	0.1	5.8	32	21.45
Former USSR	na	na	na	3.66	8.99	–	over 10	na	42.55
Japan	10,952	88	0.32	2.34	8.72	–	1.0	8	0.07
Germany (FRG)	6,890	109	0.41	2.94	9.73	–	2.8	22	1.92
UK	3,348	58	0.32	2.65	7.65	–	4.0	36	6.12
Italy	2,517	44	0.30	1.82	5.00*	–	2.4	22	1.15
France	7,484	132	0.62	1.74	6.81	–	3.7	24	10.34

Note: na = not available
Description of variables:
(1)–(3) Development assistance received by developing countries and given by developed ones, (1) in millions of US dollars in 1991, (2) in dollars given per capita, (3) assistance as a percentage of total GNP in 1991
(4)–(6) Environment: (4) emission of carbon dioxide (CO_2) per inhabitant in tonnes of carbon per inhabitant in 1990, (5) consumption of CFCs and halons as a percentage of the world total, (6) percentage annual rate of deforestation in 1980s (world rate was 0.4 per cent) (– negligible change and/or reforestation and deforestation in balance)
(7)–(9) Military expenditure as a percentage of total GDP: (8) military expenditure as a percentage of combined education and health expenditure during 1988/90, (9) for the developing countries, arms imports as a percentage of the value of all imports; for the developed countries, arms exports as a percentage of global arms exports to the Third World in 1989
Sources: (1)–(3) *World Development Report 1993*, Table 20; (4)–(6) *UNSYB 1990/91*, no. 38, Tables 108 and 109; (7)–(9) *UN Human Development Report 1992*, Tables 20, 41

the state of the environment and the natural resources at global level, although anti-pollution measures at local level have been taken. Among the first explicit studies of the possible effects of pollution on the human population and the rest of the natural world were those produced by Forrester (1971) and Meadows *et al.* (1972). The pollution ingredient appeared again in the more recent Meadows' book *Beyond the Limits*, published in 1992.

One of the most prestigious groups of experts to express concern recently and issue a dire warning about the critical stress suffered increasingly by the environment is the Union of Concerned Scientists (1993, p. 3), in whose report are the following lines:

> Human beings and the natural world are in a collision course. Human activities inflict harsh and often irreversible damage on the environment and on critical resources. If not checked, many of our current practices may put at risk the future that we wish for human society and for the plant and animal kingdom, and may so alter the living world that it will be unable to sustain life in the manner that we know. Fundamental changes are urgent if we are to avoid the collision our present course will bring about.

According to the contributors to this report, the following parts of the natural world are at risk: the atmosphere, water resources, the oceans, soil, forests and living species. Essential tasks concern environmental protection, human welfare, stable population, elimination of poverty, sexual equality. The fact that the warning in the report was signed (on 1 February 1993) by 1,680 scientists from seventy countries and included 104 Nobel laureate scientists may enhance the cause of those concerned about the future of the environment and sceptical about sustainable development, but scientists generally carry much less weight in the making of decisions that affect the environment than politicians or company directors.

The developed countries are responsible for about three-quarters of all the pollution caused in the world and, indirectly, are blamed for at least some of the environmental degradation in developing countries. It is therefore appropriate that OECD (1991) should have produced a comprehensive inventory entitled *The State of the Environment*, in which the state of the main components of the environment is considered and the impact of human activities discussed. The 1991 (no. 38) volume of *The Statistical Yearbook of the United Nations* contains for the first time data on pollution levels in the countries of the world. These data have been used in Table 4.12 columns (4)–(6) to illustrate three selected aspects of pollution and environmental impact.

Column (4) shows that the emissions of carbon dioxide in tonnes per inhabitant from fossil fuel combustion and cement manufacturing are many times higher in the developed countries than in the developing ones. They reflect fairly closely the consumption of fossil fuels, for which there are data in Table 4.1. Fossil fuel consumption levels have changed only slightly in the developed countries in the 1980s and early 1990s, and population has not increased much, but even so, each year roughly the same amount is still accumulating. The point to note is that zero growth in consumption does not mean that the problem has gone away (see Box 3.5). In the next few decades, with both increasing population and higher per capita consumption of fossil fuels, particularly in some developing countries, the amount of carbon dioxide emitted will rise enormously unless recent trends are checked. The developing countries also account for most of the

PLATE 4.5 'Exporting' pollution to the developing world

a Economic and environmental problems of the developing world. Non-ferrous metal ores are smelted at La Oroya, in the central Andes of Peru, but the refined metals then go to developed countries for manufacture. Meanwhile, the pollution is produced in Peru, damaging the vegetation and entering the river, a tributary of the Amazon

b Smoke from a copper smelter in central Chile, near Valparaiso, misses the weekend visitors to the beach. Most of the refined copper is exported to the developed countries, who in return are 'exporting' their pollution

carbon dioxide emissions from the clearance of forest, either through burning on the spot or by the removal of fuelwood for use in settlements. Of 6.4 billion tonnes estimated to have been produced from fuelwood in 1989, Asia accounted for 2.6 billion, South America 1.8 and Africa 1.5.

Column (5) shows the consumption of chlorofluorocarbons (CFCs) and halons, this data set giving the proportion of the world total. With less than one-fifth of the total population of the world, the seven developed countries included in the table accounted for about 70 per cent of consumption.

Column (6) shows that deforestation is most rapid in certain developing countries. The rate of clearance, given as the percentage of existing forest cleared each year, is related to the pressure on tropical rainforests. The absolute amount must also be taken into account. Much of the original forest in India has been cleared, while the parts that remain are fragmented, and are encircled and interpenetrated by agricultural populations that have easy access to them. In Brazil, vast expanses of forest in the north and west of the country are still very thinly populated, although the frontier of the settlement is now gradually moving northwestwards into the forest (see Chapter 11). Thus India is estimated to be losing 2.3 per cent of its 700,000 sq km (270,000 sq miles) of forest a year, Brazil only 0.7 per cent of its much larger 4,900,000 sq km (1,890,000 sq miles), but that is roughly double the absolute amount lost in India. In the developed countries of the world, most of the forest is in temperate and cold latitudes, where both environmental conditions and the quality and quantity of forestry services make it easier to replant, while the use of fuelwood is very limited in Japan and most countries of Western Europe.

Environmental issues and problems exist at various levels of scale from global to very local. The medium in which polluting substances diffuse most quickly and widely is the atmosphere where, in principle, carbon dioxide emissions from a particular source can eventually spread around the globe. If they are caught in winds, other emissions such as sulphur dioxide, nitrous oxides and radioactive materials can be carried through the atmosphere thousands of kilometres before being deposited, but most fall near the sources of the emissions. In cities such as Los Angeles and Mexico City, with very large populations, many industries, a large number of motor vehicles in use, and frequent periods in which weather conditions keep the pollution from dispersing, conditions are particularly bad.

The United Nations Development Programme's (UNDP) 1992 Report (*HDR 1992*, p. 83) contains proposals for achieving policies for environmental protection and sustainable development. The Global Environment Facility (GEF) had funds for its pilot phase 1991–4 from developing as well as developed countries. Its purpose is to assist developing countries to explore ways of protecting the global environment. Four priority areas of concern are identified: global warming, destruction of biological diversity, pollution of international waters and depletion of the ozone layer. Up to half the resources are allocated to the question of global warming. The GEF is jointly administered by UNDP, UNEP (United Nations Environment Programme) and the World Bank.

In principle, it is of interest to all countries that the economic growth of the twentieth century should continue in the twenty-first century both to cater for the increasing population and to raise material standards, particularly in the poorest countries. Many argue that economic development, if properly conducted, can be sustained without being in conflict with environmental degradation. In theory that may be so but, in practice, much economic growth is organised locally or nationally, either by private sector enterprises or by the public sector. Is there a difference in attitude and performance between the two sectors?

At first sight it might seem that the state planners controlling production in the public sector would give high priority to conservation and the prevention of environmental degradation and pollution. Increasing evidence since the late 1980s of Soviet and East European practices has shown that, in spite of professing concern over the environment, the state did little to deter pollution or to clean up badly affected sites. As Ziegler (1987, p. 165) states: '... the accident (Chernobyl) demonstrates the Soviet propensity to sacrifice public safety and environmental quality in the interest of cutting short-term costs'. By imposing sanctions and fines on polluters, the state would in practice be fining itself.

The situation is not all that different in countries in which the private sector prevails. Even in cases when a pollutant can be identified, as when the cargo of an oil tanker spills into the sea or a chemical plant produces lethal emissions, the fine imposed tends to be minute when compared with the assets of a large company. To abandon practices that can lead to pollution could cost far more than the payment of fines and of compensation to sufferers. In the last resort, the whole population benefits materially from low-cost production and (comparatively) low prices, and the imposition of high taxes on such polluting products as motor fuel to cut consumption and to increase the efficiency of use is not generally popular. Following the 'limits to growth' debate and oil price rise in the early 1970s, for a time in Western Europe and North America the owners of 'gas-guzzling' Cadillac, Mercedes and Rolls Royce cars were heavily criticised, and the use of motor fuel for car racing deplored. The 'conspicuous' consumers did not change to small cars, bicycles or public transport.

The problem of exhaustion of non-renewable or exhaustible natural resources is not difficult to appreciate. When a tonne of coal or oil is burned, it literally goes up in smoke, and a quantity of topsoil can be blown or washed away from a field. What impact the emission of carbon dioxide, nitrous oxide, methane and other gases into the atmosphere as a result of burning fossil fuel or raising livestock will have on human life and activities is a matter of speculation. The atmospheric concentration of carbon dioxide (CO_2) has increased by about 25 per cent since modern industrialisation began. According to OECD (1991, pp. 20–8), between 1860 and 1984 a cumulative total of about 183 billion tonnes of carbon was emitted as a result of fossil fuel combustion and about 150 billion tonnes from deforestation and land-use changes. In 1988, OECD countries accounted for 45 per cent of the emissions, the USSR and Eastern Europe for 24 per cent and China for 11 per cent.

Carbon dioxide and other gases affect the temperature of the earth's atmosphere. During the period 1900–40 the average global surface temperature increased by about 0.25° Centigrade but, discounting year-to-year fluctuations, did not change between 1940 and 1980. During the 1980s there was a sharp increase, a change leading many to anticipate a continuation of the trend into the twenty-first century, giving an increase of about 0.3°C per decade. This so-called 'greenhouse effect', producing global warming, has not yet been proved beyond doubt. Nor, according to OECD (1991, p. 27), is it possible 'to predict with any reliability the future emission rates of greenhouse gases, their future concentrations in the atmosphere or the global and regional changes that would follow changes in atmospheric composition'.

One change expected by many scientists as a result of global warming is a change in sea level, to which both the thermal expansion of the sea water and the melting of land ice would contribute. Even a rise in sea level of 0.5 to 1 metre (2 to 3 feet) would affect sea defences in many countries and threaten both rural and urban settlements in low-lying coastal areas. On the other hand, an increase in the temperature of the atmosphere could result in greater quantities of ice accumulating in Antarctica, Geenland and elsewhere, thereby lowering sea level.

The greenhouse effect exemplifies a number of big environmental issues. There is a means of preventing further global warming: ban the burning of fossil fuels and fuelwood and also certain agricultural practices. At present, such a path of action is unthinkable. Alternatively, prevent the accumulation of greenhouse gases from actually causing the heating of the earth's atmosphere. How? It has been shown that in the not too distant future it would be technically feasible (albeit at enormous cost) to put in orbit outside the earth's atmosphere shields that would reflect some of the sun's energy back into space. Humans create problems and science and technology find answers. Coping with the greenhouse effect will be more difficult than controlling the emission of nuclear fallout into the atmosphere. In this case it was accepted almost universally that nuclear tests in the atmosphere should cease.

If global warming is taking place, then it is a striking example of a shared change that affects all the twelve major regions, some adversely, some perhaps favourably. Each region also has its own particular environmental problems. In very cold latitudes, environmental pollution is particularly serious because low temperatures and the lack of precipitation make it difficult for waste to be degraded. In Saharan Africa there is concern about the encroachment of the desert on grazing and arable land. In mountain areas in which forest is cleared, heavy rains may pass quickly into rivers, increasing flooding and soil erosion in lower reaches, a trend noted in the Himalayas and thought also to have influenced flooding in the Rhine basin in 1995.

In conclusion, it is undeniable that concern over the environment has grown rapidly in many countries since the 1960s through various 'green' and other movements. In contrast, so far action to confront environmental problems with curative and preventive measures has been very limited. Whether this contradiction will be a feature also of the twenty-first century remains to be seen.

4.8 Military expenditure

Expenditure on defence has several related functions. Ostensibly since the Second World War it has been justified in most countries as a necessary insurance against attack from one or more other, usually neighbouring, countries. Other functions are the use of military equipment and personnel to supplement the work of the police during times of internal disorder, to run emergency services in times of industrial action, and to help when natural disasters occur. Virtually every country in the world is a member of the United Nations, one of the main aims of which is to maintain peace in the world. It is therefore to be expected that since 1945 few countries have explicitly built up armed forces in preparation for the conquest of other countries, as Germany, Italy and Japan did in the 1930s.

With hindsight it may be argued that the advent of nuclear weapons, with their capacity by the 1960s to wipe out much of the population of the world in a very brief 'all-out' war, deterred the major industrial countries from attacking each other. Nevertheless, since 1945 there have been numerous international conflicts in the developing world, in some instances with the superpowers involved, usually indirectly, but at times directly, as when US forces were engaged in fighting in Korea and in Viet Nam and Soviet forces were in combat in Afghanistan. Both restricted their armoury to conventional weapons. The situation is summarised in *HDR 1993* (p. 10) as follows:

> Worldwide there have been more than 100 major conflicts in the past four decades, taking the lives of some 20 million people. The United Nations was often powerless to deal with these conflicts – paralysed by vetoes by major powers on both sides of the East–West divide. Since May 1990, however, no such vetoes have been cast.

Many developing countries still allocate a large proportion of their budgets to defence expenditure, while the major industrial powers have devoted massive resources and research efforts to perfecting new weapons of destruction, the former Soviet Union in particular, it is now admitted, having set up huge military industrial complexes in many cities.

Although the protagonists in the Cold War have now renounced the use of nuclear arms, they have only committed themselves to actually decommissioning a small part of their nuclear weaponry. The future attitudes to and role of defence in the twenty-first century are therefore unclear. Will the use of military force again become publicly accepted as a means to achieve political ends, as it was before the Second World War? In his book *Vom Kriege* (*On War*), published in 1832, Carl von Clausewitz viewed war as a rational instrument of national policy. During the following hundred years, military force was used at some time or other to conquer territory by all the major powers of the time, whether they were engaged in taking new territory in North America, Africa and Asia, or in fighting each other in the confined space of Europe.

The US naval officer A. T. Mahan (1840–1914) propounded a global strategy based on sea power. In the inter-war period, the German geographer Karl Haushofer edited a journal, *Zeitschrift für Geopolitik*, in which territorial expansion was justified, a matter in which the Nazis took great interest. In 1923, J. F. Lee published *Imperial Military Geography*, the grandiose title of a book with facts and ideas for examinees entering British military establishments. Since 1945 discussion of military matters has been more muted, with politicians tending to stress the need for military expenditure in terms of defence. In recent years, however, influential political leaders, including Saddam Hussein of Iraq (see Chapter 13) and Vladimir Zhirinovsky of Russia (see Chapter 7) have made statements or written in a style similar to that used by Adolf Hitler in his book *Mein Kampf*.

During the Second World War, US and British geographers, like other scientists, were engaged as advisers to help the war effort in their respective countries. The careful study of the beaches in France chosen for the D-Day landings was made with the assistance of geologists. In recent decades many North American and West European geographers have avoided studying and researching military matters and some have criticised earlier geographers and cartographers for their involvement in the imperial process. It is unlikely, however, that matters of defence will suddenly melt away and that conflicts will cease. An appreciation of the role of defence and the probability of conflict in the major regions of the world is an essential part of their political, economic and social future. In the rest of this section, attention focuses on three particular aspects: expenditure on defence, the continuing role of the military in the post-Cold War period and the probability of conflict in certain parts of the world.

The Second World War was followed by a brief period of arms reduction, including the demobilisation of forces and the decommissioning of equipment, but after a few years most industrial countries embarked on massive rearmament programmes, Germany and Japan for a time being exceptions. Even newly independent, usually very poor, ex-colonies in Asia and Africa, duly spent heavily on arms. The 'official' end of the Cold War came in the late 1980s and has resulted in some reduction in military spending round the world. The 'peace dividend' is not, unfortunately, a simple transfer of manpower and productive resources from one kind of production to another. As has already become evident in the USA, Western Europe and the former USSR, the establishments producing and servicing military equipment may be inappropriate for producing civilian goods. Such establishments may be badly located to assemble materials and distribute products for civilian needs, production costs often being less of a consideration in choosing locations for military production than strategic advantages. Job losses in localities where military production has ceased or has been reduced, and also in places where large military bases have been closed, have given rise to serious local economic problems in the 1990s.

While the developed countries are able to trim their defence expenditure and to redeploy their forces to defuse tension, as the Russians have done by withdrawing their forces from Central Europe, many developing countries appear determined to carry on as if the Cold War had not ceased, had never affected them, or indeed had never taken place. Large sums are spent on the import of arms from the industrial countries, but some so-called developing countries are themselves capable of producing sophisticated weaponry and of exporting it, among them China and India.

Columns (7)–(9) in Table 4.12 compare aspects of the military sector in the largest developing and developed countries. Column (7) shows large variations in military expenditure as a percentage of GDP in 1989. Low levels (e.g. Mexico, Japan) and high levels (e.g. Ethiopia, USSR) are found among both developing and developed countries. It might be expected that a very large country would need to spend less per inhabitant on defence than a small one. The situation among the larger countries of the world in the table is, however, such that global and regional location seems to be a stronger influence, with for example Brazil, Mexico and Nigeria having very low levels of expenditure (Mexico 'protected' by its superpower neighbour in the north) while countries in the hotspots of world affairs or engaged in internal conflict, such as Pakistan and Ethiopia, have relatively large military establishments.

Healthcare and education (column (8)) are the two aspects of social development given most prominence in the UN Human Development Programme. The evidence in column (8) is that relative to these two civilian sectors of the economy, military spending is much higher in most of the developing countries listed than in the developed ones, with more being spent on the military sector than on healthcare and education combined in such countries as Pakistan, Ethiopia, Indonesia and Turkey. The extremes are even further apart between smaller developing countries, with military/social spending ratios (%) ranging in 1987–9 from 511 in Iraq and 500 in Somalia to 5 in Mauritius and 4 in Costa Rica.

In *Human Development Report 1992*, a section is devoted to new structures of peace and security. It is noted (p. 85) that 'The peace dividend opens a window of opportunity for both rich and poor countries.' Global military spending has declined in developing countries between 1984 and 1990 and in industrial countries since 1987. In the former, expenditure declined by about 20 per cent from 155 billion US dollars to 123 billion, in the latter by almost 10 per cent between 1987 to 1990, from 838 to 762 billion. If growth had continued after 1984 and 1987 respectively, instead of declining, the peace dividend can be seen as having saved some 325 billion dollars already by 1990. Should the decline continue, then by the year 2000 more than 1,000 billion dollars could be saved. Ironically, by 1990, the two poorest major regions of the world, South Asia and Africa south of the Sahara, were the only ones in which no decline in military spending was recorded.

The authors of the United Nations Development Programme (UNDP) make tentative proposals that follow from arms reduction and should be implemented as a result. Aid donors should take the military spending levels of recipients into account when allocating aid. Nevertheless (p. 86), 'they should also recognise that the recipients do have legitimate security needs'. The big powers should curtail military assistance and close military bases. Industrial countries still subsidise arms exports to developing countries. It is argued that it would be more helpful if they directed subsidies to helping arms producers to switch to more peacefully oriented products. The United Nations has begun to assume a greater role in political and security matters. If this trend is to continue, the UN would have to be reformed in several respects, and to have a larger and more reliable source of finance for its peace-keeping efforts.

Military expenditure may have been declining since the mid- to late 1980s in many countries, but no major power, developed or developing, is going to do away with its military establishment altogether. Indeed, the emergence of almost twenty new countries in the former USSR and Central Europe since 1990 (fourteen in the USSR, four in Yugoslavia and one in Czechoslovakia) may be accompanied by an increase in spending for a time.

In the next century the military will still be used for the following purposes:

- To ensure internal stability, in some cases maintaining a civilian government or a military government in power or suppressing unrest in a region. Examples in the 1990s are the use of the army in Mexico to suppress the Zapatista movement in Chiapas state, in Peru to combat the Sendero Luminoso movement, in China to pacify and maintain control of Tibet, in Russia to prevent the Republic of Checheniya from gaining independence, and in the UK to support the police in attempting to prevent violence in Northern Ireland.

Throughout the developing world, the military has frequently exceeded its brief by actually taking over the government from civilian leaders, themselves not always elected. Changes between military and civilian governments were frequent occurrences in many Latin American countries until recently:

- To protect national boundaries. In the interior of tropical Latin America and in parts of Africa, boundaries run for long distances through virtually uninhabited forest or desert. Such borders are not manned but are loosely monitored. Some large developing countries share borders with several neighbours. Brazil and Zaire, for example, each have ten neighbours, with mutual boundaries totalling several thousand kilometres in length.

- Actually to occupy part or all of another country or territory held by another power, or to intervene to maintain or restore stability, usually not without an element of self-interest. The Iraqi occupation of Kuwait in 1990 was justified on the grounds that Kuwait was previously a region of Iraq. Soviet intervention in newly independent Georgia was not a surprise.

- To help to maintain peace at the request of the United Nations. In Yugoslavia the peace-keeping forces have come from several developing nations as well as from developed ones. In Somalia and Haiti the USA has contributed.

Finally, it is appropriate to speculate briefly about the likely trouble spots in the decades to come. It is appreciated that no country is immune from the possibility of some kind of conflict. The expectation of the authors of *HDR 1993* (p. 10) is that 'future conflicts may well be between groups of people rather than states'.

- High probability of conflict, in some instances with considerable loss of life.

(a) Africa, where both interstate and internal conflicts have been numerous since the 1950s, when many colonies became independent. In the whole of Africa there are about sixty countries, all except Madagascar with anything from one to ten neighbours, inviting cross-boundary conflicts, especially where tribal areas were split by colonial boundaries. Internal conflicts have devastated the economies of the two former Portuguese colonies of Angola and Mozambique, while Sudan and Ethiopia have also suffered greatly. The situation has been deteriorating in Algeria since 1993.

(b) The former USSR, with three possible types of conflict in the future: between Russia and the other fourteen newly independent republics (e.g. Russia and Ukraine), between pairs of the fourteen non-Russian republics (e.g. Armenia and Azerbaijan), and between Russia and republics within the Russian Federation (e.g. Russia and Checheniya).

(c) Southwest and South Asia (e.g. Yemen, ethnic conflict in India and Pakistan in dispute over Kashmir).

(d) China, where the leadership succession and the handover of Hong Kong could be the catalysts for internal conflict, with the possibility of attempts to break away from central power either by the non-Han regions or by economically sophisticated provinces such as Guangdong.

(e) Southeastern Europe, where linguistic, religious and territorial diversity and fragmentation provide conditions for the spread of conflict from Bosnia and Croatia into the rest of the former Yugoslavia and beyond.

- Probability of some conflict in Latin America (e.g. Peru–Ecuador) and Southeast Asia, with continuing internal struggles (e.g. Myanmar, Cambodia, Indonesia–Timor) and possible border clashes.

- Probability of minor internal conflict but no interstate conflict in North America (riots), the EU (nationalist movements such as the Basques within countries, and tensions due to increasing migration), Oceania, Japan.

The defence sector may be seen as an insurance service, producing nothing material *per se* and with a destructive potential that may not be used. In the more active sense, a recent example being Iraq's war with Iran and later its occupation of Kuwait, the logic of using military force is such that the pay-off is worth the outlay, the resources gained being perceived as greater in value to the conquering side than the lives and material lost in acquiring them. The negative by-product of military conflict is that, instead of the resources used to maintain the military machine being used to create producer and consumer goods, as they would in a civilian context, loss of life and the destruction of means of production are caused.

KEY FACTS

1. The production and consumption of goods and services is very unevenly distributed among the people of the world. The general consensus is that the gap has increased rather than decreased since the Second World War.

2. International trade is dominated by the developed countries, but the traditional exchange of manufactured goods for primary products between developed and developing countries is no longer very marked. Many developing countries now export substantial quantities of manufactured goods.

3. Development assistance flows from developed to developing countries, but the quantity involved is well below 1 per cent of the total GDP of the developed countries and it is disbursed unevenly.

4. Large transnational companies dominate many sectors of mining, manufacturing and services in developed countries and they also operate extensively in developing countries. By their nature, transnational companies are concerned with satisfying their shareholders rather than maximising benefits in countries hosting their enterprises.

5. Concern has grown in many quarters about the present state of the natural environment. What is less certain is the extent to which the environment has been damaged so far.

MATTERS OF OPINION

1. Do the rich countries have any obligation or incentive to help the poor countries other than possibly self-interest?

2. If development assistance is given, should it be with 'strings'? For example, that none should be given unless family planning measures are implemented, the defence budget is cut, the poorest families are targeted, a democratic political system is in place?

3. If it is shown convincingly that a particular practice (e.g. forest clearance) or device (e.g. the motor car) is causing excessive/irreparable damage to the environment, should coercion be used to halt/slow the process?

4. With reference to Table 4.1, examine carefully the differences on each relevant variable between and within the groups of developing and developed countries and think about the saying 'development is indivisible' (i.e. is it realistic to expect great improvements in one variable when little change occurs in others?). In some countries indivisible might be replaced by invisible.

SUBJECTS FOR SPECULATION

1. How can the development gap be narrowed?

2. How can the terms of international trade, frequently described as unfair, be made more equable?

3. Can development assistance be made both larger and more effective than it has been over the last few decades? Can it ever be 'enough'?

4. Will the natural environment and natural resources be capable of supporting an increasing and more affluent population indefinitely?

5. Is it physically, logistically, geographically possible in practice to relocate enough people or productive resources to ensure reasonably similar living standards throughout the world?

Chapter five

The European Union and EFTA

In 1957 the heads of state of the six founder members of the European Economic Community

Determined to lay the foundations of an even closer union among the peoples of Europe,

Resolved to ensure the economic and social progress of their countries by common action to eliminate the barriers which divide Europe,

Affirming as the essential objective of their efforts the constant improvement of the living and working conditions of their people.

The first three paragraphs of the Treaty establishing the European Economic Community, 1957 (Treaties 1987)

5.1 Introduction to Western and Central Europe

The traditional western, northern and southern limits of the continent of Europe are the coasts of the Atlantic Ocean and the Mediterranean Sea. The eastern and southeastern limits, however, are mostly on land, and are therefore less obvious. The Ural Mountains and Ural River mark the eastern limit of Europe, but the former run through the centre of one of the major economic regions of Russia. The Caspian Sea, the boundary between former Soviet Transcaucasia and Iran, and the Black Sea mark the limit of Europe in the southeast. Both Russia and Turkey are divided in such a way that each is partly in Europe and partly in Asia. In this book the former Soviet Union, which accounts for nearly half of the area of Europe and a third of its population, is the subject of Chapter 7, while Turkey is included in Southwest Asia, in Chapter 13.

Until the Second World War, geographers wrote books about the whole of Europe. Some of the books were still used after the war in spite of the fundamental political and subsequently economic restructuring of the continent. After 1945, Europe became sharply divided between the US-backed western part and the Soviet-dominated central and eastern part, the Soviet bloc, commonly referred to as Eastern Europe. A few neutral countries formed a discontinuous buffer zone in between. Following the break-up of the Soviet Union in 1991 and the withdrawal of Soviet forces from Central Europe, the pattern of political allegiances and economic transactions has been changing rapidly (see Table 5.1).

In the early 1990s, Europe excluding European USSR could be subdivided on political and economic grounds into three subsets of countries, the twelve of the European Union (EU), the six countries of EFTA, and an increasing number of countries in what is now widely referred to as Central Europe (refer to Figure 5.1). Data for the EU and the EFTA countries are given in Table 5.2. The data are mainly for 1993, before Sweden, Austria and Finland left EFTA and joined the

TABLE 5.1 Calendar for Western and Central Europe since 1815 (see also Table 5.2 for details of European Union)

1815	End of the Napoleonic War
1820s	Spain and Portugal lose colonies
1860–70	Italian unification
1870	German unification. Franco-Prussian War
1914–18	First World War
1917	Russian Revolution
1928–33	World economic depression
1936–9	Spanish Civil War
1937	Germany annexes Austria
1938	Germany occupies Czechoslovakia
1939	Germany invades Poland
1939–45	Second World War
1945–8	Soviet influence in Central Europe grows
1947	Benelux Customs Union signed
1948	Czechoslovakia drawn into Soviet orbit but Yugoslavia withdraws
1951	European Coal and Steel Community signed
1957	Treaty of Rome, EEC and Euratom signed
1960	European Free Trade Association Convention
1989–90	Unification of Germany
1991	Fourteen Soviet Socialist Republics given independence from Russian Federation
1991	Start of break-up of Yugoslavia
1993	Maastricht Treaty ratified

EU. This chapter starts with an overview of Western and Central Europe. The general aspects of the EU and EFTA are the subject of the rest of the chapter. Western Europe consists of these two supranational units. In Chapter 6, individual countries of both Western and Central Europe are described.

At the time of writing, Europe and the USSR were subject to greater political change than anywhere else in the world. Three EFTA countries joined the EU in January 1995, while all Central European countries are in the process of reorientating their economic life towards Western Europe and adopting features of a market economy. The contents of Chapters 5 and 6 are therefore organised to take into account this transition. Apart from brief summaries of each of the twelve pre-1995 Member States of the EU in Chapter 6, these countries are dealt with collectively in this chapter as parts of an increasingly integrated supranational unit. Thus the EU and EFTA are discussed topic by topic rather than country by country.

Western and Central Europe have a combined population of about 510 million people, about 9.3 per cent of the world total, but their total area, less than 5 million sq km (1.93 million sq miles), is less than 4 per cent of the world total. The physical environment is relatively benign by world standards, experiencing neither extremes of cold, dry conditions nor rugged terrain except fairly locally. Where soil and other physical conditions are suitable, cultivation is possible almost everywhere, even north of the Arctic Circle in Norway and Finland. Europe's climate is modified by warm ocean currents and air masses, so that Scandinavia has a less harsh climate than the parts of North America and Siberia at similar latitudes. The arid conditions that prevail in North Africa not far south of its Mediterranean coast are experienced in a limited way only in parts of southern Spain. Even the most mountainous areas of Western and Central Europe, the Alps, Pyrenees, Carpathians, and other mountains of the Balkans, together with the mountains of Norway and Sweden, do not reach the altitude or extent of the Himalayas in Asia or the Andes in South America. They have a number of positive economic features, providing water, hydroelectric power, timber and pastures as well as favourable conditions for tourist resorts.

The search for and use of economic minerals in Western and Central Europe goes back to Roman times and earlier, and many deposits of both fossil fuels and non-fuel minerals have been worked out. With its 9.3 per cent of the population of the world, the region has only about 6 per cent of the world's fossil fuels, in particular the coal of Germany and the oil and natural gas of Norway, the UK and the Netherlands. Western and Central Europe also have about 7 per cent of the world's non-fuel mineral reserves, although there are few very large deposits. Given the high level of industrialisation in the region, however, and the virtual absence of some key minerals (bauxite, copper, phosphates), large quantities of non-fuel minerals are imported, as well as fuel, mainly oil.

The population of Western and Central Europe is

FIGURE 5.1 The countries and major cities of Western and Central Europe. The population size is indicated in the key on the map. See Figure 6.2 for details of divided Yugoslavia

now changing only very slowly and it has the lowest rate of population growth of any major region in the world. About 75 per cent of the population is defined as urban compared with the world average in 1993 of 42 per cent. About 14 per cent of the population is over the age of 64, compared with the world average of only 6 per cent, while about 20 per cent is below the age of 15, compared with the world average of 33 per cent. The prospect of having to shift resources from education for the young to healthcare for the elderly is a

TABLE 5.2 Country-by-country data set for Western Europe, mainly for 1993

	(1)	(2)	(3)	(4)	(5)	(6)	(7)	(8)	(9)	(10)	(11)	(12)	(13)	(14)
	Population				*Real GDP (US dollars)*									
	rank in world	*millions*	*Area in thous. sq km/(miles)*	*Density persons per sq km/(mile)*	*total (billions)*	*per head (thousands)*	TFR	*% over age 64*	*infant mortality*	*% in agriculture*	*% urban*	*Steel output per head (kg)*	*Energy conservation per head (kg)*	*Cars per 1,000 popn*
European Community														
Germany	12	81.1	356.9 (137.8)	227 (588)	1,177	14.5	1.4	15	7	4.4	85	603	5,610	415
UK	18	58.0	244.9 (94.6)	237 (614)	796	13.7	1.8	16	7	1.9	90	328	4,980	361
Italy	19	57.8	301.3 (116.3)	192 (497)	787	13.6	1.3	14	8	6.7	68	443	3,770	450
France	20	57.7	551.5 (212.9)	105 (272)	817	14.2	1.8	14	7	5.0	73	334	3,800	399
Spain	27	39.1	504.8 (194.9)	77 (199)	341	8.7	1.3	14	8	10.2	78	299	2,480	293
Netherlands	55	15.2	37.3 (14.4)	408 (1,057)	203	13.4	1.6	13	7	3.5	89	375	6,490	353
Greece	64	10.5	132.0 (51.0)	80 (207)	71	6.8	1.4	14	10	23.6	58	95	2,970	152
Belgium	69	10.1	30.5 (11.8)	331 (857)	134	13.3	1.6	15	8	1.7	97	1,119	5,640	351
Portugal	74	9.8	92.4 (35.7)	106 (275)	61	6.3	1.4	13	11	15.6	30	71	1,900	239
Denmark	102	5.2	43.1 (16.6)	121 (313)	72	13.8	1.7	16	8	4.5	85	135	4,370	313
Ireland	123	3.6	70.3 (27.1)	51 (132)	27	7.5	2.2	11	9	13.1	56	0	3,720	217
Luxembourg	–	0.4	2.6 (1.0)	154 (399)	7	16.5	1.6	13	9	1.7	86	9,250	11,500	443
European Free Trade Association														
Sweden[1]	80	8.7	450.0 (173.7)	19 (49)	129	14.8	2.1	18	6	3.7	83	552	4,910	411
Austria[1]	85	7.9	83.9 (32.4)	94 (243)	103	13.1	1.5	15	8	5.5	54	595	3,850	367
Switzerland	90	7.0	41.3 (15.9)	169 (438)	130	18.6	1.6	15	7	3.9	60	143	3,460	414
Finland[1]	104	5.1	338.1 (130.5)	15 (39)	74	14.6	1.8	14	6	7.8	62	569	5,690	372
Norway	111	4.3	323.9 (125.1)	13 (34)	72	16.8	1.9	16	8	5.0	72	209	7,020	375
Iceland	–	0.3	103.0 (39.8)	3 (8)	4	14.2	2.2	11	6	6.5	91	0	4,670	413

Note: [1] Sweden, Austria and Finland joined the EU in 1995 but the data in the table are for 1993, when they were still in EFTA

Description of variables:

(7) TFR – Total fertility rate, average number of children born per female

(9) Deaths of infants under the age of one year per 1,000 live births

Sources: WPDS 1993 for columns (2), (7), (8), (9), (11); *HDR 1992* for columns (5), (6); *FAOPY 1991* for column (10); *UNSYB 90/91* for columns (12), (13), (14)

PLATE 5.1 Signs of the times in the European Union

a In this sign Belgium's French-speaking region of Wallonia is announced by the twelve stars of the EU

b In Catalonia, Spain, speakers of both Spanish and Catalan are warned of the danger of entering a river bed on account of sudden variations in the level of the water. Arguably the Catalan version is redundant because all Catalonians can understand and read Spanish

major issue and a problem for the whole region.

Economic development has been uneven in Western and Central Europe. For example, less than 2 per cent of the economically active population is engaged in agriculture in the UK and Belgium, compared with almost a quarter in Greece and parts of the former Yugoslavia, and almost half in Albania. Throughout the region, the agricultural sector has been losing workers but, until recently, the economically active population in industry was increasing. Since the early 1970s, and particularly in the 1980s, the total labour force in industry has been diminishing in several West European countries, and in the 1990s jobs have also been lost in some branches of the service sector, such as banking and communications.

For several decades the whole of Central Europe has been under command economies, dominated by investment and production in the public sector. Officially, unemployment here has been kept down almost to nothing, but this has been achieved at the expense of extensive underemployment and overmanning. The transformation to market economies is already causing unemployment in parts of Central Europe, above all in East Germany (the former German Democratic Republic) where, now unified with West Germany in the EU, it has undergone massive economic restructuring. Throughout Western and Central Europe, therefore, unemployment is now regarded as a major issue and problem, possibly only to be reduced by changes in definition, such as a rise in the age at which it can first be recorded, a reduction in the official retirement age, or the use of a shorter working week. Any drastic tampering with the situation, such as job creation schemes or an increase in the length of national service by the state, would run against the current thinking in a market economy. Likewise, the creation of a new lower age limit for pensioners below present levels, which vary among countries, would presumably be an unacceptable financial burden for the state.

5.2 The European Union and EFTA

Although almost two-thirds of the area and half the population of Europe is in Central Europe and the former USSR, the supranational units of Western Europe, the EU and EFTA, have appropriated the name of the continent for themselves. From a geographical point of view as much as from economic and political ones, the EU and EFTA countries must be considered both as Member States of their groups and as individual countries. It should, however, be appreciated that in spite of four decades during which the countries which first joined the EU have renounced part of their sovereignty to supranational bodies, the individual Member States are still very much alive, as are many regions within them, such as Scotland in the UK and Catalonia in Spain. The limited extent of the political powers of the institutions of the EU is reflected in the

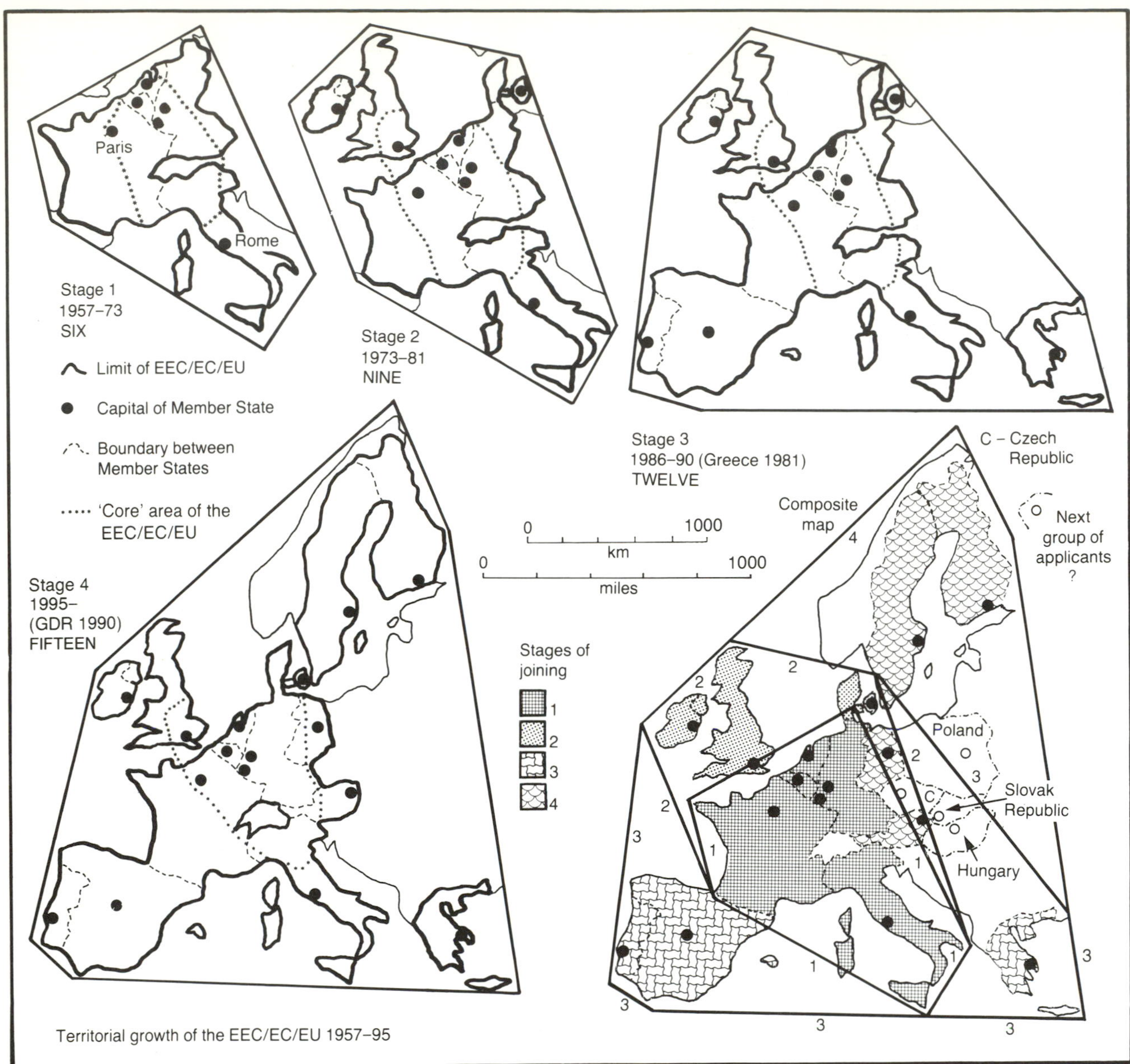

FIGURE 5.2 The territorial growth of the EU showing its changing shape. The outer lines join extremities of the EU space. The shading on the lower right-hand map indicates the duration of membership

fact that its budget is equivalent to no more than about 1 per cent of the total GDP of the EU or 3 per cent of the combined budgets of the EU Member States. The EFTA countries have been even less strongly bound together, their association being primarily economic. Until the 1990s, on the other hand, all the countries of Central Europe apart from Yugoslavia and Albania were tied politically, economically and militarily to the Soviet bloc and the CMEA. They are now changing in the opposite direction to the EU countries.

During the Second World War, Germany briefly occupied and organised much of Europe for its own purposes. For a time France, Italy, the Netherlands, Belgium and Luxembourg were drawn into the German war orbit against their wishes. Those same countries later voluntarily became the founding members of the European Coal and Steel Community (ECSC) in 1952 and of the European Economic Community (EEC) and the European Atomic Energy Community (EAEC or Euratom) in 1958. To counterbalance the EEC, EFTA was established in 1960. Subsequently, six additional countries joined the EEC – three of them, the UK, Denmark and Portugal, defecting from EFTA to do so – while in 1990 East Germany (the former communist GDR) was absorbed. Table 5.3 gives key dates in the development of the EU and Figure 5.2 shows its territorial expansion. One of the main reasons for creating and developing the EU was the fact

TABLE 5.3 Key dates in the formation of the European Union

1951	Signing of the Treaty of Paris establishing the European Coal and Steel Community (ECSC)
1957	Signing in Rome of Treaties establishing the European Economic Community (EEC) and the European Atomic Energy Community (EAEC)
1968	The Community becomes a customs union (abolition of all import and export duties between the Member States)
1973	Accession of Denmark, Ireland and the UK to the European Community
1979	The European Monetary System (EMS) comes into operation
1979	First direct elections to the European Parliament
1981	Greece accedes to the European Community
1984	Second direct elections to the European Parliament
1986	Portugal and Spain accede to the European Community
1986	The Single European Act amends basic Treaties
1989	Third direct elections to the European Parliament
1990	The Federal Republic of Germany embraces the new Länder of the former GDR after the collapse of the communist regime there
1992	Signing of Treaty of Maastricht
1993	Maastricht Treaty ratified
1994	Fourth direct elections to the European Parliament
1994	Applications for membership of the European Union from Austria, Finland, Norway and Sweden accepted
1995	Entry of Austria, Finland and Sweden (Norwegians voted against entry, November 1994)

Between 1987 and 1991 Cyprus, Malta and Turkey also applied for membership

that trade between the countries of Western Europe has grown enormously since the Second World War. Thus an economic reason rather than a strategic one for joining the EEC is shown by the data in Table 5.4, which demonstrate, for example, the change of direction of UK trade between 1951 and 1990.

The ultimate aims of the EU are not agreed on by all the Member States. Since its foundation, much of the debate and legislation has been devoted to the gradual establishment of a customs union, the outstanding feature of which has been the creation of a common tariff for the twelve Member States in their trade with the rest of the world and the elimination of all barriers to trade within the Community. By contrast, a Free Trade Area focuses only on the reduction of internal barriers. The external tariff barrier has been particularly influential in protecting agriculture in the EU through the Common Agricultural Policy, but quotas and tariffs have also been variously applied to imports of manufactured goods. Although freer trade worldwide would seem to be in the interest of the EU, some Member States have been reluctant to accept measures proposed by GATT to cut support to some sectors of the economy. In December 1993, however, the Uruguay Round was concluded, leading to further liberalisation of trade world-wide in a wide range of goods and services, a development regarded by many as of particular benefit to the EU.

By the mid-1990s, theoretically all barriers to the movement of people, goods, services and capital between EU Member States should have been removed, thanks to the creation of the Single Market. In practice, however, customs and immigration posts remain in place at many border crossing-points. The final ratification of the Treaty of Maastricht in 1993 after its signing at the end of 1991 marks another step towards greater integration in the EU, including greater conformity in social matters such as working hours, and a move towards monetary union and a common currency. Individual Member States have, however, opted out of certain parts of the Treaty.

TABLE 5.4 UK's trading partners compared, 1951 and 1990

	Percentages of value of			
	1951		*1990*	
	imports	*exports*	*imports*	*exports*
USA	10	5	12	13
Canada	7	5	2	2
Australia	6	13	1	1.5
New Zealand	4	4	0.4	0.3
India	4	4	0.6	1
Japan	0.5	0.4	5.4	2.5
EC (11)[1]	20	18	53	53
World total	100	100	100	100
Absolute values	£3,904m.	£2,580m.	£225 bn	£186 bn

Note: [1] EC as of 1990, excluding UK itself (of course)
Sources: *UNYITS 1953*, p. 456; *ITSY* (1990), vol. 1, p. 944

In the mid-1990s, then, two issues dominate thinking and policy in the EU, first greater integration, cohesion, and the reduction of disparities in economic level between regions in the Community itself and, second, enlargement. One of the features of the Treaty of Rome establishing the EEC in 1957 has been the provision for other states to apply to join. Among the conditions for membership are location in Europe, a system of democratic government, and a predominantly market-based economy, broadly compatible in economic level with that of the EU.

The next steps towards the enlargement of the EU, following German unification in 1990, have been the applications to join by four EFTA countries, Austria, Sweden, Norway and Finland. The last three are commonly referred to as the Nordic countries, a more appropriate term than Scandinavia(n), which excludes Finland. Switzerland (with Liechtenstein) and Iceland have not applied, Switzerland being concerned about the loss of sovereignty over financial and residential decisions, Iceland over the prospect that it would lose fishing grounds to other EU countries. In 1994 the EU agreed to the membership of Austria and the three Nordic countries. Austria, Sweden and Finland became full members in 1995 but near the end of 1994 the Norwegian referendum produced a negative response from the electorate.

Although the entry of Austria and two of the three Nordic countries into the EU does not pose major economic or financial difficulties, since they would all be net contributors to the EU budget, there are specific sectoral problems to resolve, such as fisheries and agriculture, as well as political issues, such as neutrality. The major drawback for the EU in accepting more members lies in its own internal decision-making and administrative activities. Larger Member States such as France and Germany have reservations about the addition of three 'smaller' countries, whose weight in the Council of Ministers' decision-making procedures would exceed their size, in both population and economic terms. Existing smaller Member States, such as Luxembourg, Ireland and Denmark, already have more MEPs in the European Parliament per head of population than the larger countries, and each has a Commissioner in the EU Commission. The three new 'smaller' members of the club could slow down the decision-making procedures, with the increased risk of the formation of blocking minorities. The addition of two new official languages to the existing nine will further increase the already high costs of translation and interpretation incurred by the EU institutions.

The EFTA countries, through their location (Switzerland, Austria), natural resources (the Nordic countries) and high levels of economic development, have much to offer the EU. Such is not the case with the countries of Central Europe, all of which are economically well below the level even of the poorest EU countries, Portugal and Greece. The most likely countries to join the EU after the EFTA countries are absorbed are the former CMEA countries Poland, the Czech Republic and Hungary, while Malta and Cyprus may also be considered for entry. After that, Slovenia (the extreme north of Yugoslavia), Slovakia (the other 'half' of Czechoslovakia), Bulgaria and Romania might expect consideration.

5.3 The population of the EU and EFTA

Table 5.2 contains information about the population and area of the countries of the EU and EFTA. In total population size Germany has about 200 times as many inhabitants as Luxembourg, while in total area, France is more than 200 times as large. Problems of representation in the EU caused by such disparities in population size between Member States have been overcome. The drawback remains, however, that in the various bodies of the EU organisation the smaller sovereign countries such as Ireland and Denmark are well, if not over, represented, whereas 'nations' of similar size such as Scotland and Catalonia are submerged in the UK and Spain respectively. A comparable situation is found in the USA, where all states send two senators to Washington, but California, the largest in population, with about 30 million inhabitants, has sixty times the population of the smallest, Wyoming. Thus in the EU, as in the USA, at the top level in the political-administrative hierarchy, tradition is respected, and a particular arrangement of units has frozen, albeit not in an ideal form for the running of the whole Union.

A distinguishing feature of the EU among the major regions of the world is its stable population. The total population of Western and Central Europe is expected to increase only very slightly between the 1990s and about 2020 but then it should decline if present fertility rates continue. In Table 5.2, column (7) shows that only Ireland, with an average of 2.2 children per female, remains above the figure of about 2.1, regarded as necessary for replacement and 'zero growth'. Since Spain and Italy now have the lowest fertility rates of any large countries in the world it is perhaps time to stop crediting the Roman Catholic religion with responsibility for large families in all Catholic countries.

As birthrates fall and life expectancy increases, the average age of the population continues to grow, as does the proportion of citizens over various threshold ages regarded as the start of an elderly status, old age and senior citizenship. In some EU and EFTA countries the proportion of total population over the age of 64 is expected to reach a quarter in a few decades' time. A demographic consequence for East Germany of joining the EU in 1990 has been a sharp decline in birthrate, from 12.5 per thousand to under 5 per thousand in 1994, far below the replacement level.

Population change in the countries and regions of the EU is influenced not only by natural growth or decline but also by migration. Since 1945 there has been considerable migration within individual EU countries, for example from southern to northern Italy. Migration has also occurred between pairs of EU countries, such as from southern Italy to Germany, France and Belgium, from Portugal to France and from Ireland to the UK. Finally, migrants have entered from countries outside the EU. For example, French expatriates and Algerians have migrated to France, and citizens of British Commonwealth countries to the UK. West Germany in particular received a very large influx of Germans from its pre-war eastern territories when they were transferred in 1945 to Poland and the USSR and, much more recently, from East Germany. In general the people moving into the EU are aged between about 20 and 40 and are therefore likely to bring young children with them or to produce them once they are settled, thus increasing population size, or at least delaying a decrease. Further migrants into the EU are expected, especially from Central Europe and the former USSR, but since most are economic 'refugees', they are likely to be of all ages.

Whether or not a stable or even a declining population, accompanied by ageing, is an advantage to the EU, there is little that can be done to change the trend. Advances in medicine still continue to lower mortality rates in all EU and EFTA countries, and life expectancy thus continues to rise inexorably. Improvements in healthcare, and the availability of pensions for the elderly, have made it unnecessary for people to have large numbers of children to support them, while family planning and the improved status and job prospects for women, together with economic depressions and recessions, have resulted in many women having no children or only a single child.

One of the main factors to which the reduction in fertility is attributed is the increasing proportion of population living in cities. The definition of urban varies considerably from one country to another, usually reflecting the size, status and/or function of settlements. With about 80 per cent of its population classed as urban, the EU is one of the most highly urbanised regions of the world. It has several of the earliest cities in the world to pass 1 million in population, although now there are many cities elsewhere in the world that have overtaken Paris (9 million) and London (8 million).

Figure 5.1 shows the distribution in Western and Central Europe of cities with over 1 million inhabitants as well as all capital cities and other selected centres. Problems of inner-city decline, increasing crime, the emergence of minority ghettos, traffic congestion, and pollution have been increasing over the decades, although improvements have also been made in recent years in some conditions. Some of the larger urban agglomerations have also become obstacles to the movement, particularly of road traffic, between various pairs of regions and centres in the EU. Paris, London and the Rhine-Ruhr cluster of cities in Germany stand out in this respect. Efforts have been made to 'bypass' larger cities with ring roads (e.g. M25 round London, a ring road round Berlin). In contrast, large areas of the EU remain comparatively empty, including much of the interior of Spain, the Massif Central in France, and northern Scotland, a reason for concern in the past but perhaps one for satisfaction now. The Nordic countries are even more 'empty' in this respect.

The sovereign states of Europe, whether those in place at the start of the twentieth century, those that emerged between the two world wars, or those since 1945, have tended to be characterised by strong central governments, whether unitary or even federal, like Switzerland. In the interests of cohesion, economic growth or sheer survival, minority groups, usually so defined on account of their language, have been forced to conform culturally in their states and have been granted little autonomy. Such minorities include the Basques in Spain and the German-speakers of Alto-Adige in Italy, as well as Scotland, distinct in its location and history from England, but no longer in language. The development of the EU has not only already reduced to some extent the powers of the twelve national parliaments, but has also opened the prospect for direct liaison between regions within the countries of the EU and the supranational entity of the Union, symbolically associated with Brussels (see Box 5.1 on the Committee of the Regions).

5.4 Natural resources and agriculture of the EU and EFTA

At first sight Western Europe appears to be fairly well endowed with natural resources. The EU and EFTA together have rather less than 7 per cent of the total population of the world, 4–5 per cent of the productive land and 5–6 per cent of the mineral reserves of the world. What is more, in contrast to all the developing regions of the world, population is not likely to grow much in the future. On the other hand, Western Europe consumes fuel and raw materials at about twice the world average rate, and it has been industrialised longer than most other regions, with the result that the exploration for minerals has been more intensive. There is therefore less chance that further major deposits of minerals will be found than in most other regions of the world, while the agricultural land, pastures and forests cannot be greatly extended into present non-productive land.

In the early 1990s, OECD Europe (Western Europe plus Turkey) consumed about 1,400 million tonnes of oil *equivalent* of primary energy a year. About 200 million was from nuclear and hydroelectric power stations, the rest from oil, natural gas and coal, half of which was imported. Oil alone accounted for over half of the fossil fuel total, and 500 million tonnes were imported in 1991, including almost 200 million from the Middle East and 100 million from North Africa. Western Europe's dependence on imported oil has been a matter of concern since the 1950s, especially from the first Middle East crisis in 1956, when the Suez Canal was closed for over a year.

Western Europe is even more heavily dependent on imports of many of the non-fuel minerals it uses than it is for fuel. These include almost all of the bauxite, copper and ferro-alloys needed as well as high-grade iron ore. With regard to food and beverages, in the last three decades, thanks to increases in yields, the output of cereals and livestock products has risen very sharply in the EU, thereby reducing imports. The region depends, however, on other parts of the world, particularly the tropics, for beverages and for raw materials for industry, such as cotton and vegetable oils.

The four largest countries of the EU in population, Germany, the UK, Italy and France, are in general poorly endowed with natural resources relative to their population size and high levels of industrialisation, but the first two have considerable reserves of fossil fuels, while France has good bioclimatic resources. The entry of Sweden and Finland into the EU brings useful forest, hydroelectric and mineral resources into the Union. Meanwhile, especially since the rises in world oil prices in 1973–4 and in 1979, substantial advances have been made in the more efficient use of energy and other primary products, thanks to fuel-saving devices and the recycling of materials.

With the increasing use of steamships in the nineteenth century it became possible for West European countries to import large quantities of basic foodstuffs, agricultural and pastoral raw materials, and timber from distant parts of the world for their growing populations and expanding industries. In the 1930s, however, attempts were made by Germany and Italy to move as near as possible to self-sufficiency in agricultural products, while the neglected agricultural sector in the UK was a primary focus of attention during the Second World War. One of the priorities in the early years of the EU was to ensure a reliable supply of food, even though some products could be obtained more cheaply from other regions of the world, particularly from North America and Australia.

Since 1945, the agricultural sector in Western Europe has been characterised by a number of clear trends, all interconnected in a complicated way:

- In the region as a whole, the area under field and tree crops has hardly changed since 1945, although there have been considerable local increases or decreases.
- Yields of both crops and livestock have increased

BOX 5.1

The Committee of the Regions (reference Parlement Européen 1993)

The twelve Member States of the EU are based on the political situation in Europe following the Second World War. Earlier political maps of Europe were very different. Several of the twelve contain 'lost' national groups, whose aspirations and voices are not adequately catered for in the present organisation of the EU. The twelve countries of the EU are subdivided into smaller areas at three levels, the existence of which is primarily for administrative purposes. To consider regional problems and means of expressing regional issues, in 1993 (Article 198a of the Treaty of European Union): 'A Committee consisting of representatives of regional and local bodies, hereinafter referred to as "the Committee of the Regions" is hereby established with advisory status.'

The creation of the above body is intended to implement the principle of subsidiarity, whereby regions at various levels exercise powers over issues and decisions more appropriately handled locally than centrally. There are to be 189 representatives from the twelve countries, the numbers determined by degressive proportionality, with twenty-four members from the four largest countries, and at the other extreme six from Luxembourg. Each country is using its own criteria to compile the membership. Belgium, for example, is allocating representatives scrupulously according to languages, with five of its twelve to Dutch-speaking Flanders, four to French-speaking Wallonia, two to Brussels and one to the small German-speaking minority in the east. The German allocation of twenty-four is distributed on the basis of the Länder, themselves entities with considerable federal powers and long cultural traditions dating from before the time when Germany was unified in 1870.

two- to threefold, depending on the region, with much of the growth achieved by the greatly increased application of artificial fertilisers. Although there are extensive areas of natural pasture in some parts of Western Europe, much of the livestock population is supported by fodder from the arable areas, supplemented by imported animal feed.

- Employment in agriculture has declined sharply throughout Western Europe, but is still much higher in some regions than in others (see Figure 5.6, p. 123). Increased mechanisation has been the main cause of the decline in the labour force, but it has also had other effects, including the move towards larger farm units and fields.

- A stable total population in Western Europe, and greatly increased production from the land, have reduced imports of basic foods, and have resulted in surpluses of some products. These have led in the 1980s to a policy of reducing the area cultivated by creating set-aside areas on which farmers are paid not to grow commercial crops.

The effect of the EU's Common Agricultural Policy has broadly been to enable crop production to continue even in areas where labour productivity is lower than in other parts of the world, and production costs higher. Parallel with the Common Agricultural Policy have been the continual reafforestation in many parts of the EU and a policy aimed to regulate fishing activities, allocating fishing quotas 'equitably' among EU members and preserving fish stocks, though not to the satisfaction of many in the industry itself. By 1994, however, it had become evident that the whole fishing industry of the North Sea could be threatened by overfishing and pollution.

5.5 Industry and the environment in the EU and EFTA

The broadest definition of the industrial sector includes all economic activities (e.g. agricultural industry, tourist industry, as well as manufacturing). Here and elsewhere in this book industry refers more specifically to extractive (mining), processing and manufacturing and not to agriculture or services. Although the value of industrial output in the EU is about ten times as great as the value of agricultural output, much less has been done in the Union to support the sector. Some sectors have been protected from foreign competition by the common tariff barrier of the EU, but cheaper production costs in many parts of the developing world have affected EU industries, especially those producing consumer goods. Meanwhile, both the USA and Japan, with superior research facilities in many sectors, very large home markets, and a positive attitude towards quality, export both capital and consumer manufactures to Western Europe.

At any given time, the industrial sector of the EU can be subdivided according to performance into three broad categories.

- The first category consists of branches in which production has been falling, and employment usually falling even faster, including coal, iron and steel, textiles and clothing, and some branches of engineering, particularly shipbuilding.

- The second category includes sectors in which there have not been great changes in output in the last two decades, such as chemicals, transport engineering, and food but, again, with employment tending to decline.

- In the third category are mainly newer industries, in which both production and employment have been growing, including, for example, information and communication technologies, office equipment and aerospace industries.

The UK, France, Belgium, Germany and Spain have experienced most of the decline in traditional sectors, in coal-mining for example, to such an extent that by the early 1980s none was mined in Belgium, and very little in France. In 1991 the EU inherited the extensive lignite quarries of the former East Germany, a particularly dirty fuel, supplying about 70 per cent of all the energy used there. Areas in which both heavy industry and textiles flourished in the past, such as Lancashire and Yorkshire in the UK, Nord and Pas-de-Calais Departments in northern France and the Rhine-Ruhr cluster of cities in Germany, have suffered greatly. More locally, shipbuilding centres have lost most of their capacity and jobs. The most recent change in the industrial sector has been a dramatic shift in the fortunes of the defence industry.

One of the purposes of the structural funds of the EU has been to identify declining industrial regions, referred to as Objective 2 regions (see Figure 5.3). These include many coalfields, on which, in the early decades of modern industrial development, heavy industrial growth was rapid. All the coalfield areas in the EU have experienced the closure of mines and the decline of associated heavy industry.

On the positive side, some regions of the EU have maintained their industrial capacity thanks to the presence of growth industries or through structural changes. In southeast England, the Rhône valley in France, and many areas in southern Germany, industrial problems are less severe, while in some regions of northern and central Italy, and in Portugal, textiles and clothing sectors have remained competitive. In many

PLATE 5.2 European Union: mining in decline

a Remains of buildings at a tin mine at Morvah in Cornwall, England, UK. Cornwall once produced a third of the world's tin. Now only one small mine is still working

parts of the EU, however, modern industrial development has been limited and often of local impact. Such has been the situation in many regions of Greece, Portugal, southern Italy, Spain and Ireland. These areas have attracted some industries because of their low labour costs, while southern Italy has been assisted as part of national policy for four decades.

Industry is the greatest single contributor to pollution of the environment in Western and Central Europe. Given the high density of population over much of the region, the heavy use of fossil fuels and chemicals, and the production of massive quantities of waste, it is not surprising that pollution of the environment has become a major issue and concern, often at the transnational level. Many regulations and much advice on acceptable practices have been proposed at both EU and national level, but on account of the cost and difficulty of monitoring and measuring pollution, and of enforcing sanctions and penalties on offenders, many environmental problems remain.

Pollutants such as nitrogen from fertilisers can filter below the ground and affect underground water, a process now affecting extensive areas in northern France and other areas of intensive arable farming in northern and Central Europe. To reduce or even ban

b Imposing entrance to a now closed coal-mine in southern Belgium, near La Louvière. In the mid-1950s Belgium produced about 30 million tonnes of coal a year. By the 1990s production had ceased altogether

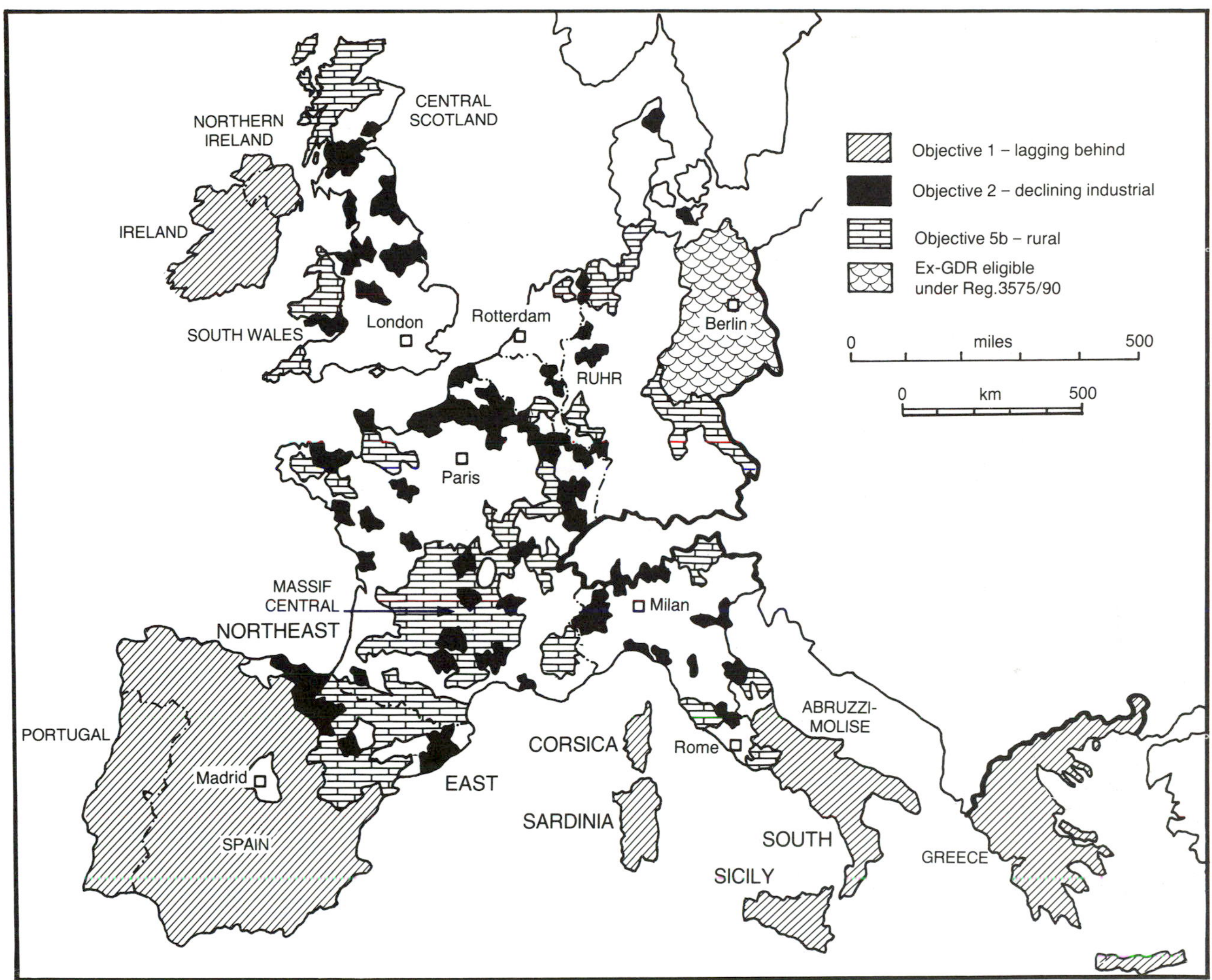

FIGURE 5.3 Regions eligible for structural assistance under the EC structural funds, 1989–93. A new Objective 6 type of region is being introduced to cater for the needs of very thinly populated, cold regions in the north of Sweden and Finland
Source: Commission of the European Communities (1993). The map in the source has been modified slightly. See also Cole and Cole (1993), p. 248

the use of fertilisers would reduce crop yields. Sulphur dioxide emissions, particularly from coal-fired power stations, are considered to be one of the main causes of acid rain, causing friction between heavy users of coal for generating electricity, including the UK, Germany, Poland and the Czech Republic on the one hand, and countries such as Norway and Sweden, where very little coal is burned, but where it is claimed acid rain produced elsewhere affects the forests. Motor-vehicle emissions, also responsible for extensive pollution of the atmosphere, are less 'conspicuous' than power station emissions and less obviously concentrated in some areas of Western and Central Europe.

In the UK the move since the late 1980s to switch to cleaner, gas-fired power stations may be more 'friendly' environmentally but it has resulted in the closure of most of the remaining UK coal-mines, with heavy job losses. Between 1984 and 1994 more than 150 out of 170 mines were closed with a reduction in jobs from 190,000 to 10,000. Germany is also likely to follow this course, although more slowly, since the large subsidy provided by the German government to keep German mines open is to be cut.

Where pollution is an international and even a global problem, co-operation between various countries is essential, and the EU and EFTA countries are gradually coming to terms with the need to combine their efforts. Very few regions of Western and Central Europe are now entirely unaffected by pollution, whether on the land or in the adjoining seas, but some localities are much worse than others. The older industrial regions of the EU, with their traditional coal-mining and metallurgical industries, have suffered badly from pollution. The larger cities also have problems of pollution, in the nineteenth century through the widespread use of horse-drawn traffic on

PLATE 5.3
Industry in the European Union

a The main plant of the motor vehicle company Volkswagen at Wolfsburg in northern Germany

b A Japanese transplant: the motor vehicle company Toyota opened its factory to reach the EU market in 1993 on a greenfield site near Derby, East Midlands, England, UK

the streets and the burning of coal for industrial and domestic purposes, and more recently with the rapid increase in the numbers of motor vehicles.

As time passes, cleaner fuels, such as natural gas in place of coal and lignite, nuclear energy, and some wind and solar power, could bring gradual improvements. The situation is, however, complex, and improvements in some areas may cause problems elsewhere. For example, the extensive use of electric battery-powered cars would undoubtedly result in cleaner streets in cities, but the electricity still has to be generated somewhere. The nuclear power industry offers clean electricity, but even in its short history of half a century it has given rise to new problems. For example in the southern part of the former East Germany there are areas affected by uranium-mining which require massive outlays to clean up and safe places for the processing and disposal of waste. The high cost of decommissioning the first generation of West European nuclear power stations is a problem that has to be faced in the next century. Even the generation of clean electricity from wind power has raised concern since clusters of wind-generating towers are too conspicuous in the rural landscape and spoil the views.

5.6 Services and transport links in the EU and EFTA

While most of the agricultural and industrial production in the EU in any given locality is sold over a wide area, even throughout the entire EU market and in countries beyond, much of the service sector is provided to satisfy particular local or regional needs. Although there are considerable regional variations in the EU, every community should have access to such services as schools, healthcare facilities, basic retail outlets and a postal service. Many of the jobs in the local services are poorly paid, however, compared with manufacturing jobs. At regional level, work in higher education and specialised hospitals is to be found. At national and international level, above all in the national capitals, services are provided for the entire nation and sometimes for the whole of the EU and beyond.

Only in a few regions in Greece and Portugal, out of a total of 170 regions in the EU as a whole at the second level of regional subdivisions, does the percentage of employment accounted for by services fall below 40 per cent. In contrast, in a few regions it exceeds 70 per cent, as in Brussels in 1987 with 80 per cent, Lazio, Italy, containing Rome, 76 per cent, and Ile-de-France, with Paris, 73 per cent. Brussels, Luxembourg and Strasbourg contain most of the EU institutions and Brussels, in particular, is increasingly attracting representatives from private and public firms and institutions world-wide. Financially, London, Paris, Frankfurt and Milan are of international importance. The transport and tourist sectors also reach above national level in some centres and areas. The ports of Rotterdam and Antwerp serve not only the Netherlands and Belgium but also Germany, Switzerland and eastern France. Hamburg seems destined to recover an extensive lost hinterland in eastern Germany and beyond. Various coastal zones in France, Italy and Spain attract an international tourist clientele.

In the late 1980s services employed almost 60 per cent of the economically active population of the EU and accounted for roughly the same proportion of GDP. In contrast, the proportions in agriculture were 7–8 and 4–5 per cent respectively and in industry 33 per cent of employment and rather more of production. While there have been some services in all but the simplest of economies throughout history, the great increase in productivity per worker in agriculture and industry in Western Europe in the last two centuries has enabled a decreasing proportion of the population to produce the food and other material needs, if not all the wants, of the community. That has left room for the expansion of the service sector, providing not only essential services such as healthcare, education, transport, distribution and communications, but also leisure and luxury ones.

There are now many regions in the EU in which the agricultural sector is virtually non-existent and manufacturing is also hardly represented. Such regions pay their way by 'exporting' services beyond their limits in exchange for goods. Such a situation can hardly be healthy for a whole country, and in the 1990s concern has been expressed over the decline of the industrial sector in various EU countries. It is worth pointing out that in 1987 West Germany had over 40 per cent of its employed population in industry, while France with 30 per cent, Italy with 32 per cent and the UK with 33 per cent (also the EU average) had appreciably less. This situation has been considered partly responsible for the dominant economic position of West Germany in the EU.

Transport is a key part of the service sector because it serves all branches of the economy. The EU does not have a comprehensive policy for transport as it does for agriculture, but broad guidelines are set out at EU level. Encouragement is given to the revival of rail transport, to take pressure off the roads. At regional level, assistance is given to support the improvement of existing roads, especially in more thinly populated peripheral areas, and to construct bridges to link islands to the mainland.

In the study of transportation a distinction should be made between the various modes of transport, some of which, such as pipelines and electricity transmission lines, are specialised and others, such as road and rail, which are general. A distinction must also be made as to what is carried or transmitted: people, goods, or information.

1. Sea and waterways Although the layout of the land of Western and Central Europe is such that nowhere is very far from the coast, in the new political situation, with the break-up of Czechoslovakia and Yugoslavia, there could eventually be eight land-locked states. Western and Central Europe have had sea links with the rest of the world since the sixteenth century, whereas Russia has depended heavily on land transport to integrate its vast empire, which extends right across Asia to the Pacific coast.

Since the Second World War great changes have taken place in the direction of traffic by sea and in the types of ship and cargo carried, while there has been a massive transfer of passengers from sea and rail to air. The famous Peninsular and Orient (P & O) Steamship Company, serving India and the Far East in its heyday, now concentrates on ferry services between the UK and the mainland of Europe, while Cunard now runs expensive holiday cruises instead of carrying millions of passengers across the Atlantic as it did earlier this century. Many of the merchant ships now in use are

BOX 5.2

New fixed links for the EU and EFTA

Since the 1950s the motorway network of Western Europe has grown impressively in extent, reducing travel times by road by about half between many major centres. Two types of area still cause delays on journeys in the region: large urban agglomerations and physical obstacles. These areas also affect rail transport. The impact of urban obstacles is gradually being overcome either by the construction of ring roads avoiding built-up areas (London's M25, Paris, Milan) or by improvement of roads through the built-up areas (the Rhine-Ruhr conurbation, several cities in the Netherlands). Figure 5.4 shows details of four key existing or proposed links on the EU/EFTA transport system. They are discussed roughly in order of progress.

1. The Channel Tunnel was opened in 1994, a year behind schedule. It cost about 15 billion US dollars, twice the original estimate for its construction. Such delays and increased costs are not unusual in construction works of this kind. The 'short sea crossing' has been favoured throughout history as the shortest ferry link between England and France. The various links between Dover

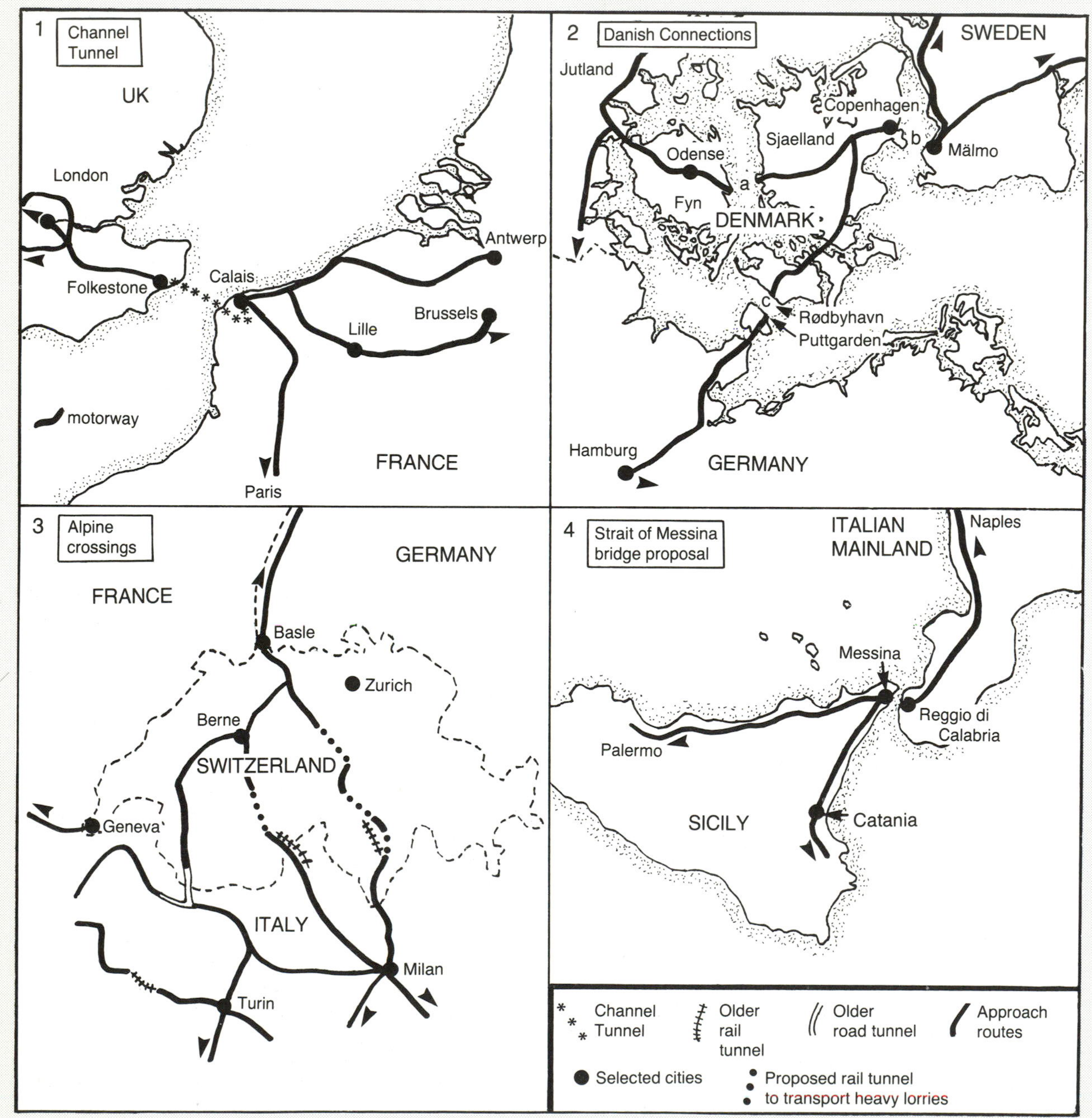

FIGURE 5.4 Existing and proposed fixed links at three places in the EU and across Switzerland

and Folkestone in England, and Calais and Boulogne in France, carry a very heavy flow of goods and passengers, taking foot passengers who have reached the ferry ports by public transport as well as cars, buses and lorries with their drivers and passengers. The crossing by conventional ferry usually takes between 75 and 90 minutes although there are faster hovercraft and hydrofoil services.

The Channel Tunnel is a rail link capable of taking fast trains at short intervals. Four kinds of traffic are carried: through passenger and through goods trains operating direct services between places in Britain and places on the mainland of Europe, and Shuttle trains carrying passenger cars and coaches, and similar trains carrying lorries. The latter trains currently run only between the tunnel terminals at Cheriton near Folkestone and Sangatte near Calais.

The journey through the tunnel saves varying proportions of journey time according to the total distance of the journey and according to the type of journey. When the ferry crossing and tunnel are compared, the saving by tunnel for cars and lorries is between about 1 and 1½ hours. Such a saving is a considerable proportion of a journey between, for example, London and Lille, cutting the time by car from about 5 hours to 3½ hours or 30 per cent. On the other hand, a journey between Glasgow in Scotland and Cologne in Germany, which might take 15 hours, would be reduced only by 10 per cent. The tunnel offers greater advantages to the movement of both goods and passengers travelling by train because it avoids transshipment at the ferry ports. Thus, for example, the rail journey between Paris and London is reduced from 7–8 hours to under 3 hours, and will be even quicker once fast tracks are in place in Kent to complement those already in use in France.

Cross-Channel passenger numbers have been growing and are expected to continue increasing in the 1990s. In 1993, ferries on all routes between Britain and the mainland carried about 30 million passengers (Butler 1994). That number is expected to drop to little more than 20 million, while the Channel Tunnel should reach 20 million by the year 2000. Air services take 40 million passengers, and it is unlikely that the tunnel will compete here on any but the shortest journeys such as London–Paris and London–Brussels. Eurotunnel shareholders may have to wait well into the first decade of the next century before they get dividends.

2. There are ambitious plans to build three fixed links between the Danish island of Sjaelland, on which the capital, Copenhagen, is situated, and (a) the rest of Denmark, (b) Sweden and (c) Germany.

Link (a) is to join Sjaelland to the island of Fyn (already linked by bridge to Jutland) and thus to the rest of Denmark. This crossing is under construction and will carry road and rail by a combination of bridges and tunnels. International navigation between the Baltic and North Seas must not be interrupted.

(b) There are many different ferry links between the Copenhagen area and the southern extremity of Sweden and one is served by a train-carrying ferry. It is proposed to build a bridge and tunnel between Copenhagen and Mälmo carrying both road and rail. Technically the problem is not great, but the cost is high, the Oresund Channel is sensitive as the main Baltic–North Sea passage, and the project has been condemned by some as environmentally suspect.

(c) Motor vehicle and train-carrying ferries link Puttgarden in Germany and Rødbyhavn in Denmark, but the waiting time and loading and unloading slow down journeys. The type of fixed link most appropriate here is still under study, but with two new Nordic countries now in the EU, the reduction of journey times between southern Sweden and Germany is a major concern.

3. Switzerland and Austria are both located astride busy road and rail links between Italy to the south and Germany, France and countries beyond to the north. Great improvements have been made to road crossings of the Alps since the Second World War and there are also several important rail links. By entering the EU, Austria has implicitly accepted, even if as a necessary evil, that the very heavy traffic between Italy and Germany will pass through it, using the Brenner Pass route. On the other hand, the Swiss have put restrictions on the passage of heavy goods vehicles through their territory. Their ultimate solution is to construct, at a current estimated cost of 20 billion US dollars, a special rail link, with a 49-km (30-mile) tunnel, to carry heavy goods vehicles between the Italian and German borders, with links to France and Austria. For several decades now, special wagons carrying passenger cars have run through some of the Swiss Alpine tunnels, but the vertical and horizontal dimensions of the tunnels do not allow the transport of lorries. The new route would involve extending the existing St Gotthard Tunnel.

4. The island of Sicily is separated from the region of Calabria on the Italian mainland by the Strait of Messina, which is less than 4 km (2.5 miles) across. The sea here is too deep for a tunnel to be built, so the state-owned Stretto di Messina Company has proposed the construction of a suspension bridge. The cost would be about 6.5 billion US dollars, considerably more than the original estimate for the Channel Tunnel. Japan, with much experience in bridge construction (see Chapter 8), is interested.

The technical problems are great. The longest bridge span in the world at present is the main span of the Humber Estuary Bridge in England, 1,410 metres (4,626 feet) in length. The Messina Strait Bridge would need a main span of 3,300 metres (10,830 feet) and towers 370 metres (1,214 feet) high, and would use 600,000 tonnes of steel. Earthquakes and very strong winds are environmental hazards of the area. The economic 'risks' are also great. The existing ferries, carrying both motor vehicles and through trains, cross the strait in about 30 minutes. With little more than 5 million inhabitants, Sicily does not have a large exchange of goods of great weight or urgency with the mainland. The existence of the bridge would give satisfaction to Sicilians, who are on the very periphery of the EU, and to the European Commissioners, one of whose priorities is to strengthen the integration of the Union.

either specialised carriers of bulk commodities such as oil and ores, or are container carriers. The traffic in some major West European ports, including Liverpool, London, Marseilles and Genoa, has declined, but others, especially Europoort near Rotterdam in the Netherlands and Antwerp in Belgium, have continued to expand. Inland waterways also serve considerable hinterlands in parts of the EU and Central Europe, with the Rhine-Main in Germany now linked by canal to the Danube, thus providing a through route between the North Sea and the Black Sea.

2. **Rail transport** became the main mover of goods and passengers in Western and Central Europe from about the middle of the nineteenth century and a dense network developed in the more heavily populated areas. Since the Second World War there has been little change in the amount of goods traffic carried by rail in the EU and EFTA countries, but in Central Europe, in the centrally planned communist countries, rail traffic increased greatly. A comparison of journey time for both goods and passengers shows that the 'door-to-door' capability of both road freight services and private passenger cars often makes road more attractive than rail for short journeys, while above a certain time-distance, air travel is more attractive for passengers than road or rail. With improvements in the speed of trains, both goods and passenger, it is hoped to restore to the rail system some of its 'lost' traffic, thus helping to relieve congestion on both road and air systems.

3. **Roads** The first motorways in Europe were built in the 1930s in Italy (*autostrade*) and Germany (*Autobahnen*), but motor transport almost everywhere in Europe was restricted by narrow roads, which passed through numerous settlements, and by vehicles ill-equipped to make long journeys. In what is now the EU there were about 4 million passenger cars in use around 1930 compared with 20 million in the USA for a smaller population. The number of cars in the EU is now about thirty times as large as it was in 1930, while commercial vehicles have increased correspondingly in number as well as gaining in size and speed. Most major centres of population and economic activity in the EU are now on or close to the motorway system or are served by other adequate roads. Improvements in routes across the Alps, the opening of the Channel Tunnel in 1994 and, later in the 1990s, the planned completion of fixed links between Sweden, Denmark and Germany will help to integrate an impressive network of fast roads (see Box 5.2). The motorway system is not, however, supported by good secondary links. To the east and southeast of Germany and Austria, road transport in Central Europe has been neglected, but the region is likely to trade much more with the EU in the future and improvements will have to be made there (see Figure 6.3).

4. **Air transport** plays a major role in the movement of passengers on journeys of more than about 300 km(180–90 miles) in Western and Central Europe, but fares for similar distances are generally much higher than in the USA, and many cities of considerable size have no local airport or have infrequent services on links into the airline system. Regular and holiday flights together congest some routes, and air traffic control, still partly organised at a national level, has to handle very heavy traffic in some air lanes.

The distribution of centres of population in the EU is such that the busiest interregional road and rail links are highly concentrated along certain corridors and bands, in an area extending from England through northeast France, Belgium, the Netherlands and West Germany to North Italy. These interregional links are shown in Figure 5.5. On account of their distance from the central regions of the EU and the presence of intervening obstacles, the Balkans for Greece, the Irish Sea for Ireland, access to the core of the EU is difficult for these countries.

5.7 Regional economic disparities in the EU

Marked cultural differences within several countries of the EU have been referred to earlier in this chapter. Regional disparities in income per inhabitant are also found in all but the smallest countries of Western and Central Europe. One of the aims of the EU policy makers is to reduce economic differences within the Union as a whole, a process referred to sometimes as convergence. The resources available to produce major changes in regional disparities at the EU level are very limited, however.

In order to distribute assistance equably, avoiding conflicts between Member States over resources, systems of regions at three levels have been used for the collection of data on economic and social aspects. Several criteria have been used to determine where regional development funds should go. Among regions eligible for assistance are those with Gross Regional Product levels well below the EU average at a given time, and with unemployment levels well above it. Location within the EU may also be taken into account, in particular the isolation of many smaller islands in, for example, Greece and Scotland, and of mainland areas at the extremities of the Union such as northern Denmark, Portugal and the extreme south of Italy. Regions adjoining international boundaries, both within the EU and, more importantly now, at its margin with non-EU countries, have also been regar-

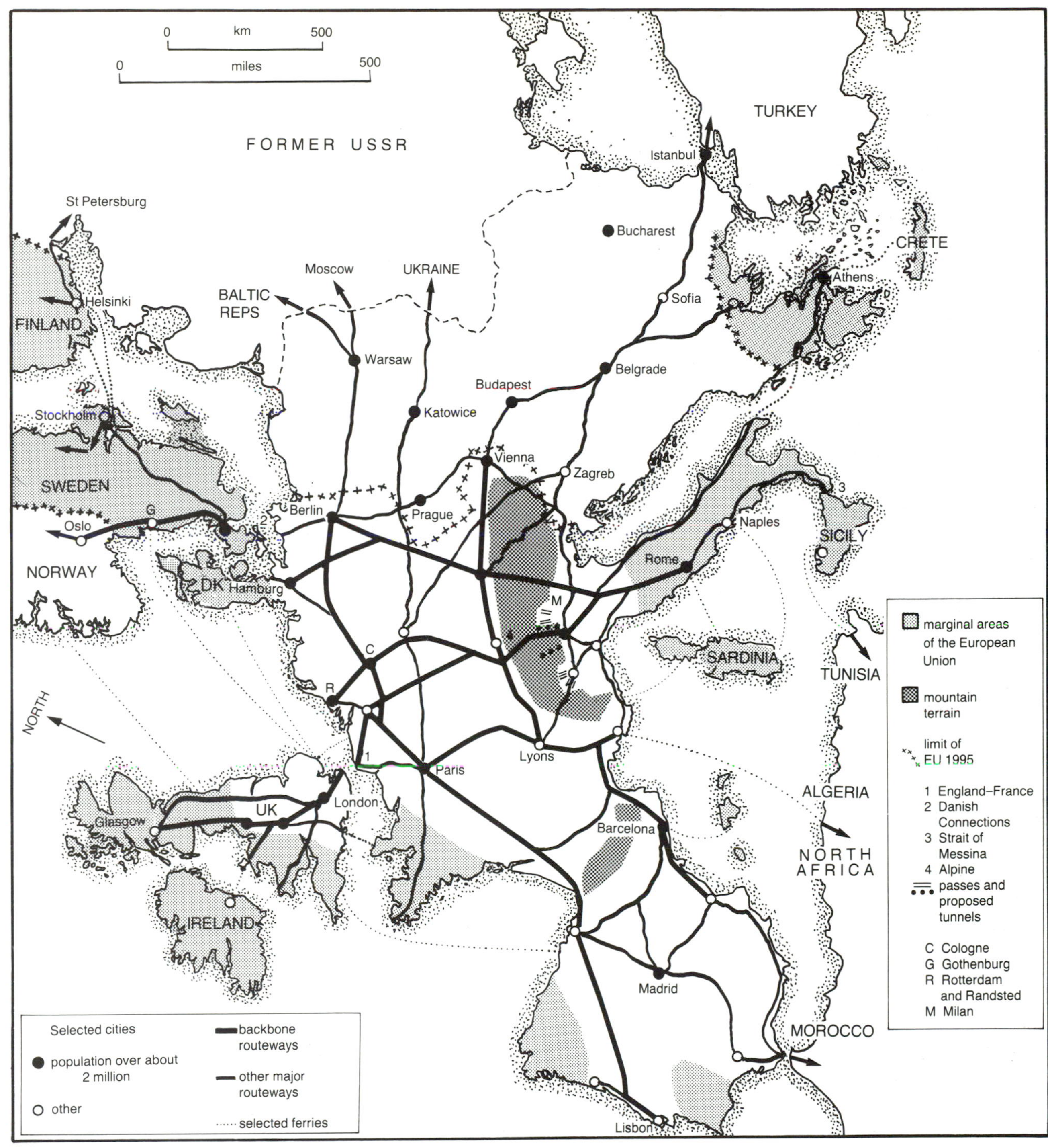

FIGURE 5.5 Major routeways in Western and Central Europe, a view from an unfamiliar angle. Spain, Italy, Greece, the UK and Scandinavia stand out clearly as projections from the spatial core of the region. Now that two more Nordic countries have joined, Central Europe is adjoined on three sides by EU countries

ded as disadvantaged and therefore eligible for assistance. Table 5.5 (pp. 124–5) contains some of the data sets used in the EU to study regional differences.

In a country or a supranational entity run largely under market economy conditions it is to be expected that new economic activities will be located in places where the lowest production costs and greatest profits are anticipated. For enterprises producing goods or services for all or a large part of the EU, a central location would generally be preferable to a peripheral

PLATE 5.4 European Union: contrasts in German Autobahn quality

a Former East Germany, near Chemnitz, built before the Second World War. Virtually no major repairs were carried out on the Autobahn system of East Germany between 1945 and 1990 and no new links were built

b Viaduct on a newly built Autobahn in former West Germany, near Kassel

one. A location in or near a large cluster of population, near other economic activities, and with good transport links to other regions would be attractive. Without measures to 'correct' this tendency, areas of the EU that are already more attractive economically and prosperous would therefore stand to gain. Their progress would tend to be at the expense of many areas in which run-down industries have become a liability, a heavy dependence on agriculture a burden, or the lack of a well-trained and skilled labour force a major drawback.

The designation of areas eligible for EU regional/structural funds is related to the above considerations (see Figure 5.4). Thus Objective 1 regions extend over much of the southern regions of the EU, characterised by a large labour force in farming, low productivity per worker in that sector, and limited industrial development. Objective 2 regions, referred to already, consist mainly of old coal-mining, textile, iron and steel and engineering centres, many active since the early decades of the Industrial Revolution, but declining in some cases since the inter-war depression. Objectives 3, 4 and 5a do not have regional implications, but Objective 5b regions extend over thinly populated, often remote, areas with difficult physical conditions and poor employment prospects. In such areas of rural decline the maintenance of adequate levels of services provision is difficult.

Since the EU budget is roughly equal to only 1 per cent of the total GDP of the EU and only about one-quarter of it is explicitly targeted to backward regions, the assistance is very limited and, even if it is used to

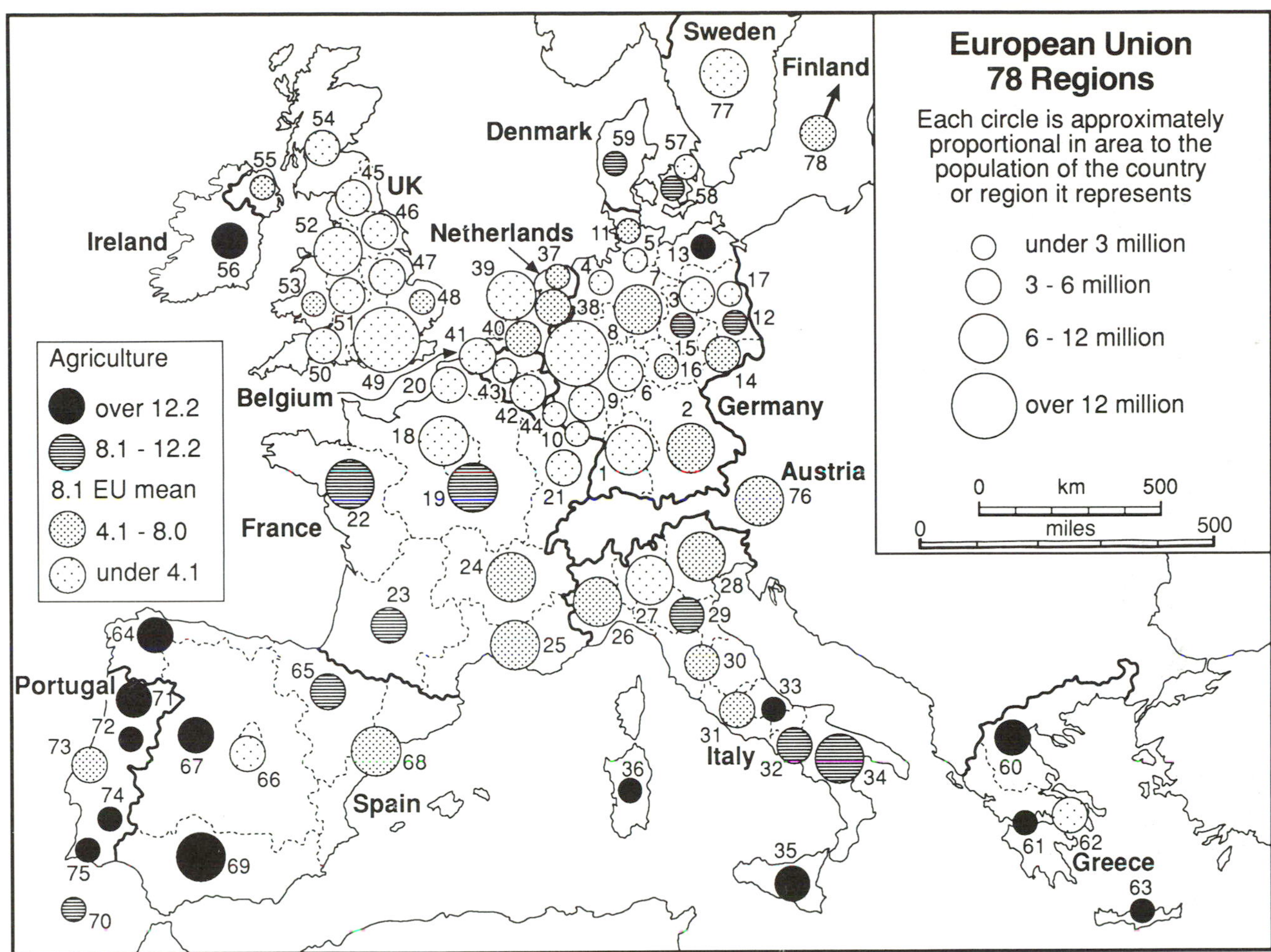

FIGURE 5.6 Economically active population in agriculture as a percentage of total economically active population in 1990 in the regions of the EU
Source: European Union (1993)

maximum advantage, is unlikely to reduce existing disparities quickly. Indeed, in the early 1990s, the gap was widening. Ironically, the distribution of Common Agricultural Policy funds to support prices of agricultural products tends to benefit farmers with the lowest production costs, many in comparatively prosperous regions in Denmark, the Netherlands and the UK.

An underlying reason why the reduction of regional disparities has been difficult is the fallacy of the concept of comparative advantage, at least in the context of such a large piece of territory as the EU, managed from the centre. Theoretically, every region should concentrate as far as possible on what it is best suited to produce. Thus the French Riviera has long exploited its outstanding climatic and landscape attractions to develop tourism and more recently to attract new high technology and research enterprises. London should continue to develop its business and financial expertise. Denmark's agriculture is highly organised. But what can the mountains of Greece and southern Italy or the semi-arid areas in Spain and Portugal offer that cannot be done better somewhere else? In some respects, investing large amounts in the least promising areas of the EU is wasteful, but persuading or forcing some of the population to leave such areas is unacceptable on both humane and political grounds.

Economic indicators for the regions of the EU show that in some Member States the disparity in income per inhabitant has increased in the early 1990s. Until the 1980s, southern Italy and (after 1973) Ireland were the only two areas in the EC that lagged far below the average level. The addition of Greece, Spain and Portugal in the 1980s and of East Germany in 1990 has greatly increased the number of relatively poor areas to be assisted. In the 1950s and 1960s, economic growth in the EEC was rapid and it was more easy to transfer funds to poorer areas. With economic uncertainty in the 1970s and 1980s and a major recession in the early

TABLE 5.5 Data for the EU level 1 regions

	(1) Popn in thous.	(2) Popn change	(3) Migration	(4) Participation in work	(5) Unemployment 1990–2	(6) Agriculture	(7) Industry	(8) Services	(9) GDP in thous. pps per head
1 Baden-Württemberg	9,726	0.5	3.1	51	2.7	3.5	46.5	50.0	16.7
2 Bayern	11,335	0.4	1.7	53	3.0	6.5	41.4	52.1	16.1
3 Berlin (W)	3,420	6.1	29.1	53	6.8	1.0	30.2	68.8	15.5
4 Bremen	679	−0.2	0.0	47	8.6	0.7	30.7	68.6	19.9
5 Hamburg	1,640	−0.1	4.4	51	6.3	1.4	25.6	73.0	24.7
6 Hessen	5,717	0.2	1.9	50	3.7	2.7	38.2	59.1	18.2
7 Niedersachsen	7,340	0.1	0.0	47	5.8	4.9	36.8	58.3	13.5
8 Nordrhein-Westfalen	17,244	0.1	1.5	46	5.9	2.3	42.0	55.7	15.0
9 Rheinland-Pfalz	3,734	0.3	2.6	48	3.8	3.9	40.5	55.6	13.9
10 Saarland	1,070	0.0	2.5	45	6.4	1.6	40.2	58.2	14.5
11 Schleswig-Holstein	2,614	0.0	−1.6	51	5.1	4.7	28.5	66.8	13.2
12 Brandenburg	2,664	0.0	−22.8	54	12.7	11.4	42.4	46.2	5.9
13 Mecklenburg-Vorpommern	1,976	0.9	−20.9	53	14.7	15.0	30.4	54.6	5.4
14 Sachsen	4,979	−0.5	−24.1	53	11.5	5.7	51.1	43.2	5.4
15 Sachsen-Anhalt	2,995	−0.3	−28.4	54	12.9	9.1	46.7	44.2	5.8
16 Thuringen	2,713	−0.1	−25.3	53	12.9	7.5	51.0	41.5	5.0
17 Berlin-Est	1,279	1.1	−23.0	55	13.3	0.8	32.7	66.5	7.5
18 Ile-de-France	10,692	0.7	−0.2	49	7.6	0.5	25.1	74.5	22.5
19 Bassin parisien	10,284	0.4	−0.6	43	9.3	8.2	33.2	58.6	13.7
20 Nord-Pas-de-Calais	3,967	0.1	−6.0	38	12.0	3.9	32.8	63.3	12.3
21 Est	5,031	0.2	−3.7	44	7.0	4.0	36.2	59.8	14.0
22 Ouest	7,461	0.5	0.8	44	9.4	11.2	29.2	59.6	12.6
23 Sud-Ouest	5,963	0.5	4.7	43	9.4	11.1	25.8	63.1	12.9
24 Centre-Est	6,689	0.6	1.9	45	8.5	5.2	33.9	60.9	14.3
25 Mediterranée	6,649	1.0	7.7	41	12.0	5.5	21.8	72.7	13.0
26 Nord-Ouest	6,196	−0.4	0.8	44	7.0	7.8	32.9	59.3	16.1
27 Lombardia	8,926	0.0	2.1	45	3.7	3.2	43.6	53.1	18.2
28 Nord-Est	6,483	0.1	1.7	44	4.3	7.3	38.3	54.4	15.9
29 Emilia-Romagna	3,925	−0.1	2.9	47	4.3	9.8	37.0	53.2	17.2
30 Centre	5,816	0.0	2.5	44	7.4	7.1	35.3	57.7	14.4
31 Lazio	5,181	0.4	2.6	43	10.4	6.3	18.8	74.9	15.5
32 Campania	5,831	0.7	1.6	38	20.4	11.4	23.8	64.8	9.4
33 Abruzzi-Molise	1,605	0.4	2.8	42	11.1	16.6	25.9	57.5	11.8
34 Sud	6,852	0.5	−1.2	38	17.1	11.4	23.8	64.8	9.3
35 Sicilia	5,185	0.6	0.6	37	21.6	15.1	21.1	63.8	9.0
36 Sardegna	1,661	0.5	1.3	40	18.5	12.8	21.8	65.3	10.0
37 Noord-Nederland	1,596	0.2	−2.1	44	9.2	6.3	26.9	66.6	13.6
38 Oost-Nederland	3,050	0.9	3.4	46	7.1	5.7	28.4	65.6	11.4
39 West-Nederland	6,997	0.5	2.0	48	6.7	3.3	20.3	75.8	14.7
40 Zuid-Nederland	3,306	0.5	0.7	47	6.6	5.1	32.8	61.8	12.6
41 Vlaams Gewest	5,754	0.2	0.8	41	5.3	2.9	32.5	64.6	14.2
42 Région Wallonne	3,251	0.1	1.7	39	10.4	3.3	25.4	71.3	11.3
43 Bruxelles	962	−0.4	−2.0	40	9.5	0.0	13.4	86.6	22.0
44 Luxembourg (GD)	381	0.4	4.0	43	1.6	3.3	30.6	66.2	16.7
45 North	3,075	−0.2	−2.0	48	11.0	2.0	31.7	66.3	11.9
46 Yorkshire and Humberside	4,952	0.1	−0.5	50	9.3	2.0	31.0	67.0	12.4
47 East Midlands	4,019	0.5	3.5	52	7.6	2.6	34.5	62.9	13.1
48 East Anglia	2,059	0.9	8.5	51	6.3	4.5	26.6	68.9	13.8
49 South East	17,458	0.3	0.3	53	7.5	1.0	22.2	76.7	16.6
50 South West	4,667	0.7	8.0	51	7.4	3.7	25.3	71.0	13.0
51 West Midlands	5,219	0.1	−1.3	51	9.0	2.0	35.1	62.9	12.6
52 North West	6,389	−0.1	−2.2	50	10.2	1.0	30.4	68.6	12.5

	(1) Popn in thous.	(2) Popn change	(3) Migration	(4) Participation in work	(5) Unemployment 1990–2	(6) Agriculture	(7) Industry	(8) Services	(9) GDP in thous. pps per head
3 Wales	2,881	0.2	3.0	47	9.1	4.4	29.6	66.0	11.5
4 Scotland	5,102	−0.1	−2.7	50	10.3	2.8	26.9	70.3	12.7
5 Northern Ireland	1,589	0.2	−7.4	44	16.7	6.6	24.7	68.7	10.2
6 Ireland	3,503	0.3	−8.6	38	16.4	14.8	28.5	56.7	8.8
7 Hovedstadsregionen	1,713	−0.2	0.1	57	7.9	0.9	21.6	77.5	17.2
8 Ost for Storebaelt	587	0.0	3.4	57	9.8	8.7	27.6	63.6	12.1
9 Vest for Storebaelt	2,842	0.2	1.5	57	9.2	8.3	31.4	60.2	13.5
0 Voreia Ellada	3,314	0.6	0.8	42	6.8	31.5	27.2	41.3	6.3
1 Kentriki Ellada	2,402	0.6	0.8	43	6.4	43.4	20.6	36.0	6.1
2 Attiki	3,506	0.6	0.8	38	9.4	1.4	30.3	68.2	6.9
3 Nisia	986	0.6	0.8	42	4.0	34.0	17.3	48.7	6.1
4 Noroeste	4,457	0.1	−0.6	40	14.7	27.7	27.1	45.3	8.4
5 Noreste	4,123	0.0	−0.4	40	14.9	8.4	38.7	52.9	11.8
6 Madrid	4,878	0.3	0.6	38	12.3	1.0	30.4	68.7	12.6
7 Centre	5,467	0.6	−0.7	37	17.2	20.8	30.7	48.5	8.3
8 Este	10.477	0.2	0.2	42	13.7	5.8	41.4	52.9	11.4
9 Sur	8,072	1.0	0.9	36	24.8	15.6	28.1	56.3	7.8
0 Canarias	1,485	0.6	1.3	40	24.2	8.3	21.9	69.9	10.0
1 Norte	3,456	0.0	−10.4	50	3.0	19.7	43.8	36.3	6.5
2 Centre	1,727	−0.3	−7.9	49	2.4	31.9	32.1	36.0	5.4
3 Lisboa	3,309	0.0	−6.2	50	5.4	7.8	27.9	63.8	10.0
4 Alentejo	546	−0.8	−9.5	44	9.2	24.6	24.5	50.6	4.6
5 Algarve	339	0.4	1.7	44	3.6	14.3	19.4	66.3	6.5
6 Austria	8,000	0.1	0.0	50	5.8	5.5	30.5	64.0	15.0
7 Sweden	8,800	0.2	0.0	55	3.3	3.7	31.3	65.0	18.5
8 Finland	5,100	0.3	0.0	50	7.5	7.8	32.2	60.0	18.0
EU only	344,149	0.8	1.0	45	8.7	8.1	32.3	59.7	13.5

)etails of variables in Table 5.5, data for the EU level 1 regions and for Austria, Sweden and Finland. Some of the data for the three former EFTA ountries are estimates by the author:
1) Population in thousands at the beginning of 1993
2) Rate of annual average change of population per thousand total population
3) Average net migration rate in per thousand for last available 5 years, the difference between the total change and the natural change in population
4) Participation rate of the total labour force as a percentage of total population
5) Unemployment as a percentage of total labour force, harmonised to be comparable between Member States, average for 1990–1–2
6), (7), (8) Percentage of total employment in agriculture (6), industry (7) and services (8)
9) Average of GDP per inhabitant indices for 1988–89–90 in purchasing power standards. The average for the European Union = 100
ource: Data derived from *European Union* (1993)

1990s the equalisation process is evidently faltering. An adequate goal would be to maintain conditions above a given 'poverty' line. As already noted in this chapter, the future of the regional problem is, however, greatly complicated by the prospect that new members may join the EU. Any country in Europe is theoretically eligible for consideration for membership provided it fulfils certain political and economic conditions. In 1993, however, the Commissioner responsible for External Affairs, Hans Van Den Broek, expressed the view informally that none of the former Soviet Socialist Republics of the USSR, with the exception, presumably, of the Baltic republics, could ever join the EU.

The outstanding difference in Western Europe between the half century before the Second World War and the half century since 1945 has been the complete absence of any military conflict between the countries of the region apart from the use of force to resolve fishing disputes. One can only speculate whether such a situation would have occurred without the European Union or without a perceived threat from the Soviet Union. With time, it seems that the EU will become increasingly integrated and also larger, thus continuing and consolidating the conflict-free situation in the

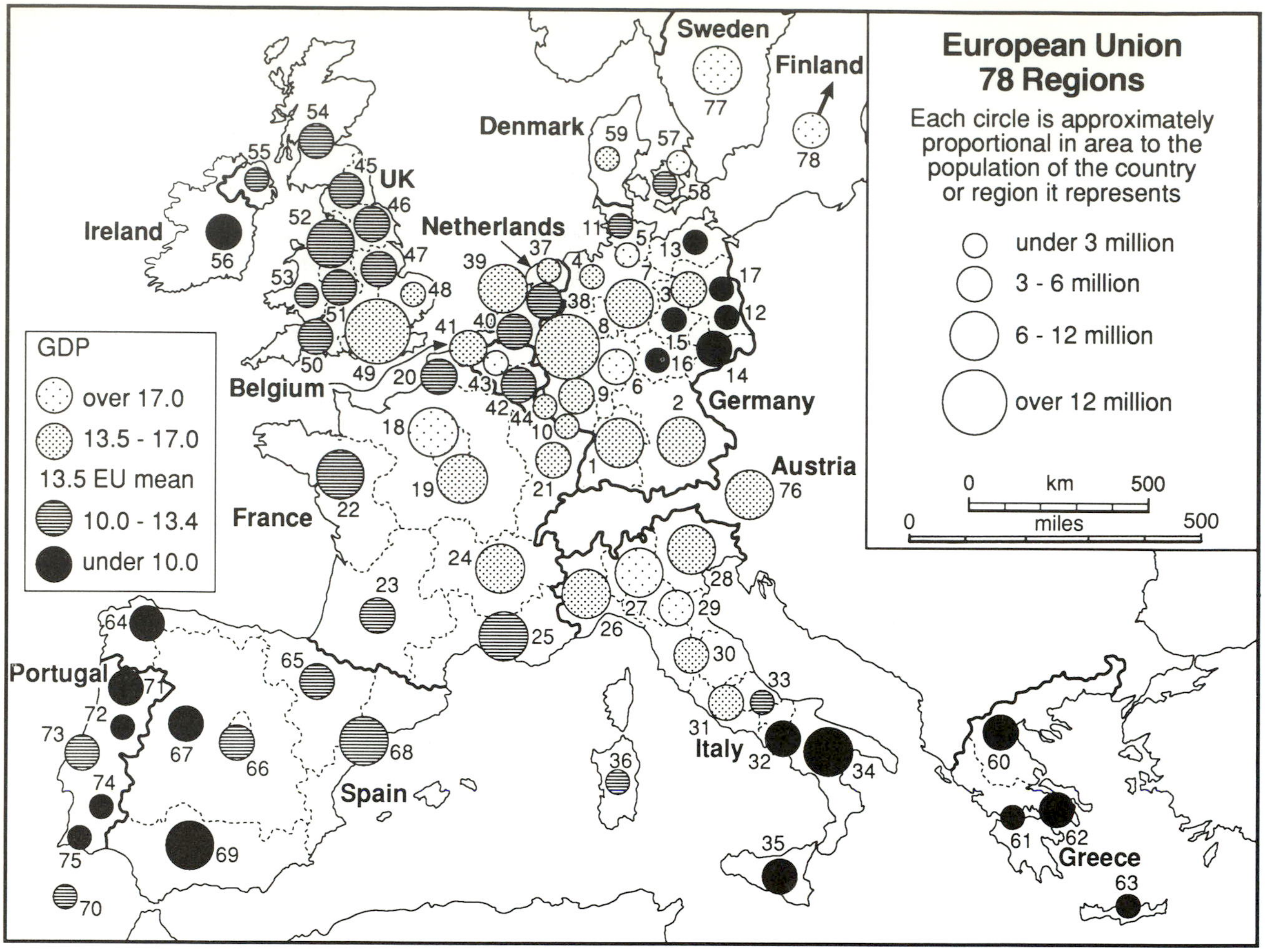

FIGURE 5.7 Per capita GDP in the regions of the European Union, in relation to the EU mean of 13,500 ECU per inhabitant in 1988–90

Source: European Union (1993)

PLATE 5.5
European Union: Sunday morning in one of Bologna's squares. Bologna, capital of the region of Emilia-Romagna, northern Italy, is one of Europe's most affluent cities. It is a stronghold of the Italian Communist Party. But where are the women? Preparing Sunday dinner?

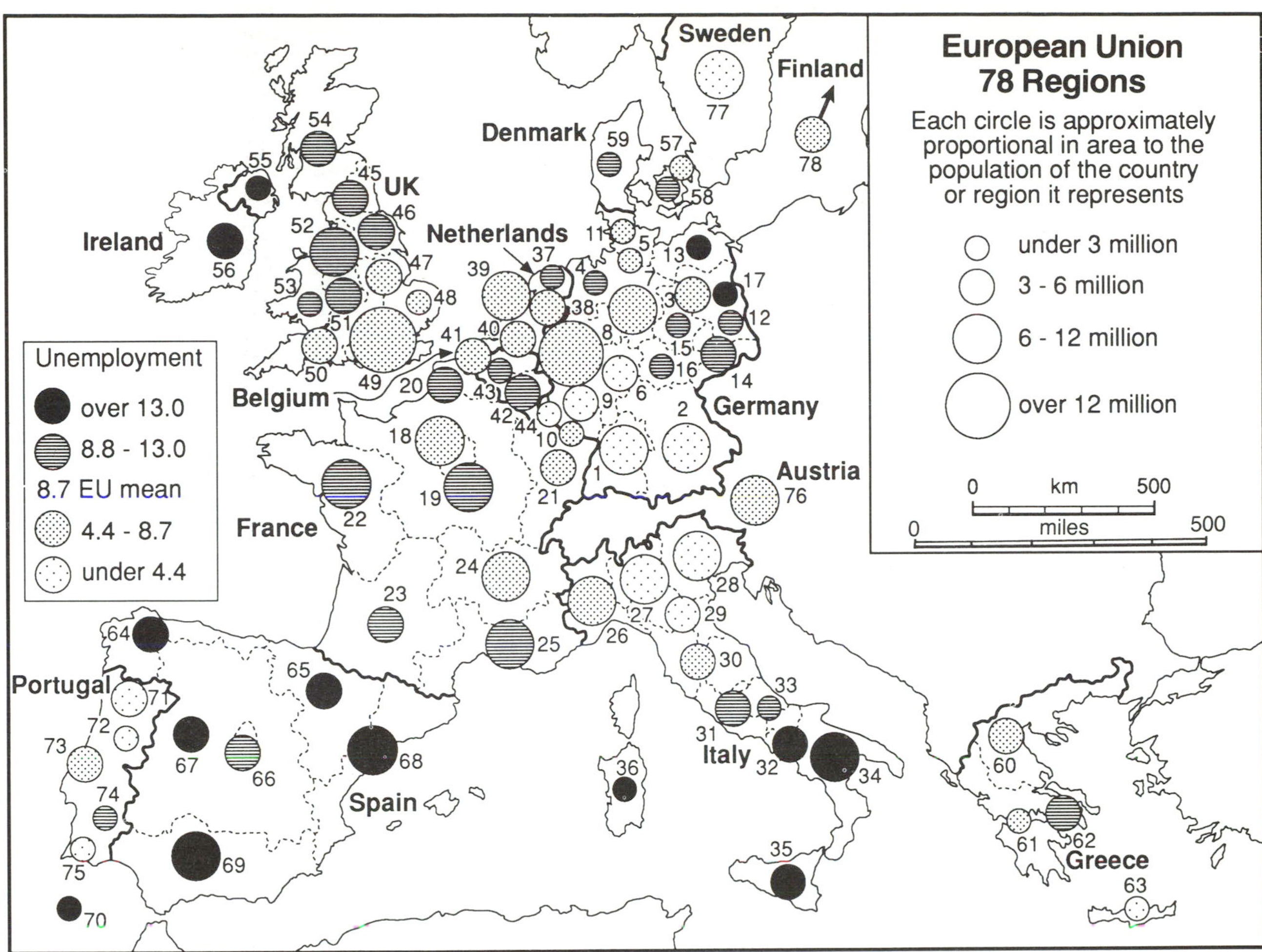

FIGURE 5.8 Unemployment as a percentage of economically active population in the European Union in 1989–91
Source: European Union (1993)

future. Western Europe differs from most other parts of the world demographically, with concern greater now over the ageing of population, the declining fertility rate, and inward migration than over population growth. Another problem in Western Europe is the heavy dependence on the rest of the world for fuel and raw materials. No matter how efficiently energy and materials are used, Western Europe cannot become self-sufficient for decades without the unlikely discovery of large new reserves of minerals or unthinkably draconian measures to cut the consumption of many products. In comparison with other major regions of the world (see Table 17.10) Western Europe scores highly in terms of both living standards and political stability.

PLATE 5.6 The European Union: the historic and picturesque. How much can survive? The number of buildings and other structures of historical value and interest in Western and Central Europe probably exceeds a million. For example, England alone has about 10,000 parish churches dating from the Middle Ages

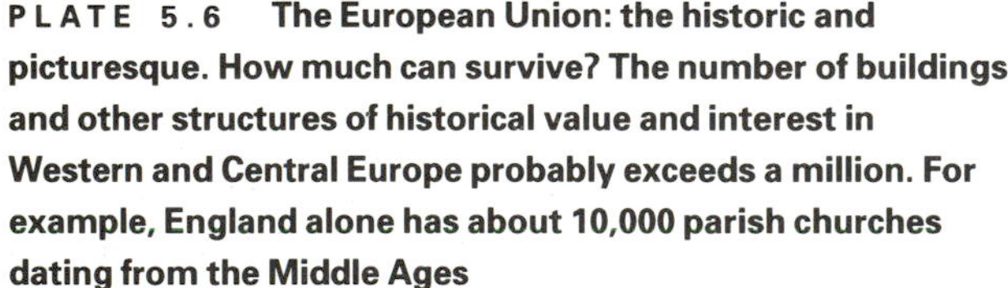

a The village of St Jean de Bueges in the Massif Central behind Montpellier, southern France. Many of the inhabitants have left and in less accessible mountain areas numerous buildings are in poor repair or already in ruins

b The village of Alberobello, Apulia, southern Italy. The local domestic architecture is so unusual that the district has become a tourist attraction. Many equally attractive villages in the remoter areas of Italy are rarely or never visited by outsiders

KEY FACTS

1. Until the end of the Second World War, Europe was characterised by the presence of a large number of political units, frequently in conflict. After 1945 there was a polarisation into western and eastern blocs, with very limited contact between the two.

2. The western bloc has gradually evolved into the European Community, but since 1989 the eastern bloc has become fragmented; the repercussions and implications of this change are not yet entirely clear.

3. The population of Europe is growing only very slowly. There is considerable immigration into Western Europe from various parts of the world. An increasing proportion of the population is elderly.

4. Western Europe is a net importer of agricultural products and it also imports about half of its energy needs, mostly in the form of oil, as well as most of its raw material needs.

5. The agricultural sector now employs only a small proportion of the economically active population, and industry is also declining in relative importance in the economy, giving ground to services.

6. Although no areas of Western Europe are desperately poor, there are regional disparities in living standards which the EU policy of convergence and cohesion is slow to reduce.

MATTERS OF OPINION

1. How tightly should the EU restrict the inflow of migrants from poorer neighbouring regions (Northwest Africa, Turkey, Central Europe and the former USSR)?

2. The EU provides considerable development assistance (under the Africa-Caribbean-Pacific Agreement – ACP) to developing countries, particularly to some of the ex-colonies of France and the UK. What proportion of this assistance, if any, should be diverted to Central Europe and the former USSR?

3. At present the general consensus in the EU is that restrictions on international trade are negative in effect. Should the EU tighten controls on the import of manufactured goods in the future in the way it has done for agriculture in order to protect its remaining industries from the competition from low-wage newly industrialising countries?

4. Western Europe has a heritage of tens of thousands of archaeological sites, listed buildings and other distinguished relics of earlier centuries as well as millions of more humble dwellings of historical interest. Should this remarkable heritage be preserved as fully as possible at the expense of funding current and future social needs?

5. Should restrictions on private motoring be enforced in the EU and greater outlays be made in improving public transport? If so, to what extent can a revival of rail transport provide the impetus needed to reduce the present dependence on road transport?

6. The EU, like the USA, depends heavily on the importation of oil, much of it from the Middle East. Is this situation likely to continue far into the future? Is western interest in protecting Kuwait, Saudi Arabia and other Gulf states more economic than political or humanitarian?

SUBJECTS FOR SPECULATION

1. The regional contrasts in production and in living standards per inhabitant in the EU are frequently attributed to the location of regions, a central location being more favourable than a peripheral one, both within the larger Member States and within the EU as a whole, in the latter case especially since the 1970s with the addition of Greece, Portugal and now Finland and Sweden. How can central and peripheral be meaningfully measured? Can the 'location' gap be narrowed by improvements in communications?

2. The enlargement of the EU to include three former EFTA members brings in affluent countries, which will be net contributors to the EU budget. The proposed inclusion by 2005 of six Central European countries will greatly reduce the present flow of resources to the existing poorer EU countries. How should the EU deal with this problem?

3. How soon can inexhaustible (renewable) sources of energy (wind, tidal and solar power) begin to reduce the dependence on fossil fuels, and what contribution can nuclear power make?

Chapter six

Countries of Western and Central Europe

Eastern Europe's conversion from communist central planning to democratic market economies is one of the most difficult undertakings imaginable. As the Carnegie Endowment's National Commission referred to it in a report last July, 'You can make fish soup out of an aquarium, but you can't make an aquarium out of fish soup.'

Bruce W. Nelan (1992)

6.1 Founder members of the European Union

In Chapter 5 Western and Central Europe were considered in general terms, both as major world regions, and on the basis of the supranational EU. In this chapter, outstanding geographical features, issues and problems of individual countries are discussed. In Section 6.1 the original EU countries are covered in the order of joining, in Section 6.2 the remaining EU countries and the EFTA countries are examined, and in Sections 6.3 and 6.4 those of Central Europe.

1. **Germany** West Germany, the Federal Republic of Germany, was a founder member of the EEC. After 1957 it emerged as the leading industrial power in the Community. Having little contact with East Germany, formerly the German Democratic Republic, or with the rest of Eastern Europe at the end of the Second World War, West Germany focused on Western Europe for increased trade. Along the eastern side of West Germany, areas close to the border were disadvantaged. In the northern part of the country, Hamburg and Bremen flourished as the main ports of the country, having, however, lost the eastern part of their pre-war hinterlands. Netherlands and Belgian ports also served much of Germany. The industries of the Rhine-Ruhr conurbation were restored to their former prominent place, being the main energy base following the loss of the other pre-war lignite and coal deposits to East Germany and Poland. The Rhine valley became the main north–south routeway in the country. Cities such as Stuttgart and Munich in the south are now among the most sophisticated and heavily industrialised places in the EU.

Following the unification of Germany in 1990 a new geographical layout came into being. East Germany had only about a quarter as many inhabitants as West Germany. By the standards of East Europe and CMEA it was one of the most advanced industrial regions in the bloc. Nevertheless, an estimated 700 billion US dollars coming largely from increased taxes in West Germany, needs to be spent over about a decade to bring East Germany up to West German standards of efficiency and environmental requirements, to upgrade rail, road and electricity supplies, and to improve other aspects such as housing. The cost of restoring Berlin to the status of national capital, replacing Bonn, is daunting. Most importantly for Germany in the longer term is the prospect of greatly improved access to markets in Central Europe, with the former East Germany the best located part of the EU to reach Poland, the Czech Republic and beyond.

2. **France** When the EEC was created in 1957, one of the reasons why France joined was to prevent Germany from becoming an independent military force in Europe again. Most of France's heavy industry was in the east and northeast of the country, and ECSC policy was to integrate the iron and steel industries of France, Germany, Belgium and Luxembourg. As new countries joined the EEC, France's central position became increasingly stronger. The centre of gravity of population of the EU is not far northeast of Dijon, and the most dynamic part of the EU includes eastern France.

After the Second World War, and particularly after 1958, when France had finally given up control of Indo-China and Algeria, a strong central government was concerned with the planning and 'management' of the territory of the country. A prime concern initially was to foster the infrastructure and industrial development of a number of regional centres to counterbalance the dominant role of Paris in the affairs and industrial activities of the country. The rail and highway systems were modernised to facilitate interregional links and to integrate the more peripheral areas such as Brittany in the northwest and the Mediterranean coastlands in the south into the national economy. Currently the more dynamic part of France is the northeast and east of the country, closest to Germany, Belgium, Switzerland and North Italy, a situation likely to be strengthened as new countries, all to the east of France, join the European Union.

3. **Italy** The elongated and fragmented nature of the territory of Italy leaves the southern part of the peninsula and the islands of Sicily and Sardinia much further from the centre of the EU than North Italy. In addition to their locational advantage North Italy, and to a lesser extent Central Italy, have positive attributes not shared by the south. The best agricultural land is in the north, most of the hydroelectric potential is there, as well as the largest reserves of natural gas, important in the 1950s and 1960s. The long tradition of manufacturing, research and good educational facilities all favour the northern regions. Milan is the main financial centre of Italy, as influential in its way as the capital, Rome.

In the inter-war period, Italy was very inward-looking economically, but after 1945 large numbers of Italian workers migrated north, some permanently, some temporarily, to work mainly in mining and heavy industry in France, Belgium and Germany. Italy's modern links with the founder members of the EU go back to that time. An increasing share of Italian foreign trade is with other EU members and in some branches of industry, especially textiles, clothing and footwear, it has become very competitive, thanks to the adaptability and flexibility of its organisation.

North and Central Italy are in a comparatively healthy economic state but attempts to develop the south through both Italian and EEC funding have had only mixed results. The choice of coastal locations in the south for new iron and steel, oil-refining and chemicals plants to stimulate local and regional economies has had disappointing results in the 1970s and 1980s because steel production in Western Europe has been reduced and there is excessive oil-refining capacity. Italy's north–south gap remains, a lesson in miniature for those who aim to narrow the gap between the richest and poorest regions in the EU as a whole and, even more ambitiously, hope for something similar to happen in the world as a whole.

4. **Benelux** Even before the creation of the EEC, Belgium, the Netherlands and Luxembourg established links which reduced some of the barriers between the three countries. An east–west line through Brussels separates southern Belgium and Luxembourg, with a long tradition of heavy industry, from northern Belgium and the Netherlands, more orientated towards agriculture and port activities. Coal production has now ceased in Belgium, and the iron and steel industry has declined, leaving this French-speaking region with severe economic problems. Antwerp and Rotterdam (with Europoort) continue to attract shipping and industries, while the Netherlands has been fortunate to have large deposits of natural gas in the extreme north of the country and offshore in the North Sea. Brussels and Luxembourg together have most of the institutions of the EU, and of all its cities, Brussels has the strongest claim to being capital of the Union, even if Belgium itself divides into two sub-nations. The central location of the Benelux in the economic and political life of the EU seems to ensure its international importance in Europe for the foreseeable future.

6.2 Members joining the EU later, and the residue of EFTA

1. **The UK** (1973) Widely regarded as one of the more reluctant members of the EU, perhaps because it has never been invaded since 1066, the UK has nevertheless frequently participated in the affairs of mainland Europe. The UK has indeed had strong overseas links with its colonies and former colonies, and in the inter-war period was heavily engaged in trade within the Empire. Since the Second World War, UK trade with other countries of Western Europe has grown rapidly, an incentive to move into the EEC customs union (see Table 5.4).

The most highly industrialised regions of Britain in the eighteenth and nineteenth centuries were in the coalfield areas, mainly in central and northern England, South Wales and Central Scotland. The political capital, London, and other cities in the southeast of the country, attracted many new industries in the twentieth century, while traditional industries such as textiles and metal-working in the coalfield areas declined, leading to very high unemployment in the 1930s.

In the decades after the Second World War, a 'north–south' divide, in practice a greatly simplified description of the situation, was recognised. Both nationally and more recently within the framework of the EU, the gap has narrowed, albeit for the negative reason that unemployment has been rising most quickly in the south of England. London itself, however, stands well above the UK average of GDP per inhabitant and, at the other extreme, Northern Ireland remains one of the poorest regions in the northern half of the EU. Whether the increasing economic and transport links (such as the Channel Tunnel) between the UK and the rest of the EU will continue to favour the south of England on account of its greater proximity to the continent (as the British call it) remains to be seen.

2. **Spain (1986)** After the death of Franco and the emergence of a more liberal system in Spain, for both economic and strategic reasons it made sense to think of membership of the EEC. Much of Spain's foreign trade was with Western Europe in the 1970s and Spain was the destination of the largest number of foreign tourists in Western Europe, most coming from the northern countries. Spain itself is a country of great cultural diversity and economic contrasts, having strongly nationalist Basque and Catalan minorities, very backward agricultural areas and modern manufacturing sectors.

Spain's peripheral position in the EU has made the improvement of road and rail links both within the country and with the rest of the EU a high priority. Much assistance from EU funds has already gone to Spain, most of which is defined as an Objective 1 area, eligible on the grounds of its economic backwardness and heavy dependence on agriculture. Few farming areas in Spain have such favourable climatic and soil conditions as those in France, North Italy or the Netherlands, but EU policy is to maintain rural communities, even if their material conditions are adequate rather than flourishing. Unemployment is very high throughout Spain, however, and the future is uncertain economically.

3. **Denmark (1973), Ireland (1973), Greece (1981) and Portugal (1986)** These four small countries are all later joiners of the EEC. They have in common a peripheral location and their heavy dependence on agriculture. Denmark, however, differs from the other three in having a very well-organised farming system, with mixed arable and livestock production, in addition to a well-developed industrial sector. The agriculture of Ireland depends heavily on livestock farming and on the UK as the main outlet for its exports. Greece and Portugal are the poorest Member States of the EU and, in the context of the EU, they compete with southern Italy and Spain in supplying fruits, vegetables and wines to a stagnating market.

All four countries are characterised by the dominant size of the capital, although in Salonika and Oporto, Greece and Portugal each have second cities of some distinction. Portugal and Ireland are at the end of the line in the EU transportation system, at great distances from the central concentration of population and economic activity. Denmark's location on the main link between the Nordic countries and the EU makes it less peripheral than Portugal or Ireland. Although Greece is very peripheral in relation to the EU, it shares a boundary with four countries, Albania, Macedonia (the existence of which it has refused to recognise), Bulgaria and Turkey. The Balkans and eastern Mediterranean are likely to experience many changes in the next decade or two, ensuring that Greek politicians will not lack problems to solve or decisions to make, and that Greece will be a major concern in the centres of power in the EU.

4. **Austria (1995), Sweden (1995), Finland (1995), Norway, Switzerland and Iceland** The countries of EFTA before its breakup in 1995 had a combined area half as large as that of the EU, but had less than a tenth of the population. The overall density of population was 25 per sq km (65 per sq mile), but it was much higher in Austria and Switzerland (about 120 per sq km, 310 per sq mile) than in the Nordic countries (about 16) and Iceland (about 3). Switzerland and Austria are poorly endowed with natural resources, but, by European standards, the Nordic countries have good bioclimatic and mineral resources. The average EFTA real GDP per head was higher than the EU average and was fairly evenly distributed over each national area. There are very few areas that would be classed as seriously backward agriculturally and almost no old and declining problem areas of heavy industry, one reason being that none of the countries has any coalfields of consequence.

After 1945 Finland and Austria had a strictly neutral status in European affairs, a situation required by the USSR as the price of independence. Sweden, in support of Finland, also remained neutral, while Switzerland chose that status voluntarily. Both Norway and Iceland were of great strategic interest to the USSR, given their presence by the only Soviet open sea outlet to the Atlantic Ocean, that from the White Sea and Murmansk. Unlike Norway and Iceland, the four neutrals did not join the North Atlantic Treaty Organisation (NATO). At times they did, however, assume roles of international importance comparable with those of larger countries.

The centre of population of the original six EEC countries was very close to Basle, arguably one of the most international cities in Europe, since its built-up area extends into three countries, Switzerland, France and Germany, underlining Switzerland's central location in Western Europe. About two-thirds of the value of Switzerland's imports come from the EU. Swiss roads and railways carry very heavy traffic between Germany, France and other EU countries to the north, and Italy – and eventually Greece – to the south. The Swiss use three EU languages and most of the population are Roman Catholics or Protestants, the predominant religious groups of the EU. Swiss companies have branches throughout the EU. Nevertheless, the fear of losing independence and special democratic features if Switzerland became a full member of the EU have made the Swiss public hesitant about joining.

In Austria, there appears to have been widespread support for joining the EU once the Cold War and the need for neutrality ended, but there have been misgivings on the part of some EU Member States over Austria's former links with the German Nazi Party. Over two-thirds of Austria's imports come from EU countries, 45 per cent from Germany alone. Austria does not have the financial resources and expertise of Switzerland and indeed has very little new to offer the EU, but its entry would be geographically desirable if not essential should the EU extend into Hungary and the Balkans in the future. Like Switzerland, western Austria is crossed by road and rail links between Germany and Italy, the busiest through the Brenner Pass, an inescapable result of its location. In the past, Austria has influenced the Balkans and traded extensively in that area. With the end of communism, its location makes it a stepping-stone between Germany and southeast Europe.

The Nordic countries and Iceland have some environmental and cultural features in common with Austria and Switzerland: mountainous areas, good water and hydroelectric resources, and extensive forests. The density of population is much lower, however, and they all have long coastlines. Sweden has been a major political, cultural and economic influence in the Nordic region in recent centuries, in terms of both territorial conquest and industrial development. Finland, however, was in the Russian Empire in the nineteenth century. After 1917, when it became independent, and even after 1945, the USSR remained a strong influence there. Each of the Nordic countries has much to offer the EU. Trade is heavy, especially with the northern EU Member States and now that two of them have joined the EU, there will be a net flow of funds to poorer EU countries.

Sweden has been characterised by its affluence and a fairly equitable distribution of wealth and income in recent decades. Its industrial development was not greatly affected by the two world wars this century and it has been an exporter of timber and high-grade iron ore as well as manufactures. In spite of a small home market, it has been a successful producer and exporter of ships and motor vehicles. Now, however, with the increasing size of industrial companies in Western Europe, it is increasingly forced to consider linking up with other countries.

Finland has relatively few industries, but its extensive forests, which cover almost 70 per cent of its total area, and various economic minerals (e.g. vanadium) provide exports. In the early 1990s its trade with the USSR dropped sharply, one good reason for its interest in the EU. In the context of an EU enlarged to absorb part of EFTA, Finland has a very peripheral location, but it is well placed to assist in the rehabilitation of the Baltic republics and the North and Northwest regions of the Russian Federation.

Only 3 per cent of the total area of Norway is cultivated, a quarter is forest and the rest of the land is not productive at all. In compensation it has magnificent scenery, good water resources, extensive fishing grounds and large oil and natural gas deposits in its sector of the North Sea, discovered in the 1960s. These fossil fuels are of interest to the EU, given the heavy dependence of the Union on imported oil, much from areas of the world regarded as politically unstable. Norway's oil reserves of about 1,000 million tonnes (0.8 per cent of the world total) would not last much more than a decade, however, at the 1991 rate of production. It is in a much better position with regard to natural gas reserves (1.7 trillion cubic metres, 1.4 per cent of the world total) which would last over sixty years at the 1991 rate of production.

Iceland depends very heavily on its fishing industry and, given its small population and the large adjacent zone in which it claims fishing rights, it would lose out in any reallocation of fishing catches to other EU countries if this was made consistent with previous EU fishing policy. With the increasing number of special opt-out clauses granted to EU Member States following the establishment of new targets and eventual aims of greater integration, it is likely that all newcomers would have to accept all existing EU conditions without reservation, after an appropriate period allowed for changes to take place.

6.3 General features of Central Europe

As indicated in Chapter 5, Central Europe includes everywhere in Europe not in the EU, EFTA or the Commonwealth of Independent States (CIS). Figure 6.1 shows the location of the countries, including the three Baltic states, and Table 6.1 contains data for the countries of the region. Since 1945 the whole region has been strongly influenced by the Soviet Union, but socialism (or communism) has been applied with

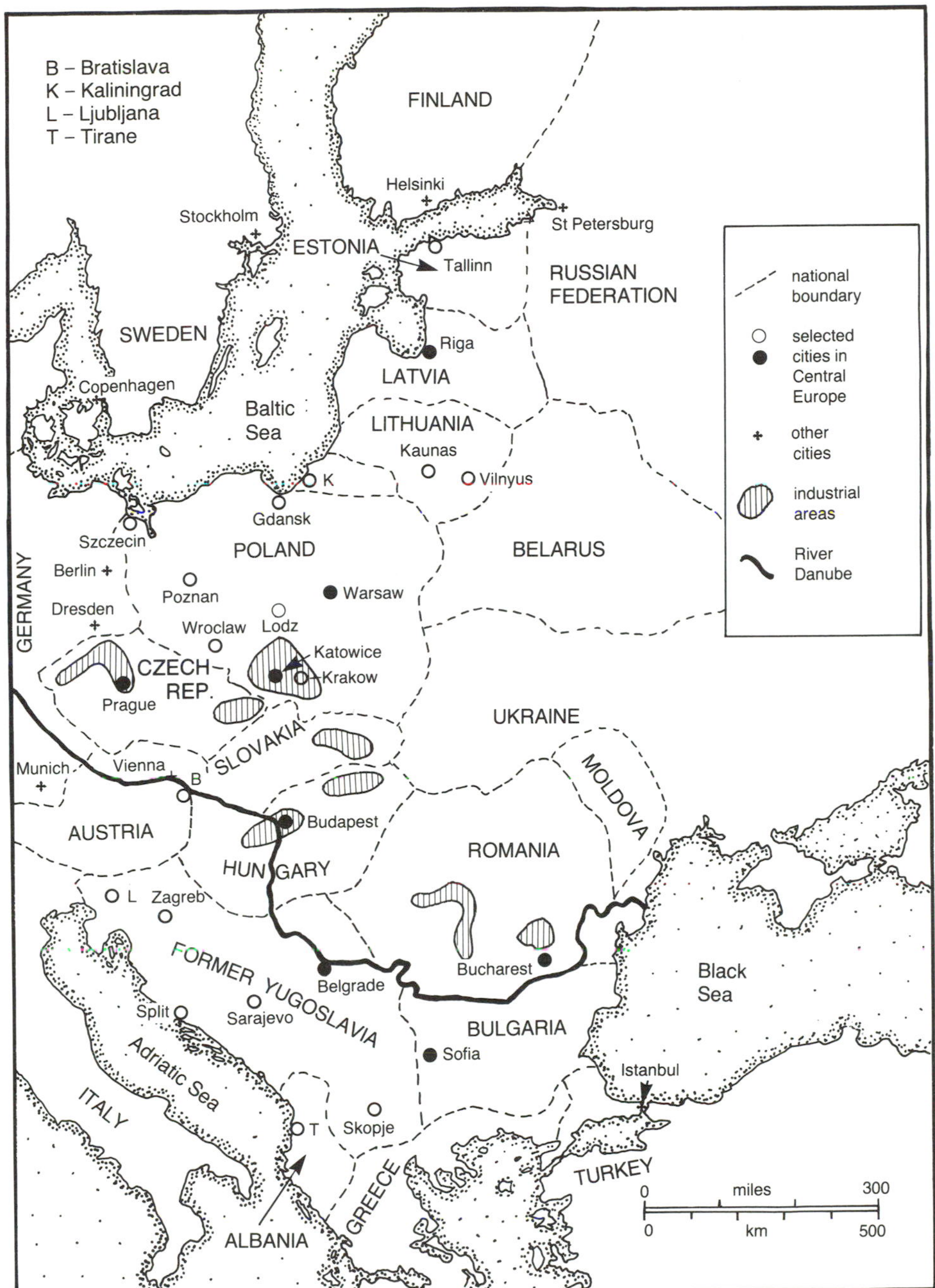

FIGURE 6.1 General location map of Central Europe

varying degrees of intensity during that period.

The Baltic states have been transformed economically and ethnically, following their reoccupation by the Soviet Union in 1939, and the settlement of many ethnic Russians there. Poland, Czechoslovakia (since 1948), Hungary, Romania and Bulgaria have been dominated by the USSR, both militarily as members of the Warsaw Pact and economically as members of the Council for Mutual Economic Assistance (CMEA or COMECON). They were each governed by a party sympathetic to the Communist Party of the Soviet Union (CPSU). Although run by socialist governments and centrally planned, Yugoslavia and Albania remained outside the Soviet bloc. Yugoslavia professed neutrality and its rulers gave greater freedom to its citizens to work in Western Europe than was enjoyed in the CMEA countries. Albania did not allow their citizens to work in Western Europe but remained a hermit state.

Soviet-style communism has apparently been abandoned in Central Europe, but the future is uncertain on account of at least two trends that could affect changes. First, the political, economic and social legacy of Soviet influence cannot be changed suddenly. Many of the influential professional and managerial personnel running the economies were members of communist or

PLATE 6.1
The Czech Republic in Central Europe

a Ostrov in the northwest part of the Republic, a mining and manufacturing centre with typically crumbling buildings dating from early in the twentieth century

b Post-Second World War housing at Chomutov, constructed during the communist period

cognate parties. Most of the foreign trade of each country was either with the USSR, as part of the system of regional specialisation in CMEA, or between the CMEA partners themselves. Second, according to Marxist-Leninist theory, it would be expected that workers would feel solidarity once they no longer toiled for profit-orientated capitalist owners of the means of production. Thus they would quickly overcome or forget their various nationalistic, religious and ethnic prejudices. This assumption has not proved true in Central Europe. The cultural history of Central Europe before 1945 must be studied if post-communist issues and problems are to be understood, since the recent changes have revealed various layers of cultural and political features beneath the communist facade.

For centuries Central Europe has been influenced by the aspirations and territorial acquisitions of powers located outside its limits as here defined. The greatest influence has come from Germany and Austria to the west, from Russia to the east and from Turkey to the south, combining to produce pressures on the peoples of the region. Now that Germany has united and few Germans remain in Central Europe, most languages spoken in the region are Slavonic rather than Germanic. Hungary and Romania, however, have distinct languages, while Estonia has a language close to Finnish, and Latvian and Lithuanian are distinguished as Baltic. No less than fourteen different languages,

TABLE 6.1 Country-by-country data set for Central Europe, mainly for 1993

	(1) Population	(2) Population	(3)	(4)	(5) Real GDP (US dollars)	(6) Real GDP (US dollars)	(7)	(8)	(9)	(10)	(11)	(12)	(13)	(14)
	rank in world	*millions*	*Area in thous. sq km/(miles)*	*Density per sq km/(mile)*	*total (billions)*	*per head (thousands)*	TFR	*% over age 64*	*Infant mortality*	*% in agriculture*	*% urban*	*Steel output per head (kg)*	*Energy conservation per head (kg)*	*Cars per 1,000 popn*
Poland	29	38.5	312.7 (120.7)	123 (319)	184	4,780	2.0	10	14	20.1	62	325	4,500	126
Romania	38	23.2	237.5 (91.7)	98 (254)	70	3,020	1.6	11	23	19.4	54	603	4,480	65
Czech Republic	67	10.3	78.9 (30.5)	131 (339)	81	7,860	1.8	13	10	7.0[2]	70[2]	1,019	6,000	216
Hungary	68	10.3	93.0 (35.9)	111 (287)	64	6,210	1.9	13	15	11.0	62	320	3,800	168
Yugoslavia[1]	73	9.8	102.2 (39.5)	96 (249)	40	4,080	2.0	11	15	20.0[2]	47	255	2,800	150
Bulgaria	78	9.0	111.0 (42.9)	81 (210)	46	5,110	1.6	13	17	11.8	68	322	5,060	67
Slovakia	99	5.3	49.0 (18.9)	108 (280)	35	6,600	2.0	10	13	10.0[2]	60[2]	943	6,000	170
Croatia	109	4.4	56.5 (21.8)	78 (202)	26	5,910	1.7	12	11	15.0[2]	51	455	4,000	200
Bosnia-Herzegovina	116	4.0	51.1 (19.7)	78 (202)	20	5,000	1.6	6	15	25.0[2]	34	0	2,000	100
Lithuania	119	3.8	65.2 (25.2)	58 (150)	23	6,050	1.9	11	14	15.0[2]	69	0	2,630	40
Albania	126	3.3	28.8 (11.1)	115 (298)	14	4,240	3.0	5	28	47.7	36	0	1,240	1
Latvia	131	2.6	65.2 (25.2)	40 (104)	16	6,150	1.7	12	17	12.0[2]	71	0	2,310	40
Macedonia	138	2.0	25.7 (9.9)	78 (202)	11	5,500	2.1	7	35	30.0[2]	54	0	2,000	100
Slovenia	139	2.0	20.0 (7.7)	100 (259)	16	8,000	1.7	11	9	15.0[2]	49	0	2,800	200
Estonia	142	1.6	45.1 (17.4)	35 (91)	10	6,250	1.8	12	14	8.0[2]	71	0	2,500	40

Notes:
[1] On 27 April 1992 Serbia and Montenegro formed a new state
[2] Author's estimate
Description of variables:
(7) TFR – Total fertility rate, average number of children born per female
(9) Deaths of infants under the age of one year per 1,000 live births
Sources: *WPDS 1993* for columns (2), (7), (8), (9), (11); *HDR 1992* for columns (5), (6); *FAOPY 1991* for column (10); *UNSYB 90/91* for columns (12), (13), (14)

each with over a million speakers, are used by a total population of about 130 million.

Religious diversity is no less striking. Three branches of Christianity, together with Islam, share the religious affiliations of almost all of the religiously committed population. The Greek and Russian Orthodox faith predominates in Romania, Bulgaria and most of the former Yugoslavia, whereas Lithuania, Poland, the Czech Republic, Slovakia, Slovenia, Croatia and much of Hungary are predominantly Roman Catholic. Protestants are the majority in Estonia, Latvia and parts of Hungary and Romania. Finally, a considerable number of Muslims in Bosnia and Albania form the main legacy of Turkish occupation of the Balkans. With the end of socialism and concomitant atheism, the new governments of the 1990s are coming to terms with a revival of religious feeling in Central Europe as well as the obligation, if acceptance by and assistance from Western Europe are to be obtained, to ensure human rights and to install democratic institutions.

Economically, Central Europe is by no means badly off in many respects. The hothouse industrialisation programmes of the 1950s and 1960s took most countries of Central Europe from a predominantly agricultural economy to one led by industry. With hindsight it is evident that the direction of industrial development led to an excessive emphasis on heavy industry while light industry was neglected. What is more, by world standards Central Europe is poorly endowed with natural resources, especially water, raw materials and fossil fuels. Fuel and raw materials were therefore imported, albeit on generally favourable financial terms, from the Soviet Union, which itself provided a captive market for the industrial products of its various CMEA partners, particularly East Germany, Poland and Czechoslovakia. It is possible to generalise to some extent about the common characteristics and problems of the various countries of Central Europe, given their particular location and their specialised production in the international system of CMEA. After four decades, however, national feeling is still strong and many issues and problems are peculiar only to one individual country or to some subset of all the countries.

6.4 Individual countries of Central Europe

1. **Poland** has the largest population of the countries of Central Europe. Almost half of the total area, most of which is lowland or hill country, is cultivated, while over a quarter is forested. Labour productivity in agriculture is low compared with that in Western Europe and until recently yields were lower than in nearby parts of the EU. With changes in management, Poland could produce considerably more from the agricultural sector, but such an achievement would be a liability rather than an asset if Poland joined the EU. By EU standards much of the countryside in Poland would be classified as lagging agricultural, Objective 1 in EU terminology.

After Poland was partitioned between Russia, Germany and Austria late in the eighteenth century, the western part became more industrialised under Germany than the eastern part held by Russia, although Russian Poland was one of the main light industrial regions of the Russian Empire. In the nineteenth century Poland's main coalfield was developed, and after 1945 it formed the basis for the expansion of heavy industry under Soviet-style planning policy. In the 1980s Poland's industrial growth ceased, but even in the late 1980s it was proposed by communist planners that more investment should be made in heavy industry. In the event of Poland's entry into the EU, it would add more regions to the existing large number of old/declining industrial areas (Objective 2) in the EU.

Issues of a more political and social nature also complicate the reorientation of Poland towards the West. The Catholic Church takes a rigid attitude towards such issues as family planning, and the population is expected to grow considerably in the next few decades. Whether or not Poland joins the EU, the movement of goods and passengers into, out of and across Poland is likely to increase in the next decade or two and the road and rail networks will have to be greatly improved.

2. **Czechoslovakia** was amicably divided in 1993, leaving the more industrialised and sophisticated Czech Republic in the west and the more rural and backward Slovakia in the east. Czechoslovakia was considered one of the showpieces of CMEA but, on closer examination, like East Germany and Poland, it depends very heavily on its own reserves of coal and lignite, together (until recently) with imports of oil, natural gas and raw materials from the USSR. Immediate problems are the organisational one of moving towards a market economy, the economic one of changing the sectoral structure of economic activities, and the environmental one of coming to terms with the high cost of cleaning up in many areas and cutting down pollution in the future.

3. **Hungary** has many of the features of the Czech Republic and Poland but is less highly industrialised. The concentration of about a quarter of the total population in Greater Budapest is itself a matter of concern, if not an urgent problem. Hungary's distinctive ethnic background also highlights the situation of the two million ethnic Hungarians embedded in the present Romania and of half a million in the former Yugoslavia.

PLATE 6.2
Central Europe: Hungary

a The gracious Parliament building on the bank of the River Danube, Budapest. Since the late 1980s Hungary has been working towards a democratic system and a market economy

b Fashionable office and shopping area of Budapest, near the Parliament building. In the early 1990s the smaller towns and the countryside did not have such a drab appearance as that characterising many areas in other Central European countries

4. **Romania** is also distinctive in the region, having a language of Latin origin, surviving through Slavonic and subsequent Turkish influences. Romania has good-quality agricultural land, but its once famous oil deposits have now largely been used up, and it has few minerals. At first sight the most promising line of development seems to be in agriculture, but as in Poland, increased output would embarrass the EU should Romania join.

5. **Bulgaria** In the Soviet orbit, Bulgaria was regarded as more developed than Romania, with a high level of production in some industrial sectors, built up from almost nothing in 1945. The country has a problem of shortage of fuel and depends heavily for its energy supply on a suspect, potentially unsafe, nuclear power station.

6. **Yugoslavia** was regarded world-wide as a successful experiment in socialist planning and ethnic harmony. It was one of the original Third World countries, seen as neutral in the ideological struggle between communism and capitalism. Strategically it was presumably satisfactory for there to be a buffer zone between East and West, in which the military confrontation was less tense than along the Iron Curtain to the north and along the interface of CMEA with Greece and Turkey to the east. By European

standards, however, Yugoslavia is a comparatively poor country, but with very marked regional disparities between the better-off north and the backward south.

The credentials for EU membership of all the five new states that have proclaimed sovereignty in the former Yugoslavia since 1991 could hardly be worse: three languages, Slovenian, Serbo-Croat (divided by two distinct alphabets) and Macedonian, and three religions, Roman Catholic in the northwest, Orthodox in the southeast and Islam in many small pockets of territory (see Box 6.1). The heavy loss of life incurred in former Yugoslavia's civil war and the massive destruction of property add to the complexity of this unfortunate part of the world. Beside Albania, however, Yugoslavia looks comparatively modern and developed. Albania has been characterised by atrocious material conditions and a growing, predominantly Muslim population, fed by a total fertility rate of three children per female, more than double that in neighbouring Greece and nearby Italy.

7. **The Baltic states** The final elements in the mosaic of Central European countries as defined in this book are the Baltic states. Out of a total population of 8 million altogether in the three republics in 1989, about 1.5 million were ethnic Russians, making up 30 per cent of the total population of Estonia, 34 per cent in Latvia but under 10 per cent in Lithuania. Almost all the Russians were settled there after 1939, not because the Baltic republics were underpopulated, compared with, say, Siberia, nor because they have natural resources of more than modest significance, but to satisfy the Soviet obsession of consolidating frontier areas in the vast Soviet Union, in their case forming a useful frontage on the Baltic coast. Considerable industrial development has taken place in the three states

BOX 6.1

Ethnic nightmare: the case of Yugoslavia

The conflict and chaos in Yugoslavia since the summer of 1991 seemed at the time of writing unlikely to end satisfactorily for a long time. Nevertheless, enough has happened to deserve attention here. Although Yugoslavia is culturally the most diverse country in Europe outside the former USSR, its problems illustrate the general situation in the continent. To appreciate the ethnic tension and conflict and the political structures anywhere in the world, it is useful to make a checklist of possible influences and causes of troubles. With Yugoslavia as a case study, these include the following:

- Physiological characteristics, such as skin colour (not a major issue in Europe)
- Language (e.g. Romanic (or Latin), Germanic, Slavonic, or other in origin)
- Religion (various branches of Christianity, plus a substantial Muslim influence)
- Location, generating national feelings of cohesion

Some of the 'nation' states of Europe were established at least several centuries ago (e.g. Spain, France, England, Russia), some were consolidated in the nineteenth century (particularly Italy and Germany) and some are of more recent origin. Few coincide precisely with the territories of distinct ethnic groups. Table 6.2 shows the distribution of ethnic groups in Yugoslavia

TABLE 6.2 Ethnic composition of the Yugoslav republics

	Area in thous. sq km/(miles)	*Population (millions) 1991*	*Density persons per sq km/(mile)*	*Population 1994 estimate*	*% of each ethnic group*
New Yugoslavia[1]	102 (39)	10.3	101 (262)	10.5	62 Serb, 17 Albanian, 5 Montenegran, 3 Muslim, 3 Hungarian
Slovenia	20 (8)	2.0	100 (259)	2.0	91 Slovene, 3 Croat, 2 Serb
Croatia	57 (22)	4.8	84 (218)	4.8	78 Croat, 12 Serb, 1 Muslim
Bosnia-Herzegovina	51 (20)	4.4	86 (223)	4.6	44 Muslim, 31 Serb, 17 Croat
Macedonia	26 (10)	2.0	77 (199)	2.1	65 Macedonian, 21 Albanian, 5 Turkish, 3 Muslim
Old Yugoslavia	256 (99)	23.5	92 (238)	24.0	

Note: [1] Consisting of the former Republics of Serbia and Montenegro
Source: Calendario Atlante de Agostini (1994)

since the Second World War, but much of the labour force in industry has been made up of Russians. Although the three republics are in theory totally independent of the former USSR, it is politically impossible for them to change their economic dependence on the Russian Federation, because they must remain outlets for Soviet foreign trade. There seems to be little that they can do in the near future to develop new lines of economic activity.

The sudden break-up of the USSR in 1991 illustrates how even without a world war political, economic and strategic situations that have changed little in decades can suddenly give rise to completely new alignments of countries. The entry of any Central European country into the EU, not least East Germany, would have been seen in the mid-1980s as a strategic triumph for the West, the removal of a key piece on the global chessboard of world affairs. In the mid-1990s the admission of a Central European country into the EU would be an act of magnanimous generosity since such a country would be a net receiver of whatever assistance is available for the increasingly large intake of poorer neighbours, starting with Ireland in 1973 and continuing with Greece in 1981, Spain and Portugal in 1986, and East Germany in 1990/1.

Assuming no further drastic changes in Europe, the countries of Central Europe will probably develop closer ties with the EU. These will be primarily of an economic nature, through increased trade and restructuring assistance via the PHARE programme of the EU, initially for Poland and Hungary: assistance for the restructuring of economies (hence the acronym) but subsequently extended to all of Central Europe and the Baltic republics. Cultural and educational links are also being reinforced through specific EU programmes.

as it was in the 1980s. Even by European standards the great diversity of the six republics formed after the Second World War is evident. The following events and influences have contributed to the current problems of Yugoslavia.

1. From 1948 when Yugoslavia broke away from the Soviet-dominated Cominform, to the late 1980s, the country managed to maintain a 'neutral' position in the Cold War under the leadership of Marshal Tito (1892–1980, Josip Broz, born in Croatia) until his death in 1980. It had a flexible socialist system and was not tied, as were the other countries of Central Europe, to the Soviet bloc and to the directives of the Soviet Five-Year Plans. The USSR failed to establish military bases in the interior of Yugoslavia or a naval base on the Adriatic coast. The latter would have been a useful psychological location to counter NATO naval domination of the Mediterranean. When the Cold War ended, the need for solidarity among the Yugoslavs was no longer a factor holding the country together.

2. Yugoslavia is a 'young' country by European standards. It was created on 24 November 1918, shortly after the end of the First World War. The United Kingdom of the Serbs (including Macedonia), Croats and Slovenes was proclaimed in Zagreb, with King Peter of Serbia the monarch; Montenegro joined shortly after. During the Second World War Yugoslavia was occupied by Germany and Italy, and while the Croats accepted the German presence, the Serbs ran one of the most successful resistance movements against Nazi occupation. The country was liberated by the USSR in 1945. Yugoslavia is not therefore a cohesive and long-standing national entity.

3. Yugoslavia was part of the Roman Empire for several centuries, but the use of local languages and of Latin at that time was superseded by Slavonic languages. Yugoslavia actually means 'South Slavia'. The two main ethnic groups, the Serbs and the Croats, have the same Serbo-Croatian language, although there are differences of dialect and different alphabets are used, the Cyrillic and the Latin respectively.

4. Religious differences are more marked than linguistic ones (cf. Northern Ireland). The oldest religious division is between the Catholic north (Slovenia and Croatia) and the Orthodox south, resulting from the split in the Christian Church in 1054. A second influence on religious allegiance came in the Orthodox south, when many people in the Balkan region were converted to Islam following the invasion of the area by the Turks of the Ottoman Empire, whose base was Anatolia, now modern Turkey. The Muslims are mostly in Bosnia-Herzegovina.

5. The area that became Yugoslavia in 1918 had previously been influenced, and parts of it held for long periods, by outside powers: Italy (Venice), Austria and Hungary from the north, the Ottoman Empire from the south. In the nineteenth and early twentieth centuries the Adriatic coast was the only seaboard held by the large Austro-Hungarian Empire.

6. Partly depending on outside influences, partly on location and partly on natural resources, by the present century there were striking differences in economic levels between the various parts of the country. According to Fisher (1966), there was a 'north–south' difference in per capita income in the 1960s, as shown by the figures for per capita income in 1963 (thousand

PLATE 6.3
Central Europe: basic means of transport in the Carpathians of Romania, near Bicaz. In the Balkan countries there were still few motor vehicles in use in the 1980s and motor traffic on secondary roads was frequently delayed by slow-moving carts and herds of livestock

BOX 6.1 *continued*

dinars): Slovenia 420, Croatia 270, Serbia 200, Bosnia-Herzegovina 150, Macedonia 150, Montenegro 140. According to Munchau (1991) the divergence was even more marked in 1989 (GNP per capita in US dollars): Slovenia 12,520, Croatia 7,110, Serbia 4,640, Montenegro 3,970, Bosnia-Herzegovina 3,590, Macedonia 3,330. Slovenia and Croatia were the most highly industrialised of the republics, while Croatia had the popular Adriatic tourist coast. The density of the road network and quality of road surfaces, like other features of the infrastructure, was also superior in the northern part of the country. One reason for the independence movement in 1991 was the resentment in the more affluent Slovenia and Croatia that these republics were subsidising the weaker economies of southern Yugoslavia.

One by one the smaller republics opted out of the Yugoslav Republic, in effect rejecting Serb dominance. Thanks to its location, Slovenia's withdrawal was fairly painless. When Croatia's turn came, Serbia proceeded to attack Croat cities near the mutual border, while Serb minorities within Croatia declared their separation from the Republic and their allegiance to Serbia. In due course, Macedonia also declared its independence, but the small Republic of Montenegro, with little of economic consequence except one of Yugoslavia's largest iron and steel plants in Titograd, stayed with Serbia.

The most painful declaration of independence was that of Bosnia-Herzegovina, the main reason being that there are substantial numbers of Muslims, mainly in the central area, with Serbs on all sides but the southwest, and Croatians on the Adriatic side. In many parts of Bosnia-Herzegovina, however, two ethnic groups live side by side in the same settlements, and in some places all three do. There is a tendency for the Muslims to predominate in the larger urban areas and the Serbs to populate the countryside.

In wars, civil or between states, the conflict is usually between two sides. The experience of Bosnia-Herzegovina has been unusual in history. At any given time there could be simultaneous conflicts between Serbs and Croats, Serbs and Muslims or Croats and Muslims. In 1993 the Muslims and Croats agreed to form an alliance. During the conflict, thanks to the support of Serbia, and profiting from their superior supply of arms, the Serbs made inroads into predominantly Muslim and Croat areas, consolidating their territorial gains by 'ethnic cleansing', a term that has in some cases become widely used in the 1990s, referring to the extermination of enemies, usually to their removal elsewhere from their home areas. The other two ethnic groups in their turn used similar methods.

But for the presence of United Nations peace-keeping forces, loss of life and suffering might have been greater than it was. In the event, the crucial outcome of the conflict has been that by mid-1994 Serbs held about 70 per cent of the total area of Bosnia-Herzegovina, Croats about 10 per cent and the Muslims 20 per cent. The international consensus is that the Serbs are entitled to only about 50 per cent, and that Bosnia-Herzegovina should keep its pre-conflict form, with the three parties working together in the new state. Any subdivisions of Bosnia-Herzegovina would result in the creation of a Muslim state, a Bosnian Serb state, with parts of the former Yugoslav Republic of Croatia annexed, and the Croatian parts of Bosnia-Herzegovina cutting off the other two areas from the coast.

Their actual membership of the EU is a prospect that is far more remote. Typically, however, the policy makers and planners of the EU have now drafted a long-term blueprint (see Figure 6.3) for the improvement of transport corridors between Germany/Austria and Central Europe and also within Central Europe itself. The creation of a comprehensive high-speed rail network in Western and Central Europe, a dream of Hitler in the Second World War, remains an intention, as yet barely started except in France. Figure 6.4 shows the proposed links together with some non-high-speed lines. Hopeful arrows point towards the former USSR, lines to be negotiated.

Closer integration through new or improved transport links is needed if Central Europe and the western part of the former USSR are to become more closely associated with Western Europe. A transformation is also expected in the economic structure through the reduction of state ownership of means of production and the removal of rigid national plans. Table 6.3 is a useful guide to who is where on the path of transition to a market economy. Figure 6.5 shows a general northwest–southeast trend from more to less privatisation, with Belarus standing out as a reluctant laggard in the process.

Apart from Yugoslavia and Albania, all the countries of the region defined here as Central Europe have been strongly influenced by the Soviet Union since the Second World War. Politically they were controlled by Communist Parties or closely associated Parties and economically they were closely integrated into the Soviet planning system. During several centuries before the Second World War virtually all of Central Europe has for long periods been ruled by one or more stronger powers outside: Germany and Austria in the west,

New small states such as those in Yugoslavia and the Baltic republics now have international recognition, are represented in the UN as sovereign states, have their own national currencies and football and sporting teams, new features they are unlikely to give up in a hurry. Allcock (1990) argued in 1990 that 'despite discord dividing Yugoslavia there are deep imperatives for unity'. In 1994 it seemed that his interpretation of the situation was completely flawed, but he covered himself by concluding: 'In spite of its apparently inexorable movement towards disintegration, there are nevertheless good reasons for supposing that this event, if it did occur, would be a case of death by misadventure.'

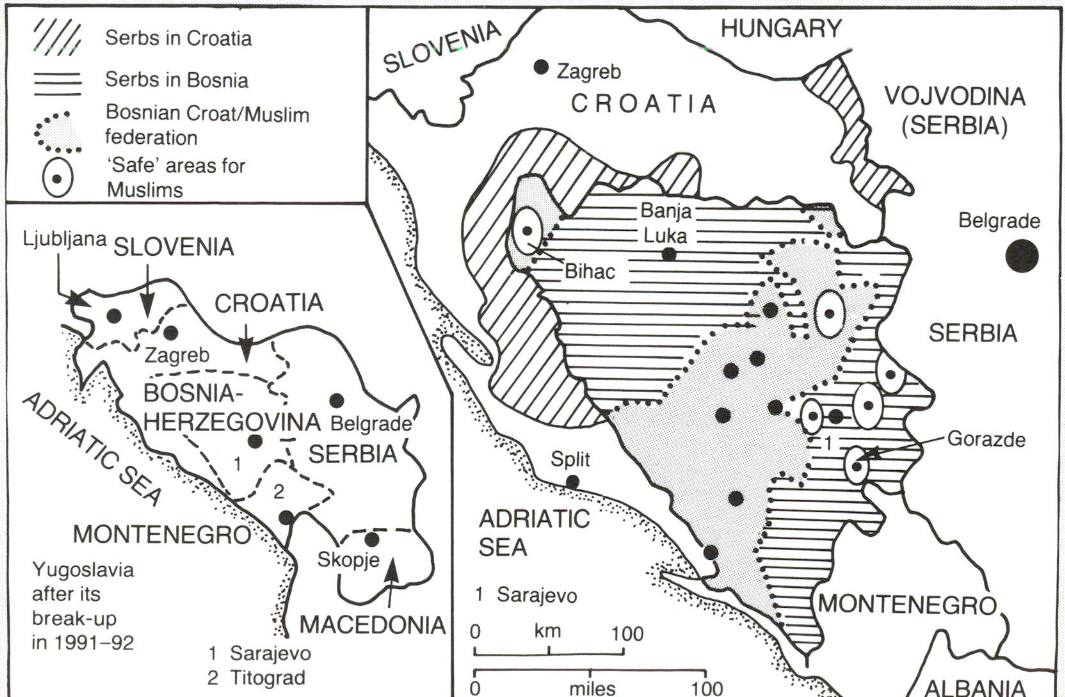

FIGURE 6.2 The break-up of Yugoslavia and the fragmentation of Bosnia-Herzegovina and adjacent areas. In 1995 the Serbs in most of Croatia were 'ethnically cleansed', expelled by force to Bosnia or Serbia, while Bosnian Serbs captured two Muslim 'safe' areas. In September 1995 much of the area hitherto occupied by Serbs to the south of Banja Luka was captured by Bosnian and Croat forces with the (unrelated?) assistance of NATO air attacks

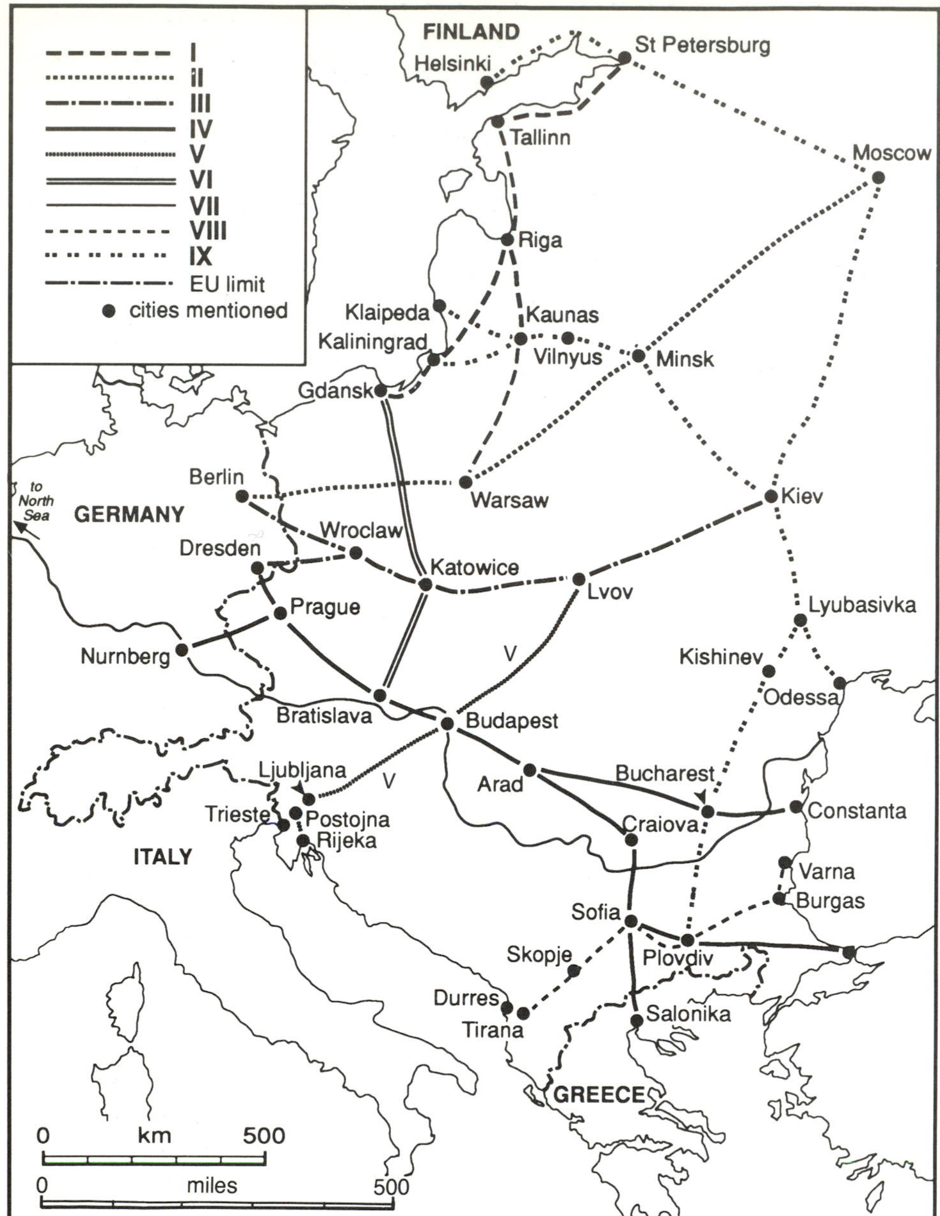

FIGURE 6.3 Proposed improvements to existing transportation links in Central Europe and new links. Note that routeway VII is the Danube and its link with the Rhine. The other routes include both road and rail

Sweden in the north, Russia in the east and the Ottoman Empire in the south. The countries are now settling down to complete political independence, while at the same time switching economic links and transactions from the former Soviet Union to Western Europe. In due course the advantages and disadvantages of the countries of the region will become more evident in the new competitive world of Europe and Russia, with more affluent neighbours to emulate in the EU and more abundant natural resources in the Russian Federation, now less accessible than in Soviet times.

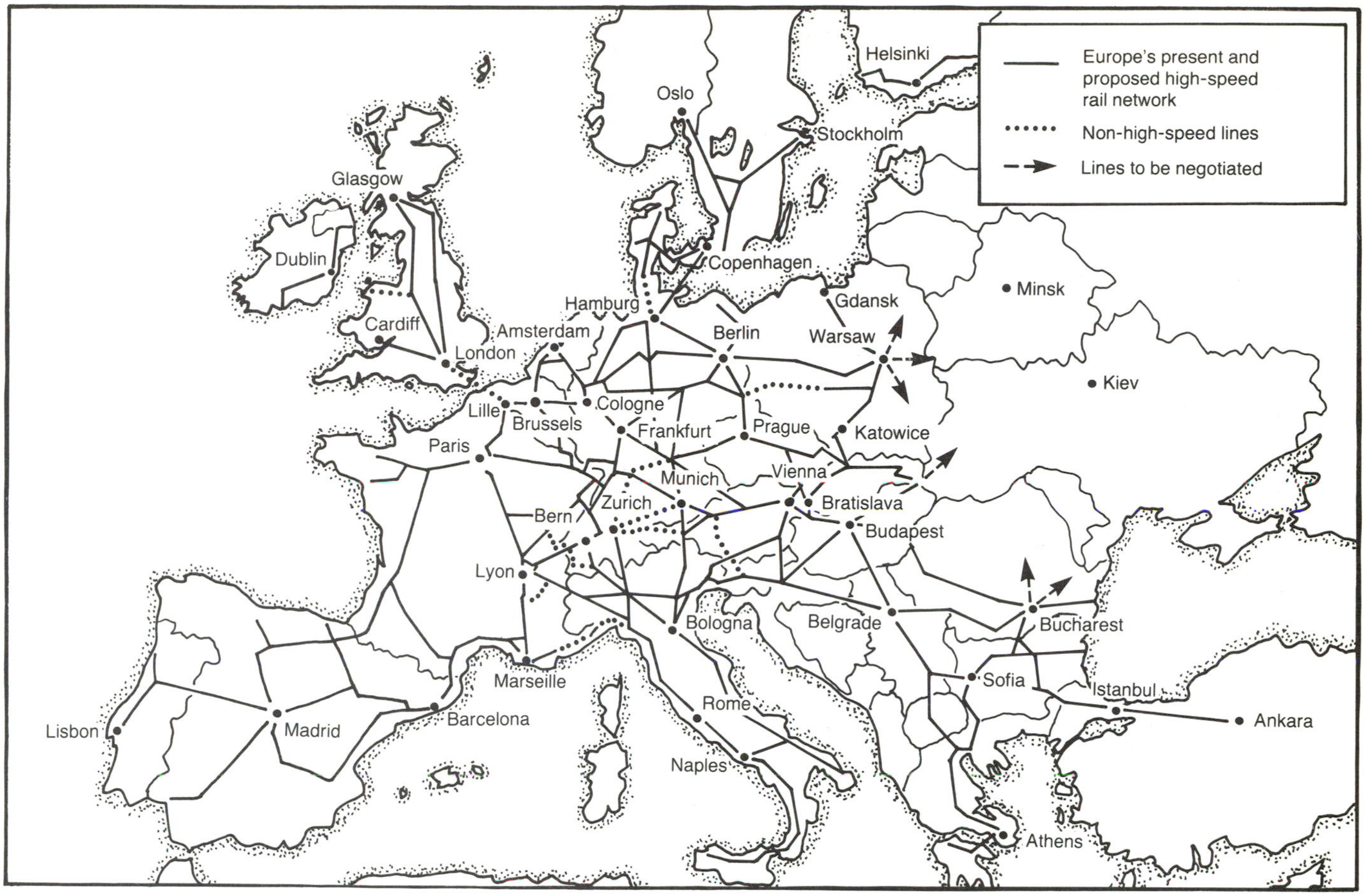

FIGURE 6.4 The present and proposed high-speed rail network for Western and Central Europe, with non-high-speed link lines also shown

Source: Healey and Baker/International Union of Railways, cited in Branson (1994)

PLATE 6.4 Berlin: Checkpoint Charlie shortly after the removal of the Wall between East Berlin and the United States zone of West Berlin. The introduction of free movement between the sectors of the city followed quickly

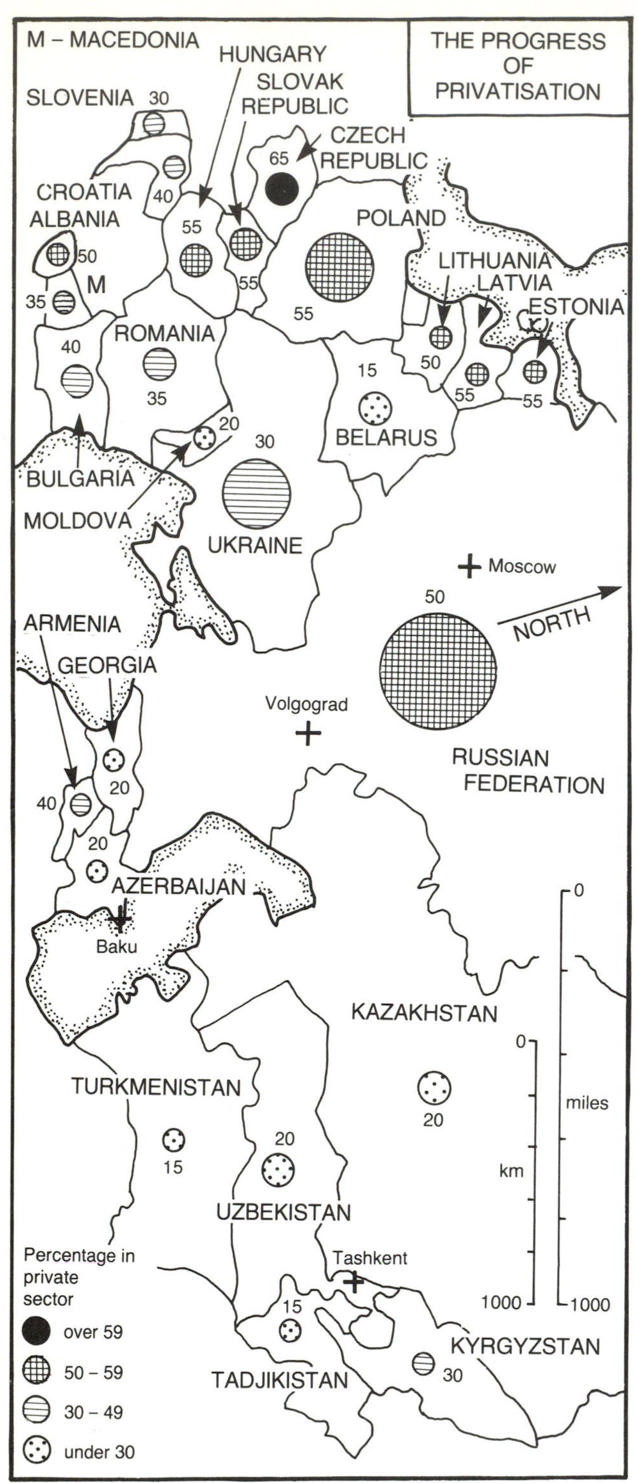

FIGURE 6.5 The transition to the market economy in the countries of Central Europe and former USSR according to the percentage share of GDP in the private sector in mid-1994
Source: Naudin (1994)

TABLE 6.3 The percentage share of the private sector in GDP, mid-1994, in Central European and former Soviet countries

1	Czech Republic	65	14	Romania	35
2	Estonia (R)[1]	55	15	Kyrgyzstan (R)	30
3	Hungary	55	16	Slovenia	30
4	Latvia (R)	55	17	Ukraine (R)	30
5	Poland	55	18	Azerbaijan (R)	20
6	Slovak Republic	55	19	Georgia (R)	20
7	Albania	50	20	Kazakhstan (R)	20
8	Lithuania (R)	50	21	Moldova (R)	20
9	Russian Federation (R)	50	22	Uzbekistan (R)	20
10	Armenia (R)	40	23	Belarus (R)	15
11	Bulgaria	40	24	Tadjikistan (R)	15
12	Croatia	40	25	Turkmenistan (R)	15
13	Macedonia	35			

Note: [1] (R) = former Soviet Socialist Republic
Source: Naudin (1994)

KEY FACTS

1. Central Europe is experiencing rapid political and economic changes. The political tendency is for the number of countries to multiply while the economic tendency is for private ownership to replace state ownership in many branches of the economy.

2. Compared with the richest countries of Western Europe, most countries in Central Europe are economically backward, with inefficient agricultural sectors, industry biased towards the output of producer goods, and most services at a low level.

3. If Central Europe is to become more closely associated with the EU transport links must be improved both within the countries of the region and between them and Western Europe.

4. Environmental pollution is a serious problem in many parts of Central Europe and the cost of reaching levels acceptable to EU standards is very great.

MATTERS OF OPINION

1. Now that Soviet domination of Central Europe has collapsed, eliminating the strategic satisfaction still felt when East Germany was removed from the Soviet bloc, is there any good reason why the EU should bother about developments in the region?

2. Should outside countries intervene in Yugoslavia to attempt to restore the country to its integrated form? If not, why not?

3. Should Western Europe welcome immigrants from Central Europe in the immediate future, an inevitable prospect if and when in due course Central European countries join the EU?

SUBJECTS FOR SPECULATION

1. How quickly can the transformation of Central Europe to a market economy take place? Are some countries more adaptable than others? Is the short experience of East Germany relevant, given that, unlike the rest of the former CMEA partners of the USSR, it has had virtually unlimited support from West Germany?

2. Is the 1994 EU proposal that six countries of Central Europe should be in the EU by 2005 realistic?

3. Does the Yugoslav conflict have lessons for other countries of the world? Can the UN continue to support large peace-keeping forces for long periods in such circumstances?

Chapter seven

The former USSR

In the longer term, after 1990, given current and prospective development efforts, Soviet natural resources could conceivably be more important in the world economy – or less.

R. G. Jenson et al. *(1983)*

7.1 The formation of the USSR

Great political upheavals in the former USSR resulted in its break-up in 1991 after more than seven decades of existence. Fifteen new sovereign states replaced the former Soviet Socialist Republics. Political changes and more gradual economic developments have, however, left many of the main features of the human geography unchanged. Economic activities are modified or cease in some places while new ones emerge elsewhere, but even if the dominance of the public sector is reduced by privatisation, the distribution of such features as cities, railways and factories will alter only gradually. It is therefore appropriate to start this account of the former USSR by describing the main geographical features of the country as it was around 1990, for which period data still referred to the whole of the USSR. Table 7.1 contains basic data for the Soviet Socialist Republics of the USSR and Table 7.2 for the regions of the Russian Federation.

In 1991 the USSR had a population of 292 million, 5.4 per cent of the total population of the world, making it the third largest country, after China and India. The total area of the USSR was 22,400,000 sq km (8,649,000 sq miles), 16.6 per cent of the total land area of the world. It was by far the largest country in the world in area, considerably more than twice the size of the next largest, China, Canada and the USA. The former USSR was characterised by its great latitudinal extent, reaching from 20° East of Greenwich in the west at Kaliningrad to 170° West of Greenwich at the Bering Strait – still a feature of the Russian Federation. It shared an international boundary with twelve other countries. The most southerly point was 35° North, but about half of the national area was situated north of about 55° North.

To the north, the USSR was bounded by the coast of the Arctic Ocean, frozen for varying periods of the year, and navigable only for a short period in the summer. The USSR claimed as its territorial waters all of the Arctic Ocean between its north coast and the North Pole. To the east, the USSR bordered the Pacific Ocean, with rugged mountain ranges behind the coast for most of its length, and again a sea frozen for up to several months in the year. On its southeast side the USSR shared a long boundary with China, interrupted by the landlocked state of Mongolia (formerly the Mongolian People's Republic). The major world region to the southwest of the USSR was Southwest Asia (or the Middle East), with which in part it shared a land boundary and from which it was elsewhere separated by the Caspian and Black Seas. Finally, in the west the boundary of the USSR ran through relatively densely populated lands, bordering Central Europe, the Baltic Sea and Finland.

The former USSR has a great diversity of physical environments (see Figure 7.1). In general, average temperature increases from north to south, but while that is true in summer, temperatures in winter are lower in the east of the country than in the west, the latter being kept less extreme by air masses from the Atlantic Ocean (see Figure 7.2). Snow remains on the ground for at least five months over the north and east of the Russian Republic and, over large areas, the

TABLE 7.1 Selected data for the fifteen Soviet Socialist Republics of the former USSR on the eve of its break-up

	(1)	(2)	(3)	(4)	(5)	(6)	(7)	(8)	(9)	(10)
	Popn		*Popn change*	*%*	*% Russian*	*Monthly pay 1990*	*Monthly pay 1990*	*Infant mortality*	*% in industry*	*Number in higher education 1989/90 (per*
	1970	*1990*	*1970–90*	*urban*	*1989*	*workers*	*collectives*	*1990*	*1989*	*10,000)*
USSR	241.7	288.6	119	66	50.8	275	241	22	57	147
Russia	130.1	148.0	114	74	81.5	297	265	17	59	158
Estonia	1.4	1.6	114	72	30.3	341	352	12	56	126
Latvia	2.4	2.7	113	71	34.0	291	301	14	58	142
Lithuania	3.1	3.7	119	69	9.4	283	303	10	58	139
Belarus (Belorussia)	9.0	10.3	114	67	13.2	265	247	12	56	143
Moldova (Moldavia)	3.6	4.4	122	48	13.0	233	252	19	53	120
Ukraine	47.1	51.8	110	68	22.1	248	220	13	55	150
Georgia	4.7	5.5	117	56	6.3	214	233	16	58	81
Armenia	2.5	3.3	132	68	1.6	241	273	19	61	142
Azerbaijan	5.1	7.1	139	54	5.6	195	239	23	49	86
Turkmenistan	2.2	3.6	164	45	9.5	244	246	45	40	97
Uzbekistan	11.8	20.3	172	40	8.3	215	203	35	43	136
Tajikistan	2.9	5.2	179	31	7.6	207	185	41	43	79
Kyrgyzstan	2.9	4.4	152	38	21.5	219	220	30	49	105
Kazakhstan	13.0	16.7	128	58	37.8	265	246	26	60	153

Description of variables:
(1), (2) Population in millions
(3) Change 1970–90 (1970 population = 100)
(4) Urban population as percentage of total population
(5) Ethnic Russians as percentage of total population
(6), (7) Average monthly pay in roubles per worker in industry and state farms (6) and in collective farms (7)
(8) Deaths of infants under the age of 1 year per 1,000 live births
(9) Employment in industry as a percentage of all types of employment
(10) Number of students in higher education per 10,000 total population
Sources: Goskomstat SSSR (1991) (1)–(3) p. 67; (4) pp. 68–73; (6) p. 38; (7) p. 39; (8) p. 92; (5) Harris (1993); (9), (10) Cole (1991)

moisture in the subsoil remains permanently frozen. In the south of the former USSR, very high summer temperatures are reached. With regard to precipitation (rain, snow, sleet), the southern part of the former USSR is extremely dry, while, because of the low temperature, precipitation is generally low in the north of the country (see Figure 7.3).

On the basis of relief most of the former USSR can be divided into two regions: to the west, rather more than half is low-lying while to the east there is a series of mountain ranges and a very extensive plateau. The western half is subdivided by the Ural Mountains into the East European Plain in European USSR, actually undulating terrain, with hill country and intervening broad valleys, and to the east of the Urals, the almost flat West Siberian Lowland. Along its southern margins are the highest mountains of the former USSR.

Russian and Soviet scientists have made major contributions to the study of soils and vegetation, especially as related to climate, a relationship more evident at continental than local level. Many events and processes in Russian/Soviet history, and the past and present distribution of bioclimatic resources and rural population, relate closely to the distribution of different types of soil and natural vegetation (see Figure 7.1), much now modified by economic activities. In the extreme north of the former USSR is an extensive zone of tundra (cold desert), which merges southwards into coniferous forest. Further south the forest thins out and disappears, to make way for steppe (grassland). Further south again are areas of semi-desert and desert. Wherever they occur, the higher mountain regions interrupt the broad west–east zones of vegetation, bringing colder conditions south.

In addition to its large and varied endowment of bioclimatic resources, the former USSR has abundant reserves of many economic minerals, but at present many deposits are at great distances from centres of population and industry, from railways and/or from coasts. The former USSR has just over 5 per cent of the population of the world, but its approximate share of natural resources is well in excess of 5 per cent: about

TABLE 7.2 Selected data for the twelve regions of the Russian Federation

	(1)	(2)	(3)	(4)	(5)	(6)	(7)	(8)	(9)	(10)	(11)
									Industrial sectors contributions (%)		
	Area in thous. sq km/(miles)	*Popn 1992 in millions*	*Density of popn per sq km/(mile)*	*% urban 1992*	*Natural change per 1,000*	*Retail sales 1991*	*Doctors per 10,000 popn 1991*	*Students per 10,000 1991/2*	*fuel energy*	*heavy industry*	*light industry*
Russian Federation	17,075 (6,593)	148.7	8.7 (22.5)	74	0.7	3,150	44	186	10	53	37
1 North	1,466 (566)	6.1	4.2 (10.9)	77	2.3	3,260	43	97	8	65	27
2 Northwest	196 (76)	8.3	42.3 (109.6)	87	−3.3	3,650	55	307	6	60	34
3 Centre	485 (187)	30.4	62.7 (162.4)	83	−3.4	3,690	52	252	4	46	50
4 Volgo-Vyatka	263 (102)	8.5	32.3 (83.7)	70	−0.3	2,710	38	155	5	60	35
5 Blackearth Centre	168 (65)	7.8	46.4 (120.2)	61	−3.3	2,520	36	158	2	51	47
6 Volga	536 (207)	16.6	31.0 (80.3)	74	1.2	2,710	43	180	7	60	33
7 North Caucasus	355 (137)	17.2	48.5 (125.6)	57	3.9	2,490	42	145	6	40	54
8 Ural	824 (318)	20.4	24.8 (64.2)	75	1.9	2,730	39	146	11	67	22
9 West Siberia	2,427 (937)	15.2	6.3 (16.3)	72	2.7	3,090	44	188	32	43	25
10 East Siberia	4,123 (1,592)	9.3	2.3 (6.0)	72	4.8	3,040	40	174	13	58	29
11 Far East	6,216 (2,400)	8.0	1.3 (3.4)	76	5.1	3,840	51	153	7	64	29
12 Kaliningrad	15 (6)	0.9	60.0 (155.4)	79	1.8	3,350	40	169	2	47	51

Notes on variables:
(6) Retail sales in roubles per capita
(9) Electricity and fuel
(10) Heavy industry (metals, chemicals, engineering, materials)
(11) Light, food

Source: Goskomstat Rossii (1992a): (1), (2), pp. 5–10; (3)–(5) pp. 82–6; (6) pp. 171–5; (7) pp. 289–92; (8) p. 257; Goskomstat Rossii (1992b): (9)–(11) pp. 35–8

10 per cent of the freshwater resources, 16 per cent of productive land, 17 per cent of non-fuel minerals and 23 per cent of fossil fuels (see Tables 3.4 and 3.5, pp. 64, 65).

The borders of the former USSR in 1990 were established as the result of several centuries of colonial expansion from the Russian heartland around Moscow in the fifteenth century (Table 7.3). An initial northward thrust extended Russian control to the Arctic coast. In the sixteenth and seventeenth centuries a massive push eastwards took the Russians to the Pacific coast around 1650. In the eighteenth century smaller, but more strategically significant, areas were gained to the west. Finally in the nineteenth century extensive areas were conquered in the Caucasus, Soviet Central Asia and the Soviet Far East (see Figure 7.4 and Table 7.3). Box 7.1 contains a small selection of views expressed about Russia over the centuries.

Around 1900 the Russian Empire extended over an area roughly equal to that of the Soviet Union in 1990. During the process of empire-building, vast natural resources were acquired, many little used until the Soviet period. Altogether about 100 non-Russian ethnic, in effect linguistic, groups were absorbed. These included the very thinly scattered population of Siberia, with simple economies, and denser clusters of population with more sophisticated economies, some with irrigation-based agriculture, in the south. In European USSR itself, Ukrainians and Belorussians, culturally very similar to the Russians, at least in Russian eyes, came under Russian control. When the Bolshevik Party, subsequently the Communist Party of the Soviet Union, gained power in 1917, it inherited this vast empire, the antiquated practices and injustices of which provided a reason or a pretext for revolutionising the political, economic and social set-up of tsarist Russia. The theoretical and ideological framework for the transformation was provided by an appropriately modified and updated set of assumptions based above all on the works of the German historian Karl Marx (life 1818–83).

7.2 The Soviet economy

In the Soviet/Leninist version of marxism, the economy would initially be run centrally, under the guidance of the Communist Party. Private ownership of most means of production and of natural resources was abolished in the 1920s in industry, transport and other non-agricultural sectors, and between 1928 and 1932 in the large agricultural sector. Wealth could not

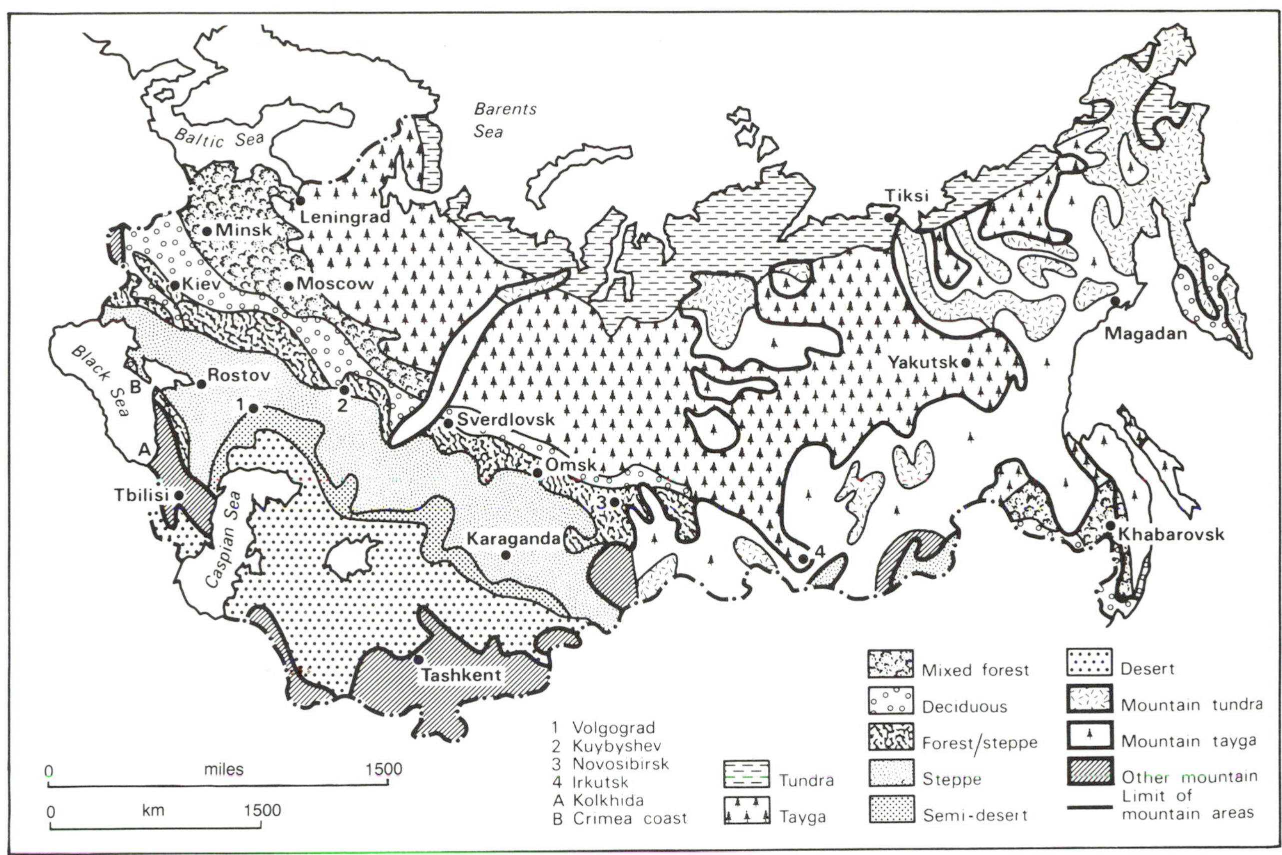

FIGURE 7.1 Natural zones of the USSR
Source: based on *Geograficheskiy atlas SSSR dlya 7–8 klassov*, Moscow 1966, pp. 6–7

therefore remain in or become concentrated in a few hands, and the working population would be happy to work for the state, the great community, of which they were the key members. Industrial development was given priority in the 1930s at the expense of agriculture and services. Emphasis was put on heavy industry, especially the expansion of the existing coal and steel sectors and the development from almost nothing of engineering. Strategic considerations made it desirable to develop resources and place new industrial investment in areas to the east of the two main existing industrial regions of the USSR in the 1920s, the Centre (around Moscow) and the Donbass (eastern Ukraine). The Ural region, southern Siberia and northern Kazakhstan in particular were opened up between about 1930 and the 1950s. The main mode of interregional movement of goods and passengers was the rail system, the route length of which (in contrast to those of Western Europe and North America) has continued to grow up to the 1990s.

Under a series of Five-Year Plans, virtually all sectors of economic life were organised centrally by Gosplan (State Planning) in Moscow, with investment allocated through ministries to different sectors of production, and through planning regions to different parts of the country. A strong political-administrative structure was necessary to determine the needs and priorities of different parts of the country. In the 1920s, such a structure was developed, creating at the top level of the hierarchy a number of Soviet Socialist Republics, each formed as closely as geographical irregularities allowed around given clusters of people with a common ethnic/linguistic identity. After various modifications in the 1920s and 1930s, changes resulting from the annexation of the Baltic republics in 1939, and boundary changes after the Second World War, there emerged the fifteen Soviet Socialist Republics (SSRs) that lasted until 1991, with the Russian Soviet Federal Socialist Republic by far the largest in area and population.

In theory at least, the above changes were intended to mark the end of the Russian Empire and the granting of equality to everyone in its former limits. The SSR level was only granted to fifteen national groups, leaving the remainder with lower political-administrative status. Some twenty were defined as Autonomous Soviet Socialist Republics (ASSR), most of them in the Russian Republic, and there were still

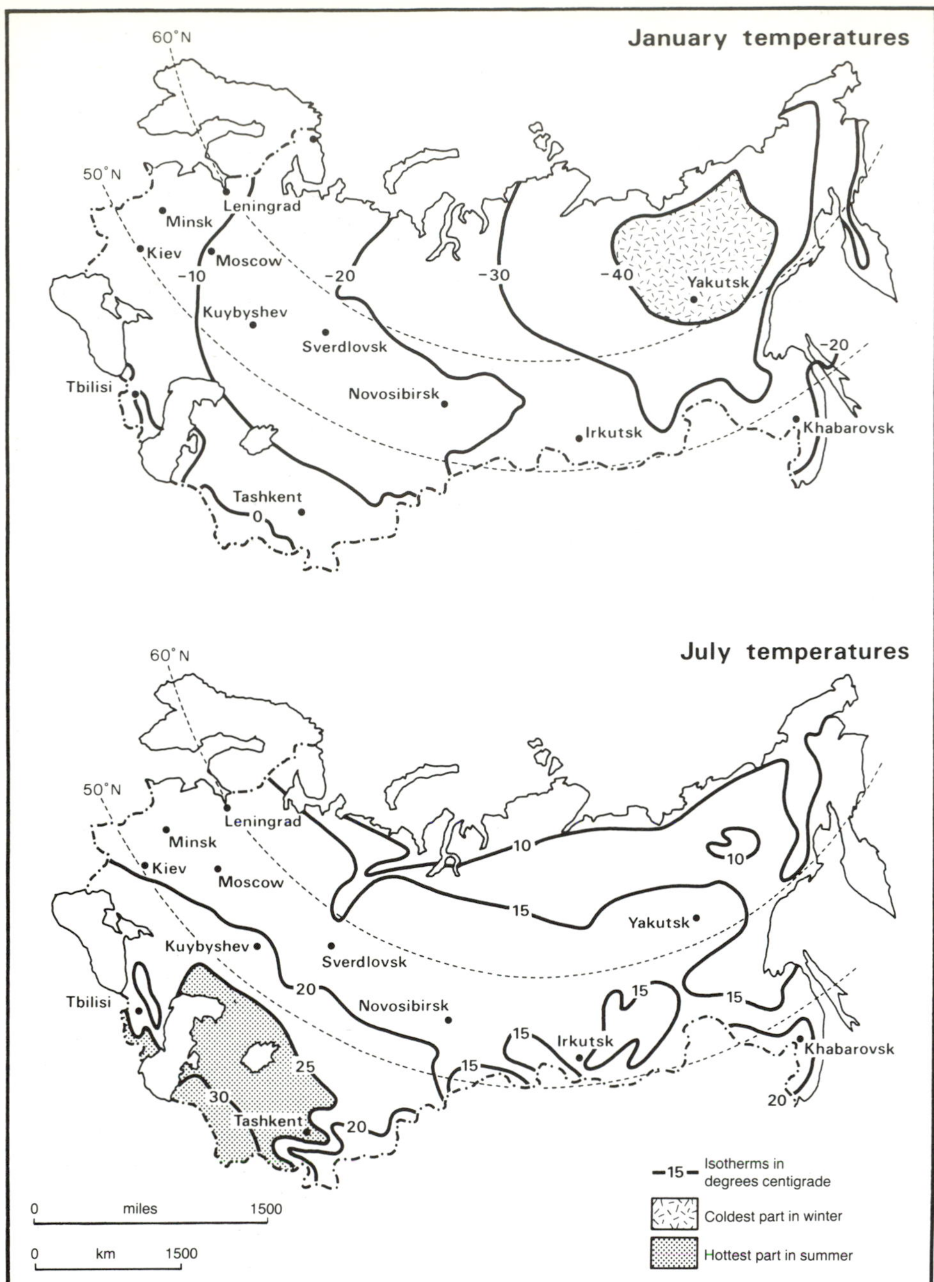

FIGURE 7.2 Mean January and July temperatures in the former USSR. For simplicity, isotherms are shown only for every 10° Centigrade (18° Fahrenheit) for January and every 5° for July

others at appropriate lower levels. When the USSR broke up in 1991 the fifteen new sovereign states kept virtually intact the boundaries of the SSRs. In a similar fashion, when the Spanish Empire in Latin America broke up early in the nineteenth century the ten new countries kept closely to the boundaries of the major subdivisions of the eighteenth century. More recently, almost all of more than fifty former African colonies of France, the UK, Portugal, Belgium and Italy, mostly given independence around 1960, again kept exactly or closely their colonial boundaries. The USSR experienced seven decades of communist rule, with the establishment of republic Communist Parties working with the Russian Communist Party, and constant propaganda extolling the historical role of the Soviet state in transforming society. Nevertheless, as soon as the opportunity came, each of the fourteen non-Russian republics quickly developed or reactivated a strong sense of nationalism, with many former Communist Party members in prominent positions being quick to adapt to the new situation. Figure 7.5 shows what 'Russia' would look like without not only the fourteen non-Russian republics (the SSRs) but also the twenty or so Autonomous Soviet Socialist Republics (ASSRs) and other related, predominantly non-Russian units at present, like for example the Chechen Republic, considered by Russians to be part of the Russian Federation.

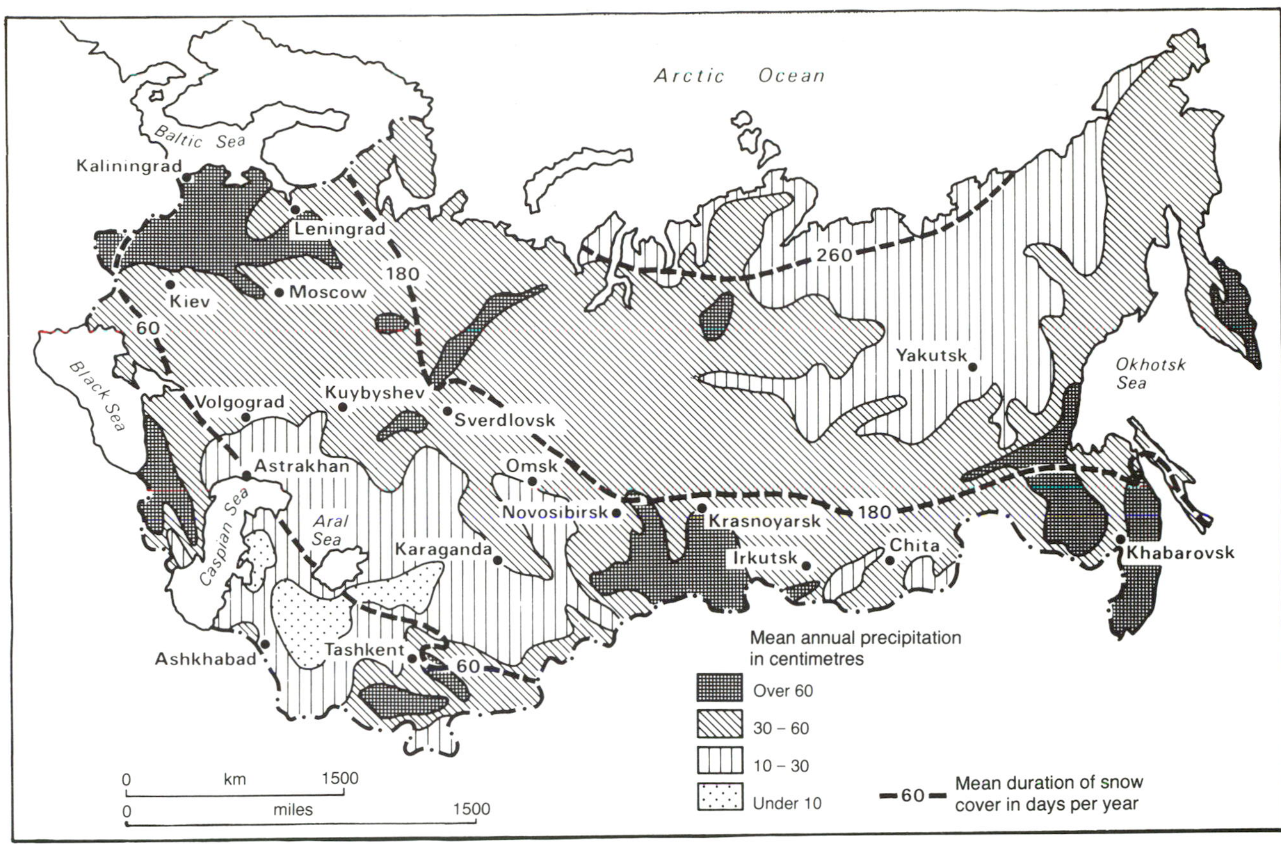

FIGURE 7.3 Mean annual precipitation and duration of snow cover in the former USSR

The policy of Soviet politicians and planners was to maintain a high degree of self-sufficiency within the country, one justifiable reason in the 1920s and 1930s being the widespread hostility towards the socialist experiment in Russia, especially in the industrially more advanced 'capitalist' countries, particularly Germany. In the 1980s, Soviet imports were equal to only about 6 per cent of the total value of production and about half of that 6 per cent came from the six 'friendly' CMEA partners of Central Europe.

Within the USSR itself, a considerable degree of self-sufficiency in basic, often bulky, products, such as building materials and vegetables, was encouraged at a local level. Regional specialisation, the 'international' division of labour, was, however, also encouraged and, in the case of some products, unavoidable. Thus, for example, more than half of Soviet oil output came from the single region of West Siberia in the 1980s, almost half of the pig-iron was smelted in the eastern Ukraine, and most of the textiles were manufactured in the Centre region. Similarly, some primary products were obtained from very small areas, for example tea mainly from Georgia, diamonds from the Yakut ASSR in Siberia, and cotton from the oases of the Central Asian republics. Thus the various republics and regions of the USSR were economically highly interdependent, much of the flow of goods being handled by the rail network or on inland waterways and coastal shipping links. Each year an average of one tonne of goods per inhabitant was carried some 10,000 km (6,200 miles).

One of the outstanding problems of the break-up of the USSR, therefore, is the fact that each of the fifteen former Soviet Socialist Republics is dependent economically in some way on every other one, but the main flows of goods are between the Russian Federation and each of the other fourteen. The situation is complicated by the fact that some key products, including oil and natural gas, were delivered from Russia to the other republics at prices well below those on the world market.

Just as various primary and manufactured products were traded between the republics, so also large numbers of people migrated within the USSR. Before 1917, and especially after the building of the Trans-Siberian Railway in the 1890s, Russians migrated far across the country from European Russia to the Urals and beyond, to settle in thinly populated areas of Siberia, some of them to penal settlements. Relatively few migrated to the more densely populated non-Russian parts of the empire in the west (Ukraine) or south (Transcaucasia and Central Asia). In contrast, after the Revolution, large numbers of Russians were

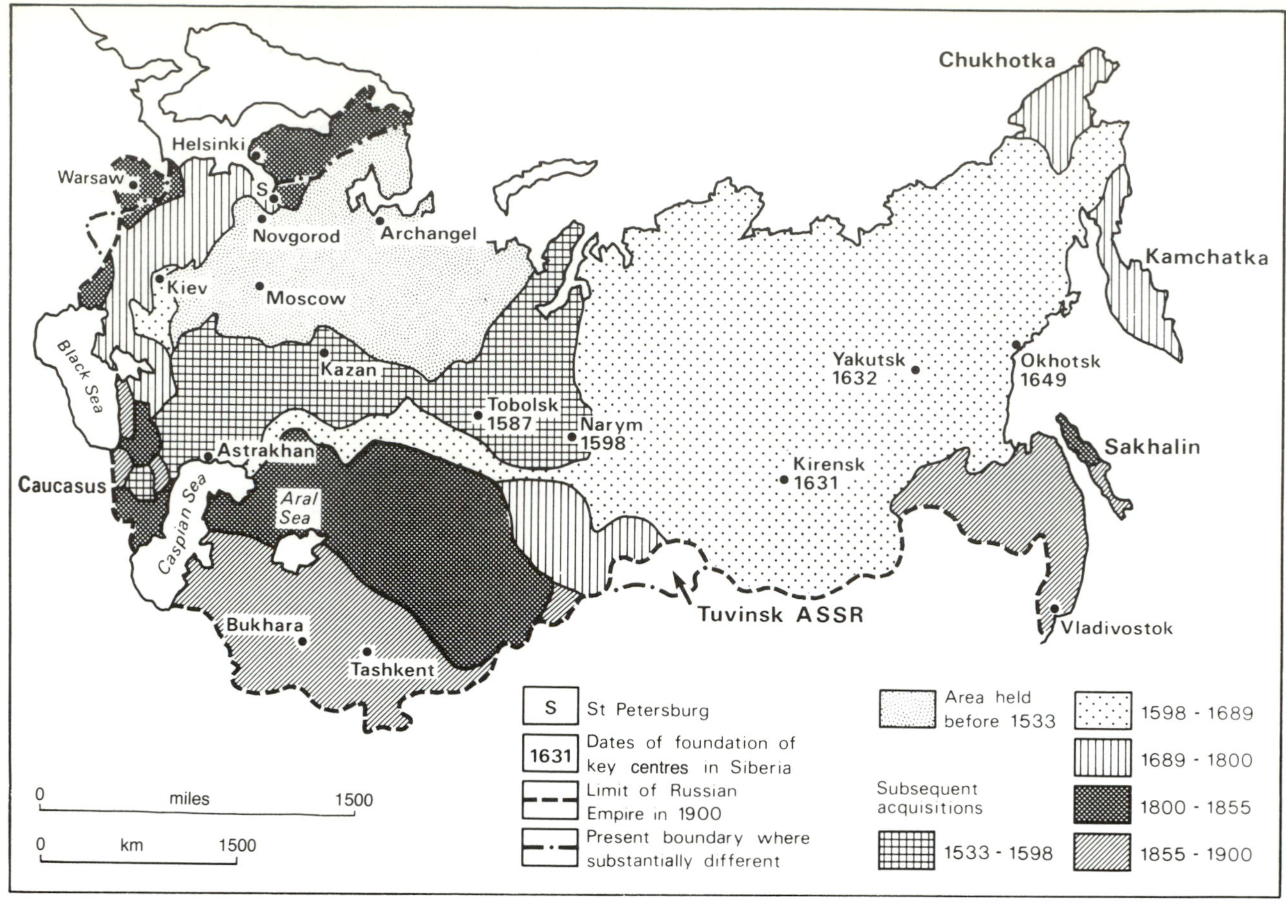

FIGURE 7.4 Territorial growth of the Russian Empire since the fifteenth century. Note that Alaska is not shown as it was held for only a few decades and was sold to the USA in 1867

settled in some of the non-Russian republics, often in administrative, managerial or manufacturing jobs. According to the 1989 census, some 25 million Russians, about one in six, lived outside the Russian Republic (RSFSR), while roughly the same number of non-Russians lived within the RSFSR, many of them in the ASSRs referred to above. Box 7.2 (p. 165) lists some of the present issues and problems that have arisen from the momentous changes in Gorbachev's USSR and Yeltsin's post-Soviet Russia and Box 7.3 (p. 170) shows a small selection of cartoons illustrating problems of a geographical nature identified in the Soviet satirical magazine *Krokodil* (Crocodile).

7.3 The reorganisation of the Russian Federation

In spite of 'losing' its fourteen fellow Soviet Socialist Republics, the Russian Federation (hereafter Russia) is still the largest country in the world in area (17,075,000 sq km – 6,600,000 sq miles) but it is now sixth in the world in population (after China, India, the USA, Indonesia and Brazil). In 1993 it had a mere 2.7 per cent of the total population of the world and in 2025, after increasing from its present 149 million to about 152 million, it will then have about 1.8 per cent of the world total. Its importance in the world economy and in world affairs lies more in its natural resource endowment than in its population size or present economic strength.

Russia has well over half of the water and bioclimatic resources, the fossil fuel reserves and non-fuel minerals of the former USSR and, as it contains most of the area as yet only partially explored for minerals, it could be even better placed in the future. A major problem for Russia, however, is the fact that about 70 per cent of its population is concentrated in the western quarter of the territory, whereas most of the natural resources are located to the east of the Ural Mountains. Figure 7.6b shows the distribution in Russia of some of the basic geographical features.

The division of the population of Russia into urban and rural is based on a definition related to the function of settlements, but size is also taken into account. In 1992, 74 per cent of the population was defined as

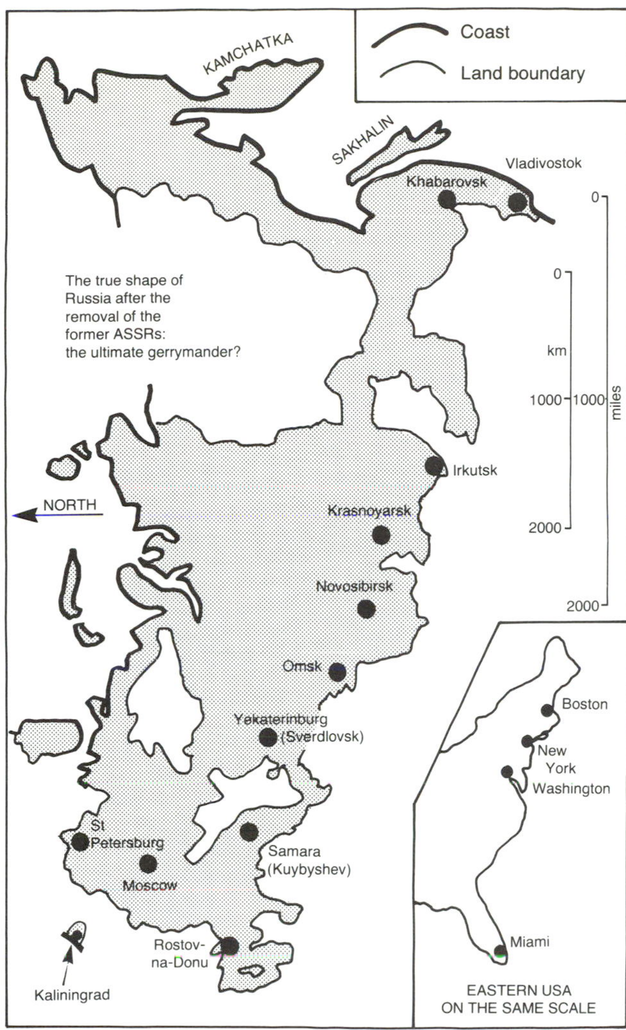

FIGURE 7.5 The ultimate gerrymander. The heartland of the Russian Federation without the former Autonomous Soviet Socialist Republics, most now with 'republic' status in the new Russia. Even here there are many small non-Russian national groups

TABLE 7.3 Russian and Soviet calendar

1547–84	Ivan IV ('The Terrible') Tsar (life 1530–84) initiates Russian eastward expansion
1649	Russian conquest of Siberia reaches Pacific coast at Okhotsk
1689–1725	Peter ('The Great') Tsar (life 1672–1725) Russia expands westwards and western technology is adopted
1703–18	St Petersburg built (capital of Russian Empire 1713)
1812	Napoleon invades Russia
1861	Emancipation of Russian serfs
1900	Trans-Siberian Railway completed
1904–5	Russia suffers defeat in war with Japan
1905	Revolution in Russia, followed by Tsarist concessions
1914–17	First World War
1917	Revolution in Russia, Tsar abdicates (March), Bolsheviks take power (November) under Lenin. Socialist state established, with guidelines based heavily on ideas of Karl Marx (life 1818–83)
1918–22	Civil war and foreign intervention
1924	Death of Lenin (life 1870–1924). Stalin takes power soon after
1928	First Soviet Five-Year Plan, industrialisation and collectivisation
1939	German–Soviet non-aggression pact
1941	Germany under Hitler invades USSR
1941–5	Germany devastates much of western USSR
1945	Germany defeated. Second World War ends
1947–8	Soviet influence consolidated in Eastern Europe. Cold War intensifies
1949	USSR explodes atom bomb
1953	Death of Stalin, Khrushchev later emerges as leader
1955	Warsaw Pact signed
1956	Unsuccessful revolts in Poland and Hungary
1957 and 1959	Soviet Union launches first space satellite and puts man in space
1964	Khrushchev removed from power. Brezhnev succeeds him
Late 1970s	Soviet economic expansion falters
1979	Soviet forces invade Afghanistan
1985	Gorbachev becomes General Secretary of the Communist Party
1986	(14 January) Gorbachev expresses wish for a future of peace to the citizens of the USA
1986	Nuclear catastrophe at Chernobyl nuclear power station
1991	Gorbachev removed from power, succeeded by Yeltsin. The fourteen non-Russian former Soviet Socialist Republics of the USSR given independence from the Russian Federation

urban. All centres with over about 900,000 inhabitants are shown in Figure 7.6a. The urban population of Russia has grown in this century both absolutely, from about 15 million around 1900 to 110 million in 1992, and also relatively. Urban growth accompanied rapid industrialisation in the Soviet Union, especially between 1928 and the 1970s, and many new cities were founded during that time. The overall size of the rural population has changed little and has even declined in some areas. Its distribution reflects broadly the distribution of arable land, but the density is higher in the older areas of cultivation in European Russia than in newer areas of settlement east of the Volga, especially in West Siberia.

Future population change in Russia and in its various regions will depend on natural increase or decrease and on migration. According to the 1989 census, the rate of natural increase has been small in most parts of Russia where Russians form the main

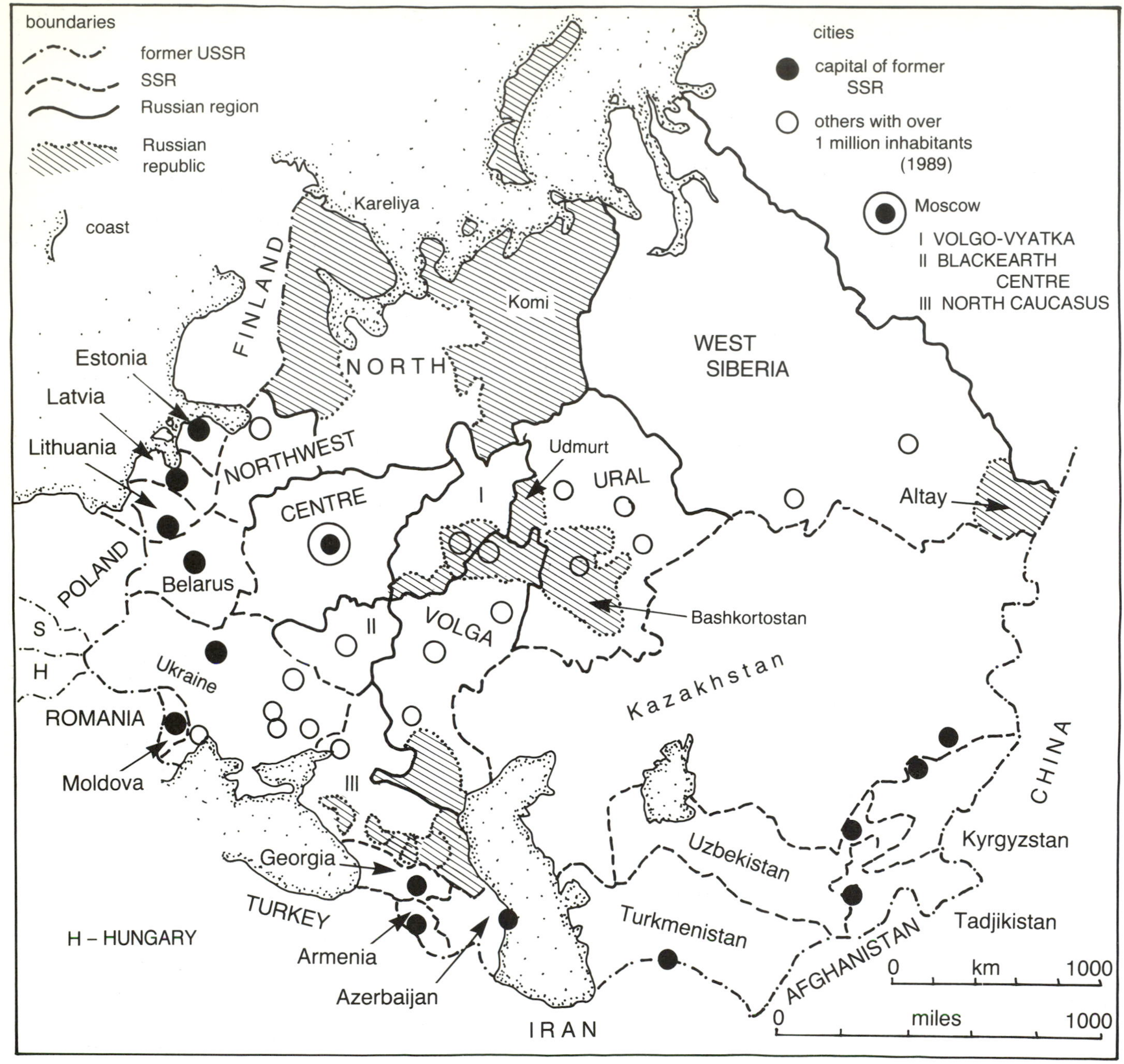

FIGURE 7.6

a Location map of former republics, Russian regions and large cities in the western half of the former USSR

component of the population. However, Specter (1994) reported a dramatic reduction in male life expectancy in the last five years, and, even more striking, a decline in total fertility rate from 2.17 to 1.4, matching a change that took three decades in Italy. In many predominantly non-Russian republics of the Russian Federation the rate of natural increase of population is considerable, especially in the republics in the region of North Caucasus. At the time of writing it is difficult to judge the effects of economic restructuring on migration, but in an economy that is more market-orientated, and with greater freedom for people to move between regions, a gradual shift of population from poorer to more prosperous parts of Russia could be expected. Moscow and St Petersburg can be expected to attract many migrants if migration patterns elsewhere in the world are followed.

From the point of view of Russian policy makers

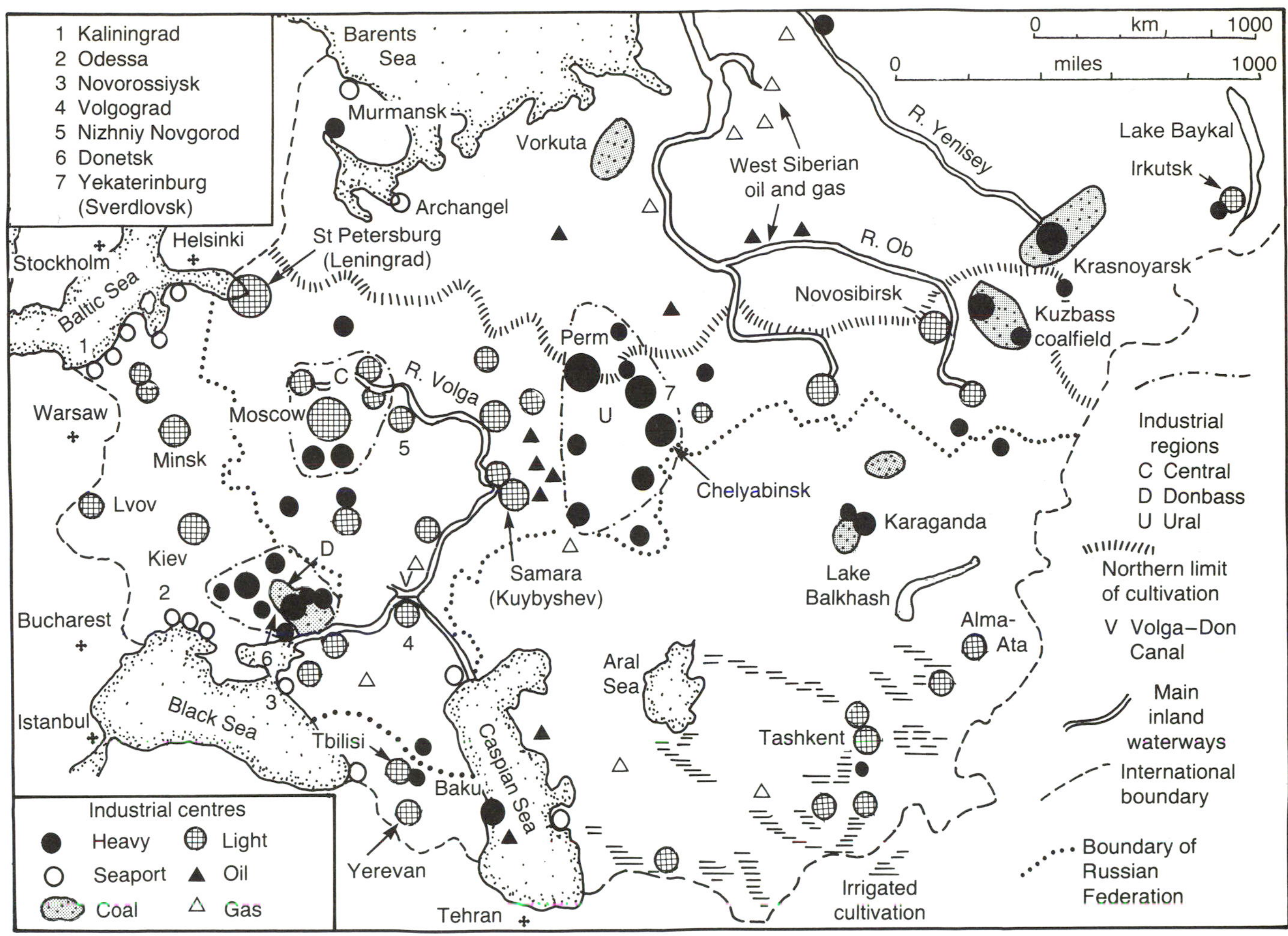

b Main economic features of western two-thirds of the former USSR

and planners, one demographic 'threat' to Russia has been reduced. The prospect that large numbers of people from Transcaucasia and Central Asia might migrate to places in the Russian Federation, a distinct likelihood in the former USSR, has receded, always assuming that migration control is enforced at the new international boundary of Russia. On the other hand, it would be difficult for any future government of Russia (assuming Russia remains intact) to ban the return of up to 25 million now 'expatriate' Russians from the other republics. The return since 1990 of about a million military service personnel from Central Europe, especially East Germany, has caused enormous problems. A decade ago labour shortages were predicted for the USSR, yet now the prospect of expatriates who are returning to Russia getting a house or a job quickly seems bleak. In the longer term, and under whatever mix of state and private enterprises and investment comes about, Russia will have to continue using and developing its natural resource base. A gradual shift of economic activity and population eastwards therefore seems likely.

7.4 The economy of the Russian Federation

Soviet agriculture has always proved difficult to plan. One reason is that over most of the main belt of arable land both temperature and precipitation are characterised by variations from year to year, with greatly reduced yields in dry years, and sowing and harvesting affected by late spring and early autumn frosts. Organisational problems, partly self-inflicted by the system of state and collective farms, have resulted from bad management and a lack of incentives to individual farmers. The quality and general shortage of facilities for the harvesting, storage and marketing of agricultural products have been blamed in many Soviet sources, often with amusing frankness in the Soviet journal *Krokodil* (see Box 7.3).

Since the late 1980s, attempts have been made to reorganise some farms, giving greater responsibility to individual farmers. The enthusiasm with which farmers cultivated their small private plots compared with the collective or communal land led some to hope that yields could be substantially increased. A decline in

BOX 7.1

What they said about Russia and the USSR

Circa 1290, Marco Polo

Russia is a very large province lying towards the north. The people are Christians and observe the Greek rite. They have several kings and speak a language of their own. They are very simple folk, but they are very good-looking, both men and women; for they are fair-skinned and blond. Their country is strongly defended by defiles and passes. They are tributary to none, except that from time to time they pay tribute to a Tartar king of the West, whose name is Toktai, but not a heavy one. It is not a country of much commercial wealth. It is true, however, that it produces furs – sable, ermine, vair, ercolin, and foxes in abundance the best and most beautiful in the world. There are also many silver mines, yielding no small amount of silver. But there is nothing else worth mentioning; so we shall leave Russia and speak of the Black Sea.

1810, J. Bigland (1810)

If we extend our views into futurity, and imagine a period when Russia shall attain to that complete population which she is endeavouring by a multiplicity of means to acquire, and to which, according to the most authentic documents, and the evidence of visible circumstances, she is continually approximating in an ascending ratio, this immense empire presents a distant and dazzling prospect, which opens a wide field both for political and moral speculation. Such a period, whenever it shall arrive, whether we suppose the continued union or the division of the empire, cannot but be productive of extra-ordinary revolutions both in Europe and Asia. Russia in a united state, with a complete population, must sway the destinies of these two quarters of the globe. What the consequences may be in regard to India, is a question too remote for present examination.

1836 (19 February) Lord Dudley, Stuart History in Hansard (1952)

What is the character of the population over which Russia rules? It is a population completely devoted to the Sovereign who sways the sceptre, whom they view and reverence as the chief of their race and the head of their Church, and to whom they are bound by the triple tie of race, language and fate. No property is held in Russia that is not subject to the disposition of the autocrat. So supreme is his power that one stroke of his pen can banish to distant countries any of his subjects, no matter what the rank, birth or property of that subject might be. There is no career open to any man but one connected with the state. No matter what his riches, if he is not in the service of the state he is as nothing. The very clergy are known to wear military orders. That organization disposed them to look for acquisitions and aggrandizement. But one enthusiasm pervades the entire population – that of advancing the pre-eminence of their country and its superior power over the rest of the world.

1914 W. C. Showalter (1914)

But as full of possibilities as the wheat-growing industry of the United States may be, they are few in comparison with those of Russia. That wonderful country, possessing more latent agricultural resources, perhaps, than any like area in the world, has 288,000,000 acres [117,000,000 hectares] of excellent wheat land. Even at our present standard of production, which is less than half of that of western Europe, Russia alone could produce more wheat than is raised on the entire globe today.

As matters now stand, the Russian crop is only about ten bushels per acre. That her lands are as fertile and her climate as well suited to the growing of wheat as those of England and Germany, are facts well known to all those who have considered her relation to the world's future food problem. Even today, in spite of her small per-acre production of every principal crop, Russia is the greatest exporter of grain in the world.

1937 S. Webb and B. Webb (1937)

In 1933 when settling the title of the book-to-be, we chose 'Soviet Communism' to express our purpose of describing the actual organisation of the USSR. Before publication, in 1935, we added the query, 'A New Civilisation?' What we have learnt of the developments during 1936–37 has persuaded us to withdraw the interrogation marks. We see no sign in the USSR of any weakening on the stern prohibition of private profit-making; meaning by this either the buying of commodities with the object of selling them at a higher price (termed speculation), or the hiring of workers for the purpose of making pecuniary gain out of their product (termed exploitation). Moreover, fifteen years' experience of three successive Five-year Plans has demonstrated the practicability of what the Western world declared to be beyond human capacity, namely the advance planning of the wealth-production and the cultural activities of an immense population; together with the deliberate organisation of the whole for the supply and service of the community without the guidance of 'price in the market' arrived at by the chaffering of buyers and sellers.

PLATE 7.1 Clash of cultures in the former USSR

a St Basil's (Orthodox) Cathedral, Moscow, Russia, not used during the Soviet years but a conspicuous landmark and tourist attraction, on the edge of Red Square. It was built in the sixteenth century for Tsar Ivan 'The Terrible'

b Temple in Bukhara, Uzbekistan, formerly Soviet Central Asia. Even before the break-up of the USSR, some of Central Asia's mosques were being restored. Now their restoration will be connected also with freedom to use them for religious purposes

Soviet exports of fuel, raw materials and manufactured goods in the early 1990s has made it more important than ever to increase agricultural production. One effect, hardly a propitious one, has been the reversion of food production to a near-subsistence situation in many areas, with attempts to ensure that basic needs can be met locally.

Only about 10 per cent of the total area of the Russian Federation is arable, (i.e. sown area plus fallow). To extend that area requires considerable investment in improving the land through drainage in some areas, improved water supply and infrastructure in others. The better prospect for Russian agriculture is to increase yields in existing areas of cultivation. Partly for environmental reasons, yields are much lower than in Western Europe for the same crops and rather below those in regions of Canada and Australia with comparable environmental conditions.

Russia has inherited a massive industrial capacity, but with an imbalance in emphasis due to the policy of supplying the military machine and of achieving self-sufficiency in all basic and strategic production. In the 1930s and again during 1950 to 1980, following a period of post-war restoration of destroyed capacity, there was pressure from planners to expand industrial capacity and output rapidly, one reason being the attempt to catch up with the USA. Thanks to the wide range of natural resources in Russia, most of the inputs of raw materials for industry are available somewhere in the national territory. There were, however, times when, for example, tin, copper and rubber were imported, as well as nuclear fuels, the latter provided mainly by East Germany and Czechoslovakia. With the break-up of the USSR, Russia has lost direct control of

PLATE 7.2 Former USSR – the cold and the empty east

a Icebreaker *Arktika* opens the way for convoys on the Northern Sea Route through the Arctic Ocean

b Baykal – Amur Magistral Railroad opens a new link across southern Siberia. Like many projects devised by politicians and planners during the Soviet period, this line took longer to complete than originally estimated, cost more, and is of doubtful economic value since it only taps new resources that exist elsewhere in Russia in abundance. It is also regarded increasingly as a threat to the natural environment

some natural resources and products from other republics, including for example the Central Asian cotton-growing lands, and sources of iron ore, coking coal and grain in Kazakhstan. It still has abundant resources of water, a great hydroelectric potential, several major coalfields, considerable proved oil reserves and further potential, about a third of the world's natural gas, and almost all of the forest area of the former USSR.

Even after the loss of Ukraine and Kazakhstan, Russia still has most of the steel-producing capacity of the former USSR, concentrated mainly in the Ural region and West Siberia, but with several plants elsewhere. The engineering industry is widely dispersed, but St Petersburg and the Moscow region are prominent in various branches of light engineering, and the Urals and Kuzbass specialise in heavy engineering. The chemicals industry has developed especially since about 1960, with many oil refineries either in the older Volga-Ural oilfields or in markets for refined products and petrochemicals. The manufacture of various types of textiles is concentrated in the Centre region. Clothing and footwear industries are more widely dispersed as, also, is the production of such bulky items as building materials and furniture.

At least a decade passed after the Revolution of 1917 before Soviet/communist planners began to reshape the industrial map of the USSR. How quickly changes can again be made depends on how ready the government is to shift emphasis from heavy industrial production in general and military production in particular to consumer industries, and also from industry to services. In the late 1980s the USSR lacked a whole range of services necessary for the running of a market economy, such as financial and legal services, sophisticated banks, stock exchanges, and facilities for ensuring the diffusion of information about privatisation processes. To shift about 10 per cent of capitalisation in industry and services from the public to the private sector in the UK took over a decade, even with the mechanisms in place to organise such a transfer. As shown in Table 6.3, by mid-1994 half of Russian GDP was accounted for by the private sector.

PLATE 7.3
Former USSR and Japan

a People waiting for taxis, Leningrad (now St Petersburg), Russia. Queuing was an essential part of life, in a system that managed to produce, some say even deliberately encouraged, shortages of goods and services

b Taxis waiting for people by the Shinkansen train station in the centre of Sendai, north Honshu, Japan. Perhaps the sheer shortage of flat and not too steeply sloping land in most of Japan has induced a sense of orderliness and meticulous organisation, especially in the transport sector

Some insights can now be gained into privatisation in Russia. Data for 1993 show, for example, that almost half of all accredited accountants were based in Moscow, as well as over a third of all dealers in shares. A few other centres, including St Petersburg (formerly Leningrad), Nizhniy Novgorod (Gorky), Yekaterinburg (Sverdlovsk) in the Ural region (Box 7.4, p. 172), and Omsk in West Siberia, are emerging as regional 'financial' centres. In the absence, so far, of a national financial system and network, and given the huge size of Russia, it seems likely that many private investors will prefer to or even be forced to invest in new enterprises in their own regions, a situation common in the earliest decades of the Industrial Revolution in the much smaller industrialising countries of Western Europe. Because some regions have much larger amounts of industrial capacity up for privatisation than others, the prospects for investors seem more favourable in some regions than in others.

Whatever historians say about Russia and the USSR with the wisdom of hindsight, it is a fact that modern industrial development took place there during a time-scale very similar to that in Japan and Italy, roughly 120 years. The USSR had the advantage over Japan and Italy of vast natural resources, but the disadvantage of having to move goods about in a huge territory, mainly by rail. It is a matter of opinion whether or not Soviet citizens were fairly or adequately remunerated

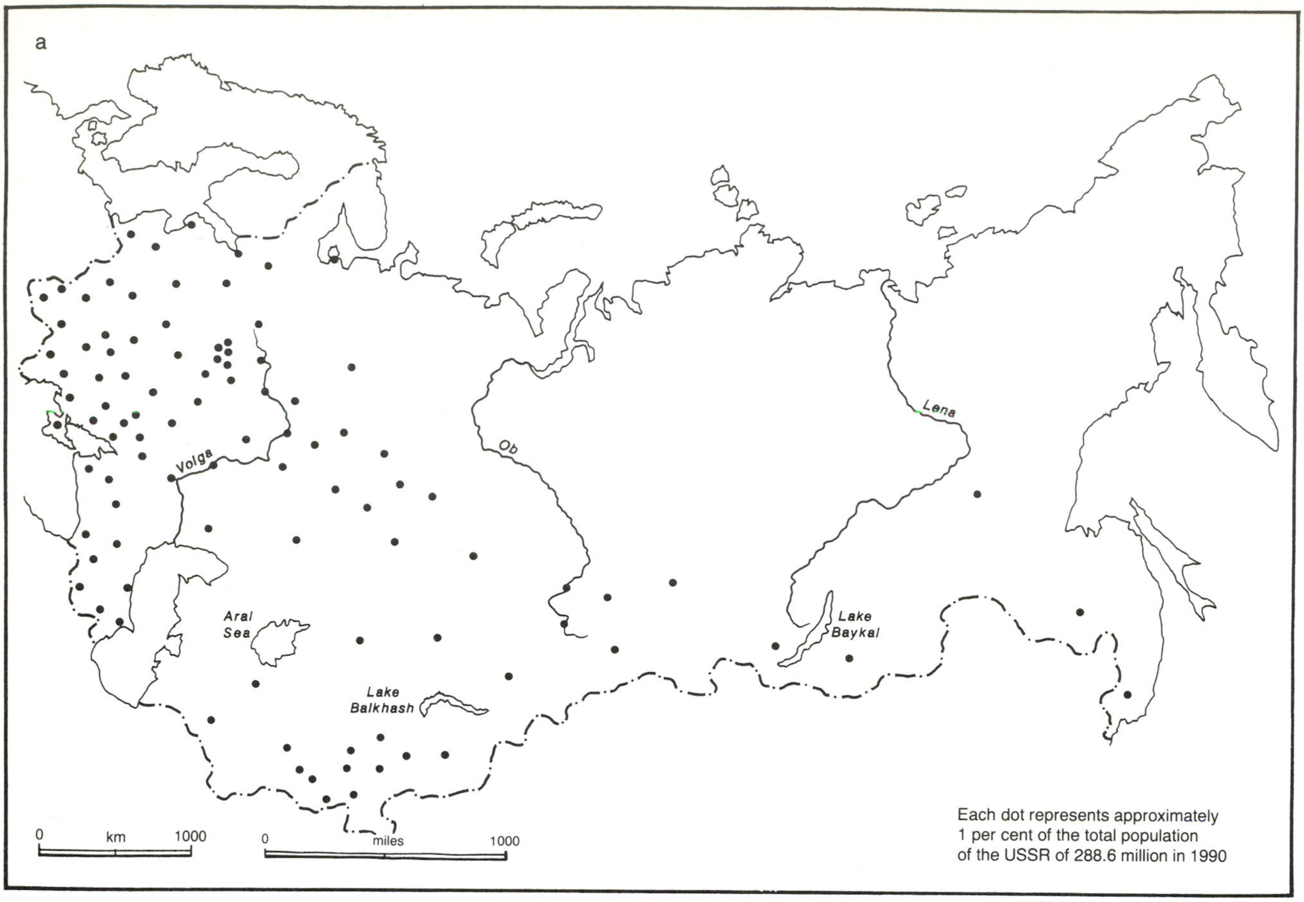

a
Lena
Ob
Volga
Aral Sea
Lake Baykal
Lake Balkhash
0 km 1000
0 miles 1000
Each dot represents approximately 1 per cent of the total population of the USSR of 288.6 million in 1990

b
St Petersburg
Moscow
Samara
Sverdlovsk
Omsk
Novosibirsk
0 km 1000
0 miles 1000
253,000 or approximately 1% of Russians outside RSFSR
250,000 Russians in ASSRs in RSFSR

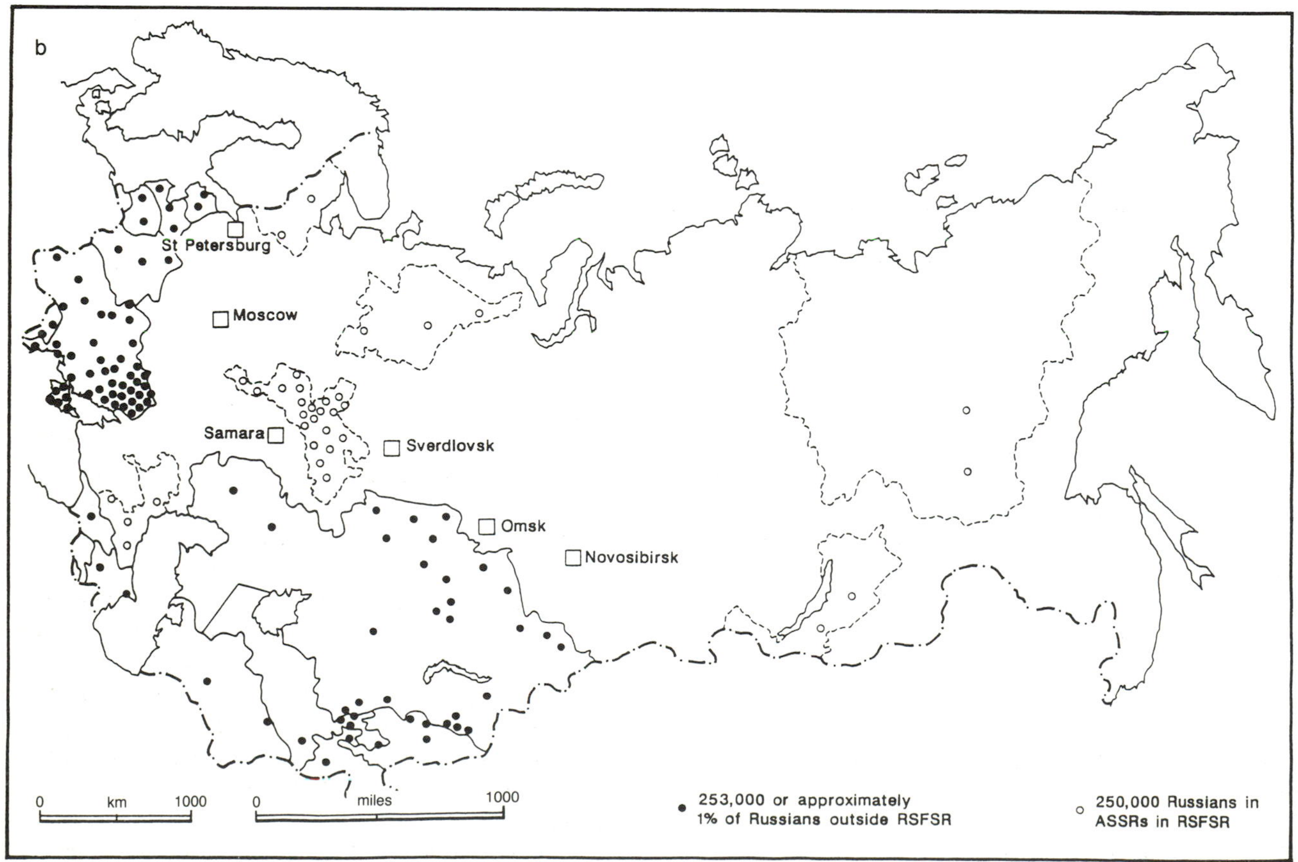

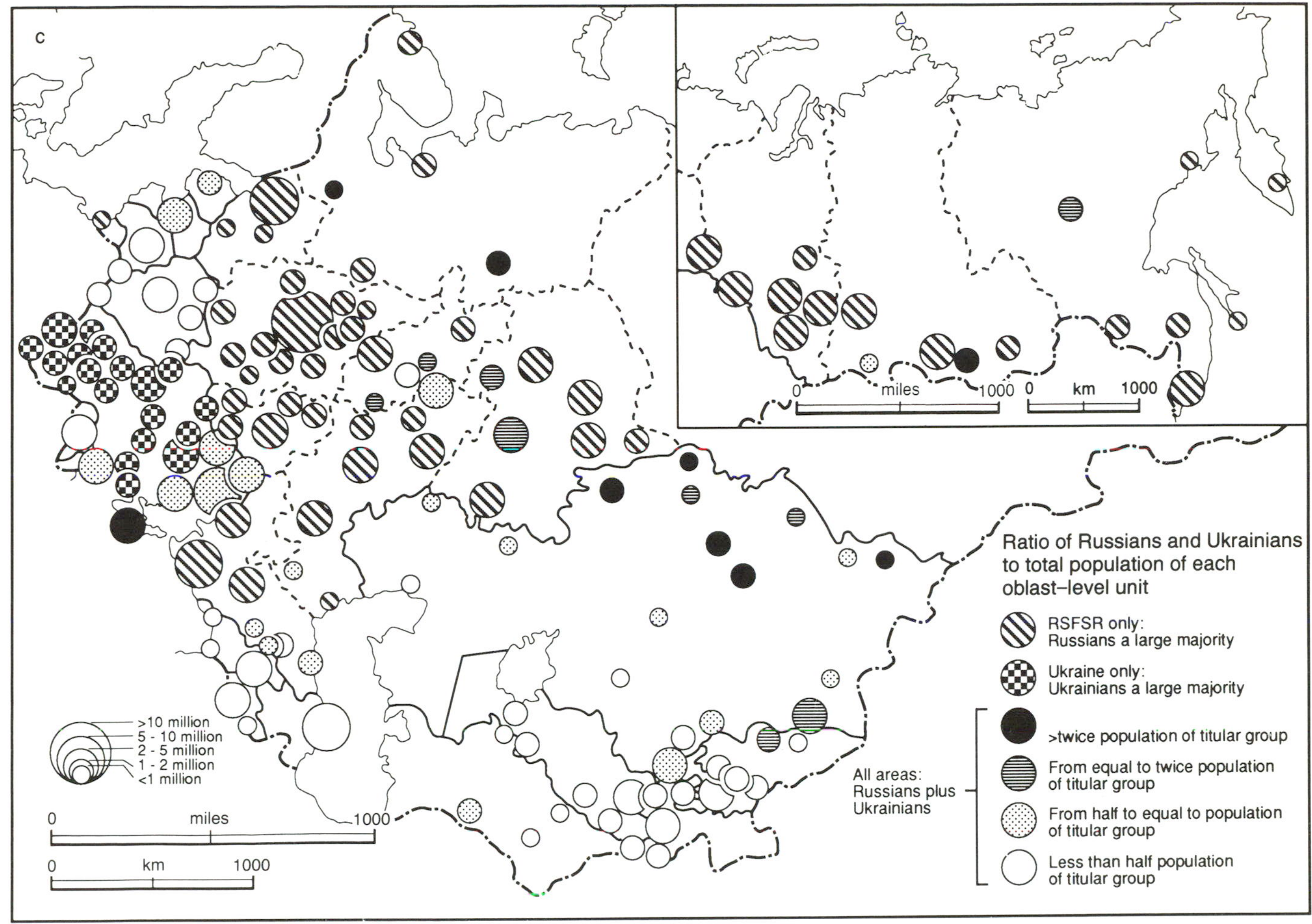

FIGURE 7.7 The population of the USSR on the eve of its break-up

a Distribution of total population, 1990

b Distribution of Russians living outside the Russian Republic, 1989

c Ratio of Russians and Ukrainians to total population (Russia, Ukraine) or titular nationality in each oblast-level unit, 1989

Source: Cole and Filatochev (1992)

for their efforts. In spite of never-ending professions of equality between regions and classes, marked regional and class disparities are a feature inherited by Russia in the 1990s from six decades of central planning (Figures 7.7–7.9). Table 7.2 includes a selection of indicators showing differences at the level of the eleven economic planning regions of Russia, while Figures 7.7 and 7.8 show variations at the more detailed scale of 75 oblast-level units.

7.5 Transport, the environment and foreign relations

Without a drastic restructuring of the existing transportation network, Russia will not be able to implement other changes needed to create a new market-based economy. Between 1985 and 1990, each year about 8 trillion (8 x 10^{12}) tonne-kilometres of goods were handled by all means of transport in the whole of the USSR. In the RSFSR the amount was about 4.5 trillion (the rest being between the RSFSR and other republics or between and within other republics). In 1991 the total in Russia had dropped to 4.1 trillion, an uncharacteristic admission of a marked decline. The rail system took 56.7 per cent, 25.6 per cent was moved through pipelines, 16.1 per cent by sea and inland waterway, but only 1.6 per cent by road and a negligible amount by air.

The movement of goods by road on journeys other

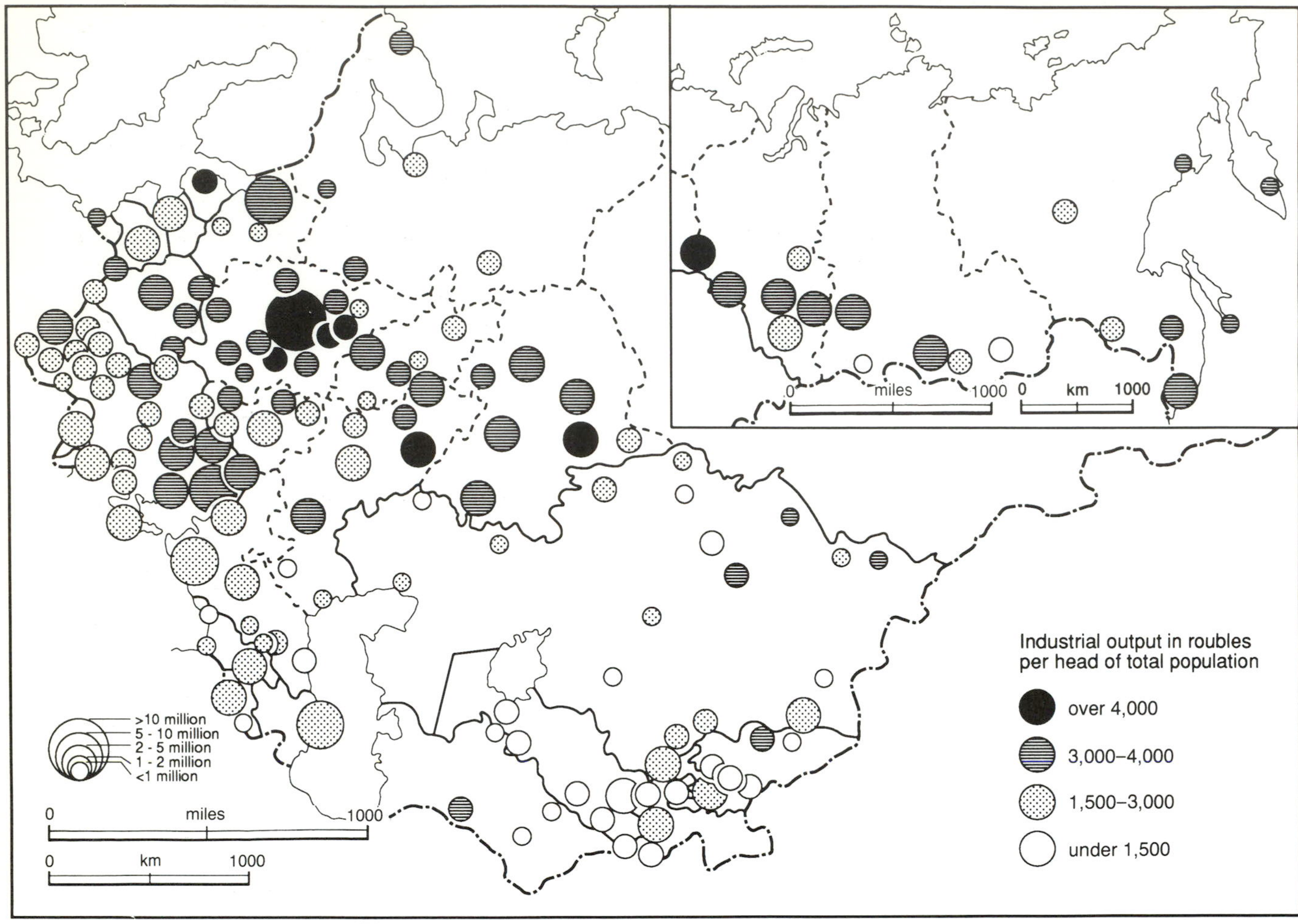

FIGURE 7.8 Industrial output per capita in the former USSR in roubles per head of total population in 1989 by republics and by economic regions of the Russian Republic (RSFSR). The limits of the latter are indicated by the broken lines

than local ones is largely confined to taking them between places of production and the nearest convenient railway station, except for the 'empty half', which includes most of Siberia, where there are no railways at all. Although there is a very extensive road network in Russia, much of it is badly surfaced, while facilities for the servicing of motor vehicles are notoriously poor. The rail system will continue to dominate land transport in Russia, but for the movement of both goods and passengers, road transport must increase, if sufficient flexibility is to be achieved to ensure fast delivery between enterprises at least within a few hundred kilometres of each other, as happens now in Western Europe and North America. If Russia forms stronger links with Western Europe, and if border-to-border motorways are built between Germany and Russia through Poland and Belarus, then corresponding new road arteries will be needed, to extend at least over much of European Russia.

Misgivings were expressed publicly in the USSR about environmental pollution at least as far back as the 1960s, and protection of the natural environment was built into the Soviet constitution. It was not until the late 1980s, however, that data on natural resources, conservation and environmental protection began to be included explicitly in Soviet statistical sources. One estimate made shortly after German unification of the cost of cleaning up East Germany to bring the environment up to EU requirements or at least expectations was about 100 billion US dollars. Russia has almost ten times as many inhabitants and at least ten times the amount of industrial activity causing heavy pollution as East Germany, including extractive, metallurgical and chemical industries. In addition, some areas of the former USSR have suffered nuclear accidents (Chelyabinsk oblast in the Urals in late 1950s, Chernobyl in Ukraine in 1986 – see Box 7.5, p. 174) and others have been the scene of nuclear weapons testing (Kazakhstan) and explosions for economic purposes (Kola Peninsula). To clean up the former

Soviet Union could cost 1,000 billion US dollars – not a task that the new generation of private investors and entrepreneurs could expect to make any profit from. Figure 7.11 (p. 176) shows a selection of local areas and cities in Russia associated with various kinds of pollution. Fortunately, the territory is so large that only a small proportion is seriously affected.

Soviet foreign trade was growing in the 1970s and 1980s and an increasing proportion was with 'non-socialist' countries, mainly the developed 'capitalist' countries of Western Europe and Japan. In the 1980s the USSR mainly imported capital manufactures and food from the West, while obtaining both capital and consumer manufactures from its CMEA partners in Europe, and sugar from Cuba. To pay for the essential imports from the West it had to export oil, natural gas, minerals, timber and timber products, but production of these items has fallen. Russia still needs to trade with the West, while its economic links with its former CMEA partners cannot stop suddenly. In addition it now has to grapple with obstacles to its trade with the fourteen other now independent republics against new local currencies and various overt restrictions to the movement of goods and people.

Russian leaders still seem unaware of the need to take a less aggressive role in world affairs. Pride alone seems to prevent Russia from returning a few small islands in the southern Kuriles to Japan, taken from it at the end of the Second World War. Japanese investment in Russia, especially the Russian Far East, could increase to the mutual benefit of both countries, following a goodwill gesture over the islands. In 1994 Russia was emphatic that it should have a major role in determining UN policy in Bosnia, while it will expect to carry weight in the deliberations of the North Atlantic Treaty Organisation. It is still involved in the affairs of several of its former republics. In this respect, it is following the path of the UK and France, which frequently became involved in disputes with their former colonies, especially in Africa.

Few practising political leaders also write books about their intentions. While still an obscure figure in German politics, in 1925 Adolf Hitler wrote *Mein Kampf*. The Second World War in Europe was the result of the application of some of his intentions. It is misguided to expect to find exact parallels in history, but in some respects Vladimir Zhirinovsky resembles Hitler in his younger years. Frazer and Lancelle (1994) have written a book entitled *Zhirinovsky – the Little Black Book: making sense of the senseless*. The authors have grouped the pronouncements of Zhirinovsky under a number of headings: the Near Abroad (the

BOX 7.2

A selection of post-Soviet issues and problems

1. Politically the Communist Party is banned, but it continues under another name. Presidential and parliamentary elections had been held by the end of 1993. Much power is in the hands of the President.

2. The power of central planners to manage the economy and allocate investment for future developments has been considerably curtailed. It is, however, difficult to achieve the level of privatisation of western countries over a short period, because virtually all branches of the economy have been in state hands and many of them, including the massive military-industrial establishment and, for example, the railways, do not in any western sense make a profit, which would be necessary for privatisation.

3. However much their new leaders would like to be independent economically, the other fourteen republics cannot do without fuel, raw materials and manufactures from the Russian Federation in the immediate future. Compromise will be needed.

4. Many of the Russians living outside the Russian Federation may choose to return to Russia from the other republics or may even be persuaded or forced to leave the republics in the same way that almost 1 million French left Algeria of their own accord or were eased out or driven out after 1958. Dunlop (1994) argues, however, that many of the new elite groups in the Russian Federation would prefer the Russian 'expatriates' to stay where they are, thus keeping alive the prospect of eventually reassembling the Soviet Union. Another ethnic complication is that within the Russian Federation there are clusters of non-Russians, such as the Bashkirs and Yakuts, each with their own ASSR, some restless after a long period of Russian domination. All of the other fourteen republics have a similar problem, in some cases with several other national minorities in addition to the statutory quota of Russians already referred to.

5. In the transition from a command economy to a mixed one, a few people have quickly become rich but many, including pensioners, are materially worse off than in the Soviet system. There are many homeless, including military personnel withdrawn from Central Europe, victims of ethnic conflict in Transcaucasia, and victims of the earthquake in Armenia. An undisclosed number of Belorussians have suffered from the fall-out from the Chernobyl nuclear disaster.

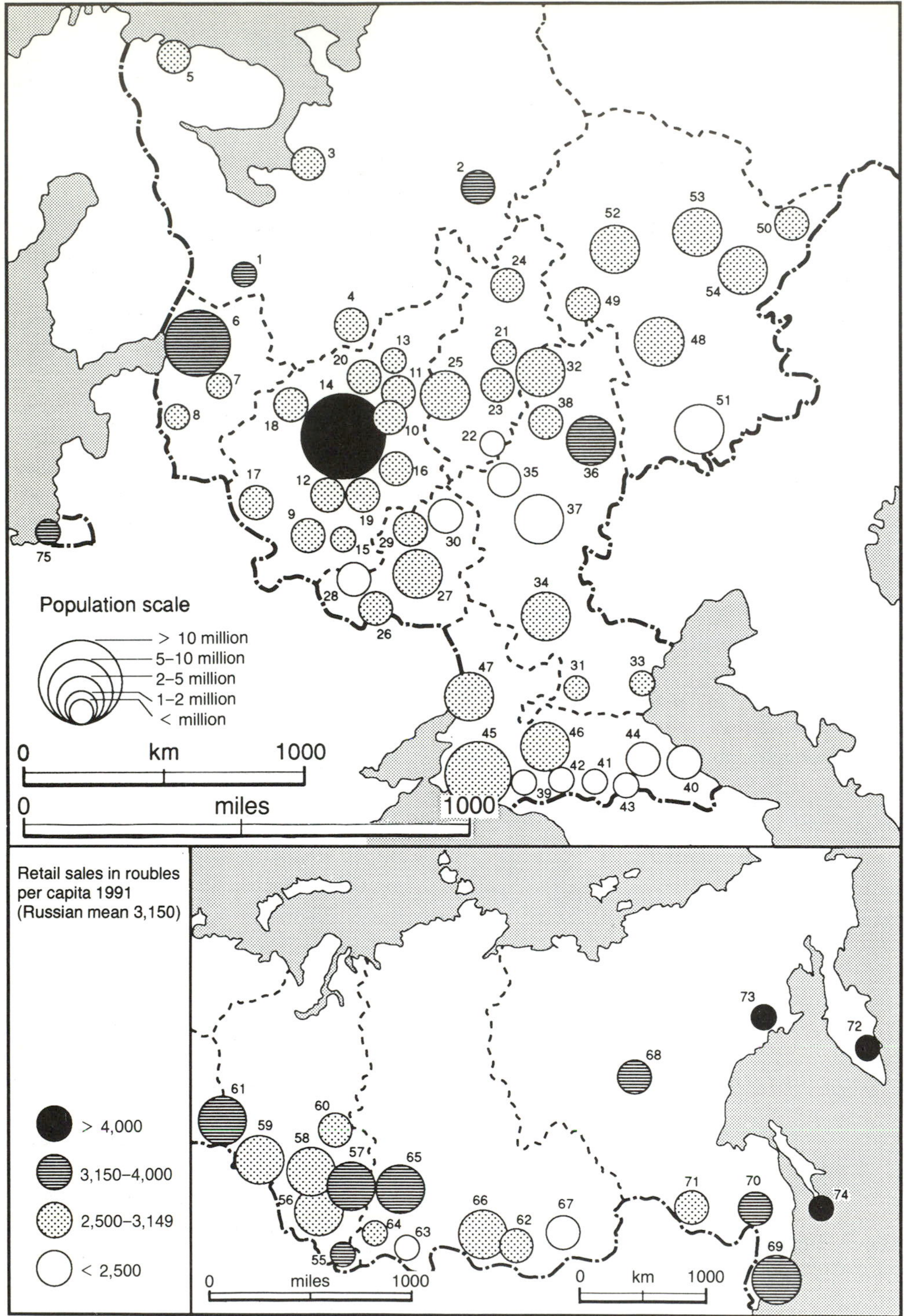

FIGURE 7.9

a Retail sales per inhabitant in Russia, 1991

b (*see facing page*) Doctors per ten thousand total population in Russia

former Soviet Union outside Russia), the Russian–Serbian axis, the Far Abroad, and Russia itself.

Zhirinovksy is of relevance to the present book because (Frazer and Lancelle, p. 149):

> He prides himself on being an expert in geography, which, as he put it, is his favourite occupation. He once gave the former French diplomat Rolf Gauffin a map with the new frontiers of Central Europe drawn on it. He signed it – just as Stalin signed a map showing the partition of Poland in the German–Soviet pact of 1939.

Figure 7.12 (p. 177) shows Zhirinovsky's sketch map.

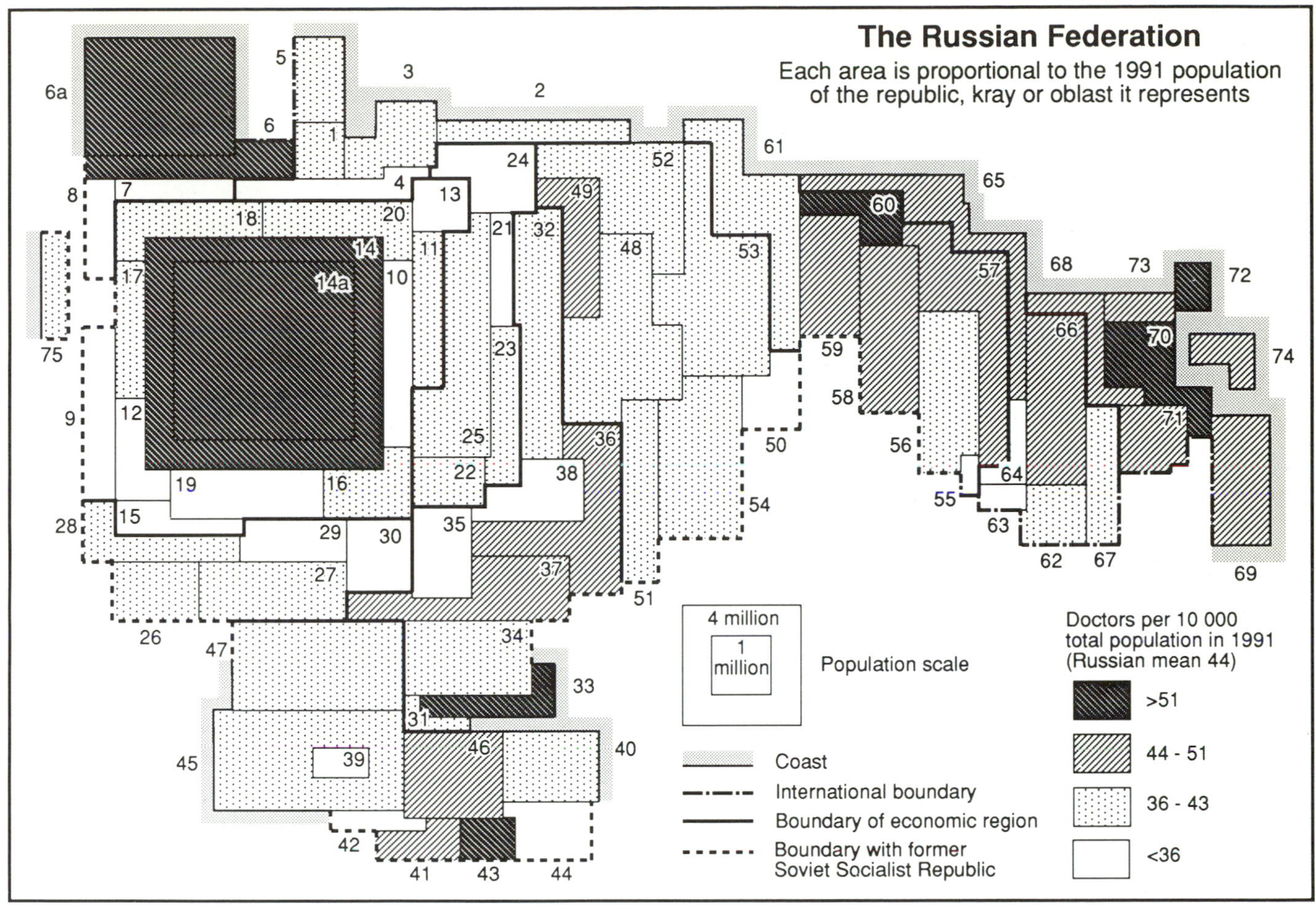

Zhirinovsky won a quarter of the votes in the Russian presidential election of 1992. Next time his Liberal Democrat Party could be more successful.

From a geographical point of view, Zhirinovsky's intentions are of particular interest. As interpreted by Frazer and Lancelle, they include the following:

- to reclaim all or most of the fourteen other former Soviet republics for Russia
- to regain other territories once part of the Russian Empire, including Finland, Poland and Alaska
- to have a frontage on the Mediterranean through association with Serbia
- to have a frontage on the Indian Ocean by controlling Afghanistan and Baluchistan (now part of Pakistan)

These territorial aspirations would have to be achieved by the Russian armed forces, using nuclear weapons if necessary. It is chilling, and worth knowing about, even if it reflects the hopes of only a very small minority in Russia.

7.6 The other fourteen former Soviet republics

The final sections of this chapter consist of an account of the geographical features of the remaining fourteen former Soviet Socialist Republics. These republics were formed after the Bolshevik Revolution in Russia in 1917–18 and the collapse of the Russian Empire. Each is based on an ethnic group. Owing to the presence on the ground in many areas of two or more ethnic groups living side by side, often the boundaries of republics could not be made to include areas with only one ethnic group, without drastic gerrymandering and/or the resettlement of large numbers of people. In addition, by the 1989 census, some 25 million Russians resided in the other fourteen republics, many of them in urban centres. For this reason it is unrealistic to argue that the fourteen new countries are straightforward nation states, even if national consciousness is very strong in most of them.

When the ten republics pre-dating the Second World War were established in the 1920s, one stipulation was that each should adjoin a stretch of coast or share part of its boundary with a foreign country. This condition

PLATE 7.4 Former USSR: high technology

a Space Pavilion, Moscow Exhibition Park, showing a portrait of Yuri Gagarin (1934–68), the first human in space (1961), presiding over a Soviet spacecraft

b High-rise flats on the southern side of Moscow. Dwellers prefer the top floors because the air is thought to be cleaner at a distance from the ground

was satisfied also by the three Baltic republics and Moldavia when they were later brought into the Soviet Union. Such a condition would make it more realistic if a republic chose to secede from the USSR. It was assumed that with time, assisted by the progress of socialism, such a step would not be taken. In the event, in the late 1980s several of the non-Russian republics were becoming restless under Soviet domination and in 1991 all fourteen took the opportunity offered by Yeltsin to gain independence.

The fourteen newly independent countries vary greatly in area and population size, in their natural resource endowment and in the sectors of the economy in which they specialise. In addition, they differ in their respective locations, both in relation to the Russian Federation, and to the possession of a stretch of coast, some having access via the Baltic Sea or Black Sea to the world's oceans without the need to cross the territory of another country. In broad terms, the fourteen countries fall – by their cultural background and their locations – into two groups, referred to here for convenience as the western group and the southern group.

The western group consists of Ukraine, Belarus,

TABLE 7.4 A comparison of human development levels in the republics of the former USSR with those of countries in other parts of the world

Rank in world		*$ per head*	*HDI score*	*Rank in world*		*$ per head*	*HDI score*
29	Lithuania	4,910	881	30	Uruguay	5,920	881
34	Estonia	6,440	872	33	South Korea	6,730	872
35	Latvia	6,460	868	36	Chile	5,100	864
37	RUSSIAN FEDERATION	7,970	862	38	Belarus	5,730	861
40	Bulgaria	4,700	854	45	Ukraine	5,430	844
46	Argentina	4,300	832	47	Armenia	4,740	831
48	Poland	4,240	831	49	Georgia	4,570	829
50	Venezuela	6,170	824	54	Kazakhstan	4,720	802
53	Mexico	5,920	805	62	Azerbaijan	3,980	770
61	Colombia	4,240	770	64	Moldova	3,900	758
65	Surinam	3,930	751	66	Turkmenistan	4,230	746
69	Jamaica	2,980	736	80	Uzbekistan	3,120	695
81	Syria	4,760	694	83	Kyrgyzstan	3,110	689
84	South Africa	4,870	673	88	Tadjikistan	2,560	657
89	Ecuador	3,070	646				

Moldova and the three Baltic republics. The latter are discussed in Chapter 6, since they have moved further into the sphere of influence of the Nordic countries and the European Union than the others. The southern group can be subdivided into the three republics situated in Transcaucasia (Georgia, Armenia and Azerbaijan), the four republics of Central (or Middle) Asia, and finally Kazakhstan. According to both their location adjoining the Baltic Sea and Central Europe and their cultural features, the six western countries are essentially European. The remaining eight are bordered by Asian neighbours, although Transcaucasia is conventionally considered to be part of Europe, notwithstanding the formidable barrier of the Greater Caucasus mountain range to its north, separating it physically from Europe.

The UNDP *Human Development Report 1993*, Table 1, contains estimates of the real GDP per capita in ppp US dollars for 1990 for each country. The latest HDI, based on life expectancy, educational attainment and adjusted real GDP per inhabitant, is also given for the same year. In Table 7.4 the levels of the fifteen former Soviet republics are shown alongside levels in other selected countries. The rank of each country in the world on the HDI is shown. The highest score, 983 out of a maximum possible of 1,000, is achieved by Japan, the lowest score by Guinea (West Africa), 45 out of 1,000, leaving it last of 173 countries considered. Most of the fifteen new countries of the former USSR mingle with the lowest scoring countries of Western and Central Europe (e.g. Portugal, Poland) and with Latin American countries.

The official Soviet figures (see Table 7.1) on the eve of its break-up credited all fifteen republics with favourable scores by world standards for both healthcare provision and educational attainment. Subsequent revelations about the quality of healthcare services and the high infant mortality rates have, however, cast some doubt on standards.

The regional differences in the HDI levels of the former Soviet republics in Table 7.4 tally broadly with those shown in Table 7.1. One achievement of the Soviet period (whether it is positive or negative is a matter of opinion) was the implementation of a planning policy that prevented the existence of a large disparity in living standards between different parts of the USSR. Soviet data in the 1970s and 1980s have consistently shown that the standard of living in the most prosperous republic, Estonia, has been about 2.5 times as high as in Tadjikistan, the poorest republic on most indices of economic and social indicators. The estimate of real GDP in Table 7.4 credits Estonia with 6,440 dollars per head, Tadjikistan with 2,560, a disparity very close to the 2.5:1 ratio. Across the Baltic Sea and Gulf of Finland from Estonia are Sweden and Finland, with a GDP per head of 17,010 and 16,450 dollars respectively, while across the southern border of Tadjikistan lies Afghanistan, with 710 dollars per head. These economic indicators point to a disparity of about 25:1 between the two Nordic countries and the poorest part of Southwest Asia. Whether Estonia would have been better off economically and Tadjikistan poorer if they had not been part of the USSR for five and seven decades respectively is a matter of speculation.

BOX 7.3

Crocodile tears

One of the features of Communist Party rule of the USSR was the policy of self-criticism. One of the main outlets for this anomalous policy was the periodical *Krokodil*, more or less weekly. The shortcomings of many aspects of Soviet life were criticised, although the system itself was not questioned. The reader is invited to enjoy four of the thousands of cartoons shown in *Krokodil* over seven decades since 1922. The captions have been translated into English and a commentary added by the author.

PLATE 7.5 Former USSR

a A cartoon from the satirical periodical *Krokodil* showing the unfortunate result of poor planning. Bureaucrats are trying to save the grain harvest from the rain with sheets of paper. Inadequate provision was made in the 1950s for harvesting, transporting and storing the grain in the newly cultivated Virgin and Long Fallow Lands

b
Caption: They call desert that part of the land on each side of our pipeline
Comment: There is growing concern in the USSR about the threat to wildlife and the natural environment by mining activities, the creation of large reservoirs and other construction work. About one-third of the area of the USSR is forested
Source: *Krokodil* no. 18, June 1976

c

Caption: At the nail factory. Sorry, we do not make nails of that size

Comment: Commentary on wrong incentives in Five-Year Plan directives. The nail production target is met more easily by producing a few large nails than by making a range of sizes

Source: Ekonomicheskaya gazeta 1971

d

Caption: Go past the *dacha* (villa) of our skilled craftsman, then past the foreman's and the next is Ivan Ivanovich's

Comment: Commentary on the luxurious nature of some country villas, probably 'owned' by a high-up member of the Communist Party

Source: Krokodil no. 11, April 1985

7.7 Ukraine, Belarus and Moldova

1. **Ukraine** Ukraine is 604,000 sq km (233,000 sq miles) in area, rather larger than France, and it has a population of 52 million, rather less than that of France. Although it is regarded by Ukrainians as culturally distinct from Russia, much of it has been part of the Russian Empire for several centuries, and in Russian eyes has no grounds for separation. In the 1920s it was given the status of Soviet Socialist Republic. As a result of the westward shift of the national boundary following the Second World War the USSR was extended to include much of inter-war southeastern Poland. As a result, eastern and central

BOX 7.4

Yekaterinburg (alias Sverdlovsk)

Yekaterinburg is situated in the centre of the Ural region of the Russian Federation. Its latitude is 56° North. In North America the parallel 56° North passes through Labrador and the Alaskan panhandle and in Britain near to Edinburgh. The city was founded early in the eighteenth century in the newly industrialising iron-making region. Much of the countryside around is forested, winter conditions are harsh, soil is mostly of poor quality and yields in agriculture are low by world standards. The population of Yekaterinburg at the 1989 census was 1,370,000, making it the fifth largest city in the Russian Federation and the tenth largest in the former USSR.

Like Nizhniy Tagil and Serov to the north and Chelyabinsk and Magnitogorsk to the south, Yekaterinburg's economy is still based on ferrous and non-ferrous metal production and metal-working, but since the 1930s it has become an outstanding centre for heavy engineering. One of the Soviet military industrial complexes is located near Yekaterinburg. It specialises in the production of high-tech and optical equipment, and among other products makes missile guidance systems. The largest enterprise in Yekaterinburg is Uralmash (Ural Machinery). The end of the Cold War has forced many factories producing military equipment in the former USSR to turn to civilian products. For example, Uralmash is switching from the production of guns and mortars to making, among other things, potato cultivators. Uralmash has been partially privatised, with participation in its ownership of its 33,000 workers, who have formed a workers' collective, and a 20 per cent share in the hands of a Georgian financier, Gaga Bedukidze.

Soviet regional planning policy encouraged some degree of regional and local self-sufficiency in agricultural products even in areas with poor conditions. The approximately 300 former state and collective farms in the oblast of Yekaterinburg have been transformed into joint stock companies and co-operatives. Agriculture was heavily subsidised by the state in Soviet times, so the restructuring and withdrawal of support has been painful. Given the environmental conditions and large urban populations in the northern and central part of the Ural region, two-thirds of the food needed has to be 'imported' from other parts of Russia and from now independent Kazakhstan.

Conditions for the citizens of Yekaterinburg have been changing in the last few years. There is no shortage of food and indeed the range of products is more varied than before, with greater quantities of sausages and butter, and more food products imported from abroad. The quantity and variety of household goods have also improved. Prices are no longer controlled, however, while the range of incomes has greatly widened. As a result, such groups as pensioners and students fare badly, while the newly well-off run Mercedes cars. There has been an increase in crime, some of it organised in 'protection' businesses.

In the centre of Yekaterinburg there are a few blocks of wooden or brick buildings from the eighteenth and nineteenth centuries, but almost everyone lives in high-rise apartments built since the 1920s by the state, the local council or individual industrial enterprises. The maintenance of these public-sector dwellings has been neglected. Similarly, the streets have many potholes and traffic is badly regulated. There are no lines to mark lanes on the wider streets and it is accepted that, if they were marked, drivers would ignore them, while at the same time giving way to the affluent drivers of large modern cars. Cinemas are frequented less than in Soviet times; the state-produced films have been superseded by old American ones. One of the most fashionable places of entertainment now is the newly opened casino, where the Mafia and the mobsters have a hand in organising the gambling. Only the *nouveaux riches* can afford to eat in the good restaurants.

According to Russian sources, Yekaterinburg is one of the most polluted cities in the Federation. The burning of fossil fuels, especially coal, is the main cause of pollution in the city, but a large area of the countryside to the south was seriously contaminated by a nuclear accident in 1959, and the local river carries large quantities of poisonous chemicals.

Ordinary people in Yekaterinburg do not see a brighter future just round the corner. They have lost the relative stability and the shared austerity, if not poverty, of ten years ago, and find themselves in a new situation, in which the contrast between the well-off and the poorest is widening by the year.

Ukraine have been more sovietised than the more recently added western part. Another difference is the intensity of industrialisation. The eastern Ukraine has the first coalfield in the Russian Empire to be exploited on modern lines and the oldest iron and steel industry to use coke for smelting. Donbass coking coal, and iron ore from the coalfield area and then from the large deposit at Krivoy Rog, and the availability of the ferro-alloy manganese ore at Nikopol, have been the main ingredients for the industry. In contrast, the western Ukraine had very few industries when it was incorporated into the USSR after the Second World War. Partly for strategic reasons, partly because it lacks sources of fuel and raw materials, it has remained one of the most rural and backward parts of the USSR.

Ukraine is often credited by both Russians and other outsiders with a very impressive natural resource base and a thriving economy. Over half of its territory is arable land and it has some of the most fertile soils in the former Soviet Union, with reasonably favourable climatic conditions, but with the constant threat of a lack of precipitation in the drier southeast. It has produced up to half of the pig-iron and one-third of the steel for the USSR in the post-war period, but, thanks to its superior strategic location and more varied minerals, the Ural region of Russia has been responsible for the production of most of the higher grade steel. Apart from the steel and heavy engineering industries, Ukraine does not have a strong manufacturing sector.

Of all the former republics of the USSR, apart from Russia itself, Ukraine is arguably the most viable as a sovereign state. The only rugged areas are the Carpathian Mountains in the west and the Crimean Mountains in the extreme south. The Polesye (Pripyat Marshes) in the extreme northwest have still been only partially drained. There are considerable areas of forest in the north. Elsewhere, almost everywhere is cultivated or used for pasture. Average yields of grain in Ukraine are much higher than those in West Siberia and Kazakhstan. Ukraine has usually had a surplus of grain, as well as supplying other parts of the former USSR with meat, sugar and sunflower oil.

In addition to its bioclimatic resources, Ukraine has large reserves of high-grade coal and more modest reserves of natural gas. For a time the hydroelectric station at Zaporozhye on the Dnepr River, completed in the early 1930s, was the largest in the world. Before 1991, however, almost all of the oil consumed in Ukraine came from the Russian Republic. In the extreme east of Ukraine is the heavy industrial area of the Donbass, with several of the largest iron and steel mills in the former USSR. To the northwest, Kharkov was one of the principal centres of heavy engineering in the former USSR, while along and close to the Lower Dnepr are other industrial centres. In contrast, industry in the western part of Ukraine is limited and is largely based on food processing, textiles and light engineering. The southern coast of the Crimea is one of the principal resort areas of the former USSR.

Paradoxically, the favourable image of Ukraine as it appears in the Soviet context, with its benign climate, good soils and prestigious heavy industrial base, contrasts with the reality of a newly independent state, attempting to achieve political independence and economic restructuring and growth. Ukraine is similar to Poland in many respects, but it is not as well placed to break away from Russian influence. It includes most of the Black Sea coast of the former USSR and its possession of the Crimea is a military and strategic setback for Russia. In 1989 there were 11.4 million Russians in Ukraine, making up a substantial minority in the eastern region, and actually outnumbering Ukrainians in the Crimea.

Ukraine's only obvious escape route from continuing Russian economic domination would be membership of the EU, but its estimated GDP per inhabitant in 1991 of 2,340 US dollars (*WDR 1993*, Table 1) compares unfavourably even with those of Portugal (5,930) and Greece (6,340), and even more markedly with Ireland (11,120) and Spain (12,450), the four poorest EU Member States and, with southern Italy, the main recipients of EU regional funds. Against the entry of Ukraine into the EU are its main products, virtually all of which, including high-cost coal, steel, grain, meat and sugar (from beet) are all sectors in which cutbacks in production are policy in the EU itself. It has horrendous problems of pollution in the heavy industrial regions and around Chernobyl, scene of the world's worst civilian nuclear disaster (see Box 7.5). It would arguably bring two new languages into the EU, the 11.4 million Russians (22 per cent of the population) having a case to make for Russian as well as Ukrainian to be made official. Its location is peripheral, even in an EU with its centre of population and economic life shifting gradually eastwards. Its rail system has a different gauge from the standard European gauge, and it has very poor road links at present with its Central European neighbours.

2. **Belarus** Like Ukraine, the western part of this former Soviet Socialist Republic was in Poland during the inter-war period. The new country of Belarus (referred to formerly as Belorussia, mistranslated as White Russia) has about 10 million inhabitants and a territory of 207,000 sq km (80,000 sq miles), an area roughly equal to the average size of one state in the USA (cf. Kansas, 213,000 sq km – 82,000 sq miles). It is landlocked, but in the west is only 250 km (155 miles) from the Baltic coast, reached through neighbouring Lithuania and the outlying Russian oblast of Kaliningrad to the port of that name. Unlike most of the other former Soviet Socialist Republics, Belarus

lacks a strong national identity. It has been greatly influenced by Russia in recent centuries, although its language comes from Belo-Ruthenian, another branch of Slavonic.

Until recent decades Belarus has been heavily dependent on agriculture. Soils over most of the area are, however, low in fertility, ill drained in the south, sandy in the north and extensively forested. The agricultural sector is dominated by livestock raising, with both dairying and meat production. Apart from the capital Minsk, the urban centres are relatively small. Given the very limited reserves of fossil fuel and non-fuel minerals, and the lack of a metallurgical base, central planners chose to develop light industry in the Republic, using energy and materials from other republics, mainly Russia.

BOX 7.5

The Chernobyl accident

In the mid-1980s there were twenty large and medium-capacity nuclear power stations in the former USSR and the CMEA partner countries of Eastern Europe. The fifteen in the USSR accounted for about 10 per cent of Soviet electricity-generating capacity. On 26 April 1986 an accident took place in Reactor 4 of Chernobyl nuclear power station near Kiev in Ukraine. The power station was built between 1978 and 1982 and had a generating capacity of 3,000 MWe. It was located less than 150 km (95 miles) to the northwest of the capital of Ukraine, Kiev, in a predominantly agricultural area, in part of the USSR with no low-cost fossil fuels.

The cause of the accident was attributed to operator error. An unnecessary turbogenerator experiment resulted in overheating of the reactor core, a power surge and an explosion that blew off the cap of the reactor. The event was initially given little publicity in the Soviet press, but the atmospheric dispersion of radio nuclides quickly carried material evidence of the explosion in a prevailing northwesterly wind across Belorussia (now Belarus) and the Baltic republics and over Scandinavia, where radioactivity dose rates had risen dramatically by 28 April. In due course wind changes dispersed radioactive material more diffusely. It reached the western Ukraine and Poland five days later and Germany and even the UK a week later. North Italy and the Balkans to the southwest were also affected (see Figure 7.10).

The official casualties resulting immediately from the accident were 31 deaths and 299 injured, while 135,000 people were evacuated, including the whole population of the city of Pripyat. A 10 km (6.2 mile) exclusion zone was established round the plant as well as a 30 km (18.6 mile) outer zone. Kiev to the southeast was apparently little affected by the Chernobyl disaster as the wind did not blow in that direction.

Since the accident the remaining three reactors of Chernobyl have continued to function but there has been growing concern about their safety. When the Soviet Union broke up in 1991, the legacy of Chernobyl became a Ukrainian problem, while Belarus inherited some of the worst affected agricultural land, with much of the oblast of Gomel seriously affected. By 1994 the EU and USA had come to the view that the whole nuclear power plant at Chernobyl should be closed down. Since its contribution to the power supply of Ukraine is indispensable, three extra nuclear reactors would have to be constructed at other Ukrainian nuclear power stations. Ukraine, however, would not itself have the financial means to carry out this work. Finally in October 1994 the Ukrainian government confirmed its readiness to close down the reactor. The EU committed an initial sum of $600 million to start the process.

The Chernobyl issue is, however, part of a much greater problem. It is possible that the safety of all or most of the Soviet and Soviet-designed nuclear plants is suspect so that the future of Chernobyl itself must be seen in a much broader context. Chernobyl has also thrown light on a potential future problem of the former USSR. Russians and Ukrainians living in non-Russian regions of the south are feeling increasingly insecure, so much so that some 2,000 settlers have moved from Central Asia into Gomel in Belarus, to areas vacated or depopulated after the Chernobyl accident, where levels of caesium-137 are 1,000 times what they were before the accident. The health of these and many other people has been or will be affected by the Chernobyl explosion. One estimate is that altogether 55 million people world-wide will be affected in some way.

Chernobyl was not the first serious civilian nuclear accident. One occurred, but was hushed up, in an area in the Russian Ural region in 1957 or 1958, leaving a legacy of pollution. In 1957 a fire occurred in the UK Atomic Energy Authority's plant at Windscale in northwest England. In 1979 the US nuclear industry was greatly influenced by the unplanned single release of radioactivity into the atmosphere from Three Mile Island nuclear power station at Harrisburg, Pennsylvania, located in a densely populated area. The official numbers of immediate deaths and injuries from civilian nuclear accidents have not been large (compare with Table 1.6), but the long-term effects may be much greater.

In a world in which comparative advantage is to be exploited, Belarus does not appear to have any economic activity at which it might excel, one reason, perhaps, for the slow shift from a state economy to privatisation since 1991. If, however, this part of Europe is to be revived economically by contact with the European Union, then the location of Belarus is relatively favourable. In particular, any improvement in transport links between Germany and the Moscow area of Russia would benefit Belarus, already crossed by the main railway between Moscow and Warsaw and by various oil and gas pipelines from Russia into Central Europe.

3. Moldova This new sovereign state has a population of about 4.5 million, but its 34,000 sq km (13,000

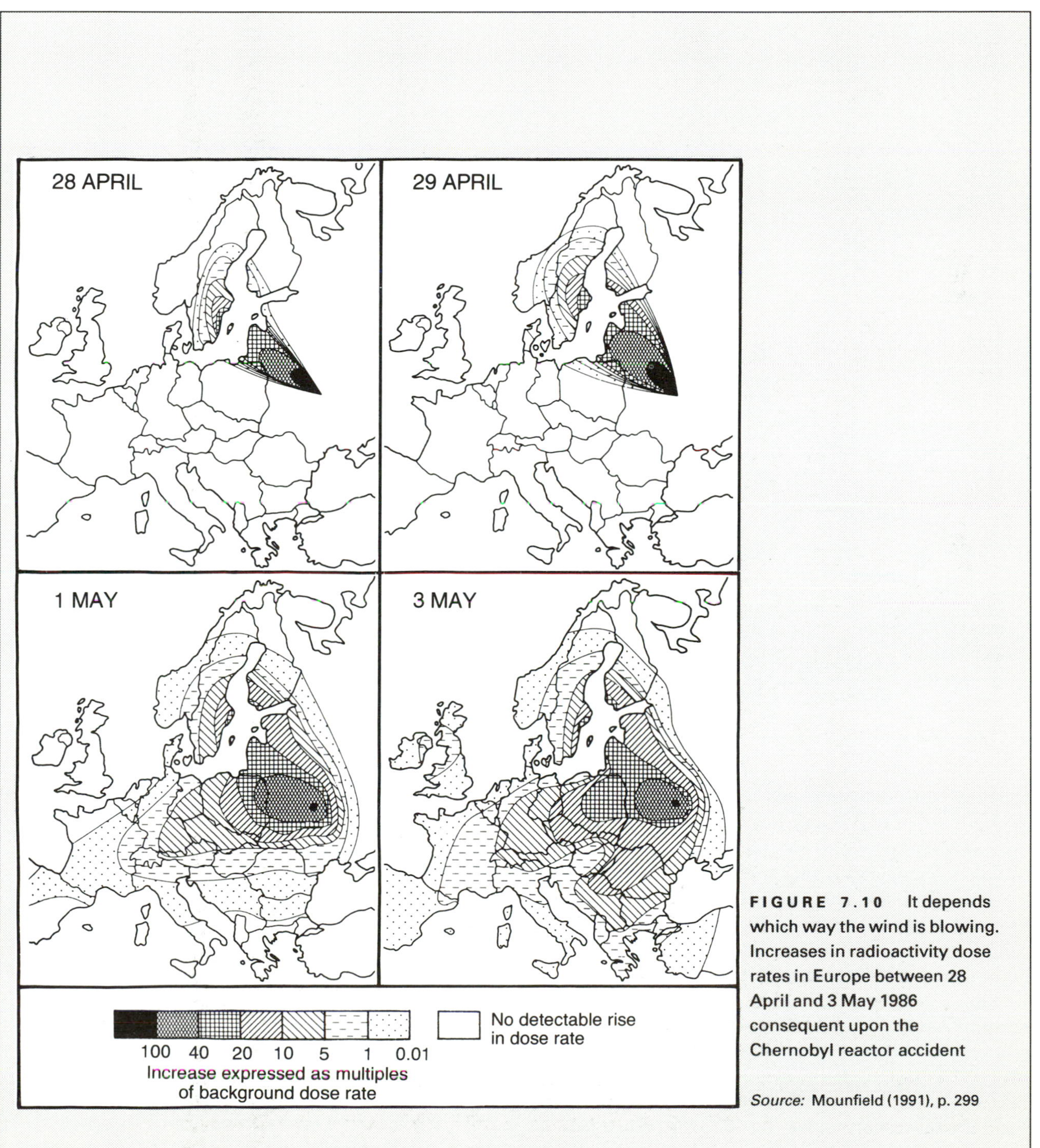

FIGURE 7.10 It depends which way the wind is blowing. Increases in radioactivity dose rates in Europe between 28 April and 3 May 1986 consequent upon the Chernobyl reactor accident

Source: Mounfield (1991), p. 299

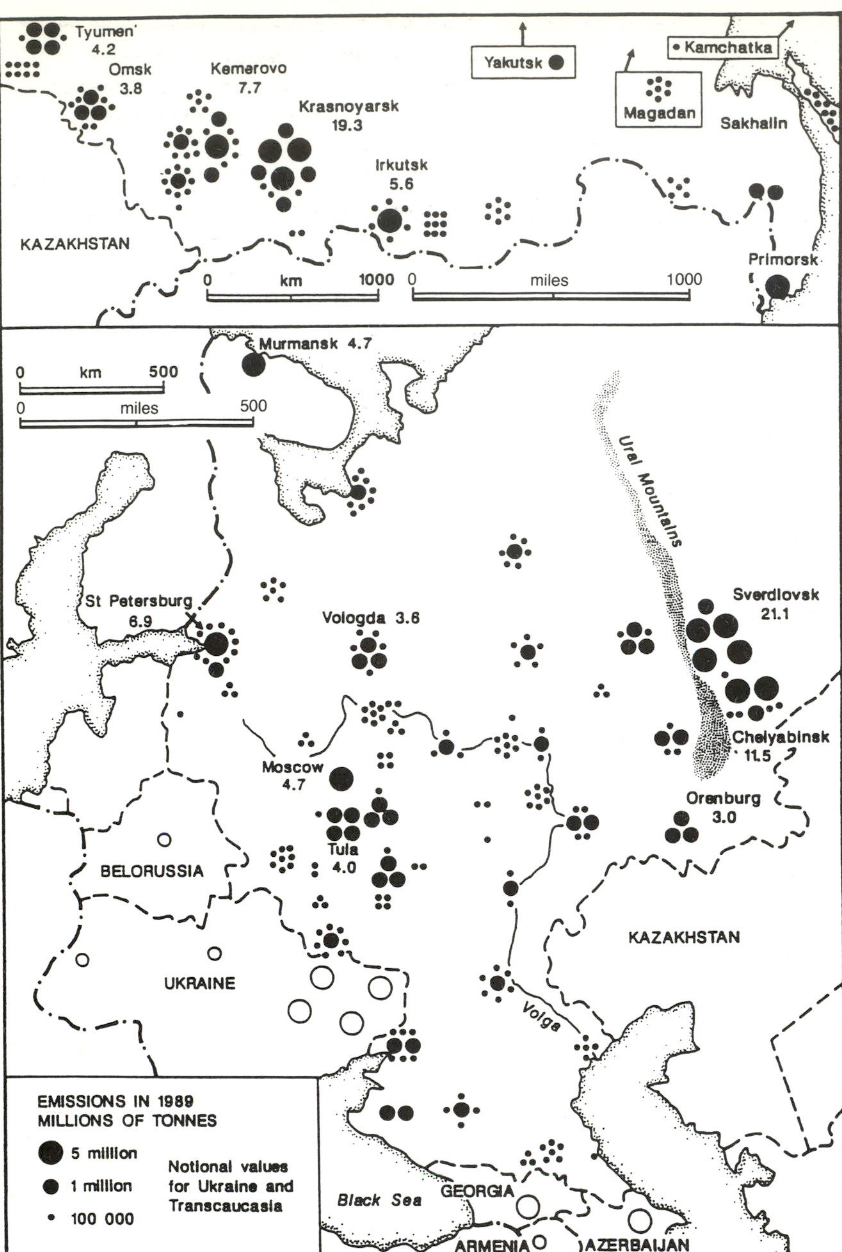

FIGURE 7.11 Main concentration of emissions of pollutants in the Russian Federation in 1989. Goskomstat, material provided for the author, 1991

sq miles) of territory makes it only slightly larger than the US state of Maryland. Moldova (formerly Moldavia) only became part of the USSR in 1945. The Moldavian language is very similar to Romanian and is a member of the Romance (or Romanic) language group, but it is written in Cyrillic script (the Latin script is used for Romanian). Russians and Ukrainians make up over a quarter of the population of the Republic. National feeling is, however, very strong, and unification with Romania has been rejected in a referendum. Although very close to the coast and to the major Ukrainian port of Odessa, Moldova, like Belarus, is landlocked. With virtually no fuel or non-fuel mineral resources, Moldova has little prospect of expanding its modest industrial base. Conditions are generally good for agriculture, and it may continue to specialise in the cultivation of vines, vegetables and sugar beet, as it has done in Soviet times.

7.8 The southern republics

1. **Transcaucasia** The three former Soviet Socialist Republics of Georgia, Armenia and Azerbaijan are located to the south of the Greater Caucasus mountain range. This barrier is crossed by only three main ('military') highways along its whole length of about 1,000 km (over 600 miles) and is skirted at either end by rail links. This region was conquered by Russians during tsarist times and consolidated during the Soviet period in

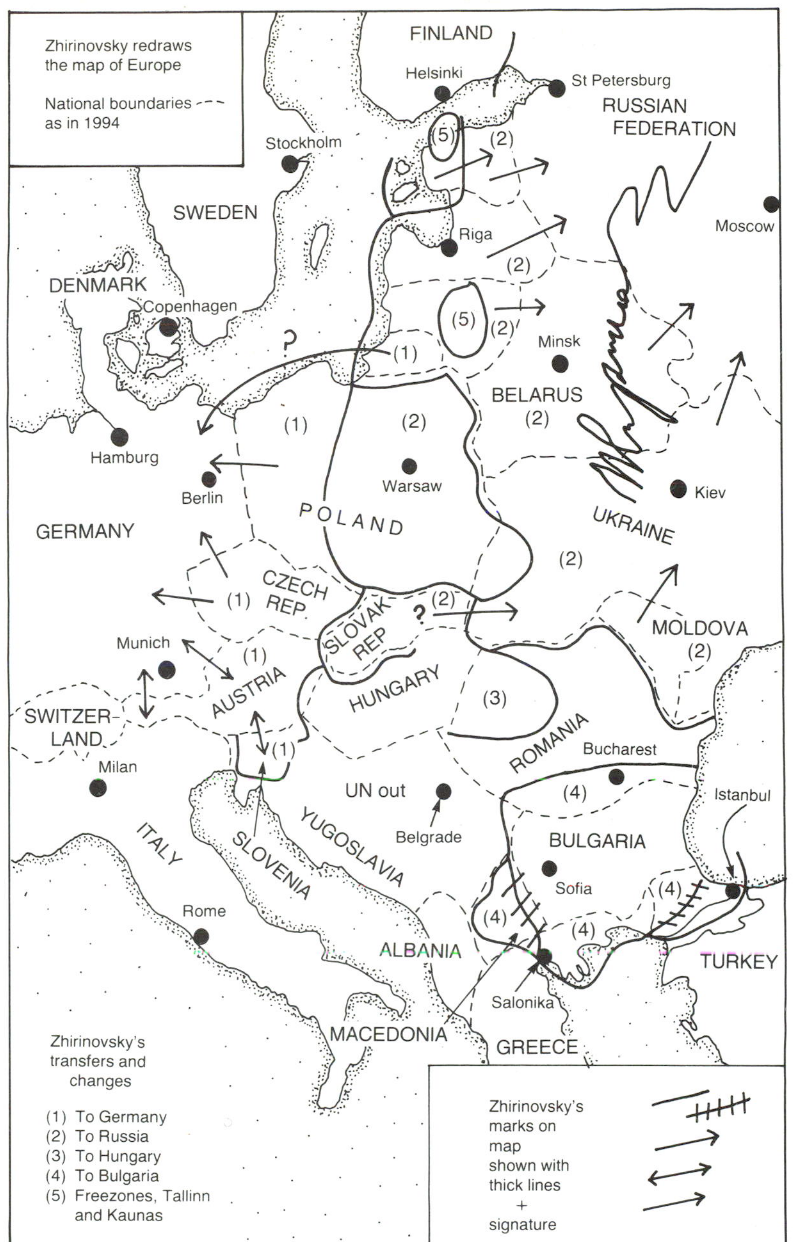

FIGURE 7.12 Zhirinovsky's aspirations in the West: Poland to be divided between Germany and Russia; Germany to take over Austria, the Czech Republic and Slovenia; Russia to take over Ukraine, Moldova and probably Slovakia; Russia to take over the Baltic republics (apart from Tallinn and Kaunas); Bulgaria to take over Macedonia and parts of Greece, Turkey and Romania

Source: *The European*, 4–10 February 1994, p. 1

the form of three republics into a region with both strategic and economic significance beyond its territorial size, 186,000 sq km (72,000 sq miles), a mere 0.8 per cent of the total area of the former USSR. Its southern boundary is shared with Turkey, a member of NATO, and with Iran, in the northern part of which there is a large Azerbaijani element. Economically the region has been of particular interest to Soviet planners as the source of almost the whole of Soviet tea production, of various metallic ores, and of oil and gas.

Once granted independence, Georgia broke away completely from Russia, declining membership of the new Commonwealth of Independent States (CIS), which to some extent filled the economic gap left by the break-up of the former USSR. Armenia and Azerbaijan engaged in an ethnic conflict arising from the difficulty of defining territories to the mutual satisfaction of the two new countries. Georgia has itself been torn apart by the presence of ethnic minorities, each seeking some degree of autonomy within Georgia, if not independence from it.

The three new Transcaucasian countries come between the European and Central Asian republics of the former USSR on the scales of real GDP per inhabitant and the HDI (see Table 7.4). They are fairly similar to Turkey on the HDI and above it on the scale of real GDP. They are the most distant from Western Europe of the European republics of the former USSR and, apart from the Central Asian republics, the least likely to feel the effect of closer links with Western

BOX 7.6

The vanishing Aral Sea

The drying up of the Aral Sea in the former Soviet Union has been described by Micklin (1992, p. 274) as one of the planet's most serious environmental and human tragedies. With the break-up of the USSR in 1991 the inland drainage basin of the Aral Sea has become the scene of an international problem, directly or indirectly affecting the four former republics of Soviet Central Asia as well as Kazakhstan, while also involving a small part of Afghanistan. The boundary between Kazakhstan and Uzbekistan actually subdivides the Aral Sea (see Figure 7.13).

Reference to Table 7.5 shows that in roughly three decades from 1960 to 1991 the surface area of the Aral Sea has halved, diminishing through desiccation from 67,000 sq km (26,000 sq miles), a size between those of Lake Superior (84,000 sq km – 32,400 sq miles) and Lake Huron (60,000 sq km – 23,200 sq miles) to less than 34,000 sq km (13,100 sq miles). This change has affected local climatic conditions. During that time the level of the Aral Sea has fallen 16 metres (52 feet), the volume of water has diminished to one-quarter and the level of salinity has increased threefold. Like the much larger Caspian Sea, in which the level has also dropped during the present century, the Aral Sea has no outlet to the oceans. Its volume of water and surface area depend almost entirely on the water reaching it from two rivers, the Syrdarya and the Amudarya (see Figure 7.13), the headwaters of which are mainly fed by rain and snow falling on the mountains of Tadjikistan, Uzbekistan and Kyrgyzstan. After leaving the mountains, these rivers flow across deserts, where the average annual rainfall is a mere 10–20 cm (4–8 inches), a negligible contribution to the water in the Aral Sea. During the long summers, evaporation is very high throughout the basin.

Several reasons have been given to account for the drying up of the Aral Sea. Climatic change may be leading to a gradual reduction of rainfall in this part of Asia, but the difference over a few decades would be negligible, certainly not enough to account for the drastic changes in conditions shown in Table 7.5. These have already led to the virtual disappearance of a flourishing fishing industry, the loss of most of the native wildlife species that existed in the sea and on its shores, and the drying up of large areas, from which lethal doses of salt and dust are blown over nearby populations. The Aral Sea is rapidly becoming another 'Dead Sea'. Growing population pressure in the basin of the Aral Sea and changes in the organisation of farms, in the system of irrigation practised, and in the type of crops grown have led to the use of a far greater amount of water than earlier this century.

TABLE 7.5 Data for the Aral Sea

	Average level in metres/(feet)	*Average area in thous. sq km/(miles)*	*Average volume in cubic km*[2]	*Average salinity in grams per litre*
1960	53 (174)	67 (26)	1,090	10
1971	51 (167)	60 (23)	925	11
1976	48 (158)	56 (22)	763	14
1991[1]	37 (121)	31 (12)	270	about 30
2000[1]	32 (105)	20 (8)	131	65–70

Notes: [1] Estimates
[2] 1 cubic km = 0.24 cubic miles

Before this part of the Russian Empire became subject to Soviet Central Planning, the population was largely rural and the economy was largely agricultural. It has been argued (Micklin 1992, p. 270) that 'a well-developed and environmentally stable irrigation system existed in Central Asia at the time of the October 1917 Revolution'. Farms were small, and small irrigated fields were carefully managed, protected by earth walls and sheltered by trees. Water withdrawals were much smaller than now and salinisation was not a problem. During and following the first Soviet Five-Year Plan (1928–32), large state-run enterprises (state or collective farms) were established, fields were greatly enlarged and irrigation efficiency was reduced. The campaign to extend the area under the cultivation of cotton and rice thus led to the 'modernisation' of agriculture on the irrigated lands of Central Asia. At the same time, increasing quantities of water were withdrawn by the growing number of urban and industrial users upstream, while pollution increased in the rivers. Commercial agriculture has been pushed particularly since the 1950s, with Central Asia accounting for about 90 per cent of all the raw cotton harvested in the USSR. In the last three decades there has been a considerable increase in the area irrigated. The USSR became a net exporter of raw cotton to its CMEA partners in Eastern Europe, while much of that used internally was manufactured in the Moscow region, not locally.

One legacy of the Russian and Soviet occupation of Central Asia and southern Kazakhstan has been a population explosion. Three republics, Turkmenistan, Uzbekistan and Tadjikistan, are situated almost entirely in the basin of the Aral Sea. Their populations totalled 13.4 million in 1963 but reached 32 million in 1994. The population is expected to reach 55 million in 2025. Parts of Kyrgyzstan and Kazakhstan, with several million inhabitants, also depend on water from tributaries of the Syrdarya River. The small Russian element is characterised by small families, while the Muslim majority is fast growing and not currently

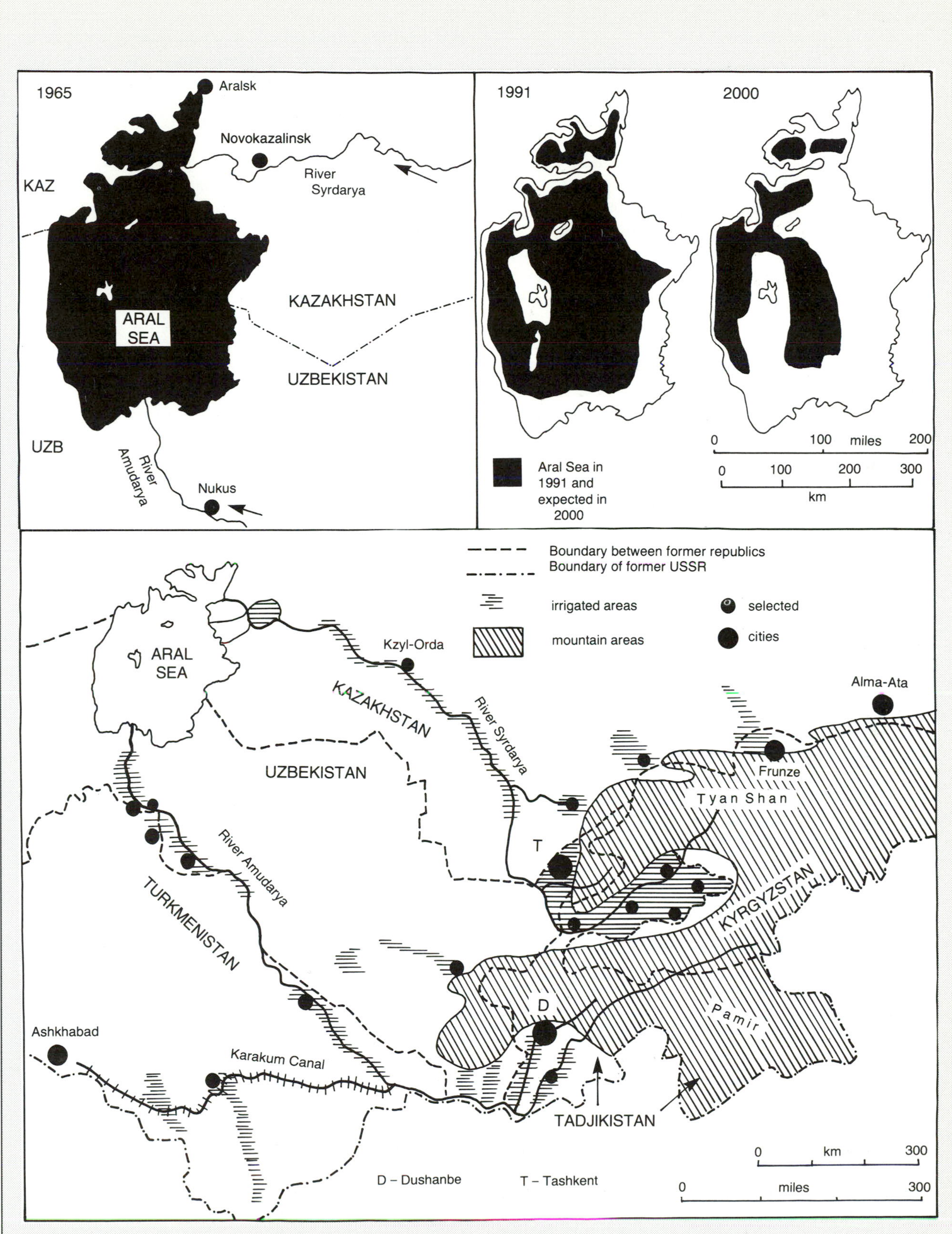

FIGURE 7.13 The shrinking Aral Sea. See Box 7.6 for details of sources

BOX 7.6 *continued*

greatly concerned about family planning and the lowering of the fertility rates. So, what are the prospects for the future of the Aral Sea?

1. To carry on more or less as now, withdrawing far more water than the amount needed to keep the level of the sea from falling, let alone enabling it to rise. The emphasis may shift from the cultivation of cotton for sale outside the region to a diversification of cultivation, while the large state and collective farms may be broken up. It is unlikely that land would be taken out of cultivation in order to reduce the demands of irrigation on water supply and thus increase the flow to the Aral Sea. Agricultural production depends very heavily on irrigation, and the population is growing. Other activities include the extraction of oil and gas, grazing and light manufacturing, but these can only supplement, not replace cultivation. The supply of water from the River Amudarya, currently diverted into Turkmenistan via the wasteful Karakum Canal (see Figure 7.13), might be reduced.

2. At great cost it would be possible to use the water of the two rivers more efficiently by reducing water losses through exfiltration and evaporation, but the present 7 cubic km of water reaching the Aral Sea annually would have to be increased to 35 cubic km merely to stabilise the level at 40–1 metres (131–4 feet).

3. For decades there has been talk of diverting water into southern Kazakhstan and Central Asia from the basin of the River Ob, which flows into the Arctic Ocean. The construction costs of such a project would be enormous and environmental damage would result in West Siberia. The Arctic Ocean would lose several tens of cubic km of relatively warm water a year, with unknown effects on conditions there. The long canal carrying the water from Siberia into the Aral Sea could lose much of the water flowing along it. In the late 1980s the diversion option was officially abandoned.

4. Population could be shifted from the basin of the Aral Sea, but where to? Now that Russia is no longer in control of Central Asia and Kazakhstan, it is even less likely than in the Soviet period that migration would be encouraged or permitted out of Central Asia into the Russian Republic.

Problems similar to those afflicting the Aral Sea are found in many arid areas of the world where irrigation is practised and traditional methods have been modernised and commercialised. Whether Russia and other countries should or will contribute to the cost of stabilising the volume of water in the Aral Sea, if that is deemed to be an environmental priority in a global context, remains to be seen. Both Russia itself and the former communist countries of Central Europe have acute environmental problems. Given their location, in some cases relatively close to Western Europe, Central Asia may be too distant to gain attention and assistance.

Further reading on the Aral Sea issues

- *Post-Soviet Geography*, May 1992, vol. 33, no. 5, special issue.
- Ellis, W. S. (1990), 'A Soviet sea lies dying', *National Geographic*, February, vol. 177, no. 2, 73–92.
- Smith, D. R. (1994), 'Change and variability in climate and ecosystem decline in Aral Sea basin deltas', *Post-Soviet Geography*, vol. 35, no. 4, pp. 142–65.

Europe. Only exceptional economic circumstances, such as the rejuvenation of the oil industry in Azerbaijan, thanks to new oil discoveries put at 3 billion tonnes (about 2 per cent of the world total), are likely to excite investors from the industrial powers of the world.

The leaders and the population of the three Transcaucasian republics may feel that their independence is worth the termination of the Russian economic assistance received in the past. In a region characterised more by physical and ethnic diversity than by stability and great economic promise, life could be difficult in the decades to come.

Much of the terrain is rugged, the eastern lowlands are very dry, and the non-fuel mineral reserves are mostly small in scale. Industrialisation has been considerable in the Soviet period, mainly to provide token self-sufficiency, as with the integrated iron and steel works at Rustavi in Georgia, which uses local ingredients. Oil has been produced onshore for more than a century and later also offshore in the Baku region of Azerbaijan, but output has declined in recent decades. Western investment is now forthcoming to exploit newly found reserves. Only about 15 per cent of Transcaucasia is under field and tree crops. With the prospect of further ethnic hostility if not open conflict, and an increase in population from about 16 million in the early 1990s to 20 million by 2025, most Transcaucasians can hardly expect great economic improvements in the near future.

2. Central Asia (Middle Asia is actually a more appropriate translation of Srednyaya Aziya) There were four Soviet Socialist Republics in the Central Asian economic region of the former USSR: Uzbekistan, Turkmenistan, Kyrgyzstan and Tadjikistan. Each of these is now an independent country, although there is still a considerable Russian civilian presence in the cities, engaged in the industrial and service sectors, as well as a military presence in Tadjikistan.

The four new countries of Central Asia had a combined population estimated to be 36.6 million in 1994 and expected to grow to over 60 million by 2025 (see Table 7.6). Central Asia is about six times as large in area as Transcaucasia, 1,277,000 sq km (493,000 sq miles), making it almost twice as large as Ukraine or the state of Texas, USA. On most economic and social indicators the four republics were the poorest and least developed of the former USSR. In three respects, the new countries resemble those of Southwest Asia (see Chapter 13): water is in short supply, oil and natural gas reserves are large and the potential good, and Islam is the main religion of the non-Russian population. For seven decades Soviet policy was to repress the religious life of its citizens, whether Christian or Muslim. The languages of Uzbekistan, Turkmenistan and Kyrgyzstan are members of the Turkic group, but Tadjik belongs to the Iranian group.

Most of the former Soviet Central Asia was absorbed into the Russian Empire in the second half of the nineteenth century, with key rail links built to tie it into the growing rail network of European Russia and Siberia. After 1917 Soviet pressure increased the influence of Russia as Russians and Ukrainians were settled there, the nomadic way of life of livestock herders was suppressed, and the Communist Party recruited local citizens. Central planners brought the economy into national Plans, achieving increasing economic specialisation and the exchange of products with other parts of the country. For example, Central Asia 'exported' raw cotton, livestock products, oil and natural gas to other regions of the Soviet Union, and 'imported' food products, machinery and equipment, coal and timber from regions to the north.

Although the four new countries of Central Asia vary considerably in area and population size and in many details, they have a number of features in common that point to possible future problems. These include the following:

- The bioclimatic resources of the countries are restricted by the lack of water. Each country has a mountain range in which precipitation is quite heavy, but the rest of its territory is desert or semi-desert, described mostly as natural pasture but in some areas nothing but blown sand. Two large and several smaller rivers originate in the mountain area (see Box 7.6 – on the shrinking Aral Sea). Most of the cultivated land in the region depends on water from these rivers as they flow through mountain valleys and across the lowlands to the north and west. Without bringing water from rivers of West Siberia and northern Kazakhstan that drain into the Arctic Ocean, a possibility considered by Soviet planners, but abandoned in the late 1980s by Gorbachev, there is no scope for irrigating more land. The cost of a worthwhile river diversion would have been very great and would have been borne mainly by Russia, without direct benefit to the Russian Federation.

- As in Transcaucasia, industrial development, mainly of light industry, was the policy of planners both before and after the Second World War. None of the four countries has more than a limited industrial base on which to build further in the future, should they aspire to join the Newly Industrialising Countries of Asia.

- Population is growing fast among the indigenous peoples of the four countries. Tadjikistan has a total fertility rate of 4.5, while Uzbekistan and Turkmenistan have rates around 4.0, although modest by comparison with Iran's TFR of 6.6, and one of 6.9 for Afghanistan.

- Ethnic tensions and conflicts have occurred both within the countries of Central Asia and between them, the situation aggravated by the complex and devious layout of the national boundaries and the considerable presence of members of each country, particularly Uzbeks, in all the other countries.

- A glance at the new map of Asia is enough to show that the four countries are landlocked. Turkmenistan is bordered in the west by the Caspian Sea, but access to the Black Sea via the Volga, Volga–Don Canal and Don is only available for small sea-going vessels. Given the intervening mountains, and deserts crossed only by poor-quality roads, time and cost distances from Central Asia to the Persian Gulf and Arabian Sea are considerable. The governments of Iran, Pakistan and (when stability returns) Afghanistan may be favourably inclined towards the emerging Muslim countries

TABLE 7.6 Population of the former Central Asian republics and Kazakhstan in millions

	1963	*1994*	*2025 (expected)*
Turkmenistan	1.8	4.1	6.3
Uzbekistan	9.7	22.1	37.2
Tadjikistan	2.3	5.9	11.2
Kyrgyzstan	2.4	4.5	6.7
Kazakhstan	11.5	17.1	19.3

to the north. However, bureaucracy as well as inadequate transport links could force the former Central Asian republics to resort to outlets across Russia to the Black Sea and Baltic or to ports on the Pacific in the Soviet Far East in the event of trade building up with the 'outside world'.

3. **Kazakhstan** In area Kazakhstan is the second largest of the republics of the former USSR. Most of the population of some 17 million is settled either in the north, where steppe and even semi-desert lands have been extensively cultivated by Russian settlers since the late 1950s, or in the south, where the Kazakh people, still in the majority, depend on irrigation to cultivate a desert area continuing northwards from adjoining Central Asia. Kazakhstan was an important supplier of grain and other agricultural products to other parts of the Soviet Union and of coking coal, iron ore and other minerals to Russia, particularly the heavy industrial region of the Urals.

Although the new leaders of Kazakhstan have stressed its good prospects as an independent state, it is difficult to envisage a complete break with the Russian Federation. The new political map of Europe and Asia since the break-up of the USSR is unfamiliar and the implications of the changes for the future are difficult to assess. A locational reason why Kazakhstan is still of great importance to Russia is the fact that it has to be crossed by any link between Russia and former Soviet Central Asia. It is also crossed by the shortest rail link between European Russia and China. In addition, there is a strong ethnic reason why Russian leaders in the Russian Federation may wish to retain some influence over Kazakhstan, if not control. In 1989, the number of Kazakhs just exceeded the number of Russians in the Republic, 6.5 against 6.2 million.

The presence of low-cost coal reserves and iron ore in central Kazakhstan, of large oil and gas reserves in the west, and of non-ferrous metals in the east, all relatively close to the Russian regions of Ural and West Siberia, is a good economic reason for continuing interest at least in the northern two-thirds of the area. One of the main areas of nuclear testing in the former USSR was Kazakhstan. On the other hand, cleaning up the environment after decades of nuclear explosions and of pollution from fuel, metallurgical and chemical industries set up in the Republic since the First Five-Year Plan (1928–32) is a financial obligation the Russian Federation would no doubt be glad to avoid.

Not since the 'winds of change' swept through Africa around 1960, giving birth to some thirty new independent countries, has such a political transformation as the break-up of the Soviet union occurred anywhere in the world. The underlying features of the physical and human geography remain in place in the former Soviet Union, but there will be changes in the way the natural resources are used and production is organised. As the fourteen former Soviet Socialist Republics reduce their transactions with Russia, they may become self-sufficient. At the same time, the hitherto almost non-existent trade with countries outside CMEA should grow, whether through the export of oil and gas to western industrial countries or the exchange of food and consumer manufactures with countries of Southwest Asia and even with western China. For Russia itself, the removal of the military and economic burden of keeping CMEA together and supporting various associated countries elsewhere in the world could be a positive change. Resources are being released for use internally to take greater advantage of the large and varied natural resource endowment of the country and to achieve a more balanced industrial structure.

KEY FACTS

1. The former USSR was the end product of four centuries of empire building in Europe and Asia by the Russians. It was the largest country in the world in territorial extent and, given its population size, one of the most generously endowed in terms of natural resources per inhabitant.

2. Much of the population of the former USSR was concentrated in the western and southern parts of the country whereas many of the natural resources were in the eastern regions of Siberia and the Far East.

3. The economy of the country was managed through central planning and state ownership of means of production, with policy determined by the Communist Party. The top level of regional administrative subdivisions consisted of fifteen Soviet Socialist Republics.

4. In 1991 communist control of the USSR was broken and each of the fifteen republics was given independence, leaving the very large Russian Republic politically separate, still very large territorially, but with weakened economic links with the rest of the country.

MATTERS OF OPINION

1. Was the seventy-year Soviet 'experiment' in the USSR a success or a failure?

2. Should the policy in Russia with regard to the problem of environmental degradation and pollution be to allocate some of the limited investment available to improving environmental conditions at the expense of neglecting modernisation of agriculture, industry and transport?

3. Do the Russians in the Russian Federation have a right to expect concessions such as dual nationality for the 25 million ethnic Russians 'stranded' in the other now independent republics?

4. Are the western industrial countries justified in diverting development assistance from the countries of the developing world to Russia to help to keep in power politicians and political parties considered to favour the market economy and democratic processes current in the West?

5. Are the EU, the IMF and the World Bank justified in allocating financial aid to Russia, which is in many respects a 'developed' country, if 'maldeveloped', at the expense of far poorer regions of the world? As Frazer and Lancelle (p. 168) note: 'After all, no-one has ever had the experience of making a capitalist giant out of a Communist superpower.'

SUBJECTS FOR SPECULATION

1. What might the former USSR have been like in 1991 if, instead of taking a 'socialist path', after 1917 it had remained predominantly a market economy?

2. What prospect is there for the emergence of a 'reassembled' USSR with most of the former republics rejoining Russia in the Commonwealth of Independent States?

3. Although it was a major industrial power in the late 1980s, the USSR was still primarily an exporter of fuel and raw materials. What policy should be followed to widen the range of industrial products and to improve their quality?

4. How long might it be before Siberia is extensively opened up and its natural resources exploited for the benefit of Russia itself and the rest of the world?

Chapter eight

Japan and the Korean Republic

The ancient history of Japan is shrouded in the mists of mythology. Though its history is old, Japan is endowed with youthful energy and a vital spirit and is constantly seeking to improve the old and discover the new. Throughout its history, Japan has shown an aptitude for the assimilation of foreign ideas and the adaptation of foreign techniques to its own tradition. Today Japan is a meeting point between the new and the old, as well as the East, and the West, which are fused into a unique harmony.

Statistical Handbook of Japan 1987 p. 2

For the purposes of this book, Japan and the Republic of Korea (hereafter South Korea) are treated together because environmentally and economically they have much in common. On account of its ambivalent position internationally and its strong cultural links with Communist China, Taiwan is grouped with China and discussed in Chapter 16, although in many ways it resembles South Korea.

8.1 A brief history of Japan

Around the middle of the nineteenth century the term 'development' and the concept of a development gap were not explicitly expressed on a global scale. With hindsight one can propose a division of the world at that time into the industrialising countries (Western Europe, eastern USA), the colonies (e.g. India, Australia) and former colonies (e.g. Latin America) of European powers, and other regions, which included both China and Japan. Both the latter were beginning to feel pressure from the industrial powers of the world to trade more widely, but while Japan quickly began to adopt western technology after an initial push by US forces in the Pacific, China resisted change. Around 1870, 85 per cent of the economically active population of Japan was still in agriculture compared with only 15 per cent in the UK in that year, but Japan already had in place more than a simple subsistence economy.

Economically, Japan has come a long way since 1870. At current exchange rates the per capita GNP of Japan in 1991 was 26,930 US dollars, while that of the USA was only 22,240. Apart from some very small countries, only Switzerland surpassed Japan's level, with 33,610 (*WDR 1993*, Table 1). The United Nations Human Development Programme (*HDR 1992*) ranked Japan second in the world after Canada on the basis of three criteria in 1990, this very high position being achieved by its very high life expectancy and educational levels rather than by real GDP. Based on purchasing power parity, the calculation gave Japan

TABLE 8.1 Japanese and Korean calendar

1853	Commodore Perry forces Japan (treaty 1854) to open trade with USA
1868	End of Tokugawa Shogunate, and Meiji Restoration in Japan. Capital transferred from Kyoto to Tokyo. Explicit policy to unify and modernise the state
1873	Population passes 35 million
1894–5	Chinese–Japanese War. Japan occupies Formosa
1904–5	Russo–Japanese War; Japanese sink Russian Baltic fleet (May 1905)
1910	Japan annexes Korea
1923	Tokyo and Yokohama devastated by earthquake (1 September)
1931	Japan occupies Manchuria
1937	Population passes 70 million
1937	Beginning of full-scale war between Japan and China
1941	Japan attacks Pearl Harbour
1942	Japan overruns Southeast Asia
1945	US drops atom bombs on Hiroshima and Nagasaki. All Japanese colonies lost.
1945	Korea partitioned between US (South) and Soviet (North) spheres of influence along latitude 38° North
1950–3	Korean War following communist North's invasion of South
1952	Japan and USA sign peace treaty
1961	Japanese coal production starts to decline
1966	Population passes 100 million
1968	Japanese rice production starts to decline
1973	Peak year for steel production
1991	Coal production down to 8 million tonnes compared with post-war peak of 55 million in 1961

14,310 US dollars per head in 1989, very close to the levels of France (14,160) and Germany (14,510), but well below the US level at 21,000. In material living standards as well as in its poor natural resource base, Japan closely resembles the larger EU countries.

Japan reached its present position in the 'development' league by achieving in about 120 years what took more than twice as long in Britain, one and a half times as long in France and Germany, but only slightly longer in Italy. The later developers have been able to build on the experience of the earliest developers, and they can see the mistakes that should be avoided. Japan was, however, very different in its global location and its culture from the North Atlantic world of Western Europe and eastern USA. Its present situation can be more fully appreciated with reference to certain events in its recent history (see Table 8.1).

In the sixteenth century Portuguese traders were granted a very tenuous foothold in Japan on Kyushu Island, but neither Portugal, Spain (which colonised the Philippines), nor other European colonial powers later tried to take on Japan militarily, a country that in the thirteenth century had seen off a Mongol invasion fleet or, according to other versions, had been saved by bad weather conditions. Over the last few centuries, Japanese traders acquired trading techniques from China, while its pirates harassed the coastal provinces of the mainland. In 1854 Japan was eventually forced by the USA to open its ports to trade with the colonial powers of the time and quickly acquired technological know-how and equipment from the USA and Europe. By 1904–5 it was sufficiently advanced economically and militarily to defeat Russia on the Asian mainland and later to sink the Russian Baltic fleet, which had been dispatched from St Petersburg and passed half way round the world to reach East Asia.

Although Japan was able initially to industrialise in the decades following the 1850s mainly on its own natural resources, in 1895 it occupied China's offshore island of Formosa, which gave it a modest subtropical agricultural base, and in 1910 it occupied Korea, which had useful minerals. Japan steadily increased its industrial and military strength and its foreign trade until 1930, when it embarked on the occupation of Manchuria, the northeast of China, a region less densely populated than the rest of eastern China, with considerably more bioclimatic resources, fossil fuel and non-fuel minerals than Japan itself had. In establishing colonies, Japan was following the example of the West European powers and of Russia, gaining control of resources lacking at home. Its attempt in the late 1930s to occupy the rest of China, and after 1941 to occupy Southeast Asia, turned out to be over-ambitious. With the end of the Second World War in 1945 came the end of Japanese colonial aspirations and the loss of all its possessions.

In due course Japan's limited natural resource base proved inadequate for the needs of its growing population and expanding industrial output and it has become heavily dependent on foreign trade and investment to provide most of the food, raw materials and fuel it uses. Nevertheless, thanks to its large home market and its ability to add greatly to the value of its many imports through manufacturing and duly re-exporting some of the products, Japan's foreign trade is equivalent to only about 15 per cent of its GDP. It has also had a tradition of high tariff and strong non-tariff barriers against imports, the cause of continuing friction with other countries, especially the USA.

8.2 Natural conditions and population

Japan is situated on the northwest side of the Pacific Ocean, forming part of the so-called Pacific Rim in two senses: a ring of mountains with active volcanoes and

PLATE 8.1 These are two pictures of the same sign, which changes from Japanese characters to the Latin alphabet about every 10 seconds. Here the imminent departure of Shinkansen trains from Tokyo to northern Honshu is announced. Chinese symbols, Japanese phonetic characters, Latin letters and Arabic numerals all appear. The trains will depart virtually on the second. This extraordinary punctuality matches the mental agility of Japanese school-children, who have to familiarise themselves with the diversity of symbols, including thousands of Chinese ones, a problem they share with the Chinese themselves

frequent earthquakes and, more recently, a fashionable concept of a region of rapid economic growth. Figure 8.2 in Box 8.1 shows the location of Japan in relation to the whole of the Pacific Rim, Figure 8.1 the location of Japan and South Korea in relation to eastern Asia. In area (373,000 sq km – 144,000 sq miles) it is half as large again as the UK (244,000 sq km – 94,000 sq miles) but considerably smaller than France (544,000 sq km – 210,000 sq miles). It has slightly more than twice as many people as either of these countries. In comparison with states in the USA, it is about three times the area of New York State, but little over half the area of Texas.

Compared with almost any other densely populated region of the world, Japan is extremely rugged, with the result that only about one-eighth of its surface is cultivated, almost exclusively the many small plains around the coasts and the narrow interior valley floors. About two-thirds of the land is forested, however, while precipitation everywhere is high enough for water resources to be adequate so long as the monsoon arrives. Rainfall was lower than average in the summer of 1994 and many areas of Japan, especially on the island of Shikoku, suffered acute water shortages. Offshore and further away there are excellent fishing grounds. Japan's mineral resources are characterised by their variety rather than their quantity, and many reserves have now been exhausted. Coal production has fallen in recent decades (see Figure 8.8), oil and natural gas reserves are negligible, and almost all non-fuel mineral requirements are imported.

The Japanese economy is analogous to the tropical rainforest. A dense, flourishing ecosystem exists on a flimsy base of soil thanks to generous inputs of the sun's energy and a heavy rainfall. The Japanese economy is fed by massive inputs of food, energy and materials from elsewhere. A comparison with Switzerland also comes to mind, but the idea that difficult natural conditions and a poor natural resource base will always produce a strong positive response and spectacular economic growth cannot be supported by experience in many other parts of the world, including Nepal, Bolivia and Haiti, to mention a few. Anyone designing a country for a great industrial power would have made Japan much less rugged, as well as more compact in shape. Nowhere is more than 110 km (68 miles) from the coast, but the inconvenience of moving along and across Honshu, the main island, and linking it to the three other large islands, has given transport engineers and bridge- and tunnel-builders plenty to do, as is shown in Box 8.2.

Three aspects of the population of Japan will be stressed here: the age–sex structure of the total population and long-term changes in it, the distribution of population over the national area, and cultural aspects. The total population of Japan was around 30 million in 1850. It doubled from 35 million in 1873 to 70 million in 1937. Since 1950 the rate of increase has dropped slowly (see Figure 8.3) as birthrate and deathrate have converged. Figure 8.4 compares the structure of the population of Japan in 1991 with that of the RSFSR (now the Russian Federation) in 1989. Population is expected to increase from the 123.6 million in 1990 to a peak around 2010 of about 130 million, thereafter

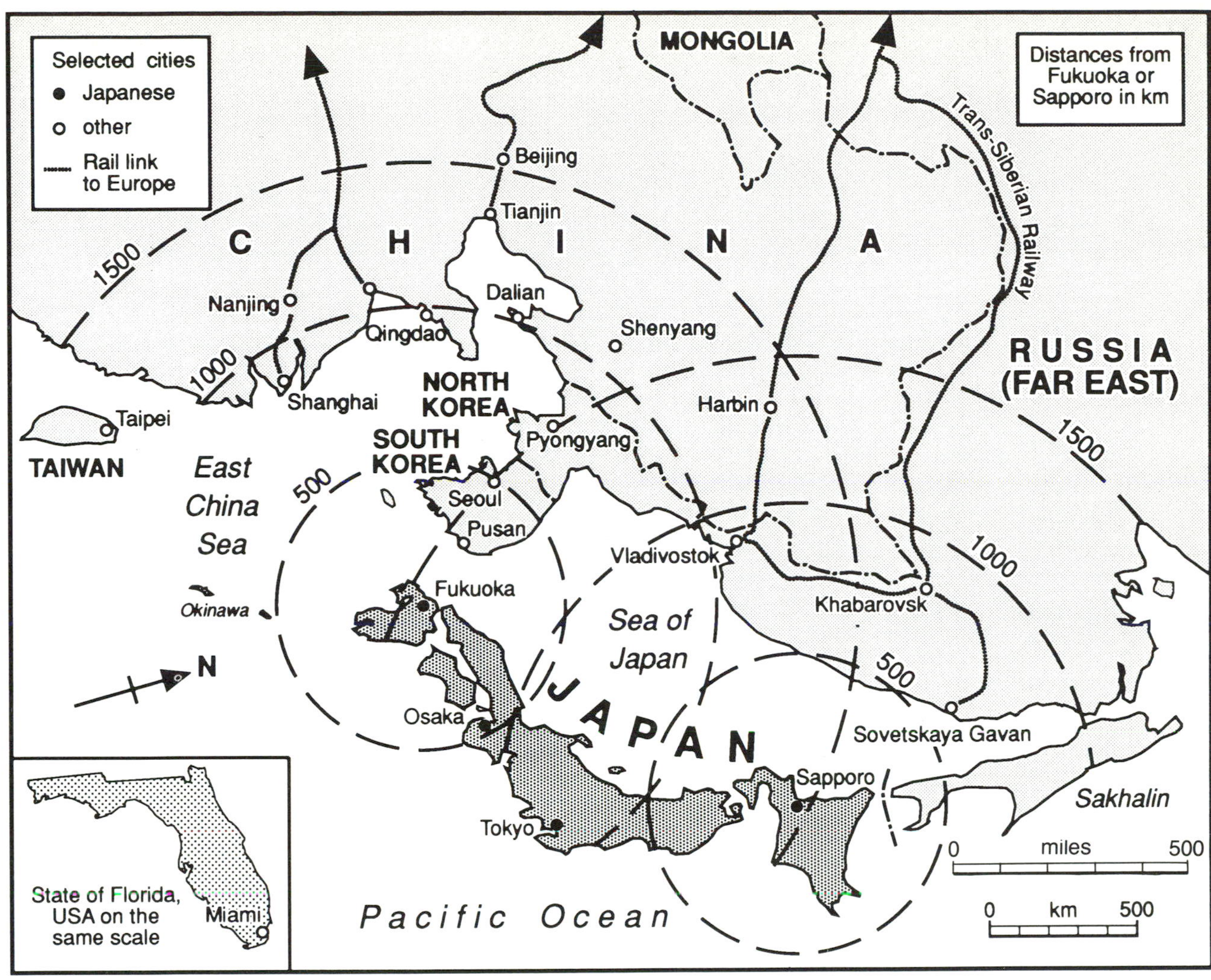

FIGURE 8.1 Location of Japan in relation to the mainland of Asia

declining again to 124 million in 2010–15 and 114 million in 2045, according to a long-term projection (see *Nippon*, 1993).

The irregularities in the age structure of the population of Japan will continue as the children of the post-1945 baby boom, which followed the war and the return of expatriates, themselves produce another bulge early in the twenty-first century. A matter of great concern already is the prospect that the proportion of population of 65 years and over will grow from about 12 per cent in 1990 to about 21 per cent after 2010, reaching possibly 28 per cent by 2045. Indeed, Japan has the highest life expectancy in the world and a total fertility rate of only 1.5. The prospect that the population trend could be affected by emigration or immigration seems highly unlikely. Japan is distinguished among developed countries by its tight restrictions on immigrants, including refugees, presumably on the grounds of lack of space but also through a desire to protect the racial and cultural integrity of its population.

The growth of Japan's population has been accompanied by internal migration from rural areas to urban areas and from smaller urban centres to larger ones. The outstanding feature of the distribution of population in Japan is the great concentration in two huge urban agglomerations, centred on Tokyo and Osaka (see Figure 8.7). Various criteria have been used to delimit these areas, and several different figures for population size have been calculated. Four adjoining prefectures can be taken to represent an extended Greater Tokyo, and three to represent Osaka. By these definitions, with 31.5 million people, Tokyo has 25.5 per cent of Japan's population on 3.6 per cent of the national area while Osaka's 16.7 million inhabitants account for 13.5 per cent of Japan's population, living on less than 4 per cent of the national area. When the prefectures of Nagoya (Aichi) with 6.6 million and

Fukuoka with 4.8 million are added to Tokyo and Osaka, the nine prefectures in question have almost half the population of Japan on a tenth of the national area. Considering the size of these urban agglomerations and centres, problems of transport, congestion and pollution are overcome surprisingly effectively, if comparisons are made with giant cities in China, Brazil and Mexico. The problem of dealing with the growth

BOX 8.1

What Pacific Rim?

Like the eastern and western hemispheres and the North–South rich–poor economic areas, the Pacific Rim is a subdivision of global proportions. The Pacific is the largest of the world's oceans, covering over a third of the earth's total surface (180 million (69.5 million sq miles) out of 510 million sq km (197 million sq miles)). Adjoining land areas that might justifiably be attributed to it bring the proportion to about 40 per cent of the world's area. The purpose of the box is to introduce a note of caution about some of the characteristics attributed to the Pacific Rim.

In a physical sense, the Pacific Rim refers to a discontinuous ring of mountains associated with volcanic and earthquake activity at the edges of tectonic plates in the Pacific Ocean, the so-called ring of fire. In a human sense, the Pacific Rim has been credited with exceptional economic growth, particularly in the 1970s and 1980s. What is the Pacific Rim, and is it experiencing particularly fast economic growth? Figure 8.2 shows the hemisphere centred on the centre of the Pacific Ocean. Trans-Pacific distances are greatest between the northwest and southeast sides (China–Chile), and least between the southwest and northeast sides (Australia–USA).

The sheer size of the Pacific Ocean must be appreciated. From Shanghai, China, to Santiago, Chile, the

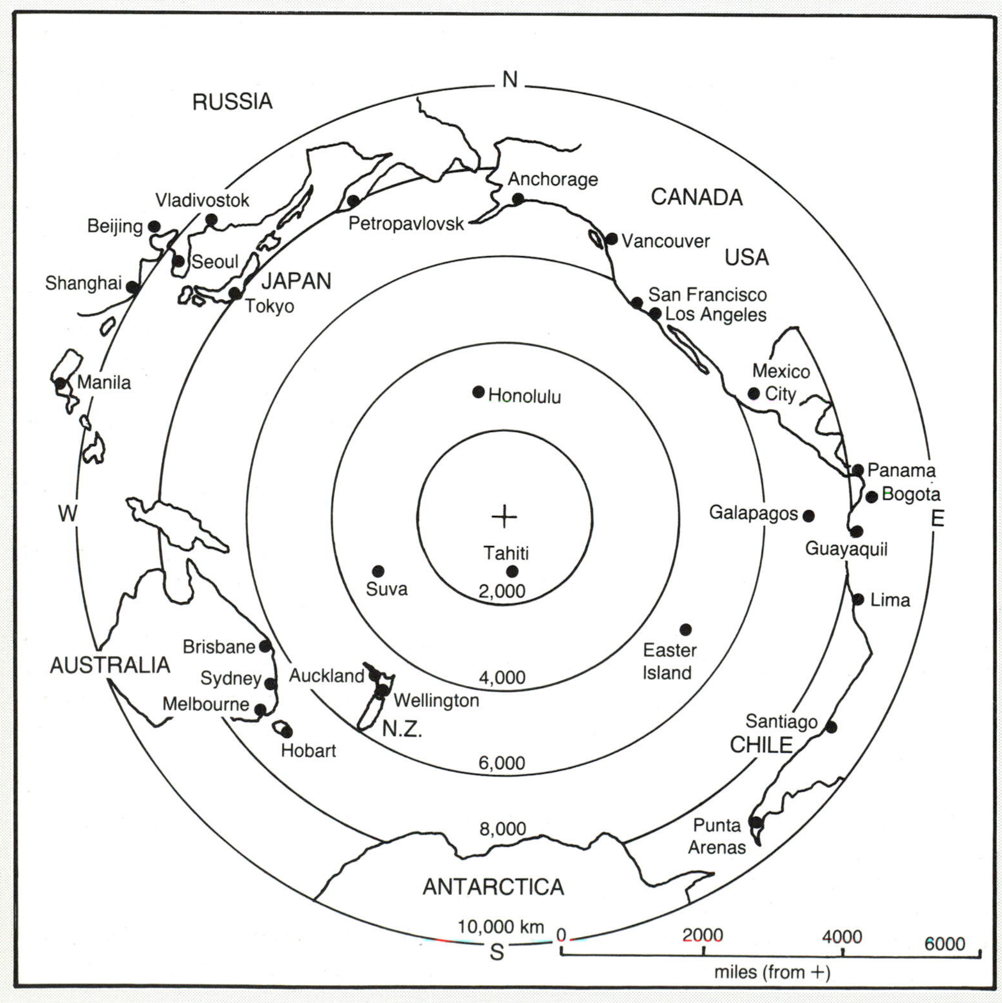

FIGURE 8.2 The Pacific Rim in relation to the Pacific Ocean. The oblique Zenithal Projection made by the author is centred on the centre of the Pacific Ocean and the outer circle contains almost half of the surface of the globe

of population and the expansion of the built-up area has been tackled to some extent through the construction of artificial islands, especially in Osaka-Kobe. Here, for example, the new Kansai International Airport, opened in 1994, is located on a man-made island several kilometres from the shore in Osaka Bay. According to Robinson (1994) the cost was some 16 billion US dollars, almost double the cost of

TABLE 8.2 Leading industrial and financial companies in East and Southeast Asia in 1993 or nearest available year (turnover in billions of £s)

Country	*Company*	*Turnover in billions of £s*	*Main activity*
Japan	Itochu	89.1	Conglomerates
	Sumimoto	86.2	"
	Marubeni	81.0	"
	Mitsubishi	78.4	"
	Mitsui	77.1	"
South Korea	Samsung[1]	15.2	Electronics
	Daewoo	6.8	Various industrial
	Korea Electric Power	5.6	Electricity
	Pohang Iron and Steel	5.2	Metals
	Hyundai	5.1	Transport equipment
Hong Kong	Jardine Matheson	5.2	Conglomerates
	Swire Pacific	3.3	Transport
	First Pacific	1.8	Business services
	Hutchison Whampoa	1.8	Conglomerate
	HK Telecommunications	1.4	Communications
Singapore	Singapore Airlines	1.9	Transport
	Keppel	0.6	Transport
Malaysia	Sime Darby	1.3	Miscellaneous
	Telekom Malaysia	0.9	Communications
Thailand	Thai Airways	1.2	Transport
	Siam Cement	0.9	Building materials

Note: [1] Two companies
Source: Extel Financial (1993), various tables

distance is about 19,000 km (11,800 miles), almost half the circumference of the globe. To pretend that the Pacific Ocean void is a unifying influence for the region is highly optimistic. The 35 km (21.8 miles) of sea separating England from France is credited with forming a major barrier, enhancing Britain's isolation from the Continent. Average distances across the Pacific Ocean are about 300 times as great.

The diversity of countries with shores along the Pacific Rim should also be appreciated. Six developed countries qualify, with New Zealand and Japan entirely in the Pacific region, only the western parts of the USA and Canada, and the eastern parts of Russia and Australia. Twelve developing countries in Asia and twelve in Central and South America also border the Pacific. Of these, China is by far the largest in population. Of the thirty countries, only eight have experienced fairly continuous fast per capita economic growth since the mid-1960s: Japan, Malaysia, Singapore, Thailand, Hong Kong, Taiwan, South Korea and parts of China. In the remaining developed countries growth has been slow but fairly steady, while in most of the Latin American countries it has been sporadic, with periods of actual decline. Fast economic growth has also been achieved by a number of countries bordering the Atlantic and Indian Oceans, the 'rims' of which have not caught on as geographical regions. Economically, the exceptionally fast growth on the Pacific Rim has largely been confined to countries in East and Southeast Asia, from Malaysia, Singapore and Thailand in the south to Japan and South Korea in the north. Table 8.2 shows, however, that the largest industrial and financial companies outside Japan in the newly industrialising countries are mostly modest (refer also to Tables 4.7 and 4.9) and are based more on services than on manufacturing.

By a remarkable coincidence, the Pacific, Atlantic and Indian Ocean 'rims' each have about one-third of the world's total population, about 1.8 billion in the early 1990s. On average, however, the Atlantic area, which includes all but the western part of the USA and the whole of Europe, is much more affluent than the Pacific region, with almost 70 per cent of the total GDP of the world (per capita average about 6,000 US dollars), compared with just over 25 per cent for the Pacific countries (2,300 per capita) and 5 per cent for the Indian Ocean region (500 per capita).

Further details and reservations about the Pacific Rim are provided in Cole (1992). Points to consider include the lack of natural resources in many of the Pacific Rim countries (e.g. less than 5 per cent of the world's reserves of oil and natural gas) and the experience of many countries that fast economic growth does not continue indefinitely. The prospect is that with population slowing down dramatically in most of the developing countries of the Pacific Rim the region will have a gradually diminishing share of the world's population in the decades to come, with the Indian Ocean region's share growing.

Final thought: Refer to a globe to appreciate geographical aspects of the Pacific Rim phenomenon.

Further reading

Cole, J. P. (1992), 'What Pacific Rim?' *Geography Review*, March, vol. 5, no. 4, pp. 25–30.

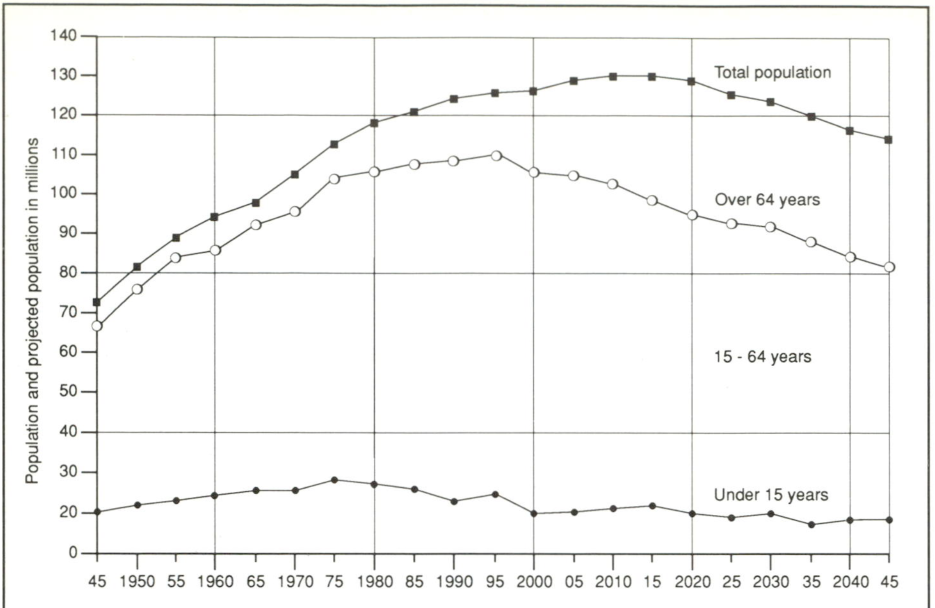

FIGURE 8.3 Population change in Japan, past and projected, 1945–2045
Source: Nippon (1993), p. 44

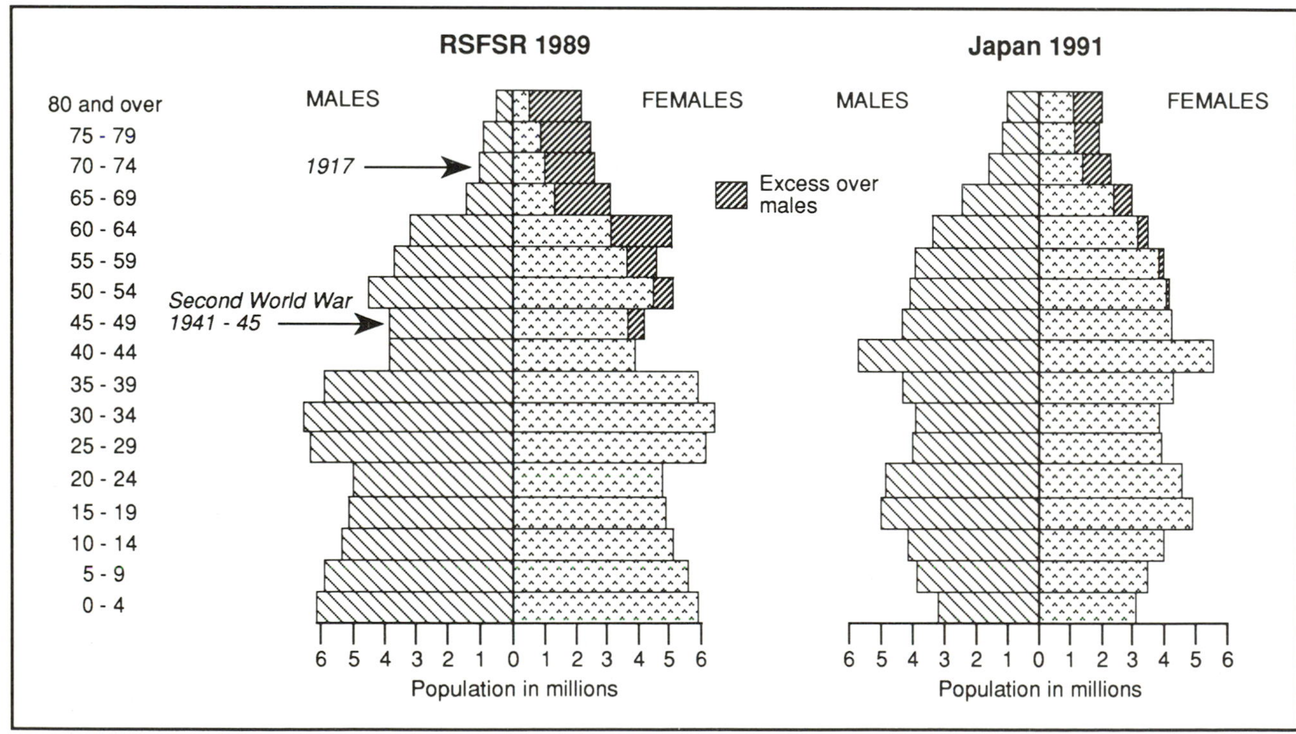

FIGURE 8.4 A comparison of the population structures of the Russian Federation (1989) and Japan (1991)

constructing the Channel Tunnel, but on a par with the cost of building Hong Kong's new airport.

The present size and distribution of population in Japan are unlikely to change greatly in the next few decades, but the Japanese people themselves are changing. Up to now they have been very insular both literally and figuratively and, apart from forays into Korea, China, and briefly Southeast Asia, particularly from 1930 to 1945, Japan has kept itself very isolated. To the outsider, the Japanese appear remarkably homogeneous. There are only about 200,000 foreigners, mostly Koreans, in the country. The original inhabitants of Hokkaido in the north do, however, differ from the rest of the Japanese. There is also a class, the Bukharamin, whose situation resembles the Indian caste of untouchables. The shift of employment from agriculture to

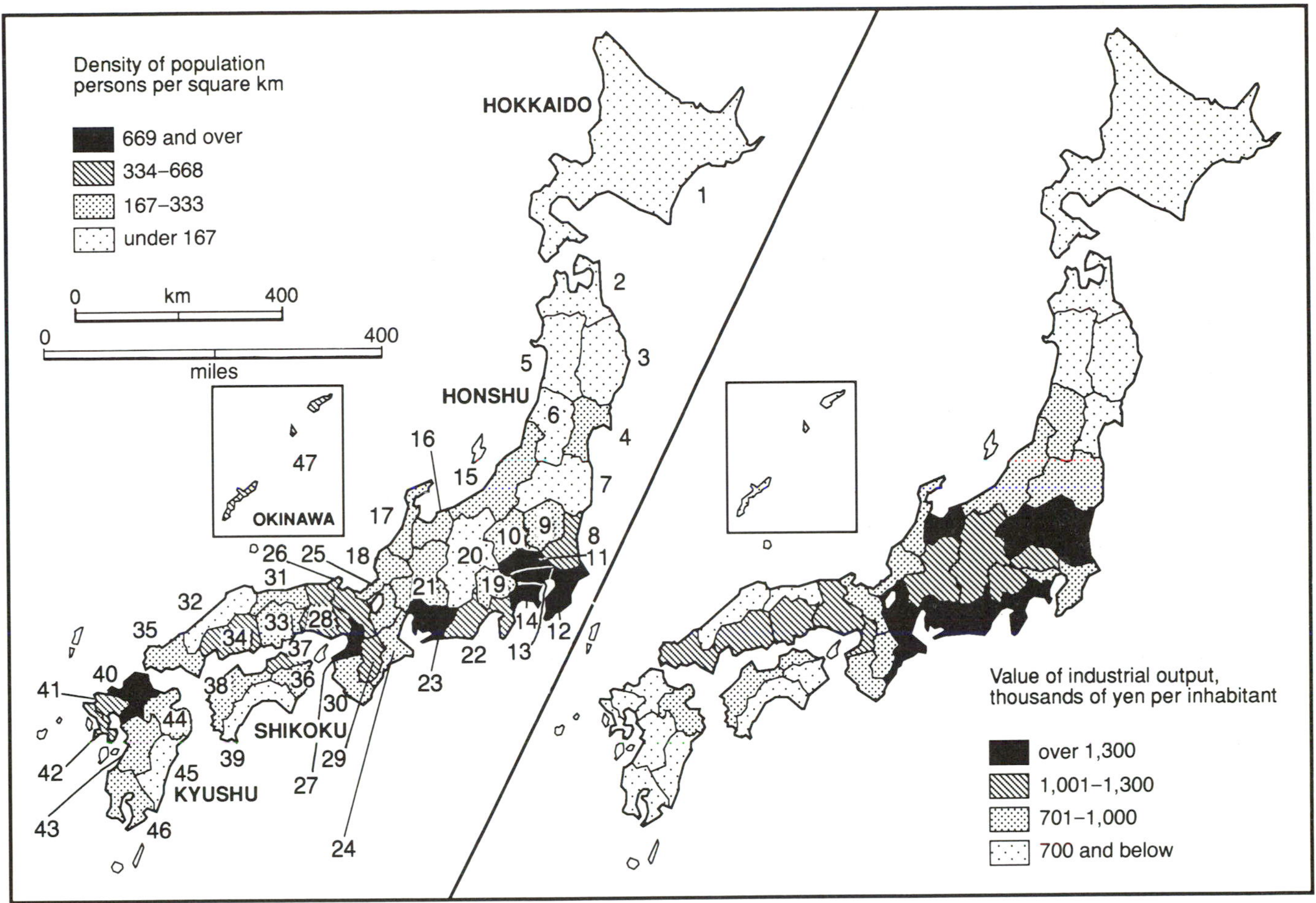

FIGURE 8.5 Density of population in 1991 and per capita value of industrial output in 1990 in Japan by prefectures
Source: Nippon (1993), Tables 1 and 5. Key to numbering is in Table 8.3

industry and services has occurred fairly smoothly in Japan; because of the physical conditions, the cities and shrinking cultivated areas share the same plains.

Since the introduction of elections on parliamentary democracy lines, the Liberal Party has run the country until 1993, with socialists in the minority and communists largely confined to the larger and more affluent industrial centres. A new political scene has emerged in the 1990s as younger leaders attempt to modify traditional economic and social concepts and practices. Trends that are eroding the Japanese way of life in the 1990s are the influence of western culture, resulting from the increase in global communications, the pressure to import consumer manufactures from other countries, the increase in visits abroad by Japanese tourists, and the need to invest and manage factories and services throughout the world.

The perceived need to reduce Japan's trade surplus, especially with the USA and the EU, has worried US politicians. Japan's own spectacular post-war economic growth has faltered in the 1990s. E. W. Desmond (1994) argues that Japan's situation, complicated by its overvalued yen, making Japanese exports relatively expensive, is forcing it to set up manufacturing firms overseas, thus becoming an even more formidable international competitor. Developing countries, especially those in East and Southeast Asia, mostly welcome Japanese branch factories. The attractiveness of such countries to Japanese investment is above all cheap labour. The value of production from Japanese transplantings overseas rose from 3 per cent of the value of its total production in 1985 to 6.5 per cent in 1992. Japan's Second World War dream of a Great East Asia Co-Prosperity Sphere is being realised peacefully. China, Thailand, Malaysia and Indonesia are among the main destinations of new developments not only by the very large Japanese companies but by many small ones as well.

8.3 Economic activities

In 1990 almost exactly half of the total population of Japan was economically active (62 million out of 124 million). Agriculture accounted for 7 per cent of employment, industry for 33 per cent and services for 60 per cent, proportions very similar to those in the EU as a whole. The relatively low level of productivity of workers in the agricultural sector is evident from the fact that it accounts for only about 2 per cent of GDP; a similar imbalance occurs in Germany and Spain, but not in the USA. The fact that many workers in agriculture are part-time may partly explain the disparity. The service sector is the largest employer of labour in every prefecture of Japan, ranging between a little under 50 per cent in three of the prefectures and 71 per cent in Tokyo. While every prefecture needs a wide range of basic local services, Tokyo, and to a lesser extent other big cities, have services of regional, national and international significance.

Compared with most of the countries of the EU, industrial employment is fairly evenly distributed according to the population of prefectures (see Table 8.3), only falling below 20 per cent in Okinawa, while exceeding 40 per cent in some prefectures located between Tokyo and Osaka, including Nagoya (Aichi). Given the layout of Japan, with good access to ports from anywhere in the country, combined with high educational levels everywhere and a versatile labour force, it is economically realistic to locate factories in or near any sizeable city. In absolute numbers, however, there is a very high concentration in Tokyo prefecture and three adjoining prefectures, with about 5 million of Japan's total of 20.5 million industrial workers. Productivity varies according to the size and age of factories. Many of the largest firms and plants are in the zone between Fukuoka in the west and Tokyo in the east, Japan's industrial heartland, long and narrow in shape.

Agriculture is the branch of the economy in which the greatest regional variations occur in Japan, with over 20 per cent still employed in the sector in two northern prefectures and less than 1 per cent in Tokyo and Osaka. Differences in per capita income between prefectures reflect, among other things, the relative importance of the poorly paid agricultural sector. In Japan, as in the EU, maintaining as high a level as possible of food production has been a national priority, and Japanese rice production, in particular, has been subsidised. In 1989 Japan had 5.3 million hectares (13.1 million acres) of cultivated land; there were therefore 23 people per hectare (9.3 per acre), compared with about 12 (5) in China but under 2 (1) in the USA.

The total cultivated area and the area planted with rice have been reduced since the 1950s, partly due to the need to release the plains for residential, commercial, industrial and transport uses (see Figure 8.6). The greatest pressure on rice-growing land has come since the late 1960s: the area has decreased from 3.3 million

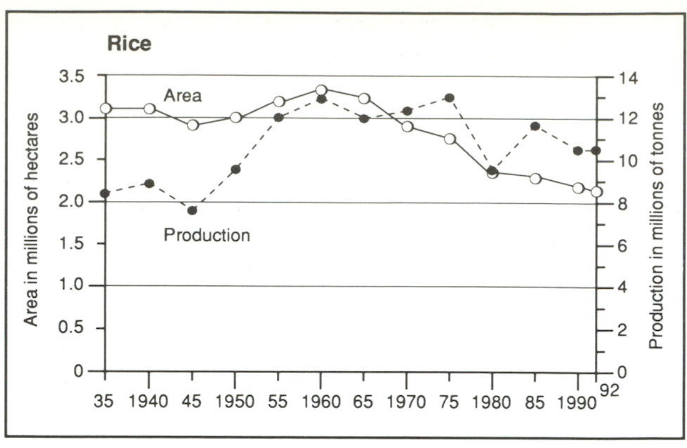

FIGURE 8.6 Japanese production of rice, 1935–92
Source: Nippon (1993), Appendix: 'Statistics by prefectures'. Annual Transition of Selected Statistics

hectares (8.2 million acres) to 2.1 million (5.2 million acres) around 1990. Yields have increased considerably, however, doubling between the late 1930s and the late 1980s. The production of other cereals is small and declining, but the number of cattle, pigs and poultry, while still very small in relation to population compared with levels in northwest Europe, has risen in response to diet changes. The fish catch has doubled in the last thirty years, but thanks only to the use of fishing grounds far from Japan itself, even in the Indian and Atlantic Oceans.

8.4 The Japanese industrial sector

Japan's rise to economic prominence in the world has been achieved mainly through the growth of its industrial sector. Traditionally Japan was largely self-sufficient economically, but towards the end of the nineteenth century the development of modern industries and the growth of population put increasing pressure on the limited natural resources of the country. Virtually all the value of merchandise exports now comes from the industrial sector. The experience of Japan's industrialisation as the first major country outside Europe and North America to industrialise on modern lines is of particular interest to many developing or 'emerging' countries today.

Before the Second World War, Japan's industrial development was strongly influenced by the expansionist policies of the state. By successfully occupying Formosa, Korea and Manchuria (northeast China), Japan gained control of raw materials and food supplies. At the same time its own coal, hydroelectric, forest and non-fuel mineral resources were intensively used. From 1937 to 1945 the ambitious 'guns before butter' policy of occupying at least eastern, if not all, China, and then Southeast Asia and possibly Australia,

TABLE 8.3 Selected regional data for Japan. The key to the location of the prefectures is in Figure 8.5

	(1)	(2) Population		(3)		(4)	(5)
Prefecture	*total in millions*	*per sq km/(mile)*		*per ha/(acre) of arable*		*Mf value*	*Retail sales*
1 Hokkaido	5.6	72	(186)	5	(2)	351	1,238
2 Aomori	1.5	154	(399)	9	(4)	304	997
3 Iwate	1.4	93	(241)	8	(3)	521	962
4 Miyagi	2.3	311	(805)	15	(6)	527	1,108
5 Akita	1.2	105	(272)	8	(3)	504	1,007
6 Yamagata	1.3	135	(350)	9	(4)	779	1,012
7 Fukushima	2.1	153	(396)	12	(5)	859	1,004
8 Ibaraki	2.9	471	(1,220)	15	(6)	1,345	1,062
9 Tochigi	1.9	304	(787)	14	(6)	1,481	1,148
10 Gumma	2.0	310	(803)	21	(9)	1,406	1,102
11 Saitama	6.5	1,707	(4,421)	68	(28)	996	966
12 Chiba	5.6	1,089	(2,821)	38	(15)	771	1,029
13 Tokyo	11.9	5,444	(14,100)	1,081	(438)	833	1,587
14 Kanagawa	8.0	3,334	(8,635)	309	(125)	1,324	1,079
15 Niigata	2.5	197	(510)	13	(5)	795	1,016
16 Toyama	1.1	264	(684)	17	(7)	1,439	1,131
17 Ishikawa	1.2	279	(723)	23	(9)	914	1,145
18 Fukui	0.8	197	(510)	18	(7)	968	1,140
19 Yamanashi	0.9	192	(497)	28	(11)	1,175	1,118
20 Nagano	2.2	159	(412)	16	(6)	1,128	1,165
21 Gifu	2.1	196	(508)	31	(13)	1,110	1,064
22 Shizuoka	3.7	474	(1,228)	42	(17)	1,619	1,122
23 Aichi	6.7	1,306	(3,383)	75	(30)	1,816	1,209
24 Mie	1.8	312	(808)	25	(10)	1,416	1,105
25 Shiga	1.2	307	(795)	21	(9)	2,071	1,032
26 Kyoto	2.6	565	(1,463)	72	(29)	983	1,190
27 Osaka	8.7	4,631	(11,994)	485	(196)	1,145	1,310
28 Hyogo	5.4	649	(1,681)	62	(25)	1,118	1,099
29 Nara	1.4	376	(974)	51	(21)	733	876
30 Wakayama	1.1	228	(591)	27	(11)	701	935
31 Tottori	0.6	176	(456)	15	(6)	521	1,029
32 Shimane	0.8	117	(303)	17	(7)	541	990
33 Okayama	1.9	271	(702)	23	(9)	1,145	1,048
34 Hiroshima	2.9	337	(873)	39	(16)	1,077	1,169
35 Yamaguchi	1.6	257	(666)	26	(11)	1,086	1,036
36 Tokushima	0.8	200	(518)	22	(9)	699	949
37 Kagawa	1.0	546	(1,414)	26	(11)	787	1,265
38 Ehime	1.5	267	(692)	22	(9)	800	956
39 Kochi	0.8	116	(300)	23	(9)	300	974
40 Fukuoka	4.8	973	(2,520)	46	(19)	635	1,070
41 Saga	0.9	360	(932)	13	(5)	636	943
42 Nagasaki	1.6	381	(987)	26	(11)	331	854
43 Kumamoto	1.8	249	(645)	13	(5)	431	938
44 Oita	1.2	195	(505)	17	(7)	790	993
45 Miyazaki	1.2	151	(391)	16	(6)	401	973
46 Kagoshima	1.8	195	(505)	13	(5)	339	891
47 Okinawa	1.2	543	(1,406)	26	(11)	143	731
Total	124.0	333	(862)	24	(10)	977	1,134

Description of variables:
(1) Total population, 1 October 1991 in millions
(2) Persons per sq km
(3) Persons per hectare of cultivated land, 1992
(4) Value added by manufacturing per inhabitant in 1990 in thousands of yen (or approximately tens of US dollars)
(5) Retail sales in thousands of yen per inhabitant
Source: Nippon (1993)

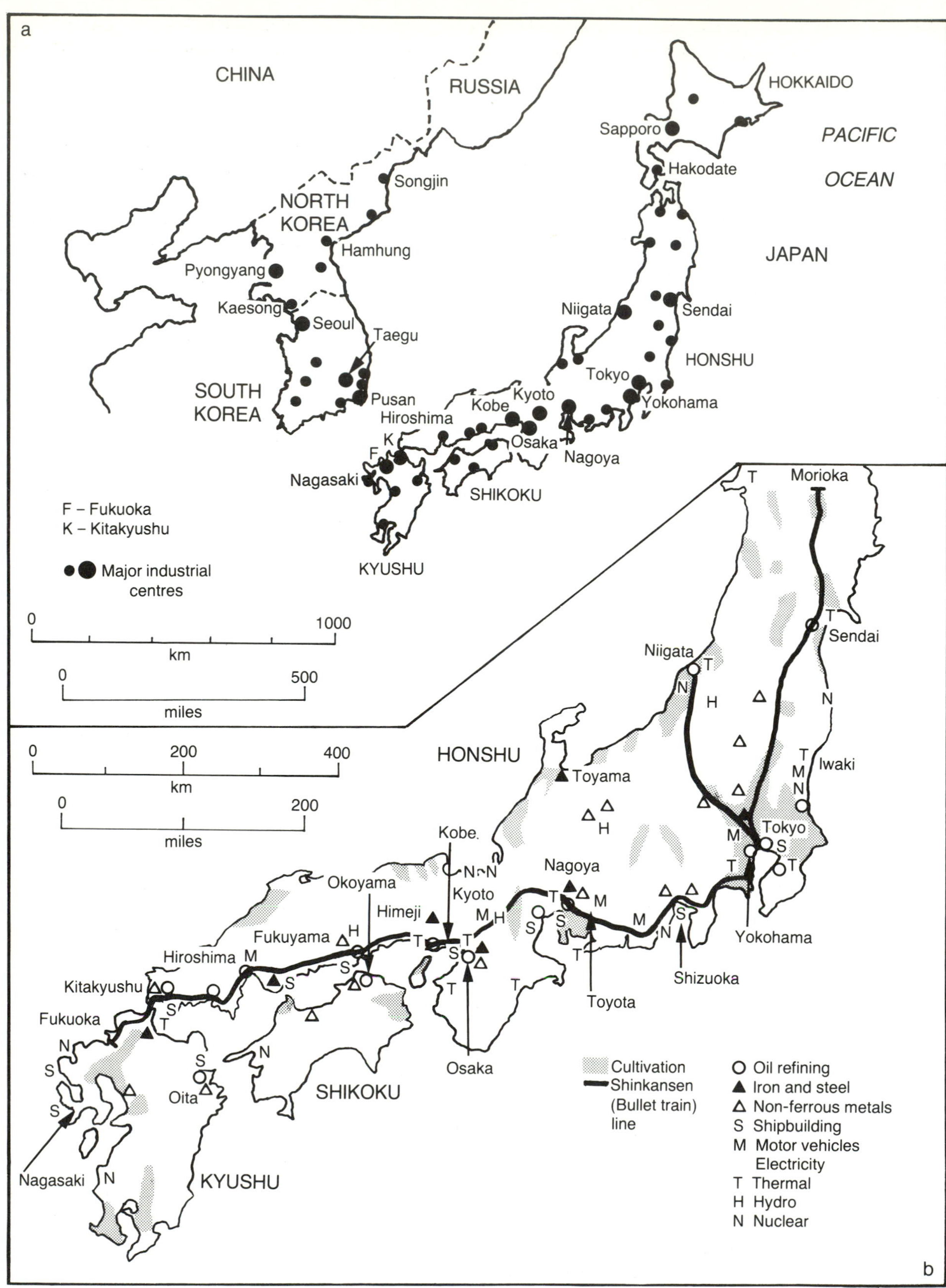

FIGURE 8.7 (a) The location of major industrial centres in Japan and Korea (b) Economic map of Japan

required the maintenance of a formidable war machine. The raw material and industrial base of the country was stretched to the limit and, uncharacteristically, large numbers of women were brought into the labour force. Between the 1870s and the start of the Second World War, Japan was already an exporter of manufactured goods, mainly from the light industrial sector, including textiles, but it was not distinguished then as an exporter of high-quality goods.

By 1945 many of the factories of Japan had been destroyed by bombing, as also had much of the urban housing. The natural resource base was very modest. A new approach to industrialisation was needed if Japan was to become a country with high levels of consumption. Development assistance from the USA following 1945, and the Korean War (1950–3), are considered to have had a major role in the initial revival of the Japanese economy. Subsequent development has been attributed to a number of indigenous features of the country. The state and the banking system of Japan have relied on the continuing patriotism of the population (in spite of defeat in the Second World War) to save, participate positively in the development of the companies employing them, work hard for limited remuneration and, with the help of a strong protectionist policy, to buy Japanese manufactures in preference to imported goods. Indeed, for many of its manufactured goods, the home market is the main market. As well as benefiting from state support and guidance, Japanese manufacturers, large and small, have themselves carefully watched trends in the world economy, switching emphasis from products that the country is not well suited to produce and export, such as silk fabrics and aluminium, to ones such as computers and robots, in the production of which it excels.

Japan is best known internationally for its very large companies and large industrial establishments. In reality, only about one-quarter of the labour force in industry is employed in plants with over 300 workers. The latter account, however, for about half of the total value of output, underlining the high level of automation and robotisation in large-scale undertakings. In spite of having over 60 per cent of the industrial robots in the world in the early 1990s, Japan has a much lower level of unemployment than other major industrial countries, one reason being the tendency to employ people 'for life' in many enterprises. In February 1995 the percentages unemployed in the economy as a whole in selected countries were: Japan 2.9, the USA 5.5, the UK 8.6, France 12.6 and Spain 23.9

One of the major preoccupations in the Japanese economy has been the dependence on imported energy and raw materials. Only about 15 per cent of Japan's primary energy consumption now comes from the contributions of home-produced coal, hydroelectric power and nuclear energy. Oil, coal and liquefied natural gas are all imported in large quantities. Over 70 per cent of Japan's oil comes from the Middle East, some from Indonesia. The oil price rises in 1973–4 and 1979 were absorbed without economic disaster, although industrial output fell by more than 10 per cent in 1975. Japanese energy consumption per inhabitant, at about one-third the US level, is low for a country with such high material standards, and it is used efficiently. Coal production has declined almost to nothing (see Figure 8.8), the hydroelectric potential is fully utilised, and nuclear power is suspect on safety and environmental grounds. It therefore seems inescapable that Japan will need to maintain an unobtrusive profile in world affairs in order to keep on good terms with Middle East countries and with other suppliers of its huge fuel needs. Virtually all the requirements of many raw materials are also imported. The amount of timber cut has fallen by about 50 per cent between 1960 and 1990 (see Figure 8.8). Japan now imports virtually all the iron ore it consumes, almost half from Australia and a quarter from Brazil.

Although nowhere in Japan is far from the sea, much of the industrial capacity is located on or very close to the coast and to facilities for handling cargoes from ocean shipping. Figure 8.7a shows the major industrial centres of Japan and the Koreas and Figure 8.7b shows the location of heavy industrial, processing and manufacturing establishments. Japan's major oil refineries, chemicals plants, iron and steel mills and thermal power stations all have coastal locations, as do nuclear power stations, most of which are tactfully located at some distance from the largest cities. The main industrial zone of Japan extends from Tokyo in the east to Fukuoka in the west. Given the shortage of

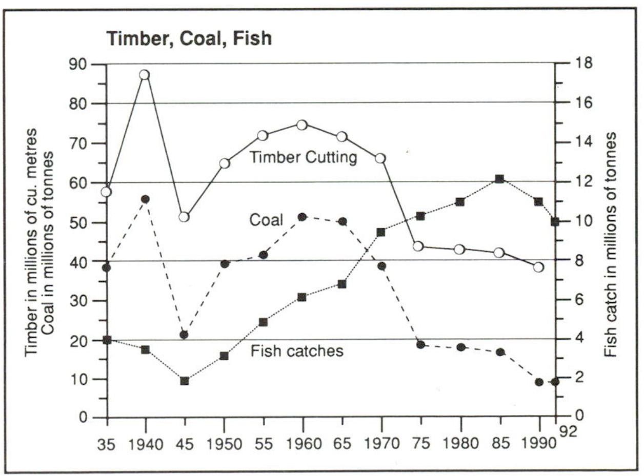

FIGURE 8.8 Japanese production of timber, coal and fish, 1935–92

Source: Nippon (1993), Appendix: 'Statistics by prefectures'. Annual Transition of Selected Statistics suitable flat land along this zone a number of areas, both adjoining the land and offshore on islands, have been reclaimed. The extensive damage to buildings and transport links on the artificial island off Kobe in the earthquake of January 1995 may raise doubts about the wisdom of such developments. By world standards, every prefecture in Japan is highly industrialised. Figure 8.5 (right-hand map) shows that there are, however, considerable relative differences in the value of industrial output per inhabitant, with prefectures in central Honshu attracting much of the expansion of industry in the post-war period. Many smaller centres throughout Japan do, nevertheless, have key industrial functions.

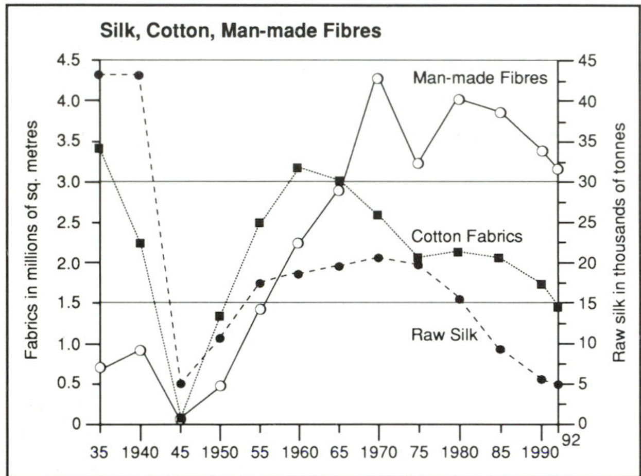

FIGURE 8.9 Japanese production of silk, cotton and man-made fibres, 1935–92

Source: Nippon (1993), Appendix: 'Statistics by prefectures'. Annual Transition of Selected Statistics

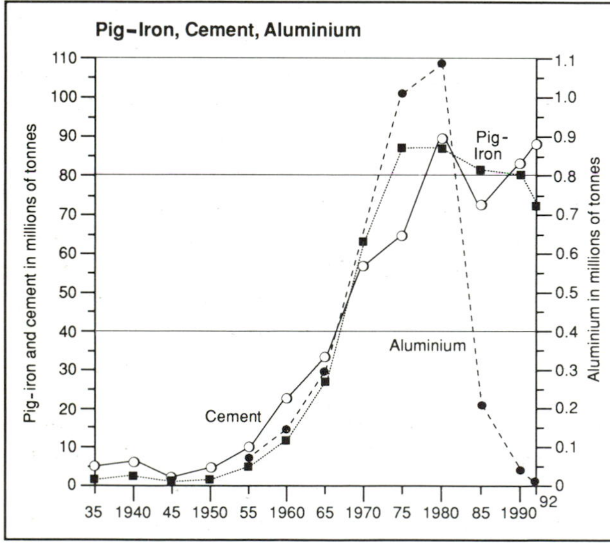

FIGURE 8.10 Japanese production of pig-iron, cement and aluminium, 1935–92

Source: Nippon (1993), Appendix: 'Statistics by prefectures'. Annual Transition of Selected Statistics

Figures 8.9 and 8.10 focus on the performance and fortunes of selected sectors of Japanese industry since the 1930s. Between 1935 and the early 1990s the population of Japan almost doubled, a trend to be taken into account in interpreting the data in these diagrams. The decline in the production of silk and cotton textiles (see Figure 8.9) reflects to some extent the competition from China and other Asian countries, as does more recently the decline in the output of man-made fibres. The great expansion of production of cement, pig-iron and aluminium between 1950 and the late 1970s (see Figure 8.10) reflects the importance to Japan of having a heavy industrial base. All three sectors are large users of energy. For this reason the first stage of smelting of aluminium ores was dropped in the 1980s. Further expansion of the iron and steel, and cement, industries has been curtailed.

Japan's position in the world with regard to five representative manufactured products is shown in Table 8.4. The first year, 1953, marks the end of the Korean War, while 1973 is the last year before the first

TABLE 8.4 The post-war performance of five selected branches of Japanese manufacturing

	1953	*1963*	*1973*	*1980*	*1990*
Steel (millions of tonnes)	7.7	31.5	119.3	111.4	110.3
% of world production	3.3	8.1	17.2	15.5	14.5
Merchant vessels (million GRT)	0.6	2.4	15.7	8.2[1]	6.6
% of world production	10.9	27.7	49.7	47.4	44.3
Passenger cars (millions)	–	0.4	4.5	7.0[2]	9.9
% of world production	–	2.5	14.9	25.1	28.5
Radio receivers (millions)	1.4	18.0	24.5	15.3	11.0
% of world production	5.4	28.8	20.0	8.2	7.3
TV receivers (millions)	–	4.9	14.4	15.2	15.1
% of world production	–	19.4	26.6	21.0	11.9

Notes: – negligible
[1] 1982
[2] 1981
GRT = gross registered tonnes
Sources: UNSYB 1971, Tables 121, 129, 130, 131, 132; *UNSYB 1979/80*, Tables 114, 122, 123, 124, 125; *UNSYB 1990/91*, Tables 89, 97, 98, 99

rise in world oil prices and 1980 the first year after the second round of oil price rises. Between 1953 and the early 1990s the population of the world roughly doubled whereas that of Japan grew by only 30 per cent.

- **Steel** The peak year for Japanese steel output was 1973. Since 1974 the output of steel has fluctuated between 97 and 111 million tonnes. In contrast, EU and US steel output has declined appreciably in the last two decades, while output has risen sharply in several developing countries.

- **Merchant vessels** At the peak of their production, Japanese shipyards accounted for almost half of the ships built in the world. The tonnage launched globally has fluctuated sharply over the last two decades but Japan has retained a large share, while in the USA and Western Europe the industry has declined drastically, virtually disappearing in some countries. On the other hand, South Korea has greatly increased its share of the world's shipbuilding industry, building its first ships in 1972 and gaining 22 per cent of world production by 1991.

- **Passenger cars** Car ownership in Japan has risen from 9 million in 1970 to 35 million in 1991, and is now near to saturation. The growing, potentially large, home market, was a major factor in enabling Japan to account for a quarter of world car production by 1980 and to become the world's leading exporter. The production of commercial vehicles has a similar history. A rejuvenation or revival of the motor-vehicles industry in the USA and the EU, the rise of new producers in the developing world, particularly South Korea, and the policy of Japanese car firms to open plants in Western Europe and North America, as well as in developing countries, points to future stagnation in production in Japan itself.

- **Radio and TV receivers** These were among the first industrial products of Japan to figure prominently among exports in the post-war period. By the 1970s, with labour costs well below those of Japan, Hong Kong was producing 40 per cent of the world output, virtually all for export. In the 1980s, both China and South Korea outstripped Japan in the production of TV sets. Japanese production of radios and TV sets has settled at a level where output is mainly for the home market.

The Japanese industrial sector usually responds quickly to changes in the world economy. As far as possible, the home market is satisfied, while special attention focuses on sectors turning out products likely to be suitable for export in given periods. Since Japan has to import large amounts of primary products from such countries as Saudi Arabia and Australia, which take only small quantities of Japanese exports, it has to maintain a favourable balance of trade with both North America and the EU and with as many developing countries as possible. The newly industrialising countries of East and Southeast Asia increasingly expand sectors of manufacturing that compete with Japan, the whole of the population of South Korea, Taiwan and Hong Kong having been in competition for more than two decades, the large city populations of Thailand, Indonesia, the Philippines and Malaysia more recently. Already some of the coastal cities of China are shaping up to becoming future miniature Japans. It is easier to trace the way these places have followed the path of Japan than to see the future path of Japan itself, especially if there is another sharp rise in world oil (and natural gas) prices and if North America and the EU make it more difficult to export Japanese-manufactured goods to them.

Of the twelve regions of the world covered in this book, Japan seems to have the most straightforward future, particularly because its population size is unlikely to change much in the next few decades and, unless it recreates a non-colonial union on the lines of the European Union in the western Pacific, its modest natural resources will not change. It is a 'mature' country in the sense that there is little room for manoeuvre, and the future should not contain great surprises. A number of 'problems' will keep Japan busy, but they can be dealt with or at least mitigated.

The concentration of the population in certain small areas, with the need even to reclaim new spaces off the coast at great expense, has been noted. A very severe earthquake centred on or near Tokyo could, however, disrupt the organisation of Japan itself and have world-wide repercussions. The speed of travel for passengers and goods continues to increase, with occasional lapses in planning, such as the 'white elephant' rail tunnel between Honshu and Hokkaido (see Box 8.2). Regional disparities in living standards are no higher than those within some larger EU countries, notably Italy, the UK and Spain. Excessively competitive conditions in the educational system lead to problems, such as stress among children in school and students in higher education, but Japan's economic success is widely attributed primarily to the high level of participation in higher education. The healthcare services have been so successful that the size of the elderly population is increasing noticeably, a situation requiring perhaps a redefinition of 'old' from 65 and over to 70 and over. Internally, Japan's problems are not as great as those in most other regions of the world, but the future of its foreign trade, investments and development assistance are more uncertain and less easy to control.

BOX 8.2

Japan's inter-island fixed links

Japan consists of one main island, Honshu (231,000 sq km, 89,000 sq miles, 62 per cent of the area of Japan), three sizeable islands, Hokkaido (21 per cent), Kyushu (11 per cent) and Shikoku (5 per cent), and a very large number of smaller islands. Numerous ferry services link the various islands of Japan, while regular shipping services link ports on the 'mainland', Honshu. The development of Japan's rail and road networks has been accompanied by a perceived need to provide fixed links between Honshu and the other three largest islands. The locations of the fixed links can be seen on the map of the whole of Japan in Figure 8.7, while their sites are shown in Figure 8.11.

A At its narrowest, the strait between the western extremity of Honshu and the north of Kyushu is only a kilometre (0.6 miles) wide. It is crossed by an older rail tunnel (Kanmon), a newer very long rail tunnel (Shin–Kanmon) carrying the fast Shinkansen tracks, and a road bridge.

B Shikoku is separated from Honshu by the so-called inland sea (Setonaikai). With the help of smaller islands serving as stepping-stones, it is proposed to provide three fixed links, two carrying both road and rail, to link the two islands. The cost of the whole project, which (if and when completed) will include nine of the twenty longest suspension bridges in the world, was estimated in the late 1980s at 16–17 billion US dollars, probably an underestimate. The prefectures of Shikoku are among the poorest in Japan and it is hoped that they will benefit from the links.

C The main reason for constructing the bridges and tunnels described above was to cut down journey times between Honshu and the two islands in question. In the case of the Hokkaido link, an additional consideration made its construction desirable: the vulnerability of ferries to inclement weather conditions on the Tsugaru-Kaikyo Strait; a typhoon once sank five ferries, drowning 1,430 people. Several have sunk there since the Second World War. The Seikan Tunnel was opened in 1988. Its total length is 54 km (33.5 miles), of which less than half (23 km, 14 miles) is under the sea. The outstanding engineering feature of the tunnel is the depth below sea level to which it has to descend, 240 metres (790 feet), because of the comparatively deep water of the strait. The cost of construction was about 70 billion US dollars and the work took two decades, the plan having originally been put forward in 1956. The rail tracks through the tunnel carry conventional trains but not the Shinkansen bullet train services. Much more traffic goes by road in Japan than in the 1950s when the tunnel was designed, but the tunnel does not carry lorries. Again, most passengers travelling from central and southern Japan to Hokkaido now go by air. In short, the tunnel now has little socio-economic value. Watts (1988) noted that it would lose money at a rate that will make the Channel Tunnel look like a bargain. It is perhaps unkind, but also reassuring, to find that Japanese politicians are as expert at sponsoring white elephants as other governments.

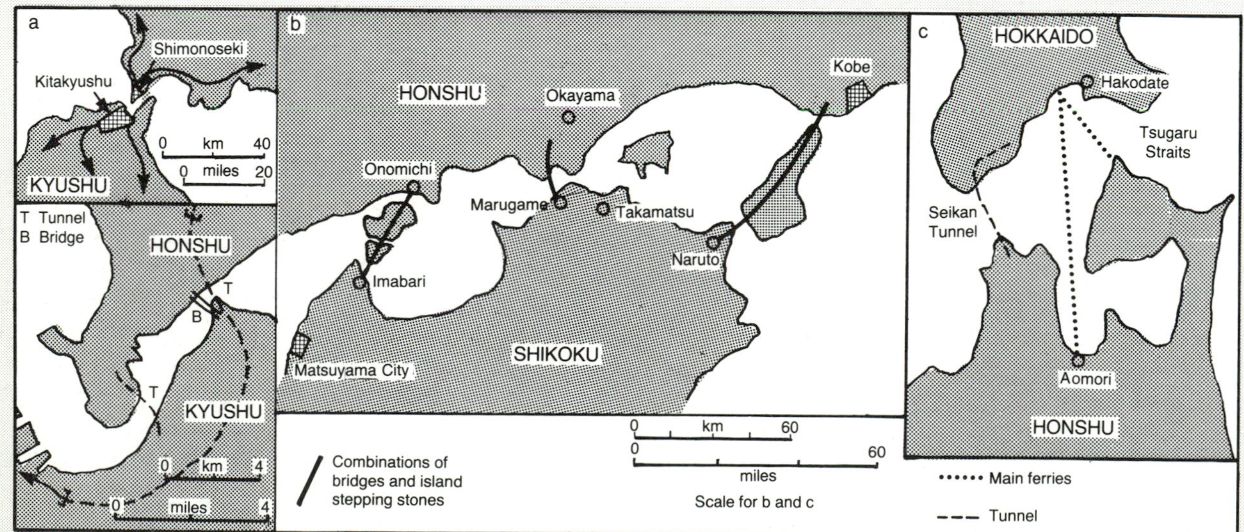

FIGURE 8.11 Japan's existing and proposed fixed links between 'mainland' Honshu and the three other large islands

PLATE 8.2 Old and new in Japan

a Nara, historic city in western Japan near the former capital Kyoto. Since many older buildings in Japan were constructed of wood not many remain today on account of the occurrence of fire. Many have been greatly restored or completely rebuilt

b Ginza, Tokyo, dazzling shopping streets and heavy traffic. Tokyo's subway is crucial for keeping the flow of passengers from seizing up. Cyclists use the pavements, not the roads. The traffic drives on the left, a feature of no significance to the Ministry of Transport since the Japanese do not take their cars abroad

In conclusion, the data in Table 8.3 allow a study of some aspects of regional diversity and regional homogeneity in Japan.

Column (1) shows the intense concentration of population in a few prefectures already referred to, notably nos. 11–14, 23 and 26–8. The relative importance of the extreme west of Japan has declined since the loss of colonies, while Nagoya and Tokyo have grown in relative importance.

Column (2) shows the marked contrasts in density of population among the prefectures, reflecting the extent of plains, the basis of the predominantly agricultural population in the nineteenth century, and the concentration in ports and industrial centres in the twentieth.

Column (3) highlights the lack of agricultural land in Japan as a whole and the uneven distribution of population.

Column (4) shows marked differences in the average manufacturing output per inhabitant, an indication of the dependence on the industrial sector.

Column (5) shows the difference in retail sales per inhabitant and, allowing that the cost of living is higher in the larger cities than in the countryside and smaller ones, at this level of regional breakdown the difference is not very marked.

8.5 South Korea

Korea was a colony of Japan from 1910 to 1945. At the end of the Second World War it was subdivided into the communist North, supported by the USSR, and the non-communist South, supported by the USA. The division was along the parallel of 38° North. South Korea was proclaimed a republic in 1948 but its new status seemed doomed when in 1950 it was almost overrun by forces from the North. After three years of war, in which much damage was caused in both parts of the country following United Nations (mainly US) and communist Chinese military intervention, a new boundary was established between the two countries.

After 1953 South Korea was governed first by a civilian dictatorship then by a military government. Both Koreas embarked on a programme of industrialisation, the North following the Soviet model of central planning, the South based more on the Japanese model, with considerable state influence but also a private sector, which is now dominant. Heavy industry was the leading industrial sector in North Korea, but more diversified industrial development was favoured in South Korea. North Korea is covered in more detail in Chapter 16.

South Korea is 99,000 sq km in area (38,000 sq miles), which compares with the areas of the state of Virginia in the USA and with Portugal in the EU. The total population was 44.5 million in mid-1994, seven times the population of Virginia. South Korea extends only about 400 km (250 miles) in a north–south direction and 250 km (160 miles) across in an east–west direction. Nowhere is more than about 120 km (75 miles) from the coast. In such a small area it is to be expected that natural resources are very limited. As in Japan, much of the surface is mountainous, and only 21,000 sq km (8,000 sq miles) are under arable or permanent crops; two-thirds of the total area is forested. Coal is the main fossil fuel in South Korea, but it satisfies only about 10 per cent of the country's energy needs. There are few non-fuel minerals.

Agricultural production in South Korea is still growing, thanks to increasing yields, but the sector's share of the total economically active population has fallen sharply in the last three decades, from over half in 1965 to less than a fifth in the early 1990s. The absolute number of about 4 million employed still means that the sector is very labour-intensive, with traditional farming methods used and high-cost rice produced thanks to protection from imports, a situation to be changed according to new trade agreements. There are about 20 persons per hectare (8 per acre) of cultivated land in South Korea and, as might be expected, the country is a large importer of food, as well as of both agricultural and mineral raw materials.

The industrial development of South Korea has followed the Japanese model closely in some respects. As a Japanese colony, Korea was organised to supply Japan with primary products, and after the Second World War industry was largely based on processing. After 1953 various sectors of manufacturing were developed, with consumer manufactures, steel production and shipbuilding growing fast. More recently, as earlier in Japan, emphasis moved to light engineering and to electrical and electronic products.

With the incentive of a fast-growing home market for cars and commercial vehicles, South Korea has built up one of the largest motor vehicles industries in the world in less than three decades. The first cars were assembled by Hyundai in 1967. In 1994, 2.3 million vehicles were produced, while current expansion should provide capacity for the production of 6 million units a year by the year 2000. In the highly protected home market of South Korea, growth in domestic sales is now slowing down. The 'Big Three' car manufacturers, Hyundai Motors, Daewoo Motors and Kia, to be joined by Samsung, the largest South Korean conglomerate, in 1997, will need to win a considerable share of the world's motor vehicles market to dispose of such a large output.

Other sectors of manufacturing are growing in parallel fashion to the motor vehicles sector. For example, Samsung claims world-wide recognition for advances in consumer electronics, semiconductors, computers and information systems. In the service sector South Korea has also spread its influence globally. For example, the important Korea Exchange Bank has twenty-eight overseas branches or representatives, six each in North America, Western Europe and Southwest Asia, three each nearer home in Japan and China, two in Latin America and one each in Australia and the Middle East. These branches are located in eight out of the twelve major regions in this book.

Level and gently sloping land is limited in extent in South Korea and offshore reclamation for industrial purposes is taking place. The internal movement of goods and passengers depends very heavily on road transport and there is a good system of motorways. Nowhere is far from the coast. On the other hand, close proximity to the border with North Korea is a negative influence on the choice of a location for new industries, while distance from the capital, Seoul, the largest concentration of population, has both positive and negative considerations. Harvard (1995, p. 12) points out that:

> Seoul lies perilously close to the de-militarised zone, or DMZ – the border with North Korea.... Taxi drivers say that if North Korea invades, Seoulites know what to do. Before fleeing southwards, drivers will leave their cars parked on the roads to block the military advance. Actually this would be nothing

new: Seoul's traffic gets snarled up every day, and the rush hour seems to last from dawn to dark.

From the rubble left by the Korean War, with a population of less than 1 million, Greater Seoul has grown to about 11 million and has within commuting distance about half of the population of the country. Recognition of its growth and potential congestion was shown with the opening of the first underground line in 1974.

Just as the shadow of communist China falls over Hong Kong and Taiwan, so South Korea is haunted by the spectre of North Korea. The North has the prospect of developing nuclear weapons, and has leaders like Zhirinovsky of Russia still using Cold War rhetoric, last year threatening to turn Seoul into a 'sea of flames'. According to UN data (*HDR 1993*, Table 21), 10 per cent of North Korea's GDP goes on military expenditure but the smaller proportion in South Korea has diminished from 6 to 4 per cent of GDP between 1960 and 1990. The possibility of reunification has been considered in recent years, but the process is unlikely to be as rapid as it was with the two Germanies. For example, the transformation of the North from a state to a market economy would be a complex undertaking, while the development of road transport to the level reached in the South would be slow and costly. In 1990 the real GDP per inhabitant in South Korea was 6,700 US dollars, and rising fast, that in North Korea was 2,000 and stagnating.

South Korea's already impressive position in the world economy is matched by its aspirations. It is expected that the country will become a member of the Organisation for Economic Cooperation and Development by 1996 or soon after. That would make it the first country in the developing world to 'graduate' to the developed world. It is hoped that the country will soon join the Security Council of the United Nations for a term of office. A South Korean is in the running for the post of head of the World Trade Organisation (WTO), the successor to GATT. At home President Kim Young Sam, the first civilian leader for over thirty years, is attempting to stamp out corruption, extend democratic freedoms and ensure the functioning of a critical press. Like Japan, South Korea needs to protect a favourable image round the world because it depends so heavily on many sources of both primary products and capital manufactures for its existence and to achieve its goal of having by 2010 or 2015 a larger economy than those of Germany or the UK.

By far the largest trading partners of South Korea are the USA (24 per cent of imports, 30 per cent of exports) and Japan (27 and 19). Exports consist almost exclusively of manufactures, notably means of transport, electrical and electronic goods, textiles and clothing. The sizes of the largest industrial companies in East and Southeast Asia are compared in Table 8.2. The scale of Japan's largest companies is matched only by those of the USA, but in East Asia, South Korea comes second. The scale of the largest companies in some of the Newly Industrialised Countries of East and Southeast Asia is compared with the giant companies of Japan. Except in South Korea, the emphasis is on services rather than manufacturing. Some of the manufacturing is organised in small companies and some is owned by transnational companies based outside the region.

A few decades ago few would have predicted that there would be rejoicing in Northeast England at the promise of investment there by South Korea's Samsung conglomerate, planning to manufacture microwave ovens and computer monitors. According to Narborough (1994), 6,000 new jobs will be created directly or indirectly in the Cleveland area, the scene of industrial

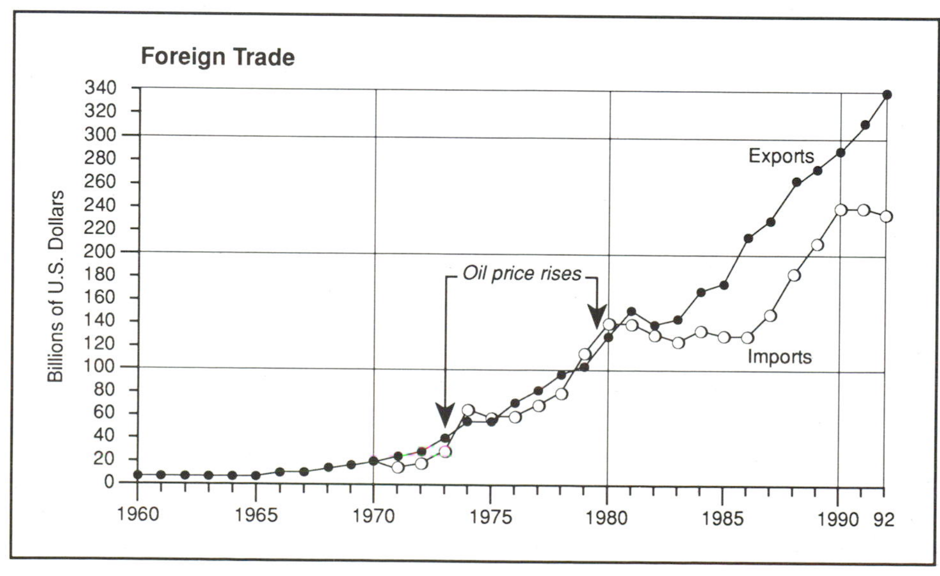

FIGURE 8.12 The value of Japan's foreign trade, 1960–92
Source: Nippon (1993), Appendix: 'Statistics by prefectures'. Annual Transition of Selected Statistics

developments in the earliest decades of the Industrial Revolution. Understandably, the British Government pledged generous grants and loans to support the operation. The story is that Samsung's representatives were attracted by the proximity of one of England's most distinguished football clubs, Newcastle United.

Since 1945 Japan's prominence in the world economy has been achieved initially through industrial growth, more recently also because of the way it has established itself as a centre of international finance. It depends heavily on the import of primary products, arguably an advantage so long as world trade favours exporters of manufactures. At present fuel and raw materials are forthcoming from developed countries (e.g. Australia, Russia) and from developing countries (e.g. Brazil, Malaysia). In the last resort, however, the Japanese economy is vulnerable for two different reasons. Catastrophic events, such as serious earthquakes, or further sharp rises in the price of oil, produce an unstable situation, making planning for the future difficult. Second, slower processes of change cause concern. Some, such as the ageing of the population and pressure on areas suitable for settlement, are internal. Others, such as the rise of new industrial powers, are external. South Korea so far has followed the path of Japan, with a time lag of two to three decades, but its future is less predictable, given the unresolved problem of uniting (or not uniting) North and South.

KEY FACTS

1. Japan and South Korea are rugged, densely populated countries with little agricultural land and few natural resources.

2. Since the 1850s Japan's leaders have deliberately sought to develop modern industries. Today it is second only to the USA, industrially and financially. In the 1950s South Korea set out on a path similar to that taken a century earlier by Japan.

3. The remarkable development of Japan has been attributed to the dedication and high educational levels of its population.

4. Japan's success has been achieved at a price: heavy dependence on other parts of the world for fuel, raw materials and some types of manufactures, as well as pollution, congestion and stress at home.

5. South Korea's political and economic future is uncertain, as the prospect of the reunification of North and South Korea remains unresolved.

SUBJECTS FOR SPECULATION

1. The Japanese situation raises a question of global relevance. How long can the net importers of primary products, including Japan and South Korea, as well as Western Europe, count on other parts of the world to supply their needs?

2. Put another way, will many countries of the world be able to use Japan as a model in their development aspirations in the next decades by following its experience?

3. If the two Koreas are unified, will the cost of rehabilitating the North interfere with the relative prosperity and growing international influence and status of the South?

Chapter nine

The USA

There are, at the present time, two great nations in the world which seem to tend towards the same end, although they started from different points: I allude to the Russians and the Americans. Both of them have grown up unnoticed; and while the attention of mankind was directed elsewhere, they have suddenly assumed a most prominent place among the nations; and the world learned their existence and their greatness at almost the same time.'

Alexis de Tocqueville (1835)

9.1 Introduction

As a physical unit North America conventionally extends south to the isthmus of Panama, but the USA and Canada are usually regarded as culturally distinct from Latin America, from which they are separated by the US–Mexican border. The USA and Canada were commonly referred to some decades ago as Anglo-America, but usage has changed, one reason being the growing nationalist feeling in Canada's French-speaking Quebec, another the increasing number of Hispanics and other 'non-Anglo' peoples in the USA.

The USA and Canada (henceforth referred to as North America), occupy 14.4 per cent of the total land area of the world (19,350,000 sq km – 7,470,000 sq miles), making them together roughly comparable in area to Latin America, but considerably smaller than the former USSR. In contrast, the 287 million inhabitants (in 1993) of North America account for only 5.2 per cent of the world's population, a share expected to decline to about 4.4 per cent in 2025, in spite of an increase of about 85 million people in the next three decades.

Like Latin America, North America is characterised by its great north–south extent, with the northern part of Ellesmere Island in the north of Canada reaching to within a few degrees of the North Pole and the state of Florida in the USA extending to 25° North of the Equator, close to the Tropic of Cancer. In relation to its population size, North America is generously endowed with natural resources: with 5.2 per cent of the world's population, it has 14 per cent of the productive land of the world, 11.7 per cent of the freshwater resources, 11.4 per cent of the fossil fuels and 17.5 per cent of the non-fuel minerals.

After the start of the Industrial Revolution in Britain in the eighteenth century, the northeast of the USA was one of the first regions outside Britain to which the technology and organisation of modern industry spread. The natural resources of North America have been used intensively ever since, both for use in the region and for export. Population spread westwards

TABLE 9.1 United States calendar

Year	Event
1607	First permanent English settlement in America (Jamestown, Virginia)
1620	Puritans land in New England (*Mayflower*)
1625	Dutch settle New Amsterdam (New York)
1684	Louisiana claimed for France
1728	Bering reaches Alaska from Siberia
1775	American Revolution begins
1776	American Declaration of Independence, recognised by British 1783
1789	George Washington first president of USA
1803	Louisiana Purchase (from France) doubles area of USA
1819	USA purchases Florida from Spain
1823	Monroe Doctrine: US declaration that there should be no further colonisation or recolonisation of the Americas by Europe
1838	Cherokee Indians removed from Southeast USA and marched to reservations in Oklahoma (the Trail of Tears)
1845–8	USA annexes Texas and conquers New Mexico, Arizona and California
1846	Oregon Treaty delimits US–Canadian boundary
1849	California Gold Rush
1861–5	US Civil War. Slavery abolished
1867	Russia sells Alaska to USA
1868	Completion of first transcontinental railway in USA
1898	Spanish–American war: Spain cedes Guam, Puerto Rico and the Philippines to the USA, which also guarantees Cuba's independence
1917–18	USA in the First World War
1920	USA withdraws into isolation, choosing not to join the League of Nations
1921	USA restricts immigration
1929	Wall Street Crash. World depression starts
1934	Good neighbour policy in Latin America: no interference
1941–5	USA in Second World War
1945	'Atom' bombs dropped on Hiroshima and Nagasaki in Japan
1950–3	Korean War
1959	Alaska and Hawaii become states
1961–73	Heavy US involvement in Viet Nam
1962	Cuba missile crisis: Kennedy–Khrushchev confrontation
1963	Assassination of President J. F. Kennedy
1964	Civil Rights Act goes into effect
1969	First US astronauts land on the moon
1989	California earthquake destroys parts of San Francisco and surrounding cities
1990–1	Iraq invades Kuwait. Gulf War

into the best farming lands, coal and then oil production grew, and non-fuel minerals were extracted. Well before the end of the nineteenth century the eastern part of the USA already had a dense railroad network, while several transcontinental lines linked east and west across the mountain states. The fact that North America has been a major industrial region for almost two centuries, and now accounts for over a fifth of the total production of goods and services in the world, is an indication that many of its non-renewable natural resources have already been used up. Increasingly the USA has to import fuel and raw materials for its industries. Initially Canada was a major source, but more recently it has traded extensively also with regions further afield.

Many excellent accounts of the territorial growth and of the demographic and economic history of North America are readily available (see Table 9.1 for a selection of key dates in US history). Attention in this chapter therefore focuses largely on current features

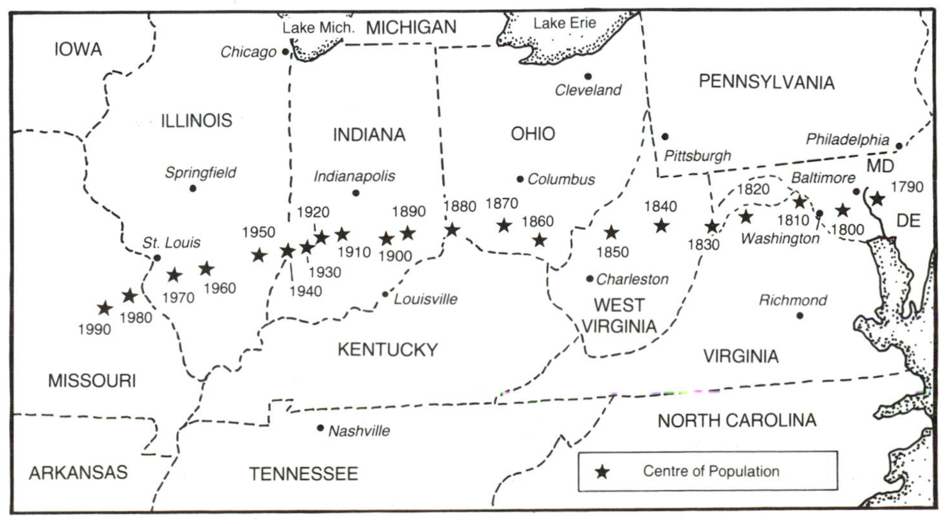

FIGURE 9.1 The centre of United States population, 1790 to 1990

Source: Statistical Abstract of the United States 1992, p. 9

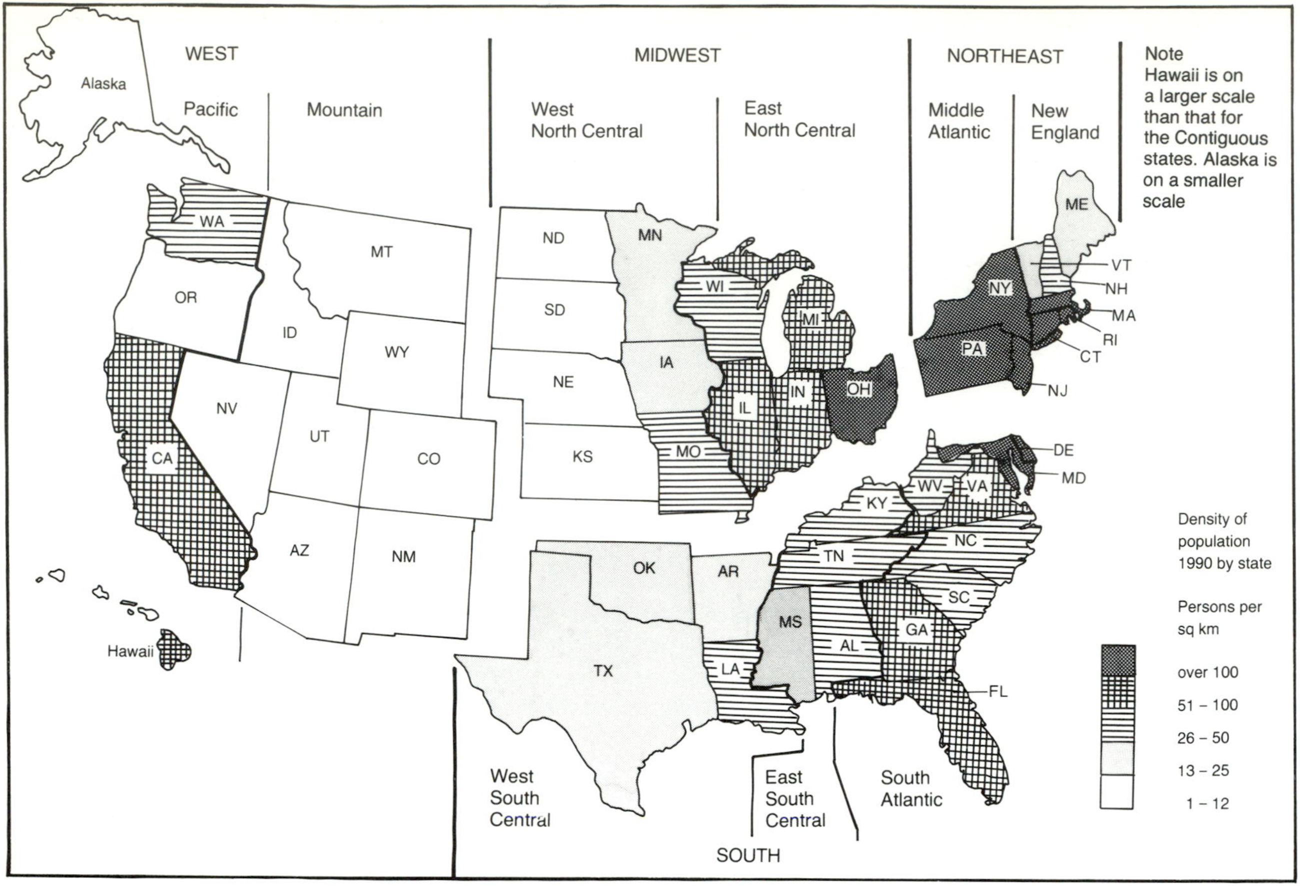

FIGURE 9.2

a Density of population, persons per square mile

Source: Statistical Abstract of the United States 1992

b (*see facing page*) Cartogram of US population by states

and problems of the geography of the USA, on its relations with other major world regions, and on issues that can be expected to arise in the future. The USA has passed through periods of isolationism and periods of involvement outside its borders. Although the Russian Federation remains an industrial and military power of great importance, and convergence in the EU is creating a new economic superpower in Western Europe, in the 1990s the USA has emerged as the single strongest power in the world both economically and militarily, giving it inescapable commitments outside North America.

The influence of the USA seems likely to increase, at least for a time, and its various transactions with foreign countries, including trade, investment, and development assistance, will grow. The 'police-keeping' role of the USA in the world, whether conforming to the policy of the UN, or in its own interests, ensures that the US presence is felt very widely round the world. Even the former Soviet bloc and China, with which friendly contacts before the 1990s were sporadic, are now increasingly influenced by US policy.

Although the establishment of NAFTA will strengthen the already very close economic ties between the USA and Canada, the sovereignty of the two countries is not to be diminished in the way that convergence in the EU is eroding the national independence of its Member States. Canada will be covered in Chapter 10, where it fits appropriately with Australia and New Zealand.

9.2 Population

Among the developed regions of the world, the USA resembles Canada and Australia, and differs from Europe, the Russian Federation and Japan, in experiencing a considerable rate of natural increase in population, enhanced by considerable net in-migration from other countries. This growth trend is expected to continue well into the twenty-first century. Population Reference Bureau estimates anticipate an increase in

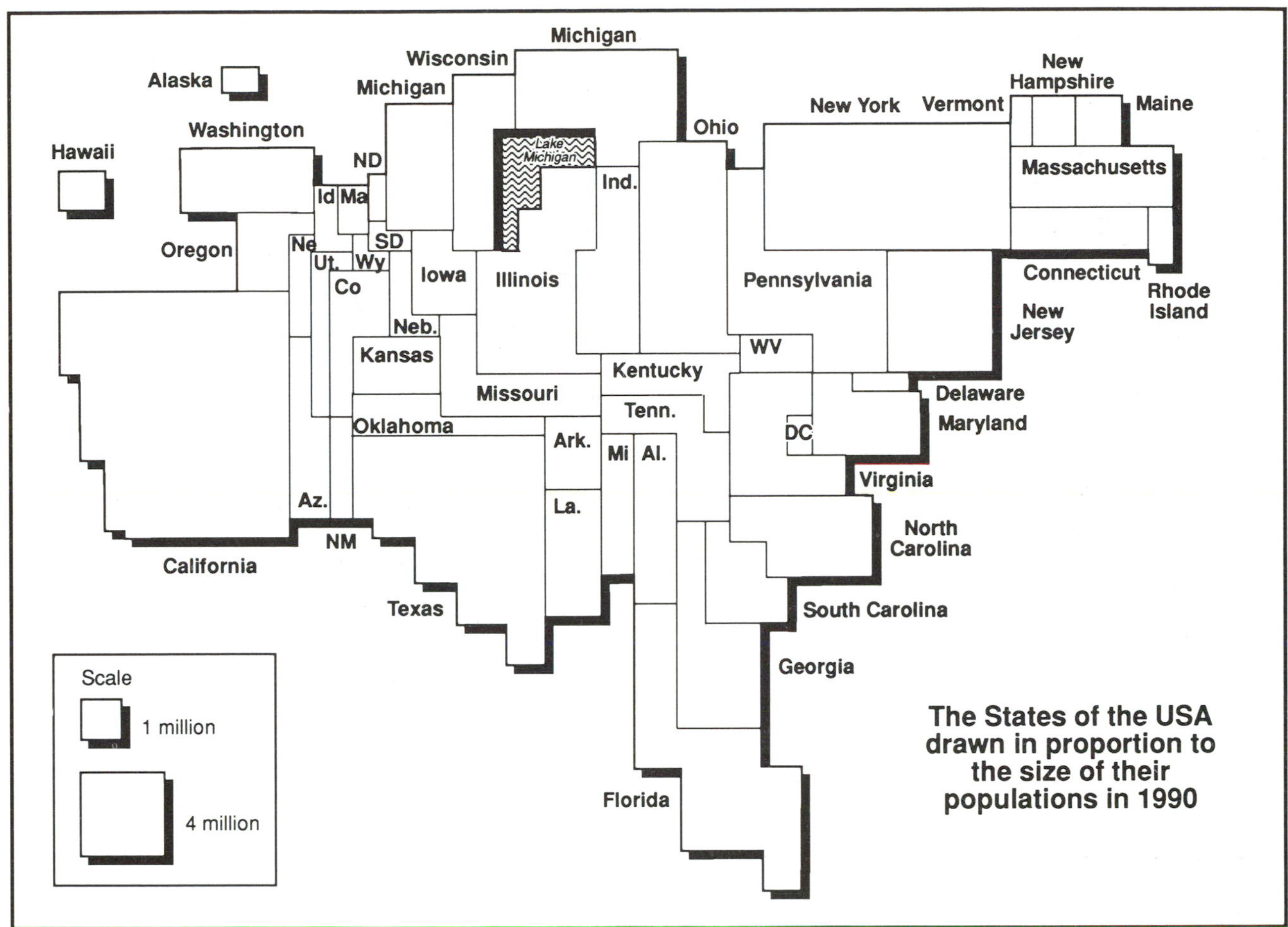

the US population from 258 million in 1993 to 299 million in 2010 and 335 million in 2025, an average annual increase of 2.4 million during both 1993–2010 and 2010–25. The effect of the total fertility rate of only 2.0 in 1993 should eventually slow growth down after 2025, producing a near rectangular-shaped population pyramid of under-70s by about 2050 (*Population Bulletin* 35, 1 April 1980, pp. 18–19). With the continuing increase in life expectancy and a net immigration constant at 750,000 a year, a total of 350–60 million is possible by 2050.

The concept of 'over-population' has rarely been used with reference to the USA. If current high levels of consumption per inhabitant of energy and raw materials continue unchanged or indeed increase, then the US economy will have to expand greatly not only in the service (or non-goods) sector but also in the agricultural and industrial sectors, in which employment has declined in recent decades. This is one reason why US involvement with other parts of the world is likely to increase. Two demographic features and trends in the USA will be discussed in the rest of this section, the changing distribution of population over the national area and the changing ethnic composition.

Distribution of population

Since the foundation of the USA over two centuries ago, the centre (of gravity) of population has shifted westward (see Figure 9.1), reflecting the continuing westward expansion of settlement since independence. In 1990 the centre was in Missouri, having moved roughly two-fifths of the way across the continent. A gradual southward bias can be noted from the 1950s, reflecting also the southward movement of population, especially to Florida, Texas, Arizona and California.

The centre of gravity of population is a point on the map and does not itself give an indication of the overall distribution of population in the USA. This is shown in Figure 9.2 by density of population in persons per sq km, the US average in 1990 being 26. Above-average densities are found in all but three states east of the Mississippi-Missouri Rivers (Maine, Vermont, Mississippi), below-average in all but four states west of it (Missouri and Louisiana, which adjoin the eastern area of above-average density, and California and Washington on the Pacific coast). The current great disparities in the density of population between different regions and states of the USA are the result largely of

PLATE 9.1 North America

a Political heart, Capitol, Washington, DC

b Financial heart, Stock Exchange, Wall Street. In fund investment capital, New York now comes third in the world after Tokyo and London

spontaneous demographic developments in recent decades, although earlier, people were encouraged to settle the interior and the west. The average US density of population of 26 persons per sq km (in 1990) is far below the average of 145 for the EU. Even the 37.6 million people in the three states of the Middle Atlantic (New York, New Jersey and Pennsylvania) have a combined density of 142 per sq km. (The density of population in Japan (1990) was 331 per sq km.)

A feature of US demography since early in the nineteenth century has been the growth of cities. According to the US definition of urban, the proportion of urban dwellers is now 75 per cent. Rural–urban migration has been a feature of US population for at least a century, as the agricultural sector has lost jobs. Especially since the 1920s, however, the urban population itself, especially in the larger cities, has tended to become more dispersed, with a decline in density in city centres. Because of the proliferation of automobiles since the 1920s and, by world standards, the development of an excellent highway system, suburbs extend great distances out from city centres, and ribbon development is common along major routeways radiating from the cities. Such large concentrations of population produce not only congestion and stress but also pollution. Some US cities, most conspicuously San Francisco and Los Angeles, are badly located in relation to the risk of serious earthquakes. Others are afflicted by flooding (e.g. 1993 Mississippi basin floods) or hurricanes, the latter especially along the Gulf of Mexico and South Atlantic coastlands.

In order to place the demographic features of the USA in their NAFTA context, Figure 9.3 shows the population structure of Canada, the USA and Mexico. The population 'pyramids' have been drawn on the same scale to allow comparability of absolute size. To make the comparison completely valid, Mexico should be moved one cohort up the scale and the cohort of 1986–90 births inserted. The abrupt reduction in fertility in Mexico in the 1970s, producing equal-sized

c Public housing, Harlem, New York. Such housing partially replaces nineteenth- and early twentieth-century tenements, many very run down, some undergoing gentrification

d Private housing of the well-off in New York, some of whom have good-sized houses and pieces of land to the north of Manhattan. Others prefer to be nearer the centre of things on East Side

cohorts for the under-15s, was not anticipated in the 1960s, when population was still growing fast. The prospect, relished by a former Mexican President, that the population of Mexico would one day exceed that of the USA, seems highly unlikely now. Figure 9.4a is a cartogram showing the distribution of population in major administrative divisions of Canada (provinces), the USA and Mexico (states). It brings out the great population size of the eastern USA compared with that of the territorially extensive West North Central and Mountain groups of states. With 90 million inhabitants in 1993, Mexico forms a large and still fast-growing southern portion of NAFTA, whereas Canada's 28 million people are stretched out along the northern side in a form cruelly described by one geographer as a footnote along the top (*sic*) of the USA.

Columns (1)–(4) in Table 9.2 contain data relating to the area and population of the states of the USA. The density of population is mapped in Figure 9.2 and change in Figure 9.5. Between 1980 and 1990 the population of the USA grew by 9.8 per cent. Only two states (New York and Iowa) and the District of Columbia actually decreased in population. In almost all the states of the Northeast and of the East North Central region the increase was below the US average. The states with the fastest growing populations were

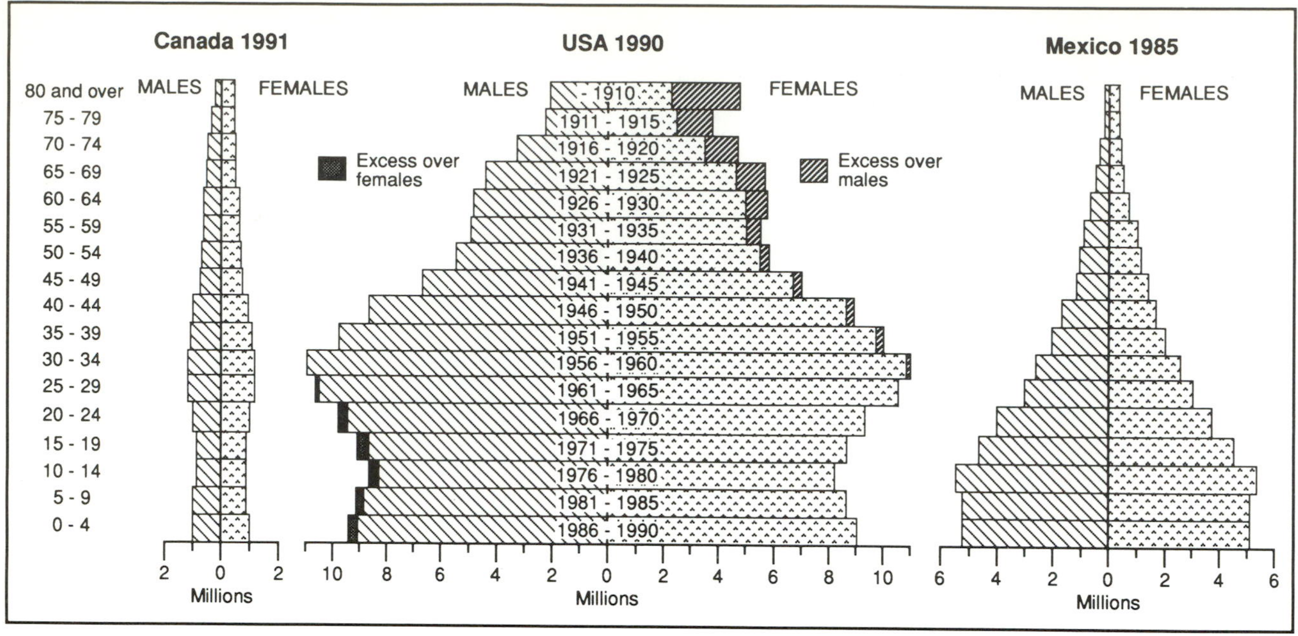

FIGURE 9.3 A comparison of the population structures of Canada, the USA and Mexico. Note that the horizontal scales show absolute numbers (not percentages) and are comparable

all in the west except New Hampshire and Florida. Since the fertility rate does not vary appreciably between states, the changes are largely due to migration, either internal or from foreign countries (column (8) shows the percentage of foreign born). The most striking broad feature has been the net migration from the northern 'snowbelt' states to the southern 'sunbelt' states. In the 1980s, only about 10 per cent of all legal immigrants into the USA came from Europe, whereas almost half were from Asia and over a third from Latin America. In the 1990s many of the non-Europeans have settled in the west or the south of the USA.

The ethnic composition and distribution of population (see Table 9.2, columns (5)–(7))

While it is not politically correct to refer to the USA as an empire, its territorial expansion since independence has in practice led to the acquisition, partly by purchase, partly by conquest, of large, thinly populated areas, inhabited predominantly by indigenous peoples. Those now in the Southwest were already controlled by Spain after its conquest of Mexico (New Spain) in the sixteenth century. The Spanish, Portuguese, British, French and Dutch overseas empires in due course disappeared, almost without exception, one reason being that they were far-flung, and in many areas inhabited originally by large populations of non-Europeans. In contrast, as the USA grew, the English-speaking Americans were able to maintain cohesion in their country through a common language and the 'melting-pot' effect, whereby after a generation or two, if not instantly, immigrants became loyal US citizens. Even so, some of the more recent European immigrants and their descendants have succeeded in maintaining ties with their countries of origin, for example the Poles and the Irish, and in particular the Jews who, since the Second World War, have strongly supported the newly formed state of Israel. There is currently concern that some of the newest immigrants, those from South Asia, have no interest in assimilation at all.

In the 1990s the US ethnic scene is changing, particularly through the large-scale immigration from Latin America, mainly from neighbouring Mexico and from various Caribbean islands. New York in the east, Florida in the south, and the Southwest are absorbing the bulk of the Hispanic migrants, but the melting-pot process can be slow in taking effect. The Asian immigrants from various parts of East and Southeast Asia in the 1970s and 1980s have settled mainly in the West of the USA. In these parts of the country the European basis of US culture is being eroded. Additionally, the African element links the USA to Africa, albeit tenuously, since the ancestors of many of the African migrants have been in the USA longer than most Europeans, but are still not completely integrated.

It is not surprising that US politicians are now more outward looking and aware of world affairs than they were earlier in the century. The country is becoming a complex and cosmopolitan microcosm of most of the

TABLE 9.2 Data set for the states of the USA (data are for 1990 unless noted at foot of the table)

	(1)	(2)	(3)	(4)	(5)	(6)	(7)	(8)	(9)	(10)	(11)	(12)	(13)	(14)
		Population			Percentage of total population									
	Area thous. sq km/(miles)	in thous.	per sq km/(mile)	% change 1970–90 (1970 = 100)	Blacks	Asians	Hispanics	% foreign born	% farming	% manufacturing	Foreign investment (%)	1990 export contributions ($ per cap.)	% with degree	1989 income per cap.
1 ME Maine	86 (33)	1,228	14 (36)	124	0	1	1	3.0	0.9	8	7.9	708	18.8	13.0
2 NH New Hampshire	24 (9)	1,109	46 (119)	150	1	1	1	3.7	0.5	8	10.1	877	24.4	16.0
3 VT Vermont	25 (10)	563	23 (60)	127	0	1	1	3.1	2.1	8	6.1	2,050	24.3	13.5
4 MA Massachusetts	21 (8)	6,016	280 (725)	106	5	2	5	9.5	0.2	9	8.7	1,579	27.2	17.2
5 RI Rhode Island	3 (1)	1,003	319 (826)	106	4	2	5	9.5	0.1	10	6.6	593	21.3	15.0
6 CT Connecticut	13 (5)	3,287	253 (655)	108	8	2	6	8.5	0.2	10	10.0	1,325	27.2	20.2
7 NY New York	127 (49)	17,990	191 (495)	99	16	4	12	15.9	0.5	6	8.1	1,227	23.1	16.5
8 NJ New Jersey	20 (8)	7,730	383 (992)	108	13	4	10	12.5	0.2	8	13.8	987	24.9	18.7
9 PA Pennsylvania	117 (45)	11,882	101 (262)	101	9	1	1	3.1	1.0	9	10.2	715	17.9	14.1
10 OH Ohio	107 (41)	10,847	101 (262)	102	11	1	1	2.4	1.8	10	10.2	1,233	17.0	13.5
11 IN Indiana	94 (36)	5,544	59 (153)	107	8	1	2	1.7	3.4	11	9.9	951	15.6	13.1
12 IL Illinois	146 (56)	11,431	78 (202)	103	15	2	8	8.3	1.8	9	10.5	1,134	21.0	15.2
13 MI Michigan	152 (59)	9,295	61 (158)	105	14	1	2	3.8	1.3	10	7.0	1,988	17.4	14.2
14 WI Wisconsin	145 (56)	4,892	34 (88)	111	5	1	2	2.5	4.0	11	7.5	1,054	17.7	13.3
15 MN Minnesota	219 (85)	4,375	20 (52)	115	2	2	1	2.6	4.8	9	7.5	1,164	21.8	14.4
16 IA Iowa	146 (56)	2,777	19 (49)	98	2	1	1	1.6	9.2	8	8.0	788	16.9	12.4
17 MO Missouri	181 (70)	5,117	28 (73)	109	11	1	1	1.6	3.5	8	7.0	612	17.8	13.0
18 ND North Dakota	183 (71)	639	3 (8)	103	1	0	1	1.5	9.4	3	6.7	563	18.1	11.1
19 SD South Dakota	200 (77)	696	3 (8)	105	0	0	1	1.1	10.9	4	5.1	295	17.2	10.7
20 NE Nebraska	200 (77)	1,578	8 (21)	106	4	1	2	1.8	7.5	6	6.3	439	18.9	12.5
21 KS Kansas	213 (82)	2,478	12 (31)	110	6	1	4	2.5	4.4	8	5.2	853	21.1	13.3
22 DE Delaware	5 (2)	666	126 (326)	122	17	1	2	3.3	1.0	10	16.0	2,018	21.4	15.9
23 MD Maryland	27 (10)	4,781	176 (456)	122	25	3	3	6.6	0.7	4	12.0	542	26.5	17.7
24 DC Dist. of Columbia	0.2 (0.1)	607	3,409 (8,829)	80	66	2	5	9.7	–	2	2.2	527	33.3	18.9
25 VA Virginia	106 (41)	6,187	59 (153)	133	19	3	3	5.0	1.3	7	10.4	1,508	24.5	15.7
26 WV West Virginia	63 (24)	1,793	28 (73)	103	3	0	0	0.9	1.3	5	20.1	864	12.2	10.5
27 NC North Carolina	136 (53)	6,629	48 (124)	130	22	1	1	1.7	1.8	13	11.7	208	17.4	12.9
28 SC South Carolina	81 (31)	3,487	43 (111)	135	30	1	1	1.4	1.4	11	13.3	894	16.6	11.9
29 GA Georgia	153 (59)	6,478	42 (109)	141	27	1	2	2.7	1.2	9	11.9	890	19.3	13.6
30 FL Florida	152 (59)	12,938	85 (220)	191	14	1	12	12.9	0.4	4	7.6	899	18.3	14.7
31 KY Kentucky	105 (41)	3,685	35 (91)	114	7	0	1	0.9	4.7	8	12.7	862	13.6	11.2
32 TN Tennessee	109 (42)	4,877	45 (117)	124	16	1	1	1.2	2.3	10	12.7	768	16.0	12.3
33 AL Alabama	134 (52)	4,041	30 (78)	117	25	1	1	1.1	1.5	9	6.7	701	15.7	11.5
34 MS Mississippi	124 (48)	2,573	21 (54)	116	36	1	1	0.8	2.2	9	5.5	624	14.7	9.6
35 AR Arkansas	138 (53)	2,351	17 (44)	122	16	1	1	1.1	2.7	9	7.6	391	13.3	10.5
36 LA Louisiana	124 (48)	4,220	34 (88)	116	31	1	2	2.1	1.0	4	11.3	3,365	16.1	10.6
37 OK Oklahoma	181 (70)	3,146	17 (44)	123	7	1	3	2.1	2.6	5	8.1	523	17.8	11.9
38 TX Texas	691 (267)	16,987	24 (62)	152	12	2	26	9.0	1.1	6	9.2	1,939	20.3	12.9
39 MT Montana	381 (147)	799	2 (5)	115	0	1	2	1.7	5.7	3	6.4	287	19.8	11.2
40 ID Idaho	216 (83)	1,007	4 (10)	141	0	1	5	2.9	4.5	6	6.6	884	17.7	11.5
41 WY Wyoming	253 (98)	454	2 (5)	137	1	1	6	1.7	3.5	2	8.9	581	18.8	12.3
42 CO Colorado	270 (104)	3,294	12 (31)	149	4	2	13	4.3	1.4	5	5.5	690	27.0	14.8
43 NM New Mexico	315 (122)	1,515	5 (13)	149	2	1	38	5.3	1.0	3	6.3	164	20.4	11.2

TABLE 9.2 *continued* Data set for the states of the USA (data are for 1990 unless noted at foot of the table)

	(1)	(2)	(3)	(4)	(5)	(6)	(7)	(8)	(9)	(10)	(11)	(12)	(13)	(14)
		Population			*Percentage of total population*									
	Area thous. sq km/(miles)	*in thous.*	*per sq km/(mile)*	*% change 1970–90 (1970 = 100)*	*Blacks*	*Asians*	*Hispanics*	*% foreign born*	*% farming*	*% manufacturing*	*Foreign investment (%)*	*1990 export contributions ($ per cap.)*	*% with degree*	*1989 income per cap.*
44 AZ Arizona	295 (114)	3,665	12 (31)	206	3	2	19	7.6	0.2	5	7.0	1,017	20.3	13.5
45 UT Utah	220 (85)	1,723	8 (21)	163	1	2	5	3.4	0.7	6	7.0	926	22.3	11.0
46 NV Nevada	286 (110)	1,202	4 (10)	246	7	3	10	8.7	0.4	2	5.6	328	15.3	15.2
47 WA Washington	176 (68)	4,867	27 (70)	143	3	4	4	6.6	1.2	8	5.4	5,020	22.9	14.9
48 OR Oregon	251 (97)	2,842	11 (28)	136	2	2	4	4.9	2.4	8	4.9	1,430	20.6	13.4
49 CA California	411 (159)	29,760	72 (186)	149	7	10	26	21.7	0.5	7	9.0	1,496	23.4	16.4
50 AL Alaska	1,831 (707)	550	0.3 (0.8)	182	4	4	3	4.5	0.2	3	17.5	5,182	23.0	17.6
51 HI Hawaii	17 (7)	1,108	66 (171)	144	2	62	7	14.7	0.6	2	8.0	162	22.9	15.8
USA	9,373 (3,619)	248,710	26 (67)	122	12	3	9	7.9	1.6	7.6	9.3	1,267	20.3	14.4

Fuller description of variables:
(9) Farm population as a percentage of total population
(10) All employees in manufacturing as a percentage of total population
(11) Foreign direct investment in US manufacturing as a percentage of all investment 1989
(12) Contribution of each state to the total value of US exports 1991 in dollars per capita
(13) Educational attainment: percentage of persons 25 years old and over with a bachelor's degree or higher
(14) Per capita income in thousands of dollars (to nearest hundred) in 1989

Sources: (1) *Calendario Atlante de Agostini* (1994); all remaining variables: *Statistical Abstract of the United States 1992*, US Department of Commerce, Bureau of the Census: columns: (2) p. 22; (4) p. 22; (5)–(7) p. 24; (8) p. xv; (9) p. xv; (10) p. 741; (11) p. 787; (12) p. 796; (13) p. xv; (14) p. xvi major regions of the world: Europeans from Western and Central Europe and the former USSR, Africans, Latin Americans, immigrants from Southeast Asia (Viet Nam, Philippines) and Hong Kong, as well as Chinese and Japanese from earlier decades.

Columns (5)–(7) in Table 9.2 show the number of African American (formerly referred to as Black), Asian and Hispanic citizens as percentages of the total population in each state. Their distribution is shown in Figure 9.6. The states of New York and New Jersey have above national average levels of all three groups, Florida above-average levels of African American and Hispanic people, California of Hispanic and Asian. The large African American element in the Southeast, the old Deep South, reflects the position before the Civil War (1861–5), the substantial proportion of African Americans in the East North Central and Middle Atlantic states, the migration to expanding industrial centres after the Civil War.

In 1990, almost 8 per cent of the US population was foreign born. The main distinction between the members of the three ethnic groups discussed above and the element of the population defined as foreign born is that very few African Americans are foreign born, whereas a considerable proportion of Europeans are, as well as most Asians and Hispanics. Figure 9.7 shows the distribution of foreign-born US citizens in 1990. Apart from Illinois, all the states with an above-average proportion of foreign born are on the coast or adjoin the Mexican border. Greater New York (New York, New Jersey) and the three southern states of New England have above-average numbers of migrants, as do Florida, Texas and California. The interior and the Southeast of the USA have comparatively few foreign born. The distribution reflects the slow process of filtering into other states from the main places of entry into the USA. Five of the seven states of the West North Central region have less than 2 per cent foreign born. The prominence or lack of foreign born partly at least reflects the presence or absence of economic developments. The presence of a large city (e.g. Chicago, Atlanta, New Orleans) also appears to raise the level in certain states markedly above those in the adjoining ones.

The large presence of Hispanics in New York and New Jersey is enough to require for the first time in US history recognition that a second language is necessary, as signs and advertisements in the New York subway demonstrate. In 1984 I visited Paterson (NJ) with a colleague to savour the industrial archaeology, including once-flourishing silk mills converted into apartments. During a taxi ride to a suburb the driver, a Hispanic, was asked on his phone if he could divert to pick up another passenger. His reply was that it would not be appropriate because, he said in Spanish, 'los pasajeros son americanos'!

9.3 Economic features

The service sector of the US economy accounts for over three-quarters of all employment in the country and of total GDP. Its rise to such prominence in the USA reflects not only the growth of essential services in a sophisticated market economy but also the enormous increase in the productivity of workers in the agricultural (Figure 9.8) and industrial sectors. The material production side of the economy still underpins the service sector and, in my view, for that reason deserves some attention in this chapter.

Column (9) in Table 9.2 shows the farm population as a percentage of total population in each state of the USA (*Statistical Abstract of the United States 1992*, p. xv). Out of the total population of the USA in 1990 of 248.7 million, a mere 1.6 per cent, or 3.9 million, were defined as farm population. According to *FAOPY 1990*, 2,880,000 persons were actually employed in agriculture, 2.3 per cent of the total economically active population of 123.2 million in the USA. Yet this modest number of workers produces enough for the USA still to be a net exporter of agricultural products, after importing such products as beverages, fruits and agricultural raw materials that cannot be cultivated at home for climatic reasons. The prominence of agriculture in the Midwest and the three northern Mountain states shows clearly in Figure 9.9. South Dakota is, however, the only state in which the farm population exceeds 10 per cent of total population.

In 1990, 1,900,000 sq km (734,000 sq miles) of the USA were used as arable land or for permanent crops, out of a total area of 9,370,000 sq km (3,620,000 sq miles), that is, almost exactly 20 per cent, an area that has changed little in the 1970s and 1980s. Permanent pasture accounts for another 26 per cent, while 31 per cent is forest and woodland. Compared with almost all the other major regions of the world, the USA is well endowed with bioclimatic resources, and the potential output from the agricultural sector is far above present output, implying its future importance in supplying other regions of the world with food products, whether through trade or in the form of aid to cover emergencies.

The USA has been the largest single producer of manufactured goods in the world throughout this century, although the total output of the EU is greater. In general its industries have been developed primarily to serve its very large home market, whereas those of the countries of the EU, and especially Japan, have to focus seriously on exports to other regions of the world in order to be able to import raw materials, fuel and food products. The industrial sector of the USA has been stimulated to expand and diversify by the two world wars, especially during and after the second, when civilian industrial output flagged in the other

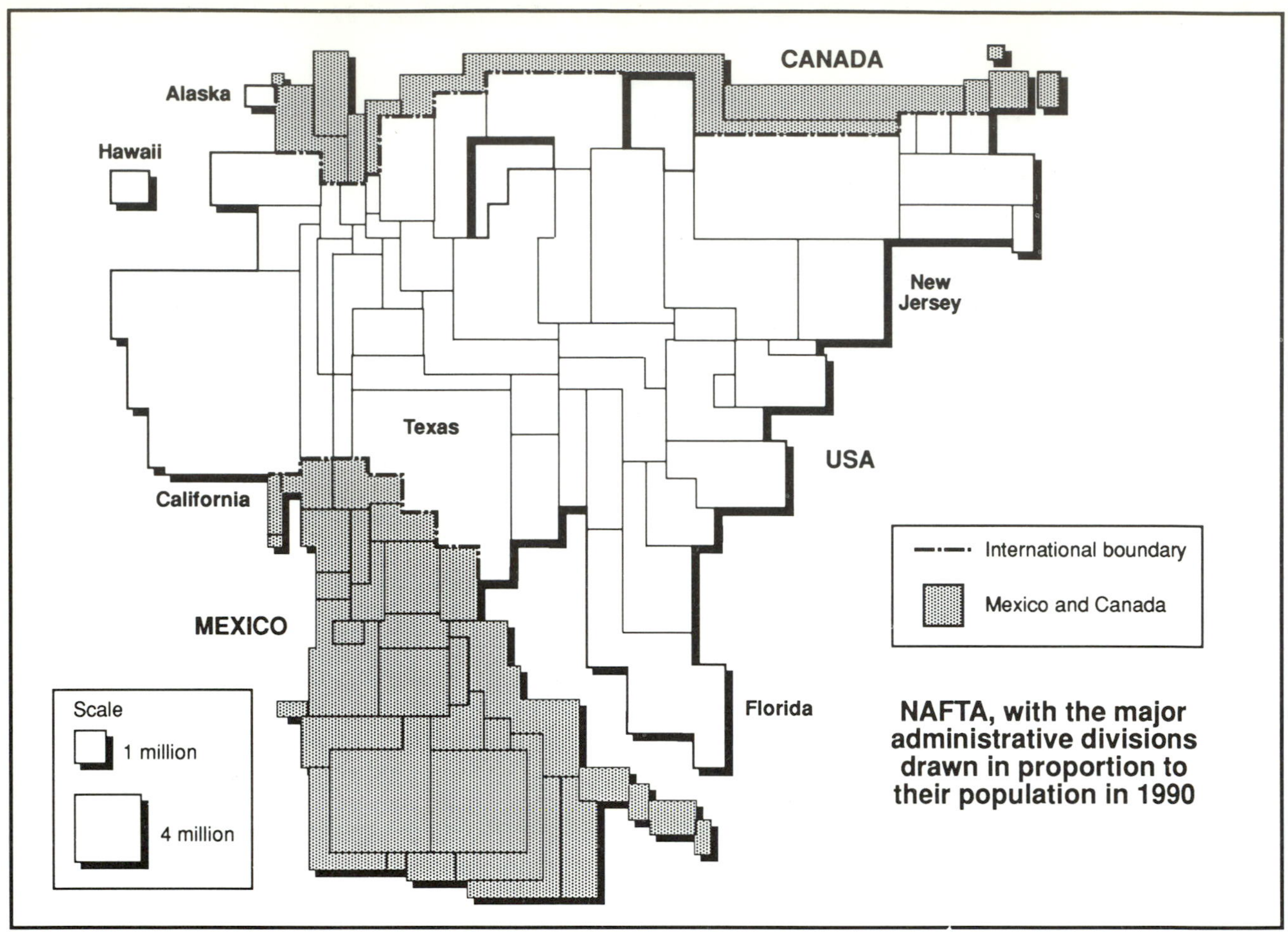

FIGURE 9.4

a The population of the NAFTA countries. Each major administrative division is drawn proportional to its population size

b (*see facing page*) Areas of USA and Mexico compared

leading industrial countries. Since the 1940s, however, it has increasingly been the target of exports of manufactured goods from Western Europe and Japan and also of both primary products and cheap manufactures from many developing countries. Some sectors of US industry, such as textiles, clothing, steel, motor vehicles and electronics, have suffered badly from foreign competition. Some, such as the motor vehicles industry, have adapted and survived, while in other sectors, including the aircraft and computer industries, the USA has dominated world markets. As in the EU, the coalfield-based industries of eastern USA have suffered particularly with the decline of coal and steel production. Textile manufacturing has likewise declined. Figure 9.10 shows the location of the main cluster of older manufacturing centres.

The level of dependence on manufacturing varies greatly from state to state in the USA. In 1990, 18,840,000 employees were on the manufacturing payroll, 7.6 per cent of the total population of the USA. Of that total, only 12,130,000 were actually production workers. Column (10) in Table 9.2 shows the employees in manufacturing in each state as a percentage of the *total* population of the state. All the states of the Northeast region except New York and all the states of the East North Central region had above the US average in manufacturing, although in most the level was not far above that average. The traditional industrial base of the USA thus remains strong in spite of the decline of many sectors there. In the South there is greater variation, but here also most of the states have above-average levels. By the criteria used, the Dakotas, Florida, and most of the states of the West are the least dependent on manufacturing, with (in addition to DC), Wyoming, Nevada and Hawaii having only about 2 per cent.

With increasing globalisation of the world economy the transactions between the USA and the rest of the

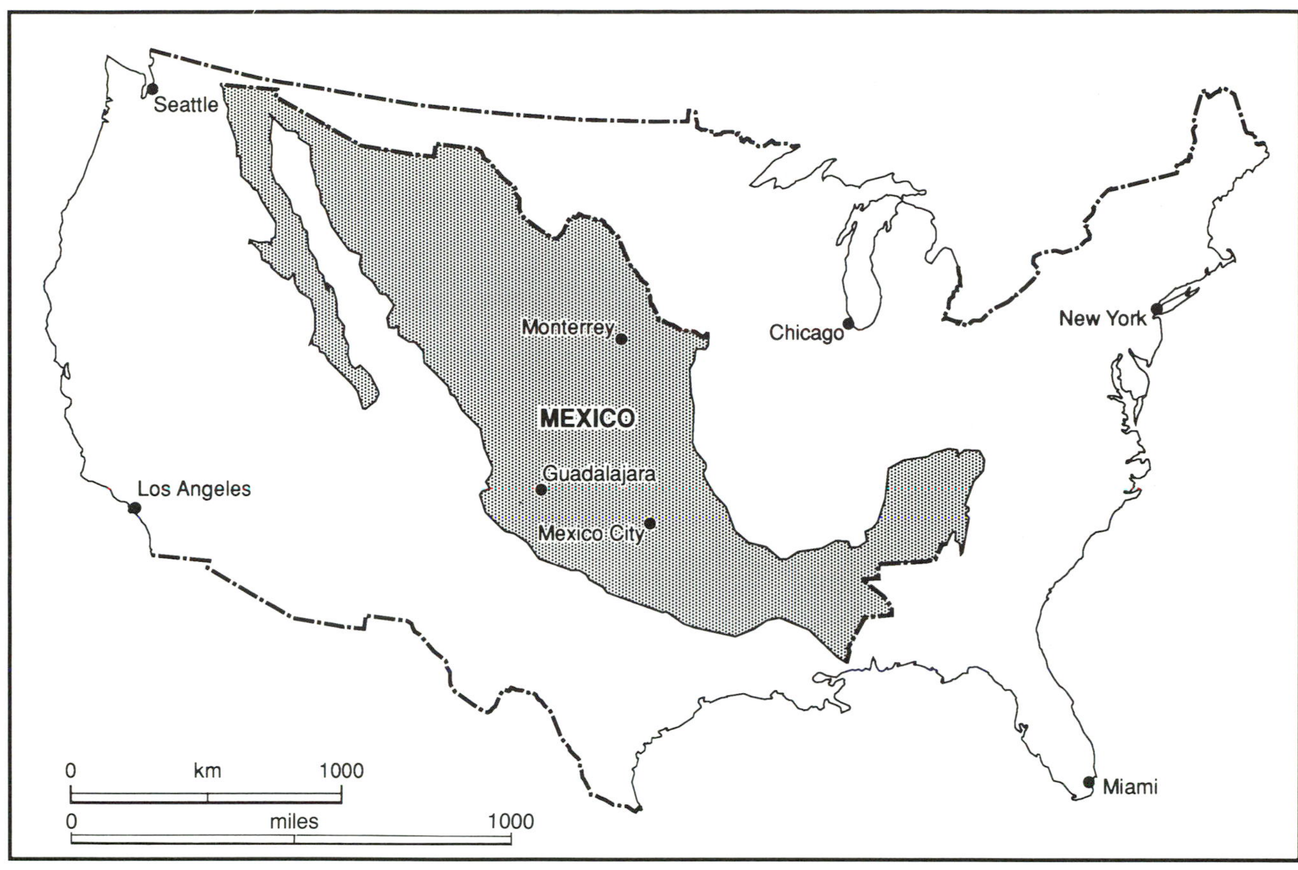

world are themselves growing. For many decades now US private investment has spread to most parts of the world, financing such diverse activities as oil extraction in Venezuela (from the 1920s to the 1970s), motor vehicle manufacturing in the UK (Ford and General Motors), and fruit growing in Central America. In recent decades, and especially in the 1980s, foreign companies have moved into the USA, a process regarded with mixed feelings there, negatively as a threat to US independence, positively especially in localities benefiting from jobs in new enterprises. The wide variations from state to state must partly be accounted for by conditions of infrastructure and labour, and partly by location, but they also reflect the initiative of some states in publicising their positive attributes abroad. Box 9.1 introduces some aspects of industrial location in the 1990s.

In 1989 (*Statistical Abstract of the United States 1992*, p. 787), 1,816,000 persons were employed in manufacturing in US affiliates of foreign companies, 9.3 per cent of all employment in manufacturing. Manufacturing accounted for 41 per cent of all employment in foreign companies in the USA. The increase in such employment in the 1980s is striking, the total rising from 2.4 million as recently as 1981 to 4.4 million in 1989. Column (11) in Table 9.2 shows the marked variations between states in foreign investment in the manufacturing sector in 1990 through the proportion employed by foreign companies expressed as a percentage of total manufacturing employment. Compared with the US average of 9.3 the five highest and lowest (excluding DC) were:

Highest		*Lowest*	
West Virginia	20.1	Nevada	5.6
Alaska	17.5	Mississippi	5.5
Delaware	16.0	Washington	5.4
New Jersey	13.8	Kansas	5.2
South Carolina	13.3	South Dakota	5.1

The broad distribution of states with high and low employment in foreign companies, shown in Figure 9.11, reveals a marked preference on the part of these companies for the eastern two-fifths of the national area, with a continuous region of above-average scores here. The states that are least attractive to foreign investors are mostly in the western half of the country where, apart from Alaska, the level is mostly well below average (Box 9.2). Thanks to its large population, California had almost 200,000 persons employed

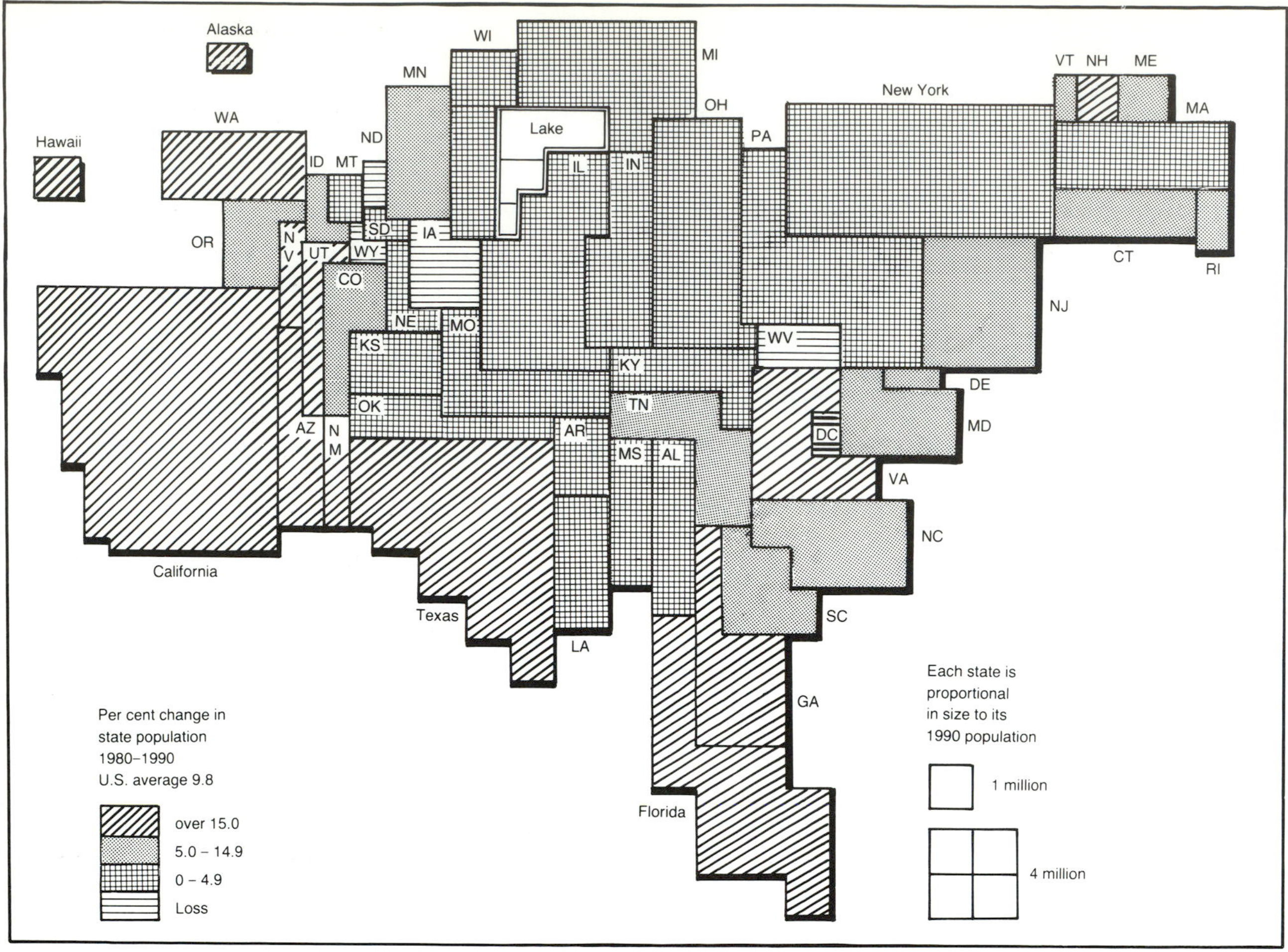

FIGURE 9.5 Population change 1980–90
Source: Statistical Abstract of the United States, 1992, p. 7

in foreign manufacturing companies and almost 500,000 altogether in foreign companies. Nevertheless, the evidence demonstrates the strong attraction of the Atlantic face of the USA to foreign investors compared with the Pacific side, casting doubt on the significance of its frontage along the much-publicised Pacific Rim (see Box 8.1).

The involvement of the US economy in the world economy is illustrated by another set of data (see column (12) in Table 9.2). The data show the contribution in dollars per inhabitant of each state to US exports (Figure 9.12, p. 225). In 1990, the average value of exports per inhabitant for the whole of the USA was 1,267 dollars. Multiplied by the number of people in the USA, that gives a value of 315 billion dollars (in 1990), which does not include another 75 billion that could not be traced to individual states. For various reasons the contribution per inhabitant varies greatly among the states. The highest and lowest five are as follows:

Highest		*Lowest*	
Alaska	5,182	Nevada	328
Washington	5,020	South Dakota	295
Louisiana	3,365	Montana	287
Delaware	2,018	New Mexico	164
Vermont	2,050	Hawaii	162

The top and bottom places are occupied by Alaska and Hawaii respectively. The former has vast natural resources in relation to its population size, while the latter is a cluster of densely populated small islands with few natural resources. Washington state comes top by a long way in contiguous USA thanks to the presence of the Boeing Company (see Box 9.1), but it has one of the lowest rankings in foreign investment in manufacturing. Louisiana and Delaware have useful coastal locations. In contrast, the four lowest, apart from Hawaii, have interior locations.

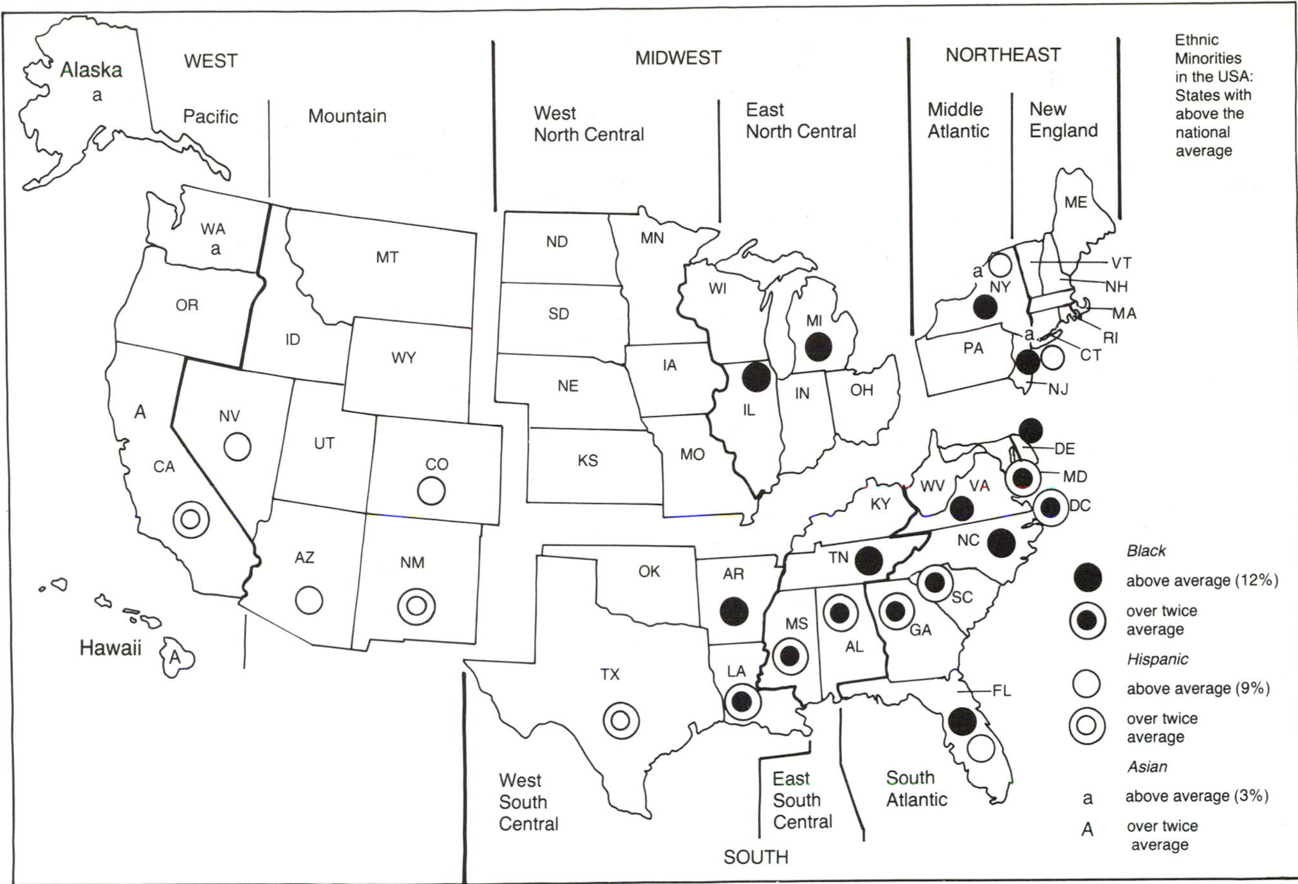

FIGURE 9.6 Ethnic minorities in the USA: the proportion of Black, Hispanic and Asian

Source: Statistical Abstract of the United States, 1992

9.4 Overseas transactions and commitments

In this section several aspects of US involvement abroad are discussed.

Foreign trade

Before the break-up of the Soviet Union it could be argued that the USSR valued self-sufficiency and conducted foreign trade reluctantly, primarily to obtain products not available in adequate quantities at home (e.g. at times tin, rubber, grain, equipment). At the other extreme, producers in West European countries trade to increase turnover and profits from foreign countries. As already noted, the US home market is so large that foreign markets have mostly been of secondary importance.

Table 9.3 shows the largest sources of US imports. The twelve countries listed accounted for just below 75 per cent of the value of all US merchandise imports. The countries are ranked according to the favourability

TABLE 9.3 US balance of merchandise trade with its twelve largest suppliers of imports in 1991 (in billions of dollars)

	US imports from	*US exports to*	*Balance*	*Exports/ imports*
1 UK	18.5	22.1	+3.4	119
2 France	13.4	15.4	+2.0	115
3 Mexico	31.2	33.3	+2.1	107
4 Canada	91.1	85.1	–6.0	93
5 South Korea	17.0	15.5	–1.5	91
6 Hong Kong	9.3	8.1	–1.2	88
7 Germany	26.2	21.3	–4.9	81
8 Italy	11.8	8.6	–3.2	73
9 Saudi Arabia	11.0	6.6	–4.4	60
10 Taiwan	23.0	13.2	–9.8	57
11 Japan	91.6	48.1	–43.5	53
12 China	19.0	6.3	–12.7	33
Total[1]	488.1	421.9	–66.2	86

Note: [1] Includes all other partners not listed above

Source: US Statistical Yearbook 1992, Table 1335, pp. 800–3

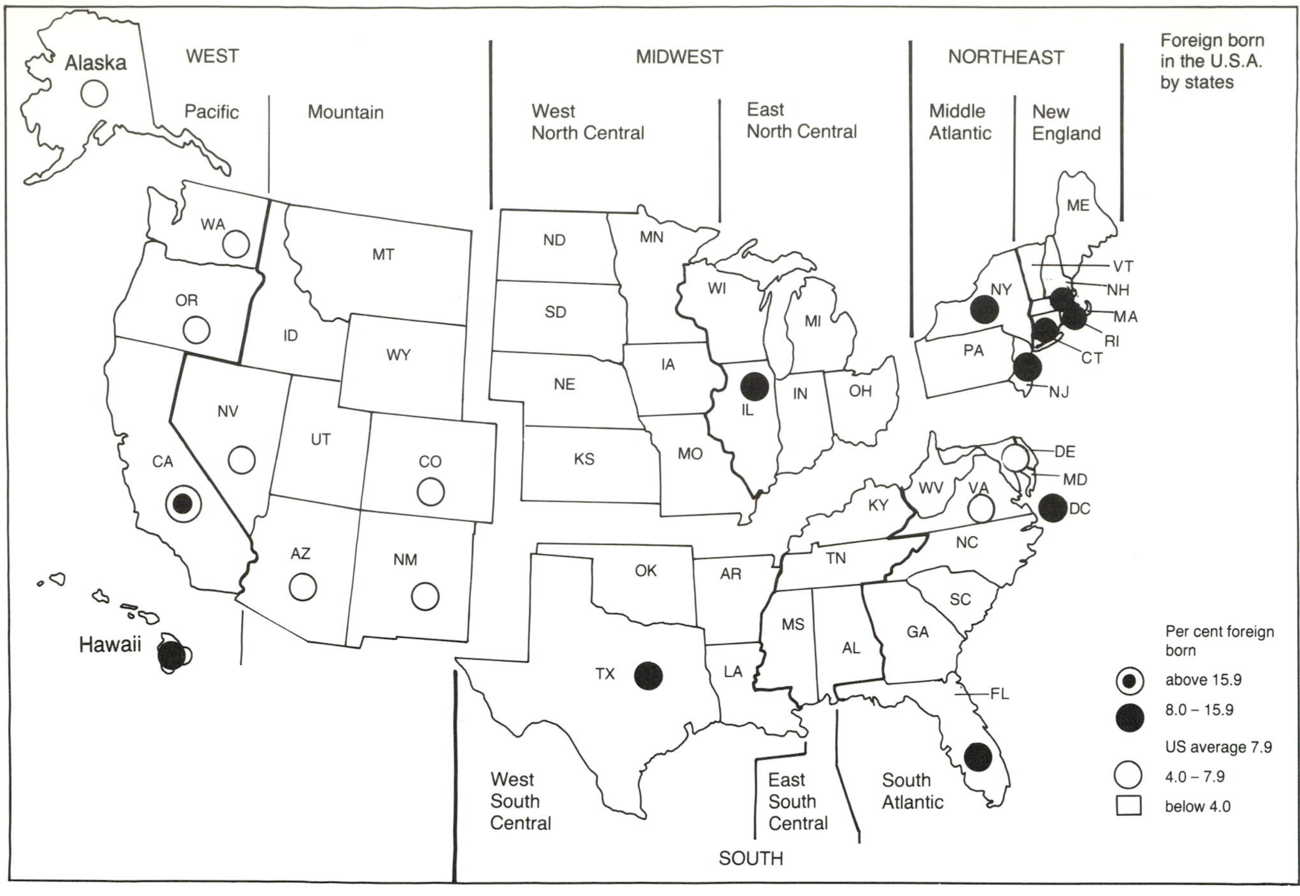

FIGURE 9.7 The foreign-born population of the USA
Source: Statistical Abstract of the United States, 1992

of the balance of trade of each with the USA. Thus, for example, US exports to the UK were 19 per cent higher in value than its imports from the UK, whereas in its trade with Japan, its imports were nearly twice the value of its exports. Almost exactly 25 per cent of US imports came from its immediate neighbours, Canada and Mexico. Another 14 per cent came from the four largest economies of the EU, compared with 19 per cent from Japan. A comparatively recent trend has been the increasing US trade with South Korea, Hong Kong (an outlet for Chinese goods as well as its own), Taiwan and China. These four countries accounted for another 14 per cent of US foreign trade.

TABLE 9.4 Direction of US foreign trade, 1951 and 1990

	1951		*1990*	
	imports	*exports*	*imports*	*exports*
Canada	21	17	18	21
UK	4	6	4	6
Rest of EU (12)	10	15	14	19
Japan	2	4	18	13
Latin America	31	24	13	14
Malaya[1]	4	5	3	3
India	3	3	0.7	0.6
World total	100	100	100	100
Absolute amounts	$US 10,817m.	$US 14,879m.	$US 515,635m.	$US 389,860m.

Note: [1] In 1990 Malaysia and Singapore
Sources: UNYITS 1953 (1954), vol. 1, p. 466; *ITSY 1990* (1992), vol. 1, p. 956

Table 9.4 provides a broad picture of changes in the direction of US foreign trade between 1951 and 1990. During that time, US foreign trade has increased greatly, even after inflation is taken into account, so percentages are used to facilitate comparisons. The composition has also changed, with increasing imports of oil and of manufactures. Since the Second World War the positions of Canada and the UK have not changed much, but the relative importance of the rest of the EU has increased. The most marked increase in trading has been with Japan and the other countries of East Asia referred to above. On the other hand, the contribution of Latin America to US foreign trade has diminished substantially, following an increase during the Second World War, when trade with most other areas in the tropics was difficult, if not impossible. India and Malaya were important sources of US

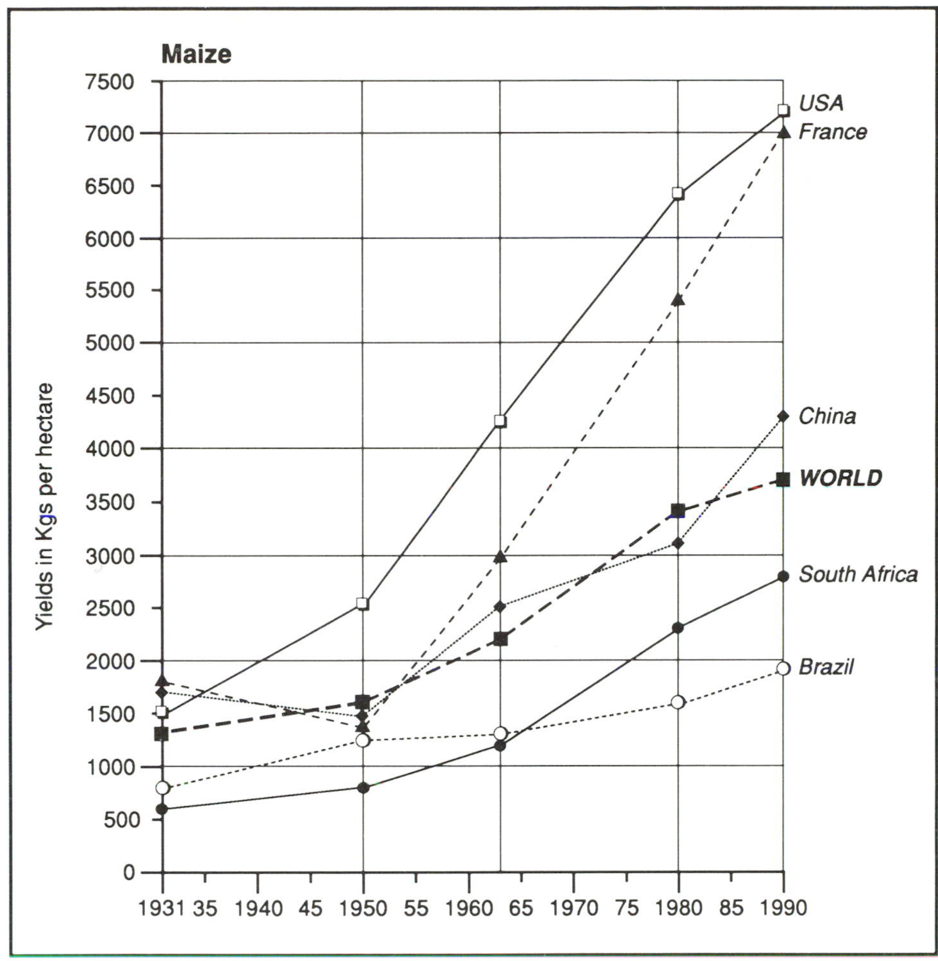

FIGURE 9.8 Maize yields in selected countries, 1931–90

imports in 1951 but, given the great size of India's population if not of its economy, US trade with it is now very modest indeed.

The US investment position abroad

In 1990, US investment abroad totalled 421.5 billion dollars (compared with 215.4 billion in 1980). About 40 per cent was in manufacturing, 14 per cent in petroleum (oil) and over 23 per cent in finance. Table 9.5 (p. 227) shows the distribution by major regions of the world. The distribution differs markedly from the distribution of foreign trade. In spite of the growing US interest in East Asia, almost half of all US investment abroad was in Western Europe, where the share actually increased between 1980 and 1990. As membership of the EU increases and as the economy grows, bringing greater integration and cohesion, it has become increasingly important for both the USA and Japan to manufacture within the customs union in order to be able to reach the home market of the EU. Canada and Latin America together receive another third of US overseas investment, but Canada's share has actually declined, whereas there have been increases in several other major regions of the world.

Western Europe, Canada and nowadays Latin America are comparatively stable areas of the world for US investment and seem likely to remain the preferred regions; on the other hand, instability in Africa and Southwest Asia, resistance of varying intensity from China and Japan, and the small size of the market in South Asia leave many regions of the world on the margin of US investment preferences. The picture could be complicated by an increase in the current very small amount of investment in Central Europe and the former USSR, although no great change seems likely in view of the unstable political, economic and ethnic situation there.

US development assistance

In relation to its very large GDP, among the developed countries the USA is one of the least 'generous' contributors of development assistance. The situation

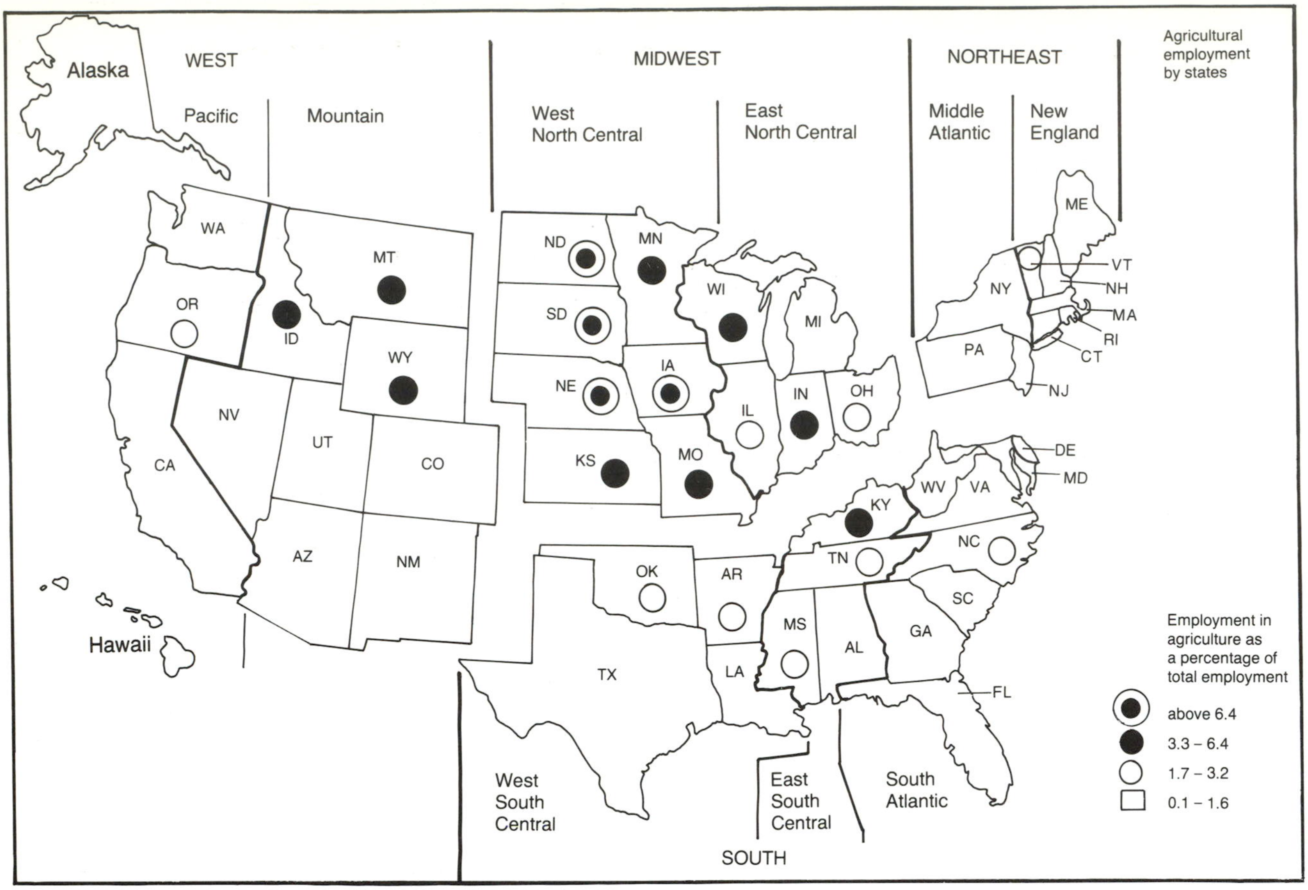

FIGURE 9.9 Employment in agriculture in the United States
Source: Statistical Abstract of the United States, 1992

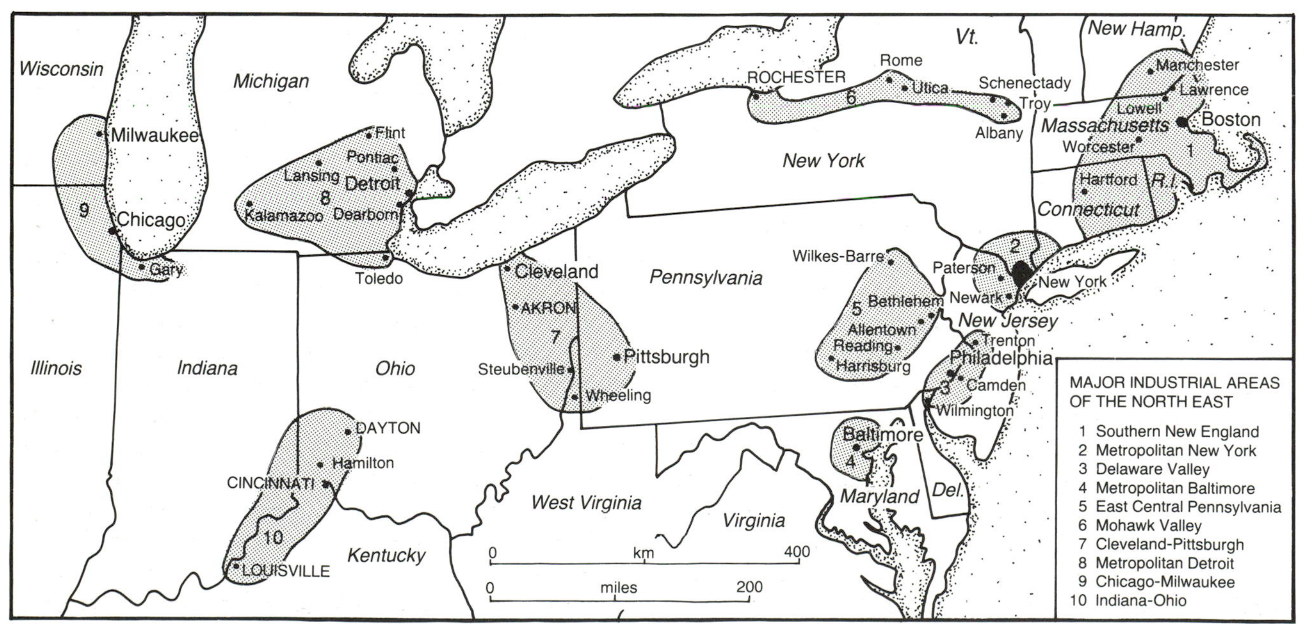

FIGURE 9.10 The US manufacturing belt
Source: Clark (1984), p. 82

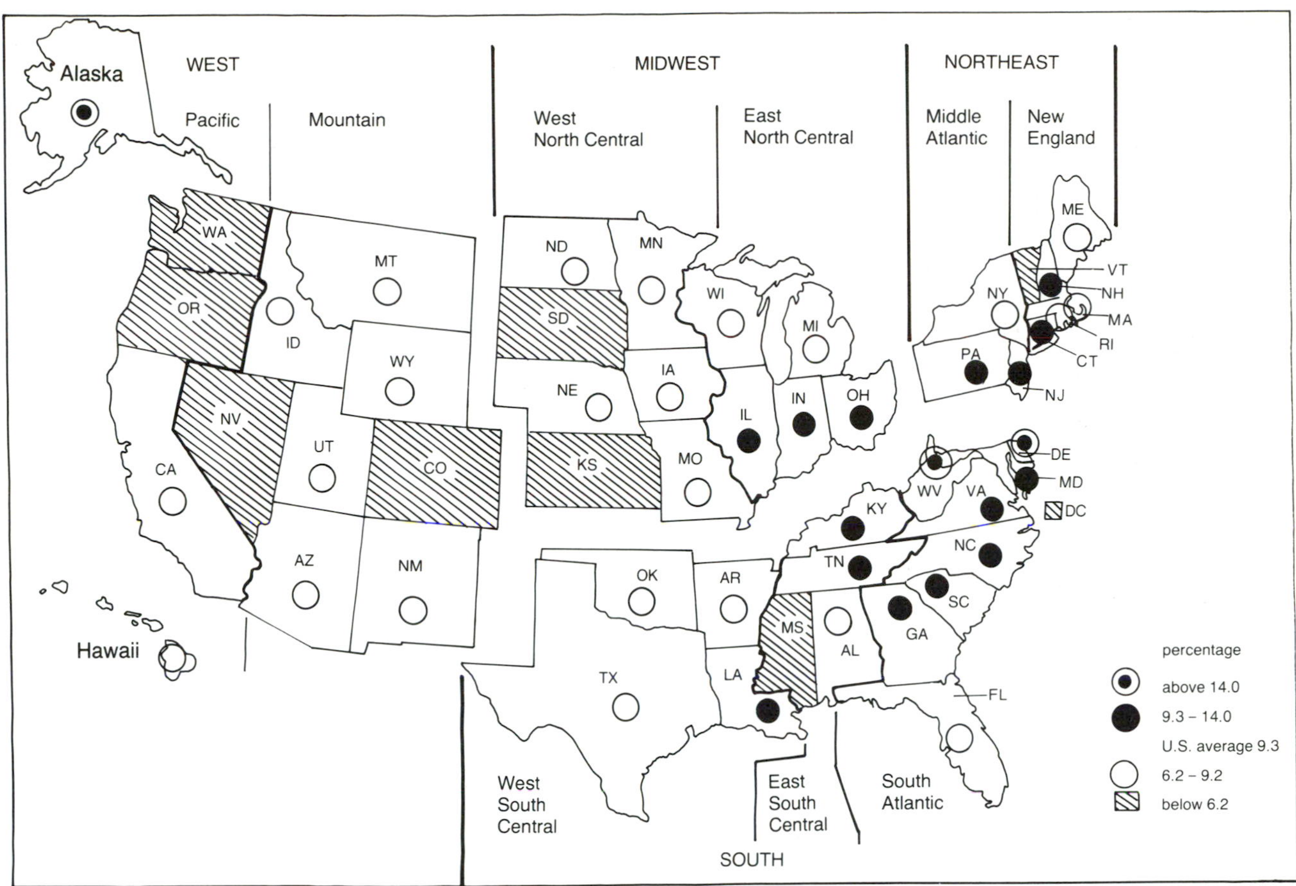

FIGURE 9.11 Foreign direct investment in the manufacturing sector of the United States. Per cent of total employment in manufacturing

Source: Statistical Abstract of the United States, 1992

was very different in the decade 1946–55, following the Second World War. Table 9.6 shows the direction of US government foreign grants and credits during the period 1946–55 and in 1990. Between 1946 and 1955, over two-thirds of the aid went to Western Europe, another 9 per cent to Japan and South Korea. In other words, the aim was to assist the war-damaged economies of countries that in the 1930s were among the richest in the world. Although no less deserving of such assistance, given the massive material losses inflicted on it from 1941 to 1944 by the mutual enemy Germany, the USSR received very little, while Central (at that time Eastern) Europe received a moderate amount for a short time. After some years, US aid to Europe and Japan diminished to negligible proportions and attention focused on the developing countries, initially on many in Latin America.

The situation in 1990 reflects US preoccupation within Southwest Asia and North Africa. Although Israel hardly qualifies on grounds of underdevelopment, in 1990 it received about 4.4 billion dollars of US assistance, while Egypt received almost 5 billion dollars. In contrast, Bangladesh, with twenty times as many inhabitants as Israel, and far poorer, received a mere 140–80 million a year between 1986 and 1990, while India received nothing, and the whole of Africa, excluding Egypt, about 1.8 billion. Even without further detail it is evident that US government foreign grants and credits are not targeted specifically at the poorest countries of the world. The USA is under no obligation, other than a moral or philanthropic one, to provide any assistance at all outside its own borders, but it is transparently clear that political as well as economic considerations determine the placing of the modest amount it does disburse. The same could be said of EU and Japanese assistance and also of that from countries of Eastern Europe and from the USSR before their efforts virtually ceased. Central Europe and the former USSR are now hoping to be net receivers of assistance rather than net donors, thereby diverting part of the traditional flows from developed to developing countries, which is bad news for the latter.

BOX 9.1

Post-industrial America?

The decline in employment in manufacturing in the USA has led to much speculation about the changing nature of production and job prospects. In 1984 Clark wrote a book entitled *Post-industrial America*. The term is misleading because it implies that manufacturing has declined absolutely or even has ceased to exist. The purpose of this box and of Box 9.2 is to examine briefly two aspects of US manufacturing, the Ford Motor Co., and foreign transplants in Southeast USA. First, a crucial fact must be recorded. Total employment in manufacturing industries in the USA fell from 21.9 million to 20.4 million between 1980 and 1991, declining from 22.1 per cent of all employment to 17.5 per cent during that period. However, employment in the agricultural sector has also been declining and in 1991 was only 2.9 per cent of total US employment. No one questions the loss of jobs in the US agricultural sector. In practice, output per worker in manufacturing in the USA has risen, in contrast to the decline in employment.

PLATE 9.2
Paterson, New Jersey

a Textile mills, some now converted to apartments. The silk industry of this city declined and collapsed in the 1920s and 1930s. Most of the population of Paterson is Hispanic or non-white

b Water power was an essential ingredient of industrial development in the oldest states of the USA before steam power prevailed early in the nineteenth century

9.5 Regional inequalities in the USA

Regional variations in average income, in living standards and in the quality of life within countries of the world, especially the larger ones, is a subject that concerns national governments and has been the subject of much research by geographers. Large disparities have already been referred to within the EU, while even in Japan there are marked differences. In spite of the Soviet policy of attempting to narrow regional differences in the vast empire of the former USSR, large disparities also occur there. The settlement of the USA was such that a relatively empty area 200 years ago could be settled progressively by waves of migrants enjoying broadly equal opportunities to make something out of nothing. The slave-based economy of the Southeast states was the exception.

At the highly aggregated level of the fifty-two states (plus DC), the 2:1 difference in per capita income between the richest and the poorest states is not large when compared with the 3:1 gap in the EU at the highest level of regional divisions. A similar gap occurred also in the former USSR, between the Baltic republics and Moscow on the one hand and some parts of Central Asia on the other. In anticipation, it may be noted that regional differences within large *poor* countries – notably Brazil, India and China – are also more marked than they are in the USA. In the last six decades

The Ford Motor Company

Although the Ford Motor Company's annual turnover in the early 1990s of around 100 billion US dollars was 25 per cent less than that of the world's largest motor vehicle company, General Motors, Ford is special because of its association with the innovation of mass-production technology. Henry Ford (1863–1947) was born near Dearborn, Michigan, site of the main works today. Ford's first successful automobile was the Model T. It was first produced in 1908, and eventually 15 million were sold. Ford aimed to keep production costs low by using completely interchangeable parts, having a high degree of division of labour in the plant, and adopting the assembly line, the latter idea described, but in the manufacture of pins, by the economist Adam Smith in the eighteenth century.

About a century after the start of the mass-production of cars by Ford, the company is on the threshold of developing a series of 'world' cars, with the Mondeo model already on the market in 1994. This model cost 6 billion US dollars to develop and was launched a year late. According to Lorenz and Randell (1994), the chairman of Ford in 1994, Alex Trotman (actually British born), considers that for a motor company to survive far into the next century operations must be on a global basis. The point is made that it costs less to develop one type of engine for a production run of 1 million cars than types for two production runs of 500,000 cars each. Ford expanded into Europe in the 1920s and also into Canada to cater for these regional markets.

A current question is the extent to which the company should manufacture in developing countries, where labour costs are much lower than in Western Europe or the USA. Other transnational companies run integrated global operations. For example, Coca Cola and McDonalds do so, but they are essentially one-product concerns. The Japanese car manufacturer, Toyota, has global products, but almost all their design and engineering is done in their home base.

Ford continues to extend its operations abroad. For example, it bought the British car-making firm Jaguar in 1989. Thanks to the launch of a successful new Jaguar model, Ford are considering developing a new small car in Britain, rather than in the USA. State aid to ensure the car is made in Britain is under discussion. The project would create some 10,000 jobs, directly and indirectly. The situation exemplifies the struggle for new jobs in manufacturing, which in both the USA and Western Europe do not offset the continual loss of jobs which has taken place since the 1970s.

In 1945 Detroit was producing half of the world's cars. Greig (1994) describes what Detroit was like in the 1960s: 'Detroit was the celebrated birthplace of the factory assembly line, the high industrial wage, the affordable car and Motown Music. . . . This energetic city lived in the fast lane as the motor capital of the world.'

Fifty years after the end of the Second World War, the financial leaders of the western world attended President Clinton's Jobs Summit:

> These leaders of the seven richest industrialised nations flew into Motown, only to ride through America's worst urban decay: whole swathes of Detroit are now burnt-out ghetto land. It is a city of broken glass and abandoned buildings. The wind whistles through gutted blocks of flats, scrubby weeds grow in empty lots and seagulls feed on the debris of a dying city.

'Post-fordism' is a term that has recently come into fashion to describe new, more flexible industrial operations and labour relations. The Ford Motor Company has survived the onslaught of European and Japanese car manufacturers. Will its global operations help it to cope with competition from the growing motor industries in the NICs, or the 'emerging markets', as it is becoming fashionable to call them?

the gap between states of the USA has been reduced from about 3:1 in the 1930s to 2:1 now. The gap is wider at county level in the USA, while the contrast between even smaller districts, and ultimately between families, is enormous.

Only one of several different variables that can be used to measure regional differences in income in the states of the USA has been chosen, per capita income. In 1989 the US average was 14,420 dollars. The data set is shown in Table 9.2 (column (14)). The five states with the highest and lowest per capita income were as follows:

Highest		*Lowest*	
Connecticut	20,189	South Dakota	10,661
District of Columbia	18,881	Louisiana	10,635
New Jersey	18,714	Arkansas	10,520
Maryland	17,730	West Virginia	10,520
Alaska	17,610	Mississippi	9,648

In general, the states of the Southeast fall in the lowest quartile in income per capita, but depressed industrial states such as West Virginia and stagnating agricultural

BOX 9.2

Foreign investors in US manufacturing

In Box 9.1 it was stressed that in 'post-industrial America' the presence of a strong, innovative manufacturing sector is vital if the USA is to maintain its position in the world economy. US politicians lament the unavoidable flow of investment into industrial enterprises in foreign countries, indirectly reducing jobs at home. They also have misgivings about foreign investments in their own country, although these create jobs at home and are the cause of satisfaction in the states where much of the flow is concentrated. In Table 9.2, column (11) shows how great are the variations among the states in the scale of foreign investments in manufacturing per inhabitant.

One region of the USA in which there has been considerable foreign investment since the 1970s is the Carolinas. According to Brodie (1994), one particular zone in this part of the South Atlantic region attracted some 800 foreign companies in recent years, the area extending southwestwards from Raleigh and Durham in North Carolina, through Charlotte, to Greenville in South Carolina. Some of the largest industrial companies of the EU and Japan have come to the area. These include, for example, companies connected with the motor industry such as BMW (Germany) and Michelin (tyres, France); chemicals – BASF (Germany); electronics – Hitachi (Japan); and pharmaceuticals – Glaxo (UK). This trend mirrors the shift of textile manufacturing from Northeast USA to the Southeast many decades ago. One change that has taken place since then is that the attractions of particular locations are different now. If hydroelectric power, locally grown cotton and cheap labour attracted cotton manufacturing a century ago, many new considerations are now taken into account.

A number of reasons why this particular part of the USA has proved attractive in the 1990s have been proposed.

- Labour costs are low in comparison with those in many other states and only two-thirds of the German level. Unions are generally weak.
- Land is readily available and is cheap or may be given away. Tax relief is available. Congestion common in the older industrial regions is not a problem.
- The region has the advantage of being in the 'sunbelt', without the risk of serious interruptions from snow, forcing closedowns. Nor does it have the nagging prospect of earthquakes, a small but perhaps significant consideration.
- Charlotte is now the third largest banking centre in the USA. Charlotte and Atlanta have major international airports, while seaports on the coast of the Carolinas are not far away.

The USA is still by far the largest single market in the world, only the combined market of the fifteen countries of the EU together rivalling it. Foreign companies need to have a considerable presence in the USA in a useful part of the country. Although the Detroit area has much of the American manufacturing capacity of the motor vehicles industry, the choice of the Carolinas by BMW and of Alabama by Mercedes Benz shows that traditional locations are no longer necessarily favourable in the way they once were.

The preference for the South Atlantic region by some of the world's largest transnational companies outside the USA is thought-provoking. In comparison with America's Pacific states' window on the Pacific Rim, the Carolinas are much closer to the larger part of the home market than California is. They are also better located to reach the bulk of the Canadian and Latin American markets and the EU.

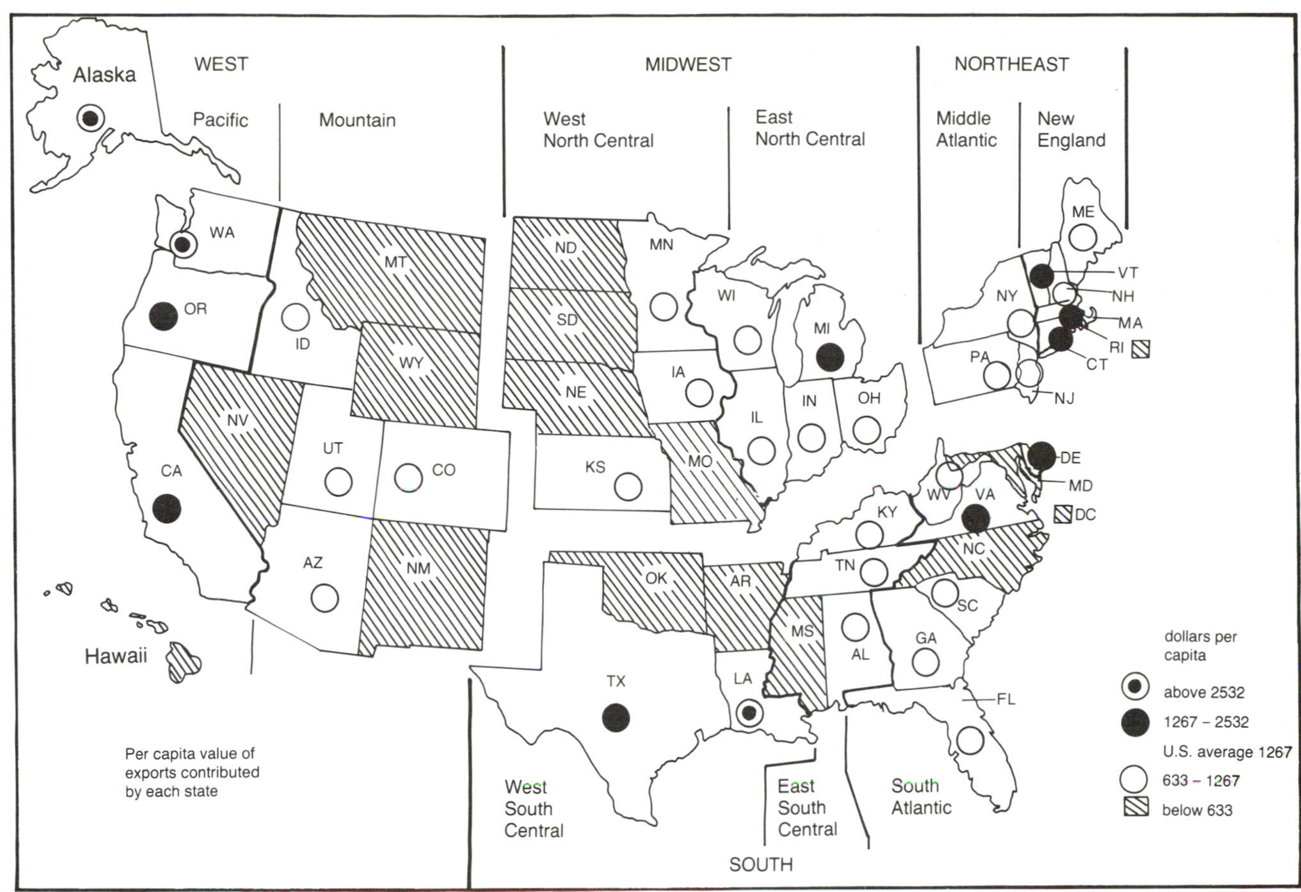

FIGURE 9.12 Contribution of states to US exports

Source: Statistical Abstract of the United States, 1992

states such as the Dakotas also come low in the league table.

In 1989, 10 per cent of the families in the USA were below the official poverty level. The situation was much worse for female householder families, over 40 per cent of which were below the poverty level. The disparity between states in the percentage of families below the poverty level is close to 5:1, compared with the 2:1 disparity for income noted above. The states with the lowest and highest percentages of families below the poverty line are:

Lowest percentage		*Highest percentage*	
New Hampshire	4.4	West Virginia	16.0
Connecticut	5.0	Kentucky	16.0
New Jersey	5.6	New Mexico	16.5
Maryland	6.0	Louisiana	19.4
Hawaii	6.0	Mississippi	20.2

The states at the extremes for families below the poverty line include several also in that position on per capita income.

In the United Nations' *Human Development Report 1992*, the USA is ranked sixth among the countries of the world on the scale of the human development index, achieving 976 points out of a possible maximum of 1,000. Its 1990 life expectancy of 75.9 years is slightly exceeded by those of Canada, Japan and several West European countries, but it had the highest mean years of schooling in that year of any country in the world. Its real GDP per capita of almost 21,000 dollars in 1989 was exceeded only by that of the United Arab Emirates (23,800). On the combination of the three criteria, life expectancy, mean years of schooling, and a drastically adjusted real GDP level, its final human development index score of 976 was exceeded only by those of five countries: Canada (982), Japan (981), Norway (978), Switzerland (977) and Sweden (976). Such minute differences are insignificant when compared with the lowest scoring countries, notably 52 out of 1,000 for Guinea and 62 for Sierra Leone, both in West Africa, and 65 for Afghanistan.

On the principle that it is easier for a small region or country to be near the top or bottom of a scale that measures such aspects of economic development as income or energy consumption, the USA's position almost at the top is a remarkable achievement. The fact that most of the ethnic minorities, for various reasons, tend to predominate among the families below or near the poverty line is, however, a matter of concern, if not surprise, as also is the need to identify a poverty line at all. The 1,878,000 American Indians are widely dispersed over the states of the USA, but are most prominent in the Southwest. As with the indigenous populations of Australia, Canada and Siberia, their collapse before an unending flow of technologically better equipped Europeans is one of the saddest aspects of the US success story. No less unfortunate is the failure of the 'Whites' in the USA to overcome more than partially their antipathy towards the 11 per cent who have misleadingly been referred to as Blacks if they have some trace of African physiological characteristics.

What, then, do the Whites think of the present situation in the USA and its future prospects? Being a citizen of the richest large country in the world is not enough to keep people happy all the time. Box 9.3 shows the consensus of views of a small number of students asked in 1993 to record ten negative features or problems of their country. However bad things are perceived to be, very few Americans among either the older ethnic groups or the many new immigrants are rushing to leave their country. The occasional American migrates permanently to savour the relaxed affluence of Canada or Australia or the cultural heritage of Western Europe. The main exception is the population of Mexican origin, officially 13.5 million in 1990, not counting illegal immigrants. Many send remittances back to Mexico, but not a few eventually return there to the places of their birth, taking back both material benefits and ideas about different life styles, contributing presumably among other things to the sharp drop in fertility rates in parts of Mexico.

The interface between Mexico and the USA is a zone of unique interest in the world, the only long land frontier between a developing and a developed country. It makes the exclusion of migrants from Mexico or via Mexico impossible, without the massive protection of an 'iron curtain' type of barrier. Southern Europe attempts to keep its former North African colonial population at bay with the help of the Mediterranean. Japan's insular location enables it to control and in practice prevent entirely the arrival of any aspiring but unwanted migrants from the developing countries of Southeast Asia.

9.6 Metropolitan areas of the USA

In 1990, 197.5 million people lived in areas defined as metropolitan, almost 80 per cent of the total population of the USA, compared with 51.2 million who lived in non-metropolitan areas. The metropolitan population is divided among 250 Metropolitan Statistical

BOX 9.3

How they saw their country

Maximum points for a topic 23

A major problem in the USA is pollution (15), of air, water, cities. Natural resources (2) are diminishing, water (2) is short, especially in the Southwest, and the wilderness (2) is disappearing.

Education (14) is generally unsatisfactory at all levels and many people are not interested in it. Healthcare (10) is inadequate and inequitable.

Unemployment (11), related to decline of heavy industry (2), is growing, together with homelessness (6) and crime (12), manifested in violence/gangs (8) and drug abuse (6). Lax firearm control (5) is related to the above. Poverty (1), hunger (1), poor urban housing (2) and a rich/poor gap (3) are other negative features.

The politicians and government (9) have much to answer for, including the debt/deficit/depression (9) and an unsatisfactory foreign policy (7), with heavy dependence in the energy sector (4) on imported oil.

Racism/ethnocentrism (8) and illegal immigrants (2) may be connected with overpopulation (3). The decline of family farms, overproduction and excessive subsidies are negative aspects of agriculture (6), while public transport (3) leaves much to be desired. The role of the military (3) is ambivalent. Television/the media (4) are generally a negative influence and materialism/greed (6) are common features. Working parents (2) and AIDS (3) are negative aspects. The welfare system (8) is a shambles and is too lucrative to be good. Socialism is creeping in (2) and abortion rights (2) are a controversial issue.

The following were also mentioned once each:

(a) 'Places': NAFTA, South America, China, Japan, 'sunbelt' cities, Los Angeles, Euro Disney

(b) 'People': Michael Jackson, Mrs Clinton

(c) 'Other': the car, inflation, loss of values, loss of world dominance (*sic*), cancer, civil rights, family values, lack of role models, lack of parking, drink, rape, income tax, child abuse, lack of recycling, religious indifference, foreign companies, narrow view of world, multilingual society, federal government too strong.

TABLE 9.5 US investment position abroad by major world regions, 1980 and 1990 (billions of US dollars in columns (3) and (4))

	(1)	(2)	(3)	(4)	(5)
	%		Total		1980–90 change (1990 = 100)
	1980	1990	1980	1990	
European Union[1]	37.4	41.0	80.5	172.9	215
EFTA and others	7.3	7.4	15.8	31.3	198
Japan and S Korea	3.2	5.5	6.8	23.1	340
Canada	20.9	16.2	45.1	68.4	152
Australia and NZ	3.8	4.2	8.2	17.7	216
Latin America	18.0	17.2	38.8	72.5	187
Africa S of Sahara	2.1	0.7	4.5	3.0	67
N Africa and SW Asia	1.9	1.7	4.0	7.0	175
S Asia	0.2	0.3	0.5	1.1	220
SE Asia	2.2	2.9	4.8	12.4	258
China and others	1.2	2.2	2.6	9.1	350
International[2]	1.8	0.7	3.8	3.0	79
Total	100.0	100.0	215.4	421.5	196

Notes: [1] As of 1990 (i.e. in 1980 includes Greece, Spain, Portugal, not yet members)
[2] Also includes some small countries not accounted for above
Source: *US Statistical Abstract* 1992, p. 789

TABLE 9.6 Direction of US government foreign grants and credits, 1946–55 and 1990 (billions of dollars)

			Percentage	
	1946–55	1990	1946–55	1990
Western Europe	34.5[1]	0.3	68.2	2.5
Central Europe[2]	2.2	0.8	4.3	6.8
Japan and S Korea	4.7	−0.8	9.3	–
Latin America	1.2	1.9	2.4	16.1
Africa S of Sahara	0.1	1.6	0.2	13.6
N Africa and SW Asia	2.3	5.7[3]	4.5	48.4
South Asia	0.6	0.7	1.2	5.9
Southeast Asia	3.7	0.7	7.3	5.9
China plus[4]	1.3	0.1	2.6	0.8
Total	50.6	11.8	100.0	100.0

Notes: [1] Of which EU (as in 1990) 31.2, EFTA 3.3
[2] Includes USSR, 1946–55
[3] Largely accounted for by Egypt and Israel
[4] Mainly Taiwan, 1946–55

Areas (MSAs) and eighteen Consolidated Metropolitan Statistical Areas (CMSAs). The eighteen CMSAs are themselves made up of groupings of Primary Metropolitan Statistical Areas (PMSAs). For example 'Greater' New York consists of fifteen PMSAs, and extends beyond the State of New York into New Jersey, Connecticut and a corner of Pennsylvania. In 1991 New York itself (the five boroughs) had only 8.6 million (44 per cent) out of the total CMSA population of 19.6 million. Most of the eighteen CMSAs of the USA comprise only two or three PMSAs.

Before the metropolitan areas are reviewed in this section it is appropriate to note problems in defining large urban agglomerations and obtaining or creating a set of data in which the population sizes can be compared both within countries and internationally. The problem is exemplified by the disparity in population noted above between New York 'proper' and New York CMSA. Similarly, for example, in the San Francisco-Oakland-San Jose CMSA in California, San Francisco PMSA has only 1.6 million people out of the total of 6.3 million. The problem occurs world-wide. In his book *1400 Governments*, Wood (1964) drew attention to the particular problem in the USA of managing large, administratively highly fragmented urban areas. He wrote (p. 1):

> A vigorous metropolitan area, the economic capital of the nation, governs itself by means of 1,467 distinct political entities (at latest count), each having its own power to raise and spend the public treasure, and each operating in a jurisdiction determined more by chance than design.

In the USA metropolitan areas have proliferated and expanded in both area and population over the last few decades. Compared with urban agglomerations in other parts of the world, those in the USA mostly have low densities of population outside the older central areas. Some extend over several counties and many adjoin neighbouring metropolitan areas. There is a continuous zone of counties in metropolitan areas from southern Maine to Virginia, while much of the eastern part of Florida and the southern part of California are now also metropolitan. In the eastern two-thirds of the USA, counties are relatively small, but in the Southwest, where they are generally larger, the metropolitan areas of some small cities extend over territories the size of small European countries. Whatever their shortcomings, the metropolitan areas of the USA are a major influence on the human geography of the country and a key aspect of the gradual transformation from a predominantly agricultural and rural economy and society to one based on manufacturing and services, with an urban life style.

In the middle of the nineteenth century almost two-

TABLE 9.7 Decline in employment in the US agricultural sector, 1965–92

	Population in millions				
	US total	*US agricultural*	*Total*	*Economically active in agriculture*	*Percentage of economically active in agriculture*
1965	194.3	9.9	79.4	4.1	5.1
1980	227.7	8.9	109.9	3.8	3.5
1992	254.9	6.2	125.1	2.7	2.1

Sources: FAOPY vol. 31, 1977, Table 3; *FAOPY* vol. 46, 1992, Table 3

thirds of the economically active population of the USA was in the agricultural sector. The percentage was still 50 in the 1870s and 25 in the 1920s. In 1850 the US percentage of 64 compared with 52 in France but only 22 in Great Britain. Table 9.7 shows how the decline in employment in the agricultural sector in the USA has continued in recent decades to reach a mere 2.1 per cent in 1992 (cf. Japan 5.8, France 4.8, Germany 4.3, UK 1.9, Belgium 1.7). About a quarter of the population of the USA is still defined as rural, but only about a tenth of that population (including dependants) is now in the agricultural sector.

The proportion of the total population living in metropolitan areas varies greatly from state to state. It exceeds 90 per cent in three of the six states of New England, two of the three in the Mid-Atlantic division, and three of the nine (including Washington DC) in the South Atlantic division, reaching 93 per cent in highly

BOX 9.4

Buffalo Commons proposal (see also Box 16.5)

In most of the major regions of the world, the prospect is that in the decades to come pressure will increase on the natural environment as greater quantities of food and other agricultural products are needed. Forests and pastures will be cleared and many species and their habitats will disappear. Thanks to both its generous bioclimatic resources and the impressive development of agricultural technology, the USA is fortunate that for some decades the area under cultivation and pasture has not had to be extended appreciably. There has been a debate since 1987 about the proposal put by Frank and Deborah Popper to take land from farmers in order to recreate the natural grassland habitat of the buffalo, which came very near to extinction through hunting and farm development in the nineteenth century. They justified their proposal by describing it as a way of avoiding an inevitable disaster. They assume that human interaction with the environment, as it has existed since European settlement, cannot continue in the Great Plains.

The so-called Buffalo Commons thesis, which developed from an exercise in land-use planning, involves parts of ten states, Montana, Wyoming, North and South Dakota, Colorado, New Mexico, Nebraska, Kansas, Oklahoma and Texas. The Great Plains extend over the area roughly between 100° West and the front range of the Rocky Mountains. Precipitation diminishes westwards from about 20 inches (51 cm) in the north and 30 inches (76 cm) to the south in an area badly affected in the inter-war period by mechanised cultivation, soil erosion and the formation of the 'dust bowl'. The ten states cover more than 30 per cent of the area of the USA but the combined population of 31.6 million in 1990 is less than 13 per cent of the US total. Many counties have declined in population since the 1930s and some now have very low population densities. Many farmers in the Great Plains region have been receiving agricultural subsidies.

After examining economic and social aspects of the various counties in which decline has occurred, the Poppers proposed that some 413,000 residents should be encouraged to abandon their homes and farms. They would be resettled elsewhere, leaving former cultivated farmland and range land to revert in two or three decades to prairies, with some help such as tearing down fences and planting short grass, to be grazed by buffalo. Such land would be referred to as the Buffalo Commons. Their key paper on the subject appeared in *Planning* in 1987. De Bres and Guizlo (1992) critically examined the proposal of the Poppers and noted controversial aspects of both a practical and an ethical nature.

The counties selected as 'eligible' for transformation and depopulation were examined by the Poppers on the basis of 1980 census data and now some no longer

TABLE 9.8 The ten largest Consolidated Metropolitan Statistical Areas (CMSAs) in order of population size, 1991

	(1) Population	(2)	(3)	(4)	(5)	(6)
	in thous. 1991	*annual % change 1980–90*	*% 65 years and over*	*Unemployment rate (%)*	*% Black*	*% Hispanic*
1 New York – N. New Jersey – Long Island	19,592	0.3	13.1	8.7	17.7	14.6
2 Los Angeles – Riverside – Orange County	14,818	2.3	9.8	9.1	8.5	32.9
3 Chicago – Gary – Kenosha	8,340	0.2	11.4	7.4	19.0	10.9
4 Washington – Baltimore	6,830	1.5	9.9	6.0	25.2	3.9
5 San Francisco – Oakland – San Jose	6,332	1.5	11.1	6.7	8.6	15.5
6 Philadelphia – Wilmington – Atlantic City	5,925	0.4	13.5	7.5	18.3	3.8
7 Boston – Worcester – Lawrence	5,431	0.6	12.7	8.3	4.8	4.4
8 Detroit – Ann Arbor – Flint	5,215	−0.2	11.5	8.9	20.5	2.0
9 Dallas – Fort Worth	4,135	2.8	8.3	6.9	14.0	13.0
10 Houston – Galveston – Brazoria	3,859	1.8	7.3	7.3	17.9	20.7

Sources: U.S. Metro Data Sheet (1993), Washington, Population Reference Bureau Inc., prepared by Carl Haub and Anne Lang; *Statistical Abstract of the United States 1993* for 65 years and over

receive aid. A methodological flaw was the choice of variables measuring land-use distress, based on population change, density and age, plus poverty, rather than considering indicators of land use itself. Moreover, measuring features at county average level misses contrasts within counties and also rural–urban differences. The counties eventually targeted were widely dispersed through the region, some singly, others in clusters. Buffalo ranges returned to a natural state as opposed to large reservations would need larger areas than most of those proposed.

In addition to questions of the practicality of the proposal, there was a discussion of human issues involved in forcing people to abandon their homes. Deprivatisation (a more palatable word than nationalisation?) and displacement of population are not general practices in late twentieth-century American society, although in the nineteenth century ethnic cleansing was for some decades not uncommon as Indian tribes were displaced westwards. Many of the counties targeted by the Poppers are inhabited by descendants of the pioneers of the period 1850–80. The region is their home. Some of them commute to adjoining counties that form metropolitan areas. The assumption of the Poppers that environmental changes such as the 'greenhouse' effect would lead to a deterioration of climatic conditions for farming is based on speculation at present.

The Poppers are based in New Jersey, so their proposals for changes in the Great Plains could be seen as meddling by outsiders. Only time will prove whether or not this proposal to put the clock back and carry out ethnic cleansing by removing humans to replace them by wildlife will materialise in any form. Over the centuries, the process has been applied on many occasions in Europe and Asia to clear settled areas and create lands for hunting. Only in China at the time of the Mongol invasions in the thirteenth century have such large areas been affected.

In 1800, before the coming of the White settlers to the west there were, according to Hodgson (1994), 30 million bison in the prairies, a number reduced in 1870 to 20 million and by 1889 to a mere 1,000. Thanks to buffalo farming, there are now some 130,000. The more deprived and thinly populated parts of the Great Plains may be the preserve now of the Whites, but according to Burton (1994) the very rare event of the birth of a white buffalo on a farm (of a White farmer) in Jamesville, Wisconsin (not in the Great Plains), has revived the Indian legend that astonishing historic events will occur. At the moment Buffalo Commons is more of interest as a subject that raises many issues and problems worth discussion rather than a concrete geographical fact. For the present it seems realistic to ensure the survival and expansion of buffalo herds in large ranches, while leaving the human inhabitants in their existing settlements, many of which are satisfactory, some thriving.

Further reading

Hodgson, H. (1994), 'Buffalo back home on the range', *National Geographic*, November, vol. 186, no. 5, pp. 64–89.

TABLE 9.9 Extreme percentage values of population change, people over 65, unemployed, Black and Hispanic

(a) Average annual percentage population change 1980–90

Five with the greatest increase		
Punta Gorda	FL	6.4
Naples	FL	5.7
Fort Pierce – Port St Lucie	FL	5.1
Fort Myers – Cape Coral	FL	4.9
Las Vegas	NV–AZ	4.8
Five with the greatest decrease		
Waterloo – Cedar Falls	IA	–1.1
Decatur	IL	–1.1
Steubenville – Weirton	OH–WV	–1.4
Wheeling	WV–OH	–1.5
Casper	WY	–1.6

(b) Percentage of population 65 years old and over

Highest five		
Sarasota – Bradenton	FL	30.4
Fort Myers – Cape Coral	FL	24.8
West Palm Beach – Boca Raton	FL	24.3
Fort Pierce – Port St Lucie	FL	23.6
Daytona Beach	FL	23.0
Lowest five		
Austin – San Marcos	TX	7.8
Boulder – Longmont	CO	7.6
Houston – Galveston – Brazoria	TX	7.3
Provo – Orem	UT	7.0
Fayetteville	NC	6.1

(c) Unemployment rate (percentage) 1992

Five with the lowest unemployment		
Iowa City	IA	2.0
Sioux Falls	SD	2.3
Lincoln	NE	2.4
Columbia	MO	2.9
Madison	WI	2.9
Five with the highest unemployment		
Modesto	CA	16.0
Merced	CA	16.5
McAllen – Edinburg – Mission	TX	17.0
Yuba City	CA	18.1
Yuma	AZ	22.4

(d) Percentage Black 1990

Five with the highest percentage Black		
Albany	GA	45.8
Sumter	SC	43.2
Pine Bluff	AR	43.1
Jackson	MS	42.5
Rocky Mount	NC	41.9
Five with the lowest percentage Black		
Eau Claire	WI	0.2
Provo Orem	UT	0.1
Laredo	TX	0.1
Bismark	ND	0.1
Wausau	WI	0.1

(e) Percentage Hispanic

Five with the highest percentage Hispanic		
Laredo	TX	93.9
McAllen – Edinburg – Mission	TX	85.3
Brownsville – Harlingen – San Benito	TX	81.9
El Paso	TX	69.6
Las Cruces	NM	56.4
Five with the lowest percentage Hispanic		
Owensboro	KY	0.4
Wheeling	WV	0.4
Altoona	PA	0.3
Gadsden	AL	0.3
Parkersburg – Marietta	WV–OH	0.3

metropolitan Florida. In the West, the proportion of metropolitan population reaches 97 per cent in California. At the other extreme, seven states have less than 40 per cent of their population in metropolitan areas: Maine, Vermont, South Dakota, Mississippi, Montana, Idaho and Wyoming.

Unless some major demographic or administrative change takes place in the USA, the metropolitan population should continue to grow both absolutely and relatively. It increased by 20.8 million during 1980–90 while the non-metropolitan population increased by only 1.37 million. If the population of the USA increases from 249 million in 1990 to an expected 338 million in 2025, the metropolitan population

TABLE 9.10 A comparison of the largest urban agglomerations in six parts of the world: population in millions

USA		*European Union*		*Former USSR*	
New York	19.6	Paris	8.7	Moscow	9.0
Los Angeles	14.8	Rhine–Ruhr	8.0	St Petersburg	5.1
Chicago	8.3	London	7.7	Kiev	2.7
Washington – Baltimore	6.8	Madrid	4.1	Tashkent	2.2
San Francisco	6.3	Rome	3.1	Baku	1.8
Japan		*Latin America*		*Asia excluding Japan*	
Tokyo	16.0	Mexico City	20.0	Shanghai	13.3
Osaka – Kobe	13.0	São Paulo	17.0	Seoul – Inchon	11.0
Nagoya	7.0	Buenos Aires	11.5	Beijing	10.8
Shimonoseki – Kitakyushu	2.7	Rio de Janeiro	11.0	Bombay	9.9
Kyoto	2.6	Lima	6.4	Calcutta	9.3

could rise from 197 million in 1990 to about 280 million.

The rate of population growth (and decline) in the 1980s varied greatly from state to state, so while the population of some metropolitan areas may change little in the next decades, the population of others could even double. To appreciate some of the characteristics of the metropolitan areas of the USA and the factors influencing population change it is necessary to examine a number of variables.

The USA has some of the largest urban agglomerations in the world in size of population and indeed also in the extent of the built-up area. Table 9.8 contains data for the ten largest CMSAs in population size in 1991. These top ten in size – out of 278 metropolitan areas altogether – have about 80 million inhabitants, about 40 per cent of the metropolitan total. The data in column (2) show a marked difference in the rates of population change between the five located in the old industrial belt of the country (Areas 1, 3, 6, 7 and 8 in Table 9.8, with 4 marginal) and the remaining five. Marked contrasts also occur in the proportion of Black and Hispanic to total population. As would be expected, the scores of the largest urban agglomerations on the five variables (2)–(6) in Table 9.8 are not at the extremes of the range of values for all metropolitan areas. These are shown in Table 9.9a–e, which contains the highest and lowest scoring MSAs; none of the eighteen CMSAs actually falls at the limits of the range. There follow brief comments on each of the subdivisions (a) to (e) in the table. The reader whose interest has been sufficiently aroused to examine the data more thoroughly can consult either *U.S. Metro Data Sheet* (1993), produced by the Population Reference Bureau, or the section on metropolitan areas in the *Statistical Abstract of the United States* (e.g. pp. 37–41 in the 1993 volume).

(a) There are very marked contrasts in the rate of population change in the metropolitan areas of the USA. Ten of the fastest growing SMAs are in Florida, the remainder in the four states bordering Mexico. The differences are due partly to natural change (birth against deaths) and partly to migration, both within the USA and between the USA and other countries. Many of the fastest growing MSAs are located close to the border of Mexico or, in the case of Florida, to the Caribbean islands. They are all in the 'sunbelt', and attract population, particularly the retired, from other parts of the USA.

(b) The proportion of the population aged 65 and over is highest in some of the MSAs of Florida. The increasing share of elderly in such cities will tend to reduce the rate of natural increase. That tendency could be cancelled out by the continuing net inflow of foreign migrants. The MSAs with the smallest proportion of population 65 and over are widely scattered throughout the country, but in some cases the high contribution of foreign migrants and the absence of retired people other than local ones accounts for the young average age.

(c) The unemployment rate should give a rough idea of the attractiveness of a city to job seekers leaving their home area. In practice it is unlikely that many people would be greatly influenced by unemployment statistics, even if they had easy access to them. The unemployment rate in the USA as a whole has been around 5–6 per cent in the early 1990s, about half the average for the European Union (see Chapter 5), where its influence on migration is considerable. In the USA, some of the MSAs with an unemployment rate well below the national average are in the Midwest region. At the other extreme, the presence of large numbers of foreign migrants may account for the very high levels of unemployment in several MSAs in California and in other places on or near the Mexican border.

(d) Information about the Black (the Afro-

American) population is plentiful in US statistical sources and is provided in both the sources used here for all the metropolitan areas of the country. The extremes are very striking, with well over a third Black in many MSAs in the southeast states, 25 per cent in Washington – Baltimore, and around 20 per cent in Detroit and Chicago. In marked contrast there are virtually no Blacks in many MSAs, whether in the north and northwest of the country or along the Mexican border (e.g. Laredo).

(e) It might not be obvious to a non-American why the US Census appears to attach such great importance to obtaining and publishing data about skin colour rather than, for example, about blood groups, tall and short people, or other physiological characteristics. Given the conditions under which the original Black population arrived in the USA, most of them well before the non-British immigrants, they differ little from the White Americans culturally. On the other hand, there is a practical reason for finding out where in the USA the Hispanics live, whether permanently or temporarily. Many are living and working illegally in the USA. They have problems with language and with adjustment to a new life style. Of the thirty MSAs and CMSAs with the highest Hispanic element, twelve are in Texas, nine in California. The eastern half of the Mexican–US boundary is with Texas while the other half is shared by California, Arizona and New Mexico. The eastern half is nearer than the western half to most of the population of Mexico and is more easily reached and therefore attractive, hence the high proportion of Hispanics in many Texas MSAs.

At this point in the book it is useful to compare the largest urban agglomerations in other parts of the world with those of the USA (see Table 9.10). The data are for the early 1990s. New York and Los Angeles are much larger in population than the largest cities in the EU and the former USSR. The three largest in the EU are not likely to grow much in size. On the other hand, with restrictions on migration to cities in Russia and the other former republics now removed or reduced, it is possible that these places will grow fast. Although the population of Japan is less than half the size of that of the USA and the former USSR and about a third that of the EU, its three largest urban agglomerations match the three largest in the USA. Further population growth in Japan's cities is likely to be modest.

The population of the selection of very large cities in the developing world has doubled in the last two or three decades. The natural increase in population is still high in most cities in the developing world apart from those of China and South Korea, but even here inward migration adds substantially to growth. Elsewhere rural areas and smaller cities are growing reservoirs from which migrants make their way in to the largest cities. A doubling of the population of Mexico City or São Paulo would yield an absolute gain of up to 20 million, with awesome problems for city managers and planners. Even with the many resources available, New York and Tokyo have problems in providing adequate amenities for their populations. With a fraction of the resources available and twice as many people it is hard to imagine how the emerging mega-cities of Latin America, India, China and elsewhere in the developing world will survive, let along improve and prosper.

In Table 17.10 North America and Oceania have the highest scores among the world's major regions for material living standards and political stability, scoring more highly in particular on natural resource endowment than Western Europe and Japan, but North America is not without problems. If oil and natural gas are consumed at current rates far into the next century in spite of dwindling home reserves, then as the only country powerful enough to take effective military action, as in the Gulf War, the USA will constantly be concerned about Middle East affairs. Internally, both Canada and the USA face ethnic problems. In Canada, some autonomy or even complete independence for Quebec is considered a serious prospect. The fact that in 1994 over half the US homicide offenders and also the victims were of the Black 'race' points to a situation which, if it occurred elsewhere in the world, might draw strong criticism from many Americans. The 'melting-pot' may have been more successful in bringing European immigrants together. Even so, after several generations of life in the USA, those of Irish origin can influence attitudes and actions towards affairs in Ireland, and those of Jewish origin likewise focus attention on Israel. Whether or not the majority of Americans want it, thanks to its great economic strength and its cosmopolitan population, the USA cannot escape the 'most prominent place among the nations' anticipated in 1835 by Alexis de Tocqueville.

KEY FACTS

1. Until the end of the eighteenth century most of the present USA was thinly populated. Most of what is now very productive land was forested or natural grassland. The minerals were little used. The flood of European settlers in the nineteenth and early twentieth century had almost an economic vacuum to develop.

2. With the spread of railways and the introduction of other techniques from Europe, supplemented increasingly by American inventions and innovations, modern development could take place without the encumbrance of large numbers of traditional farmers to be absorbed, as was the case in many other parts of the world.

3. Thanks to the large size and wide range of natural resources in the twentieth century the USA has depended far less on foreign trade than most countries of the world. That situation is changing as foreign trade grows.

4. Since independence in 1885 the USA has not remained entirely isolated. Its influence has been considerable in Mexico, Central America and the Caribbean as well as in the northern Pacific. Only during and since the Second World War have large numbers of US bases and service personnel remained permanently outside North America.

MATTERS OF OPINION

1. In recent decades restrictions and quotas have been placed on aspiring migrants from the rest of the world. The 'net' has been illegally penetrated most easily by migrants from Mexico and the Caribbean islands. In 1994 it was estimated that there were 1.5 million illegal immigrants in California alone, about 5 per cent of the population of that state. Should the USA, against its basic principles, set up an impenetrable 'Berlin wall' type of barrier the whole length of its mutual border with Mexico?

2. An off-the-cuff list of issues considered to preoccupy the USA in the 1990s included the following. To what extent do you agree with the list? What other issues come to mind?

- Heavy dependence on imported oil, particularly that from the Middle East. Should fuel for motor vehicles be more heavily taxed?
- Migration policy or lack of policy
- Ethnic diversity resulting from the growing presence of migrants from outside Europe
- Natural and man-made hazards and disasters (e.g. earthquakes, hurricanes, floods, pollution, soil erosion) do not go away. They cause less loss of life than a century ago, but immeasurably more material damage
- To what extent should the natural environment be protected and even restored or recreated (see Box 9.4 on the so-called Buffalo Commons)?
- What future role should the USA have as 'global policeman', a role 'enjoyed' by Britain in the nineteenth century. How big should the military establishment be, and should foreign bases and treaty links (e.g. NATO) be maintained?
- What would be your order of priority in US involvement elsewhere in the world aimed at protecting US interests (e.g. Middle East oil), establishing democracy (e.g. Haiti) or delivering humanitarian assistance (e.g. Somalia)?

SUBJECT FOR SPECULATION

The outcome of the American Civil War (1861–5) apparently ensured once and for all that the USA would remain a single sovereign state, though with considerable power remaining in the states. Frazer and Lancelle (1994) have combined two statements Vladimir Zhirinovsky made in 1994 about the future of the USA: 'We will not gloat when California joins Mexico [he did not say "rejoins"], when a Negro republic is created in Miami and when the Russians take back Alaska, or when America dissolves into a Commonwealth of New States.' Ten years ago, few people would have predicted the collapse of the USSR. Is there any reason to anticipate some kind of break-up of the USA? If so, which states might be responsible for initiating moves towards greater autonomy?

Chapter ten

Canada and Oceania

How could a man or a people seize a vast territory and keep out the rest of the human race except by a criminal usurpation – since the action would rob the rest of mankind of the shelter and the food that nature has given them all in common?

Jean-Jacques Rousseau (1762)

10.1 Canada, Australia and New Zealand

Canada and Australia resemble one another in many ways. Throughout this chapter comparisons are made between the two countries. In two very important respects, however, they differ. First, the natural environment of Canada is characterised by extremely cold conditions for varying periods during the year whereas Australia is characterised over much of its area by hot and dry conditions. Second, Canada shares with the USA the longest land boundary in the world between two countries. Its neighbour has almost ten times as many inhabitants. In contrast Australia enjoys an isolated location between the Indian and Pacific Oceans. Table 10.1 contains comparative data for the USA, Canada and Australia. Given the very large trade between Canada and the USA, such a comparison is relevant, Canada being one of the main sources of US imports of primary products.

By world standards, both Canada and Australia have very low densities of population, Canada having 7.4 per cent of the world's land area with only 0.5 per cent of the world's population, Australia 5.8 per cent with 0.3 per cent. Figure 10.1 shows the two countries represented on the same scale with the broad types of natural environment in each. Figure 10.2 shows winter temperatures in Canada and rainfall in Australia. Although only a small proportion of the total area of each country is under arable and permanent crops, Canada has more than six times the world average amount of arable land per inhabitant, Australia more than eleven times. Australia has forty times the world average amount of natural pasture per inhabitant, although most of it is of very low productivity, while Canada has almost eighteen times its 'share' of the world's forest. In terms of bioclimatic resources, Canada and Australia are among the most generously endowed large countries in the world, a feature shared by three regions of the Russian Federation: East and West Siberia and the Far East.

The USA, Canada and Australia have about 5.5 per cent of the total population of the world but a far larger share than that of the reserves of coal and of almost all the principal non-fuel minerals. Together, however, they consume between a fifth and a quarter of most mineral products, which puts them in a weak position with regard to reserves of oil and natural gas.

The demographic data in Table 10.2 show that Canada, Australia and New Zealand are similar in many respects to the USA and West European countries. Low infant mortality rates testify to good health services, and downward trends in fertility rates to a future in which in some decades' time the replacement rate of population will not be maintained without net in-migration. The rate of natural increase of population resembles that found twenty to thirty years ago in Western Europe. The estimates of population increase between 1993 and 2025 point to a combined absolute addition in the three countries of a mere 13–14 million people to their present population, less than the

TABLE 10.1 Natural resources of the USA, Canada and Australia in percentages of the world total

	USA	*Canada*	*Australia*	*All three*
Population	4.7	0.5	0.3	5.5
Total area	7.0	7.4	5.8	20.2
Arable	13.2	3.2	3.4	19.8
Pasture	7.1	0.8	12.3	20.2
Forest	7.3	8.9	2.6	18.8
Coal	23.1	0.8	8.7	32.6
Oil	3.4	0.8	0.2	4.4
Natural gas	3.9	2.2	0.3	6.4
Aluminium	–	–	21.1	21.1
Copper	16.8	5.0	2.3	24.1
Iron ore	10.5	7.7	10.3	28.5
Lead	22.1	12.6	16.8	51.5
Tin	0.7	2.0	5.9	8.6
Silver	11.7	14.8	10.0	36.5
Zinc	12.9	15.3	10.6	38.8
Industrial diamonds	–	–	51.0	51.0
Gold	6.3	3.3	1.8	11.4
Manganese	–	–	7.5	7.5
Nickel	0.5	13.8	4.0	18.3
Sulphur	12.0	11.6	–	23.6
Tungsten	5.4	17.1	4.6	27.1
Potash	1.0	48.4	–	49.4

amount by which the population of India or China increases in a single year.

The percentage of the economically active population in agriculture has declined sharply over a long period in all three countries, while urbanisation has proceeded rapidly. As a result, a large proportion in each country is concentrated in a few cities, while a very small number of people is thinly dispersed over vast areas, engaged in agriculture, forestry and mining. In both Canada and Australia large areas are virtually untouched by human activities. Table 10.3 includes some key dates in the history of the three countries. Canada, Australia and New Zealand will now be discussed separately.

10.2 Canada

Like the Nordic countries of Western Europe and also the former USSR, Canada is characterised by its high latitude. A very small, albeit important, part of Canada extends south to 42° North, the latitude of Chicago in the USA, but almost all of the country is closer to the North Pole than to the Equator. The mountains of the west shut off the interior of Canada from much of the influence of air masses from the Pacific, while the eastern side of Canada does not feel the effects of warm North Atlantic drift waters which reach similar latitudes in Western Europe. On account of unfavourable climatic and soil conditions, less than 5 per cent of the total area of Canada is cultivated, and even in the main area of cultivation, the prairies of the western interior, the growing season is short. Cereal yields are lower than those in the USA and Western Europe (see Table 10.4). During the 1980s, Canada produced between 130 and 180 million tonnes of roundwood per year, about 5 per cent of the world total, yet, unlike most countries of the world, replanting has actually resulted in an increase in the forest area in the last three decades.

Canada's non-fuel minerals are mostly located in the geologically favourable eastern half of the country, in the Canadian Shield area (Figure 10.3). Oil and natural gas, on the other hand, are in the west, particularly in Alberta province. With its wide range of non-fuel minerals, its forest products and adequate but not spectacular reserves of oil and natural gas, Canada is well provided with the ingredients for industrial development and at the same time is a net exporter of many primary products. The small size of its national market compared with that of the USA has, however, made it difficult to establish a separate industrial base. In terms of investment per inhabitant, Canada has been by far the largest destination of US investment in the world, not only in manufacturing, but also in minerals and in services. Most of the industrial capacity of Canada mirrors in miniature the activities of the industrial belt of Northeast USA, with, for example, the manufacture of motor vehicles in Windsor, just across the border from Detroit, centre of the industry in the USA, and with steel mills on the shores of the Great Lakes. This part of Canada is well served by the St Lawrence Seaway (see Figure 10.4).

The human geography of Canada is a little difficult to appreciate on account of the presence of the USA to the south and the virtually empty lands of northern Canada to the north. About 90 per cent of Canada's population is within 200 km (124 miles) of the US border or the Great Lakes. The 'heartland' of Canada is a very long narrow strip of fairly densely populated country, roughly similar in length to Chile in South America. European penetration has come largely from the east, pushing in the nineteenth century beyond the Great Lakes, and eventually reaching the west coast. The populated zone of land has been integrated by two transcontinental railway lines, the importance of which, especially for passenger traffic, has nevertheless diminished sharply in recent decades.

From the main belt of population in Canada, numerous links cross into the USA. Transactions between different sectors of Canada and adjoining parts of the USA are as important as east–west ones within Canada itself. One example is the movement of oil by pipelines from Alberta into Northwest USA.

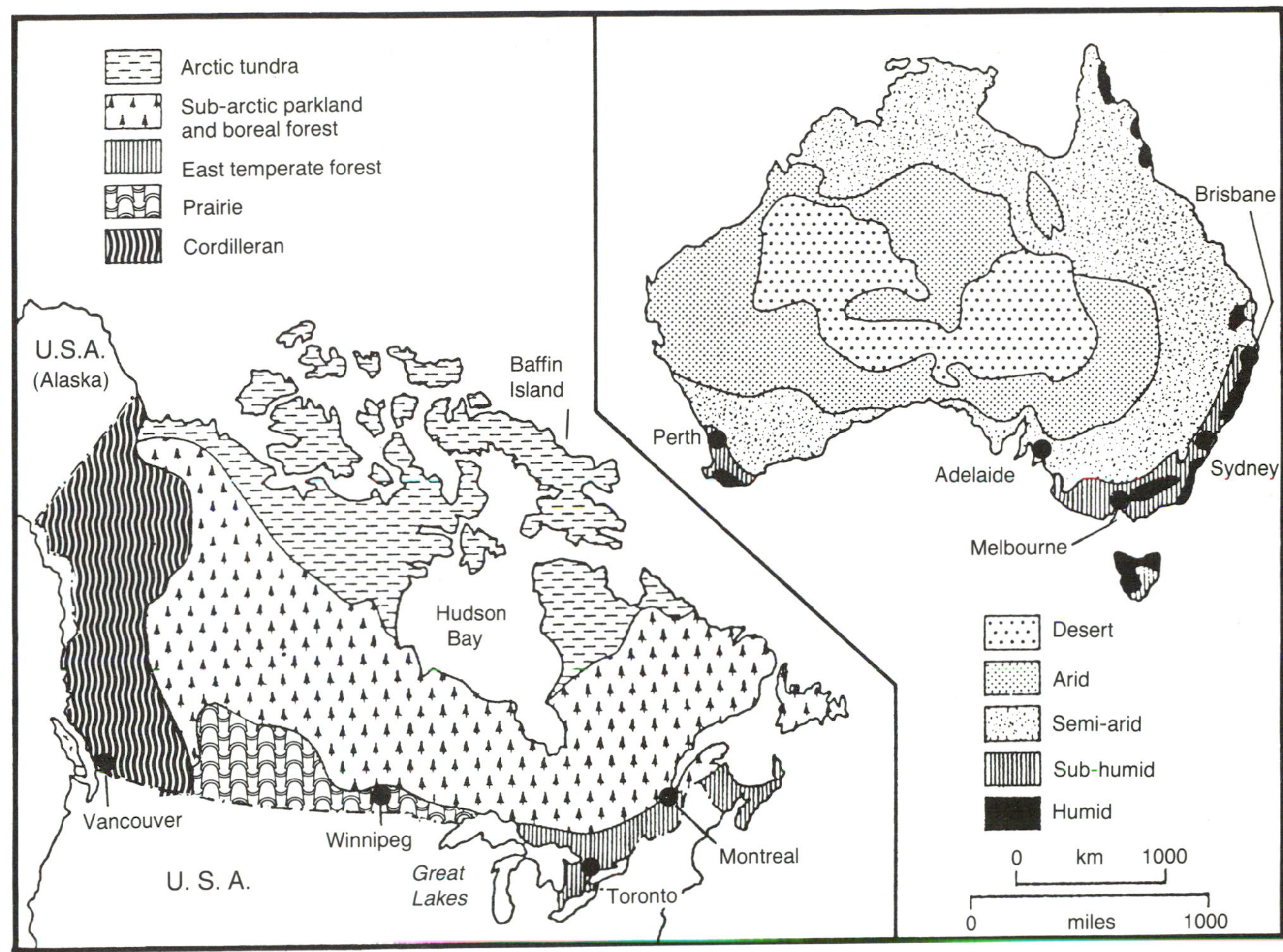

FIGURE 10.1 Natural environments of Canada and Australia compared

Eastern Canada, on the other hand, is supplied mainly by oil imported from Western Europe and elsewhere. To the north, tenuous links with distant mining and forestry centres reach into the virtually uninhabited northlands, and the Alaska Highway crosses Canada to link Alaska to the rest of the USA.

The emptiness of the north is reflected in the fact that, on the 40 per cent of Canada's area accounted for by the Northwest Territories and Yukon, there lives a mere 0.3 per cent of the total population of Canada. The population of these two divisions of Canada grew by less than 20,000 (from 68,000 to 86,000) during the 1980s while the rest of Canada grew by almost 3 million. In contrast, over 60 per cent of the population of Canada, some 17 million people, live in the two provinces of Quebec and Ontario, mainly in the

TABLE 10.2 Demographic profiles of selected countries

	USA	*Canada*	*Australia*	*NZ*	*Argentina*	*UK*
Population in millions	260.8	29.1	17.8	3.5	33.9	58.4
Birthrate per 1,000	16	14	15	17	21	13
Deathrate per 1,000	9	7	7	8	8	11
Annual natural increase %	0.7	0.7	0.8	0.9	1.3	0.2
Total fertility rate	2.1	1.8	1.9	2.1	2.9	1.8
% under 15	22	21	22	23	30	19
% 65 and over	13	12	12	11	9	16
GDP per capita 1992 in US dollars	23,120	20,320	17,070	12,060	6,050	17,760

Source: WPDS (1994)

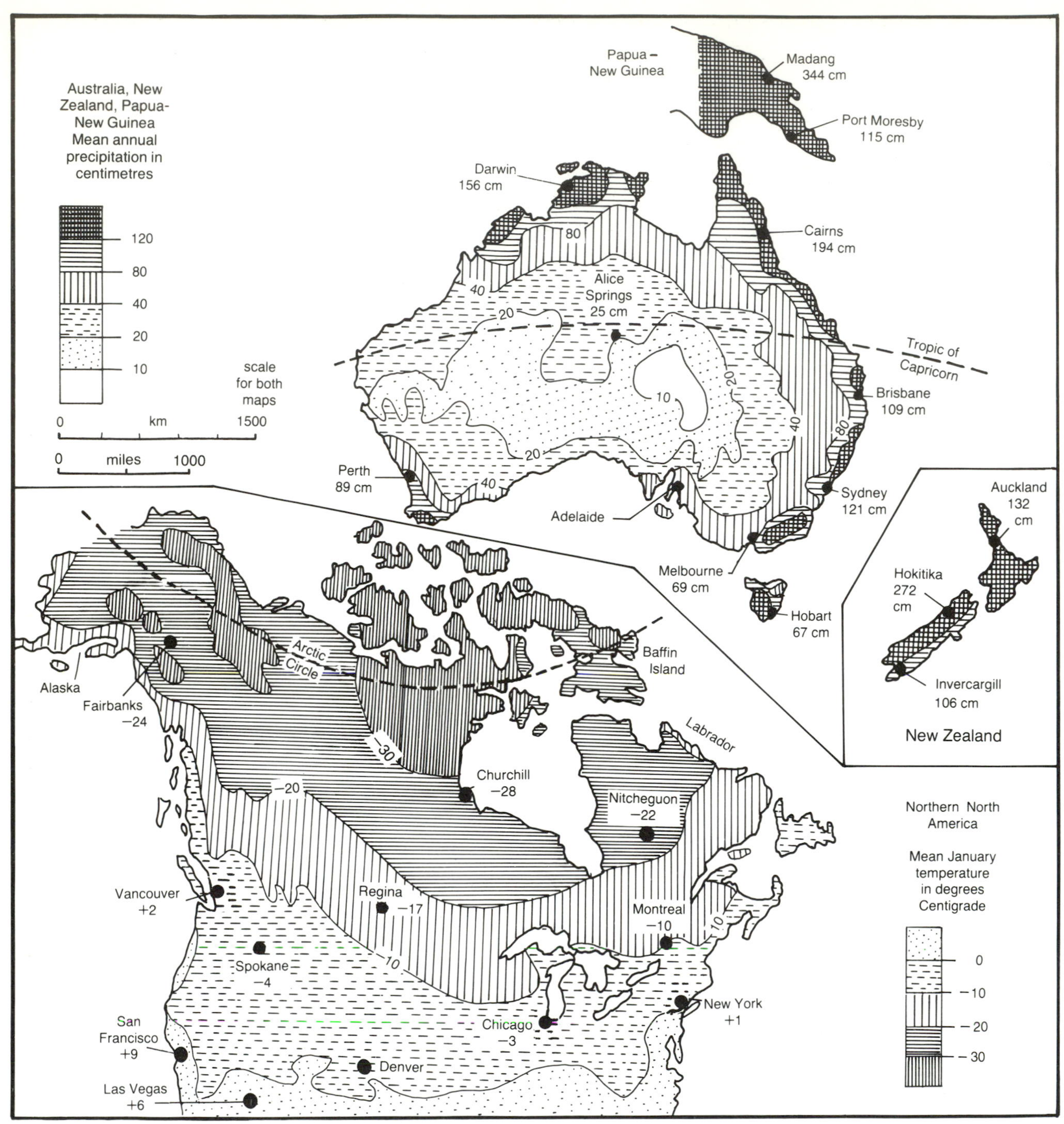

FIGURE 10.2 Rainfall in Australia and winter temperatures in northern North America. Based on *Schweizer Weltatlas*

southern parts. The Census Metropolitan Areas (CMA) of Montreal and Toronto have a combined population of 7 million, one-quarter of the total population of Canada. Seven other CMAs each have over half a million inhabitants.

Some of the positive and negative features of the geography of Canada have been noted. By world standards, Canada is extremely well endowed with natural resources, its economy is diversified, and productivity in all sectors is high. Its real GDP per capita of 18,635 US dollars per inhabitant was third in the world after those of the United Arab Emirates and the USA, while on the Human Development Index of the UN *Human Development Report 1992* it came first in the world on account of its high levels of life expectancy and mean years of schooling. To the inhabitants of most regions of the world, the life style of the citizens of Canada is far beyond their wildest aspirations or

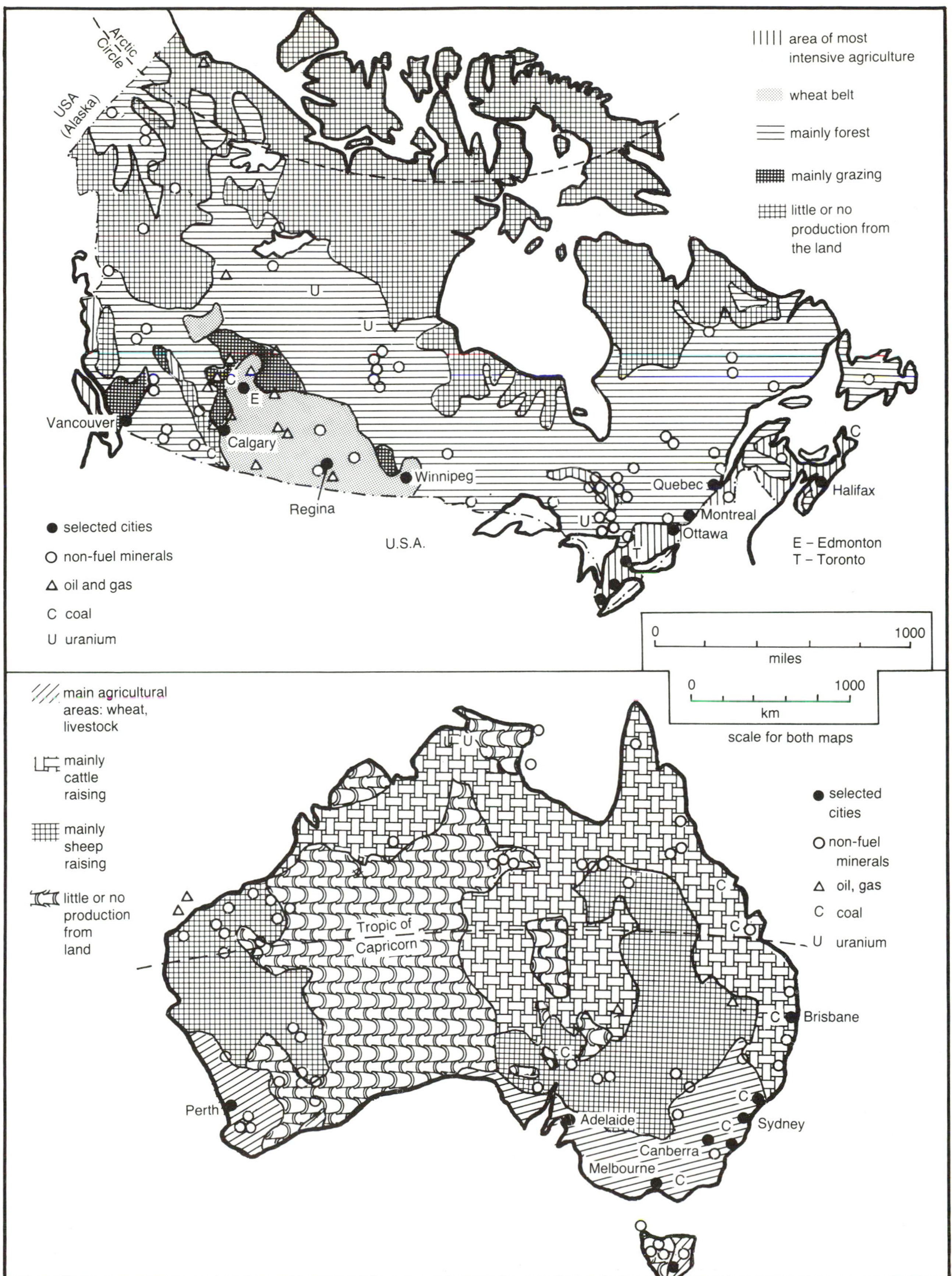

FIGURE 10.3 Economic conditions in Canada and Australia

TABLE 10.3 Calendar for Canada, Australia and New Zealand

1608	French colonists found Quebec
1645	Tasman circumnavigates Australia and discovers New Zealand
1760	New France conquered by British, Quebec captured 1759, Montreal 1760
1788	First British colony in Australia
1807	First wool exported to Britain from Australia
1836	First railway in Canada (Quebec)
1840	Britain annexes New Zealand
1850	Gold discovered at Bathurst, Australia
1867	Dominion of Canada established with Nova Scotia, New Brunswick, Ontario and Quebec
1876	Population of Australia passes 2 million
1885	Transcontinental railway completed in Canada
1901	Commonwealth of Australia formed. White Australia policy
1907	New Zealand acquires dominion status
1915	First blast furnace in Newcastle, Australia
1931	Population of Canada 10.4 million
1933	Population of Australia 6.8 million
1948	Snowy River project, Australia
1959	St Lawrence Seaway completed Canada–USA
1988	Canada joins free trade agreement with USA and Mexico (NAFTA)

prospects. Yet this remarkable country is threatened with disintegration.

The last part of the twentieth century has seen resurgence of ethnic awareness in many parts of the world, based on language, religion, physiological or other features. France was the original European colonial power in Canada, but its colonies there were ceded to Great Britain in 1763 (see Table 10.3). From a few tens of thousands of French settlers at that time, and with minimal replenishment from the mother country, the French-speaking population of Canada has risen to over 6 million in the early 1990s, most in Quebec Province, with about 1.5 million additional people of multiple origin (French and British or French and Other). It seems only a matter of time before Quebec, like many ethnic groups in Europe in the last few years, achieves sovereignty, splitting the rest of Canada into two parts. Before the end of the Cold War, such an event would have been regarded as a strategic and security setback, if not disaster, by the USA, one reason being that Quebec controls the entrance from the Atlantic to the St Lawrence Seaway system (see Figure 10.4).

Forecasts about events of a political nature are notoriously prone to failure, but there seems no obvious reason why economic life and the management of space in the northeast of North America should be more than superficially modified by an independent Quebec. One reason is that most of Canada's foreign

TABLE 10.4 Wheat yields

	Average in kg per hectare	
	1979–81	*1988–90*
Argentina	1,550	1,860
Australia	1,260	1,590
Canada	1,780	1,760
Former USSR	1,510	1,980
USA	2,290	2,380
France	4,990	6,330

trade is with the USA. In 1990, 65 per cent of its imports came from its neighbour while 75 per cent of its exports went there. In contrast, the European Union and Japan provided only 17 per cent of its imports and took the same proportion of its exports. Its trade with Mexico, the other partner in NAFTA, is very small.

10.3 Australia

If the essential Canada is a long, narrow east–west belt of settlement along the northern edge of the USA, Australia is a number of separate 'islands' of settlement spaced at considerable intervals along the southeast coast of the country, plus one (Perth) in the west. The settlement of Australia by the British is even more recent than the settlement of Canada by the French. The indigenous population of Australia was not equipped technologically to resist British conquest and was so small in number that it was easily relegated to remote areas or, as in Tasmania, wiped out. Australia was not organised fully as a single entity until 1901, when the Commonwealth of Australia was created. For example, there was no uniform gauge of railway for all the colonies. The 'emptiness' of Australia is underlined by the fact that after more than 200 years it still has only 18 million inhabitants, in spite of an almost uninterrupted though comparatively small flow of migrants from the British Isles up to the Second World War, topped up by other Europeans since then. That population is not expected to increase in the next three decades by more than a few million.

The author's calculation of the natural resources of the major regions of the world (see Chapter 3) shows Oceania to be in a league of its own. Australia has two-thirds of the population of Oceania but almost all its natural resources. With 0.5 per cent of the world's population, Oceania has 6.5 per cent of the total area of the world, 4.5 per cent of the productive land, 2.0 per cent of the freshwater resources, 3.1 per cent of the fossil fuels and 10.1 per cent of the non-fuel minerals. Much of Australia remains unexplored for minerals, yet already it is credited with more than 10 per cent of

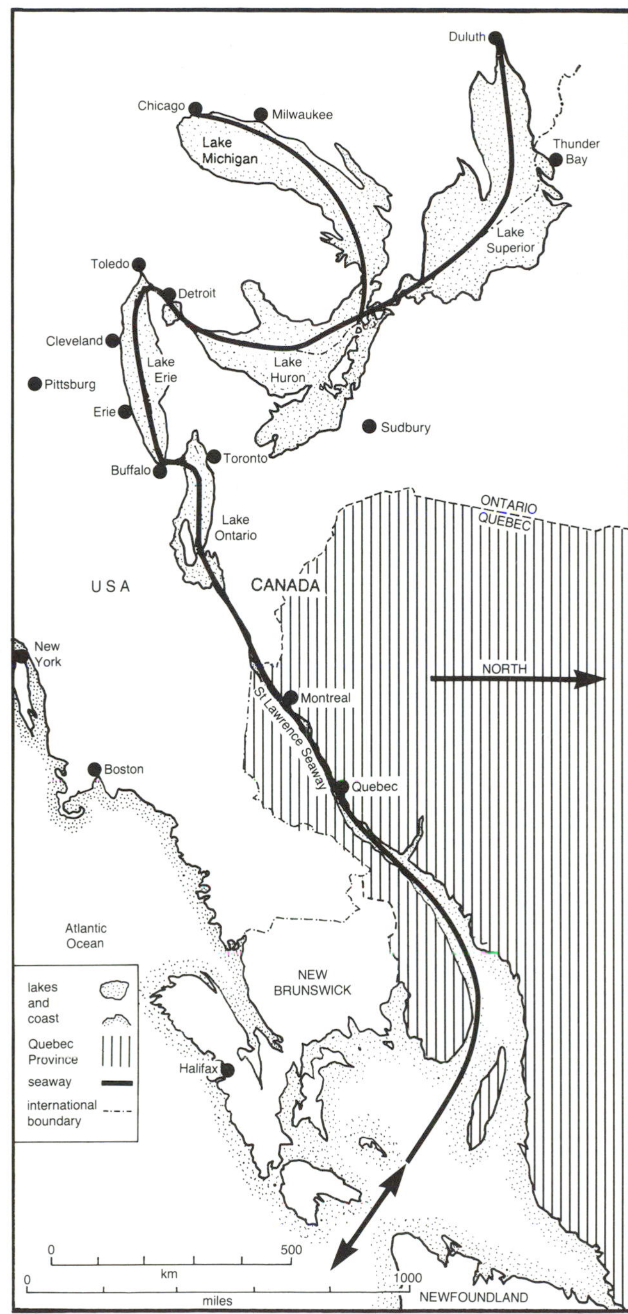

FIGURE 10.4 The St Lawrence Seaway

the bauxite, iron ore, lead, silver, zinc and industrial diamond reserves of the world. The failure to find more than modest reserves of oil and natural gas has been disappointing, making Australia a net importer of oil. On the other hand, thanks to large reserves of hard coal in deposits that are cheap to work, Australia has become a leading exporter of this fuel. Manners (1992) expects increasing problems in the exploitation of Australia's mineral resources for at least two reasons: obstacles posed by the indigenous population in some areas, and general concern over protection of the natural environment.

The concern over the Australian environment is not confined to mineral extraction, but it extends to the sensitive biological environment through the continued clearance of the remaining forest for cultivation and pasture. According to *FAOPY 92*, the area of forest and woodland in Australia has been reduced from 1,377,000 sq km (532,000 sq miles) in 1975 to 1,060,000 sq km (409,000 sq miles) in 1990, a loss of over 300,000 sq km (116,000 sq miles), with an addition of only 65,000 sq km (25,000 sq miles) of arable land. With some local exceptions, yields in agriculture are low by the standards of most of the developed regions of the world, a result of both the marginal conditions of soil and precipitation in many areas in which cultivation, especially of cereals, is carried out, and a comparatively low level of application of fertilisers (Figure 10.5). On the other hand, the economically active population of 408,000 in agriculture farm 490,000 sq km (189,000 sq miles) of land under crops, on average 1.2 sq km (0.46 sq miles) or 120 hectares (296 acres) per worker while also using over 4 million sq km (1.5 million sq miles) of permanent pasture land for grazing.

Most of the cultivated land in Australia is in a belt of country in the 'near interior' of New South Wales, Victoria and South Australia, with an outlier in the extreme southwest of Western Australia. Warm temperate crops such as vines and rice, and tropical fruits and sugar cane, are grown locally in irrigated lands in the southeast and along the coast of Queensland. Australia is virtually self-sufficient in agricultural products, apart from tropical crops such as beverages. For its population size it is a large net exporter of agricultural products, in the past mainly to Western Europe, more recently to Japan and to other countries in East and Southeast Asia, the destination of many of its exports of minerals too. Table 10.6 shows the shift that has taken place in the direction of Australia's foreign trade between 1950 and 1990. Trade with the UK has declined dramatically, especially since the UK's entry into the EEC in 1973. The natural resource bases of Australia and Japan are at the extremes of abundance and scarcity on a world scale, but Japan is able to supply many of the manufactured goods not produced in Australia.

The government of Australia this century has not been consistent with regard to its immigration and demographic policy. Earlier this century it was portrayed as a country with vast economic potential, able to support many times its small population (less than 4 million in 1901). One Australian geographer, Griffith Taylor, pointed out that much of Australia is desert. His view of the limitations of the bioclimatic resources was justified, but there is no question that Australia has enormous reserves of many minerals. Although immigration has been encouraged after the Second World

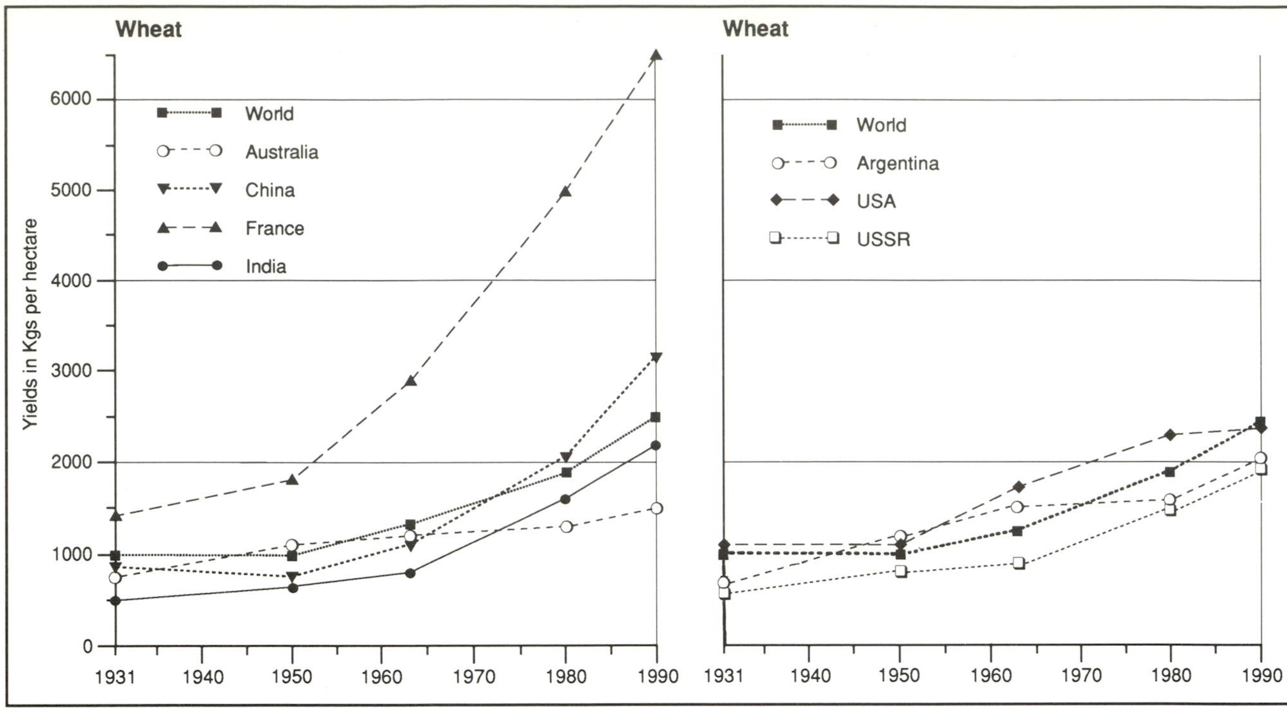

FIGURE 10.5 Wheat yields in selected countries, 1931–90

TABLE 10.5 Area and population of the provinces of Canada, the states of Australia, and New Zealand

	Area in thous. sq km/(miles)		*Population*	*Density of population per sq km/(mile)*	
Newfoundland	372	(144)	568	1.5	(3.9)
Prince Edward Island	6	(2)	130	21.7	(56)
Nova Scotia	53	(20)	900	17.0	(44)
New Brunswick	72	(29)	724	10.1	(26)
Quebec	1,358	(524)	6,896	5.1	(13)
Ontario	917	(354)	10,085	11.0	(28)
Manitoba	548	(212)	1,092	2.0	(5.2)
Saskatchewan	570	(220)	989	17.4	(45)
Alberta	638	(246)	2,546	4.0	(10)
British Columbia	893	(345)	3,282	3.7	(10)
Yukon	532	(205)	28	0.05	(0.13)
North West Territories	3,246	(1,253)	58	0.02	(0.05)
Canada	9,203	(3,553)	27,297	2.97	(8)
New South Wales	802	(310)	5,941	7.4	(19)
Victoria	228	(88)	4,439	19.5	(51)
Queensland	1,727	(667)	3,000	1.7	(4.4)
South Australia	984	(380)	1,454	1.5	(3.9)
Western Australia	2,526	(975)	1,651	0.7	(1.8)
Tasmania	68	(26)	469	6.7	(17)
Northern Territory	1,346	(520)	168	0.1	(0.26)
ACT	2.4	(0.9)	293	122.0	(316)
Australia	7,682	(2,966)	17,414	2.3	(6)
New Zealand	271	(105)	3,435	12.7	(33)

TABLE 10.6 The direction of Australian foreign trade, 1951 and 1990

	1951		*1990*	
	imports	*exports*	*imports*	*exports*
UK	48	33	7	3.5
Rest of EC (12)	9	23	16	10
USA	8	15	25	12
Japan	2	6	18	25
New Zealand	0.4	2	4	6
India	5	2	0.5	1.5
Malaya[1]	4	2	4	6
World total	100	100	100	100
Absolute values	£A 744m.	£A 982m.	US $ 39,740m.	US$ 38,662m.

Notes: [1] In 1990 Malaysia and Singapore
Sources: UNYITS 1953 vol. 1, p. 48; *ITSY 1990*, vol. 1, p. 33

War, less publicity has been given to portraying Australia as the (or a) land of the future. At a conference on its natural resources held in 1973, concern was expressed at the danger that they could be used up too quickly.

If Australia is concerned over its natural resource base, many regions of the world should consider their positions as desperate. A more immediate problem in Australia has been the need to maintain good links between the main concentrations of population and to persuade settlers to move into and remain in enough places in the interior and along the northern coastlands to provide credibility that some use is being made of these virtually empty lands. Not far away to the north lies Indonesia, where the population is growing by more than 3 million a year.

Air transport carries much of the interstate passenger traffic, although it is possible to go from Sydney to Perth by rail on a journey lasting three days and nights. Since most of the population of Australia lives on or very close to the coast, shipping services provide satisfactory links between the many ports. Various rail links penetrate far inland from points on the coast, but the problem of having different gauges in the states made interstate movement by rail difficult until certain lines in the other states were converted to the standard gauge (4 feet, 8½ inches – 143.5 cm) used in New South Wales. These lines are shown in Figure 10.6.

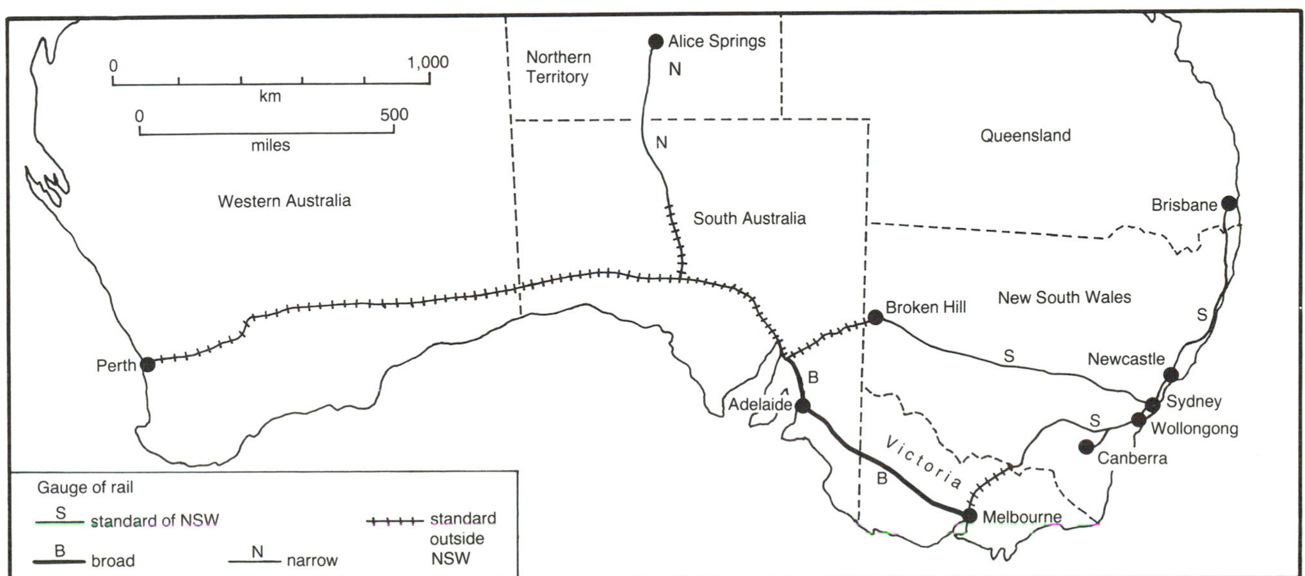

FIGURE 10.6 Transport links in Australia

Australia was ranked seventh in the world on the scale of the United Nations Human Development Index (scoring 971 out of 1,000). Like the Canadians, the Australians should have little to complain about, but nationally there is a problem because, with or without justification, most Australians prefer to live in or close to one of the big five urban concentrations. These contain over 60 per cent of Australia's total population. It is in these cities that almost all of the manufacturing in Australia is concentrated. Newcastle, north of Sydney, Sydney itself and Wollongong to the south form the main heavy industrial region of Australia, with most of Australia's coal production here. Melbourne is the second industrial centre. In contrast to most of the other major regions of the world Australia seems set to experience more of the same, with few big surprises.

Analogies are fraught with dangers. Nevertheless, in the world of comparative regional geography, they can be at least thought-provoking. For example, if Japan is cut along the middle and its two coasts put end to end, they can match almost exactly the coastline of China. If the Amazon and Lena (in Siberia) basins are compared, they have in common forest, a low density of population, the use of waterways for transport and the prospect of clearance and 'development' from the fringes and from bases in the interior. In Figure 10.7 remarkable similarities and equally notable differences can be seen in a comparison between Canada and Australia. For example, a series of large cities, linked originally by rail, now also partly by road, extends over a great distance. The two largest cities are near one end and in between is the capital city. The hinterlands are thinly populated, but inhabited largely by an increasingly vociferous indigenous population. The contrasts include the fact that on its interface with the rest of the world the Australian region is limited by the ocean while Canada borders the USA. Australia does not have the cultural division of Canada; New South Wales is not a former French colony. Even so, in many respects the problems of the management of space in the two countries are broadly similar.

10.4 New Zealand, Papua-New Guinea and the Pacific Islands

The three regions discussed in this section differ enor-

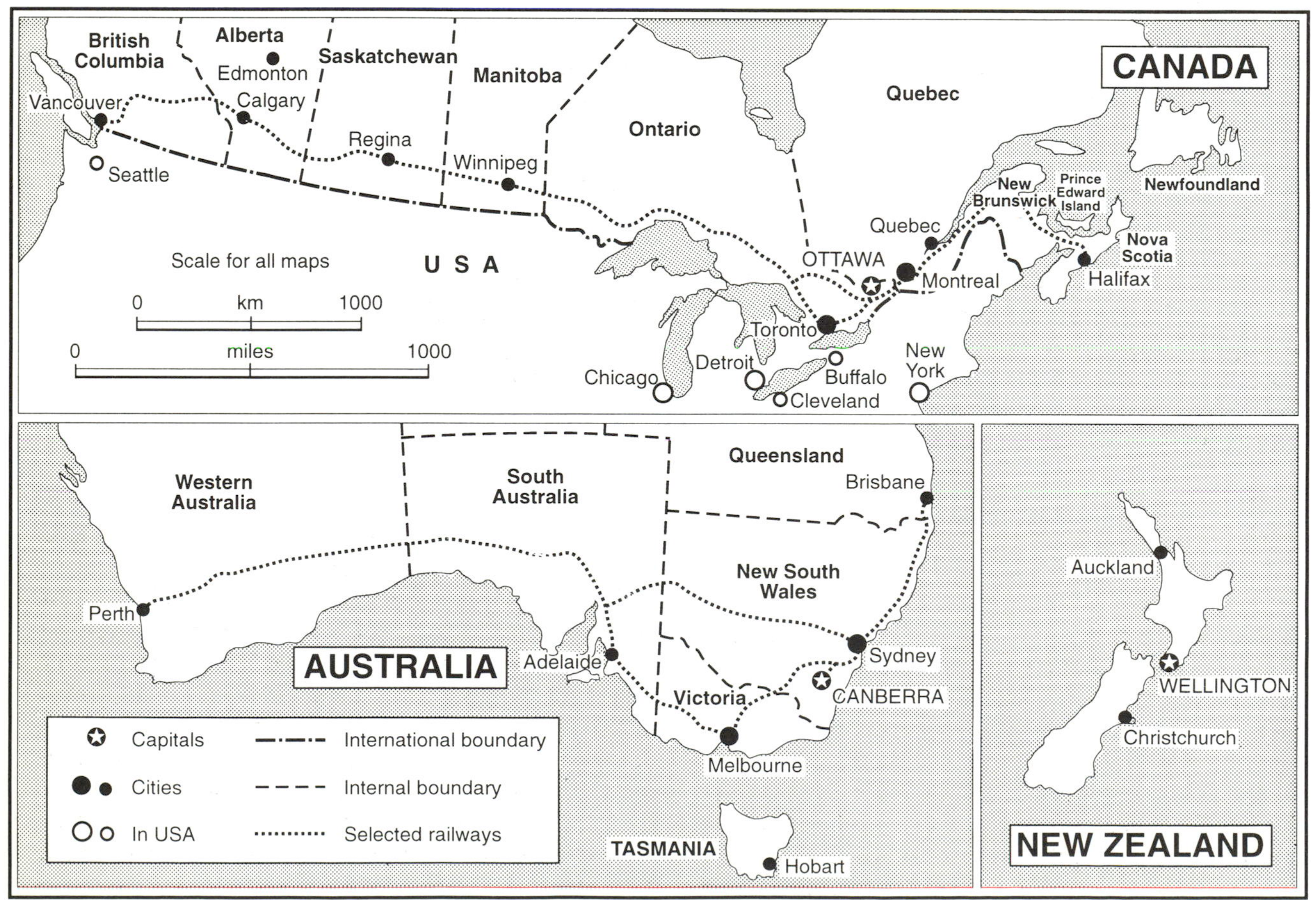

FIGURE 10.7 Similarities between Canada and Australia–New Zealand

PLATE 10.1 Oceania

a York, Western Australia, where any building from the nineteenth-century 'colonial' period is likely to be listed. The smaller cities of Australia spread comfortably over large areas and unlike the big five are still reasonably stress-free

b Logs, Rotorua, New Zealand. Very little of the land in New Zealand is under crops. Forestry and pasture make a major contribution to the economy

mously in many respects, so they will be dealt with in turn.

1. **New Zealand** The colonisation of New Zealand by the British started several decades after the colonisation of Australia. The indigenous population, however, was more advanced in technology than that of Australia and in relation to the area of the country, more numerous. In due course, as in other parts of the world colonised by Europeans, the local Maori population came under British rule. New Zealand is somewhat larger in area than the UK but the UK has about seventeen times as many inhabitants. Like Australia and Canada, New Zealand is highly urbanised, with almost half of the population in three cities, Auckland, Wellington and Christchurch.

New Zealand has no economic minerals of more than local importance and the industrial sector is largely concerned with processing. Its economy is more dependent on agriculture than that of Australia, with 9 per cent of the economically active population in that sector. Even so, less than 2 per cent of the total area is cultivated and the contribution of pastoral farming is overwhelming. Roughly half of the area of New Zealand consists of permanent pasture. In 1990, 16 per cent of the value of New Zealand's exports was accounted for by meat, 7 per cent by wool and about the same proportion by dairy products. The UK is no longer one of the main destinations for New Zealand products. Australia, Japan and the USA now take half of the exports.

In a rapidly changing world, New Zealand continues to prosper, although its real GDP per inhabitant is only half that of the USA. If Australia resembles Argentina in many ways, New Zealand can be compared with Uruguay. These two countries of Latin America are not so prosperous as their counterparts in Oceania, but their peripheral locations in the world set them all apart from most of the stresses of the Cold War and its aftermath.

2. **Papua-New Guinea** resembles much more closely the less densely populated islands of Indonesia than Australia or New Zealand, most of all west Irian, the Indonesian half of New Guinea. Papua-New Guinea became independent in 1975, Papua having been a colony of Australia since 1906 and New Guinea an Australian protectorate for almost as long. In spite of the growing influence of Australia, Papua-New Guinea is very different, with a real GDP per inhabitant of less than 2,000 US dollars, less than 20 per cent of the population urban, and great ethnic diversity.

Over 80 per cent of Papua-New Guinea is still forested, and much of the island is very rugged. Many small agricultural communities remain little changed by modern developments. According to Toth *et al.* (1992), possibly the last few makers of stone axes in the world still made these tools in one remote village, while another part of the country has been opened up to exploit one of the world's largest copper mines, which provides half of the value of the exports. Population is growing fast, and although Papua-New Guinea is considerably larger in area than New Zealand, the expected doubling of population in the next three decades could make substantial inroads into the remoter parts of the island, damaging the rainforest and modifying local ways of life beyond recognition.

3. **Islands of the Pacific** The Pacific Ocean covers about 35 per cent of the earth's surface. Its extent and shape can best be appreciated on a globe. Frequent mention has been made in recent years to the Pacific Rim with reference to economic matters but the mountain ranges and strings of islands bordering much of this ocean have been known much longer for their frequent earthquakes and active volcanoes.

The various islands on the western side of the Pacific, including the large islands of Japan and the Philippines, as well as smaller islands in the Aleutians and Kuriles, are part of Asia. In the northern Pacific, Hawaii is a state of the USA, while in the southeastern Pacific, Easter Island is a Chilean possession. Most of the remaining hundreds of inhabited islands are conventionally attributed to Oceania, and are grouped into Micronesia, Melanesia and Polynesia. The *United Nations Statistical Yearbook 1990/91* listed twenty-eight separate groups of islands in Oceania in addition to Australia, New Zealand, Papua-New Guinea, and islands directly belonging to these three countries. At the end of the Second World War, Australia, New Zealand, the USA, the UK and France held the various islands as colonies or protectorates, and they are still involved to varying degrees in the economic life of the islands, providing development assistance.

Interest by outsiders in the islands of the Pacific is on two different scales. During the Cold War, the Pacific region was basically a sphere of US strategic interest. At the same time, the economic potential of the oceans, yet to be realised to any great extent, means that the Pacific Ocean is worth attention, even though the islands are the tops of high undersea mountains and rarely have any continental shelf around them. At a more local level, the islands have various attributes and uses. For example, some have been used by the French for testing nuclear weapons, some have large deposits of economic minerals and some were vital landing places in transoceanic flights before the 1970s, when propeller planes and then the early jets were limited in range. On the other hand, the quantities of such agricultural products as coconuts, pineapples and sugar cane available for export are very limited. Above all, the future of the islanders themselves must be considered. Population is growing fast on many of the islands and food production is not adequate to support their populations.

It is difficult to calculate the area and population of the Pacific islands of Oceania with precision. Data from *FAOPY* and PRB are used here to put them into a global perspective. In 1993 Oceania had a population of 28 million, but when the combined populations of Australia, New Zealand and Papua-New Guinea of 25 million are deducted, 3 million Pacific islanders in Oceania are left. They live in islands spread over much of the Pacific, some 100 million sq km (40 million sq miles) of water or 1,000 times the combined area of the islands of 90,000 sq km (35,000 sq miles), an area equal roughly to the states of Maine or Indiana in the USA. Fiji and New Caledonia, each about 18,000 sq km (7,000 sq miles) in area, are the largest groups in area. In population Pitcairn is the smallest, with 52 inhabitants in the census of 1989.

For much of their history as British colonies and then as independent countries, Canada, Australia and New Zealand received and in due course welcomed immigrants. Almost all came from Europe, and in the case of Australia and New Zealand, largely from the British Isles until after the Second World War. Since the 1960s, it has become much more difficult to migrate permantly to these countries. Yet if the world were organised as a single political unit and its central planners were pondering over how to redeploy population more closely according to the distribution of natural resources, then large transfers of people would be required, with Australia and Canada at the receiving end. Such a thought was presumably part of the grand plan of Japanese leaders with regard to Australia in 1941. Ironically, even in the 1970s concern was expressed in Australia about the danger of over-exploiting its natural resources. Both Canada and Australia have broadly based and sophisticated industrial sectors. In the long term, however, as world population grows and more 'emerging' countries industrialise, the natural resources of Canada and Australia should become an even greater asset than they are now. Canada, Australia and New Zealand must be high on the list for people thinking of emigrating, whether from Western Europe, Hong Kong or India.

KEY FACTS

1. Canada and Australia are the two countries of the world with the largest quantity of various natural resources per inhabitant. Their shares of water, bioclimatic, fossil fuel and non-fuel mineral resources are far higher than the world average level.

2. The small population of Canada and Australia (for their territorial extent) is highly concentrated, much of it in a few large cities.

3. During the present century the industrial development of Canada and Australia has been considerable and primary products now account for a more modest share of exports than they did a century ago.

4. New Zealand is also well provided with bioclimatic resources.

MATTERS OF OPINION

1. In both Canada and Australia, a small, scattered and technically simple culture of indigenous population has been partly assimilated, partly eliminated and partly relegated to reservations. Should they be given a greater say in the running of the country and in the management of their old space?

2. The natural environment in both Canada and Australia is fragile, Canada because so much of it experiences cold conditions, Australia on account of dry conditions. To what extent should economic development be restricted in the interest of preserving areas as yet hardly touched by human activities?

SUBJECTS FOR SPECULATION

1. How long can Canada and Australia sit on their natural resources unnoticed before less well-endowed countries take an unwelcome interest in them?

2. Is the concentration of population in a few cities in Canada and Australia an advantage rather than a drawback?

Chapter eleven

Latin America

Actually Latin America is not a virgin land, awaiting the arrival of the pioneer – it is an old land, tramped over, many of its sources of accumulated treasure exploited and abandoned, many of its landscapes profoundly altered by the hand of man. Yet it is a land in which large areas remain comparatively empty of human inhabitants.

Preston E. James (1950)

11.1 Introduction (see Table 11.1)

The cultural history and background of Latin America set it apart from the USA and Canada. Spaniards often refer to Latin Americans other than Brazilians as *hispanoamericanos* and to citizens of North America as *norteamericanos*. Strictly, 'Latin' refers to peoples with a language of Latin origin, Spanish, Portuguese, French, Italian and Romanian being the main ones. However, North America has strong Latin elements in the French-speaking province of Quebec in Canada and in the large and increasing population of Hispanics in the USA, especially in the Southwest states and in Florida and New York. A large number of Italian immigrants also entered the USA, mainly between 1870 and the First World War. Latin America has English- and Dutch-speaking populations in the Caribbean and the former Guiana colonies, modest footholds in the Spanish and Portuguese domains. Even so, the countries of Latin America have much in common with one another economically and culturally.

After its 'discovery' by the Spaniards and Portuguese five hundred years ago, what is now Latin America was divided between the two Iberian powers, Portugal being allocated the eastern part of South America in the Treaty of Tordesillas (see Chapter 2). In theory the whole region was conquered by the Europeans, but even three hundred years later large areas remained beyond the effective control of Spain and Portugal, notably the vast area of tropical rainforest, roughly corresponding to the drainage area of the River Amazon, and also the southern part of the continent.

When independence was gained from Spain and Portugal, initially ten new sovereign states emerged. The large Portuguese colony remains intact to this day as modern Brazil, now the fifth country in both area and population size in the world. Spain's most important colonies were the Viceroyalties of New Spain (Mexico) and Peru, but the Peruvian Viceroyalty split in the eighteenth century. The last Spanish colonies, Cuba and Puerto Rico, were 'liberated' from Spain in 1898 by the USA. Puerto Rico has remained a dependency of the USA to the present. Spanish is used in eighteen Latin American countries, French is spoken in Haiti, and Portuguese in Brazil (see Figure 11.1).

The emergence of sovereign states from the colonies of Spain and Portugal added new countries to the international scene, following the independence of the USA from Britain and of Haiti from France. As happened in the 1950s and 1960s in Africa, and in the 1990s in the Russian/Soviet Empire, the new countries largely coincided with the major colonial administrative divi-

TABLE 11.1 Latin American calendar from 1492 to the present

1492	Columbus reaches the Americas
1493/4	Treaty of Tordesillas
1519	Cortés begins conquest of Aztec Empire (Mexico)
1532	Pizarro begins conquest of Inca Empire (Peru)
1545	Silver mines discovered in Potosí (now Bolivia), and in Zacatecas (Mexico)
1550	First university in the Americas, San Morcos, Lima, Peru
c. 1560	Sugar cultivation begins in Brazil
1693	Gold discovered in Brazil
1697	Haiti ceded by Spain to France
1808	Independence movements begin in Latin America
1823	Monroe Doctrine: US prohibits European intervention in Americas
1846–8	US occupies northern part of Mexico
1861–7	The Mexican Expedition. France attempts to establish empire in Mexico during American Civil War
1879–84	War of Pacific between Chile, Bolivia, Peru
1890s	Expansion of oil production in Mexico
1898	Spanish–American War, US annexes Puerto Rico
1903	Panama Canal Zone ceded to USA by Colombia
1906	First iron and steel works in Latin America at Monterrey, Mexico
1910–11	Mexican Revolution
1920	Panama Canal formally opened for general traffic (first opened 1914)
1920s	Expansion of oil production in Venezuela
1934	US good neighbour policy towards Latin America enunciated
1939	Assets of foreign oil companies in Mexico nationalised
1959	Fidel Castro becomes leader of Cuba
1960	Capital of Brazil transferred from Rio de Janeiro to Brasilia
1960	Latin American Free Trade Association established (now Latin American Integration Association)
1982	War in the Falkland Islands/Islas Malvinas between UK and Argentina

sions of the empires from which they emerged. They bear little relationship to the domains or territories of the pre-Columbian peoples originally conquered by the Europeans, or to the varied conditions of the natural environment, and they vary greatly in size.

Latin America has four distinct elements in its population, but there has been a great deal of mixing since the first Iberians settled there around 1500. The indigenous population, referred to misleadingly as Indians or Amerinds, was very unevenly distributed over the region, with two main areas of civilisation, one the Aztec and Maya civilisations of Mexico and Guatemala, the other the Inca civilisation centred on Peru. Elsewhere there were simple societies of hunters, gatherers and cultivators. Between 1500 and 1800 only about 300,000 settlers came to Latin America from Iberia, but they formed the dominant class, the military, the administrators and the owners of land. From the middle of the sixteenth century, large numbers of African slaves were brought across the Atlantic, mainly to work in sugar and other plantations, but also for domestic work and in some areas as labour in mines. The Africans became a large element, often the dominant one, in the population of many Caribbean islands and in Northeast Brazil. The fourth element consisted of later settlers from various European countries, many still coming from Spain and Portugal, others from Italy, Germany and, to a lesser extent, Britain. The advent of the steamship in the 1840s greatly increased the capacity and reliability of long-distance sea journeys for the movement of passengers and goods. In due course, small numbers of Indians, Chinese and Japanese also migrated to Latin America, not always voluntarily.

Given the great north–south extent of the region, the variety of natural environments in Latin America is enormous. At opposite ends of the precipitation scale are the tropical rainforests of the Amazon and the desert of northern Chile and the coastal zone of Peru. In between, the savanna and pampa consist of grassland, with scattered trees in the savanna, before their clearance for cultivation and pasture. Extending from the north of Mexico to the southern extremity of Chile are the mountain ranges of the Pacific Rim, the Sierra Madre Occidental of Mexico and the Andes of South America, with intermediate ranges in Central America.

In 1993 Latin America had a population of 460 million, compared with about half that number in 1965. The population is not expected to double again in the next thirty years, but another 220 million could be added by 2025 (*WPDS 1993*). Compared with its 8.4 per cent of the total population of the world (see Table 3.1), Latin America is large in area, accounting for about 15.4 per cent of the total land area. It has about 14 per cent of the world's bioclimatic resources, 28 per cent of the freshwater resources, 15 per cent of the reserves of non-fuel minerals, but only 6 per cent of the fossil fuels. The average GDP per inhabitant, 2,360 dollars in 1991, is more than half of the world average but, as was explained in Chapter 3, real GDP is higher in many developing countries than GDP at the official exchange rate with the US dollar. On the basis of real GDP Latin America as a whole is near the world average.

11.2 Cultural background

Almost all the countries of Latin America gained independence from Iberia early in the nineteenth cen-

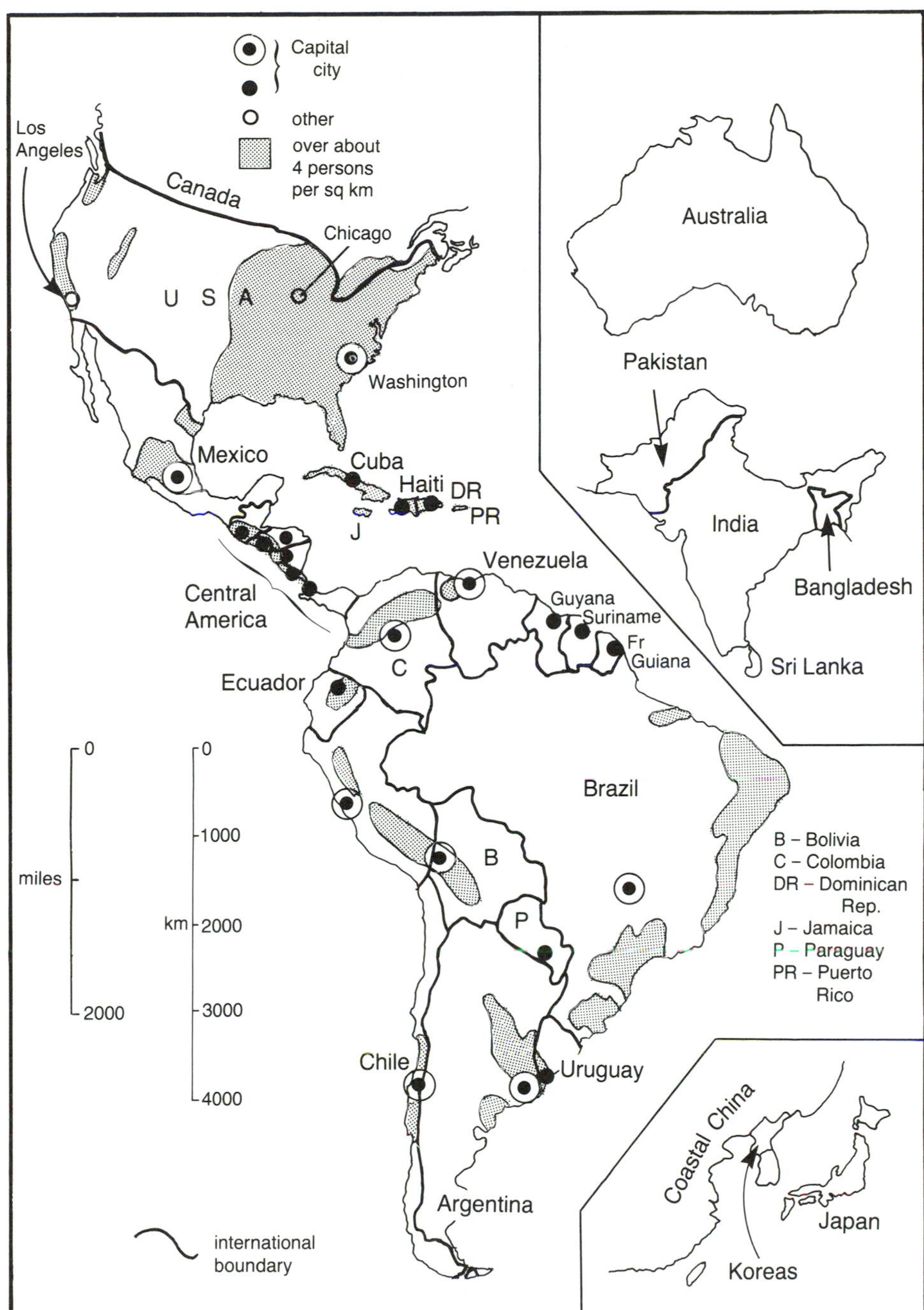

FIGURE 11.1 General map of Latin America. Other selected regions are drawn on the same scale for comparison

tury. Today they make up what may be regarded as the 'senior' developing region. They set out on the path of political independence well before parts of Asia and almost all of Africa were actually absorbed by the colonial powers of Western Europe and by Russia during the process of colonisation that continued into the twentieth century. While the Latin America of today is far from homogeneous culturally, many features are broadly similar. This section contains a summary of some of the main features.

1. **Language** In grammatical structure and in vocabulary, Castilian Spanish and Portuguese are very similar, although some words are completely different. Speakers of other languages intending to reach proficiency in Spanish and Portuguese cannot, however, 'pick up' the second from the first. For those with Spanish or Portuguese as their mother tongue, the difference in pronunciation is a particular problem. Nevertheless, the general consensus among Latin Americans is that Spanish and Portuguese are close enough not to be a barrier to communication. At regional and local level, however, completely different languages of the indigenous population are widely used in some countries. For international purposes, English is now by far the most widely studied language among

those aspiring to read foreign publications and to travel outside the region.

2. **Religion** The first areas of Latin America to be conquered by Spain and Portugal, the Caribbean, Mexico, Peru and Northeast Brazil, were colonised at a time when Christian dominance had finally been established in Iberia itself following the expulsion of the Muslims from the south of Spain. Any tolerance towards people with other beliefs, notably the Jews in Iberia, ceased when it was decreed that Christianity should be the only faith. Soon after the conquest of the Moors, Spain became involved in Europe outside Iberia, pursuing a cause initiated by the Reformation and the breakaway of Protestants from the Catholic Church. In parallel, the conversion of the indigenous population of Latin America to Catholicism was an essential part of the conquest of the Americas.

Today the presence of the Catholic Church is officially widespread if not universal in Latin America but, after independence, its influence was deliberately reduced in some countries (notably Mexico), while it is considered still to be very strong in others (e.g. Colombia). In the 1990s the influence of the Catholic Church is considerable in Latin America, although Protestant missions have now been accepted and are well established in some areas. Again, men and women of the Catholic Church are thin on the ground in many rural areas, often preferring to reside in the comparative comfort of the cities. Some of the clergy, especially in Brazil, have emphasised the role the Church should have in helping the poor in this life as well as guiding them in more spiritual matters. The influence of the Church on such moral issues as birth control seems to be modest. Better-off families readily produce a small number of children when it suits them and the rural poor have large families either because that is considered desirable or because they do not have access to effective family planning devices.

3. **Physiological features** Reference has already been made in Chapter 2 and earlier in this section to the 'racial' mixing in Latin America. The three main ingredients – indigenous, European and African – have mixed extensively over the centuries, with mestizo (*mestiço* in Portuguese) the intermixing of indigenous and European and mulato the product of European and African parents. The sharp division between White and Afro-American in the USA and between Black and White in South Africa is not usually found in Latin America. Physiological features and their cultural descriptions are more blurred and complex. The inescapable fact is, however, that economically, politically and socially those with the greatest proportion of the European ingredient tend to be the most influential and affluent. In my own view, the virtual right of the Spaniards and Portuguese colonists to power and land in the colonial period has remained an influence to this day.

4. **The economy** Much land was allocated in the colonial period to the settlers from Spain and Portugal. The more recent reform of land tenure, focusing on the allocation of small farms to the landless majority of the rural population, has been extensively carried out in some Latin American countries. Often an inadequate amount of land, located where conditions are poor, has been given to previously landless peasants with little experience of management and no capital to improve conditions. Even to this day, in areas such as the Brazilian Amazon forests, where new farms are being established for cultivation or grazing, some land still goes to large landowners, some to small farmers.

Slavery was finally abolished in some parts of Latin America only late in the nineteenth century. In the colonial period the indigenous population was often included in a package or *encomienda* along with the land. The European element in the population of Latin America has tended to continue the tradition of forming a privileged elite.

In many parts of Latin America a small percentage of the total population of the country congregates in the larger cities, keeps a large share of the wealth and has the jobs with the highest incomes. The reservoir of poor rural and urban dwellers provides a plentiful supply of domestic labour. There is no incentive to change the status quo, either by accepting the possibility of a redistribution of wealth, or by using the wealth more effectively to stimulate economic growth, a task the state takes on board sporadically.

Two true cases illustrate the situation vividly. Towards the end of the nineteenth century much of the property of the Church in Mexico was confiscated. Before this could happen, two of the sisters in a convent removed the valuables and passed them to their brother. Instead of seizing this opportunity to demonstrate entrepreneurial skills by establishing a factory, he distributed the funds among members of his family to give each a comfortable living. In January 1979 as I was returning from Chile to England by air, the plane landed in Recife, Northeast Brazil, the largest city in the poorest part of the country. A considerable number of Brazilians, some carrying fur coats, boarded the plane, clearly armed for a shopping expedition in the depths of winter in London.

11.3 Discussion of a data set for Latin America (see Tables 11.2 and 11.3)

Column (2) in Table 11.2 shows that Brazil has about one-third of the total population of Latin America, Mexico about one-fifth. Brazil is by far the largest country in area, Argentina second (see column (3)).

TABLE 11.2 Demographic features of the twenty-three largest Latin American countries

	(1) Population 1993 rank in world	(2) Population 1993 in millions	(3) Area in thous. sq km/(miles)		(4) Density of popn per sq km/(mile)		(5) Annual popn increase %	(6) Expected popn 2025 in millions	(7) % under 15 years	(8) % over 64 years	(9) % in employment
Mexico & C. America											
Mexico	11	90.0	1,958	(756)	46	(119)	2.3	137.5	38	4	35
Guatemala	71	10.0	109	(42)	92	(238)	3.1	21.7	45	3	29
El Salvador	101	5.2	21	(8)	248	(642)	2.6	9.1	45	4	32
Honduras	96	5.6	112	(43)	50	(130)	3.1	11.5	47	4	31
Nicaragua	114	4.1	130	(50)	32	(83)	3.0	8.2	46	3	31
Costa Rica	125	3.3	51	(20)	65	(168)	2.4	5.6	37	5	35
Panama	132	2.5	77	(30)	32	(83)	2.0	3.7	35	5	36
Caribbean											
Cuba	62	11.0	111	(43)	99	(256)	1.0	12.9	23	9	43
Haiti	91	6.5	28	(11)	232	(601)	2.8	12.2	40	4	44
Dominican Republic	87	7.6	49	(19)	155	(60)	2.2	11.4	38	3	32
Puerto Rico	120	3.6	9	(3)	400	(1,036)	1.1	4.7	27	10	34
Jamaica	134	2.4	11	(4)	218	(565)	1.9	3.4	33	8	50
Trinidad	147	1.3	5	(2)	260	(673)	1.4	1.8	31	6	38
South America											
Venezuela	43	20.7	912	(352)	23	(60)	2.6	32.7	37	4	35
Colombia	30	34.9	1,139	(440)	31	(80)	2.1	51.3	34	4	32
Ecuador	65	10.3	284	(110)	36	(93)	2.5	16.8	39	4	31
Peru	39	22.9	1,285	(496)	18	(47)	2.0	35.6	38	4	33
Bolivia	84	8.0	1,091	(421)	7	(18)	2.7	14.3	41	4	31
Chile	56	13.5	757	(292)	18	(47)	1.6	19.8	31	6	36
Argentina	31	33.5	2,767	(1,068)	12	(31)	1.3	44.6	30	9	35
Paraguay	112	4.2	407	(157)	10	(26)	2.7	8.6	40	4	34
Uruguay	127	3.2	177	(68)	18	(47)	0.9	3.8	26	12	39
Brazil	5	152.0	8,512	(3,286)	18	(47)	1.5	205.3	35	5	37
North America											
USA	3	258.3	9,378	(3,621)	36	(93)	0.8	334.7	22	13	49
Canada	32	28.1	9,976	(3,852)	4	(10)	0.8	35.7	21	12	50

Sources: WPDS 1993 for columns (2), (4)–(8); *FAOPY 1991* for column (9)

Some of the Caribbean islands have been developed intensively since 1500 and together with El Salvador they have the highest densities of population in the whole of Latin America (see column (4)). Within the larger countries the distribution of population is very uneven and locally high densities of rural population occur, as in the coastlands of Northeast Brazil, while the four largest cities in Latin America, Mexico City, São Paulo, Rio de Janeiro and Buenos Aires, each now have over 10 million inhabitants.

Earlier this century, accelerated by in-migration and the reduction of mortality rates, most of Latin America experienced rapid population growth. Column (5) shows that there are now large variations in the rate of population increase. Six of the smaller, poorer countries – three in Central America and three in South America – have rates of population increase that are high by world standards, whereas in contrast Cuba, Uruguay and some small Caribbean islands have low rates of increase for developing countries. The total fertility rate is now below 3.0 in Brazil, Argentina, Chile and Colombia. In comparison it is 2.0 in the USA, and only 1.3 in Spain and Italy. The population structures of Mexico (see Figure 9.3) and of the youngest age groups in Cuba, Argentina and Brazil (see Figure 11.2) show the beginning of a trend towards a stable population but (see column (6)) future growth is expected to be high, leading perhaps to an eventual doubling of the population of the region to about 900 million.

Latin American countries mostly have a large percentage of under-15s in the population and a small percentage of people aged 65 and over (see columns (7) and (8)). Percentages of about 15 for the elderly,

TABLE 11.3 Economic and social aspects of the twenty-three largest Latin American countries

	(1)	(2)	(3)	(4)	(5)	(6)	(7)	(8)	(9)	(10)
			% of economically active population in agriculture		*Quality of life*				*Non-primary products as % of all exports*	
	% urban	*Energy consumption per head (kce) 1990*	*1965*	*1991*	*life expectancy*	*adult literacy*	*real GDP*	*HDI*	*1970*	*1991*
Mexico & C. America										
Mexico	71	1,790	52	29	70	87	5,691	804	32	45
Guatemala	33	190	64	51	63	55	2,531	485	28	28
El Salvador	48	250	59	36	64	73	1,897	498	2	8
Honduras	44	180	65	55	65	73	1,504	473	9	6
Nicaragua	57	270	59	38	65	81	1,463	496	16	12
Costa Rica	45	600	48	23	75	93	4,413	842	20	26
Panama	53	630	43	24	72	88	3,231	731	4	21
Caribbean										
Cuba	73	1,460	39	19	75	94	2,500	732	na	na
Haiti	44	50	79	63	56	53	962	276	26	42
Dominican Republic	60	370	57	35	67	83	2,537	595	19	20
Puerto Rico	74	3,040	19	3	na	na	na	na	na	na
Jamaica	52	840	34	27	73	98	2,787	722	53	56
Trinidad	65	7,180	19	7	72	96	6,266	876	13	29
South America										
Venezuela	84	3,340	29	11	70	88	5,908	824	1	12
Colombia	68	810	47	27	69	87	4,068	758	8	33
Ecuador	55	740	52	30	66	86	3,012	641	2	2
Peru	72	500	47	34	63	85	2,731	600	2	18
Bolivia	51	370	65	41	55	78	1,531	394	3	5
Chile	85	1,270	26	12	72	93	4,987	863	5	15
Argentina	86	1,850	18	10	71	95	4,310	833	14	28
Paraguay	48	230	51	46	67	90	2,742	637	9	11
Uruguay	89	750	17	13	72	96	5,805	880	20	40
Brazil	76	780	48	24	66	81	4,951	739	14	56
North America										
USA	75	10,030	6	2	76	99	20,998	976	70	80
Canada	77	10,340	11	3	77	99	18,635	982	52	64

Notes:
na = not available
kce = kilograms of coal equivalent
Explanation of columns (5)–(8):
(5) Average life expectancy in years
(6) Percentage of adults literate
(7) real GDP in US dollars per head
(8) combined score of (5)–(7), possible range 0–1,000
Sources: WPDS 93 for column (1); *UNSYB 90/91* for column (2); *FAOPY 91* for columns (3), (4); *HDR 92* p. 127, Table 1 for columns (5)–(8); *WDR 93* for columns (9), (10)

common in Northern and Western Europe, are still several decades away in most of Latin America. In all the countries of the region, the dependency ratio (that of the number of people under 15 plus those 65 and over against those of 15–64) is much higher in most Latin American countries (e.g. 51:49 in Honduras, 45:55 in Bolivia) than in North America or Western Europe. As a result, the potential employed population is relatively smaller. The data in column (9) show that between 30 and 40 per cent of the total population of Latin America is economically active, with considerable numbers, mostly of women, either not working through choice, unemployed, or not registered officially as working.

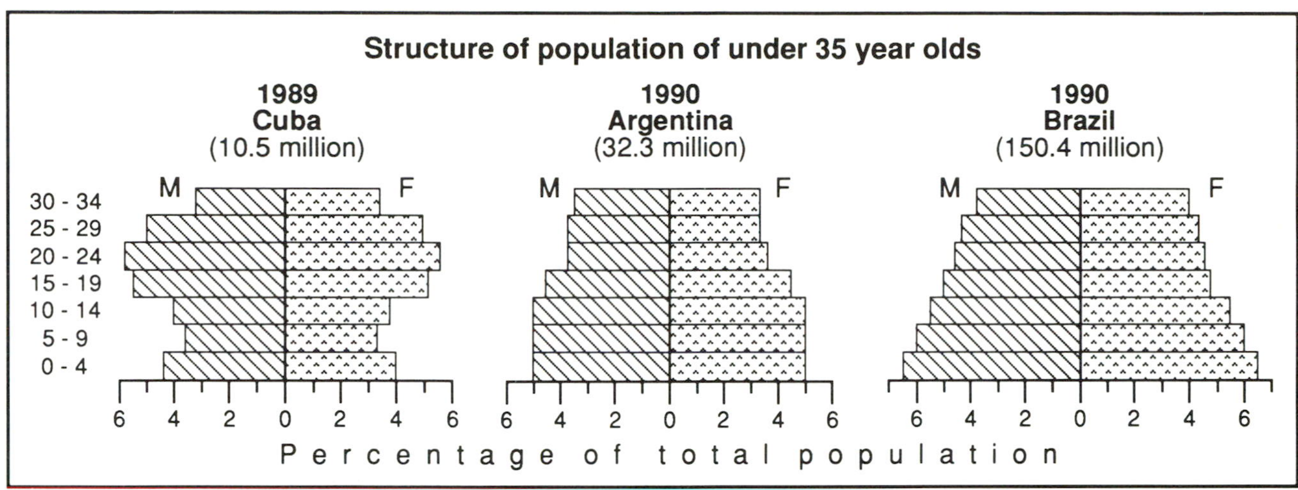

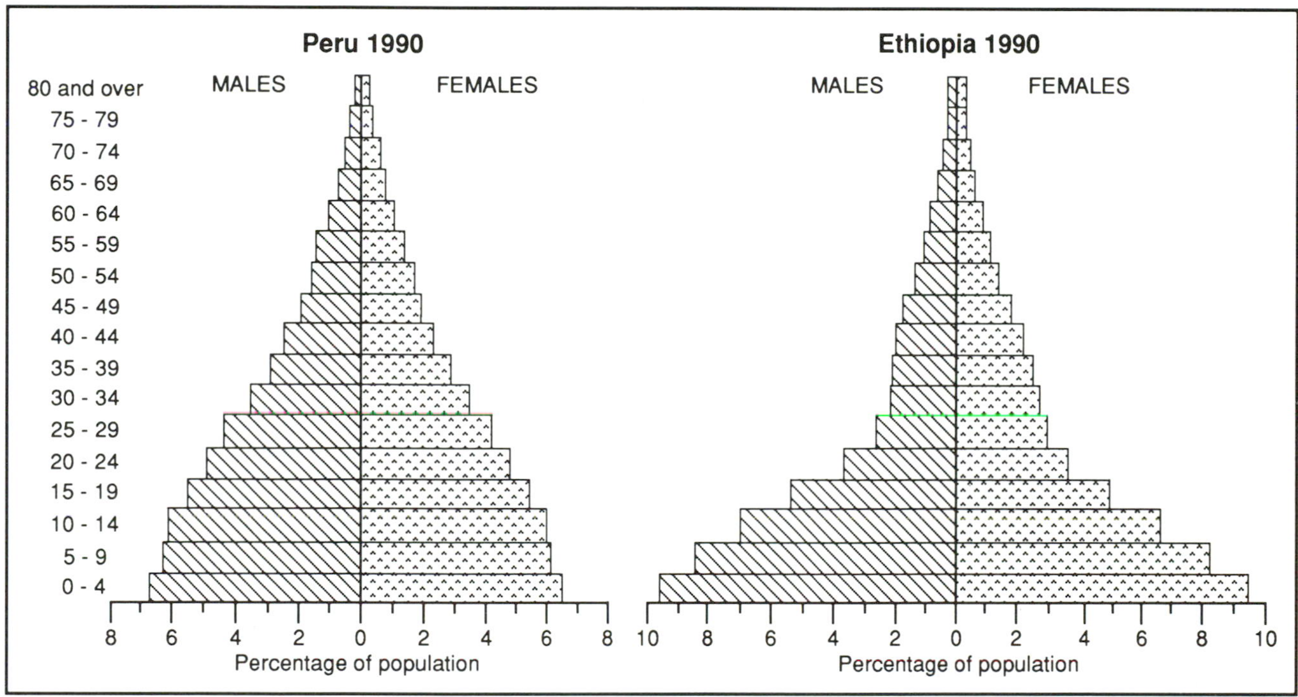

FIGURE 11.2 A comparison of population structures: (*a*) of the younger age groups in Cuba, Argentina and Brazil, (*b*) of Peru and Ethiopia

In the *WDR 1993*, Table 1, 127 countries of the world are ranked on a scale derived from 'basic indicators'. The list includes twenty-two of the twenty-three countries in Table 11.2 (Cuba is omitted). The three most backward, Haiti, Nicaragua and Honduras, are defined as 'low-income' countries, twelve are lower-middle income and seven are upper-middle-income. The GDP per capita of Haiti at the bottom end is 370 dollars, that of Puerto Rico at the top end 6,320. Towards the upper end of the basic indicators scale, Latin American countries overlap with countries of Central Europe, the former USSR and Southeast Asia. At the other end they mingle with the better-off countries of Africa. On average, all Latin American countries have improved materially since the Second World War, but the conditions of the poorest members of the population have changed little or have even deteriorated.

Column (1) in Table 11.3 shows the percentage of urban dwellers in the 23 largest countries of Latin America in population. Definitions of urban vary from country to country, but the surprising fact is that, with 71 per cent of its population defined as urban, Latin America is very close to the *average* for the more

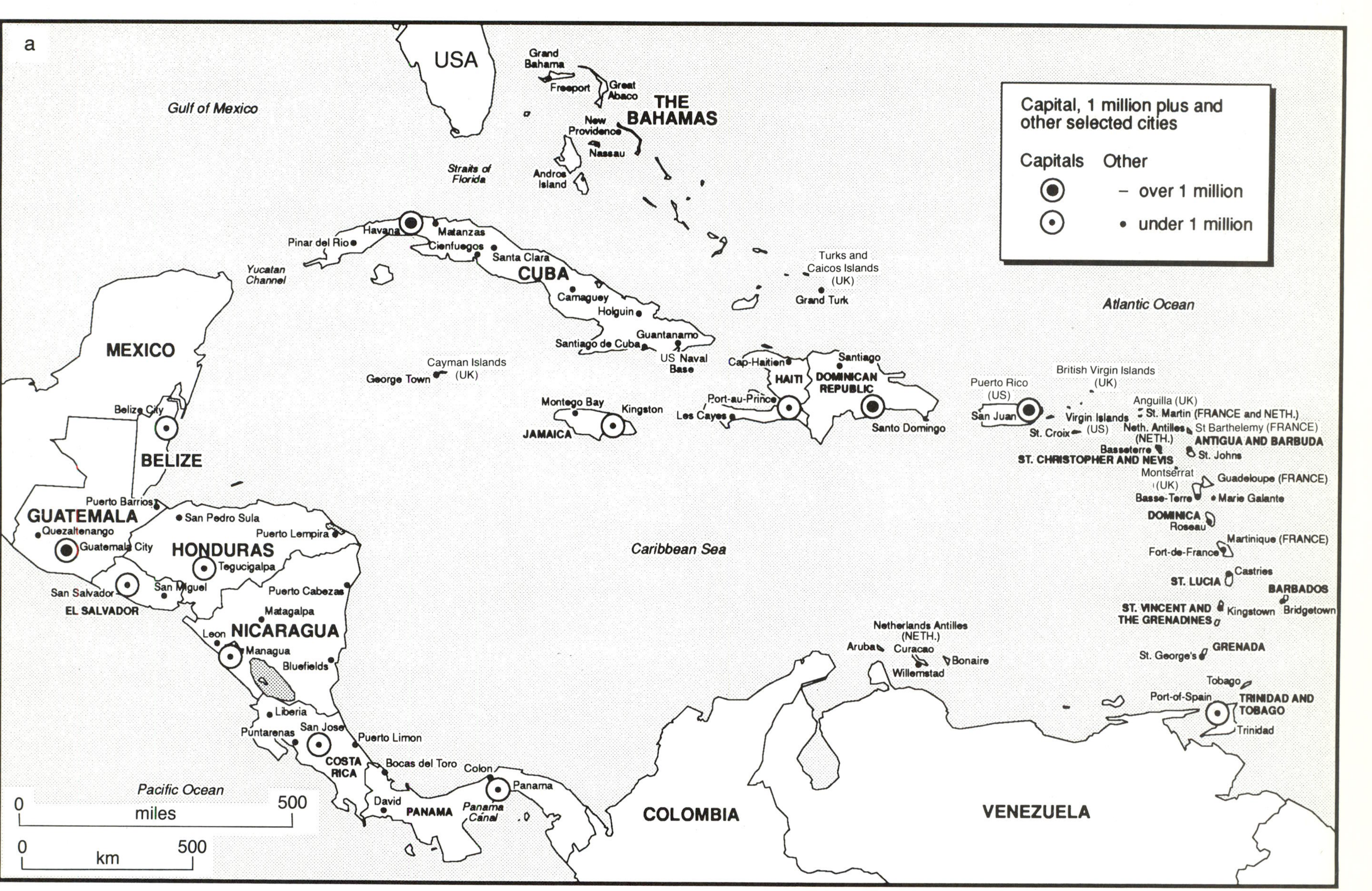

FIGURE 11.3 Countries and cities of (*a*) Central America and the Caribbean, (*b*) South America. Population sizes are indicated in the key

Caribbean Sea
Atlantic Ocean
Pacific Ocean
Maracaibo
Barranquilla
Valencia
Caracas
Ciudad Guayana
Cucuta
San Cristobal
Georgetown
Paramaribo
Cayenne
Medellin
VENEZUELA
GUYANA
SURINAME
FRENCH GUIANA
Cali
Bogota
Boa Vista
COLOMBIA
Macapa
Quito
ECUADOR
Guayaquil
Iquitos
Manaus
Santarem
Belem
Sao Luis
Fortaleza
Teresina
Piura
Natal
Trujillo
Porto Velho
BRAZIL
Rio Branco
Recife
PERU
Aracaju
Lima
Cusco
Salvador
Ica
BOLIVIA
Brasilia
Cuiaba
Arequipa
La Paz
Goiania
Cochabamba
Santa Cruz
Arica
Sucre
Belo Horizonte
Vitória
Rio de Janeiro
PARAGUAY
Sao Paulo
Antofagasta
Asuncion
Curitiba
Resistencia
San Miguel de Tucuman
Florianopolis
Porto Alegre
Cordoba
Rosario
Salto
URUGUAY
Mendoza
Valparaiso
Santiago
Buenos Aires
Montevideo
ARGENTINA
Concepcion
Bahia Blanca
Mar del Plata
CHILE
Puerto Montt
Comodoro Rivadavia
FALKLAND ISLANDS (UK)
Punta Arenas
Ushuaia
SOUTH GEORGIA ISLAND (UK)
0 1000
miles
0 1000
km
Capital, 1 million plus and other selected cities
Capitals
Other
over 5 million
1 - 5 million
under 1 million
b

developed countries of the world, which is 72 per cent. The level is 75 per cent in the USA, 74 per cent in the Russian Federation, and only 60 per cent in Central Europe. In recent decades, urban population has grown roughly twice as quickly as total population in Latin America because the larger cities, including most of the national capitals, have attracted large numbers of migrants.

Column (2) in Table 11.3 shows large differences in the consumption of energy among Latin American countries. Compared with North America, the use of energy for heating and air-conditioning in Latin America is small, while its use for transport, particularly for private cars, is also much less per inhabitant. Some countries of Latin America are much more highly industrialised than others, and in some regions considerable quantities of energy are used in the refining of oil and non-fuel minerals. There is an enormous disparity in energy consumption per head between Haiti, which uses annually an average of 50 kg of coal equivalent, at one extreme, and Trinidad, using 7,180, at the other. Venezuela and Mexico are the largest producers of fossil fuels in the region and have higher levels of consumption than most of the other countries. In Brazil there are great regional differences in the level of energy consumption per head.

Columns (3) and (4) show the relative importance of agriculture in total employment in 1965 and 1991, the period chosen covering a brief but rapidly changing twenty-six years in the recent history of the region. In almost every country in the world the percentage employed in agriculture has declined in the last few decades. Paradoxically, however, in most Latin American countries, as in many other developing countries, the labour force in agriculture has hardly changed, because total population has increased. Thus, for example, in Brazil in 1965 48 per cent of the economically active population was in agriculture, a total of 12.4 million people. By 1991 the percentage was only 24, but the number had actually risen to 13.3 million. The area under cultivation has increased during the last thirty years in most parts of Latin America, and mechanisation has been widely introduced in some regions.

In the United Nations Human Development Programme (Report for 1992 – see Chapter 4) some Latin American countries are placed fairly favourably in the world ratings. Columns (5)–(7) in Table 11.3 show the scores for average life expectancy and adult literacy as well as the raw score for real GDP (truncated for countries with over about 5,000 dollars per head in the calculation of the HDI in column (8)). Among the larger countries, Mexico, Venezuela, Chile and Argentina all score more than 800 out of a possible maximum of 1,000 for quality of life (Canada, the highest in the world, reaches 982). Costa Rica and Uruguay are highest among the smaller countries. Standards are much lower in four of the Central American countries (in the 400s), in Haiti with 270 and in Bolivia with 394, but with only 52, Guinea in Africa has the lowest score in the world.

Ever since the Spanish conquest, Latin America has been regarded as a source of primary products for Europe and later also for the USA and Japan. In the colonial period, precious metals and stones, sugar, coffee, cotton and many other products were exported to Europe in quantities commensurate with the shipping capacity of the time. Since the middle of the nineteenth century it has been possible to transport raw materials such as guano (bird droppings used for fertiliser), nitrates, copper and other non-ferrous metals, as well as agricultural products such as grain and meat. During the twentieth century Mexico and Venezuela have been among the world's leading exporters of oil.

The data in columns (9) and (10) cover only a short period of time but they show conclusively that the share of non-primary products in exports has increased substantially. During the twentieth century, partly stimulated by a reduction of trade during the two world wars, and with government support to private industries, and investment in the public sector, Latin American countries have followed programmes of import substitution. They have cut imports of manufactured goods, and increased exports of 'non-traditional' products, many of them manufactured. The direction of trade of Latin American countries has also gradually changed since the Second World War, an increasing proportion being carried out between countries of the region. Thus Venezuela supplies oil and Brazil manufactured goods to many other Latin American countries. Brazil and Mexico stand out as the leading NICs of the region, but Colombia and Peru, for example, have greatly increased exports of textiles and other manufactures since 1970. Even so, Latin American countries all remain net importers of manufactured goods.

For the remaining sections of this chapter, the countries of Latin America are divided into five groups: Mexico and Central America, the Caribbean islands, the Andean countries, southern South America, and finally Brazil.

11.4 Mexico

During the last decades of Spanish rule the Viceroyalty of Mexico covered about twice the area of the present Mexico. After independence in the nineteenth century it lost its northern, thinly populated regions to the USA, notably California, Arizona, New Mexico and Texas. The area of greatest economic interest to Spain was around and to the north of Mexico City, where the main silver-mines of the Viceroyalty were located. In the nineteenth century, Mexico's range of exports increased, but they consisted mainly of non-ferrous metals and, early in the twentieth century, of oil. Most of the population was engaged in agriculture, however, until as recently as the 1960s. Means of production in agriculture were very simple, conditions in many areas difficult for cultivation, and the rural population lacked education and modern healthcare facilities.

Given its proximity to the USA and the American policy of obtaining imports of vital primary products from 'secure' parts of the world, Mexico was particularly susceptible to US economic and political influence. In recent decades, increasing numbers of Mexicans have emigrated to the USA, legally or illegally, some permanently, others temporarily. Awareness of a different type of economy and society and, more concretely, financial remittances, have increasingly made an impact on Mexico, especially on its more northern states. At the same time, for decades millions of US tourists have visited Mexico, some seeing only the border towns but many travelling further into the country. The formation of NAFTA is a momentous step towards closer integration between the two richest countries of the world, the USA and Canada, and a Third World country. A comparable step for the European Union would be the entry of Turkey or the Maghreb countries.

The population of Mexico was only about 20 million in 1940. It doubled to about 40 million in the mid-1960s and doubled again by the mid-1990s. The urban population has risen faster than total population, to some extent reducing the rate of natural increase on account of generally lower fertility rates among urban dwellers. The rate is still 2.3 per cent per year. Emigration to the USA relieves pressure a little. Even so, a population of 138 million is expected by 2025 and a peak population of about 180 million may be reached several decades later. A quarter of the population of Mexico lives within a radius of about 50 km of the centre of the capital, Mexico City.

In many respects, Mexico is remarkably near average among the countries of the world. It has 1.63 per cent of the world's population, 1.46 per cent of its area and 1.56 per cent of the bioclimatic resources. Its 5 per cent share of world oil resources gives it about 2.3 per cent of the world's commercial fossil fuels, and it has 1.6 per cent of non-fuel minerals, with a prominent position in world reserves of non-ferrous metals. Much

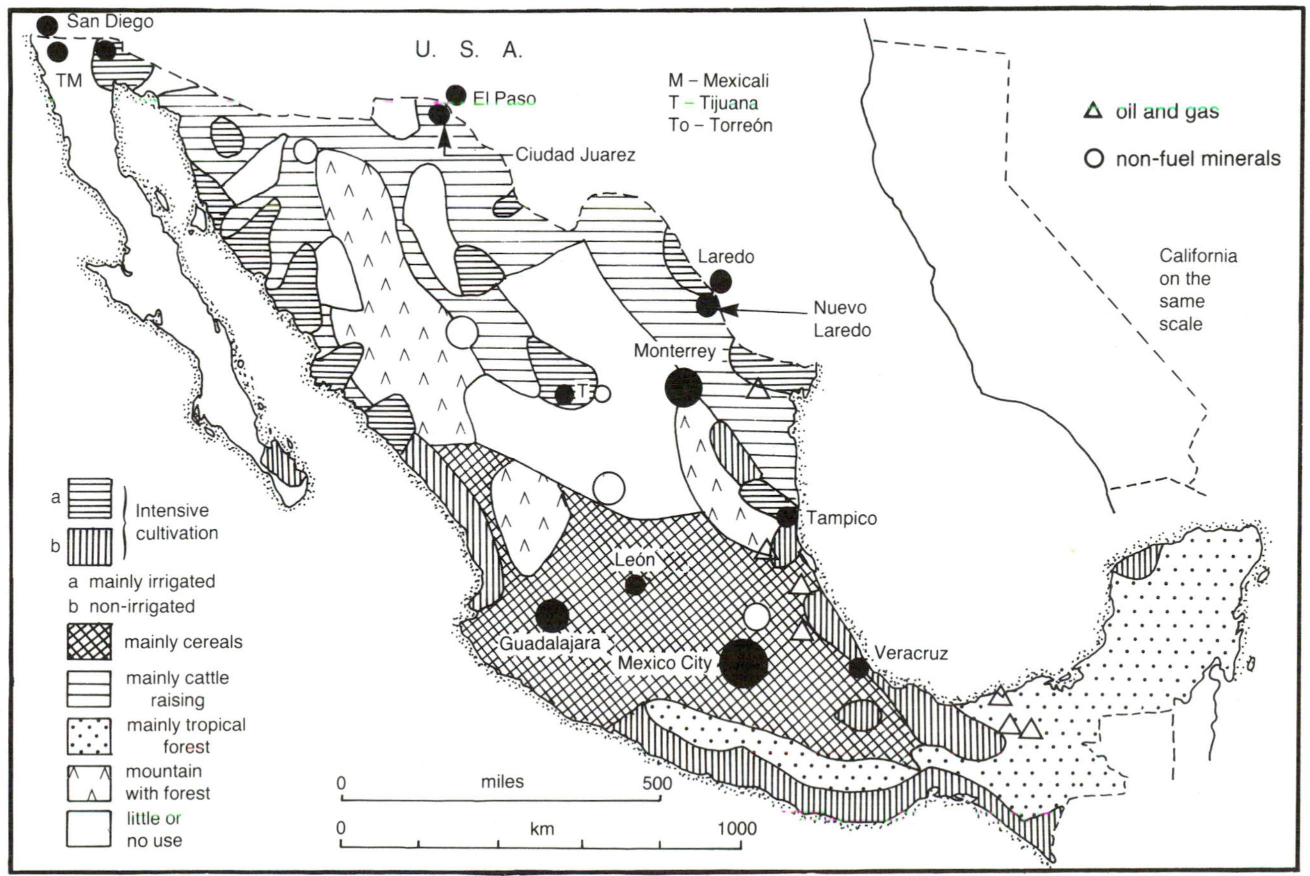

FIGURE 11.4 Economic map of Mexico

of north and central Mexico is semi-arid, and water resources are therefore restricted except in the south.

Mexico's real GDP per inhabitant was 5,700 dollars in 1989, a third that of Canada, almost half that of Australia. The distribution of GDP in 1991 (*WDR 1993*, p. 242) was made up of 9 per cent from agriculture, 8 per cent from extractive industry, 22 per cent from manufacturing and 61 per cent from services. The weakness of the agricultural sector is demonstrated by the fact that it employs 29 per cent of the economically active population but its contribution to GDP is small. In northern Mexico, especially on irrigated land, yields of cereals, cotton and other crops are high by world standards. In much of central and southern Mexico, the yields of maize, the original cereal of pre-Columbian times, are far below the world average. The cost of increasing yields substantially in the many rugged or dry areas would be high. Only 13 per cent of the area of Mexico is under crops compared with 38 per cent under permanent pasture and 22 per cent forested. There is scope for extending the cultivated area into the forest and, with irrigation, into areas that are at present pasture.

Mexico has a wide range of minerals of commercial significance. Increasing use is made of its limited coal reserves which, together with local iron ore, were crucial in the establishment in Monterrey of the first modern iron and steel plant in Latin America early in the twentieth century. Mexico currently produces about 7 million tonnes of steel a year. It is also a major producer of silver, zinc and lead. It has produced oil throughout this century and was a leading world producer in the early decades. Following the discovery of large new reserves of oil and natural gas in the south of the country in the 1970s, much of it offshore, production has risen dramatically in the last two decades from about 25 million tonnes a year at the time of the first oil price rise in 1973 to 107 million in 1980 and 155 million in 1992. Mexico does not belong to OPEC, which gives it extra flexibility in its use of oil for export. The contribution of oil to the value of exports rose from about 10 per cent in 1970 to 30 per cent in

BOX 11.1

Earthquake in Mexico City, 1985

One of the most damaging earthquakes in recent decades occurred in Mexico in 1985. Although it caused heavy damage in some rural areas and small cities, its greatest impact in human terms was in Mexico City, one of the largest concentrations of population in the world. Some outstanding features of this earthquake are described below and the impact on human life is then compared with the loss of life in serious earthquakes elsewhere in the world.

The shock started at 7.19 a.m. on 19 September 1985. It lasted between two and three minutes. It occurred at a time when most people were still in their homes. The epicentre of the earthquake was about 370 km (230 miles) southwest of Mexico City, just offshore. It was felt over an area of some 800,000 sq km (310,000 sq miles), about 40 per cent of the national area, and it affected six out of Mexico's thirty-two states. A second, weaker, shock was felt 36 hours later. Approximately 5,000 deaths were caused by the main earthquake, approximately one in 3,000 of the total population of Greater Mexico City. Damage was estimated at 5 billion US dollars.

The worst damage occurred in many of the large, mainly non-residential, buildings in and near the central business district of the capital. This area is situated on the floor of a former lake, on material that is comparatively unstable. Many of the surrounding areas of the city are on more solid ground, mainly on volcanic rock. Here damage was mostly very slight.

About 700 buildings were destroyed or badly damaged. Within the area most heavily damaged, marked contrasts occurred, related particularly to the type of construction of buildings. Many modern buildings with reinforced concrete structures were heavily damaged, while some collapsed completely. Among these were several hospitals, many office buildings and a number of hotels, factories and other non-residential buildings. In one cluster of some twenty high-rise residential buildings, one collapsed, causing about 1,000 deaths, while the others were little affected. Fire swept through the well-known Hotel Regis. Psychologically it was fortunate that most well-known buildings escaped serious damage. These included the famous cathedral in the main square (Zocalo), the Torre Latinoamericana, a miniature version of New York's Empire State Building, and a department store rather incongruously called Liverpool. There was minimal damage to the Metro (subway) system.

Immediately following the shock, the electricity supply was affected, while the central hospitals, which should have received casualties, were damaged and had to be evacuated. As rescue work continued, stories spread of the discovery of live victims up to several days after the shock. One man unintentionally did better by declining a poor breakfast in his hotel just before the earthquake struck and walking to a café

1991 while metals stayed at about 10 per cent. During that period the percentage of export value accounted for by other primary commodities, mainly agricultural, was almost half in 1970 but only 14 per cent in 1991. The export of machinery and transport equipment, like that of oil, has also expanded.

Mexico remains a country of great regional diversity, difficult to 'manage' on account of the elongated shape, rugged mountain barriers and difficult climatic conditions in many areas. Its rail network links most main centres of the country, but trains are slow and track poorly maintained. Since the Second World War great improvements have been made to the road system and much traffic is now carried by road. Only the remotest corners of the Sierra Madre mountains and the forest of the south are not adequately linked to the system. The rapid growth of the larger cities, especially the capital, brings problems of congestion, pollution and shanty towns. A serious earthquake in 1985 damaged many public buildings and blocks of flats in the centre of Mexico City, awakening politicians and planners to the danger of having many of the research and higher educational facilities of the country in a single centre (see Box 11.1).

As in many of the larger countries of the developing world, great inequalities in the average income per inhabitant at regional, local and family or individual level occur, and much of the wealth of the country is concentrated in the hands of a small proportion of the total population. In Mexico, income per inhabitant decreases from northwest to southeast, with an anomalously high level in and around Mexico City. As during the Revolution in Mexico in 1911, much of the present discontent in the country is found in the agricultural sector and involves questions of land ownership and poor infrastructure in rural areas, particularly the south, among the mainly indigenous and mestizo population. Early in 1994 unrest occurred in the southern states of Chiapas and Oaxaca. Table 11.4 shows regional differences in Mexico and Figure 11.5 the location of the states of Mexico.

along the street. His hotel was destroyed. There was also the inevitable honeymoon couple who left their hotel scantily dressed. Within 50 hours, a modest thirty flights from foreign countries had arrived with equipment and supplies. The USA contributed a third of these.

The post-mortem raised a number of issues. The most controversial issue was the fact that the reinforced concrete in many of the damaged buildings was not strong enough to take the shock. Either there had been excessive economies in their construction or some of the materials destined for construction work had been stolen. In practice building regulations in Mexico are indicative rather than mandatory. Concern for the future was expressed with regard both to the likely effect of the disaster on tourism and the danger to the economy of having so many of the research centres of the country concentrated in the one city.

Some bad earthquakes

- The greatest recorded loss of life occurred in an earthquake in July 1201, which affected many cities in the eastern Mediterranean and Near East. The death toll was about 1,100,000.

- In 1906 San Francisco was afflicted by an earthquake which caused only about 800 deaths but which left 300,000 homeless as 75 per cent of the city was destroyed by a fire storm. Richter scale 8.3. In 1989 San Francisco was again the scene of a bad earthquake, in which 100 deaths occurred, mainly through the collapse of part of an interstate highway.

- In 1923 much of Tokyo and Yokohama in Japan was destroyed, with 143,000 deaths, about 20 billion dollars' worth of damage including the destruction of over half a million dwellings. Much of Tokyo was once again destroyed during the last months of the Second World War in conventional air attacks by US bombers.

- Much of the coal-mining city of Tangshan in China, to the east of Beijing, was destroyed in an earthquake in July 1976. China declined outside assistance and the city was duly rebuilt. The story goes that the safest places to be in were the coal-mines. Reports of deaths ranged from an initial 655,000 to 750,000 and finally to only 240,000, still enough to remove a quarter of the total population.

Given the impossibility of predicting exactly where and when the next serious earthquakes will occur it is clearly not possible economically to make everywhere that is known to be at risk completely safe. There will always be casualties, damage and recriminations. On balance, the experience points to a high ratio of casualties to material damage in both urban and rural settlements in developing countries and a high ratio of material damage to casualties in developed countries (see Chapter 14 on the Indian and Los Angeles earthquakes). The overall effect on global population trends is only very slight but psychologically the inhabitants of earthquake prone cities live under just that much more stress, waiting for 'the big one'.

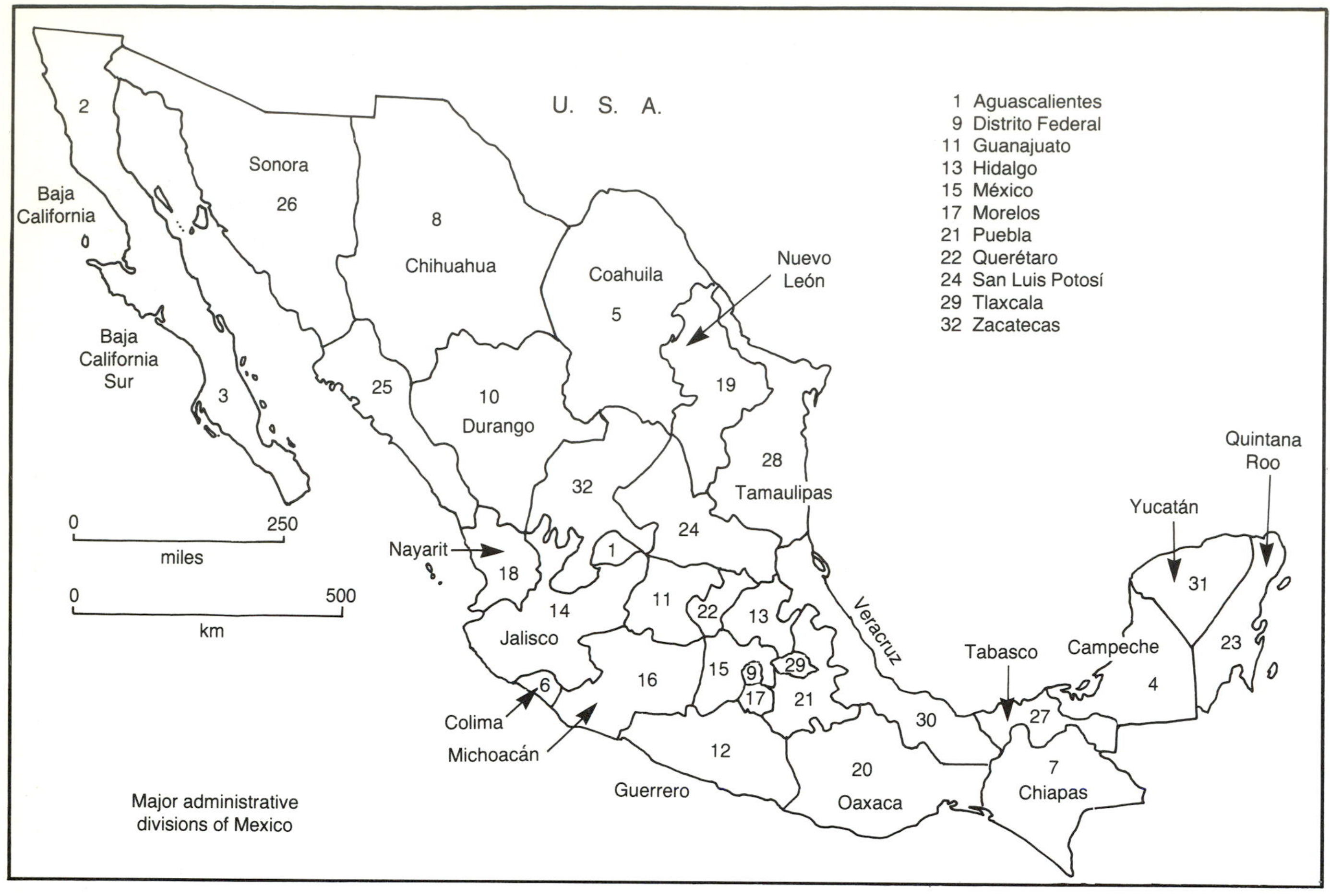

FIGURE 11.5 The states of Mexico. See Table 11.4 for data on the states

11.5 Central America and the Caribbean

Like Mexico, Central America became part of the Spanish Empire in the sixteenth century, but the virtual absence of precious metals meant that it was less intensively colonised and developed. The Panama isthmus was crossed by road early in the sixteenth century to enable the Spaniards to conquer Peru and to transfer goods between the Pacific and the Atlantic Oceans via the Caribbean. Central America eventually broke up into six ex-Spanish colonies, while British Honduras, the only non-Spanish colony on the mainland, is now known as Belize.

In 1993, Central America had 31 million inhabitants, compared with over 80 million in Mexico. In area it covers about 500,000 sq km (190,000 sq miles), making it only a quarter the area of Mexico and considerably smaller than Texas. The original vegetation of most of Central America was tropical rainforest, which only a few decades ago still covered over half of the area. Between 1963 and 1990 alone, the area of forest and woodland in Central America decreased from 280,000 sq km (108,000 sq miles) to 165,000 sq km (64,000 sq miles), with clearance taking place at roughly the same rate everywhere except in Panama and Belize, where it has been slower. Since the population of Central America is expected to double in the next three decades, the prospect is a further reduction in the area of forest. Even now, however, only about 15 per cent of the region is under crops. The agricultural sector provides basic food supplies for the growing population as well as tropical fruits and coffee for export. Agricultural commodities account for between 70 and 90 per cent of the value of exports. The dependence of the region on agriculture is underlined by the fact that there is little manufacturing, while reserves of commercial minerals are minimal. Fuel makes up a considerable proportion of the value of imports.

Central America is greatly elongated, and movement between the countries along the isthmus is difficult because of distance and poor roads. Since the products of each country are broadly similar, there is little trade between them. That is one reason why the regional association formed in the 1950s, consisting of Costa Rica, El Salvador, Guatemala, Honduras and

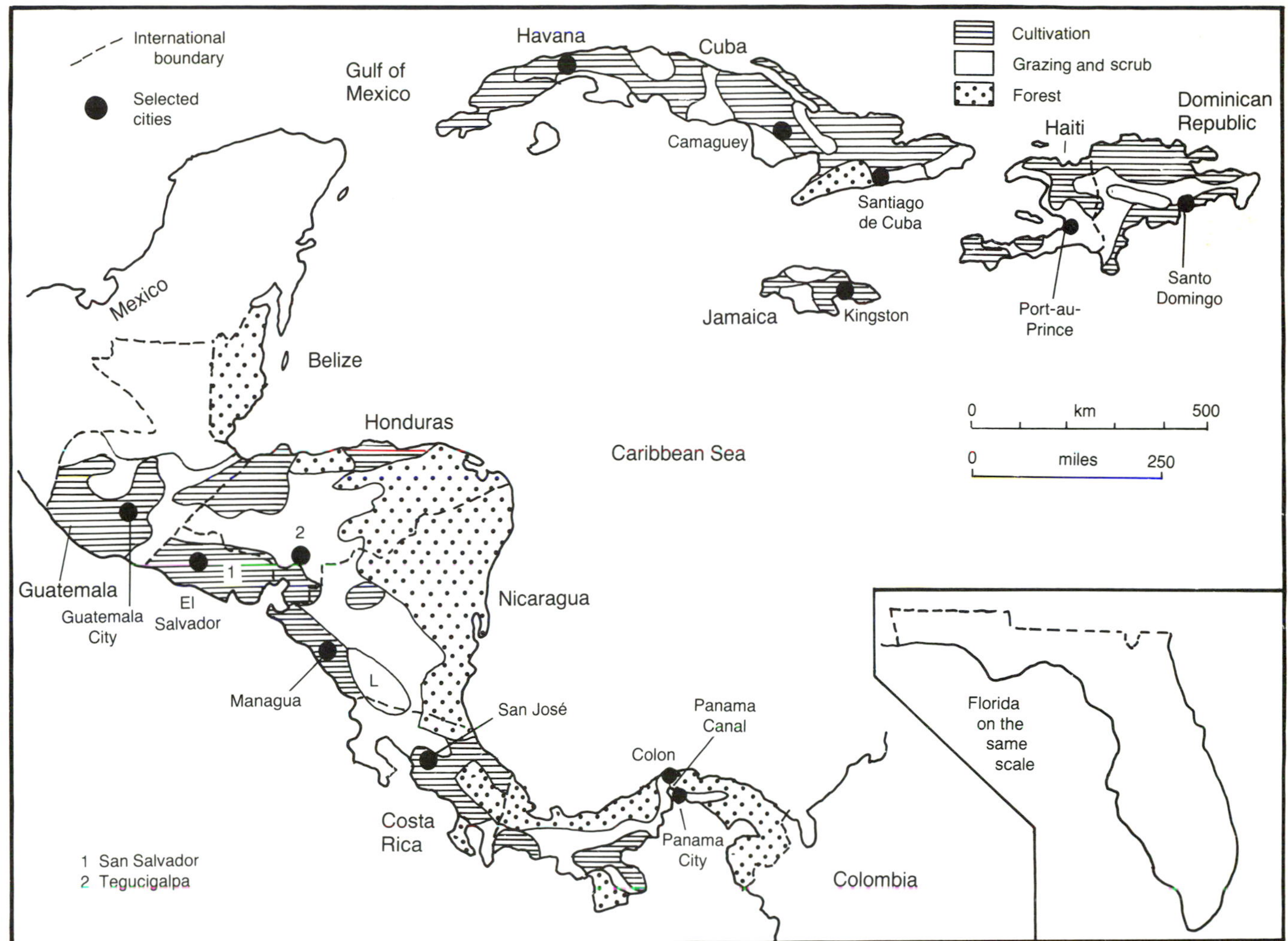

FIGURE 11.6 Economic map of Central America and the Caribbean

Nicaragua, and now referred to as the Central American Common Market (CACM), has not greatly affected the member countries. Indeed, there are marked contrasts in both the density and the racial composition of the population. El Salvador is densely populated and its citizens have moved into its less densely populated neighbours, Honduras and Nicaragua. The population of Guatemala is predominantly indigenous, that of Costa Rica is mainly of European extraction, while Panama's is more mixed, resembling that of the Caribbean islands.

Internationally, the Panama Canal is arguably the best known and most significant feature of the region and there is concern over its capacity to continue handling the growing traffic between the Pacific and Atlantic Oceans. Clearance of forest in the basins of the rivers feeding the artificial lakes that provide water for the canal locks threatens to hamper navigation. The construction of an alternative inter-ocean canal through Nicaragua has been contemplated for a long time. Such a project would require international funding, most of which would have to come from the USA. United States influence has been strong in Central America, both economically and on account of the strategic significance of the region during the Cold War. For the USA itself, the Panama Canal is a vital link between its eastern and western coasts, while socialist movements in a region so close to home, as in Nicaragua, prompted concern in Washington and considerable involvement since the Second World War.

The Caribbean islands are somewhat larger in population than Central America, with about 35 million inhabitants, but are less than half as large in area, 215,000 sq km (83,000 sq miles), of which 30 per cent is cultivated. Cuba alone accounts for half of the total area of the islands, but sixteen other smaller countries each have at least 100,000 inhabitants. The second and third largest countries, Haiti and the Dominican Republic, share the island of Hispaniola.

The fortunes or misfortunes of the Caribbean

islands have fluctuated through the centuries according to which European power held them. Spanish influence remained strong in the Caribbean for four centuries. Cuba and Puerto Rico were colonies until the end of the nineteenth century, but already in 1697 the western part of Hispaniola, Haiti, was ceded to France. Early in the nineteenth century the rest of Hispaniola became independent as the Dominican Republic together with Haiti. The Spanish used their Caribbean territories to raise cattle and to control the sea links between Spain and its mainland colonies. France in Haiti, England in Jamaica and both in their respective smaller islands, exploited their Caribbean colonies intensively, with the help of slaves, to grow sugar, coffee and other tropical crops. Minerals played virtually no part in the economic life of the Caribbean islands until the twentieth century.

Haiti gained its independence from France in 1810, but the colonial commercial system collapsed after that. Cuba became independent from Spain in 1898, but US influence remained strong until 1959, when Soviet support for Fidel Castro's Communist Party began. Puerto Rico remains in effect, if not in name, a part of the USA. The British islands have mostly become small independent countries, but Guadeloupe and Martinique remain French and are overseas territories of the European Union. The smaller islands of the Caribbean are part of the world-wide EU 'Network Cooperation Agreements' in which thirteen Caribbean countries, actually including three mainland countries (Belize, Guyana and Suriname), receive overseas development assistance.

In the 1990s the Caribbean is characterised by an intriguing variety of economic conditions. All the islands continue to export tropical agricultural products, but cane-sugar in particular has declined greatly in importance over recent decades. Cuba has considerable deposits of nickel and chrome ore, Jamaica bauxite, and Trinidad oil and gas deposits. Trinidad and the Netherlands Antilles have oil refineries. Almost all the islands have developed tourist facilities, the advent of widespread air travel since the 1950s putting them in reach in both time and cost of many people in North America and some in Western Europe.

The rate of natural increase of population in the Caribbean countries is 1.7 per cent a year, much lower than the 3 per cent in some Central American countries, and the population is expected to grow from 35 million in 1993 to only 50 million in three decades' time, still a large population, given the limited natural resources of all the islands except Cuba, the lack of industry, apart from some labour-intensive manufacturing, and the failure of the number of tourist visitors to increase regularly.

11.6 The Andean countries

Venezuela, Colombia, Ecuador, Peru and Bolivia have been grouped here because they have several physical and economic features in common related to the presence of the high ranges, basins and plateaux of the Andes. In all of the countries except Bolivia, which lost its coastal province to Chile in 1880, the Andes make penetration from the Pacific coast to the interior lowlands difficult. At the same time, the Andes have been the home of a considerable population of cultivators and, particularly in Peru and Bolivia, the source of various minerals. In the twentieth century, oil has been the main commercial source of fuel in the region, and all five countries are producers.

The Andean countries had a combined population of 97 million in 1993 (see Tables 11.2 and 11.3 for country details). That population is expected to grow to about 150 million in 2025. The combined area of the five countries is 4,720,000 sq km (1,820,000 sq miles) and the average density of population of just over 20 per sq km is low by world standards. Population is, however, very unevenly distributed, with dense rural populations locally in areas of cultivation and with eight cities of over 1 million inhabitants, the largest being the capitals of the three largest countries, Bogotá in Colombia, Lima in Peru and Caracas in Venezuela.

An indication of both the adverse physical conditions and the slow colonisation of the large areas of forest is the fact that less than 4 per cent of the total area of the five countries is under crops, about one-fifth of it taken up with permanent crops, including coffee, cacao and banana plantations. About 45 per cent of the area of the Andean countries is forest, almost all being some variety of the tropical rainforest. Most of the forest is in the interior lowlands of each country, or in the case of Venezuela, in the Guyana Highlands, but there is some mountain forest on the lower interior slopes of the Andes, and a narrow zone of forest along the Pacific coast of Colombia, characterised by very high species diversity. About a quarter of the total area of the five countries is natural pasture, while another quarter, mainly in the Andes, is unproductive.

From the data above it would appear that there is scope for extending cultivation in the Andean countries. There are, however, two problems related to the forest: the inherent infertility of the rainforest soils, making the cultivation of field crops commercially questionable, and the increasing strength of public feeling both inside and outside the region that the forest should be preserved for its own sake. The areas of natural pasture are also difficult to bring into cultivation, being mainly in the savanna environment, with unreliable rainfall, or high in the Andes. There is a possibility of increasing yields in many of the existing areas of cultivation. Both Venezuela and Peru have had

TABLE 11.4 Demographic and social data for the states of Mexico, 1990

	(1) Population	(2) % urban	(3) % immigrants	(4) Primary sector	(5) % professional and technical	(6) % literate	(7) % dwellings with electricity	(8) % dwellings with toilet
1 Aguascalientes	720	77	19	15	10	93	95	69
2 Baja California	1,661	91	47	10	12	95	90	59
3 Baja California Sur	318	78	31	18	12	94	89	57
4 Campeche	535	70	22	34	10	84	85	33
5 Coahuila	1,972	86	15	12	12	94	94	60
6 Colima	429	83	27	24	11	90	94	56
7 Chiapas	3,210	40	33	58	7	70	67	25
8 Chihuahua	2,442	77	15	17	10	94	87	59
9 Distrito Federal	8,236	100	24	1	17	96	99	74
10 Durango	1,349	57	11	29	10	93	87	45
11 Guanajuato	3,983	63	8	23	8	83	88	42
12 Guerrero	2,621	52	5	36	10	73	78	25
13 Hidalgo	1,888	45	9	37	8	79	77	31
14 Jalisco	5,303	82	14	15	10	91	93	63
15 México	9,816	84	40	9	10	91	94	51
16 Michoacán	3,548	62	8	34	9	82	87	40
17 Morelos	1,195	86	29	20	12	88	96	48
18 Nayarit	825	62	15	38	10	88	92	39
19 Nuevo León	3,099	92	23	6	12	95	96	72
20 Oaxaca	3,020	39	6	53	7	72	76	19
21 Puebla	4,126	64	9	37	9	81	84	33
22 Querétaro	1,051	60	17	18	9	85	85	46
23 Quintana Roo	493	74	6	20	9	88	85	40
24 San Luis Potosí	2,003	55	9	31	9	85	73	39
25 Sinaloa	2,204	64	12	37	10	90	91	45
26 Sonora	1,824	79	17	23	11	94	90	57
27 Tabasco	1,502	50	10	36	10	87	85	38
28 Tamaulipas	2,250	81	24	16	11	93	84	51
29 Tlaxcala	761	76	12	28	10	89	94	39
30 Veracruz	6,228	56	9	39	8	82	74	34
31 Yucatán	1,363	79	6	27	10	84	90	40
32 Zacatecas	1,276	46	8	40	9	90	87	36
Mexico	81,250	71	17	23	11	87	88	48

Source: INEGI, *XI Censo General de Población y Vivienda*, 1990
Location in the source and full definition of the variables:
(1) Table 3 Population in thousands in 1990
(2) Table 3 Urban population as a percentage of total population
(3) Table 5 Immigrants (persons born in a different state) as a percentage of total population
(4) Table 16 Persons in the primary sector (almost all agricultural) as a percentage of total economically active population
(5) Table 15 Professional and technical as percentage of total economically active population
(6) Table 9 Literate population as a percentage of total population over age of 14
(7) Table 27 Dwellings with electricity as a percentage of all dwellings
(8) Table 31 Percentage of private dwellings with a flushing toilet

to import considerable quantities of food in recent decades, the former able to do so thanks to its large oil revenues, the latter to allow its limited agricultural land to produce high-value crops for export.

Although there are coal reserves in Venezuela and Peru, the use of these has been very limited up to now, but coal production has increased steadily in Colombia, doubling between 1982 and 1992, much of it being produced for export. The Andean countries have about 7 per cent of the world's oil reserves, 6.2 per cent of them in Venezuela. The extraction of oil began in Peru around 1900 and has continued ever since, but the spectacular discoveries of oil reserves in northwest Venezuela in the 1920s by various companies, including

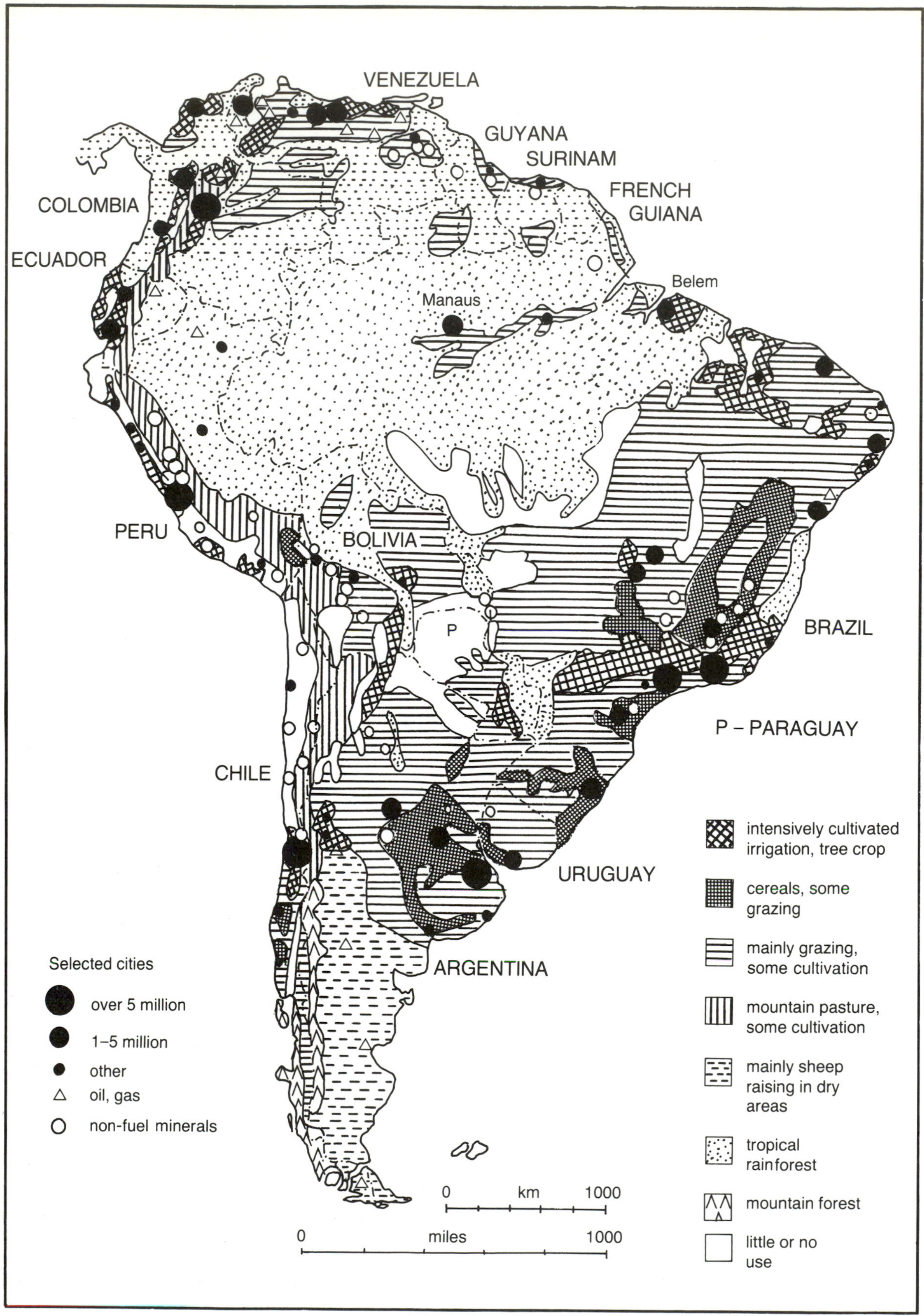

FIGURE 11.7 Economic map of South America

the now giant companies Esso and Shell, turned Venezuela into the world's largest exporter of oil for three decades. Colombia, Ecuador and Bolivia also have reserves of oil, and recent discoveries in Colombia are increasing its prospects of becoming a major exporter. Production in Colombia increased threefold between 1982 and 1992. The drop in Venezuelan oil production and exports after 1973 was made possible by the rise in oil prices, but these have declined in the 1980s, accounting partly for the rise in production from 90 million tonnes in 1985 to 129 million in 1992.

Ecuador and Bolivia have very modest industrial sectors, restricted to the processing of primary products and light manufacturing. Colombia, Peru and Venezuela have each established a heavy industrial base, with an iron and steel plant, and also have light engineering. Although its exports are dominated by oil and oil products, in 1991 over 10 per cent of Venezuela's exports were manufactures, while Colombia reached 34 per cent and Peru 10 per cent, all three attempting to diversify their exports.

In all five Andean countries the service sector accounted for about half of total GDP in the early 1990s and for a growing share of employment. Transport, healthcare and educational services are relatively poorly developed in most rural areas. The number engaged in services has grown rapidly in the cities, but there is much underemployment in retailing and other activities.

1. **Venezuela** On the United Nations Human Development Index, Venezuela ranks 44th out of the 160 countries considered, having relatively high life expectancy and literacy levels and a real GDP per capita just below that of Portugal, the poorest country in Western Europe. Until this century, Venezuela was of little importance in the world economy. The growth of the oil industry led to rapid development, first in the northwest around the Gulf of Maracaibo, then in the eastern fields. Revenue from the industry resulted in fast economic expansion and a facelift in the capital, Caracas. The oil industry has accounted for almost all the value of exports, and provided much of the government's revenue in recent decades.

Since the 1960s it has been regarded as desirable to develop other activities in Venezuela, including light industry in and near the capital, heavy industry by the lower Orinoco, and the extraction of non-fuel minerals, including iron ore and bauxite. The possibility of extracting fuel from Venezuela's large reserves of tar sands has led to a large increase in the estimates of its oil reserves (now put at 9 billion tonnes), giving a life of about 70 years at the 1992 production rate of about 130 million tonnes.

With population expected at least to double in the next few decades, it seems just a matter of time before big inroads are made into the southern forested highlands. At the other extreme, Greater Caracas could grow from its present population of 4 million to 10 million. Venezuela's progress should be watched by other developing countries, further behind on the development ladder, for both positive and negative lessons.

2. **Colombia** is dominated by the presence of three parallel ranges of the Andes, running roughly north–south and making east–west movement difficult. Although the capital, Bogotá, has a population approaching 5 million, its influence in the country is not as great as that of Caracas in Venezuela or Lima in Peru. Medellín, Cali and Barranquilla are major regional centres of influence. Population is concentrated in the western half of the country, but penetration of the eastern savanna and forest has gained momentum, stimulated by the search for oil, the clearance of the forest for pasture, and drug trafficking.

Officially the value of Colombia's exports consists of about one-third each of oil, agricultural products (mainly coffee) and manufactures, but per head of population they are only about one-third of the value of Venezuela's total exports. Colombia is the main processor of coca leaves from the coca shrub, which is grown mainly in Peru and Bolivia, but the proceeds of exports of cocaine are omitted from official trade statistics.

3. **Peru** The regional contrasts in environmental conditions in Peru are even sharper than in Venezuela and Colombia, with a desert, irrigated in a series of small oases along the coastal zone of the Pacific, the high Andes and, beyond, the Amazon lowlands, covered mainly by tropical rainforest. In pre-Spanish times the Inca Empire had extended its control along the Andes and the coast, but the forested interior was, and mostly still is, inhabited only by small tribes with simple forms of society. After 300 years of exploitation by Spain, independence was won by Peru in 1820. Today the country is dominated by Lima, with over a quarter of the total population, over half the manufacturing and many key services.

In relation to its population size, Peru has very large reserves of zinc, lead, copper and silver ore but it processes output for export and does not use these metals to any extent in its own industries. Improvements in agriculture will be slow and costly, requiring new irrigation projects along the coast and a better infrastructure in the Andes. Fortunately for the future of the interior forest, transport links are very difficult to construct and maintain over the Andes, but the search for oil and the accelerated clearance of the forest for timber and pastures, as elsewhere in Amazonia, point to the gradual disruption of the remaining indigenous communities.

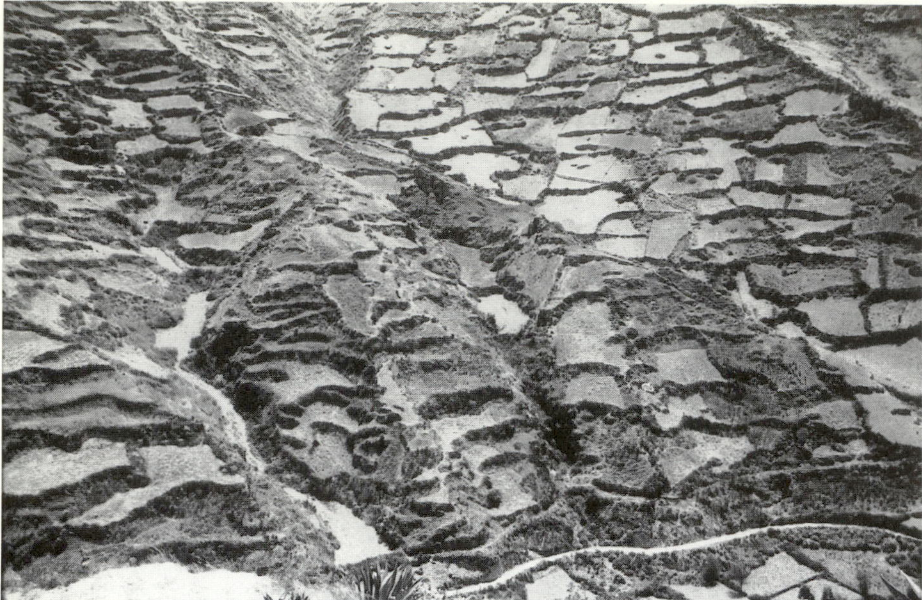

PLATE 11.1
The limits to cultivation. Areas such as these in Western Europe have been abandoned under modern agricultural conditions

a Central Andes of Peru, near Chacayan: terraces in the dry season. To take equipment and fertilisers to such fields is very difficult and costly. Soil erosion is a constant threat. Work in the fields is carried out mainly by hand or with work animals. Yields of cereals and potatoes are low

b View from the central railway of Peru, cultivation to the edge in the Rimac valley behind Lima

Peru's future prospects have been clouded by the uncertainties of the coca 'industry' and by the Shining Path insurgents. The latter movement, growing in strength in the late 1970s, was based on Maoist experience and aims as applied in China. It was led by intellectuals intent on removing social injustices in Peru, but was not generally welcomed by the rural population of the areas in the Andes and Amazon in which it operated. In the struggle between the government forces and the guerrillas some 30,000 lives were lost. On a lighter note, periodically, and belying their 'shining' image, the *senderistas* cut power lines supplying Lima with electricity, plunging the city into darkness. By the mid-1990s the activities of the Shining Path had greatly diminished. Not so the coca industry, the main difference being that the Shining Path caused human suffering at home while the export of coca leaves, albeit in theory illicit, exports the human suffering to other countries and brings in revenue to Peru itself. The coca industry and guerrilla activities pale into insignificance when compared with the demographic prospects for Peru. More fundamentally, the population of Peru has increased about threefold between 1940 and 1990 and the population of Lima about ten times. Can such a trend continue for several more decades?

4. **Ecuador** is similar to Peru in having three distinct physical environments, but it is smaller in population and area. When the Spaniards conquered this part of South America in the sixteenth century they were mainly attracted to sources of silver and gold in what are now Peru and Bolivia. Settlers from Spain became landowners in the Andean region of Ecuador, which was the main concentration of population at the time and remained so until the twentieth century. Subsequently the coastal region, once heavily forested, became the focus of commercial agriculture, producing bananas and cocoa beans for export and accounting in 1970 for as much as 97 per cent of the value of exports. Guayaquil, the main port of Ecuador, has outgrown the capital, Quito, which stagnates in the very traditional heartland of the country. More recently, in the comparatively small area of Amazon forest belonging to Ecuador, oil was discovered. Reserves of 300 million tonnes are small by Venezuelan and Mexican standards but have contributed to domestic energy supplies and also to exports (40 per cent of the value in the early 1990s).

5. **Bolivia** has two distinct physical regions. The high Andes around Potosí was the main source of silver in the Spanish Viceroyalty of Peru, while in the nineteenth century Bolivia became one of the main sources of tin in the world. In the 1950s the Andes region was linked by road to the extensive, mainly forested, eastern lowlands of Bolivia, where new settlements have been established. After Haiti, Bolivia is the poorest country in Latin America. The mining industry has brought little to the majority of the population and with world tin production stagnating since the 1960s and Bolivian production declining, the outlook is bleak.

Except in eastern Venezuela, the Andes hinder movement between the interior of the five countries and the coast. The mineral and timber resources of the interior of tropical South America have remained largely untouched. With growing awareness of the expected increase in trade round the Pacific Rim, there are moves to improve links to the coast.

11.7 Southern South America

In 1993 the four southern countries of South America, Argentina, Chile, Paraguay and Uruguay, had a combined population of about 55 million. That population is expected to grow to 77 million by 2025. The total area is just over 4 million sq km (1.54 million sq miles). About 9 per cent of that area is under crops, almost 50 per cent is natural pasture and over 20 per cent is forest. Compared with the rest of Latin America, the bioclimatic resources of southern South America, in relation to population size, are very favourable. On the other hand, fossil fuel resources are limited, although Chile and Argentina are at present largely self-sufficient in energy. The only non-fuel mineral of any great significance in southern South America is copper – there are very large copper reserves in Chile.

In contrast to the importance of the mountain environment to the Andean group of countries, the high, but relatively narrow, range of the Andes separating Chile and Argentina is almost uninhabited. The tropical and subtropical forest of the region is in northern Argentina and Paraguay, and there is temperate forest in the south of Chile. There is no distinct line between the extensive area of natural pasture and desert in southern and western Argentina, but the best grazing land is mixed with the large area of cultivation in the pampa region of Argentina and neighbouring Uruguay. Argentina and Uruguay have often been compared with Australia and New Zealand, but climatically and in the fertility of its soils, the pampa of Argentina is more favoured than its counterpart in the interior of New South Wales. Argentina has not yet made much use of chemical fertilisers, nor researched as extensively as Australia in crop and trace element improvements. In Table 10.4 wheat yields are compared in six of the world's leading wheat-growing countries.

The foreign trade of the countries of Argentina, Uruguay and Paraguay has been dominated by the export of agricultural products and the import of manufactures whereas, until the 1980s, Chilean exports consisted almost exclusively of minerals. The situation in 1991 was that Paraguay remains essentially an exporter of agricultural products whereas both Argentina and Uruguay have diversified by increasing the export of manufactures and Chile has increased its share of agricultural and forest products and of manufactures. Chile, Paraguay and Uruguay all trade extensively with Argentina, but Argentinian agricultural products are exported less than they were in the past to Western Europe (the Common Agricultural Policy of the EU is blamed) and more to Brazil and Peru.

1. **Argentina** In Spanish colonial times Argentina was a remote extremity of the Viceroyalty of Peru, and most of the population was in the northwest. Today the north and northwest of Argentina have lagged behind the pampa region economically. The influence of Buenos Aires extends far to the south of the country, where the indigenous population was deliberately exterminated in the nineteenth century.

Once Argentina became independent of Spain, the influence of other European countries grew. In the middle of the nineteenth century, with the increasing

PLATE 11.2
The driest desert in the world, the Atacama, northern Chile. It rarely rains and no rivers cross the desert here to provide water for irrigation. A mist hangs over part of the landscape but it does not produce rain of any practical use

PLATE 11.3
One of the highest mountains in the Andes, Huascaran, Peru. Three years after this picture was taken the town in the foreground, Yungay, was largely destroyed by a mudflow dislodged from the side of the mountain in the distance by an earthquake in 1970

use of steamship navigation and railways, the exploitation of the pampas region began. Settlers from Spain, Italy and Germany arrived in large numbers later in the nineteenth century. An extensive rail network served the hinterland of Buenos Aires and meat products as well as grain could be transported to Europe. Buenos Aires was already the fourth largest city in the world, the national capital, the principal port and the main manufacturing centre of Argentina.

Since the 1930s Argentina's economy has suffered through neglect of the agricultural sector, and attempts to increase industrial independence. Over three-quarters of the imports of Argentina are still, however, manufactures. The heavy industrial base of Argentina depends on imported materials; the output of steel has not expanded greatly in the 1980s (1980 2.6 million tonnes, 1990 3.6).

Today, about one-third of the total population of the country is in Buenos Aires and neighbouring cities (e.g. La Plata) (cf. Australia where about one-third live in Sydney and Melbourne). In the Federal Capital, and the province of Buenos Aires in which it is located, half of Argentina's population lives in little more than one-tenth of the total area of the country. Other large centres have attracted population and industry to the west (Mendoza) and north (Rosario, Córdoba, Tucumán) of Buenos Aires, making Argentina one of the most highly urbanised countries of the world (86 per cent) and leaving only 10 per cent of the economically active population in agriculture. All in all, however, Argentina copes with its giant capital city and wide open spaces and coexists with its smaller neighbours, albeit not always to the satisfaction of Chile. Argentina's claim to the Falkland Isles has a political motive, to foster national cohesion and identity, and an economic one, control of extensive territorial waters with fishing and offshore mineral potential. The bioclimatic resources of the islands themselves are minuscule compared with Argentina's mainland resources.

2. **Chile** The shape of Chile is a nightmare for cartographers and such a long narrow piece of territory could hardly survive as a meaningful nation if it stretched across a continent. It is, however, a viable country for two reasons: most of its population is concentrated in the central zone, and its main neighbour, Argentina, is separated from it by the Andes. A good road extends the whole length of the northern two-thirds of Chile and shipping services reach to the extreme south.

Since the middle of the nineteenth century, Chile has been a source of minerals for Western Europe and the USA, exporting nitrates, together with guano, the latter also from Peru, the first large-scale source of fertilisers not derived directly from livestock. Later in the nineteenth century, Chile became one of the main sources of copper and a small exporter of iron ore. As in the other large and medium-sized countries of Latin America, successive Chilean governments in this century have encouraged industrial growth: heavy industry, using their own coal and iron ore for steel production near Concepción, and also light manufacturing, mainly in the capital, Santiago.

In the 1980s protection and support for industry was reduced, and a return to free trade has led to an increase in the export of primary products and in the import of manufactures. Like Argentina, Chile's prospects for the future seem reasonably favourable, but the fact that about half of the population of Chile lives in Santiago and Valparaiso, on 4 per cent of the total area of the country, could be a matter of some concern.

3. **Uruguay and Paraguay** originated in the Spanish Empire but, as 'buffer' states between the more powerful Argentina and Brazil, they have features of both their large neighbours and are influenced economically if not politically by them. Materially, most Uruguayans are well-off by developing world standards, with only a small rate in increase of population and good bioclimatic resources. Paraguay's landlocked position has left it isolated compared with Uruguay, but its backwardness may also be related to the devastating effects of the Great Paraguayan War of 1864–70, in which most of the male population was killed (see Chapter 3). Paraguay's future is likely to be affected by the aspirations of Brazil and Argentina in the interior of the continent related to the hydroelectric and bioclimatic potential.

11.8 Brazil and the Guianas

Because of its size and natural resources, Brazil has gained in prominence in world affairs since the Second World War, as a Third World power with successes and potential, even if the title of one book – *Brazil, the Great Power of the 21st Century* – seems ambitious. In recent decades Brazil has been distinguished for two main reasons: first the creation of a formidable industrial base in the triangle São Paulo–Rio de Janeiro–Belo Horizonte and, second, because it has about two-fifths of the remaining tropical rainforest of the world.

With 152 million inhabitants in 1993, Brazil had almost exactly one-third of the population of Latin America. Population is expected to rise to about 200 million in 2025. Brazil occupies just over 40 per cent of the area of Latin America. About 7 per cent of the total area is cultivated, 22 per cent pasture and 58 per cent forest. Brazil is one of the few countries in which the arable area has increased greatly in the last three decades. In the early 1960s there were about 300,000 sq km (116,000 sq miles) under field and tree crops, in the early 1990s twice as many. The expansion of the cultivated area has mainly been at the expense of forest.

Brazil has about 2.8 per cent of the total population of the world, but its share of the world's freshwater resources is about five times its share of population, and its bioclimatic resources about 2.5 times the world average. The weakest aspect of its natural resource endowment is its very limited reserves of fossil fuels. On the other hand, with abundant iron ore, large reserves of manganese ore and tin, and much territory still little explored for minerals, it is well placed with regard to some non-fuel minerals.

Before the arrival of the Portuguese around 1500, Brazil's indigenous population numbered about 3 million and consisted of a large number of tribes depending on gathering, hunting, fishing and shifting cultivation. Being a country of modest size and resources, and having its principal trading interests along the coasts of Africa and in Asia, Portugal put fewer resources into its colonies in Brazil than Spain did in Mexico and Peru. Initially most of the Portuguese endeavours were concentrated on the coastal zone of Northeast Brazil, the part closest to Europe. Forest products were exported in small quantities to Portugal, but already by the end of the sixteenth century sugar was cultivated and the cane processed for shipment to Europe.

Subsequently Portuguese settlements extended

southwards, but they were still mainly confined to a narrow coastal zone until the eighteenth century, when gold and precious stones attracted exploration inland beyond the escarpment behind the coast. In the nineteenth century the cultivation of coffee expanded inland behind São Paulo. Until the 1880s northern and western Brazil were hardly touched. Interest in rubber obtained from the tree *Hevea brasiliensis* led to the penetration of Amazonia, but settlements of outsiders were mainly located along the rivers, and the indigenous population remained undisturbed over large areas, except where they were coerced into gathering rubber from trees scattered thinly in the forest.

By the end of the nineteenth century, Southeast Brazil was the dominant region of the country in economic terms. In 1890, however, there were only about 15 million people in the whole of Brazil, so the development of the country has taken place almost entirely this century. The 'empty' interior of Brazil was already a matter of concern at least a century ago for two reasons. First, it seemed exposed to possible incursions and territorial claims from the various neighbours to the west and north. Second, it was considered that great economic potential was being wasted, in spite of its poor image as a wild area, sometimes referred to as the *inferno verde* (green inferno), difficult to clear and colonise.

The symbolic transfer of the capital of Brazil in 1960 from Rio de Janeiro to Brasília, a location about 1,000 km (620 miles) inland, was one step towards opening up the so-called 'other half' (*outra metade*) of Brazil, and the implementation of regional planning policies. One direction was the establishment of a fund, SUDENE, to assist the poor Northeast. New industries were set up in Salvador, Recife and other coastal cities. Another initiative was a road construction programme to open up Amazonia. An east–west road was constructed to allow settlers from the Northeast to migrate to lands cleared in the rainforest. From Southeast Brazil another road penetrated Amazonia.

In the 1990s, the situation in Brazil and future prospects can best be summarised by looking at the five macro-regions used by the Instituto Brasileiro de Geografia e Estatística. Table 11.5 contains a selection of data for the states of Brazil and for the five macro-regions into which they are grouped.

1. **Sudeste (Southeast)** The four states of this region have 44 per cent of the total population of Brazil but account for 60 per cent of GNP, including 65 per cent

FIGURE 11.8 The states of Brazil in 1990. Tocantins is a new state recently added to the North (Norte) region

TABLE 11.5 Data set for the states of Brazil and the macro-regions

	(1) Area in thous. sq km/(miles)		(2) Density of popn per sq km/(mile)		(3) Population 1960	(4) Population 1991	(5) Popn change 1960–1990	(6) Life expectancy	(7) GRP[2] per head	(8) Infant mortality	(9) Fertility	(10) Migration emigration	(11) Migration immigration
North	3,852	(1,487)	2.7	(7)	2.6	10.3	397	64	6.8	72	6.5	7	18
Rondônia	239	(92)	4.7	(12)	0.07	1.1	1,616	na	8.9	na	6.2	10	66
Acre	154	(59)	2.7	(7)	0.16	0.4	264	na	5.4	na	6.9	13	11
Amazonas	1,568	(605)	1.3	(3)	0.7	2.1	297	65	10.5	65	6.8	7	8
Roraima	225	(87)	1.0	(3)	0.03	0.2	771	na	6.3	na	6.1	9	30
Pará	1,247	(481)	4.2	(11)	1.5	5.2	339	64	5.9	74	6.3	7	15
Amapa	142	(55)	2.0	(5)	0.7	0.3	425	na	6.8	na	7.0	10	28
Tocantins[1]	277	(107)	3.3	(9)	0.3	0.9	300	na	2.0	na	na	na	na
Northeast	1,556	(601)	27.3	(71)	22.2	42.5	191	52	4.6	119	6.1	19	6
Maranhão	330	(127)	14.8	(38)	2.5	4.9	200	55	3.0	109	6.9	13	11
Piauí	251	(97)	10.4	(27)	1.2	2.6	208	58	2.4	95	6.5	21	7
Ceará	150	(58)	42.7	(111)	3.3	6.4	193	47	3.7	132	6.1	18	4
Rio Grande do Norte	53	(20)	45.3	(117)	1.1	2.4	211	45	5.3	149	5.7	20	7
Paraiba	54	(21)	59.3	(154)	2.0	3.2	160	44	3.0	148	6.2	27	6
Pernambuco	101	(39)	70.3	(182)	4.1	7.1	174	48	4.7	145	5.4	21	7
Alagoas	29	(11)	86.2	(223)	1.3	2.5	200	47	4.2	143	6.7	24	8
Sergipe	22	(8)	68.2	(177)	0.8	1.5	198	55	6.7	108	6.0	25	8
Bahia	567	(219)	21.0	(54)	5.9	11.9	200	58	6.3	95	6.2	18	5
Southeast	924	(357)	67.9	(176)	30.6	62.7	205	64	13.3	72	3.5	14	18
Minas Gerais	587	(227)	26.7	(69)	9.7	15.7	163	63	8.7	72	4.3	24	5
Espírito Santo	46	(18)	56.5	(146)	1.2	2.6	222	67	9.4	58	4.3	24	16
Rio de Janeiro	44	(17)	290.9	(753)	6.6	12.8	193	63	13.2	75	2.9	6	22
São Paulo	248	(96)	127.0	(329)	12.8	31.5	246	64	16.1	75	3.2	7	24
South	575	(222)	38.4	(99)	11.8	22.1	188	67	10.6	60	3.6	14	14
Paraná	199	(77)	42.2	(109)	4.3	8.4	198	64	9.6	69	4.1	19	27
Santa Catarina	95	(37)	47.4	(123)	2.1	4.5	214	67	10.4	58	3.8	15	12
Rio Grande do Sul	281	(108)	32.4	(84)	5.4	9.1	170	71	11.7	45	3.1	11	2
Central	1,605	(620)	5.9	(15)	2.9	9.4	320	65	8.5	71	4.5	13	35
Mato Grosso do Sul	357	(138)	5.0	(13)	0.4	1.8	445	66	8.2	66	4.4	12	36
Mato Grosso	901	(348)	2.2	(6)	0.5	2.0	227	66	6.6	66	4.7	18	41
Goiás[1]	340	(131)	11.8	(31)	1.6	4.0	250	64	7.0	73	4.7	12	23
Distrito Federal (Brasília)	6	(2)	266.7	(691)	0.14	1.6	1,141	66	15.1	82	3.6	14	67
Brazil	8,512	(3,286)	17.3	(45)	70.1	146.9	210	60	9.7	93	4.4	15	15

Notes: [1] Tocantins was formerly the northern part of Goiás
[2] GRP = Gross Regional Product
na = not available
Description of variables:
(1) Area in thousands of sq km (p. 29)
(2) Density of population, persons per sq km
(3), (4) Population in millions in 1960, 1991 (p. 206)
(5) Population change 1960–91 (1960 = 100)
(6) Life expectancy in years in 1980 (p. 228)
(7) Gross regional product in thousands of cruzeiros per head (p. 1045)
(8) Infant mortality rate per 1,000 live births in 1980 (p. 229)
(9) Total fertility rate, average number of births per female (p. 231)
(10), (11) Emigration and immigration per 1,000 population in 1980 (pp. 232–3)
Source: IBGE, *Anuario Estatístico do Brasil 1992*, Rio de Janeiro, 1992. Page nos given above

PLATE 11.4
Mechanisation in Brazil

a Early textile machinery installed in a mill in Northeast Brazil, Salvador, dated 1908 and made in Manchester, England, by Brooks and Doxey

b Plant for processing manganese ore, Serra do Navio, Amapa, North Brazil. Such extractive enterprises do not cause extensive damage to the forest but locally their impact is devastating. In theory such mines and quarries are to be covered again with soil in the hope that the forest will return

of industrial output. Almost all of Brazilian heavy industry is located in the region, including iron and steel production, engineering and chemicals, as well as much of the production from light industry. Brazil is the leading producer of steel in Latin America, with 15 million tonnes in 1980 and a peak of 25 million in 1989 (20.5 in 1990). Hydroelectric stations near the coast were important early in the century but are limited in capacity. Most of the energy in the region is obtained from imported fuels and from large hydroelectric stations to the west, on the Paraná River and its tributaries. Southeast Brazil also has some of the best agricultural land in the country, particularly on soils formed on lava in the interior of São Paulo state.

Southeast Brazil's share of total population has risen due to immigration from other regions, especially the Northeast. Migration from the Southeast to the two most sparsely populated regions, the North and the Centre-West, has so far been slight. São Paulo and Rio de Janeiro have been the main destinations of the migrants, but the rivalry between the two to be the largest city in Brazil has now given way to concern as Greater São Paulo heads for 20 million in the year 2000 and Greater Rio for 15 million.

2. **Sul (South)** Now the main coffee-growing region

of Brazil, the South also has good cereal-growing conditions in the south and industrial capacity in Porto Alegre (Rio Grande do Sul) and Curitiba (Paraná).

3. **Nordeste (Northeast)** In 1870 the Northeast had 46 per cent of the 10 million inhabitants of Brazil but by 1990 the share had dropped to 28 per cent. Droughts have been a cause of deaths during some seasons but the decline in the proportion of Brazil's population in the region has been due to migration and also to the preference of almost all foreign migrants to Brazil for the Southeast and South. Agricultural conditions are difficult and sugar, cotton, cocoa, fibres and other export crops of the Northeast have not been in great demand outside Brazil. The region has Brazil's main oilfield, but reserves are small.

4. **Centro-Oeste (Centre-West)** consists mainly of savanna vegetation. Settlement has been spreading gradually into the region from the Southeast as roads have penetrated it, but it remains predominantly agricultural, with extensive areas of cattle-raising.

5. **Norte (North)** This region is the subject of the next section, in which the future of the tropical rainforest of Brazil is discussed.

11.9 The fragmenting forest of Amazonia

For centuries the clearance of forest for agricultural and other purposes has been regarded throughout the world as a great achievement. Paul Bunyan was a legendary figure in US history who cleared forest at a remarkable rate. A bizarre machine developed by the US Le Tourneau Company to push over large trees was introduced into the Peruvian Amazon region with much publicity in the 1950s. Why, in the last two decades, has conservation rather than clearance of the forests of the world come to be regarded as desirable, if not vital for the future of humanity? Some aspects of the debate over the tropical rainforest of South America in general and of Brazil in particular are discussed in this section.

The largest remaining area of tropical rainforest in the world is in South America. Almost all of it is in the region commonly referred to as the Amazon, or Amazonia. All forests receive some precipitation (rain, snow, hail) but the so-called rainforests of the world are characterised by heavy rain throughout the year or for most of the year. Not all are in the tropics. The tropical rainforest is characterised by its very large number of plant and animal species, the great volume (biomass) of vegetation per unit area and, generally, the poor quality of the soil on which the forest grows. Some areas of tropical rainforest are flooded each year, some are located high on the slopes of mountains. In the Caribbean islands, West Africa and parts of India and Indonesia, very little of the original tropical rainforest remains. The experience of these and other regions may be useful in ensuring that part at least of the tropical rainforest of South America can be saved.

Where is the tropical rainforest in South America? Almost all of it is in a nearly continuous area situated in nine countries, roughly, but not precisely, in the drainage basin of the River Amazon and its tributaries. Table 11.6 shows the extent of forest and woodland in the nine countries referred to above, most but not all of it in the Amazon region. Some, like the pine forest of South Brazil, is not tropical rainforest.

Figure 11.9 shows the problems of defining the 'Amazonia' tropical rainforest of South America. It extends to the Atlantic coast in the northeast and to the Andes Mountains in the west. It merges into savanna and grassland to the north and south. In Venezuela and Colombia the forest extends beyond the limit of the Amazon drainage basin into the Orinoco basin. In Ecuador, Peru and Bolivia the forest extends up only the lower slopes of the Andes, while the limit of the

TABLE 11.6 Forest and woodland in northern and central South America in millions of hectares, 1965–90

	1965	*1970*	*1975*	*1980*	*1985*	*1990*	*1990 as % of 1965*
Brazil	543	537	531	518	506	493	91
Bolivia	59	58	57	56	56	56	95
Peru	74	74	72	71	70	68	92
Ecuador	18	18	15	14	12	11	61
Colombia	55	55	55	53	52	50	91
Venezuela	35	35	35	33	32	30	86
Guyana	18	18	18	16	16	16	89
Suriname	15	15	15	15	15	15	100
French Guiana	8	8	7	7	7	7	88
Total	825	818	805	783	766	746	90

Sources: FAOPY 1991, Table 1 for 1975–90; *FAOPY 1976*, Table 1 for 1965, 1970

FIGURE 11.9 The rainforest of South America
a defining Amazonia

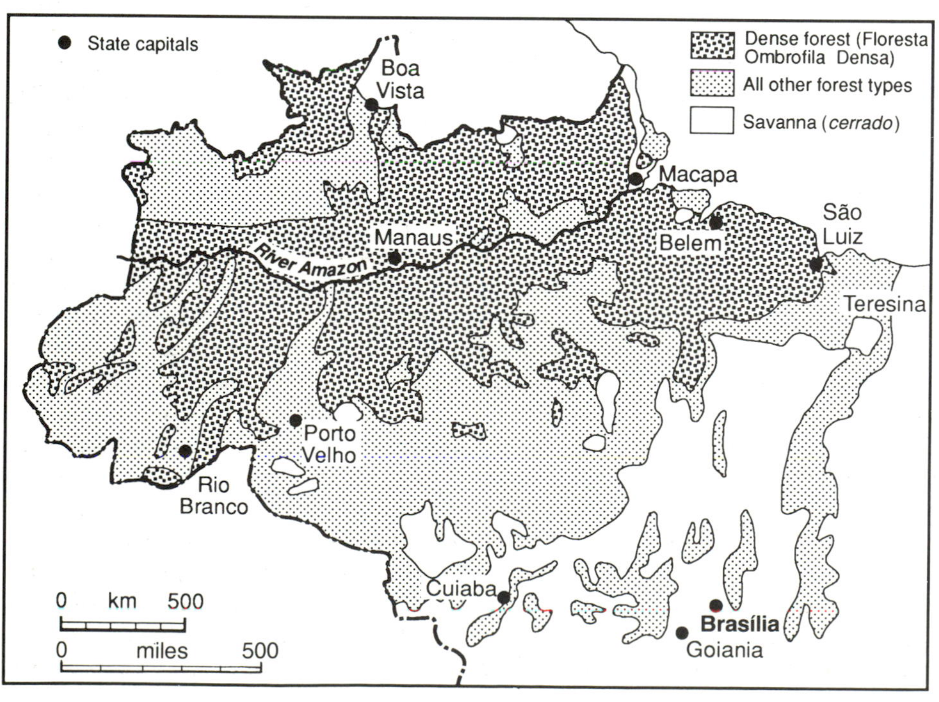

b the distribution of rainforest in Brazil

Amazon drainage basin follows the summit of the highest range. In the southeast, the Amazon basin is mainly covered with savanna (*cerrado*) vegetation. The boundaries between the various pairs of countries in this part of South America do not follow the watersheds or the limits of the forest. They mostly cut through still very sparsely populated areas, and have frequently been disputed since the countries became independent early in the nineteenth century (Guyana and Suriname more recently).

To confuse the terminology, Brazil, Venezuela, Colombia and Peru all have administrative divisions called Amazonas, the largest being in Brazil (see A in Figure 11.9a). In Brazil there is also a region called the legal Amazon, which extends southwards beyond the rainforest. Conservation of the tropical rainforest of South America is ultimately an international problem, involving nine countries of the region at continental level and of importance globally, given the influence on the composition of the atmosphere of large-scale clearance of the forest, especially by burning.

Forest may at first sight seem to be a type of vegetation that can easily be defined, but in practice the areas given in Table 11.6 are only very rough estimates. Very little of Amazonia has been mapped on a large scale on account of the size, limited economic importance and the cost of making surveys by aerial photography. Only since the 1970s have satellite images begun to reveal detailed features of the forest, including clearance by felling and by burning. Since satellites pass regularly and frequently over all parts of the earth's surface they can monitor change in the forest far more thoroughly than aircraft or ground expeditions.

Reservations should be noted concerning the prospects of monitoring forest and other environments in the less densely populated areas of the world. Technical problems remain. In areas of tropical rainforest, for example, cloud cover is frequent but images created from reflection of light depend on clear skies. The best detail or resolution from satellites used for civilian purposes is obtained in pixels, uniform rectangles or squares, 30 × 30 metres in the US Landsat system, 20 × 20 metres in the French SPOT system. Fine details may be missed, while completely different features on the ground (such as roofs of large buildings and bare rock surfaces) may not be distinguishable, but there is not such a problem, however, with vegetation. Even when abundant information about, for example, illegal forest cutting or burning (picked up by the recording of heat emissions) is passed on to authorities concerned with the protection of the environment, there is no guarantee that any steps will or even can be taken to prevent further destruction.

Until the last two decades of the nineteenth century the penetration of Amazonia by the Spaniards and Portuguese was very limited indeed. An indigenous population of several million practised a sustainable economy cultivating small areas, fishing and hunting, and using products from the vegetation, without, however, markedly modifying the ecosystem. Latex from the rubber tree (*Hevea brasiliensis*) was the first product gathered in large quantities for export to the outside world. Extraction began in the 1880s. One of the main uses of the rubber, after processing, was in the manufacture of motor tyres. Together with the extraction of Brazil nuts, timber and a few mineral products, the gathering of rubber made a considerable impact on the indigenous population of Amazonia, but little on the forest itself.

During and after the Second World War, the governments of the region and also the USA became increasingly interested in the natural resources of Amazonia. In Peru a highway was built in the 1940s down from the Andes behind Lima to the Ucayali River at Pucallpa. In the 1950s a comparable highway was built from Cochabamba to Santa Cruz in Bolivia. More recently, considerable reserves of oil have been discovered and exploited in the forested lowlands of Ecuador, Peru and Colombia, while Venezuela is beginning to look for ways to use the forest resources of its 'empty' southern area. The forest is being attacked from the north, west and southwest by these smaller countries. About two-thirds of Amazonia are in Brazil, however, and the 'declaration of war' on the empty (northwest) half of the country by the military government in the late 1960s is the theme of the rest of this chapter.

Three main means of transportation are allowing the penetration of Amazonia: the navigable waterways of the Amazon and its tributaries, the airports and airstrips, and the roads. Since the rivers are fixed, their courses and directions largely determined the accessibility of different parts of the forest until the 1970s. Airstrips can be constructed wherever the land is flat enough, but the very high cost of moving equipment into and products out of Amazonia by air has made their impact limited and local. There are only a few short stretches of special-purpose railway. Road transport is now the main means of opening up new areas of forest, hitherto protected from clearance on account of their distance from the rivers. In the late 1960s a massive road construction programme was planned for the Brazilian Amazonia. Even now they have opened up only very small areas. They are costly to construct, surface and maintain and, given the distances from Amazonia to both Northeast and Southeast Brazil, their commercial use is limited to the movement of goods that are of high value and are not perishable.

So who lives in Amazonia and who produces what? Table 11.7 shows how the population of Brazil, very approximate until the censuses of recent decades, has grown about fifteenfold since 1872, while the very

TABLE 11.7 Population in millions in the North region of Brazil in relation to the total population of Brazil, 1872–2000

	1872	*1890*	*1900*	*1920*	*1940*	*1950*	*1960*	*1970*	*1980*	*1990*	*2000*
Brazil	9.9	14.3	17.4	30.6	41.2	51.9	70.1	93.1	119.0	150.4	171.5
North	0.3	0.5	0.7	1.4	1.5	1.8	2.6	3.6	5.9	8.9	11.5
Rest of Brazil	9.6	13.9	16.7	29.2	39.8	50.1	67.5	89.5	113.1	141.5	168.0
Per cent in North	3.4	3.3	4.0	4.7	3.5	3.6	3.7	3.9	4.9	5.9	6.4

Source: IBGE, *Anuario Estatístico do Brasil 1992*, Rio de Janeiro 1992, pp. 180, 183

roughly estimated population of the North region, undercounting the indigenous population, has increased about thirtyfold. The following users and uses make different impacts on the forest.

- The indigenous population, originally numbering about 3 million, but now greatly reduced, and mixed through contact with outsiders, has largely been relegated to reserves, assimilated or in some cases even deliberately exterminated. For the original inhabitants the forest now provides not only subsistence but also some products for trading, including rubber, nuts, medicinal plants and exotic plant and animal species.

- Logging is carried out in Amazonia but, given the extent of the forest, not on a large scale. One reason is the problem of locating and extracting the few tree species of commercial value out of the many found in the forest. An alternative commercial approach is to take out the whole biomass for processing.

- Mining makes a considerable local impact, as for example at the highly mechanised manganese ore mine at Serra do Navio in Amapa, or in the quarries and river beds worked by the *garimpeiros*, small-time gold-miners, increasingly accused of polluting the rivers of the region.

- In the 1980s the Brazilian government supported the development of extensive cattle ranches, resulting in the clearance of large areas of forest, mainly in the southern part of Amazonia. Such areas generally provide only poor-quality, unsustainable grazing land.

- In some areas on the southern and eastern fringes of Amazonia, small-scale farms have been established by settlers from both Northeast and Southeast Brazil. Such farmers tend to be largely self-sufficient in agricultural products since potential urban markets in the more heavily populated areas of Brazil are very far away.

- There is little manufacturing in Amazonia at present, but some processing of primary products.

- A threat to the forest comes from the construction of hydroelectric stations with reservoirs covering areas of forest, although at present these are small and are for local needs.

Amazonia is a huge area on which these and other activities are being carried out and expanded. But who influences these developments? The following very different players are all concerned:

- The national government of Brazil in Brasília, with economic growth high on the agenda. The empty half of Brazil produces less than a tenth of the GDP of the country. Brazil has a large foreign debt to pay off.

- The governors of the states of Amazonia and their associates want both economic growth in their domain and the prestige of new developments.

- The outside world is divided in its views about the future of Amazonia. To investors and to importers of primary products in the industrial countries, here is another area that apparently has vast natural resources. In the world of international finance and trade the fate of the forest and its original inhabitants matters no more than the fate of political prisoners in communist China. To environmental groups, on the other hand, the rainforest should be preserved for several reasons, which include its intrinsic worth as a unique ecosystem, its special properties, including many species that may be useful for medicinal and other purposes in the future, its role as the home of the original inhabitants, and the prospect that its clearance would release carbon dioxide into the atmosphere.

- The indigenous inhabitants, shrinking in numbers, minimally represented politically and legally, but not averse to improving their quality of life by obtaining some of the products of modern technology: good metal tools, outboard motors, radios.

- The plant and animal species of the forest have no voice at all.

- Future generations of Brazilians and of the rest of the world's population have no voice at all.

Whatever the relative influence of the first four active players, the fact is that the rainforest in Brazilian

Amazonia is being cleared. How fast this is happening has not been easily measurable. Clearance has, however, been greatest on the eastern and southern sides of Amazonia. Fearnside (1993, p. 540) provides estimates of the percentage of original forest lost by 1991.

Percentage of forest cleared in the states of the Brazilian Amazon region

Maranhão	68.8	Acre	7.0
Tocantins	39.7	Roraima	2.6
Mato Grosso	16.4	Amazonas	1.6
Rondônia	16.1	Amapá	1.5
Pará	13.0	Brazil	10.5

The key estimates for forest clearance in the Brazilian Legal Amazon area are as follows. By 1978, 150,000 sq km (58,000 sq miles) had been cleared. Between 1978 and 1988 an average of about 22,000 sq km (8,500 sq miles) were cleared each year, bringing the total cleared to 377,000 sq km (145,000 sq miles) in 1988. Between 1988 and 1991 the rate of clearance slowed down. Fearnside argues that by clearing the forest the opportunity to use it sustainably is being thrown away. The prospects for the cultivators, ranchers, loggers and gold-miners who have populated Amazonia mainly since the 1960s are not good because their existence cannot be sustained.

Whatever the hopes of the conservationists, the prospects for preventing further clearance of Brazil's Amazonian rainforest are poor. The future for the forest ranges through the following:

- An immediate moratorium on all forest clearance could be imposed by the Brazilian government. This is highly unlikely and, even if implemented, would be difficult to monitor and enforce. If successfully implemented immediately, about 90 per cent of the forest would remain.

- Controlled clearance, where soil is fertile, minerals are to be extracted or reservoirs are needed behind hydroelectric dams. The end product would be that most of the forest would remain intact, while settlement would largely be confined to elongated zones along the roads and some of the waterways. Perhaps 70 per cent of the forest could be preserved.

- Fragmentation of the forest, with many islands of forest of varying shape and size left in areas away from the rivers and roads. Fifty per cent of the forest would remain.

- Uncontrolled development of Amazonia could result in very little of it remaining in a few decades' time. It took many centuries to reduce the forest that at one time covered most of England to less than 10 per cent of the total area of the country early in the twentieth century. Economic change will be much faster in the twenty-first century.

Clayton (1994) describes a $1 billion project to put the forest of Amazonia under surveillance from satellites, aircraft and ground radar stations. The Amazon Surveillance System (SIVAM) is planned to cover more than 5 million sq km (1.93 million sq miles). It is to be established and run by Raytheon, a US defence and electronics contractor, whose connection with the US military has worried the Brazilian military.

The great physical and human diversity of Latin America has been stressed throughout this chapter. In Table 17.10, Latin America has the highest score among the six developing regions on the consensus of natural resources, living standards and stability. There are, however, great contrasts in living conditions both within and between countries. The contrast in life styles between the indigenous population of the Amazon forest and the conspicuously rich of the big cities is now widely appreciated. Again, at international level, the Andean Indians of Bolivia and the former slaves of Haiti are among the poorest people in the world. By contrast, Argentina, early in the twentieth century classified as one of the richest countries in the world, still has a standard of living comparable with that of the poorer countries of Western Europe. Of all the major regions of the world, Latin America is the one in which change and progress should be most closely watched. If the poverty in many parts of Latin America can be eliminated, then there is the possibility that other developing regions could follow its path. If that is not possible, then the outlook is bleak for the other five developing regions.

KEY FACTS

1. Latin America is a major region of great physical and human diversity.

2. Physically it consists of extensive areas of forest and grassland and has some of the most arid conditions and highest mountains in the world.

3. Much of the region has been settled for several tens of thousands of years. Colonisation by the Spanish and Portuguese has brought profound changes in the region in the last 500 years.

4. Conditions for agriculture are difficult over most of the region but non-fuel mineral reserves and oil and gas resources are considerable.

5. Industrialisation has been rapid in some parts of Latin America this century and the region is regarded as one of the 'hottest' prospects for investment and economic expansion in the next century.

MATTERS OF OPINION

1. In many Latin American countries the rate of population growth is slowing down, although the absolute increase is still considerable: 1974–84 90 million, 1984–94 75 million, with a further increase of 210 million expected by 2025. On this evidence, should Latin American governments put more resources into reducing the rate of population growth?

2. South America has many of the remaining people in the world living in simple societies. Should attempts be made to allow them to keep their ways of life?

3. Should Latin American governments encourage investment from the transnational companies of the developed world?

SUBJECTS FOR SPECULATION

1. In spite of various attempts to achieve closer integration among the member countries of Latin America since the 1950s, trade and other transactions are mostly limited. Will the countries of the region join the global trend towards regional economic unions in the next century?

2. Latin America has few coal reserves. Has this lack of the key fuel in the earlier period of the Industrial Revolution in Europe and North America hampered development in the region?

3. As yet little concern has been expressed by governments in Latin America about the conservation of natural resources. Is there time still to save, in particular, some of the dwindling forest areas?

4. Latin America has four urban agglomerations with over 10 million inhabitants which, like most cities in Latin America, are growing fast in population. What is the threshold size of population above which the apparent advantages of scale become a liability?

Chapter twelve

Africa south of the Sahara

> **Yet there may be hope in the very instability which Africa is experiencing in the wake of this unnatural dis-Africanisation. The fate of African culture may not as yet be irrevocably sealed. With every new military coup, with every collapse of a foreign aid project, with every evidence of large-scale corruption, with every twist and turn in opportunistic foreign policy, it becomes pertinent to ask whether Western culture in Africa is little more than a nine-day wonder.**
>
> *Ali A. Mazrui (1986)*

12.1 General features

The Sahara Desert, which stretches across northern Africa from the Atlantic in the west to the Red Sea in the east, roughly between 15° and 20° North of the Equator, has a very low density of population. It separates the northern tier of African countries bordering the Mediterranean from the rest of the continent. Six countries in northern Africa – Western Sahara, Morocco, Algeria, Tunisia, Libya and Egypt – have been grouped in this book with the countries of Southwest Asia (see Chapter 13), with which they have many features in common. The boundary between these countries and their neighbours to the south is taken as the northern limit of the major world region of 'Africa south of the Sahara', a description I prefer to 'sub-Saharan Africa'. Figure 12.1 shows the countries of the region. Excluding those of North Africa, there are over fifty distinct states in the continent, including two groups of small islands in the Atlantic, as well as Madagascar and three other smaller island groups in the Indian Ocean. The number could grow. Eritrea became independent from Ethiopia in 1993 and Somaliland (formerly British Somaliland) is seeking independence from Somalia.

With 550 million inhabitants, Africa south of the Sahara had almost exactly 10 per cent of the world's population in 1993. In 2025 it is expected to have about 1,326 million, 16 per cent of the world's population in that year. The region is some 24.5 million sq km (9.5 million sq miles) in area (the whole of Africa covers 30 million), 18.3 per cent of the world's land surface. With an annual average GDP of about 500 dollars per head in the early 1990s, it has only about half of the average for all the developing countries in the world, about one-fifth of the Latin American average, and one-eighth of the world average. If South Africa is omitted from the calculation, then the average for the rest of Africa south of the Sahara drops considerably. South Asia, and Africa excluding the north and the extreme south, are the poorest regions in the world.

Given their predominantly tropical locations, Africa and Latin America share a number of features. Each has at some stage been colonised by European powers. In each, the sovereign states of the 1990s coincide

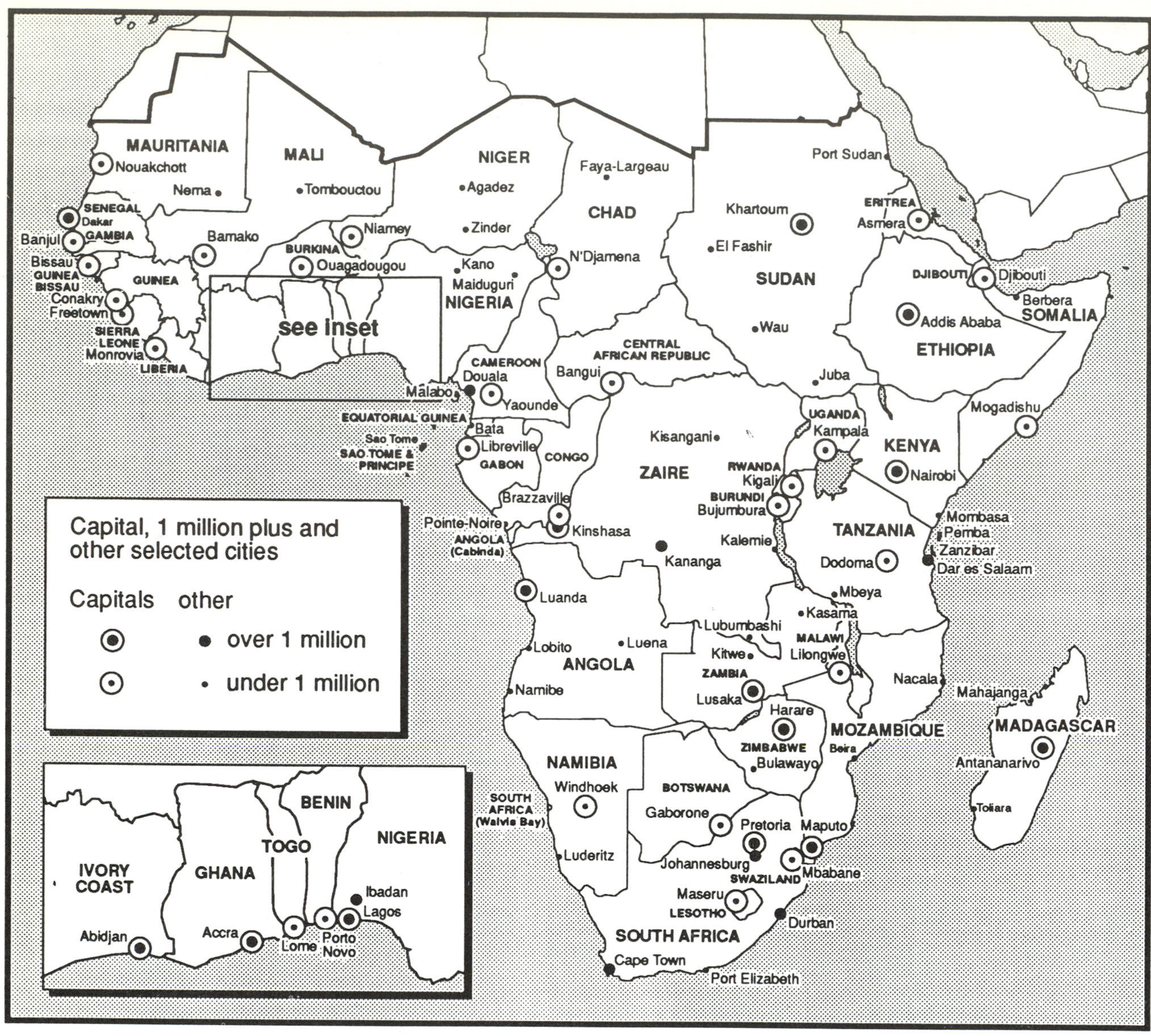

FIGURE 12.1 The countries of Africa south of the Sahara

closely with the former colonial territories. Each has been managed as a source of primary products for Western Europe and the USA. Latin America was part of the empires of Spain and Portugal between 1500 and early in the nineteenth century, after which almost all of it became politically independent. The colonisation in Africa before 1890 was largely confined to trading posts and narrow coastal strips. When the carve-up of Africa by Europe started in the 1880s, railways and steamship services already served large parts of the world and, apart from a few areas, most notably Abyssinia and Liberia, Africa was quickly taken over by various European powers and linked to the world transport and trading system.

Africa south of the Sahara is less varied physically than Latin America, with only a small southern part outside the tropics and with high mountains markedly affecting settlement and life styles only locally. Both regions have extensive areas of tropical rainforest, but the main region of the forest in central Africa is only about one-third as extensive as that of Amazonia in South America. Culturally, the indigenous population of Latin America has largely been transformed to Iberian and western ways except in the more remote mountain and forest areas, whereas except in South Africa (and Algeria in North Africa) European settlement in Africa has been limited. Although European languages are widely spoken, it is still uncertain whether, now well into the post-colonial period, European life styles will prevail or underlying African life styles will return.

The percentages in Table 12.1 show that Africa south of the Sahara is favourably placed with regard to most natural resources. Unfortunately, the cropland is

TABLE 12.1 Africa south of the Sahara: vital statistics

	% of world total		*% of world total*
Population	10.0	Productive land	13.3
Area	18.3	Cropland	6.3
Fossil fuel reserves	3.8	Forest	27.3
Non-fuel mineral reserves	18.2	Natural pasture	33.9

generally of low productivity, although in some places it is very fertile, and yields of cereals and other crops are among the lowest in the world. The usefulness of the forest is restricted, first because the soil on which it grows is of low fertility and, second, because increasing concern over the future of the rainforests of the world is restricting their exploitation. Again, most of the natural pasture yields only a limited amount of fodder, and much of it is located in areas with a very variable rainfall. One-third of the land of Africa south of the Sahara has no bioclimatic use.

The proved reserves of fossil fuels in Africa south of the Sahara are limited in quantity and highly localised in distribution. The region has only about 2 per cent of the world's oil reserves (most in Nigeria, some in Angola and Gabon), about 3 per cent of the natural gas reserves (most again in Nigeria) and about 6 per cent of the world's coal reserves (almost all in South Africa and Zimbabwe). Almost all of the oil produced is exported from Africa, as well as about half of the coal.

12.2 The colonisation of Africa

Africa is considered to be the place of origin of the first humans. In much more recent times Egyptian civilisation flourished in northeast Africa, while all of northern Africa became part of the Roman Empire. Between 900 and 1500 a number of states grew up away from the coast along the southern fringes of the Sahara Desert (e.g. Ghana eighth to eleventh centuries, Mali twelfth to fourteenth centuries). For the most part, however, Africa has been populated by a large number of tribes (see Box 12.1), with informally recognised tribal areas, variously practising cultivation, herding or hunting. Figure 12.2 shows the pre-colonial pattern of tribal areas in part of Africa.

Africa has made little impact on the rest of the world in the last several thousand years, but it has been influenced by forces from outside, particularly by Arabia and Islam since the seventh century, both in the north and along the coastlands of the Indian Ocean. Following the first Portuguese navigators in the last decades of the fifteenth century, European powers increasingly impinged on the coasts, with those of West Africa providing the main interface for European colonial powers with colonies in the Americas to collect slaves for shipment across the Atlantic. Table 12.2 may be referred to for key dates in African history since the sixteenth century. Even if events in history should not be judged by modern standards, the slave trade from the sixteenth to the nineteenth century was devastatingly inhumane in the way it was organised and in the treatment of the slaves during their transport across the Atlantic and in the new 'home' in the Americas (see Box 2.1). One negative consequence has been the socio-ethnic problems in the USA and in some Latin American countries. These seem likely to diminish only very gradually with time. Ironically, the biggest export of African culture from the continent has been through the slave trade.

TABLE 12.2 Calendar for Africa south of the Sahara

1505	Portuguese establish trading posts in East Africa
1571	Portuguese create colony in Angola
1652	Cape Colony founded by Dutch
1659	French establish trading station on Senegal coast
1806	Cape colony passes under British control
1807	Slave trade abolished within British Empire
1822	Liberia founded as colony for freed slaves from the USA
1835	Boer colonists (Dutch origin) 'trek' from Cape to found Natal (1839), Orange Free State (1848) and Transvaal (1849)
1860	French expansion into West Africa from Senegal
1884	Germany acquires Southwest Africa, Togoland, Cameroons
1885	King of Belgium acquires Congo
1886	Germany and Britain partition East Africa
1886	Gold discovered in Transvaal
1890	Beginning of colonisation of Rhodesia
1896	Italian attempt to conquer Ethiopia thwarted
1899	Boer War begins
1910	Formation of Union of South Africa
1914–15	Germany loses African colonies in First World War
1935–6	Italy invades and conquers Abyssinia (now Ethiopia)
1949	Apartheid programme inaugurated in South Africa
1957	Gold Coast (now Ghana) becomes independent
c. 1960	Many colonies of France, the UK and Belgium are granted independence
1965	Southern Rhodesia (now Zimbabwe) declares independence from Britain unilaterally
1967	Civil War in Nigeria, with attempted secession of Biafra region
1975	Portugal gives independence to Mozambique and Angola
1993	Apartheid abolished in South Africa and one-person-one-vote elections held in 1994

BOX 12.1

Tribes in Africa

The word 'tribe' is widely used in both popular and specialist publications, often without being clearly defined. One use of tribe, well known to those of the Judaean and Christian faiths, refers to the twelve divisions of the people of Israel, each claiming descent from one of the twelve sons of Jacob. Since the sixteenth century, 'tribe' has been widely used by Europeans to apply to a primary aggregate of people in a simple condition, often under a headman or chief. It has frequently been used in a derogatory sense.

When the Romans were conquering northwest Europe two millennia ago they were fighting against local tribes. In the sixteenth century, European empires in the Americas found most of the region's inhabitants in tribes, as in the nineteenth century the Europeans also did when they were exploring and conquering Africa. Generally tribal types of social organisation are found in simple societies, in which food and other basic needs are obtained locally, there is no written language, and no explicit constitution. Such a situation still prevailed in the nineteenth century over much of Africa, and also in more remote parts of Latin America and Southeast Asia. The functions and identities of tribes become eroded and superseded by more highly organised political, social and economic systems. Thus for example in South Asia and China, especially where whole river basins have been organised for purposes of irrigation, tribal features tend to disappear.

Conflicts in Africa in the 1980s and 1990s in, for example, Zaire and Rwanda, relate to some extent to the underlying tribal map, on which European colonies were imposed. They show that after a century or more of European colonial rule and subsequent independence, tribal allegiances are still strong. In an impressively comprehensive if not definitive index of tribal names in the whole of Africa, Murdock (1959) includes about 6,000 tribes. His tribal map of Africa can be contrasted with the new division of Africa following the granting of independence to the colonies. With very few exceptions, the colonial territories actually form the new countries. In a highly efficient but ruthless and arbitrary way, usually by agreement, the European colonisers of Africa used rivers, lakes and other physical features or alternatively 'geometrical' boundaries, some following parallels and meridians, to carve up most of the continent. Figure 12.2 shows how in East Africa, as everywhere else in Africa, the colonial boundaries, now the new states, often ignored the tribal areas, cutting through many of them.

One of the most cynical boundary lines is that between the former British colony of Kenya and the former German colony of Tanganyika (now Tanzania). From the coast near Tanga the boundary runs roughly northwest to Lake Victoria. Its original alignment was changed at the request of the Kaiser of Germany so that Mount Kilimanjaro would be included in Tanganyika, the justification, accepted by his cousin Queen Victoria of Britain, being that there was no mountain in any of the three German colonies in Africa.

There are two main problems with the superimposed boundaries in Africa. First, some tribal areas have been cut into two or even occasionally in three (see Figure 12.2 e.g. the Boran, Bararetta and Masai, each with a part in Kenya and each also with a part in a neighbour of Kenya). Second, in every country several, in some countries many, potentially warring tribes have been packaged into modern states, increasingly expected to be run as western-style democracies. The friction between two of the largest tribes in Nigeria, the Yoruba and the Ibo, led to a devastating war, in which the Ibo unsuccessfully attempted to secede. In Rwanda, two groups, the Hutu and the Tutsi, both part of the broader Rwandan tribe, were in conflict in 1994, and as a result, out of a total population of about 8 million, 'officially' some 500,000 people perished during that year in conflict or through deliberate extermination; an unofficial estimate puts the number at 2 million. In South Africa under the policy of apartheid the African tribes in the Union were given nominal independence, while politically, economically and militarily the 'white tribe' held on to most of the best agricultural lands and mineral deposits of global significance. Conflicts between the 'de-tribalised' supporters of the African National Congress (ANC) Party and members of the Zulu tribe continued into 1994, but have been more muted since universal suffrage was introduced in the general election.

Ferguson (1992, pp. 90–1) argues, with specific reference to tribes in North America since the sixteenth century and in the Amazon basin in the present century, that a delicate social balance usually existed between local tribes:

> Accepted wisdom even now holds that 'primitive' cultures are typically at war and that the primary military effect of contact with the West is the suppression of ongoing combat.
>
> In fact, the initial effect of European colonialism has generally been quite the opposite. Contact has invariably transformed war patterns, very frequently intensified war and not uncommonly generated war among groups who previously had lived in peace.

According to Ferguson, anthropologists have identified three main causes of social change that can destabilise the tribal zone (that area in which there is strong contact with outsiders): diseases introduced by settlers, transformation of ecosystems by alien animals and plants, and changes in the way of life made possible by new goods and technologies.

Some of the issues and problems outlined above are widely relevant today because in many parts of the world tribal systems and allegiances are still strong. There is no point, however, in trying to put the clock back. There is no way in which the present sixty-odd countries of Africa could be broken down politically and administratively into 6,000 tribes. Many decades will, however, be needed before the modern states of Africa overcome tribal allegiances and contacts.

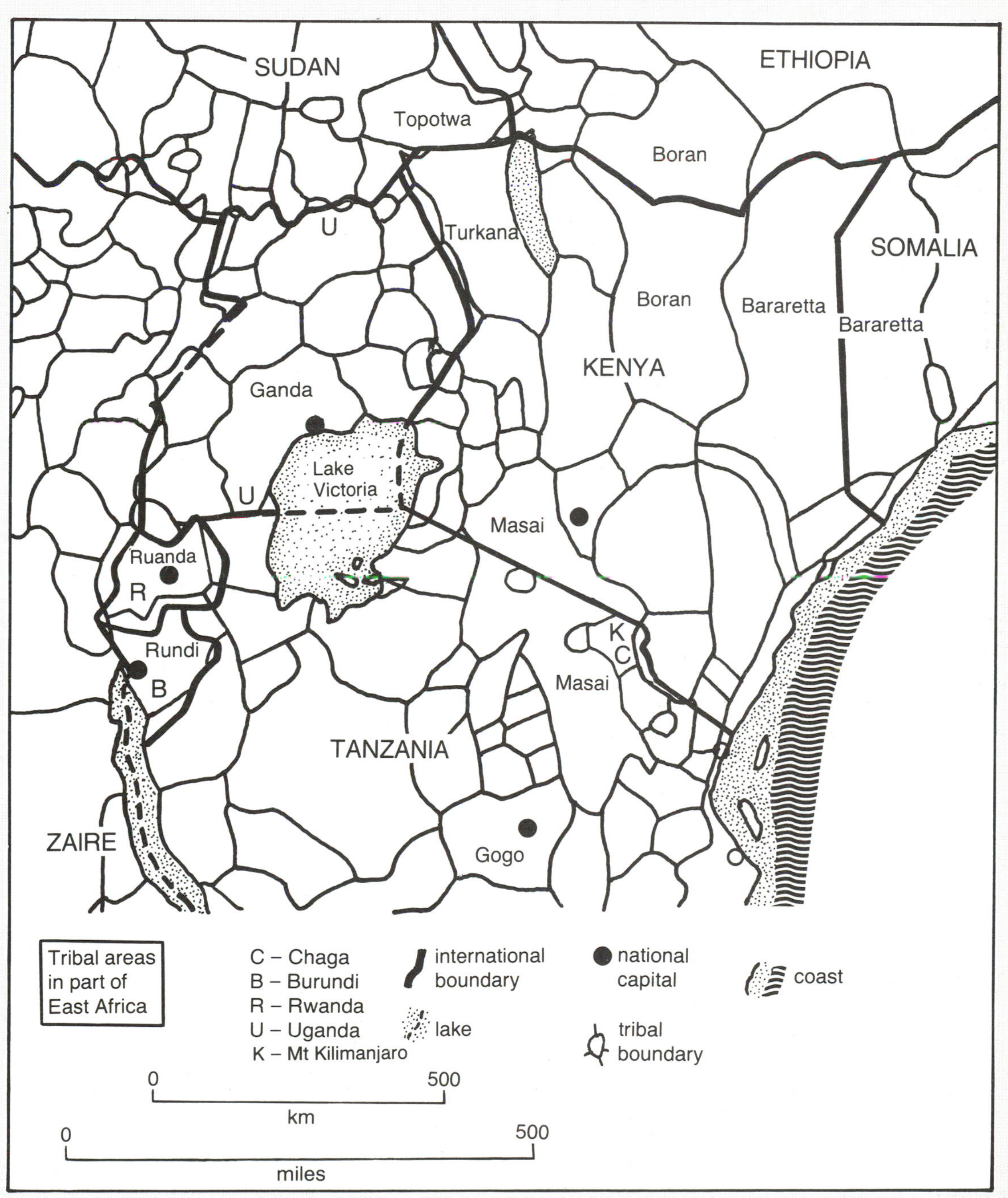

FIGURE 12.2 Tribal areas of eastern Africa before the advent of European colonisation in the 1880s

Source: Murdock (1959)

The colonisation by the Dutch of what is now South Africa, and later the arrival of British, French and Portuguese colonists in many parts of Africa have been the cause of great upheavals. As recently as the years immediately following the Second World War the prospects for moving more European settlers to the tropics was still being debated. Those already there owned much of the best agricultural land locally, and newcomers would expect additional favourable land. For example, in Tanganyika (now Tanzania) the unsuccessful Groundnut Scheme was launched by the UK during the term of office of a Labour Government in the 1940s to provide a primary product for the European market. There was speculation as to how the huge hydroelectric potential of the Congo basin could be harnessed and the electricity transmitted over the enormous distance to Europe. Africa was still regarded as an unequal partner in trade and transactions. The first quotation in Box 12.2 shows the misgivings of a geographer in 1950 about the view of Africa at that time.

Figures 12.1 and 12.3 can be consulted for a comparison of Africa on the eve of its carve-up and in the 1960s, after most countries had just gained independence. The Portuguese were first in and last out, attempting to hold on to their colonies in Angola and Mozambique into the 1970s. Cynically, it could be argued that France, the US and Belgium gave up most of their colonies when they did because by the 1950s they would have had to invest more in them than they were likely to gain from them. The most difficult colonies to dispose of were those in which substantial numbers of Europeans were settled: Algeria (North Africa), Kenya and Southern Rhodesia (now Zimbabwe).

A number of new problems and situations have arisen in Africa since the Second World War.

- Instability in many countries following the withdrawal of colonial rule, with conflicts between and within countries, tribal disputes and ethnic cleansing, together with refugee problems (e.g. Nigeria-Biafra, Ethiopia, Burundi, Zaire, Rwanda).
- A population explosion, partly the result of the

BOX 12.2

Pelzer and Fleming on white man in the tropics

The Americas and Australasia have received almost all the migrants who left Western and Central Europe since the fifteenth century. Generally the Europeans leaving their continent avoided areas that were climatically most difficult for 'white' settlers, whether for diseases that were prevalent there or for an enervating hot/humid climate throughout the year. European settlers in Asia, apart from Russians in Siberia, were few. Africa, on the other hand, tempted European settlers. It was an area in which little resistance was encountered in the face of the sophisticated military hardware and organisation of European colonial powers. In the event, white settlement in Africa was largely restricted to Algeria in the north, Kenya in East Africa and South Africa and Rhodesia in the south. In 'Geography and the tropics', Pelzer (1951, p. 341) questioned the expectations of European powers that Africa *could* be extensively settled by Europeans. He did not question whether or not it *should* be settled.

> Those who are of the opinion that the tropics will, at short notice, prove to be a panacea for the ills of the European parts of the mid-latitudes by supplying foodstuffs and raw materials which are lacking, by absorbing a greater share of industrial production than at present, and by offering settlement outlets for a large number of Europeans, are bound to be disappointed. They do not seem to realize that the settlement of tropical areas is beset with a great many obstacles inherent in the physical and cultural environment. Doubtlessly many parts of the tropics are capable of absorbing an additional population and of supplying greater quantities of food and industrial raw materials; but this will require a great deal of time, the one thing which those who are desperately seeking to bring about a rapid bonification of undeveloped tropical areas are not willing to grant. It remains to be seen, for example, whether it will be possible, at short notice, to turn an African shifting cultivator, who heretofore worked mainly with a hoe, into a permanent cultivator tilling the land with tractor-drawn implements. Would this not be too revolutionary a break with the accustomed economy and cultivation practices? It does not grow out of the African's own needs, but is something wished upon him in order to alleviate the economic distress of a far-distant people. There is danger that African leaders may arouse their followers to opposition if an attempt is made to speed the pace of enforcing more intensive cultivation practices, erosion control measures, pest control, and a reduction in the number of livestock. These reforms are needed, but unless the African peasant is willing to accept these changes and is ready to co-operate, it will be extremely difficult to obtain permanent results.

modest but effective healthcare services introduced this century.

- Old diseases remaining and the appearance of AIDS in the 1980s.

- Food shortages and famines variously attributed to natural causes, especially droughts, to the colonial/post-colonial economy in which (some of) the best land is used to grow crops for export, and to growing population.

- A new colonial era in which the old colonial powers have lost political control but, through various means, including the activities of transnational companies, continue to control economic affairs, albeit from a great distance.

One of the 'new colonial' powers was the USSR after about 1960 when it attempted to influence the Belgian Congo (now Zaire) and, after the withdrawal of Portugal, to dominate Angola and Mozambique, putting pressure on South Africa. Other newcomers are the USA, now a major trading partner of some African countries, and China, which extended its influence in Africa in the 1970s. Finally, the United Nations Organisation has become more actively involved in the distribution of food and with plans to develop the continent. The UN Report *Industrial Growth in Africa* (see Economic Commission for Africa, 1963) produced in the 1960s outlined a continental industrial strategy designed to create an iron and steel industry of considerable capacity in several African countries, various branches of engineering, and light industry. In contrast to Latin America, apart from South Africa, industrial development on modern lines has in effect been of negligible proportions in the region: for example still almost no steel is produced.

12.3 Discussion of a data set for the region

Before a brief account is given of each of the six regions into which Africa south of the Sahara has been divided for the purposes of this book, there follows a commentary on Tables 12.3 and 12.4. The thirty largest countries in population are included in the two tables.

On a more flippant note, in *The Diamond Smugglers*, Fleming (1957, pp. 88–90) describes vividly the problems of preventing diamond thefts and smuggling. The white workers get the same treatment as the black ones, but administered in a more amenable way:

> 'Supposing I was a European worker. What exactly would happen to me when I went on leave from a place like Consolidated Diamond Mines?'
>
> 'You'd be taken by the company bus to the X-ray department and shown into a fine waiting-room with plenty of magazines to read. Your luggage would be put on a conveyor belt that would carry it very slowly into a dark room and under an X-ray scanner with a man sitting above it at some levers to control the speed of the conveyor and stop it.
>
> 'He'd be looking down straight through your suitcase. He'd see the scissors and the zip-fasteners and the cuff-links, and all the other metal things in your bag. He'd recognize every black shape at once.
>
> 'If there was one he didn't recognize, you'd be asked what it was – perhaps asked to show it or open it. All very polite, like a very refined Customs examination.
>
> 'Then you'd be called through into another room – men into one and women into another – and you'd be X-rayed yourself, with particular attention to your head and your stomach and your feet. If the radiographer saw a black speck, in your stomach for instance, he might tell the mine manager and then you'd be put in hospital and very politely but thoroughly purged. On the other hand, the scanner might just make a mark on a diagram of the human body – everybody has one with his file – and wait till you went on leave again. If the speck had altered place, or if there were more specks when he looked through you again, his suspicions would have been confirmed and you'd certainly go into hospital. All very civilized and polite, as I said. But very thorough. The blacks are treated in much the same way, but without the waiting-room and magazines. Often their stomachs are full of black specks. But these generally turn out to be buttons or nails or pebbles they've swallowed just to see if the white man's magic really works or not. Astonishing what they manage to get down without hurting themselves.'
>
> 'In the average mine are there any ways of getting stones out except through the main gate?'
>
> 'It's not easy. Some of these places are like huge concentration camps. Ten feet high electrified double wire fence all round, and dogs and guards patrolling between the fences day and night – the De Beers Alsatians are the stars at the local agricultural shows, by the way. And outside the wire there are miles of flat, empty ground. No good tunnelling. They've tried shooting stones out with catapults, but they're nearly always caught looking for them. And it would have to be a pretty big stone that wouldn't just disappear in the sand.'

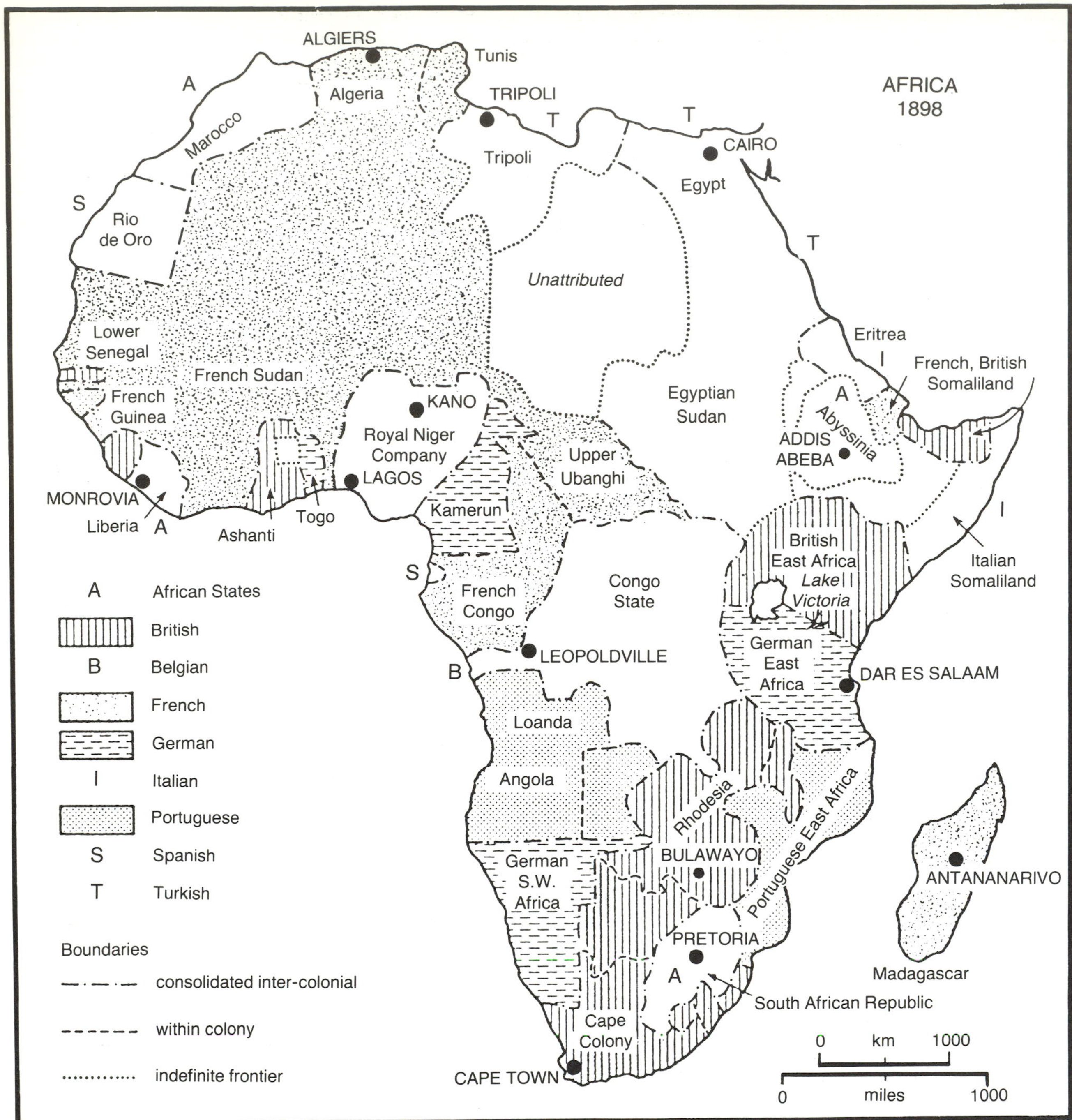

FIGURE 12.3 Africa towards the end of the nineteenth century. Based on the map of Africa in 1898 in J. G. Bartholomew's *The Citizen's Atlas of the World*, London (1898), pp. 83–4. The original spelling of the place-names is used here

The cut-off size is 3 million. Nineteen smaller countries are listed in Table 12.5.

Table 12.3

Columns (1), (2) show the world ranking in population size and the 1993 population in millions. Nigeria has the largest population in Africa, although (3) it is not the largest in area. Given the very low per capita GDP of all the countries except South Africa, the home market of the smaller countries is minute, comparable to that of a US city with about 100,000 people, or to that of Iceland in Western Europe. Variations in the density of population (4) show the contrast between the comparatively high densities in, for example, the coastal zone of West Africa and some highland areas in Central and East Africa (Rwanda, Burundi, Uganda), and the very low densities in the drier regions (e.g.

TABLE 12.3 Demographic features of the thirty African countries with the largest populations

	(1) Rank in world	(2) Popn in millions (1993)	(3) Area in thous. sq km/(miles)	(4) Density per sq km/(mile)	(5) Annual increase %	(6) Expected popn 2025 in millions	(7) % under 15	(8) % over 64	(9) % in employment
Western coastal									
Nigeria	10	95.1	924 (357)	103 (267)	3.1	246	45	3	37
Ghana	53	16.4	239 (92)	69 (179)	3.1	38	45	3	36
Côte d'Ivoire	58	13.4	323 (125)	41 (106)	3.5	38	48	3	38
Cameroon	60	12.8	475 (183)	27 (70)	2.9	33	45	3	36
Senegal	86	7.9	197 (76)	40 (104)	2.7	17	47	3	43
Guinea	92	6.2	246 (95)	25 (65)	2.5	13	44	3	43
Benin	103	5.1	113 (44)	45 (117)	3.0	12	46	3	47
Sierra Leone	108	4.5	72 (28)	63 (163)	2.6	10	44	3	34
Togo	113	4.1	57 (22)	72 (186)	3.6	12	49	2	40
Western interior									
Burkina Faso	72	10.0	274 (106)	36 (93)	3.4	26	48	4	52
Mali	79	8.9	1,240 (479)	7 (18)	3.0	24	46	4	32
Niger	83	8.5	1,267 (489)	8 (21)	3.2	21	49	3	50
Chad	98	5.4	1,284 (496)	4 (10)	2.5	10	41	4	35
Central									
Zaire	26	41.2	2,345 (905)	18 (47)	3.3	105	43	3	37
Angola	76	9.5	1,247 (481)	8 (21)	3.1	22	45	3	40
Rwanda	88	7.4	26 (10)	285 (738)	2.3	17	49	2	48
Burundi	94	5.8	28 (11)	207 (536)	3.2	15	46	4	51
C African Republic	128	3.1	623 (241)	5 (13)	2.6	7	42	3	46
Eastern									
Ethiopia	22	56.7	1,222 (472)	46 (119)	2.8	141	49	4	42
Tanzania	34	27.8	945 (365)	29 (75)	3.1	73	47	3	46
Kenya	35	27.7	580 (224)	48 (124)	3.7	62	49	2	41
Sudan	36	27.4	2,506 (968)	11 (28)	3.1	61	46	2	32
Uganda	47	18.1	236 (91)	77 (199)	3.1	50	49	3	43
Somalia	75	9.5	638 (246)	15 (39)	3.1	23	46	3	39
Southern									
Mozambique	54	15.3	802 (310)	19 (49)	2.7	36	44	3	52
Madagascar	59	13.3	587 (227)	23 (60)	3.3	34	47	3	43
Zimbabwe	63	10.7	391 (151)	27 (70)	3.0	23	45	3	39
Malawi	70	10.0	119 (46)	84 (218)	3.4	26	48	3	40
Zambia	81	8.6	753 (291)	11 (28)	3.1	21	49	2	32
South Africa	28	39.0	1,221 (471)	32 (83)	2.6	70	40	4	37

Sources: WPDS 93 for columns (5)–(8); *FAOPY 91* for column (9)

Chad) and in the tropical rainforest (e.g. much of Zaire).

(5)–(8) show in all countries a characteristic demographic structure, with over 40 per cent of the population under the age of 15, contrasting with 4 per cent or less over 64, and a rate of increase giving a doubling of population in at most three decades. Figure 11.2b compares the population structures of Peru and Ethiopia, illustrating the marked difference in the demographic situation between Latin America and Africa. The large number of females under-15 in the early 1990s will become potential mothers in five to twenty years' time. Although the ratio of dependants (under-15 plus over-64) to working population is roughly 1 to 1, such precise age limits have little meaning in most of Africa in terms of schooling and retirement. Column (9) shows that the percentage of total population in employment varies considerably among the countries. The variations depend partly on national definitions in employment statistics, and partly on the proportion of women working, which tends to be lower in the countries influenced by Islam than in the others. It is difficult to see how material conditions can improve in the future as long as many of the primary products are exported and there is very little industry other than that for basic local needs.

TABLE 12.4 Economic and social aspects of the thirty African countries with the largest populations

	(1)	(2)	(3)	(4)	(5)	(6)	(7)	(8)	(9)	(10)
			% of economically active population in agriculture		*Quality of life*				*Non-primary products as a % of all exports*	
	% urban	*Energy consumption per head kce 1990*	*1965*	*1991*	*life expectancy in years*	*adult literacy %*	*real GDP in dollars per head*	*HDI*	*1970*	*1991*
Western coastal										
Nigeria	16	190	80	64	52	51	1,160	241	2	1
Ghana	32	110	56	49	55	60	1,010	310	1	1
Côte d'Ivoire	40	220	86	55	53	54	1,380	289	6	10
Cameroon	40	250	84	60	54	54	1,700	313	na	na
Senegal	39	150	74	78	48	38	1,210	178	19	22
Guinea	26	90	85	73	44	24	600	52	na	na
Benin	38	50	84	60	47	23	1,030	111	11	30
Sierra Leone	32	80	75	62	42	21	1,060	62	63	33
Togo	29	80	79	69	54	43	750	218	6	9
Western interior										
Burkina Faso	20	30	87	84	48	18	630	78	5	12
Mali	22	20	90	80	45	32	580	81	10	7
Niger	15	60	96	87	46	28	630	78	4	2
Chad	32	20	92	74	47	30	580	88	5	2
Central										
Zaire	40	70	69	65	53	72	380	262	na	na
Angola	28	90	82	69	46	42	1,230	169	na	na
Rwanda	5	30	95	91	50	50	680	186	na	na
Burundi	6	20	95	91	49	50	610	165	2	2
C African Republic	47	40	90	62	50	38	770	159	45	44
Eastern										
Ethiopia	14	30	88	74	46	66	390	173	1	3
Tanzania	21	40	95	80	54	65	560	268	13	11
Kenya	24	120	88	77	60	69	1,020	366	13	20
Sudan	21	70	78	59	51	27	1,040	157	0	1
Uganda	11	30	89	80	52	48	500	192	1	1
Somalia	24	50	89	70	46	24	860	88	na	na
Southern										
Mozambique	27	30	69	81	48	33	1,060	153	na	na
Madagascar	24	40	84	76	55	80	690	325	7	7
Zimbabwe	29	670	73	68	60	67	1,470	397	35	32
Malawi	16	40	81	75	48	47	620	166	4	4
Zambia	49	210	81	69	54	73	770	315	0	1
South Africa	56	2,660	29	13	62	70	4,960	674	na	na

Notes:
na = not available
kce = kilograms of coal equivalent
Explanation of columns (5)–(8):
(5) Average life expectancy in years
(6) Percentage of adults literate
(7) Real GDP in US dollars per head
(8) Combined score of (5)–(7), possible range 0–1,000
Sources: WPDS 93 for column (1); *UNSYB 90/91* for column (2); *FAOPY 91* for columns (3), (4); *HDR 92* p. 127, Table 1 for columns (5)–(8); *WDR 93* for columns (9), (10)

TABLE 12.5 Smaller countries of Africa south of the Sahara (see Tables 12.3 and 12.4 for the thirty largest)

	Population (millions)	*Area in thous. sq km/(miles)*		*GNP in dollars per head*		*Population (millions)*	*Area in thous. sq km/(miles)*		*GNP in dollars per head*
1 Eritrea[1]	3.5	121	(47)	na	11 Gambia	0.9	11	(4)	390
2 Liberia	2.8	111	(43)	na	12 Swaziland	0.8	17	(7)	1,080
3 Congo	2.4	342	(132)	1,030	13 Reunion	0.6	3	(1)	na
4 Mauritania	2.2	1,026	(396)	530	14 Comoros	0.5	2	(1)	510
5 Lesotho	1.9	30	(12)	590	15 Djibouti	0.5	23	(9)	na
6 Namibia	1.6	824	(318)	1,610	16 Cape Verde	0.4	4	(2)	850
7 Botswana	1.4	582	(225)	2,790	17 Equatorial Guinea	0.4	28	(11)	330
8 Mauritius	1.1	2	(1)	2,700	18 Seychelles	0.1	0.3	(0.1)	5,480
9 Gabon	1.1	268	(103)	4,450	19 São Tome e Principe	0.1	1	(0.4)	370
10 Guinea-Bissau	1.0	36	(14)	210					

Notes: na = not available

[1] Included in Ethiopia in Tables 12.3 and 12.4

Sources for area and GNP: UNSYB 90/91, WPDS 1994

Table 12.4

(1) Variations in the level of urbanisation among the countries of Africa south of the Sahara are partly due to differing definitions, but reflect also a contrast between, at one extreme, the intensive near-subsistence agriculture of Rwanda, Burundi and Uganda and at the other the commercial agriculture of West Africa and the extractive industries of Zambia.

(2) The consumption of commercial sources of energy per inhabitant in Africa is far below the world average of some 2,000 kg of coal equivalent per year. Domestic heating (and air-conditioning) use very little fuel, and much cooking is done with fuelwood. The needs of transport and industry are very limited. Some of the energy is actually used in the extraction and processing of fuel and raw materials for export.

(3), (4) Apart from South Africa, well over half of the economically active population in Africa was still in the agricultural sector in the early 1990s in almost every country, with over 90 per cent in two (matched elsewhere in the world only by Nepal in South Asia). South Africa stands out as being more like a developed country in this respect. During the short span of twenty-six years (1965–91) the proportion of economically active population in agriculture in Africa south of the Sahara has diminished in all countries, but the absolute number has risen. In Nigeria, for example, only between 1975 and 1991 it grew from 19 to 27 million, without a parallel increase in cultivated area. The population increase is therefore resulting in both a higher density of population in existing agricultural areas and a big increase in urban populations. In the whole of Africa, including North Africa, between 1975 and 1991 employment in agriculture dropped from 72 per cent of the economically active population to 63 per cent, but the number working rose from 119 million to 156 million. In anticipation of Chapter 17, it will be shown that if such a trend continues there could be more than 250 million people on the land in Africa around 2025.

(5)–(8) show how the thirty largest African countries (apart from those in North Africa) score on the United Nations Human Development Index (see Chapters 1 and 3 for details). In most countries, life expectancy (5) is well below the average for developing countries, although it has been rising gradually in spite of famines and conflicts. Adult literacy varies greatly, reflecting both the dubious nature of relevant statistics and the greater efforts of some governments than of others to provide educational facilities. Literacy among women tends to be considerably lower than among men in countries with a large Muslim element. Real GDP per inhabitant is very low, but considerably higher than GDP when converted into US dollars at the going exchange rate (cf. Nigeria 1,160 (real) against 290, Madagascar 690 against 210). On the composite Human Development Index (HDI), out of 160 countries assessed world-wide, in the whole of Africa only South Africa (rank 70) and Libya (rank 74) came in the top half. In the third quartile were fifteen African countries and in the bottom quartile twenty-nine, the latter sharing this part of the ranking with ten Asian countries and with only Haiti in Latin America. Most of the 'poorest' countries, in greatest need of development assistance, are in Africa (see Figure 4.2).

Although data for much of its foreign trade have been kept secret in South Africa because of economic sanctions, and accurate data are not available for several other African countries, columns (9) and (10) in Table 12.4 provide enough evidence to show that non-primary merchandise accounts for virtually none of the

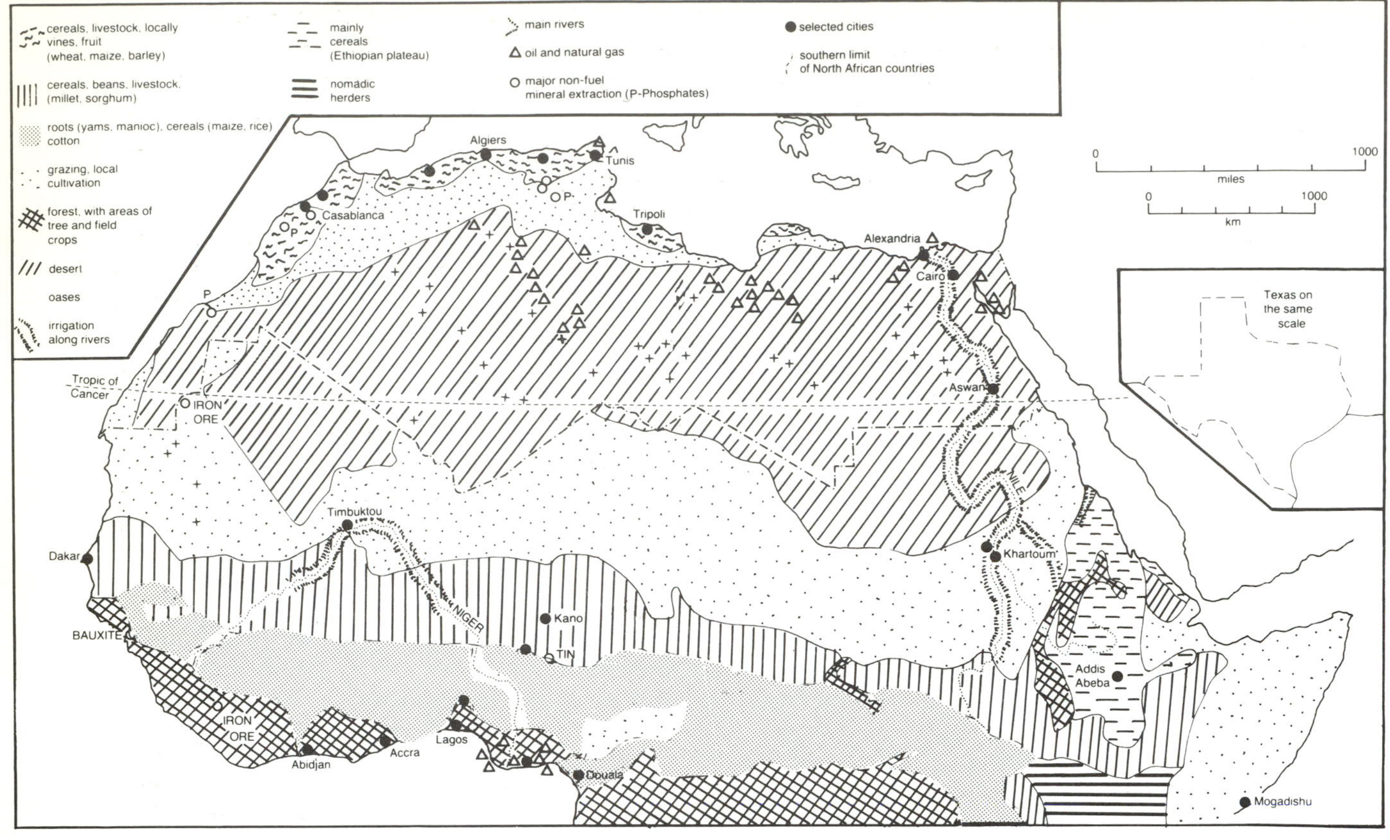

FIGURE 12.4 Economic map of the northern part of Africa, relevant also to Chapter 13

value of exports from most countries. In some cases (e.g. Sierra Leone, Central African Republic) the apparent importance of non-primary products may be due to the definition of processing as 'industrial'. There is no sign of the surge in non-primary exports between 1970 and 1991 noted in some Latin American countries (see Table 11.3) and in most countries of South, Southeast and East Asia (see Table 15.3). Figure 12.4 should be referred to in connection with sections 12.4, 12.5 and 12.6.

12.4 Western coastal countries

Nine of the countries in this region are included in Tables 12.3 and 12.4. To these can be added the following smaller countries: Cape Verde Islands, in the Atlantic off Senegal, Gambia, Guinea-Bissau, Liberia, Equatorial Guinea and São Tomé e Principe. Apart from the amalgamation of the British and French protectorates of the Cameroons, some minor territorial adjustments and a good number of name changes, the present forms of the countries reflect the colonial situation. The proliferation of states in West Africa results from the presence over several centuries of a large number of trading posts and footholds of Portugal, France and Britain, many of which formed bases for penetration of the interior in the nineteenth century. Liberia was created as a homeland for slaves from the USA and was not colonised by European powers.

After the decline of the slave trade, interest in West Africa focused on the commercial cultivation of tree crops, especially the oil palm (Nigeria), cacao (Ghana, Côte d'Ivoire), coffee, and later rubber (originally from the Amazon). Most of the existing tropical rainforest, distributed here along a coastal zone of varying width, has now been cleared for cultivation, partly for plantations, partly for food crops for the local population. Rainfall diminishes with distance from the coast, and forest gives way to tall grass savanna, generally supporting a lower density of population. The whole region has about 175 million inhabitants, over half of them in Nigeria. So far, food supply has been less of a problem here than in many other parts of Africa, but pressure on the existing bioclimatic resources continues to grow. Soon the only remaining forest will be in specially protected areas (e.g. in Côte d'Ivoire, Cameroon) and the export of timber and timber products will be very limited.

Apart from the oil reserves of Nigeria, and bauxite reserves in several countries, West Africa is poorly endowed with mineral resources. Nigeria has less than

2 per cent of the world's proved oil reserves, but in 1992 accounted for almost 3 per cent of world production; only about one-fifth is consumed in Nigeria itself. Without new discoveries of oil, reserves would last less than thirty years at the 1992 rate of extraction. Nigeria also has 2.5 per cent of the world's natural gas reserves, but so far little use has been made of these. In Venezuela, and in several countries of Southwest Asia, the export of oil over some decades has brought in revenues that have been used to develop infrastructure and to start up alternative activities. In relation to its population size, the oil and gas industry of Nigeria is small, and, even if some benefit is gained from exporting most of the production, only a small part of the country has been much affected.

Guinea is credited with 27 per cent of the world's bauxite, while Cameroon, Ghana and Sierra Leone together have another 6 per cent. Guinea is one of the world's leading exporters of alumina and bauxite, while the hydroelectric power from stations on the Volta River in Ghana is used to smelt a local bauxite. A variety of other minerals are produced, also mostly for export. West Africa does not therefore benefit fully from its mineral production, receiving export revenues but not having the greater financial advantage of adding value through manufacturing products using the materials.

The proportion of total cultivated land under permanent crops in West Africa varies greatly from country to country and is related partly to the amount of former rainforest in each. In Nigeria less than 10 per cent of the cultivated land is under permanent crops because much of the agricultural land is in the drier northern region where such crops do not grow. In Cameroon, there is 15 per cent under permanent crops, in Benin and Sierra Leone a quarter while in Côte d'Ivoire the percentage is 34, in Ghana 58 and in Liberia 66. Between 1975 and 1990 there was little overall change in the cultivated area in the seven countries referred to, and in Ghana there was actually a reduction. Oil palm, cacao and coffee exports are declining relative to exports of these products from other parts of the tropical world.

Cassava, millet, sorghum and maize are among the chief sources of food in West Africa. The data in Table 12.6 show yields for sorghum and maize in selected countries, including some of the countries in the coastal zone of West Africa as well as countries to the north in the drier grassland, semi-desert and desert areas. Sorghum is widely grown in drier areas of Africa and South Asia and also in North and South America. The average yield in Africa is well below the world average, about one-third that obtained in Argentina and one-quarter that in the USA. Yields vary among African countries themselves, depending on climatic and soil conditions and the quantity and quality of inputs. They also fluctuate from year to year, sometimes dropping in one year, or even a sequence of several years, to unacceptably low levels.

Maize yields in Africa are about half the world average and a quarter the US average. South Africa is distinguished in the continent by its considerably higher than average yields. It would seem reasonable to assume that yields of cereals and other food crops such as cassava could be increased in West Africa and elsewhere on the continent. Improvements are likely to come gradually, with attention to water supply in drier areas and greater use of fertilisers.

Under the conditions in West Africa described above, the comparatively high level of urbanisation and the rapid growth of some of the larger cities must be regarded with concern. With few industries and

TABLE 12.6 Sorghum and maize yields in kg per hectare (and kg per acre) in selected countries

	Sorghum				*Maize*			
	1979–81		*1989–91*		*1979–81*		*1989–91*	
World	1,450	(590)	1,340	(540)	3,340	(1,350)	3,700	(1,500)
Africa	900	(360)	780	(320)	1,560	(630)	1,650	(670)
Nigeria	1,260	(510)	1,020	(410)	1,350	(550)	1,260	(510)
Niger[1]	430	(170)	270	(110)	710	(290)	830	(340)
Chad	540	(220)	640	(260)	840	(340)	970	(390)
Ethiopia	1,370	(550)	1,080	(440)	1,630	(660)	1,670	(680)
Sudan[1]	730	(300)	490	(200)	580	(230)	1,110	(450)
South Africa	1,420	(570)	1,050	(430)	2,320	(940)	2,820	(1,140)
Argentina	2,930	(1,190)	2,850	(1,150)	3,160	(1,280)	3,260	(1,320)
USA	3,620	(1,470)	3,710	(1,500)	6,470	(2,620)	7,090	(2,870)

Note: [1] Total quantity grown is small
Source: *FAOPY 1991*, vol. 45, Table 24

inflated service sectors, the large population sizes of well over a million inhabitants in Ibadan and Lagos (Nigeria), Douala (Cameroon) and Dakar (Senegal) and even more in Accra (Ghana) and Abidjan (Ivory Coast), resemble the situation in Latin American countries half a century ago.

12.5 Western interior and Central Africa

The total population of the Western interior countries, Burkina Faso, Mali, Niger, Chad (see Table 12.3) and Mauritania (not in the table) was a modest 35 million in 1993, but these five countries extend over 5.1 million sq km (1.97 million sq miles), one-sixth of the total area of Africa, and about half the area of the USA. Only Mauritania has a coast. The other four countries are linked by road southwards to the coast of West Africa, but the links northwards to the Mediterranean coast are long and tenuous. The whole region was part of the French Empire, but it contained little that could be exported. Apart from the iron ore exported by Mauritania, about half of the value of its exports, minerals are of negligible importance at present, although the sheer size of the five countries should yield some after a more thorough exploration.

Of the total land area of the five countries of 5.1 million sq km (1.97 million sq miles), 3.3 million (1.27 million) is classed as of no use for bioclimatic resources, only the southern fringes of Upper Volta and Chad have forests, and almost all the rest is poor natural pasture, little more than 2 per cent (130,000 sq km – 50,000 sq miles) being cultivated. To make matters worse the yields of food crops in the region are among the lowest anywhere in the world.

The region of Central Africa consists of the former Belgian colonies of the Congo (now Zaire) and neighbouring Rwanda and Burundi, the former Portuguese colony of Angola, and the southern part of former French Equatorial Africa, namely the Central African Republic, Congo (capital Brazzaville) and Gabon. It has a population of about 70 million in an area of 4.9 million sq km (1.9 million sq miles). The outstanding natural feature of the region is the large extent of forest remaining, over 3 million sq km (1.16 million sq. miles), over 60 per cent of the total area of Central Africa, and 45 per cent of all the forest in the whole of Africa. Only 3.3 per cent of the total area is, however, cultivated (160,000 sq km – 62,000 sq miles), about 10 per cent of it under permanent crops. Agriculture is largely dedicated to growing food for the population of the region. The cultivated land is mainly concentrated in a few areas of good soil and high population density, but yields are low by world standards. There is scope for clearing some of the forest to extend the agricultural area but, as in Brazil, there is increasing pressure to proceed slowly and prudently because timber and other forest products are useful both for local needs and for export.

With regard to minerals, Zaire has 15 per cent of the world's diamond reserves and is a major exporter. It is also famous for the copper reserves in the southern part of the country (formerly Katanga), but it only has 7.6 per cent of the world's remaining proved copper reserves, and it produces only about 5 per cent of the world's copper. Gabon has a wide variety of other minerals, including iron ore, uranium and, together with Angola, modest oil reserves (about 100 million tonnes).

When independence was granted to their Central African colonies by France and Belgium, Zaire (then the Belgian Congo) became the scene of conflict between western interests, represented by Mobuto, and Soviet interests represented by Lumumba, the former winning after a lengthy period of unrest and conflict. Regional and tribal rivalries and disputes still disrupt life in Zaire, as well as in Angola, where the reluctance of the Portuguese to withdraw in the 1960s resulted in a long conflict, followed by a virtual Soviet takeover and then a conflict between Soviet and South African interests.

Central Africa remains heavily dependent on agriculture (over 90 per cent of employment in Rwanda and Burundi), having few industries beyond those processing raw materials or providing very basic local needs. Kinshasa is by far the largest city in Central Africa with a population of some 4 million, while Kananga in central Zaire and Luanda, capital and chief port of Angola, each have over 1 million.

12.6 Eastern Africa

This region consists of the six countries included in Tables 12.3 and 12.4, together with Djibouti. Eritrea has now formally seceded from Ethiopia. During the carve-up of Africa in the 1880s, Tanganyika (now, with Zanzibar, renamed Tanzania) was occupied by Germany, and most of the present Somalia, together with Eritrea, by Italy. Sudan, Kenya, Uganda and (British) Somaliland were British territories while, after an abortive Italian attempt to conquer it, Abyssinia was left alone until Italy occupied it in 1935–6. After the defeat of Germany in the First World War in 1918, Tanganyika became a British protectorate and during the Second World War, Abyssinia was taken by Britain from Italy. Eventually almost all of Eastern Africa came under British rule or protection. An attempt was made to strengthen the main area of British interests, Kenya, Uganda and Tanganyika, into the Federation of East Africa, but when the countries became independent this dissolved. The seven countries in the region have 168 million inhabitants and a combined area of

6,150,000 sq km (2,375,000 sq miles).

More than any of the regions of Africa discussed in this chapter, Eastern Africa has had serious troubles resulting from both ethnic conflict and natural disasters. The Sudan is divided both physically and culturally into a northern dry and desert region with strong Arab and Islamic influences and a southern largely forested, more 'African' and 'pagan' part. Conflict has been almost continuous since the country became independent. After the death of Emperor Haile Selassie in 1974, Ethiopia switched from being 'protected' by Western Europe and the USA to being within the Soviet sphere of influence. Internal conflicts, forced resettlement of agricultural populations, and serious droughts in the northern part of the country, have left it even poorer and weaker than it was two decades ago. Some aspects of Ethiopia's recent plight are discussed in Box 12.3. Kenya's troubles were earlier, when Britain attempted to support the British settlers against hostile tribes. Uganda suffered internal strife for over a decade, while Somalia has been the scene of both internal conflicts and a war with Ethiopia. Only Tanzania has remained largely at peace, but Julius Nyerere's attempted socialist transformation of rural Tanzania has created new problems while trying to solve old ones. The situation in Tanzania since independence is discussed in Box 12.4.

Eastern Africa is one of the poorest regions in the world, making the extensive disorganisation and damage through its numerous disputes particularly unfortunate in view of the urgent need to combat poverty and cope with periodic droughts. The region has no economic minerals of consequence. Only 400,000 sq km (154,000 sq miles) are cultivated, 7 per cent of the total area, while 21 per cent is forested and 44 per cent defined as natural pasture, much of it too dry, however, to support much livestock. Ethiopia has one of the best large areas of good soil in Africa on its central plateau, formed on trappean lava. Elsewhere there are small areas of fertile land, including the shores of Lake Victoria in Uganda and Tanzania, and irrigated river valleys of the Nile and its tributaries in Sudan.

A wide range of tropical and subtropical crops is grown in Eastern Africa for export, often occupying the best land. Coffee and tea are grown in Kenya and Tanzania, cotton and sisal in all the countries. None of these products has been in great demand, a situation that led President Nyerere to explain in a down-to-earth way to his citizens and to the world why it is difficult to be in the Third World: in the mid-1960s 17 tonnes of sisal would buy one tractor but, ten years later, 24 tonnes were needed. Since all the seven countries in Eastern Africa depend on imported oil or oil products, the price rises in the 1970s were a serious setback for very poor countries, arguably indeed more than for the rich industrial countries.

12.7 Southern Africa

In Tables 12.3 and 12.4 the Republic of South Africa has been listed separately from the other countries in deference to its very different attributes, both economic and ethnic. In addition to the six countries listed in the tables, the islands of Mauritius, Seychelles, Reunion and the Comoros form part of Southern Africa, together with Lesotho, Swaziland, Namibia (formerly Southwest Africa) and Botswana (formerly Bechuanaland). In 1993 the whole region had a total population of 105 million, expected to reach 225 million in 2025. The total area is 5.3 million sq km (2.05 million sq miles), of which about 6 per cent (325,000 sq km – 125,000 sq miles) is cultivated and 22 per cent is forested (1,150,000 sq km – 440,000 sq miles). Only in South Africa (mainly fruit) and Madagascar, do permanent crops figure prominently in agriculture. As elsewhere in Africa, forests are still being cleared for construction timber and fuelwood. The now threatened natural vegetation and fauna of Madagascar are of particular interest ecologically, because of the wide range of species peculiar to that island.

In the world economy Southern Africa is known more as a source of mineral exports than of agricultural products. Zambia has about 7 per cent of the world's copper reserves while Botswana has 13 per cent of the world's natural industrial diamonds. South Africa's prominent global position in mineral reserves is shown in Table 12.7.

Like the Sahel countries and Eastern Africa, Southern Africa has also had its share of dry conditions. World maize yields in the 1980s have ranged between 3,000 and 3,800 kg per inhabitant. During 1980–91 inclusive they averaged 6,740 in the USA. The African average was 1,430, little over 20 per cent of the US average. South African yields were considerably above the average African level, those in Namibia, Botswana and Lesotho well below it. Maize yields in Southern Africa are very low by world standards on account of unfavourable soil conditions, very limited inputs of fertiliser, poor-quality seed, and in some areas disruptions caused by frequent conflicts. They vary greatly from year to year. Sorghum and maize are still widely grown because there is nothing more suitable to grow

TABLE 12.7 The mineral reserves of South Africa: percentage of the world total

Chromite	78	Industrial diamonds	7
Gold	62	Zinc ore	7
Manganese ore	42	Coal	7
Phosphates	19	Iron ore	4
Uranium	13		

Source: US Bureau of Mines (1985)

BOX 12.3

Ethiopia's unending food supply problems

With a population of about 55 million in 1994 Ethiopia is the country with the third largest population in Africa after Nigeria and Egypt. In area 1,250,000 sq km (483,000 sq miles), it is almost twice as large as Texas and five times the area of the UK. It is unusual in Africa south of the Sahara in having been held as a colonial power for only a few years following the Italian invasion in 1935–6.

No doubt hunger and famine have struck the cultivators of Ethiopia over many thousand of years but only in the last three decades has much publicity been given to their plight. Ethiopia is still predominantly an agricultural country, with 73 per cent of its economically active population in agriculture and just under half of GDP accounted for by this sector. A number of features of Ethiopia are listed below. The reader is invited to propose ways in which the country can escape from its conspicuous poverty trap.

PLATE 12.1
Cultivators and herders

a Trappean lava plateau of central Ethiopia: patchwork of ploughed fields (dark) and land with crops, but with considerable signs of gullying. The removal of tree cover and increasing pressure from population growth threaten the stability of the soil and its productivity

b Eastern lowlands of Ethiopia: nomads herd cattle, moving from one pasture to another. Such land is classified as natural pasture but in reality usually supports only a very low density of livestock

PLATE 12.2 Agriculture

a Subsistence agriculture: a stick used to make holes in the soil to plant seeds, Ethiopia

1. The population of Ethiopia is estimated to be growing by over 3 per cent per year. In spite of the relative decline of employment in agriculture, the number of farmers is actually growing.

2. The area under arable and permanent crops in Ethiopia is about 11 per cent of the total national area. The extent of the cultivated area has hardly changed in recent decades. In the rest of Ethiopia conditions are not suitable for a quick expansion of cultivation.

3. The plateau of central Ethiopia has good soil and an average annual rainfall that is adequate for the cultivation of cereals and other food crops. The rainfall is, however, very unreliable. Beyond these natural resources, Ethiopia has virtually no fossil fuel or non-fuel mineral resources. Its manufacturing sector accounts for less than 10 per cent of GDP and produces for local and national needs.

b Commercial agriculture: sisal fibre being processed. Ethiopia exports sisal, but the commodity is not in great demand in the developed countries

BOX 12.3 *continued*

PLATE 12.3
Church hollowed out of a deposit of sandstone, Lalibela, Ethiopia

The background to this and similar structures dates back many centuries to a time when visitors to the Holy Land from this part of Africa returned with the idea of recreating scenes they encountered

4. To continue feeding its population Ethiopia depends primarily on its own resources but also increasingly since the early 1970s on foreign emergency assistance.

5. The standards of farming practices are very low indeed. Chemical fertilisers and pesticides are hardly known. Many agricultural settlements are not accessible by motor vehicles. Very little use is made of tractors and other agricultural machinery in Africa, except in South Africa and the northern tier of countries. Ethiopia has about 4,000 tractors in use, giving an average of 4,000 farmers per tractor. Facilities for storing agricultural produce are very restricted. Important sources of food include a grain crop, tef, and a tree providing edible roots, the ensete. These crops are not grown outside this part of Africa and have not been the subjects of agricultural research. Not surprisingly, therefore, agricultural yields are among the lowest in the world. Maize (corn) and sorghum are important cereals in both Ethiopia and the USA. Yields in the two countries are contrasted in Table 12.8.

TABLE 12.8 Maize and sorghum yields in Ethiopia and the USA compared (see Box 12.3)

	Area in thous. sq km/(miles)		*Yield kg per ha/(acre)*		*Production in millions of tonnes*
Maize					
Ethiopia					
Average 1979–81	750	(290)	1,630	(660)	1.2
Average 1990–92	1,000	(390)	1,610	(650)	1.6
USA					
Average 1979–81	29,660	(11,450)	6,470	(2,620)	192.1
Average 1990–92	28,050	(10,830)	7,500	(3,040)	210.7
Sorghum					
Ethiopia					
Average 1979–81	1,050	(410)	1,370	(550)	1.4
Average 1990–92	810	(310)	1,100	(450)	0.9
USA					
Average 1979–81	5,270	(2,030)	3,620	(1,470)	19.2
Average 1990–92	4,220	(1,630)	4,060	(1,640)	17.3

Source: *FAOPY* (1992)

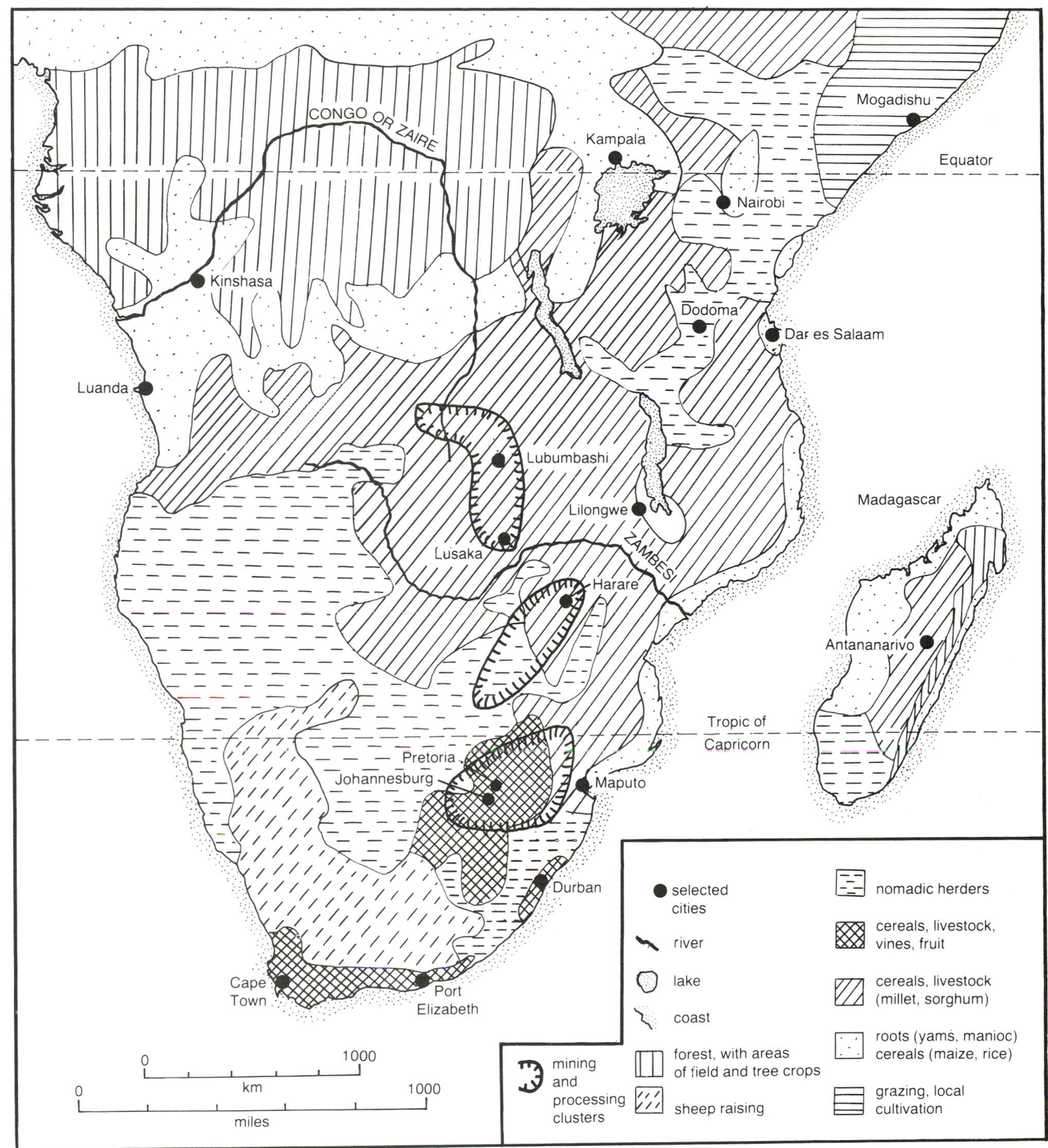

FIGURE 12.5 Economic map of the southern part of Africa

in many areas and because they are the crops that the regions are least bad at growing.

Madagascar is the only former French colony in this part of Africa apart from the island of Reunion, which is still an overseas territory of France and therefore of the European Union. Madagascar has a great variety of environmental conditions and produces most tropical and subtropical crops, but the value of its foreign trade is small, with coffee and spices the main exports and France still the main trading partner. FAO data for the last twenty years show little change in the area of arable and permanent cropland or pasture but a gradual reduction in the area of forest.

The transformation to post-colonial status has been less smooth in Mozambique than in Madagascar. After Portugal finally gave up its colony, a conflict began

BOX 12.4

Rural developments in Tanzania

Like almost every country in Africa, Tanzania consists of a piece of territory delimited during the nineteenth century through agreements between European powers. Tanganyika went to Germany. Railways were built, natural resources were duly exploited, and a small number of German settlers migrated to the country. After the defeat of Germany in the First World War, Tanganyika came under British protection. After the Second World War, in the interests of political 'economies of scale' an attempt was made to merge Tanganyika, Kenya and Uganda into a federation. In due course, however, Tanganyika became independent, and, with the island of Zanzibar coerced into a merger, became modern Tanzania. Thus the former colonial territory, including about a hundred different tribes and a great diversity of physical conditions, continued in existence. Its dynamic leader, Julius Nyerere, concentrated on improving the agricultural sector, a reasonable policy because in 1965 about 95 per cent of the population was in that sector. The ideology of the political viewpoint was socialist, considerably influenced by the misguided assumption that rural developments in China in the 1960s, where large communes had been introduced, had been a great success.

Emphasis in Tanzania was on providing basic needs such as water, food, homes, medical and educational services. It was considered that such developments could best be provided if the often widely scattered rural population was resettled in nucleated villages, a procedure termed *ujamaa*. The emphasis was on co-operation rather than on competition and individualism. Western high technology was largely ignored although the use of radio broadcasts to advise people on changes and techniques was essential for the programme. Plate 12.4 shows two of the homely illustrations included in widely circulated simple 'manuals' in Swahili.

In the 1980s enthusiasm for rural developments on the lines described faltered. Reasons given for failure in Tanzania to make much improvement in rural conditions, which were very poor in most of the country, included droughts, a conflict with neighbouring Uganda, the deteriorating position of primary products exported in relation to both manufactures and oil, and corruption among civil servants. Like most countries in Africa south of the Sahara, Tanzania faces an uncertain future. Its population in 1965 was about 12 million. This had risen to 30 million in 1994 and, with a rate of natural increase of 3.4 per cent a year and a total fertility rate of 6.3, its population could exceed 70 million as early as 2025.

It is difficult to assess the successes and failures in agriculture in Tanzania because data for the cultivated

PLATE 12.4 Tanzania

a The need to provide better housing

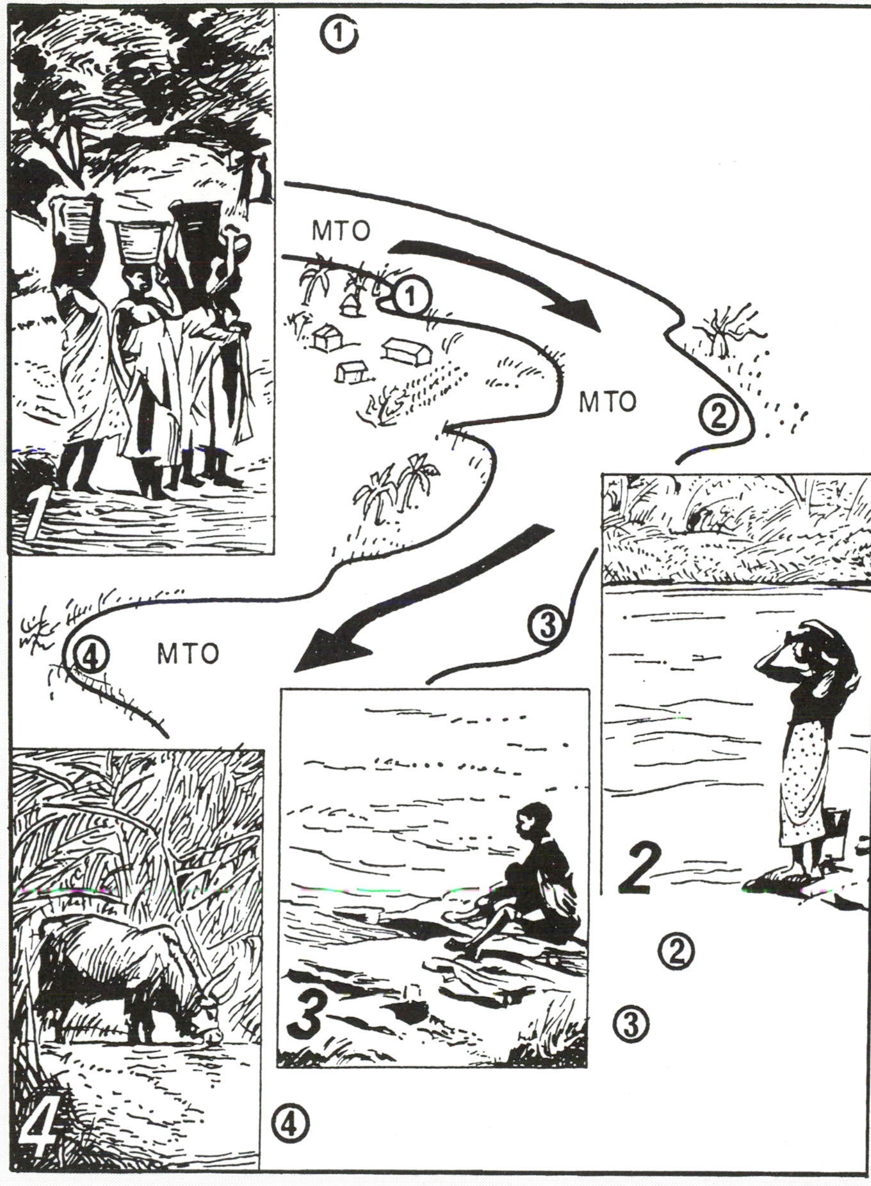

b Instructions in a Tanzanian 'manual' on health improvements *Mtu ni Afya* (published in Swahili in Dar-es-Salaam) on the use of water from a stream in a rural area:

1 Drinking water
2 Washing oneself
3 Washing clothes
4 Watering the livestock (but the next settlement downstream could be quite close to 4)

area are inconsistent. One reason is apparently the changing definitions of 'cultivated' since independence, depending on whether or not areas cultivated only at intervals as well as continuously are counted in the total. A cultivated area of 10.7 million hectares (26.4 million acres) (11 per cent of the national area), is given for 1966 in *FAOPY 1970*, contrasting with a revised 4.2 million (10.4 million acres) for that year in *FAOPY 1976*. An area of 6.1 million hectares (15.1 million acres) is given for 1975 but it drops to 3.3 (8.2) for 1976 in *FAOPY 1992* and 3.4 (8.4) in 1991. Such wildly different figures make an assessment of the present situation and future planning very difficult. In the early 1990s there were about eight people per hectare of cultivated land which, notwithstanding a large area classed as pasture, is hardly enough to support 30 million people, given the universally low yields, generally poor soils and mostly problematical climatic conditions. The reorganisational card has been played. What else is there to try?

between Soviet and South African interests. The capital Maputo is in the extreme south of the country close to South Africa and is an outlet to the coast for the northern part of South Africa. Zimbabwe, Zambia and Malawi also have links to ports in Mozambique further north. Like Madagascar it has few minerals. Since the mid-1980s it has been afflicted by droughts, conflicts and food shortages.

The rest of Southern Africa has been strongly influenced by Britain. The attempt to form a Federation of Central Africa with Southern Rhodesia (now Zimbabwe), Northern Rhodesia (now Zambia) and Nyasaland (now Malawi), to create a larger economic grouping, was abandoned, and Britain was left attempting to control Southern Rhodesia, where a powerful white minority ran the country for a time. Material conditions in the three countries have not improved in general because population has grown fast, with little increase in the area under cultivation and yields, higher in Zimbabwe than in its neighbouring countries except for South Africa, fluctuating and declining in the early 1990s.

The Republic of South Africa is the most southerly country in the continent of Africa. It is almost all situated south of the Tropic of Capricorn. It is 1,221,000 sq km (471,000 sq miles) in area, and in 1993 had a population of 39 million. Four smaller countries, Swaziland, Lesotho, Botswana (formerly Bechuanaland) and Namibia (formerly South West Africa) adjoin South Africa and are closely associated with it. In area they cover 1,471,000 sq km (558,000 sq miles), but their combined population is only 6 million.

Most of South Africa is over 1,000 metres (3,282 feet) above sea level, a plateau bounded in the southeast by the highest mountains of the country. Annual rainfall exceeds 50 cm (20 inches) only in the eastern part of the country, while there is a considerable area in the northwest which receives less than 25 cm (10 inches) and is semi-desert or desert. Most of the 11 per cent of the total area under arable or permanent crops is along the south and southeast coastlands and in the northern interior. Only a small percentage of the country is forested, about two-thirds is classified as natural pasture, and about one-fifth has no bioclimatic use.

Although South Africa is largely self-sufficient in food and is an exporter of some agricultural products, the contribution of the sector to total GDP is small (5 per cent in 1991) and declining (8 per cent in 1970). In terms of employment, the percentage in agriculture has declined sharply from 31 in 1965 to 25 in 1975 and 13 in 1991, the latter contrasting with 63 per cent in Africa as a whole.

Like Australia and Canada, South Africa is exceptionally generously endowed with mineral reserves in

PLATE 12.5 South Africa themes portrayed by Herbert Cole in the magazine *Fun* in 1901

a A confused Zulu warrior carrying his spear (then as now) ponders over the external powers of church and business. Caption of original: 'The bishop and the colonist'

b Conflict between the 'white tribes' (then as now) – the British lion confronting the Boer over the territorial cake

BOX 12.5

An ongoing tale of woe? Prospects for Africa south of the Sahara, a subject for speculation

In May 1993 the former Italian colony of Eritrea finally won independence from Ethiopia and recognition as the newest member of the Organisation of African Unity. According to Kiley (1993), President Afeworki's presentation included a criticism of the members that they belonged to a 'nominal organisation' that had failed to fulfil its objectives and meet its commitments: 'The African continent is today a marginalised actor in global politics and the world economic order.' He did not, as had been customary, blame the former colonialists or the Cold War for Africa's ills but laid most of the blame for Africa's problems at the feet of Africans.

Dowden (1994) produced a checklist of the state of affairs in African countries and for each country added a line or two on prospects. His remarks are given below for all countries apart from a few of the smallest. The occurrence of recent or current food emergencies (F) and civil wars (W) according to Kiley (1993) are added:

Angola (F, W): more war, hunger and misery

Bénin: the pleasantest and most stable place in West Africa

Botswana (F): good if South Africa stays stable

Burundi: misery

Cameroon: only African patience prevents disintegration into war

Chad: oil deposits recently discovered in south may keep this country at war for some time

Congo: continuing instability

Côte d'Ivoire: time for a power struggle

Eritrea (F, W): excellent, but could do with some aid

Ethiopia (F, W): interesting; good if weather kind

Gabon: France has too much to lose to allow much democracy

Ghana: western donors will pay highly to keep their model country on the road

Guinea: could be worse

Kenya (F): sliding on down

Liberia (F, W): might work, but chances are it might return to ugly civil war

Madagascar (F): years of peace and oppression could end in civil war

Mali: rain, and peace with Tuareg would help

Mozambique (F, W): huge potential and, coming from zero, can only improve

Namibia (F): still depends too much on the elements and South Africa

Niger: slow but orderly progress

Nigeria: huge if it gets through the economic pain barrier and its rulers stop stealing

Rwanda (F, W): difficult to envisage long-term peace

Senegal: continuing instability and stagnation

Sierra Leone (F, W): could enjoy peace and relative prosperity outside the UN-ruled areas

South Africa: could go over the edge suddenly, but it has a long way to fall

Sudan (F, W): no sign of a coherent opposition yet. War in the south continues

Tanzania (F): on the right track, but moving too slowly

Uganda: could prosper with stability and investment

Zaire (F, W): a faint chance if Mobutu goes

Zambia (F): debt and despondency

Zimbabwe (F): much depends on South Africa

relation to the size of its population, 0.7 per cent of the world total. Table 12.7 (p. 285) shows South Africa's share of a selection of minerals of which it has a prominent position in the world's reserves. It lacks oil and natural gas reserves but has large reserves of high-grade coal. While sanctions were imposed, it was able to produce petroleum products from coal, although at a high cost. The mining industry of South Africa accounts for 15–20 per cent of GDP and for most of the value of the country's exports. It gives direct employment to about 750,000, gold-mining alone employing about 400,000. About 350,000 foreign workers are

TABLE 12.9 The racial composition of the population of South Africa in the twentieth century, in millions

	Total	*White*	*Non-White*	*% White*
1904	5.2	1.1	4.1	21
1951	12.7	2.7	10.0	20
1984	26.7	4.8	21.9[1]	18
Expected 2000	46.0	5.0	41.0	11

Note: [1] Of which 18.2 Blacks, 2.8 Coloureds, 0.9 Asians
Source of 1984 data: This is South Africa, January 1987, Pretoria

legally employed in the economy of South Africa, coming mainly from the four small neighbouring countries but also from further afield.

Of all the major countries in the world, South Africa most strikingly qualifies as a 'two nations' country, principally on the basis of race. Compared with much of Africa, the extreme southern area of the continent was thinly populated in the seventeenth century when in 1652 the Dutch established their first settlement. In the nineteenth century British settlers arrived, and by the beginning of the twentieth century the whole country was a British colony, to which, however, independence was given in 1910. Table 12.9 shows that until the 1980s the whites formed about one-fifth of the total population of South Africa. The faster rate of natural increase of the non-white population was offset by considerable net immigration of whites after the Second World War, especially from Britain. The remarkable difference between the official 1984 data and the author's estimate for the year 2000 based on a PRB projection highlights the demographic situation. Explicit privilege for the white population, upheld by apartheid, no longer exists. The non-white government that came to power in 1994 will, however, have great difficulty in fulfilling the aspirations of the poor majority while also persuading enough of the white population to stay in the country and contribute to running the economy successfully. In one respect, the position of the white population may become easier because it can no longer be accused of oppressing the non-white majority.

A century ago, British domination of South Africa was established as the Boer War ended in British victory. C. Lamb (*Sunday Times*, 19 March 1995, p. 1.18) reports that the first visit of the Queen to South Africa for fifty years was marked by some bitterness and irony. 'Robert van Tonder, leader of the far-right Boerestaat party, planned to present a £100 billion bill to the Queen for compensation for damages suffered during the Boer War.' He pointed out that the armies of her great-grandmother Queen Victoria destroyed the freedom of the Boers and 'committed the infamous holocaust in which a sixth of our people were murdered in concentration camps'.

In this chapter very little positive has been said about Africa south of the Sahara. Although perhaps not politically correct to do so, it is worth pointing out that at least during the eighty years in which most of Africa was held by European colonial powers, conditions were relatively stable. It is debatable whether the European powers were forced out of Africa after the Second World War by independence movements or chose to leave because their colonies were more expensive to run than the economic gains from them. In the event, the leaders and managers of the newly independent countries had not been adequately prepared to run them in the way that the former colonial powers might have expected or hoped. Environmental disasters and human conflicts have combined to produce widespread famines as well as bitter conflicts, often between ethnic groups or tribes within countries, rather than between countries. In the 1970s there were 700,000 refugees in Africa, in 1995 about 7 million. The population of Africa has doubled between 1970 and 1995. In areas affected adversely by drought, conflict and the presence of large numbers of refugees, relief agencies can often bring in supplies of food, but to provide adequate water and also fuel for cooking is more difficult. Healthcare attention is more likely to be forthcoming when epidemics occur than for preventive purposes. Education is at best sporadic, and equipment for teachers minimal. Africa south of the Sahara has by far the lowest score of any of the major regions of the world in Table 17.10.

KEY FACTS

1. Although Asian and European navigators and traders plied the coasts of Africa for many centuries and traded in both commodities and slaves, the systematic occupation of the continent only started towards the end of the nineteenth century.

2. Before then much of African society was organised in tribal areas of varying shape and size, numbering several thousand. During the colonial period there were some fifty colonial units, some, as in French Equatorial Africa, grouped into larger territories for convenience. Since the 1950s virtually the whole of Africa has become independent, the new states based on the colonies superimposed often uneasily on the tribal areas.

3. Unlike some developing countries in Latin America and South and Southeast Asia, there has been almost no development of modern industry in Africa south of the Sahara except in South Africa. The region remains an exporter of primary products. Yields of many agricultural products are among the lowest in the world.

4. Many of the poorest countries in the world are in Africa south of the Sahara.

5. Population is growing fast in the region and if present trends continue the present 570 million would double to 1,140 million in less than twenty-five years (see Caldwell and Caldwell 1990).

MATTERS OF OPINION

1. Is it fair to blame Africans if they have not taken easily to the ways imposed on their continent by outsiders?

2. Should the United Nations, with support from the developed countries, mount an operation to restore order in Africa?

3. Should attempts now be made to implement the United Nations' proposal in the early 1960s for an industrialisation programme for the continent? A new form of colonialism?

4. In many Catholic and Muslim countries religious and political leaders are in principle against the use of family-planning procedures and devices. In Africa the attitude is more one of indifference or ignorance about the subject. Should the UN and other international institutions attach family-planning strings to the (limited) development assistance given to Africa in an attempt to reduce the rate of population growth?

SUBJECTS FOR SPECULATION

1. Is fast population growth likely to continue or will a combination of conflict, food shortages and diseases such as AIDS narrow or close the present wide gap between birthrate and deathrate? (See Anderson and May 1992.)

2. In the 1970s desertification was given a great deal of publicity when it was argued that desert conditions (from the Sahara Desert) were spreading southwards at an alarming rate into West and Central Africa (see Glantz 1987). Desertification refers to the process whereby land that was once cultivated or grazed is no longer suitable on account of dry conditions or loss of soil. It can be brought about through both climatic change and human activities. It is caused in some areas by the blowing of sand and the formation of dunes. Environmentalists now focus more on the fate of the forests of Africa than on the encroachment of the desert. Why should this be so?

3. For twenty-five years following the Second World War, Latin American countries were distinguished for their instability, frequently alternating civilian and military governments, high inflation rates and banana republic economies. Such features are less typical now. Could Africa settle down in a similar fashion?

4. What might Africa be like now if it had not been colonised by European powers in the nineteenth (or any other) century?

Chapter thirteen

North Africa and Southwest Asia

For much of the past 1300 years Europeans have regarded Islam as a menace. Even in the eighteenth and nineteenth centuries, when the tables were turned and European power was spread throughout the world, Muslims were still seen to be a danger, this time to the security of European Empire.

F. Robinson (1982)

13.1 General features

The region of North Africa and Southwest Asia extends from the Atlantic coast of Morocco in the west to the border of Pakistan in the east, a distance of more than 7,000 km (4,350 miles) (see Figure 13.1). This is roughly the same distance as from Denver, Colorado, to London, England, or from Hawaii to Washington DC. On the north the region 'faces' Europe across the Mediterranean, Black and Caspian Seas and borders former Soviet Central Asia. The five North African countries extend south into the heart of the Sahara Desert. For long periods of history, at least since the times of ancient Greece, there have been contacts to the north. During the period of the Roman Empire Europe pushed into northern Africa and Southwest Asia. At other times, particularly in the seventh century and again in the fifteenth century, the Moors and later Ottoman Turkey moved into Europe, converting Europeans to Islam. Since the middle of the nineteenth century the pressure has almost exclusively been from the European side (see Table 13.1).

North Africa and Southwest Asia cover such a vast area that it is to be expected that environmental conditions and human activities vary greatly from region to region. Nevertheless, three important themes recur throughout much if not all of the region.

- Low rainfall, and accompanying high temperatures much of the year, mean that most of the region is desert or semi-desert, and rainfall is limited in effectiveness. In many places successful agriculture depends on or is assisted by irrigation.

- About two-thirds of the world's oil reserves are located in the region, more in Southwest Asia than in North Africa.

- Islam exerts a great influence on the lives of much of the population, although the intensity of its influence varies from country to country. Most of the countries are populated by Arabs, but Turkey, most of Iran, and Afghanistan are not Arab countries, while in this century Israelis have swamped the Arab population of Palestine. Some of the outstanding characteristics of Islam are described in Box 13.1.

Like the rest of Africa, North Africa was targeted for colonisation by European powers in the nineteenth century, in this instance mainly by France and Italy, but with Britain involved in Egypt. Southwest Asia, on the other hand, was largely under Turkish control until the end of the First World War and even after that French and British interest was limited. Initially the extraction of oil in the region of the Persian Gulf before the

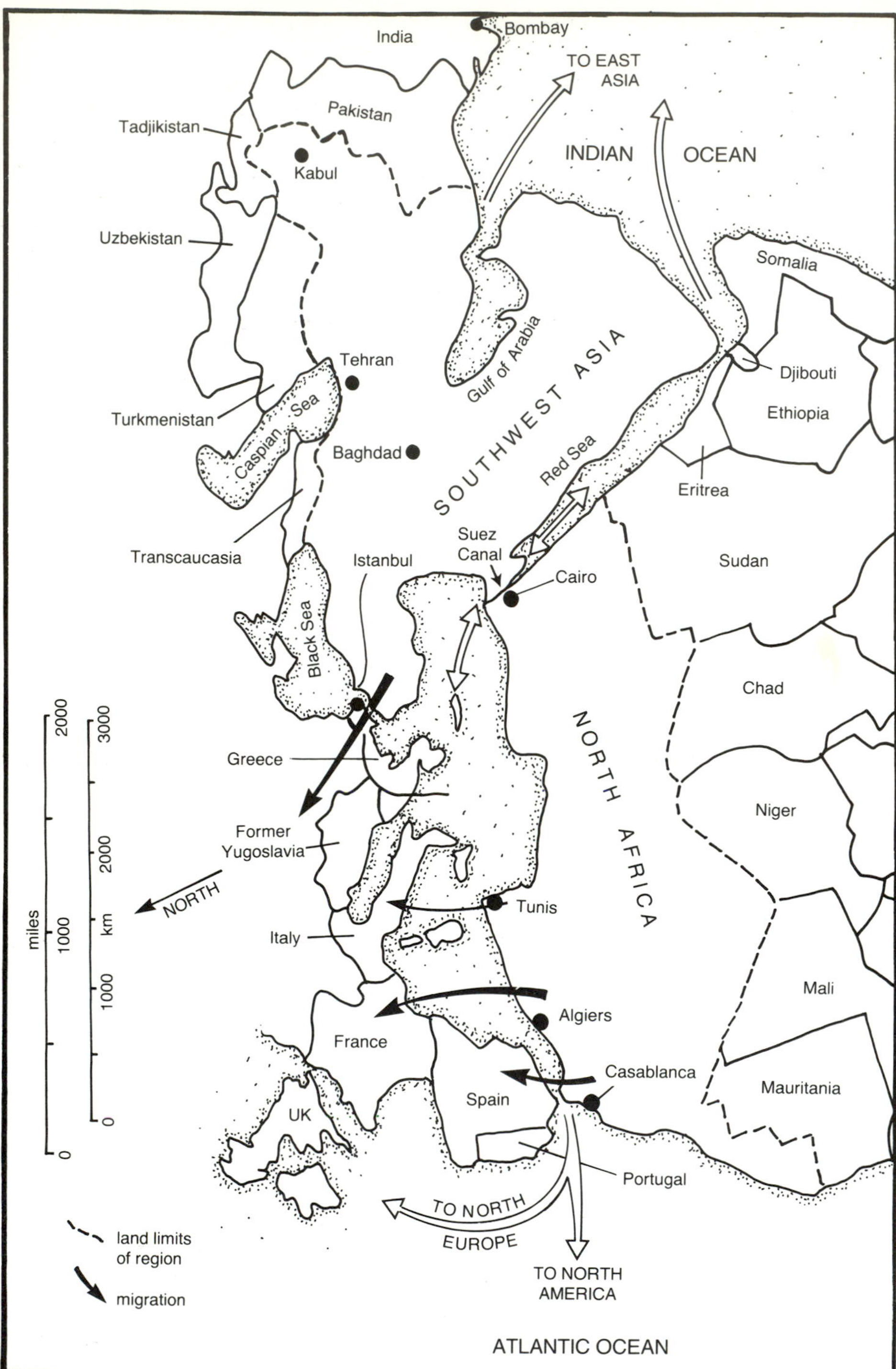

FIGURE 13.1 North Africa, Southwest Asia and neighbouring regions

Second World War was on a small scale, but rapid development of the oil industry took place in the region after 1945.

The region of North Africa and Southwest Asia is about 12,590,000 sq km (4,861,000 sq miles) in area, 9.4 per cent of the world's land area. Its population of 353 million in 1993 was 6.4 per cent of the world's total. Population has been growing very quickly in the last few decades, doubling from 80 million in the early 1930s to 160 million in the early 1960s and doubling again by the late 1980s. The total is expected to reach a staggering 750 million by about 2025 if present trends continue. The area under cultivation has not increased substantially since the 1920s and is unlikely to grow much between 1990 and 2025. Of the total area of North Africa and Southwest Asia, only 7.4 per cent (940,000 sq km – 363,000 sq miles) is cultivated, 5.1 per cent is forested (645,000 sq km – 249,000 sq miles) and 21.2 per cent (2,665,000 sq km – 1,030, 000 sq miles) is classified, much of it generously, as natural

TABLE 13.1 Calendar for North Africa and Southwest Asia

1453	Ottoman Turks capture Constantinople: end of Byzantine Empire
1492	Spaniards begin conquest of North African coast
1516	Ottomans overrun Syria, Egypt and Arabia (1517)
1830	French conquest of Algeria begins
1869	Suez Canal opened
1879	Britain interested in Afghanistan
1881	French occupy Tunisia
1882	British occupation of Egypt
1911	Italy conquers Libya
1914	British protectorate over Egypt
1917	Jews promised national home in Palestine through Balfour Declaration
1920–2	Turkey loses possessions and starts modernisation
1932	Kingdom of Saudi Arabia founded
1936	Arab revolt in Palestine against Jewish immigration
1941–3	War in North Africa, Germany versus Britain and USA. Italy loses Libya
1948	State of Israel established, first Arab–Israeli War
1950	Iran nationalises Anglo-Iranian oil company
1952–3	Military revolt in Egypt and republic formed
1956	Second Arab–Israeli War, Suez crisis, Britain and France invade Canal Zone. Canal closed 1956–7
1960	Organisation of Petroleum Exporting Countries (OPEC) established
1962	Algeria becomes independent from France after eight years of conflict
1967	Third Arab–Israeli War, Israel gains territory
1973	Fourth Arab–Israeli War
1973–4	Sharp rises in oil prices
1979	Further rises in oil prices
1979	Soviet intervention in Afghanistan
1980–8	Iran–Iraq War, ending inconclusively
1991	Gulf War, Iraq withdraws from Kuwait
1994	Arab–Israeli accord over Palestinian territories

pasture. Thus two-thirds of the total area has no bioclimatic use.

According to my assessment of natural resources, for their 6.4 per cent share of the world's population, North Africa and Southwest Asia have 4.8 per cent of the world's bioclimatic resources and 5.0 per cent of the world's non-fuel minerals, but a mere 1.5 per cent of the freshwater resources, compared with 35 per cent of the fossil fuel reserves, mostly in the form of oil and natural gas. The region therefore has a favourable natural resource to population balance (world 100: region 213) but an irregular resource profile. The distribution of oil and natural gas reserves is very uneven in relation to area (see Table 13.2) and to population. Most of the oil production is exported, a prospect also for natural gas, as the international movement of liquefied natural gas by sea increases.

The direct benefit of the oil and gas industry to the region as a whole is limited because so much is exported rather than consumed locally, and also because some of the countries have few or no reserves. The impact of the royalties from the extraction and export of oil is concentrated in a few places and a few hands except in the small states, notably Kuwait, Qatar and the United Arab Emirates, where much of their population now lives in a few centres awash with oil revenues. Thus although Algeria, Iran and Egypt are large producers of oil, most of their population is very poor.

13.2 Discussion of Tables 13.2–13.4

Table 13.2

Column (1) shows the expected availability of fresh water in thousands of cubic metres per inhabitant in the year 2000. Egypt is virtually all desert and most of its water supply comes from the Nile, a source that could be reduced by heavier use upstream in Sudan and Ethiopia. The Arabian Peninsula has a poor supply, increased in a very limited way by the costly desalinisation of sea water along the Gulf coast. The northern tier of countries in Southwest Asia is better provided with water but still has many areas of semi-desert. By comparison, the figure for the USA is 6.6, for Britain 2.0 and for India 1.5.

(2)–(4) North Africa has 4.0 per cent of the world's oil reserves and 3.8 per cent of the natural gas reserves. That amount is dwarfed by Southwest Asia's 65.7 per cent of the world's oil reserves (but 28.4 per cent of the world production) and 31.0 per cent of the natural gas reserves (but 5.7 per cent of production). So long as oil and natural gas are in demand in the world economy and huge new reserves are not discovered elsewhere, Southwest Asia is assured of a major role in the world economy and world affairs.

(5) Up to 99 per cent of the population is Muslim in some countries. Israel is the outstanding exception, with Judaism predominant, while Lebanon, Syria, Egypt and Iraq all have Christian minorities of note.

Table 13.3

(1)–(2) Iran, Turkey and Egypt are the countries with the largest populations and, with around 60 million inhabitants each, contrast with the much smaller sizes of Kuwait, Oman and the United Arab Emirates (UAE) as well as with the West Bank (1.6 million) and Gaza (0.7) adjoining Israel, the islands of Bahrain (0.5) and

Qatar (0.5) in the Arabian Gulf, and Cyprus (0.7). Algeria, Libya and Saudi Arabia are very large territorially, but contain large areas of desert, giving a low average density of population.

(5)–(6) Population is growing fast throughout North Africa and Southwest Asia, but considerable contrasts occur, with growth in North Africa, the eastern Mediterranean coastlands and Turkey generally slower than growth in much of the rest of Africa (see Table 12.3) and Southwest Asia. Some massive population increases are expected in the next few decades unless hitherto unknown circumstances, physical or social, quickly reduce fertility rates. Except in Israel, the percentage of elderly is very small, while the under-15s reach 50 per cent in Libya. The comparatively large percentage aged between 15 and 64 inclusive in Israel (60 per cent), the UAE (64) and Kuwait (57) is caused by the large numbers of recent immigrants in the case of the UAE and Kuwait, many of them males working in the oil industry and related services. Turkey, Egypt and Tunisia are the only countries not affected by in-migration in which the demographic transition has taken the population some way towards a stable structure. The very low percentage in employment (in 1991) in most countries reflects two influences: the large percentage of the dependent population, especially the young, and the lack of work outside the home for women who, traditionally in Muslim countries, remain in the home, and are not counted as employed.

Table 13.4

(1) Compared with Africa south of the Sahara and South Asia, most countries of North Africa and Southwest Asia are highly urbanised. The interiors of Oman and Yemen are still predominantly rural, as also is Afghanistan. Elsewhere there are some very large and

BOX 13.1

Features of Islam

In the introduction to his *Atlas of the Islamic World*, Robinson (1982) stresses that about one-fifth of the population of the world is made up of Muslims and that 'The pattern for existence laid down by Islam is one of the main ways men (*sic*) have had of living life.' The countries in which Muslims are in the majority or form a substantial minority extend from Morocco on the Atlantic in the west to Indonesia and the Philippines in the east. The proportion of Muslims to total population is shown in Table 13.2 for North Africa and in Figure 13.2 for all countries with a notable Muslim element.

Robinson (1982) points out that 'for much of the past 1300 years Europeans have regarded Islam as a menace'. In the early centuries, following the spread of Islam in the seventh century from its place of origin in Arabia, 'two particular aspects of Muhammad's message were singled out for Christian polemic: the support Islam was believed to give to the use of force (although Christians, too, waged holy wars), and the sexual freedom Muslims were supposed to enjoy in this life'. Old objections by Europeans to Islam have given way to new ones, such as the position of Muslim women, with less access to education than the men have and virtual confinement to the home in extreme cases.

Since very few of the predominantly Muslim countries of the world have democratically elected governments, it is likely that the states are influenced by the Islamic view of the world and its customs. Five central obligations of Muslims are: to believe in Allah and the Quran, to pray five times a day, to contribute with a 'levy' for charitable uses, to visit Mecca at least once in a lifetime, and to fast for a given period each year. As with Christians, Muslims vary in their dedication to their religion and in the thoroughness with which they carry out their duties. There are also Muslim 'fundamentalists' in many parts of the Muslim world, for example in Iran, southern Iraq, Algeria, Egypt and Sudan. Since 1979 Iran has been a theocratic state, one in which the government has been strongly influenced by Islamic mullahs.

In the countries with a large proportion of Muslims, various features of the Islamic religion affect such aspects of life as diet, dress, demographic practices and the importance of pilgrimages. For example, it is extremely important to have at least one son in Muslim families, a factor keeping fertility rates high. On the other hand, the common religion has not prevented conflicts within and between Muslim countries, while the possession of huge oil reserves and the extraction of oil in some countries has led to huge differences in living standards between neighbouring Muslim countries (e.g. Saudi Arabia and Yemen, Libya and Egypt). The antipathy between Turkey and its Arab neighbours seems far from the sympathy that might be expected between fellow Muslims.

These generalisations about Islam and its impact on Muslim countries are intended to illustrate and account for some of the cultural features of the region. In Western Europe, interest in Islam and its followers has grown greatly since the 1960s, directly through the increasing number of migrants, particularly from Turkey, Northwest Africa, Pakistan and Bangladesh into fast-growing cities, with 7–8 million inhabitants each in Tehran, Istanbul and Cairo, and at least eighteen other cities with a population of over 1 million.

(2) Different levels of energy consumption per inhabitant distinguish the highest levels of the main oil and gas producers with small populations from those of the predominantly rural Yemen and Afghanistan. Some consumption is directly related to the extraction and refining of oil, but Southwest Asia accounted in 1992 for only 6.6 per cent of the world's oil refining while producing 28.4 per cent of the world's crude oil. Virtually all the oil exported is crude, the refining capacity mainly serving regional needs, which amount to 5.4 per cent of the world's consumption.

Columns (3)–(4) show a dramatic recent fall in the percentage of the economically active population in agriculture. In contrast to Africa south of the Sahara (see Chapter 12), South Asia (Chapter 14) and China (Chapter 16), there has been only a small increase in the absolute number engaged in agriculture in North Africa and most of Southwest Asia, with no change in Turkey, Iraq or Afghanistan. Mechanisation has been more widely introduced in many parts of the region than generally in developing countries.

(5)–(8) The United Nations Human Development Index reveals a paradox in some countries of North Africa and Southwest Asia. The real GDP per inhabitant is generally high by the standards of countries in the developing world but life expectancy and literacy levels are comparatively low, literacy in particular being pulled down by the low percentage of females able to read. Around 1970 the ratio of literate females to literate males (=100) was a mere 13 in Saudi Arabia, 28 in Algeria, 40 in Egypt and 43 in Iran. On the HDI (column 8) Yemen with 240/1,000 and Afghanistan with 65/1,000 have very low scores, and everywhere except Israel (939/1,000) and Kuwait (815/1,000) scores are disappointing. The oil wealth has not

the more affluent countries of the European Union. It has also grown indirectly through contact between Muslims in their own countries and personnel from Europe and the USA connected with the oil industry and other activities in North Africa and Southwest Asia. The desperate plight of Muslims in Bosnia and the recent independence of several former Soviet Socialist Republics with large Muslim majorities have further increased awareness of the Muslim world among politicians in particular and the public in general in Western Europe and North America.

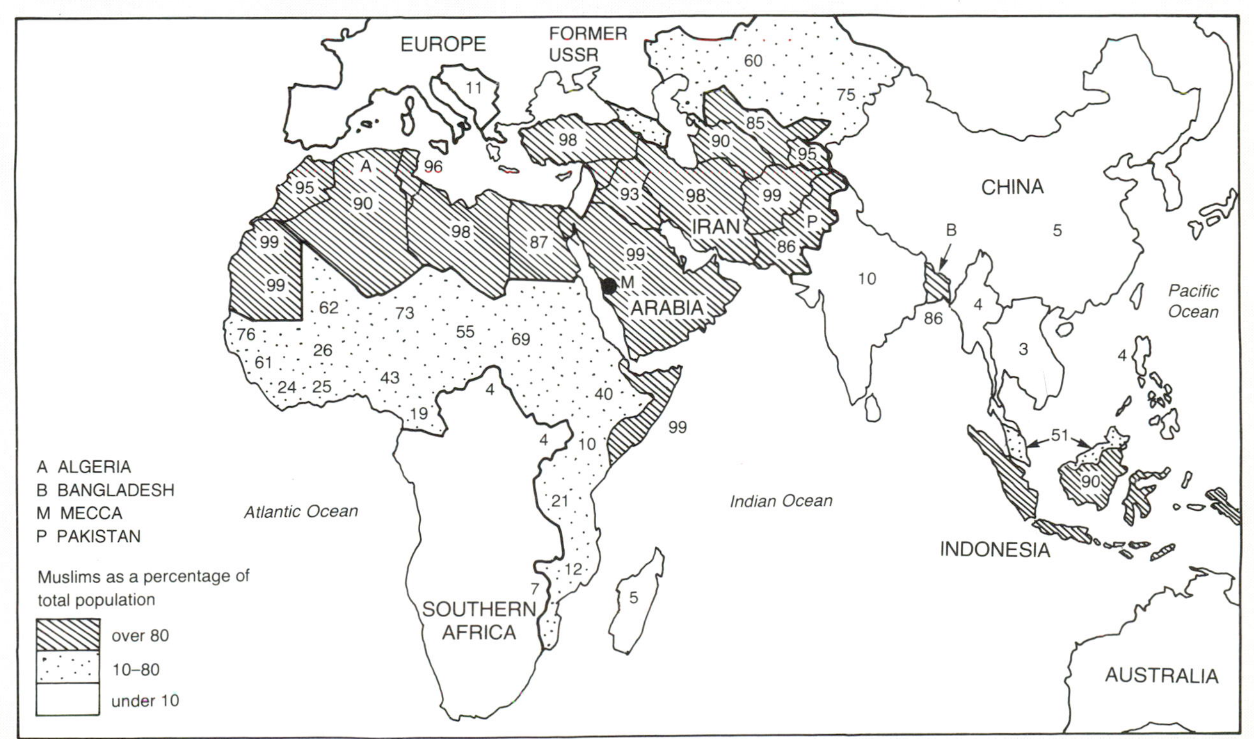

FIGURE 13.2 The world of Islam

TABLE 13.2 Water, oil and gas, and Muslims in North Africa and Southwest Asia

	(1) Water availability, in thous. cu metres per head	(2) % of world reserves 1992: oil	(3) % of world reserves 1992: gas	(4) Oil production, in millions of tonnes: 1992	(4) Oil production, in millions of tonnes: 1993	(5) % Muslims in total population
North Africa						
Morocco	0.9	–	–	–	–	95
Algeria	1.0	0.9	2.6	57.0	56.3	90
Tunisia	0.4	0.2	–	5.0	na	95
Libya	1.2	2.3	0.9	73.0	68.0	98
Egypt	0.05	0.6	0.3	46.1	47.0	87
Southwest Asia						
Israel	na	–	–	–	–	5
Jordan	na	–	–	–	–	95
Lebanon	na	–	–	–	–	50
Syria	1.0	0.2	–	25.2	30.0	82
Iraq	1.3	9.9	2.2	23.6	22.3	93
Saudi Arabia	0.3 (Saudi Arabia, Yemen, United Arab Emirates, Kuwait, Oman)	25.6	3.7	427.9	424.7	99
Yemen		0.4	–	9.0	10.6	96
United Arab Emirates		9.8	4.2	118.5[1]	115.0	99
Kuwait		9.3	1.1	54.1	97.3	99
Oman		0.4	–	36.1	39.1	99
Turkey	2.3	–	–	–	–	98
Iran	2.5	9.2	14.3	172.2	180.0	98
Afghanistan	2.2	–	–	–	–	99

Notes:
– = small or non-existent
na = not available
[1] Abu Dhabi, Dubai and Northern Emirates
Other producers: Neutral Zone 17.0, Qatar 22.6
Sources: Barney (1982) p. 156; BP (1993), p. 2 and BP (1994), p. 4

diffused or been so effectively applied as in Venezuela (824/1,000).

(9)–(10) In spite of the lack of appropriate data it is evident that primary products, almost all oil or natural gas, dominate the exports of Saudi Arabia, Iran and Algeria, as also of smaller oil producers in the Gulf. On the other hand, even between 1960 and 1991 marked changes took place in the composition of exports from Morocco, Tunisia, Jordan and Turkey. Already in 1970 Israel's exports included a large proportion of manufactures.

If changes in North Africa and Southwest Asia proceed as fast in the next thirty years as in the last thirty, the region will be very different from now. With double the population, possibly a reduction in the absolute quantity of oil exported, little addition to the cultivated area, and the growth of manufacturing, especially in the larger cities, the region could turn into a large importer of food and raw materials. If there are as many military conflicts in the next thirty years as in the last, international or internal, then the apparent advantage of having two-thirds of the world's oil reserves could have been lost.

13.3 North Africa (see Figures 12.4 and 13.3)

Throughout much of the last three thousand years, the narrow zone of economically useful land bordering the Mediterranean in North Africa has had stronger economic and cultural ties with either southern Europe or Southwest Asia than with the rest of Africa. Since the middle of the last century, French influence has been strong in Northwest Africa (the Maghreb) starting with the invasion of Algeria in 1830, followed by the French colonisation of Tunisia and Morocco in 1912 and the occupation of Libya by Italy in 1911–12. From southern Spain and the Canary Islands, Spain also gained footholds in Morocco. Britain and Egypt established a condominium over much of the Nile basin, roughly that part now occupied by Sudan.

The opening of the Suez Canal in 1869 between the

TABLE 13.3 Demographic data for the countries of North Africa and Southwest Asia

	(1) Popn rank in world	(2) Popn in millions 1993	(3) Area in thous. sq km/(miles)		(4) Density of popn per sq km/(mile)		(5) Annual % increase in popn	(6) Expected popn 2025 in millions	(7) % under 15	(8) % over 64	(9) % of total population in employment 1991
North Africa											
Morocco	33	28.0	447	(173)	63	(163)	2.4	46	40	4	31
Algeria	37	27.3	2,382	(920)	11	(28)	2.7	61	44	4	23
Tunisia	82	8.6	164	(63)	52	(135)	1.9	13	37	5	33
Libya	105	4.9	1,760	(680)	3	(8)	3.4	14	50	2	25
Egypt	17	58.3	1,002	(387)	58	(150)	2.3	105	39	4	28
Southwest Asia											
Israel	100	5.3	21	(8)	252	(653)	1.5	9	31	9	38
Jordan	118	3.8	89	(34)	43	(111)	3.6	8	44	3	25
Lebanon	121	3.6	10	(4)	360	(932)	2.1	6	40	5	31
Syria	57	13.5	185	(71)	73	(189)	3.8	37	49	4	25
Iraq	45	19.2	438	(169)	44	(114)	3.7	53	48	3	27
Saudi Arabia	50	17.5	2,150	(830)	8	(21)	3.2	49	39	2	29
Yemen	61	11.3	528	(204)	21	(54)	3.2	32	49	3	24
United Arab Emirates	137	2.1	84	(32)	25	(65)	2.8	3	35	1	51
Kuwait	141	1.7	18	(7)	94	(243)	3.0	5	42	1	40[1]
Oman	143	1.6	213	(82)	8	(21)	3.5	5	47	3	27
Turkey	16	60.7	780	(301)	78	(202)	2.2	99	35	4	43
Iran	15	62.8	1,648	(636)	38	(98)	3.5	162	47	3	28
Afghanistan	51	17.4	652	(252)	27	(70)	2.8	49	46	4	30

Note: [1] 1990 for Kuwait
Sources: for data in columns (2)–(8): *WPDS 1993*; for data in column (9): *FAOPY 1992*, Table 3

Mediterranean and the Indian Ocean via the Red Sea made Egypt of prime importance strategically to Britain, France and Holland, with their colonies and extensive trade in South and Southeast Asia. Even when most of their colonies in Asia had already achieved independence in the decade after 1945, Britain and France were still prepared in 1956 to use military force in an attempt to keep control of the canal after it was nationalised by Egypt. The closure of the Suez Canal for a time in 1956 and again in 1967 resulted in its decline as a vital strategic link, although it still carries heavy traffic. Pipelines, and supertankers taking the longer route by the Cape of Good Hope, together convey most of the oil exported from the Gulf to Western Europe and the USA.

Algeria was the main destination of French settlers emigrating to North Africa, and Tunisia also attracted French and Italian settlers. While Morocco and Tunisia achieved independence in the 1950s fairly peacefully, the French tried to hold on to Algeria, regarding it as an integral part of France, the reason being twofold: to support the French settlers, and to keep control of the already proved reserves of oil and natural gas. Italy, on the other hand, lost Libya during the Second World War.

In the three Maghreb countries the agricultural sector has stagnated, with little change in the area cultivated, continuing low yields of cereals, and little change in the size of the labour force. The mountains of the north receive the highest rainfall, but beyond them to the south, conditions rapidly become dry, and limited irrigation offers the best prospect for cultivation. Citrus fruits, wine, olive oil and dates are among the agricultural products most exported. Thus the contribution to employment and to GDP of agriculture has diminished in recent decades, while extractive activities, including phosphates in Morocco and Tunisia, oil and gas in Algeria, and service activities, including tourism in Morocco and Tunisia, have grown correspondingly. Libya is drier than the Maghreb countries, with only patches of agricultural land in the extreme north and little else but, in relation to population, large oil reserves, 2.3 per cent of the world total.

Egypt's water supply from rainfall is minimal and cultivation depends entirely on the Nile for irrigation, while settlements and industry also derive their supply from it. With only 26,000 sq km (10,000 sq miles) of land under arable and tree crops, Egypt has 22 inhabitants for every hectare of cultivated land, albeit intensively farmed and high-yielding. Even China only

TABLE 13.4 Economic and social data for the countries of North Africa and Southwest Asia

	(1)	(2)	(3)	(4)	(5)	(6)	(7)	(8)	(9)	(10)
					Quality of life				*Non-primary exports as a percentage of total exports*	
	% urban	*Energy consumption kce per head*	*% employed in agriculture*		*life expectancy in years*	*% adult literacy*	*real GDP dollars per head*	*Human Development Index*		
			1965	*1991*					*1970*	*1991*
North Africa										
Morocco	47	380	54	36	62	50	2,300	429	10	61
Algeria	50	1,520	60	24	65	57	3,090	533	7	3
Tunisia	59	820	60	23	67	65	3,330	582	19	68
Libya	76	4,890	60	13	62	64	7,250	659	na	na
Egypt	44	700	55	40	60	48	1,930	385	27	40
Southwest Asia										
Israel	90	3,240	12	4	76	96	10,450	939	70	88
Jordan	70	1,120	33	6	67	80	2,420	586	17	46
Lebanon	84	1,540	55	8	66	80	2,250	561	na	na
Syria	50	1,100	50	24	66	65	4,350	665	9	23
Iraq	70	910	50	20	65	60	3,510	589	na	na
Saudi Arabia	77	6,650	72	38	65	62	10,330	687	0	1
Yemen	29	350	84[2]	55	52	39	1,560	232	na	na
United Arab Emirates[3]	81	20,750	5[1]	2	71	55	23,800	740	na	na
Kuwait	90[1]	7,480	1	1	73	73	15,980	815	na	na
Oman	11	4,110	60[1]	39	66	35	10,570	598	na	na
Turkey	59	1,060	72	47	65	81	4,000	671	9	67
Iran	54	1,690	48	27	66	54	3,120	547	4	3
Afghanistan	18	230	87	54	43	29	710	65	na	na

Notes:
na = not available
kce = kilograms of coal equivalent
[1] Author's estimate
[2] Average of two Yemens
[3] Abu Dhabi, Dubai and North Emirates
Sources: (1) *WPDS 1993*; (2) UN (1991), *Energy Statistics Yearbook*; (3), (4) *FAOPY 1993*; (5)–(8) *UNHDR 1992*, Table 1; (9), (10) *WDR 1993*, Table 16

has about 12 inhabitants per hectare cultivated. In 1975 there were almost 5 million workers in agriculture in Egypt but in 1991 almost 6 million, resulting now in more than two workers per hectare of cultivated land. Before the construction of the Aswan Dam on the Nile in southern Egypt the waters of the river carried silt and deposited it on reaching the low-lying irrigated lands along the lower course of the river. Since the construction of the dam, much less sediment has been carried by the river, and it has been necessary to use large amounts of fertiliser in its place.

Algeria, Libya and Egypt are currently large exporters of oil and natural gas. In 1992 they produced 176 million tonnes of oil (5.6 per cent of the world total) and exported 137.4 million (78 per cent of their output). Western Europe is the main destination (83 per cent of exports). They also produce 3.3 per cent of the world's natural gas, most from Algeria, which supplies some by pipeline under the Mediterranean to Italy and some by Liquefied Natural Gas (LNG) carriers to France, Belgium and Italy. The exports of Morocco and Tunisia are more diversified, but Moroccan phosphates, unprocessed or processed, provide its main export product, with the addition of vegetables, fruits and fish. The Maghreb countries are also the source of increasing numbers of emigrants. Some have settled in EU countries legally, while others arrive illegally. A favourite zone of entry for illegal immigrants from the Maghreb is the southern coast of Spain.

All three countries of Northwest Africa depend increasingly on the EU for trade and tourism. Morocco has already sounded the EU with regard to applying for membership, although strictly its credentials are not appropriate, since it is in Africa, not Europe, and it has a much lower level of economic development even than Portugal or Greece. The close association of Algeria and Libya with the USSR and the greater strength of

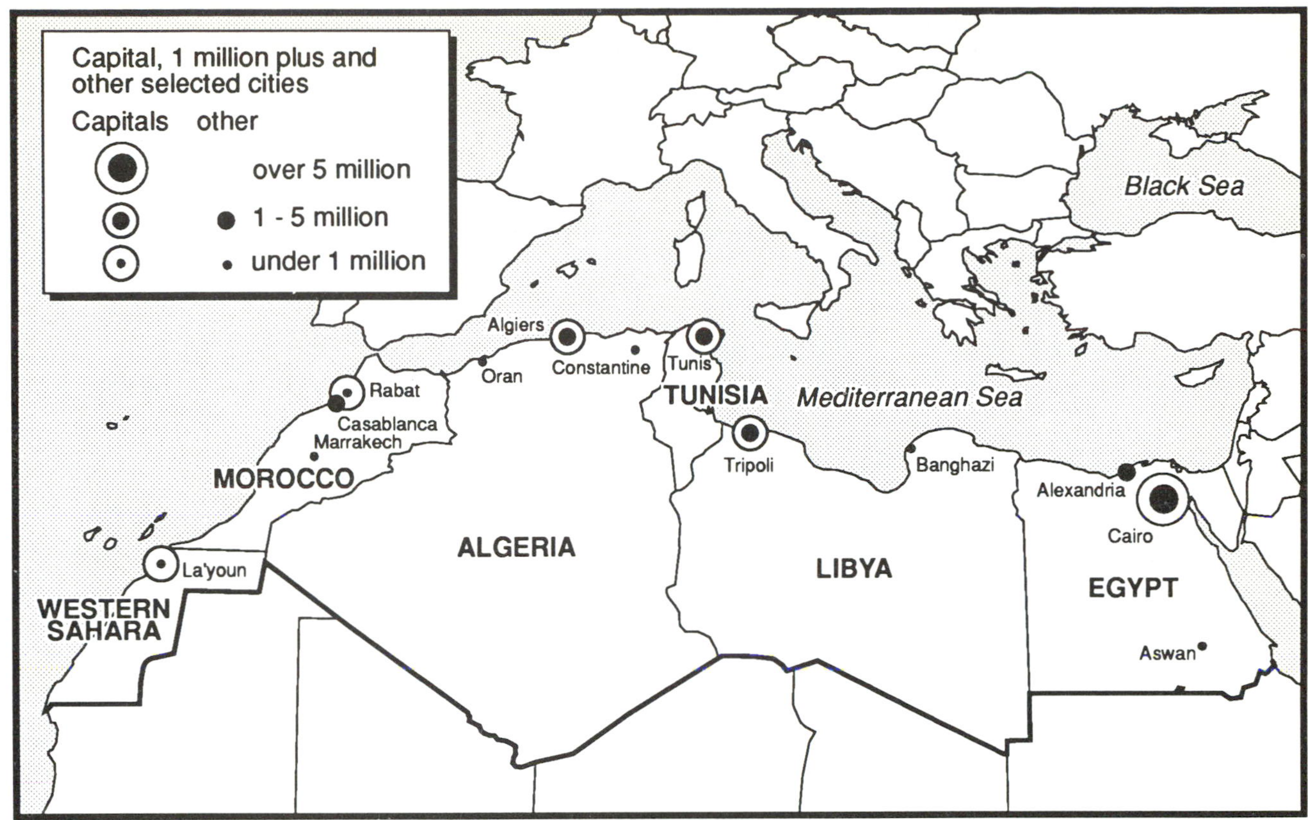

FIGURE 13.3 Countries and cities of North Africa

Muslim fundamentalism there than in Morocco or Tunisia have alienated them from Western Europe, although it is the chief market for their exports. Muslim fundamentalism appears to be increasing in Egypt, while the negative attitude towards outsiders in Algeria is growing in strength.

Although the attention of the EU is focused on possible new EFTA partners and on developments in Central Europe, the proximity of North Africa, with its oil and gas reserves, its fast growth of population, and the increasing pollution along the southern shores of the Mediterranean, mean that the region's problems

PLATE 13.1 Semi-desert landscape, with some cultivation and a date-palm, a useful and widely grown tree crop in much of the region. Matmata, Tunisia, has provided the scenery for the production of the film *Star Wars*

PLATE 13.2
Mosque in Kaiouran, Tunisia, first in status in North Africa. Over 90 per cent of the population of the region is Muslim, although western (rather than Christian) ways of life have eroded some of the strictest practices of Islam

and prospects have a prominent place in EU thinking and policy. The Italian government has proposed the setting up of a special bank and fund to help to create jobs in North Africa and thereby reduce emigration. Much will have to be done to cope with an expected increase of population from 128 million in 1991 to 227 million in 2025.

13.4 The eastern Mediterranean (see Figures 13.4 and 13.5)

Egypt stands out among the countries of the world on account of its strategic position as guardian of the Suez Canal, its remarkable historical legacy, and its extreme dependence on a single river for its water supply. Turkey's location at the interface between Europe and Asia, and between the Christian and Islamic 'worlds', is no less distinguished. The four countries at the eastern end of the Mediterranean – Syria, Lebanon, Israel and Jordan – are also located in a region of particular significance in history since here Judaism and Christianity originated, while Islam's place of origin is not far off at Mecca in Saudi Arabia.

The decline of the influence of Turkey dates back to the nineteenth century as the more distant parts of the Ottoman Empire were gradually abandoned or captured by outside powers. The presence of the Ottoman Empire in the fifteenth and sixteenth centuries in the area bordering the eastern Mediterranean, blocking trade routes between Europe and South and East Asia, was, however, an incentive to the emerging nations of Western Europe to explore the oceans and find alternative sea routes. The First World War brought Turkey's decline to an end, leaving the homeland, together with a reluctant portion of the Kurdish nation in its eastern extremity. Turkey reformed its administration, economy and educational system in the period between the two world wars and after the Second World War was powerful enough militarily to form a crucial element in the NATO defence system, intended on its southern flank to contain possible Soviet expansion into the Middle East.

With about 250,000 sq km (96,500 sq miles) of land under arable and permanent crops, almost one-third of its total area, Turkey is generously endowed with agricultural land, having only about 2 inhabitants per hectare, compared with over 20 in Egypt. Although the interior has rather low rainfall, nowhere are conditions as dry as those prevailing over most of the rest of North Africa and Southwest Asia. Yields of cereals are nevertheless much lower than in Western and Central Europe. The labour force in agriculture, between 11 and 12 million throughout the 1970s and 1980s, has declined from 65 per cent to 47 per cent of the total economically active population over twenty years, but it still exceeds the combined labour force in agriculture of all the countries of the European Union.

Turkey's mineral resources are of modest proportions, with small reserves of fossil fuels and no non-fuel mineral of outstanding importance. Its iron and steel industry has grown in the 1980s (pig-iron from 2.1 to 5.6 million tonnes, steel from 1.7 to 9.4) and various other branches of industry have expanded. Between 1970 and 1991 non-primary products increased from 9 to 67 per cent of the value of Turkey's exports.

Since the western extremity of Turkey, including most of the population of its largest city Istanbul, is in Europe and, given Turkey's alignment with the West through NATO, its leaders demonstrated their position by applying for membership of the European Community in 1989, but the application was rejected. Since the Second World War a large proportion of Turkey's foreign trade has been with Western Europe, although

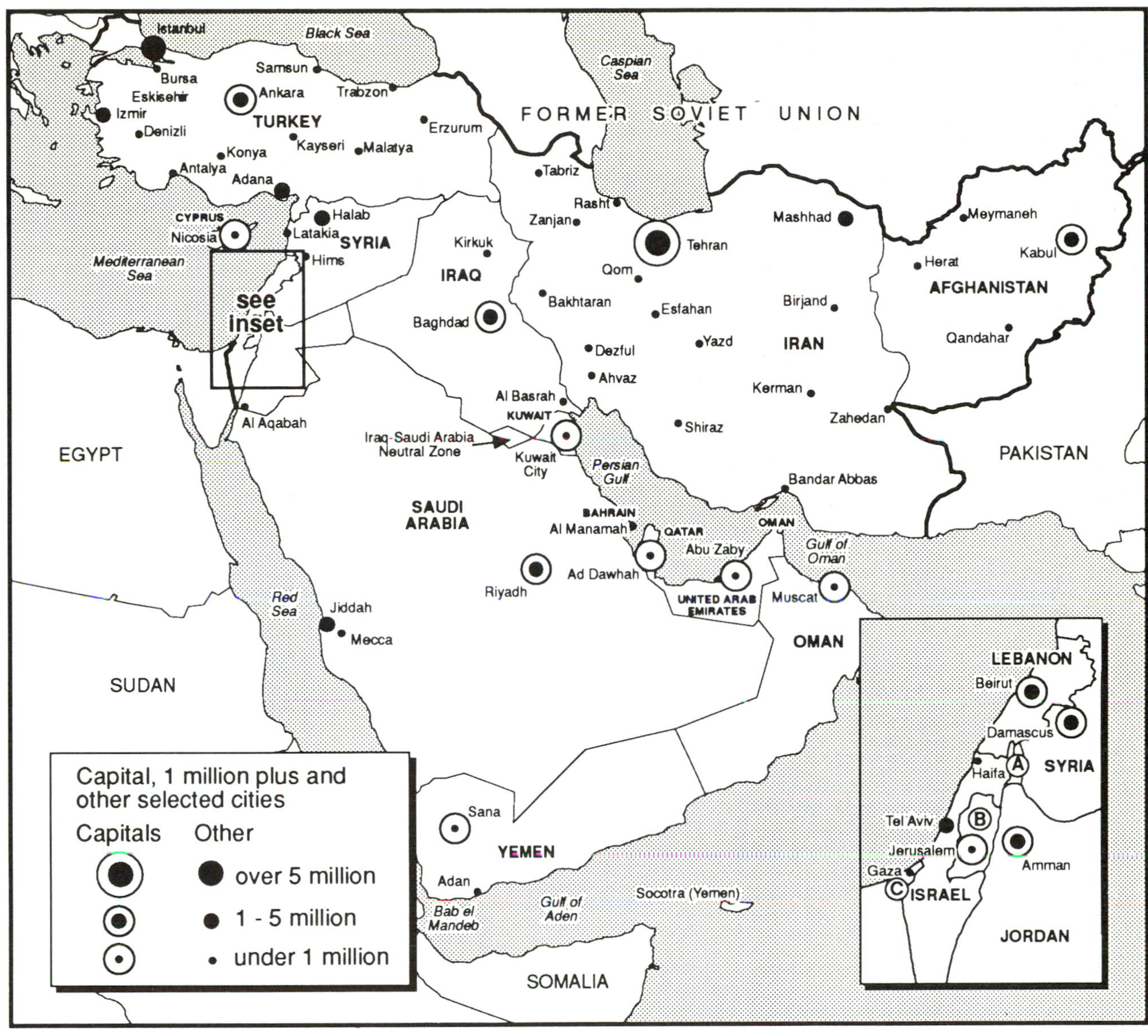

FIGURE 13.4 Countries and cities of Southwest Asia

A Golan Heights **B** West Bank **C** Gaza

the share has declined slightly in the 1980s. Unlike Morocco, Turkey does satisfy the criterion of being in Europe, but with regard to democratic institutions and human rights it has a dubious record, while economically it is far poorer even than Greece. In contrast to the Member States of the EU, it has a fast-growing population.

The antagonism between Greece and Turkey, aggravated by the Turkish occupation of northern Cyprus in 1973, makes it likely that Greece would veto its application to join the EU under present circumstances. Without oil and gas reserves of consequence Turkey does not offer anything positive economically to the EU and indeed would qualify for a large share of EU regional funds to backward agricultural areas should it enter. Once the future of the former Soviet republics of Transcaucasia and Central Asia is clearer, it is possible that Turkey may seek stronger ties with these areas. With its neighbours in Southwest Asia, namely Syria, Iraq and Iran, it has little of common interest except to keep the Kurds (see Box 13.2), divided unequally between all four countries, from forming a separate state.

Syria, the Lebanon, Palestine and Jordan became protectorates of France and Britain after the defeat of Turkey in the First World War. Gradual Jewish immigration into Palestine earlier in this century, followed by a massive flow of illegal migrants after the Second

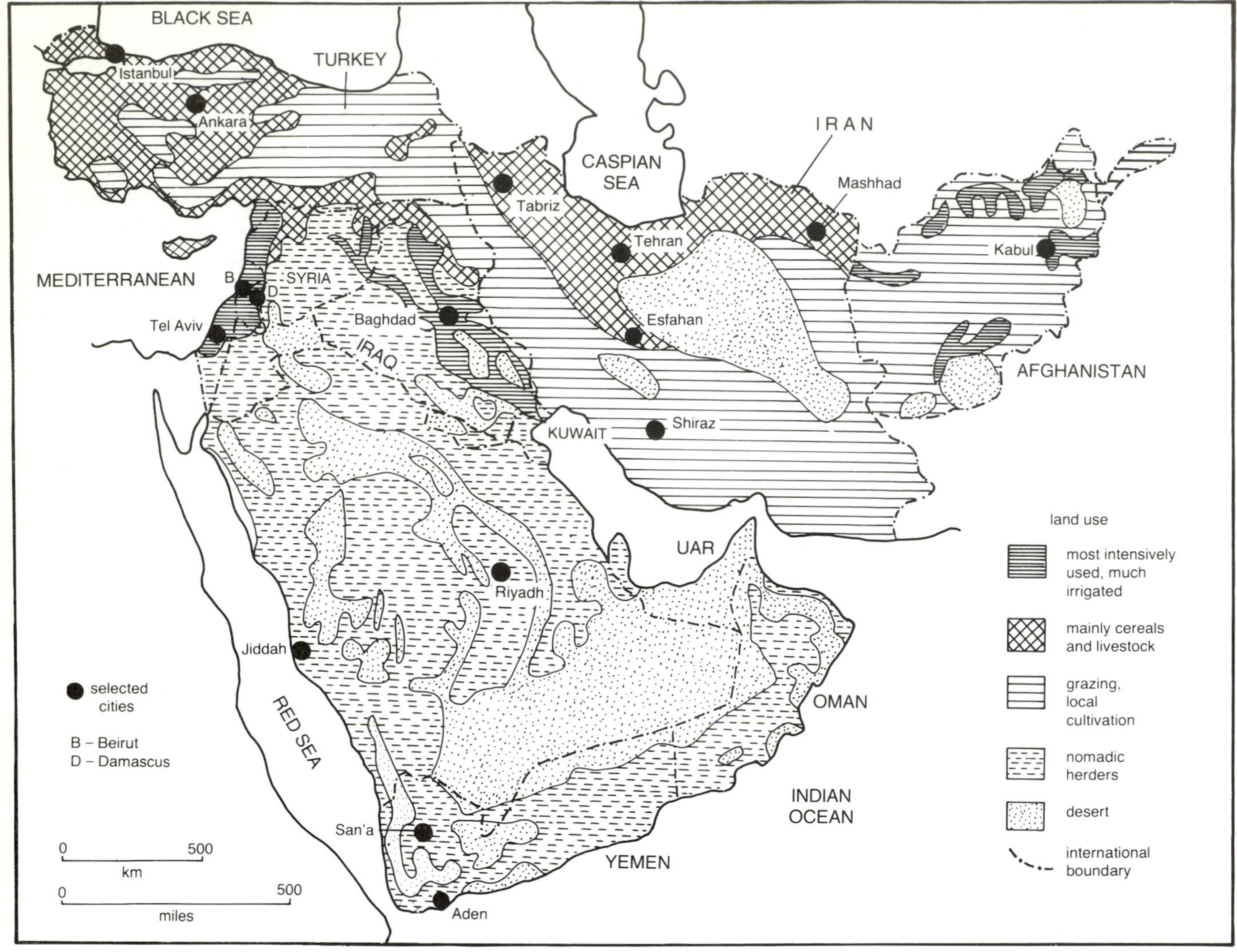

FIGURE 13.5 Economic map of Southwest Asia

World War, and the establishment of the state of Israel, led to an ongoing conflict between Israel and its neighbours. The survival of Israel has only been possible through the virtually unreserved support of the USA, provided through its large and influential Jewish population, and to the tenacity, advanced weaponry and brilliant military organisation of Israel's forces, maintaining superiority against great numerical odds and winning three wars (1956, 1967 and 1973).

The area behind the eastern coast of the Mediterranean has modest economic value but great cultural and emotional significance. Rainfall is low except in the north and in the mountains, and water supply is a problem. Syria has about 0.2 per cent of the world's oil reserves. Otherwise the natural resources of the four countries are limited. Co-operation between the countries offers a better prospect for the future than the continuation of fruitless conflict. Steps were already being taken in 1994 to establish peaceful relations between the Israelis and their Palestinian and Jordanian neighbours.

A comparison of data for the four countries in Tables 13.3 and 13.4 shows Israel to be by far the most sophisticated. It has higher educational standards, and better healthcare facilities than its neighbours. Agriculture flourishes in difficult conditions, and considerable industrial development has taken place. In spite of its lack of natural resources, primary commodities (food, fuels, raw materials) accounted in 1991 for only 25 per cent of imports, the remainder being various manufactures. In many economic respects, Israel is like an outlier of the EU, and its real GDP per inhabitant is four times that of Lebanon (set back by long conflict) and Jordan, and more than twice that of Syria and Turkey. Demographically, Syria and Jordan resemble other Arab/Muslim countries of Southwest Asia, with a large increase in population expected in the next few decades. The prospect for increasing agricultural production correspondingly seems bleak, while industrial development has been limited.

13.5 The Gulf region (see Figure 13.7)

Almost two-thirds of the world's proved oil reserves are in a zone extending from northern Iraq through the basins of the Tigris and Euphrates Rivers and along the lands bordering the Arabian (Persian) Gulf and Gulf of Oman. But for the enormous oil and natural gas reserves of this region the outside world would probably take little interest in it. Agriculture supports a low density of population, based on extensive areas of poor pasture and limited areas of irrigation, the most notable being in the Tigris and Euphrates valleys in Iraq.

Ethnically, most of Syria and Iraq and virtually all of the Arabian Peninsula are populated by Arabs, with relatively minor differences between countries. In the twentieth century this part of the Arab world has been subdivided arbitrarily, leaving the territorially extensive state of Saudi Arabia dominating the southern part, and small states along the fringes. Iraq, which has a central position in Southwest Asia, has five neighbours, and with an expansionist policy like that of Germany in the 'centre' of Europe in the 1930s it could easily reach the oil reserves of its neighbours.

In contrast to the ethnically fairly homogeneous nature of the Arabian Peninsula, from eastern Turkey through Iran and Afghanistan to Pakistan in the east, the proliferation and territorial contortions of ethnic groups is breathtaking. In the west, a 'lost' nation of some 10 million speakers of Kurdish extends unequally over southeastern Turkey (over 4 million), western Iran (over 3 million), northeastern Iraq (over 2 million) and northeastern Syria (300,000) (see Box 13.2). Some 12 million speakers of Azerbaijani are split roughly equally between the former Soviet Socialist Republic of Azerbaijan and northwest Iran. Modern Persian, with some 35 million speakers, is the main language of Iran, in which over half of the population use it, as do 4 million in Afghanistan. Baluchis (3 million speakers) and Arabs form two other important minority groups in Iran. The presence of some 2 million Arabs in the 'corner' of Iran at the head of the Gulf of Arabia has been crucial in recent conflicts in Southwest Asia. Almost all of the oil reserves of Iran are located in this small part of the country.

The protracted Iran–Iraq War (1980–8), started by Iraq, was an attempt by predominantly Arab Iraq to take over the Arab lands of Iran. In the event, Iran held on to the territory, and the Arabs of the area apparently preferred to remain in Iran. Had Iraq succeeded in taking that territory, it would have doubled its share of the world's oil reserves from about 10 to 20 per cent. In 1990 Iraq occupied Kuwait with ease, thereby acquiring 10 per cent of the world's oil reserves virtually overnight. Given the great military strength of Iraq, and in the absence of any response from the outside world, Iraqi forces could without much difficulty have continued beyond Kuwait into Saudi Arabia, taking control of perhaps another 10 per cent of the world's oil reserves in the area closest to Kuwait.

The importance of Middle East oil exports to the rest of the world, and particularly to the OECD industrial countries in 1989 on the eve of the Gulf War, is clearly shown in Table 13.6 and the subject is discussed in Box 13.3. The Middle East provided 42 per cent of the world's oil exports (664 million tonnes out of 1,577 million) (Table 13.8). It exported almost 200 million tonnes to Western Europe, 150 million to Japan and almost 100 million to North America. These quantities represented 15 per cent of the total energy consumption of Western Europe, 36 per cent of that of Japan and 4.5 per cent of that of North America. In 1992 the situation was broadly the same, but with 744 million tonnes exported from the Middle East (Iraq providing almost none). The changes in oil production and prices in the last two decades shows graphically and vividly the wild fluctuations experienced in this industry (see Figures 13.8 and 13.9). How do each of the oil-rich countries in the Arabian Peninsula fare as a result of their huge resources and position in the floodlights of world affairs?

Saudi Arabia alone has over a quarter of the world's proved oil reserves. It is over 2 million sq km (770,000 sq miles) in area, a fact that might lead to the conclusion that it could still have extensive undiscovered oil deposits. Over three-quarters of its 1993 population of 17.5 million is concentrated in urban settlements. Barely 1 per cent of the total area is cultivated (24,000 sq km – 9,300 sq miles), 40 per cent of it irrigated. About 40 per cent of the total area is classified as natural pasture. In 1970 food imports accounted for 26 per cent of the value of imports, in 1991 only 15 per cent, in spite of a big increase in population size, thanks to heavy investment in agricultural improvements, including more than a doubling of the irrigated land.

Saudi Arabia is a land of extremes. Virtually all the value of its exports is accounted for by oil. Almost all of its surface is desert, or pasture of low productivity. The most productive cultivated land and the bulk of the population are concentrated in tiny oases and modern cities. The capital, Riyadh, is growing fast, with a population of 1.5 million. A final example of the special nature of Saudi Arabia is its balance of trade. In 1991, for example, exports were valued at almost 55 billion US dollars, imports only 25.5 million. The oil industry needs some investment, but there is little else in the way of resources to invest in, so some of the funds must be invested elsewhere in the world. Repayment for the Gulf War has, however, reversed Saudi Arabia's balance of payments drastically since 1991.

While oil dominates the economic life and determines material conditions in Saudi Arabia, at a non-material or spiritual level, Saudi Arabia is the host and guardian of the birthplace of Mohammad and Islam (see Box 13.1). Mecca and Medina are the two most holy places in Islam, and every Muslim should visit Mecca at least once in his/her lifetime, financially an unlikely prospect for the mass of Muslims living in such distant and poor countries as Morocco and Indonesia.

Among the smaller states of the Arabian Peninsula there are sharp contrasts between those with large oil reserves and service activities connected with the industry, Kuwait and the United Arab Emirates each having about 10 per cent of the world's oil reserves, and Yemen, with more limited reserves and a population mostly engaged in agriculture. So long as the rest of the world needs the oil of the Arabian Peninsula, even the poorer areas can expect some material improvements. Fortunately for the rest of the world, the Arabian Peninsula is not dominated by the religious fundamentalism of Shiite Muslims, in spite of the presence of Mecca.

After its unsuccessful attempt to keep Kuwait, since 1991 Iraq has been under severe sanctions and in theory has been banned from exporting oil. After more than a decade, in which about a third of a million men were lost in the war with Iran, a period of dire austerity is the last thing Iraq needs. Even so, ethnic cleansing has proceeded both in the north against the Kurds and in the southeast against the marsh Arabs. With about 55,000 sq km (21,200 sq miles) of cultivated land, 12 per cent of the national area, almost half of it irrigated, Iraq is better able than Saudi Arabia to support its population in food, hence its ability to survive sanctions. Considerable time will be needed, even without further conflicts with its neighbours, before Iraq can restore normal conditions and make positive use of its oil wealth.

BOX 13.2

The Kurds and human rights

According to Short and McDermott (1975, p. 4), the Kurdish people, a national group without an independent state, constitute a single nation which has occupied its present habitat for at least three thousand years. They have outlived the rise and fall of many imperial races: Assyrians, Persians, Greeks, Romans, Arabs, Mongols, Turks. They have their own history, language and culture. Today almost all the Kurds live in Turkey, Iran, Iraq or Syria, but in addition about 250,000 live in Azerbaijan and Armenia (former USSR) and 50,000 in Lebanon.

The basic claim of the Kurds to national status is their language. It is a member of the Indo-European language family and is related to Persian, the main language of Iran. Turkish in Turkey and Arabic in Iraq and Syria are members of entirely different language families. Almost all Kurds are Muslims, so this cultural characteristic does not distinguish them from the rest of the population in the countries in which they live. The exact number of Kurds in Southwest Asia is not known. Table 13.5 shows the author's estimates for 1994, based on figures for the late 1960s in Short and McDermott (1975, p. 6) and for the early 1980s in Gunnemark and Kenrick (1985, p. 122). There are roughly 22 million Kurds, assuming that, in spite of attacks on them in Iraq and Turkey, their population has grown at the same rate as the rest of the population in the countries in which they live.

TABLE 13.5 The distribution of the Kurds

	Total population 1994 (millions)	*Estimate of Kurds (millions)*	*Kurds as percentage of total*
Turkey	62	9.3	15
Iran	61	7.3	12
Iraq	20	4.2	21
Syria	14	1.0	7

Like the Poles in Europe for a time, the Kurds have been partitioned among more powerful neighbours. In the seventeenth century the Kurds were split roughly 3 to 1 between the Ottoman Empire based in Turkey to the west and Persia (now Iran) to the east. Until the end of the First World War most Kurds, together with Syrians, Iraqis and other peoples of Southwest Asia, were administered from Constantinople. With the final break-up of the Ottoman Empire, new political units were formed, including Iraq, a British mandate, and Syria, a French one. The Kurds were thus partitioned between four political units. In spite of the intentions in the Treaty of Sèvres (1920), in which the creation of a Kurdish state was recommended, nothing came of the proposal. There were Kurdish uprisings in northeast Iraq in the 1920s, and in north Iraq in the 1930s. More recently, the Kurds of Iraq and Iran have found themselves on opposite sides of the conflict in the 1980s. During that period and since the Gulf War (1991), the Kurds in Iraq have come under attack from

TABLE 13.6 All international trade in billions of US dollars

	1980	*1990*
Total value of exports	2,001	3,392
Mineral fuel exports[1]	481	343
OPEC's share thereof	291	133
Developed countries' imports from OPEC	222	87

Note: [1] Commodity class 'mineral fuels and related materials'
Source: *UNSYB*, no. 38, Table 118

13.6 Iran and Afghanistan

The modern state of Iran is based on the Persian people, who inhabit the lands between the Caspian Sea in the north and the Persian Gulf in the south. There are, however, many ethnic minorities, including the Azerbaijanis and Kurds, with clearly defined homelands, Arabs widely scattered in the southern part of the country, Beluchis, separated from fellow Beluchis now in Pakistan, and Turkmeni people along the border with the newly independent Turkmenistan, formerly a Soviet (Socialist) Republic.

Iran was not systematically colonised in the nineteenth century by European powers as were North Africa, South Asia and Central Asia. Russia and Britain had spheres of influence in Persia, but agreed to leave the region independent. During the Second World War, on the grounds that Germany was hoping to reach oil supplies in the Middle East, the USSR and the UK occupied Iran (then Persia), subsequently withdrawing their troops in 1946. The first of a number of takeovers of foreign oil companies in Southwest Asia was carried out by Persia through Moussadeq in 1950. Relations between Britain and Persia were sufficiently restored for Persia to become a partner with Turkey, Iraq, Pakistan and the UK in the Central Treaty Organisation (CENTO), forming a protective screen to deter a possible Soviet invasion of the Middle East. After a royalist coup, Persia, renamed Iran, was ruled by the Shah between 1953 and 1979. During that time, changes aimed at modernising the country led to a more active role for women in society, at least in the cities. Ambitious plans were proposed to make Iran the

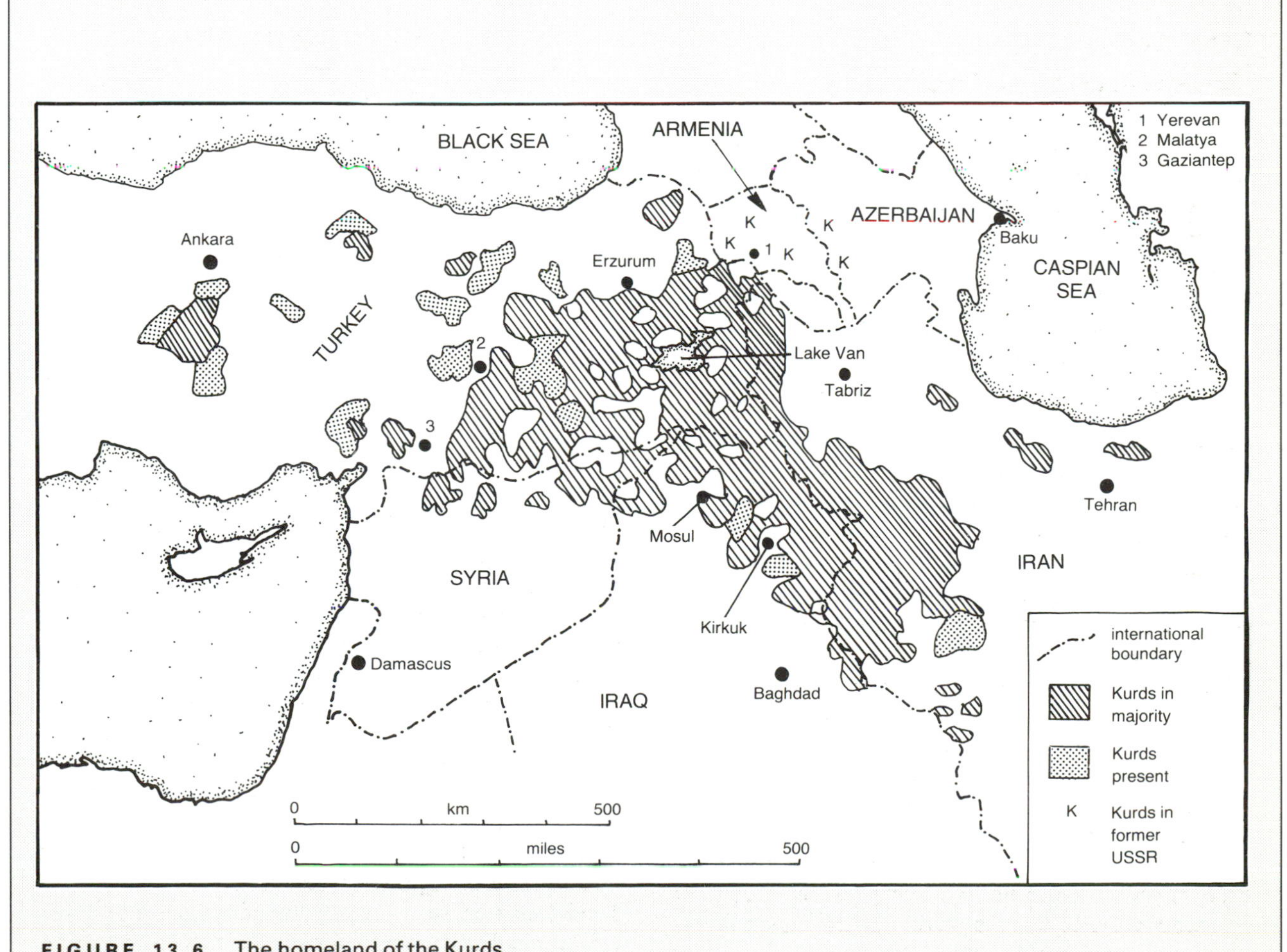

FIGURE 13.6 The homeland of the Kurds

fourth most powerful industrial country in the world by 2000 or whenever. In 1979 this process was stopped, Iran's resources were directed to preventing Iraq from occupying its oil-fields, and the USA was humiliated by the siege of its embassy in Tehran during the period 1979–81.

Little has come of the Shah's earlier plans to turn Iran into a leading industrial power. In both 1970 and 1991, oil accounted for 90 per cent of the value of the country's exports and over 80 per cent of the imports were manufactures. Iran's production of 1 million tonnes of steel in 1990 is only one-tenth as large as Turkey's, and only half of that in the newly established Saudi Arabian steel industry.

Agriculture still employed 27 per cent of the labour force in Iran in 1991 (cf. 40 per cent as recently as 1975) and the absolute number employed has actually risen slightly, whereas the area under arable and permanent crops (150,000 sq km – 58,000 sq miles – in 1990) is less than it was before the start of the war with Iraq. About one-third of the cultivated land is irrigated and highly productive, but yields of cereals and other basic crops elsewhere are low.

In the early 1990s, Iran was characterised by a generally backward agricultural sector, little modern industry, and an increasing output of oil. More of the same might seem the prospect for the immediate future, but there must be concern about the rapid growth of total population in the country as a whole, expected to grow from 63 million in 1993 to 162 million in 2025, and in the capital Tehran in particular, now with about 7 million inhabitants. Agriculture does not need more workers, the extraction of oil and natural gas accounts directly only for a tiny fraction of the economically active population, and industrial capacity is small. Growth of the service sector is therefore the hope for the future if unemployment is not to rise sharply.

Afghanistan is the poorest country in Asia and was ranked 158 out of 160 in the world on the 1992 UN Human Development Index, with only Sierra Leone and Guinea below it. Its first-ever census was held only in 1979, and with the unending internal conflict since then it is difficult to know precisely the population size

BOX 13.2 *continued*

the Iraqi army, while dissidence among the Kurds of Turkey has been severely suppressed.

Figure 13.6 shows the distribution of Kurds in Southwest Asia. They are in the majority in an area 1,000 km (620 miles) long and averaging 350 km (220 miles) in width, extending from eastern Turkey southeast along each side of the Iraq–Iran divide. The economy of the Kurds is predominantly agricultural, but the products from their mainly rugged, generally dry, environment are of little interest in world markets. Although the main oil reserves of Iraq, about 10 per cent of the world total, are in an area in which there are many Kurds, the industry is dominated by Iraqis.

The Kurds have a convincing cultural argument for the creation of an independent Kurdistan, whatever the strength of their economy. With an area of about 350,000 sq km (135,000 sq miles), Kurdistan is equal in area to Germany and almost as large as Japan. Its population of over 20 million compares in size with those of Venezuela and the newly emerged Uzbekistan. Kurdistan has the attributes of a nation state, but not the location. It would be landlocked – a drawback, but not a reason in itself for preventing its independence. More importantly, the extremely sensitive political situation and proximity to the massive Middle East oil reserves inhibit any of the more powerful countries of the world from giving more than token support to an independence movement for Kurdistan. Like many other cultural groups in the world the Kurds continue to suffer from injustices inflicted by the more powerful central governments of the countries in which they live. It is appropriate here to reproduce a selection of key Articles from the Universal Declaration of Human Rights adopted by the General Assembly of the United Nations on 10 December 1948:

Article 1

All human beings are born free and equal in dignity and rights. They are endowed with reason and conscience and should act towards one another in a spirit of brotherhood.

Article 2

Everyone is entitled to all the rights and freedoms set forth in this Declaration, without distinction of any kind, such as race, colour, sex, language, religion, political or other opinion, national or social origin, property, birth or other status. Furthermore, no distinction shall be made on the basis of the political, jurisdictional or international status of the country or territory to which a person belongs, whether it be independent, trust, non-self-governing or under any other limitation of sovereignty.

Article 10

Everyone is entitled in full equality to a fair and public hearing by an independent and impartial tribunal, in the determination of his rights and obligations and of any criminal charge against him.

of the country and to have accurate economic data. Afghanistan is only slightly smaller in area than the US state of Texas and has 17–18 million people. It is credited with the highest infant mortality rate of any country in the world, 170 per thousand, and also with one of the highest total fertility rates, 6.9. Life expectancy is very low, the proportion of under-15s very high.

Most of Afghanistan is rugged mountain or dry plateau. No less than 40 per cent of the land is classified as being of no bioclimatic use, about 45 per cent is natural pasture and only 12 per cent is cultivated. Afghanistan has few mineral or forest resources, lacks sources of water for irrigation, and depends heavily therefore on low-yielding cereal crops and livestock raising. Over half of the economically active population is still in the agricultural sector.

In the nineteenth century Afghanistan was considered by both Russia and Britain as a region that might be brought into their empires. Its remoteness and the ability to resist invaders spared it that fate. After the Second World War the USSR and the USA both took an interest in Afghanistan and competed to provide considerable development assistance. The main routes out of landlocked Afghanistan cross into former Soviet Central Asia in the north, Pakistan in the southeast, and Iran in the southwest.

The 1980s was a devastating decade for Afghanistan, with the USSR attempting to gain control of the country, possibly with a view to establishing a link between its southern border and the Indian Ocean. The USA supplied arms to rebel forces, there was heavy damage to the limited infrastructure of the country, and there was a mass exodus of refugees into neighbouring Pakistan. Like Viet Nam and many other parts of the developing world, Afghanistan thus served as a battleground for the superpowers in the Cold War, the Americans achieving some revenge for Soviet support of North Viet Nam in the 1960s and 1970s. During the Cold War Afghanistan was an area of considerable strategic significance. Regionally, on the other hand, it is a backwater, unlikely to grow rich on oil revenues in the way some other areas in Southwest Asia have, or to industrialise like many areas in South and Southeast

Article 19

Everyone has the right to freedom of opinion and expression; this right includes freedom to hold opinions without interference and to seek, receive and impart information and ideas through any media and regardless of frontiers.

Article 20

(1) Everyone has the right to freedom of peaceful assembly and association.

(2) No one may be compelled to belong to an association.

To conclude this summary of the position of one minority group, the Kurds, there follows a list of topics covered in the first twenty-seven Reports of the Minority Rights Group up to 1975. The topics are grouped into the twelve major world regions used in the present book.

1, 2. Western and Central Europe: the two Irelands, the Basques, the Gypsies of Europe

3. Former USSR: religious minorities, the Crimean Tatars, Volga Germans and Meskhetians

4. Japan: the Burakumin, Koreans and Ainu

5. USA (and Britain): race and law

6. Canada: Indians

7. Latin America: Blacks in Brazilian society, East Indians of Trinidad and Guyana, Amerindians of South America

8. Africa: Asian minorities of East and Central Africa, southern Sudan and Eritrea, Africans in Rhodesia, Namibians of Southwest Africa, genocide in Burundi

9. Southwest Asia: Israel's Oriental Immigrants and Druzes, the Kurds, the Palestinians, Arab women

10. South Asia: the Biharis in Bangladesh, India and the Nagas, the Tamils of Sri Lanka, the Untouchables of India

11. Southeast Asia: the Chinese in Indonesia, the Philippines and Malaysia, the Montagnards of South Viet Nam

The above may serve as a checklist for further topics on human rights to investigate. Some of the issues have largely been resolved in the last twenty years (e.g. Rhodesia and Namibia), but the list could be extended to include many other issues in 1975 and/or now, including, for example, Australia (aborigines), Yugoslavia (ethnic cleansing), Argentina (human rights), Haiti (military oppression), Indonesia (Timor), Tibet (Chinese attempts to eradicate many cultural features). The reader may find it illuminating to add to the list. It should be remembered, however, that while it now represents almost all of the countries of the world, the UN does not have a monopoly of truth in defining human rights, a point clearly made by the Chinese Communist Party, which rates the right to a job, for example, higher than freedom of speech and peaceful assembly.

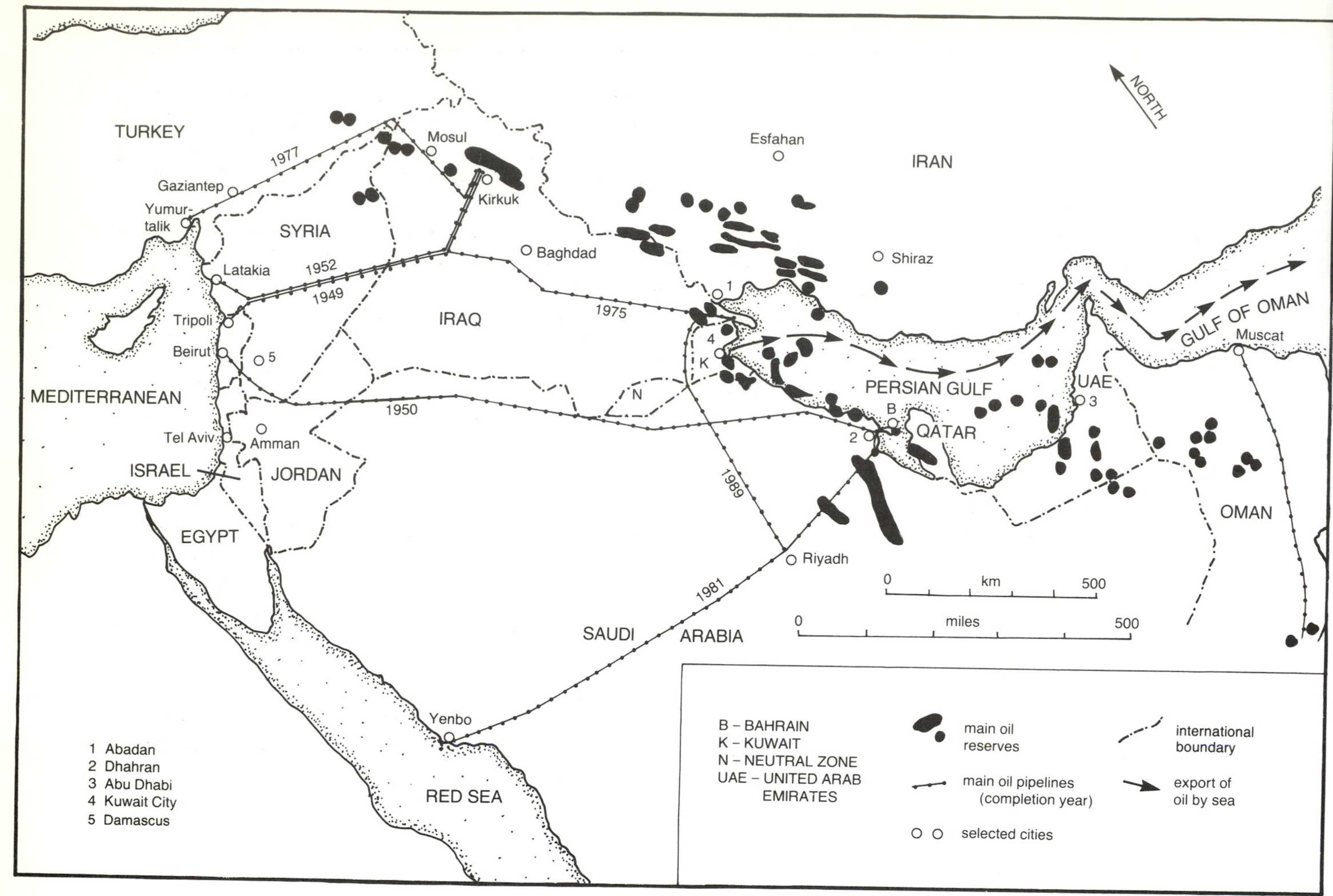

FIGURE 13.7 Oil in Southwest Asia

Asia. The emergence of three new countries, former Soviet Socialist Republics, on its northern border, may affect it economically and in other ways. On the other hand, with the greatly reduced Russian influence and presence in Central Asia, Turkmenistan, Uzbekistan and Tadjikistan seem destined to revert to conditions similar to those in Afghanistan.

The natural resource base of North Africa and Southwest Asia is distinguished by the reserves of oil and natural gas. Low rainfall makes water supply a problem almost everywhere, except in parts of Turkey and Iran. Apart from the processing of oil, there is little industry. Several of the countries depend heavily on the oil industry for government revenues and direct and indirect employment. If oil consumption is reduced in the industrial countries as a result of progress in developing renewable energy sources or through concern over the effect on the environment of the combustion of fossil fuels, the economies of the oil producers would suffer greatly. Again, the discovery of large new oil reserves elsewhere in the world could reduce the influence of the Middle East on oil prices.

BOX 13.3

OPEC versus OECD

The Organisation of Petroleum Exporting countries (OPEC), with its headquarters in Vienna, has thirteen members (see Table 13.8), of which six are in Southwest Asia and two in North Africa. The Organisation for Economic Cooperation and Development (OECD), with its headquarters in Paris, has 24 members, the 'western' industrial countries, of which 17 are in Europe. In 1992, world oil production was 3,170 million tonnes. OPEC countries produced 1,284 million tonnes, or 40.5 per cent of the total. In the same year the OECD countries consumed 1,799 million tonnes, or 57.6 per cent of the world total.

In 1992, OPEC countries consumed only 248 million tonnes of oil, 19.2 per cent of their production, and exported 1,037 million tonnes. In the same year, OECD countries produced only 772 million tonnes of oil and therefore imported 1,027 million tonnes to make up the 1,799 million tonnes they consumed. Most but not all of the trade in oil and oil products is accounted for by the flow from OPEC to OECD. There are, however, many other net exporters of oil, including developing countries not in OPEC (e.g. Mexico, Oman) as well as developed ones (e.g. Canada, Norway). An even larger number of developing countries are net importers of oil (e.g. India, Brazil, Ethiopia).

Much of the 'battle' over oil prices since the early 1970s has been fought between OPEC and OECD countries. The former pushed up prices in 1973–4 and again in 1979. The latter have at times reduced imports of oil from OPEC countries, seeking non-OPEC sources of oil, cutting consumption, and encouraging the use of other sources of fuel. Figure 13.8 shows price fluctuations since 1973 and events connected with changes in production and prices. Although in theory prices should roughly reflect production costs, in practice OPEC members have a very large proportion of the world's oil reserves (about two-thirds) and of production (about two-fifths), which means that if they agree on output and prices (not always the case), they can keep prices many times higher than the very low cost of extracting Middle East oil. The artificially engineered rises in oil prices, which actually started in the late 1960s, have halted the rapid increase in oil consumption observed in the developed countries in the 1950s and 1960s. The effect has been to prolong the life of oil reserves beyond that expected if consumption had continued to rise at the rate it was doing during those two decades.

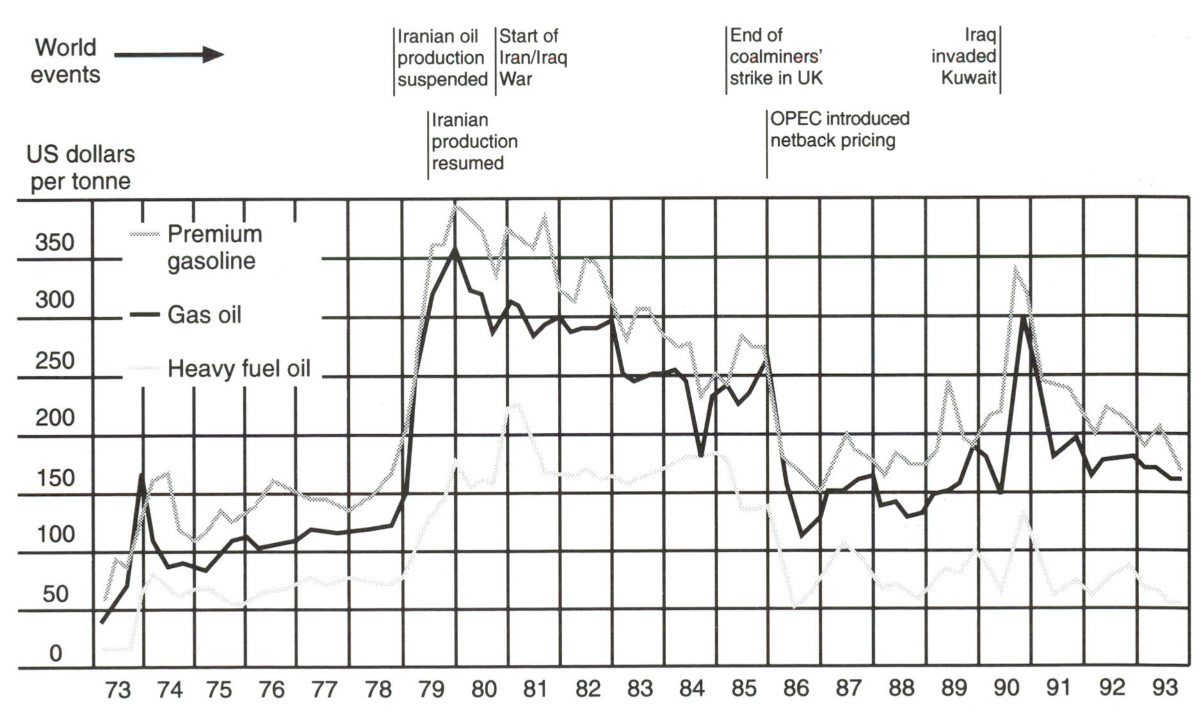

FIGURE 13.8 Fluctuations in the price of oil 1973–93

Source: BP (1991), p. 13 and (1994), p. 13

BOX 13.3 *continued*

The advantages of oil over other sources of energy have been its versatility and its transportability. A tonne of oil contains roughly 50 per cent more energy than a tonne of coal, and can be refined to make liquid fuel, essential for most modern forms of transport, far more cheaply than coal can be processed for this purpose. Oil remains in value the largest single commodity handled in international trade, but its importance relative to other commodities has fluctuated markedly in the last two decades, as will be shown by the example below.

Between 1980 and 1990 the total value of international trade grew from about 2,000 billion to almost 3,400 billion dollars. The contribution of mineral fuels to this total dropped, however, from almost 25 per cent to a mere 10 per cent, although the amount traded changed little. The year 1980 was immediately after the 1979 rise in oil prices. In 1990 prices started low, although they rose sharply after Iraq's invasion of Kuwait in the summer of that year. The commodity class considered here includes trade in coal, gas and other items, but oil was by far the largest contributor. Thus to the question of how important the oil industry is in the world economy, a precise answer is impossible. Much of the increase in the total value of international trade between 1980 and 1990, after the gradual reduction in the real value of the dollar is taken into account, was accounted for by more than a doubling in the value of manufactured goods traded.

From trends in the 1970s and 1980s it seems reasonable to propose the following prospects for the 1990s and early part of the twenty-first century.

1. The OPEC members and other developing countries exporting oil cannot hope for further massive increases in oil prices. On the other hand, it is in their interest not to exhaust their reserves too quickly. In time they should be able to make greater use of the oil they produce in their own economies.

2. The developed countries are likely gradually to reduce their need for imported oil. They can do this because (see e.g. Larson *et al.* 1986) they are constantly using fuel and raw materials more efficiently, and their populations are not growing fast. In particular, they have the resources to fund research to develop alternative energy sources and substitutes for oil. Nevertheless, as Table 13.7 shows, in North America and Western Europe, oil companies, all in practice involved also with other energy sources, figure among the largest companies.

3. The many developing countries that are importers of oil have generally suffered most in the last two decades. Oil is the main source of commercial energy used in most countries, albeit in very small quantities. Such countries have been affected adversely by the rise in oil prices and also in the price of many manufactured goods compared with the agricultural and/or non-fuel mineral primary commodities that they export.

TABLE 13.7 Major oil companies in the world, rank in value of turnover/sales

In the USA: Exxon Corpn (2), Mobil (6), Chevron (9), DuPont-De Nemours (11), Texaco (15), Amoco (17)
In Western Europe: Shell – UK and Netherlands (1), BP – UK (2), Elf Aquitaine – France (11), ENI – Italy (12), TOTAL – France (25), Petrofina – Belgium (32)

Source: Extel Financial (1993)

TABLE 13.8 OPEC oil production and consumption, 1992 and 1993

OPEC member	(1) *Region*	(2) *Oil production % of world*	(3) *Production (million tonnes)* 1992	1993	(4) *Consumption (million tonnes)*	(5) *% used at home*	(6) *GNP in US dollars per head*
1 Saudi Arabia	SWA	13.5	428	425	53	12	7,940
2 Iran	SWA	5.4	172	180	57	33	2,190
3 Venezuela	LAM	4.1	129	132	27	21	2,900
4 United Arab Emirates	SWA	3.8	119	115	10	8	22,220
5 Nigeria	AFR	2.9	92	94	15	16	320
6 Indonesia	SEA	2.3	74	73	36	49	670
7 Libya	NAF	2.3	73	68	10	14	na
8 Algeria	NAF	1.8	57	56	10	18	1,830
9 Kuwait	SWA	1.4	46	97	5	11	na
10 Iraq	SWA	0.7	24	22	15	63	na
11 Qatar	SWA	0.7	23	23	2	9	16,240
12 Ecuador	LAM	0.5	17	18	7	41	1,070
13 Gabon	AFR	0.5	15	15	1	7	4,450
OPEC total		39.9	1,269[1]	1,318	248	20	

Notes:
na = not available
[1] Discrepancy in source, with 1,284 given there as OPEC total for 1992 and 1,300 for 1993
Sources: (2) and (3): BP (1993) and BP (1994); (4) *UNSYB*, no. 38, Table 106; (6) *WPDS* (1994)

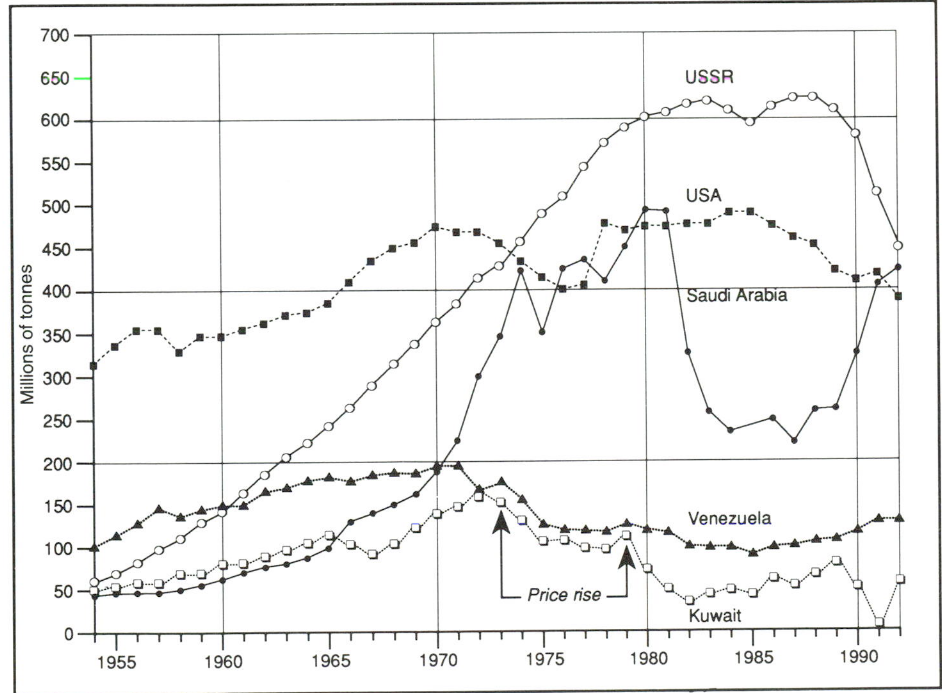

FIGURE 13.9 Oil production in the USA, USSR, Saudi Arabia, Venezuela and Kuwait, 1954–92
Source: various numbers of *UNSYB* and BP (1993)

KEY FACTS

1. Several of the oldest civilisations in the world have originated in or taken possession of the lands of Southwest Asia and North Africa.

2. Much of the region is thinly populated on account of lack of rain and, as in ancient times, population is concentrated in river valleys where irrigated agriculture is possible.

3. Population is expected to double in 25 to 30 years, increasing pressure on water supplies. Family planning is not widely publicised and in 1994 Libya was one of the countries expressing a pro-natalist attitude at the Cairo Population Conference.

4. Although there is great ethnic diversity in some areas, there are two cohesive cultural features in the region: the Arabic language and the almost universal acceptance of Islam as the religion of the region.

5. The discovery of large oil reserves in Southwest Asia since the 1920s has given the region a dominant position in the world oil industry, enabling producers to influence prices and to establish a favourable balance of trade with the rest of the world.

6. Genuine democratic elections have been rare or non-existent in most countries. Civilian or military leaders with dictatorial powers have frequently been in conflict with their neighbours (e.g. Morocco and Algeria, Libya and Egypt, Iraq and Iran, Kuwait). The major industrial countries, heavily dependent on Middle East oil, have had to deal diplomatically with the leaders of the region through self-interest.

MATTERS OF OPINION

1. Should greater equality between the sexes be a goal for the politicians in the region?

2. Should oil production be regulated more rigorously by OPEC countries than it is to extend the 'life' of the oil and gas reserves and to avoid sharp fluctuations in the price of crude oil?

3. Should the very uneven distribution of wealth in most countries be reduced by the appropriate application of taxes and stronger welfare services?

SUBJECTS FOR SPECULATION

1. What do you think will happen to the countries of North Africa and Southwest Asia when the oil reserves dry up or when other countries no longer need to import (much) oil?

2. Do you think water supply can be maintained in face of the fast-growing population?

3. Do you think the influence of Islam on political and economic affairs in the region will increase?

4. Could industrialisation usefully complement the oil and gas industry in the future as it eventually declines?

Chapter fourteen

South Asia

The Indian growth rate has been low in comparison not only with the targeted growth rates but also with the rates achieved in most other countries. . . . All that can be said about the growth rate in India is that it is more than twice the growth rate the Indian economy recorded in the final 47 years of British rule from 1900 to 1947.

Raj Krishna (1980)

14.1 Introduction (see Table 14.1, South Asia calendar)

Virtually the whole of South Asia was under British colonial rule for between 100 and 200 years, some areas having a greater degree of autonomy than others within British India. Burma, now Myanmar, adjoining India, was also a British colony, but culturally it differs greatly from the rest of South Asia and it has been included in Chapter 15 on Southeast Asia, to which it is more culturally similar. In 1947 India became independent from the UK, splitting initially into India itself, predominantly Hindu in religion, Pakistan, with West and East territories on either side of India, predominantly Muslim, and Ceylon, now Sri Lanka, a separate island, with a strong Buddhist tradition. In addition, South Asia as understood here includes the Kingdom of Nepal and also Bhutan to the north of India. Subsequently, in 1971 West and East Pakistan separated when the independence of Bangladesh in the east was recognised by (West) Pakistan. Tables 15.2, 15.3 and 15.4 in Chapter 15 include data for the main countries of South Asia.

In 1993, South Asia had a population of 1,173 million (cf. China, 1,179), increasing by about 2.3 per cent a year, and expected to reach 1,933 million in 2025. With 912 million people in mid-1994, India has over three-quarters of the population of the region, although its population is not growing as fast as the populations of Pakistan and Bangladesh. In 1750 South Asia already had a large population for that time, rather less than 20 per cent of the world total, exceeded only by China. During the period of European influence, the population increased threefold, from 150 to 450 million. By the year 2000, it could reach 1,350 million, a threefold increase in only fifty years, but with a much greater absolute number added, i.e. 900 million. The extent to which British colonial rule contributed to such a massive increase in population in South Asia is a matter of speculation. Infant mortality rates of around 100 per thousand live births are higher than in any other major region of the world except Africa south of the Sahara, while the total fertility rate of 3.9 in India, 4.9 in Bangladesh and 6.7 in Pakistan will ensure a large increase in population for many decades unless drastic reductions occur.

According to the author's assessment of the global distribution of natural resources, South Asia has a smaller amount per inhabitant than any major region apart from Japan and South Korea, with the dubious qualification that its level of consumption of food, energy and raw materials per inhabitant is also very low. With 21.2 per cent of the total population of the world, South Asia has 11.6 per cent of the productive land, 5.4 per cent of the fresh water, 2.8 per cent of the fossil fuels and 2.1 per cent of the non-fuel minerals. Further reserves of minerals should be discovered in time, but for the size of population the total area is small, being little more than half of the area of Australia while having about 65 times as many people. Furthermore, every eight months the population of South Asia increases by 18 million people, equivalent to the total population of Australia.

14.2 General features of India

Although population is growing more quickly in many of the world's developing countries than in India, it is the massive total that now draws attention to the country. The growth of India's population was slow until the 1920s, after which the rate increased from about 1 per cent a year around 1930 to over 2 per cent in the 1960s (see Table 14.2). In spite of great efforts to encourage family planning practices, in most parts of India the improved healthcare services have cancelled out family planning efforts, lowering mortality as fast as fertility. Life expectancy was 32 years in 1950–1, 59 years in 1993. There are, however, regional differences in the fertility rate among the major administrative divisions, with the states of Kerala and Tamil Nadu recording increases of 14 and 15 per cent in the decade 1981–91 (see Table 14.3), compared with 24 per cent for the whole of India and over 30 per cent in some of the peripheral units in the northeast of the country. According to Todd (1987) some of the credit for the reduction in birthrate in southern India can be attributed to the comparatively high levels of female literacy, and also to the Dravidian matrilineal descent system, giving women more influence in family matters than they have in Muslim countries or in much of northern India (see Table 14.4).

Figure 14.1 shows the population structure of India in the late 1980s. Whatever happens in the 1990s, it is clear that the cohorts of potential mothers in the next three to four decades already point to large numbers of births, even if fertility continues to fall. At the same time, the proportion of elderly is small, so there will not be a corresponding number of deaths. Thirty-six per cent of the population of India was under the age of 15 years in 1993 compared with only 4 per cent of 65 and over.

Unlike China, India does not have extensive areas of land with a very low density of population, but the data in column (3) of Table 14.3 show very marked differences in density among the twenty-five states, even without considering the small urban union territories (see Figure 14.2, map of the states of India). Low densities occur in the states located entirely or mainly in the mountain fringes of the north and northeast, and also in Rajasthan, which includes India's main area of desert. Densities are very high not only in the irrigated lowlands of the Ganges valley and delta, especially West Bengal, but also in Kerala in the extreme southwest.

Although only 25 per cent of the population of India is defined as urban, the country has three very large urban agglomerations, Delhi, Calcutta and Bombay, each with somewhere near 10 million inhabitants, the first the fast-growing capital, the other two the centres of two of the main industrial regions of the country. About twelve other cities each have between 1 and 5 million inhabitants. Although there has been a steady net flow of migrants from rural areas to urban centres, the agricultural population of India continues to grow, and 66 per cent of the economically active population was still in agriculture in 1991.

Over the last several thousand years South Asia has

TABLE 14.1 Calendar for South Asia

1175	Muizzudin Muhammad of Ghazni founds first Muslim empire in India
1398	Timur invades India and sacks Delhi
1526	Babur conquers Kingdom of Delhi and founds Mughal dynasty
1690	Foundation of Calcutta by English
1707	Death of Aurangzeb: decline of Mughal power in India
1751	French gain control of Deccan
1757	Battle of Plassey: British destroy French power in India
1796	British conquer Ceylon (now Sri Lanka)
1818	Britain defeats Marath and becomes effective ruler of India
1845–9	British conquest of Punjab and Kashmir
1853	First railway and telegraph lines in India
1857	Indian Mutiny
1877	Queen Victoria proclaimed Empress of India
1885	Foundation of Indian National Congress
1904	Partition of Bengal: nationalist agitation in India
1919	Amritsar incident; upsurge of Indian nationalism
1947	India and Pakistan independent
1948	Death of Mahatma Gandhi (1869–1948), leader of India's struggle for independence
1961	Population of India 436 million
1962	War between India and China
1964	Death of Jawaharlal Nehru (1889–1964) first prime minister of independent India
1965	War between India and Pakistan
1971	East Pakistan (now Bangladesh) breaks away from West Pakistan (now Pakistan)
1994	Population of India 912 million

TABLE 14.2 The growth of India's population

Year	*Population in millions*	*Average annual growth (%)*	*Year*	*Population in millions*	*Average annual growth (%)*
1901	238.4	–	1951	361.1	1.25
1911	252.1	0.56	1961	439.2	1.96
1921	251.3	–0.03	1971	548.2	2.20
1931	279.0	1.04	1981	683.3	2.22
1941	318.7	1.33	1991	844.3	2.11

Note: PRB estimates: 1,166 million in 2010, 1,380 million in 2025
Source: Census of India 1991: provisional population totals

TABLE 14.3 Linguistic and demographic aspects of the states and territories of India

	(1)	*(2)*		*(3)*		*(4)*	*(5)*	*(6)*
	Language[1]	*Area in thous. sq km/(miles)*		*Density per sq km/(mile) 1991*		*Population in millions 1981*	*1991*	*Change*[4] *1981–91 (1981 = 100)*
1 Andhra Pradesh	D Telegu	277	(107)	240	(622)	53.6	66.4	124
2 Arunachal Pradesh	I Miri	84	(32)	10	(26)	0.6	0.9	137
3 Assam	I Assamese	78	(30)	284	(736)	18.0	22.3	124
4 Bihar	I Bihari	174	(67)	496	(1,285)	69.9	86.3	124
5 Goa	I Konkani	4	(1.5)	316	(818)	1.0	1.2	116
6 Gujarat	I Gujarati	196	(76)	210	(544)	34.1	41.2	121
7 Haryana	I Hindi	44	(17)	369	(956)	12.9	16.3	128
8 Himachal Pradesh	I Hindi	56	(22)	92	(238)	4.3	5.1	121
9 Jammu and Kashmir	I Kashmiri	222	(86)	35	(91)	6.0	7.8	129
10 Karnataka	D Kannada	192	(74)	233	(603)	37.1	44.8	121
11 Kerala	D Malayalam	39	(15)	747	(1,935)	25.5	29.0	114
12 Madhya Pradesh	I Hindi	443	(171)	149	(386)	52.3	66.1	127
13 Maharashtra	I Marathi	308	(119)	256	(663)	62.8	78.7	126
14 Manipur	X Manipuri	22	(8)	82	(212)	1.4	1.8	129
15 Meghalaya	X Khasi	22	(8)	78	(202)	1.3	1.8	133
16 Mizoram	X Mizo	21	(8)	32	(83)	0.5	0.7	140
17 Nagaland	X Naga	17	(7)	73	(189)	0.8	1.2	156
18 Orissa	I Oriya	156	(60)	202	(523)	26.4	31.5	120
19 Punjab	I Punjabi	50	(19)	401	(1,039)	16.8	20.2	121
20 Rajasthan	I Rajasthani	342	(132)	128	(332)	34.3	43.9	128
21 Sikkim	I Nepali	7	(3)	55	(142)	0.3	0.4	128
22 Tamil Nadu	D Tamil	130	(50)	428	(1,109)	48.4	55.6	115
23 Tripura	X Tripuri	10	(4)	262	(679)	2.1	2.7	134
24 Uttar Pradesh	I Hindi	294	(114)	472	(1,222)	110.9	139.0	125
25 West Bengal	I Bengali	88	(34)	774	(2,005)	54.6	68.0	125
26 Chandigarh[2]		0.1	(0.04)	5,620	(14,556)	0.5	0.6	142
27 Delhi[2]		1.5	(0.6)	6,310	(16,343)	6.2	9.4	151
28 Pondicherry[2]		0.5	(0.2)	1,681	(4,354)	0.6	0.8	134
Total		3,288[3]	(1,269)	257	(666)	683.3	844.3	124

Notes:
[1] Language family: I = Indo-European, D = Dravidian, X = other
[2] Union Territories. Omitted from the table are Andaman and Nicobar Islands, Dadra and Nagar Haveli, Daman and Diu, and Lakshadweep
[3] Total includes small areas not listed
[4] Calculated from more precise data than those in columns (4) and (5)

TABLE 14.4 Selected demographic and social data for the states of India and for the rest of South Asia

	(1) *Age at marriage in years*	*(2)* *Female literacy*	*(3)* *Infant mortality*	*(4)* *Birthrate*
Nepal	17.1	7	na	44
Sri Lanka	25.1	85	49	29
Pakistan	19.8	20	na	44
Bangladesh	16.3	na	130	46
India	17.2	37	125	36
Andhra Pradesh	16.2	31	122	38
Karnataka	17.8	40	89	33
Kerala	21.0	83	56	28
Orissa	17.3	31	127	37
Tamil Nadu	19.6	51	110	32
Maharashtra	17.6	28	83	30
Assam	18.6	40	124	35
Bihar	15.3	18	na	34
Gujarat	18.4	45	146	40
Madhya Pradesh	14.9	24	138	42
Punjab	20.0	52	108	33
Rajasthan	15.1	14	142	40
Uttar Pradesh	15.5	23	178	45
West Bengal	17.9	43	na	32

Description of columns:
na = not available
(1) Mean age of women at first marriage in years
(2) Literacy rate of women aged 15–19 per hundred
(3) Deaths of infants under the age of 1 year per 1,000 live births
(4) Birthrate per 1,000 total population
Points to note:
(1) The later age of marriage of women in Kerala and above all Sri Lanka
(2) The high level of literacy in Sri Lanka and Kerala contrasting with the low level in Rajasthan, Bihar and (Muslim) Pakistan
(3) Infant mortality rates *three* times higher in Uttar Pradesh than in Kerala or Sri Lanka
(4) Lowest birthrates in Kerala and Sri Lanka
Final thought: Kerala has a considerable proportion of Christians, mainly Roman Catholic
Source of data: Todd (1987), Appendix I, pp. 188–91, selected variables

been invaded many times in spite of the formidable physical barriers to the north. Arabian influence from the west has long been felt in India. Conquerors and settlers from outside have brought various languages and religions into the subcontinent, but often people with different cultural features live side by side. India was unified in 1526 under the Moghul emperors. After slow beginnings following the foundation of the East India Company in 1700, Britain gradually extended and consolidated its hold over almost all South Asia, influencing India profoundly in many ways, including its administrative structure and economy.

India is subdivided into two main language families, Indo-European in the north and Dravidian in the south. Of seventeen official languages (see Table 14.3), the four Dravidian languages are spoken in the most southerly states. In the remaining large states Indo-European languages are spoken, Hindi having the largest number of speakers. The total number of languages in India, not counting dialects, is estimated at around 300 (Gunnemark and Kenrick, 1985). The Indo-European languages are similar enough for Hindi to be understood in most areas except in the states where the Dravidian language predominates. English is spoken by about 3 per cent of the population, mostly the elite, and is used for both administrative and commercial purposes (see Table 14.5). Languages with completely different origins are spoken in many of the mountain areas of the northeast. The linguistic 'partition' of India has not, however, been such a great issue or problem as the division of South Asia on a religious basis.

When South Asia gained independence from Britain in 1947 an effort was made to satisfy the dominant religious groups, the Hindus (see Box 14.1) and the Muslims. When the boundaries were finalised, large numbers of each group found themselves on the 'wrong' side of the new international boundary line, a feature not uncommon in the world at various scales. Many people died during the partition and many

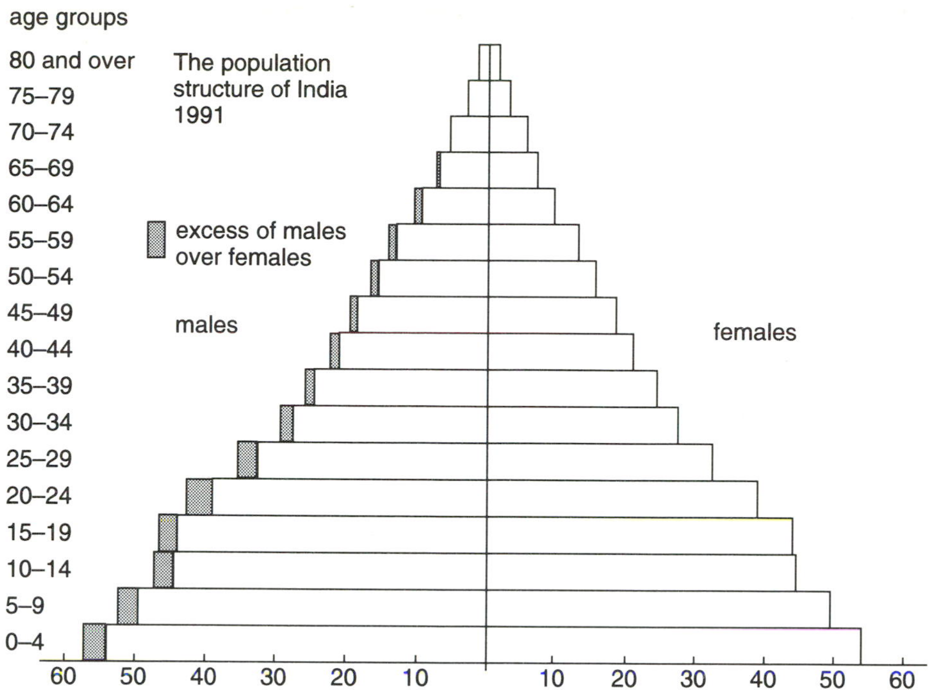

FIGURE 14.1 The population structure of India in 1991

others migrated across the new boundaries. Pakistan and Bangladesh are predominantly Muslim but India also has almost 100 million Muslims. To their credit, the leaders of India have succeeded in keeping this culturally heterogeneous mass of people in one political unit. States have considerable control of their internal affairs in a federal system and, in spite of conflicts on religious grounds, generally foster tolerance and mutual understanding between the main groups of people. There are, however, reports of suppression and persecution of minority ethnic groups, especially in the northeast, while the strong caste system has been difficult to break down, the untouchables in particular having difficulty in integrating normally in society.

Economically, India is also a country of great diversity. The range of crops cultivated and livestock raised is very large and varied, reflecting the wide variety of natural environments. There are reserves of many economic minerals, although only a few, notably coal, iron ore and manganese, are found in abundance. India satisfies almost all its needs for light industrial products and also has a considerable heavy industrial sector, capable of producing a wide range of capital goods such as transport and mining equipment. The data in Table 14.7 show, however, that in relation to its population size, India has far less than its 'share' of most natural resources and of products from mining and industry. Its per capita GNP (PRB source) of 330 dollars per head in 1992 contrasts with the average for all developing countries excluding China of 1,080, with 4,180 for the world as a whole, and with almost 16,000 for Europe. India's small total production of goods and services results mainly from the very low productivity of persons employed in the agricultural sector, a population dispersed in some 580,000 villages, the subject of the next section.

TABLE 14.5 Religion and language in India

Population by religion (%)		*Population by principal languages (%)*			
Hindus	82.6	Hindi	43	Gujarati	5
Muslims	11.4	Bengali	8	Malayalam	4
Christians	2.4	Telegu	8	Kannada	4
Sikhs	2.0	Marathi	8	Oriya	4
Buddhists	0.7	Tamil	6	Punjabi	3
Others	0.9	Urdu	6	Others	1

14.3 Indian agriculture (see Figures 14.3 and 14.4)

In the early 1990s, agriculture accounted for about two-thirds of the total economically active population of India. The proportion has declined gradually in the last two decades, by about 1 per cent every four years. On the other hand, the absolute number has been rising, and in the sixteen years from 1975 to 1991 it rose from 172 million to 218 million. During that period the area of cropland hardly altered, remaining around 1,680,000 sq km (650,000 sq miles), just over

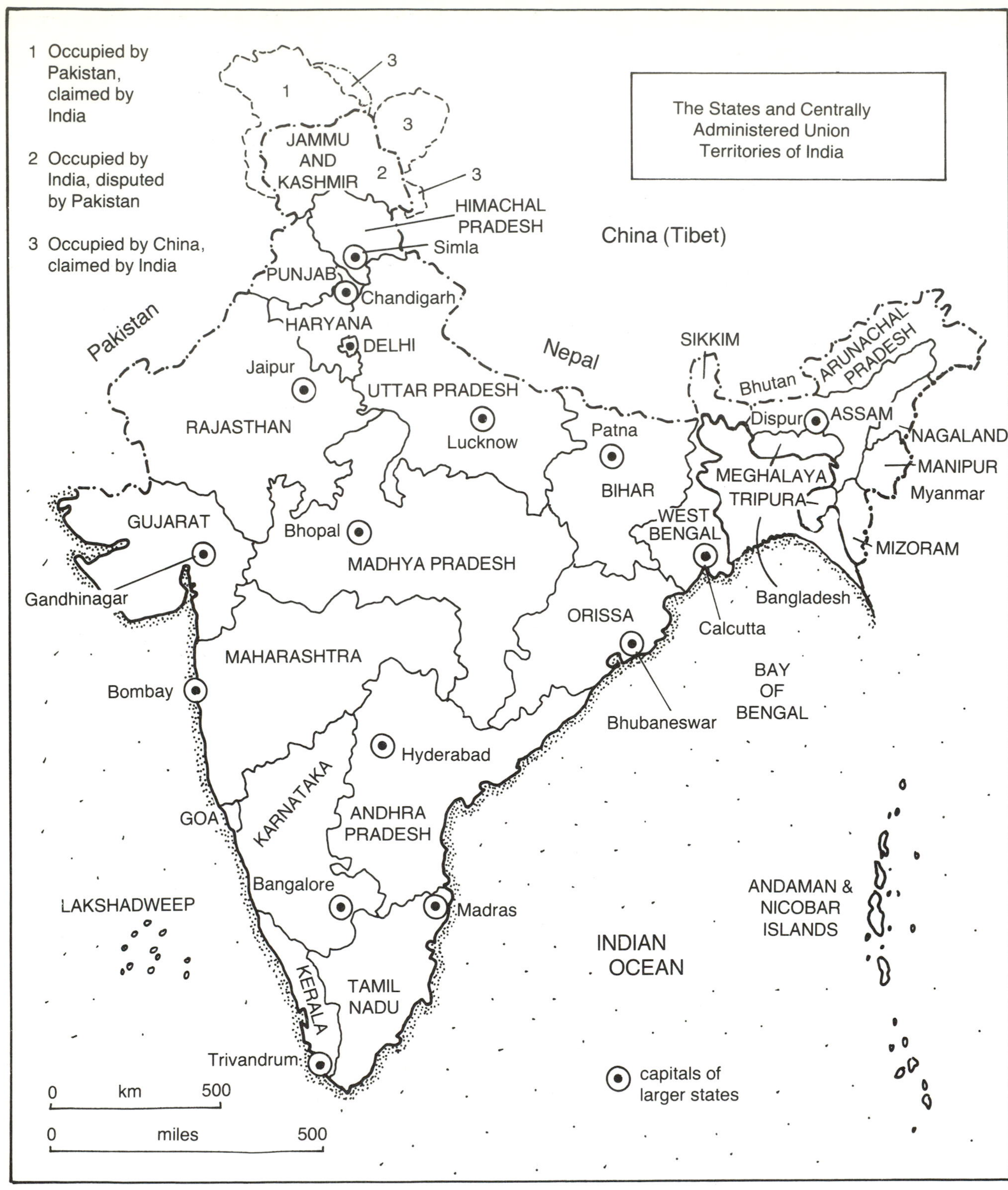

FIGURE 14.2 The states of India

half the national area. Thus in 1991, on average every 100 hectares (247 acres) of cultivated land were worked by 130 people, a figure contrasting with China at one extreme, where the number is about 400 per 100 hectares, and with Australia at the other, where each worker farms considerably more than 100 hectares. About a quarter of the cultivated land in India is irrigated and much of the cultivated land is double-cropped. Less than 4 per cent of the land in India is defined as permanent pasture, but 20 per cent is still

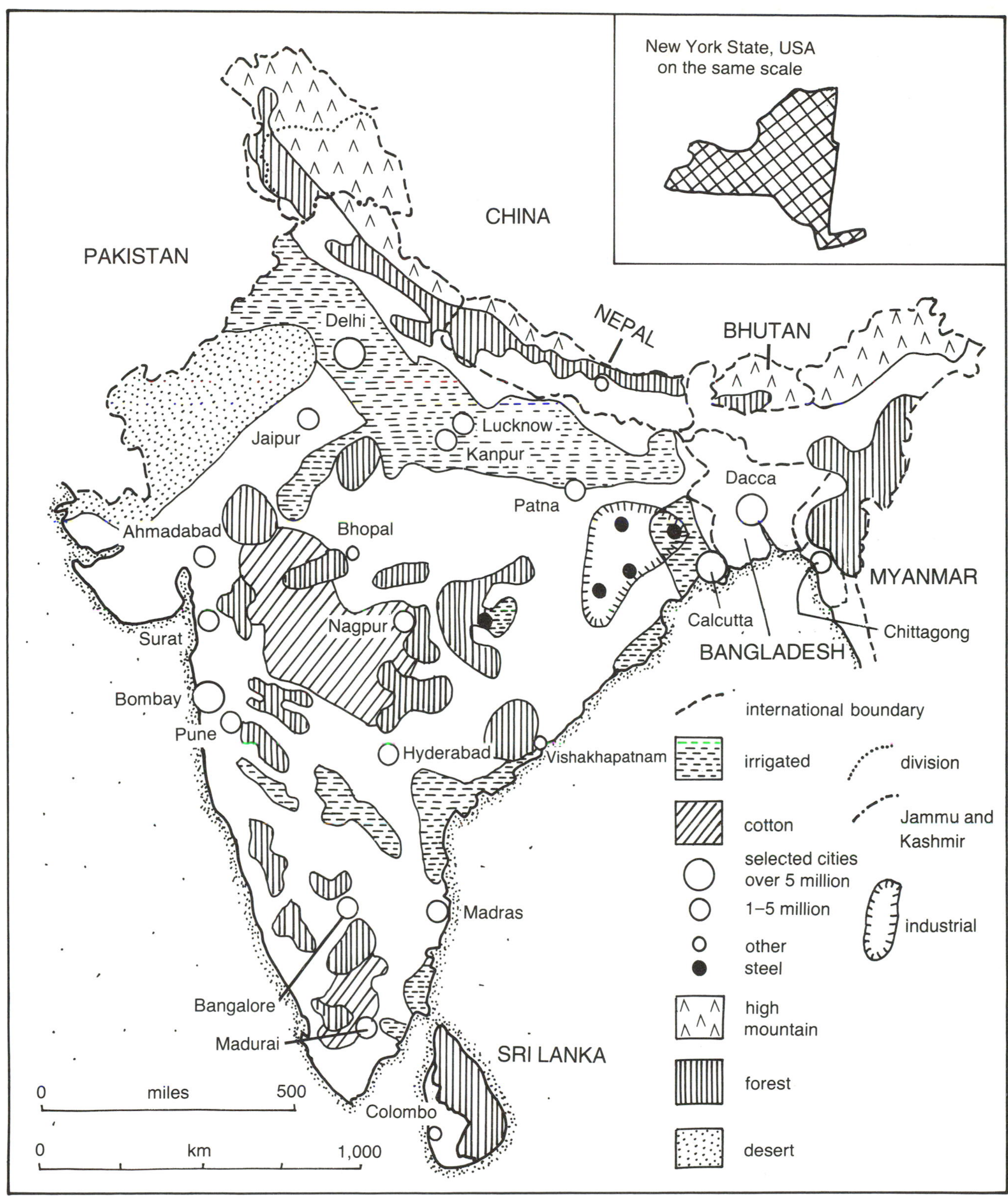

FIGURE 14.3 Economic map of India

forested, an amount that has only changed slightly in the 1970s and 1980s.

Changes in the proportions of different types of land use in India have been gradual in the last two decades and there is not much room for manoeuvre. Rugged conditions and high altitude limit the bioclimatic resources along the northern margins of the country, in the extreme east, bordering Myanmar, and to a lesser

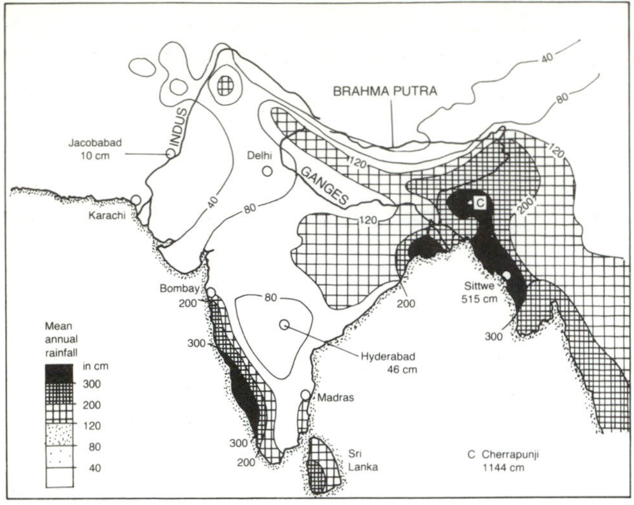

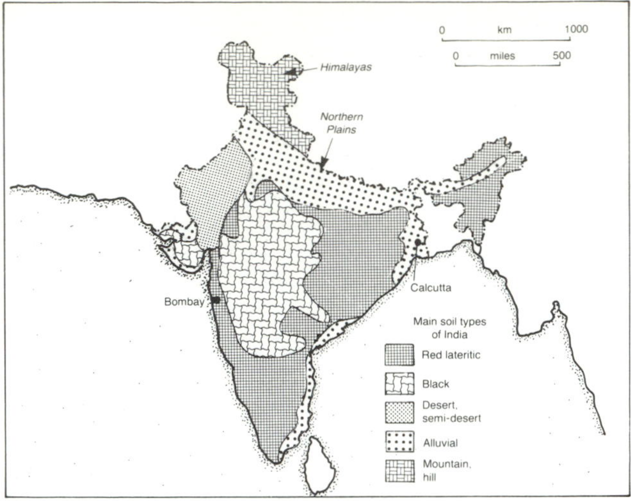

FIGURE 14.4 Rainfall in South Asia and soil in India. Soil map simplified from Johnson (1979)

extent in the mountains of the southwest. Mean precipitation varies greatly between different parts of India, with very low rainfall in the northwest (the Thar Desert) bordering Pakistan. Seasonal variations in rainfall are vary marked over most of the country, while year-to-year variations cause great fluctuations in yields. The fertility of the soil also varies from place to place. The lowlands of the many river valleys, the largest being the Ganges plains in the north, are extensively irrigated, and nutrients are deposited as irrigation proceeds. In general, soils elsewhere are not inherently very fertile except in the area east of Bombay, where they are formed on lava deposits.

India has to be virtually self-sufficient in feeding its population of over 900 million. In the past, through failures of distribution, and reluctance to accept foreign assistance, some areas have suffered serious famines. For the last twenty years this problem has largely been overcome. In some respects Indian agriculture has been successful, but in other respects results have been disappointing. The yields of the principal cereals grown – rice, wheat, millet and sorghum – are usually well below the world average and far below the highest yields achieved in the world. Although the consump-

BOX 14.1

Hinduism

From the point of view of this book two aspects of Hinduism are particularly important. First, the partitioning of British South Asia in 1947 was made primarily on the basis of religion. Second, the beliefs of Hindus have some bearing on their life styles and on their attitudes to society and the environment. Table 14.6 shows the approximate percentages of believers in the main faiths of the region of South Asia. It shows that among India's neighbours only Nepal is predominantly Hindu. Pakistan and Bangladesh are predominantly Muslim countries, Sri Lanka and Myanmar Buddhist.

TABLE 14.6 Religions in South Asia and Myanmar (percentages)

	Hindu	*Muslim*	*Buddhist*	*Sikh*	*Christian*
India	80	11	1	2	2
Pakistan	1.5	95	–	–	1.5
Bangladesh	12	87	–	–	–
Sri Lanka	–	7.5	70	–	7.5
Myanmar	2	2	88	–	2
Nepal	89	3	5	–	–

Note: Other 'religions' include animists in Myanmar

TABLE 14.7 India's share of world totals, latest available, early 1990s

	%		%
Population 1993	16.3	Energy production	2.3
Area	2.5	Energy consumption	2.5
Bioclimatic resources	7.1	Steel production	1.7
Water	4.3	Passenger cars	0.6
Fossil fuels	2.4	Telephones in use	1.3
Non-fuel minerals	2.0	World trade	0.6

tion of animal products per capita in India is very small, and food from crops is therefore mostly fed directly to humans, about 200 million cattle are kept, principally as work animals for cultivating the land and for transport. These animals are indispensable for the present rural economy of India, but they do consume large quantities of fodder, not much of it from natural pastures, in order to keep alive, while returning some as manure or for fuel.

Over half of the grain produced in India is from rice. Table 14.8 shows the area cultivated, yield and production of rice (paddy) in selected countries of the world including India (see Figure 14.5). While it would be unfair to criticise India, Pakistan and Bangladesh, their performance in raising rice yields in the last sixty years has been disappointing. China and Indonesia started from roughly the same base as India in the early 1930s but by the early 1990s were obtaining much better results. Again, from their higher base levels in the early 1930s, Japan and the USA have both achieved much higher absolute increases in the last six decades than India has. There seems to be plenty of slack to be taken up in India and, therefore, good reason for satisfaction that production has kept ahead of population growth in the 1980s.

The country's second cereal, wheat, is grown most widely in northern and eastern India. In the 1990s it

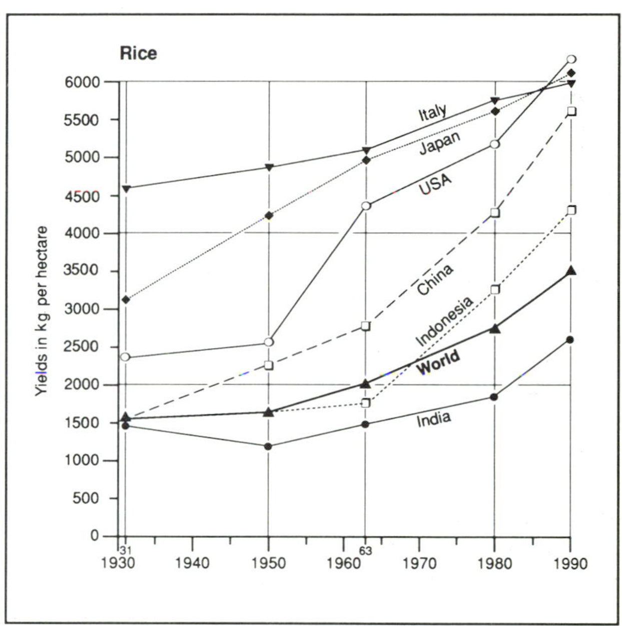

FIGURE 14.5 Rice yields in selected countries, 1931–90
Sources of data: League of Nations (1933), various numbers of *FAOPY*

Unlike Judaism, Christianity and Islam, Hinduism has evolved without reference to one great leader or teacher, or to a special book or a single god. It has developed over several thousand years, changing forms and practices with time. It is more flexible in its organisation than the other main religions in allowing members of the Hindu faith to worship a single god, several gods, or none at all. Holy scriptures, festivals and rituals differ from place to place. There is a universal belief through the idea of *Karma* that the way people live in this life affects their next life, into which they will be reborn, rather than their situation and rewards in the next world. The prospect is that those who have done well will be reincarnated in an upper echelon of society, while those who have lived badly will be reborn in a low caste or will even become animals.

The notion of the divine in Hinduism is the Trimurti, three gods, Vishnu, Shiva and Brahma, to whom, together with a number of minor deities, loyalty is given. The Vedas are the writings that support the ideology of Hinduism, while the running of temples, rituals, pilgrimages and festivals is overseen by the priestly Brahmin class of Indian society. In short, Hinduism is closely related to the rigid system of castes, which profoundly influences life in India and is characterised especially by lack of mobility through social classes.

Among the practices of Hindus is the belief that cattle are sacred and should not be slaughtered, a practice criticised as being wasteful economically. In some West European countries, however, it is customary to avoid the slaughter of horses, an equally reasonable practice, since the horse, like the bullock in India, is or was a vital work animal. The fact that the number of cattle in India has been around 200 million a year in the 1980s, roughly one per four Indians, and that arguably they consume as much food as all the humans, is a fact of life in India. Their contribution to the economy as work animals and in providing dung for domestic fuel in areas where fuelwood is scarce should be taken into consideration.

TABLE 14.8 Rice production and yields in selected countries

	Rice yields in kg per hectare/(acre)				*Area in millions of hectares/(acres) and production in millions of tonnes*			
					1979–81 average		*1989–91 average*	
	1931–2	*1961–5*	*1979–81*	*1989–91*	*area*	*production*	*area*	*production*
India	1,520 (620)	1,480 (600)	1,860 (750)	2,620 (1,060)	40.1 (99)	74.6	42.3 (104)	111.1
Pakistan	–[1]	1,650 (670)	2,470 (1,000)	2,330 (940)	2.0 (5)	4.9	2.1 (5)	4.9
Bangladesh	–[1]	–[3]	1,950 (790)	2,680 (1,090)	10.3 (25)	20.1	10.3 (25)	27.6
Indonesia	1,500 (610)	1,810 (730)	3,260 (1,320)	4,300 (1,740)	9.1 (22)	29.6	10.4 (26)	44.7
China	1,590[2] (640)	na	4,240 (1,720)	5,620 (2,280)	34.3 (85)	145.7	33.3 (82)	18.7
Japan	3,100 (1,260)	5,020 (2,030)	5,580 (2,260)	6,120 (2,480)	2.4 (6)	13.3	2.1 (5)	12.7
USA	2,370 (960)	4,370 (1,770)	5,170 (2,090)	6,340 (2,570)	1.3 (3)	7.0	1.1 (3)	7.0
World	1,540 (620)	2,040 (830)	2,760 (1,120)	3,510 (1,420)	143.8 (355)	396.4	148.1 (366)	519.6

Notes:
na = not available
[1] Then part of India
[2] Manchuria
[3] Then part of Pakistan
Sources: League of Nations (1933) and *FAOPY 1991*

TABLE 14.9 World's leading producers of cotton lint and tea

(a) Cotton lint production in thousands of tonnes and as a percentage of world total

	1979–81 average	%	*1989–91 average*	%
China	2,630	18.3	4,650	24.9
India	1,280	8.9	1,770	9.4
USA	3,000	20.9	3,280	17.5
Pakistan	730	5.1	1,740	9.3
Former USSR	2,730	19.1	2,580	13.8
World	14,340	100.0	18,720	100.0

Source: FAOPY 1991, Table 87

(b) Tea production in thousands of tonnes and as a percentage of the world total

	1979–81 average	%	*1990–2 average*	%
India	560	30.1	720	28.3
China	330	17.7	570	22.4
Sri Lanka	200	10.8	220	8.7
Kenya	90	4.8	200	7.9
Indonesia	100	5.4	160	6.3
Turkey	80	4.3	130	5.1
World[1]	1,860	100.0	2,540	100.0

Note: [1] Other producers include Japan and the former USSR
Source: FAOPY 1992, p. 190

amounted to about half the weight of rice produced, the latter usually preferred where there is a choice because it gives a heavier yield. In 1979–81 Indian wheat yields compared unfavourably with the world average, 1,550 kg per hectare against 1,890 kg per hectare. In 1989–91 the corresponding yields were 2,200 kg per hectare for India and 2,480 for the world. In France the yield of wheat in 1989–91 was 6,500 kg per hectare, three times the improved Indian level, but not a level Indian farmers could expect to achieve quickly.

Most of India's remaining cereal production is from millet and sorghum, crops grown mainly in the drier northwestern part of the country. Yields per hectare are far below the world average and even below the average for Africa. In 1989–91 the yields were 2,500 kg per hectare for the world, 880 for India and 1,090 for Africa. Indian yields of other cereals, roots and tubers, and beans, similarly compare unfavourably with those in other parts of the world.

The generally low agricultural yields in India are clearly not the result of a shortage of labour in the agricultural sector. The application of larger quantities of chemical fertilisers to the land might seem to be a solution, but India lacks the mineral resources to produce them and has to depend on imports. India's consumption of nitrogenous and potash fertilisers has actually doubled in the 1980s and the consumption of phosphate fertilisers increased 2.5-fold. Even so, per hectare of cultivated land, China uses about twice as much fertiliser as India. There is scope also for making improvements to water supply in rural areas. All in all, however, improvements in Indian agriculture can be expected to be gradual.

At present the level of mechanisation in Indian agriculture is very low indeed. Increased mechanisation would reduce the number of livestock needed for work purposes, but it would be pointless to implement it if the policy in India is to keep as many people as possible on the land. To provide enough tractors and other machinery to reduce substantially the animal and human workloads, the equipment could either be imported or, there being no technical constraints, manufactured in India. A large increase in the number of tractors and other machines such as harvesters would require the importation of additional fuel, since India is not self-sufficient in oil products.

Considerable progress has been made in the industrialisation of India in recent decades, and agriculture makes an important contribution to the raw material base. Cotton, jute and hides, for example, are produced for home consumption and also for export. In the 1980s, India accounted for rather less than 10 per cent of all the cotton produced in the world (see Table 14.9a). It is also the largest producer of jute and of tea in the world (see Table 14.9b).

14.4 Indian industry

During the period of British colonial rule India was principally regarded as a source of primary products for British industry and a market for British manufactured goods. Karl Marx even commended Britain for suppressing the traditional manufacturing processes of India and exporting goods there from British factories. Such a view is neither fashionable now nor indeed realistic, because manufactures now make up more than half of the value of India's exports. During the last decades of British rule, modern textile and metallurgical plants were already being established in India.

Shortly after the end of the British rule in 1947, India introduced what has been described as a partially planned economy, one of the principal aims of which was to promote national self-reliance in both agricultural and industrial production. The policy of substituting domestic production for imports has had considerable success in the last four decades and India's dependence on imported manufactured goods has been greatly reduced. On the other hand, the policy has been attacked for creating a sheltered, high-cost industrial structure run by the public sector. In the 1990s policy has been changing, with more emphasis on market forces, and state involvement indicative rather than mandatory. Individual sections of industry will now be discussed.

India has 16 per cent of the total population of the world but only about 2.4 per cent of the world's fossil fuel reserves, most of this being coal. It accounts for about 2.5 per cent of all the primary energy consumption in the world. Even though there is no need for domestic heating in most parts of India, and much of the transport system is based on animal power, the annual consumption of 320 kg of coal equivalent of energy per head of population is very low, compared with 840 in China. Total energy consumption in India in fact more than doubled between 1980 and 1990, but because of population growth, the amount consumed per capita rose by only 56 per cent.

Energy shortages have plagued the Indian economy throughout the last four decades and particularly in the 1980s, in spite of a doubling of coal production during that decade and almost a fourfold increase in oil production. India has a nuclear power programme, but in 1992 its eight small power stations accounted for only 0.3 per cent of the total world output. About 90 per cent of India's coal production comes from a cluster of coalfields in the states of Bihar, West Bengal and Madhya Pradesh. The oil-fields, however, are in the west of India in Gujerat, offshore near Bombay, and in Assam in the extreme northeast.

India's industrial sector is broadly based and includes both heavy and light industry. Given its

PLATE 14.1
South Asia: rural and urban scenes

a Village festival in south India, a tranquil scene, in which a bullock-cart race is taking place in the middle part of the picture

b Madras, Tamil Nadu State, south India, with a proliferation of small retail outlets, more commercial vehicles than private cars, numerous mini-cabs, some actually constructed round motor-scooters, and people everywhere. Note the extensive use of English

population size, however, production is limited in most sectors. Its 16 per cent of the world's population contrasts with 1.7 per cent of the world's steel production, 1.8 per cent of the aluminium and 4 per cent of the cement, but its share of global output in some consumer products is greater. For example, its industrial consumption of cotton is 10 per cent of the world total. Some 15 million people are employed in textile manufacturing, which accounts for about 20 per cent of the value of all industrial output. On the other hand, India's engineering and electrical sectors are very small, producing 0.6 per cent of the passenger cars in the world, no shipbuilding of consequence, and less than 1 per cent of the world production of radio and television receivers.

Much of India's factory industry is concentrated in a few regions and major cities. Bombay and Calcutta are outstanding centres of light industry, while most of the capacity of India's iron and steel industry is concentrated in a few centres to the west of Calcutta where production on modern lines began in the 1930s. The modern iron and steel industry of India was greatly expanded in the 1950s mainly in the public-sector Steel Authority of India, with financial and technical assistance from West European countries and the USSR. The location of the main steelworks from this period is shown in Figure 14.3.

In addition to the industries referred to above, many people are employed in processing and manufacturing, mainly for local consumption. Some 7.5 million people

PLATE 14.2
South Asia: agriculture

a Planting young rice seedlings by hand, south India

still produce textiles on hand looms, and leather goods, carpets and metal goods are made throughout the country. In addition to the large integrated iron and steel mills there are many mini-steel plants. A recent development in India has been the manufacture of electronics products, while the skill of Indians in creating computer software has led to the establishment of branches of European computer companies in the country.

14.5 India's problems and prospects

In the agricultural sector the prospect is that yields in India will have to be increased if output is to grow in the next few decades to keep up with the anticipated growth in population. Some of the remaining forest might be cleared to provide additional cultivated land. The large and increasing consumption of fuelwood threatens many remaining areas of forest and wildlife (see Ward 1992). India's 14 per cent share of world fuelwood consumption places it untypically near to the world average for once. Agriculture still accounts for two-thirds of employment but, in 1991, only 31 per cent of GDP (45 per cent in 1970), while industry's share is 27 per cent (22 per cent in 1970) from about 10 per cent of the total workforce. The very low productivity per worker in agriculture can only be raised slowly. Compared with the agricultural sector, therefore, industry appears to be doing well, but it has problems with energy supplies, the need to import fuel, mainly oil, and in the longer term, the lack of raw materials.

A problem that faces the Indian economy as it

b Tea gardens and tea pickers in Sri Lanka. Many of the workers in this poorly paid occupation are Tamil women originating in the southern extremity of India

PLATE 14.3
South Asia

a Pump drawing well-water, the supplier of all the domestic needs in this village in south India

b Cannibalised bicycle parts help in spinning, south India

expands is the weakness of the transport system. In this respect India has a better rail system than China, one inherited from the period of British rule. Nevertheless, the 63,000 km (39,000 miles) of route, provided now with equipment made mostly in India, is only about twice the length of the route of the French railway system, yet it covers about six times the area and serves more than fifteen times as many people. The state of India's roads is even less satisfactory in both the quality of road surfaces and the small number of commercial vehicles in use, i.e. about 4 million, the same as in France. The long-distance haulage of goods by road is increasingly important in giving flexibility in the growing Indian economy. Main roads pass through innumerable villages and the intervening countryside, where animal-drawn traffic is dense. With rather more commercial vehicles than the UK but only an eighth as many passenger cars in use, India has an annual death toll on its roads of about 60,000, compared with about 7,000 in the UK.

India has an army of over 1 million men as well as a sizeable navy and airforce. Although officially committed to a policy of peace, and with no explicit territorial aspirations outside its present borders, India's military expenditure was 3–4 per cent of its total GDP in the 1980s, its arms imports were worth

c Bombay, gateway to India, busy street and the imposing railway station, designed by the architect who designed (according to a tour guide) St Pancreas station, London

over 3 billion US dollars in 1987, and the ratio of its military expenditure to health and education expenditure was about 80 per cent – less, however, than China's 146 per cent. India has 15,000 km (9,300 miles) of land boundary, 7,500 km (4,700 miles) of coast, an exclusive economic zone offshore of over 2 million sq km (770,000 sq miles), and various island territories in the Indian Ocean. Neighbouring Pakistan and China have both been in conflict with India since 1947. The maintenance of armed forces is therefore reasonable, although it is a drain on the resources of the country.

India appears to be undergoing a major change in its economy. The power of the state monopolies is being cut, privatisation encouraged and bureaucratic regulations reduced. The Bombay Stock Exchange experienced a remarkable boom in 1992, repeated late in 1993. Foreign investors are being encouraged to move in. One estimate puts the size of the growing middle class in India at 150 million. Expectations that India could become one of the four great trading areas of the world in a few decades' time, along with NAFTA, the extended EU and East Asia, would, however, require economic growth there on a scale far greater than that achieved in the last four decades. With half the adult population still illiterate, about a third of the population below the poverty line, and two-thirds still in the agricultural sector, a massive transformation would be needed. Given the small quantities of most natural resources, the expansion of industry on the scale required would make India a net exporter of large quantities of manufactured goods and a net importer of primary products, a situation in which China now finds itself.

Whatever the future achievements of the Indian economy, natural and man-made hazards seem destined to produce disasters of great proportions in India, given the high cost of taking precautions against them and the high density of population in many areas. For example, in a list of selected accidents involving hazardous substances between 1970 and 1989 (OECD 1991), the worst by far in numbers of deaths (2,800) and injured (50,000), with 200,000 additionally evacuated, occurred at Bhopal in 1984 when there was a leakage of methyl isocyanate. The US company, Union Carbide, that owned the plant finally left Bhopal in 1994. In 1993 an earthquake with its epicentre at Latur, between Bombay and Hyderabad, was estimated to have caused 30,000 deaths as stone and mud dwellings collapsed. The earthquake of a comparable intensity in Los Angeles in the same year caused about fifty deaths, but damaged infrastructure, especially the highway system. Also in 1993, thousands died in floods in northern and northeastern India, leaving an estimated several million homeless. Such disasters occur world-wide, but partly through the great concentrations of people in both rural and urban areas and the poor quality of domestic, industrial and other structures, more people are vulnerable in India and more are afflicted when they occur. In 1994, India was again struck with a different disaster, pneumonic plague, which started in the state of Gujerat. The infection was quickly carried to other parts of India, particularly the big cities. Although successfully contained, the occurrence underlined, in particular, the poor state of hygiene in both urban and rural areas.

Should India continue with a highly self-sufficient economy or should it trade more with the rest of the world? In 1991 it accounted for only 0.5 per cent of the total value of exports in world trade and only 1.1 per cent of the world's agricultural exports (see Table

14.10) although it has 20 per cent of all the economically active population of the world in agriculture. The data in Table 14.11 show that the foreign trade of India, Pakistan, Bangladesh and China was all fairly similar in composition, depending heavily on manufactures in exports, while importing industrial supplies, fuel (except China) and capital equipment to build up, maintain and run industries.

14.6 The rest of South Asia (Pakistan, Bangladesh, Sri Lanka, Nepal) (see Tables 15.1–15.3)

1. **Pakistan** Although Pakistan differs greatly from India in territorial extent, population size, cultural background and environmental conditions, the basic features of its demographic and economic structure are broadly similar. The relative importance of agriculture

TABLE 14.10 The value of the foreign trade of selected countries in billions of US dollars, 1990 and 1991

		Total		*Agricultural products*	
		imports	*exports*	*imports*	*exports*
USA	1990	517	394	33	48
	1991	509	422	33	48
France	1990	244	222	25	34
	1991	248	229	26	34
India	1990	24	18	1	4
	1991	22	18	1	4
China	1990	108	129	10	13
	1991	126	148	11	14
World	1990	3,609	3,468	392	362
	1991	3,669	3,527	393	365

Source: *UN Food and Agricultural Organization Trade Yearbook 1991*

TABLE 14.11 The sectoral breakdown of imports and exports in India, Pakistan, Bangladesh and China

(a) Composition of imports, average percentage of total trade

	India 1986–8	*Pakistan 1987–9*	*Bangladesh 1985–7*	*China 1988–90*
Food	7.6	15.2	23.9	7.4
Industrial supplies	43.8	35.1	36.7	47.3
Fuel	17.0	16.2	16.3	2.2
Machinery	21.4	21.4	12.8	26.9
Transport equipment	3.4	8.2	5.8	9.9
Consumer goods	1.9	3.7	4.2	5.8

(b) Composition of exports, average percentage of total trade

	India 1986–8	*Pakistan 1987–9*	*Bangladesh 1985–7*	*China 1988–90*
Agricultural products	17.5	19.1	28.3	10.4
Minerals	22.8	0.3	–	8.2
Manufactures (including textiles)	59.7	80.6	71.7	81.4
Textiles	31.3	64.0	65.5	31.7

Source: *United Nations International Trade Statistics Yearbook 1990*, vol. 2

in Pakistan's GDP has declined from 37 per cent in 1970 to 26 per cent in 1991 while the share of industry has risen from 22 to 26 per cent during that time. During that period the absolute size of the economically active population in agriculture has actually risen from 10 million to 18 million, declining, however, from 59 to 49 per cent of all those employed. In contrast, the area under arable and permanent crops has increased by only a few per cent. The trends noted above seem likely to continue for some decades, since 44 per cent of the population is under the age of 15 years and only 4 per cent is 65 and over. If recent demographic trends continue, the 1993 population of Pakistan could double before 2020.

Pakistan has almost no fossil fuel or non-fuel mineral resources. Its main natural resource is the quarter of its total area that is used for cultivation. Four-fifths of this cultivated land is irrigated. Pakistan thus depends heavily on the Indus for its water supply, in the same way that Egypt depends on the Nile, and Iraq on the Tigris and Euphrates rivers. Another 10 per cent of the area of Pakistan is classed as forest and natural pasture, while almost two-thirds has no bioclimatic use at all. Grain yields in Pakistan are poor and the country has become a net importer of food, while devoting some of its best land to cotton cultivation, which supplies its textile industry. Almost 30 per cent of Pakistan's population is urban, with about 6 million people in Karachi, the chief port and industrial centre of the country, and over 3 million in Lahore in the north.

Pakistan's existence as a separate sovereign state is justified by the predominance of Muslims in this part of South Asia, Hindus and Christians making up only about 3 per cent of the population of the country. Linguistically, Pakistan is much more diversified, with Indo-European languages, the most widely spoken principally Sindhi in the south, and Punjabi (also spoken in India) in the north, as well as Iranian languages, mainly Pathani and Baluchi, in the west. With four neighbours – India, China (adjoining the Pakistani area of Kashmir), Afghanistan and Iran – and the former Soviet Republic of Tadjikistan visible across the Afghan panhandle, Pakistan's geopolitical location is similar, on a larger scale, to that of Poland in Central Europe. If the USSR had succeeded in establishing a pro-Soviet government in Afghanistan in the 1980s, Soviet influence would have extended to within 500 km (310 miles) of the Indian Ocean, a situation that could have led to Soviet pressure on Pakistan's mainly pastoral Baluchi minority to secede from Pakistan. That prospect is now dead, but assuming Vladimir Zhirinovsky never becomes President of Russia (see Frazer and Lancelle 1994), Pakistan still has to contend with India over the partition of Kashmir and, no less importantly in the longer term, to cater for perhaps twice as many people in the next twenty-five years.

2. Bangladesh In 1947, when British India became independent, what is now Bangladesh became the eastern part of the newly created, predominantly

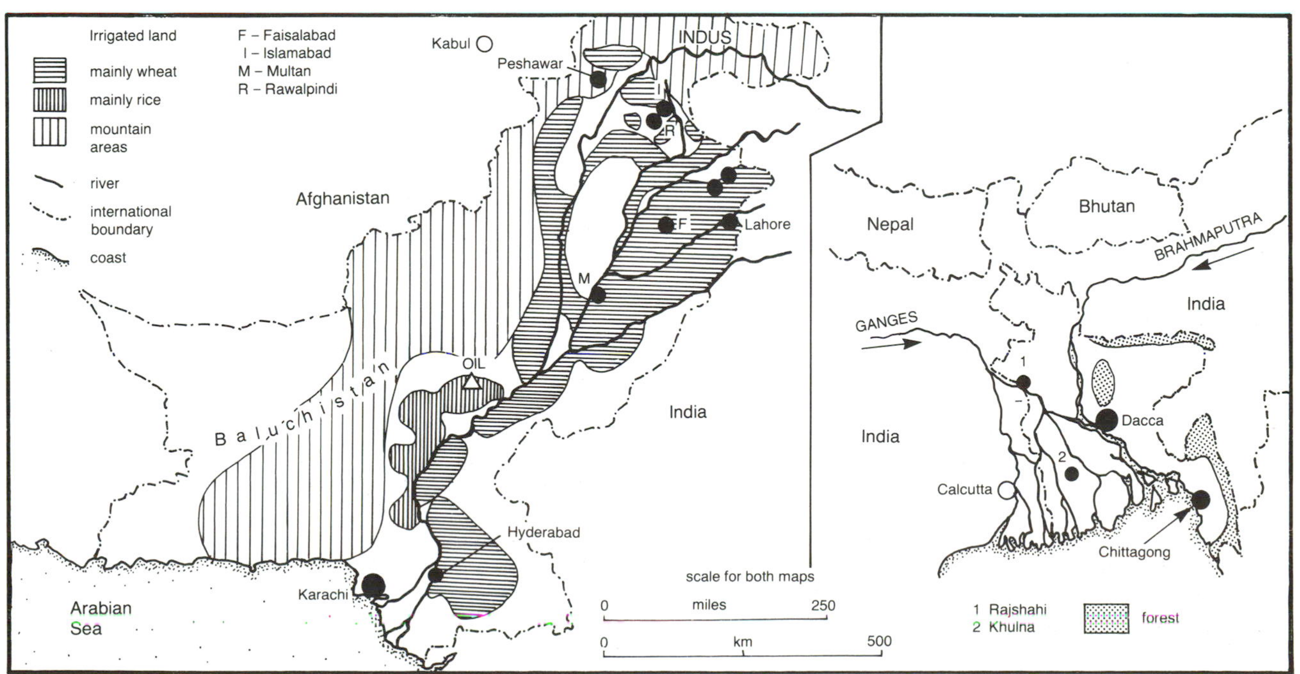

FIGURE 14.6 Location maps of Pakistan and Bangladesh

Muslim country of Pakistan. In 1971, East Pakistan broke away from the more powerful West Pakistan described in the previous subsection. Bangladesh and Pakistan have almost exactly the same number of inhabitants but Bangladesh is only about one-fifth as large in area. In Bangladesh, about 87 per cent of the population is Muslim, while almost all the rest is Hindu. The population of Bangladesh is expected to double within about three decades, rising from 114 million in 1993 to over 200 million in the early 2020s.

The relative importance of agriculture in the economy of Bangladesh has declined in recent decades. The sector still accounted for the high proportion of 68 per cent of employment in 1991, but was much lower then than in 1965, when it accounted for 84 per cent. The sector is less productive than industry and services. The contribution of agriculture to total GDP declined from 55 per cent in 1970 to 36 per cent in 1991. Even in 1991 industry accounted for only 16 per cent of GDP, compared with a mere 9 per cent in 1970. The absolute number employed in agriculture actually rose between 1975 and 1991 from 17.6 million to 23.7 million, whereas the area of arable land and permanent crops grew by only a small percentage.

BOX 14.2

Adam Smith on four regions of the world

Box 14.2 compares Bengal, now shared between India and Bangladesh, with China (see Chapter 16), Britain (see Chapter 5) and the English colonies of North America (see Chapter 9) over two hundred years ago. How closely do the features recorded by Adam Smith match circumstances now?

Today Bangladesh and the neighbouring Indian state of Bengal form one of the largest concentrations of materially poor people in the world. Evidence that only a certain amount of change is possible over a given period of time is provided by a comparison of Bengal, China, Britain and the English colonies of North America in 1776 with conditions now. Four brief extracts from *The Wealth of Nations* by the Scottish economist Adam Smith follow. They are not arranged in the order in which they appear in his book.

1. Bengal (pp. 175–6):

> But it would be otherwise in a country where the funds destined for the maintenance of labour were sensibly decaying. Every year the demand for servants and labourers would, in all the different classes of employments, be less than it had been the year before. Many who had been bred in the superior classes, not being able to find employment in their own business, would be glad to seek it in the lowest. The lowest class being not only over-stocked with its own workmen, but with the overflowings of all the other classes, the competition for employment would be so great in it, as to reduce the wages of labour to the most miserable and scanty subsistence of the labour. Many would not be able to find employment even upon these hard terms, but would either starve, or be driven to seek a subsistence either by begging or by the perpetration perhaps of the greatest enormities. Want, famine, and mortality would immediately prevail in that class, and from thence extend themselves to all the superior classes, till the number of inhabitants in the country was reduced to what could easily be maintained by the revenue and stock which remained in it, and which had escaped either the tyranny or calamity which had destroyed the rest. This perhaps is nearly the present state of Bengal, and of some other English settlements in the East Indies. In a fertile country which had before been much depopulated, where subsistence, consequently, should not be very difficult, and where, notwithstanding three or four hundred thousand people die of hunger in one year, we may be assured that the funds destined for the maintenance of the labouring poor are fast decaying. The difference between the genius of the British constitution which protects and governs North America, and that of the mercantile company which oppresses and domineers in the East Indies, cannot perhaps be better illustrated than by the different state of those countries.

Ironically, the Declaration of Independence of the North American colonies was made in 1776, precisely the year in which Adam Smith's work was published.

2. If Bengal was gradually sinking economically, Smith was cautiously less pessimistic about China (pp. 175–6):

> China has been long one of the richest, that is, one of the most fertile, best cultivated, most industrious, and most populous countries in the world. It seems, however, to have been long stationary. Marco Polo, who visited it more than five hundred years ago, describes its cultivation, industry, and populousness, almost in the same terms in which they are described by travellers in the present times. It had perhaps, even long before his time, acquired that full complement of riches which the nature of its laws

Bangladesh is located in the lower valleys, floodplains and deltas of two of South Asia's largest rivers, the Ganges and the Brahmaputra, the former with some and the latter with all of its headwaters in the Himalayas. Almost all of Bangladesh is below 200 metres (656 feet) above sea level and much of it is so low that it is frequently flooded by the two main rivers and their tributaries. In contrast to Pakistan, Bangladesh is itself in a region of very heavy rainfall. The main problems in agriculture are the need to control the excessive rain- and floodwater, and the lack of land to cultivate, given the present and expected future size of the population. Of the total area of Bangladesh (144,000 sq km – 56,000 sq miles), only 90 per cent is classed as land area, the rest being water. About another 10 per cent of the total has no bioclimatic use. Two-thirds of the total area of Bangladesh is cultivated, while there is some forest on the northern and eastern fringes. On an area similar to that of England (130,000 sq km – 50,000 sq miles) and of New York State (127,000 sq km – 49,000 sq miles), and without the recourse these two regions have to enormous quantities of products from elsewhere, Bangladesh has to support approaching 120 million people, compared

> and institutions permits it to acquire. The accounts of all travellers, inconsistent in many other respects, agree in the low wages of labour, and in the difficulty which a labourer finds in bringing up a family in China. If by digging the ground a whole day he can get what will purchase a small quantity of rice in the evening, he is contented. The condition of artificers is, if possible, still worse. Instead of waiting indolently in their workhouses, for the calls of their customers, as in Europe, they are continually running about the streets with the tools of their respective trades, offering their service, and as it were begging employment. The poverty of the lower ranks of people in China far surpasses that of the most beggarly nations in Europe. In the neighbourhood of Canton many hundred, it is commonly said, many thousand families have no habitation on the land, but live constantly in little fishing boats upon the rivers and canals. The subsistence which they find there is so scanty that they are eager to fish up the nastiest garbage thrown overboard from any European ship. Any carrion, the carcase of a dead dog or cat, for example, though half putrid and stinking, is as welcome to them as the most wholesome food to the people of other countries. Marriage is encouraged in China, not by the profitableness of children, but by the liberty of destroying them. In all great towns several are every night exposed in the street, or drowned like puppies in the water. The performance of this horrid office is even said to be the avowed business by which some people earn their subsistence.
>
> China, however, though it may perhaps stand still, does not seem to go backwards. Its towns are nowhere deserted by their inhabitants. The lands which had once been cultivated are nowhere neglected. The same or very nearly the same annual labour must therefore continue to be performed, and the funds destined for maintaining it must not, consequently, be sensibly diminished. The lowest class of labourers, therefore, notwithstanding their scanty subsistence, must some way or another make shift to continue their race so far as to keep up their usual numbers.

3. Smith is more impressed by Britain (pp. 176–7):

> In Great Britain the wages of labour seem, in the present times, to be evidently more than what is precisely necessary to enable the labourer to bring up a family. In order to satisfy ourselves upon this point it will not be necessary to enter into any tedious or doubtful calculation of what may be the lowest sum upon which it is possible to do this. There are many plain symptoms that the wages of labour are nowhere in this country regulated by this lowest rate which is consistent with common humanity.

4. Like de Tocqueville half a century later, Smith sees a bright future for North America (p. 173):

> But though North America is not yet so rich as England, it is much more thriving, and advancing with much greater rapidity to the further acquisition of riches. The most decisive mark for the prosperity of any country is the increase of the number of its inhabitants. In Great Britain, and most other European countries, they are not supposed to double in less than five hundred years. In the British colonies in North America it has been found that they double in twenty or five-and-twenty years.

In the view that population growth is a positive feature, Smith differs in principle from Thomas Malthus (see Chapter 3), but his views have been echoed by some demographers, including, for example, Colin Clarke. More importantly, after more than two centuries, much of what Adam Smith saw in the 1770s is broadly true today. Discounting the devastating impact of an asteroid or a nuclear catastrophe of unimaginable proportions, two hundred years on from now there should be humans in the four countries described by Adam Smith. Will the gap in material conditions change dramatically by then?

with England's 46 million and New York's 18 million.

There is little scope for extending the area under cultivation in Bangladesh. On the other hand, between the early 1960s and the early 1990s, rice yields have increased from 1,650 kg per hectare to about 2,700 kg, but in both periods were still well below the world average. There are plans to reduce the impact of flooding on agricultural land, but much of the cost of construction work to control the rivers has to be met by foreign sources. Given the often unsuccessful experience of attempts to reduce flooding in the USA and elsewhere with costly protection structures, it is argued that at best rivers such as those in Bangladesh can be managed, but not strictly controlled.

Bangladesh has no economic minerals of consequence apart from modest natural gas reserves, few sources of forest products, and one principal raw material from agriculture, jute, of which it produces over a quarter of the world total. It therefore apparently lacks the basis for industrialisation. Surprisingly, however, 70 per cent of the value of the exports of Bangladesh consist of manufactures. Almost all these are accounted for by textiles and clothing. Only 30 per cent of all exports are primary commodities, and Bangladesh is now a net importer of food.

Only 14 per cent of the population of Bangladesh is urban. The capital, Dacca, with about 5 million inhabitants, is the only multi-million city and, together with Chittagong (1.5 million), has much of the industry of the country. It is not surprising, then, that Bangladesh is classed among the least developed countries of the world. It scores only 185 out of 1,000 on the United Nations Human Development Index, compared with India and Pakistan, still low, with scores around 300.

3. Sri Lanka Apart from its physical detachment from India, Sri Lanka is distinct from the mainland in having a predominance of Buddhists in the population, approximately 70 per cent, but substantial numbers also of Hindus (15 per cent) and 7 per cent each of Muslims and Christians. More importantly, with regard to cultural tensions, there is a substantial minority of Tamils in northern and eastern Sri Lanka, originating in India's most southerly state, Tamil Nadu. With its strategic location in the Indian Ocean, Sri Lanka has long been on many shipping routes and its incorporation in the British Empire gave it virtually a global importance, particularly after the opening of the Suez Canal in 1869.

In South and Southeast Asia Sri Lanka stands out, with Malaysia and Thailand, in having an exceptionally high Human Development Index for a developing country. Literacy and life expectancy levels are far higher than in the rest of South Asia and real GDP per inhabitant is substantial at 2,250 dollars per head. Since independence in 1948 government policy in Sri Lanka has been to promote family planning. Even so, the population has doubled from 9 million in the mid-1950s to 18 million in 1993. With the very low total fertility rate of 2.3 and an annual natural increase in population of only 1.4 per cent, the population of Sri Lanka could cease to grow in a matter of three to four decades.

Sri Lanka is about half as large in area as Bangladesh or England, and roughly equal in area to the US state of West Virginia. Almost 30 per cent of the area of Sri Lanka is under field or permanent crops, each accounting for about half of the total. Rice is the principal food crop, but many other crops are grown, including a variety of bush and tree crops, especially tea and rubber. As recently as 1970, virtually the entire value of Sri Lanka's exports consisted of primary products, but by 1991 manufactures, mainly textiles, made up almost two-thirds of their value. Like Pakistan, Bangladesh, and many states of India, there has been a sharp increase in the production of manufactured goods for export, including goods processed or manufactured from local materials. With approaching 1 million inhabitants the capital, Colombo, is also the chief port and manufacturing centre.

4. Nepal Nepal had a population of about 20 million in 1993 and is similar in area (140,000 sq km – 54,000 sq miles) to Bangladesh. Environmentally it is completely different, consisting of the southern side of the central Himalayas and only a small area of lowland along its border with India. It is landlocked and therefore dependent on India for access to the coast, since the sole road across the Himalayas reaches Lhasa in Chinese-occupied Tibet. With a score of 168 out of 1,000 on the UN HDI it is ranked 140 out of 160 in the countries listed, being particularly low on the index of educational attainment. The population of 20 million is expected to double in about thirty years.

Nepal occupies about one-third of the southern face of the Himalayas, an area of exceptional scenic attraction. It is visited by tourists more than the Himalayas to the northwest, where India and Pakistan dispute Kashmir, and to the east, where Sikkim, Bhutan and India occupy a less accessible and scenically less dramatic mountain region. As might be expected, India is one of the main trading partners of Nepal, but Germany and the USA take half of its limited exports. Nepal lacks minerals and good agricultural land, is fragmented because of the lack of east–west links, and is remote and peripheral in South Asia. It therefore seems unlikely that Nepal will follow the rush to industrialise noted in many countries and cities in South and Southeast Asia. Because of the cost of reaching Nepal from the main source countries of international tourists, the tourist industry should remain small. Nepal's development on modern lines is therefore likely to be slow, a source of satisfaction for conservationists in the developed countries if not for those inhabitants of Nepal aspiring to improve material standards.

The population of South Asia is not growing as fast as the populations of Africa or Southwest Asia. Nevertheless, in only the next three decades the total population is expected to grow from 1.2 billion in 1995 to over 1.9 billion in 2025. Even an immediate drastic reduction in the total fertility rate would have little influence on the trend until after 2025. The expected absolute gain of 700 million people is equal to the present *total* population of Europe, including the whole of Russia. It is misleading to point out that a high density of population such as that in Japan or Germany does not prevent a country from developing and becoming rich. There is no way in which South Asia could overcome its lack of natural resources and the limitations of the productive capacity of its agricultural and industrial sectors to transform itself into an affluent country at least for half a century. Awareness of a growing 'middle class' in India has led in recent years to enthusiasm among investors and exporters in the developed countries at the prospect of an expanding market. The parallel growth of the number of very poor people in the region merits equal attention.

KEY FACTS

1. Although politically fragmented during most of its history, almost the whole of South Asia was for at least a century a colony of Britain. Although Pakistan and Ceylon became separate states after independence, against all odds India has held together.

2. In terms of natural resources per inhabitant India, like Pakistan and Bangladesh, is very poorly endowed.

3. If present demographic trends continue, the population of India will double in less than forty years. Failing the unlikely discovery of vast mineral resources or an unprecedented increase in agricultural production, the quantity of natural resources per inhabitant will decrease.

MATTERS OF OPINION

1. The evidence (see Tables 14.3 and 14.4) is that differences in the educational level and position in the family produce considerable differences in fertility levels among the states of India. Against tradition, should the central government of India attempt to make all states follow the path of Kerala?

2. Like China, India receives only a minute amount of development assistance per inhabitant from the developed countries. If they were independent countries, several of its states would be eligible to join the UN-accredited group of Least Developed Countries, to which India as a whole does not belong. Should consideration be given to this possible adjustment?

SUBJECTS FOR SPECULATION

1. In view of the grim natural resource situation, should the government of India attempt to revive and strengthen its desultory and largely unsuccessful campaign to extend family planning?

2. What is likely to happen to the 'surplus' people if the labour force in agriculture is cut over a short period to half its present size?

3. India is regarded in financial circles in the developed countries as potentially a huge consumer market. At first sight that prospect seems reasonable, even if only a fifth of the population of nearly a billion reaches a level of moderate affluence. Two hundred million 'middle-class' consumers is equal roughly to five Spains or two Germanies. But what of the remaining fourth-fifths?

4. What future do Pakistan and Bangladesh have with their 'river-dominated' economies and fast-growing populations?

5. What would South Asia be like now if it had not become part of the British or any other European empire?

Chapter fifteen

Southeast Asia

Malaysia has formulated a strategic plan to become a fully developed nation by the year 2020. We call it Vision 2020 and it requires concerted development in all areas – economic, social, political, spiritual, psychological and cultural. The balanced development of the nation, encompassing both its people and our natural environment, requires a strong capability in science and technology.

Mahathir Bin Mahamad (1994)

15.1 Introduction (see Table 15.1 for calendar)

The greatest distance across Southeast Asia is from the north of Myanmar (formerly Burma) to the southeast of the Indonesian part of New Guinea, Irian Jaya (West Irian), more than 6,000 km (3,730 miles) (see Figure 15.1). In the north of mainland Southeast Asia, Myanmar adjoins Bangladesh (a very short boundary), India and China, while Laos and Viet Nam also touch China. Indonesia has a common boundary with Papua-New Guinea. Southeast Asia has a long 'frontage' on the Indian Ocean, while Australia is not far from Indonesia across the Timor and Arafura Seas (Figure 15.2). Of the total area of 4.5 million sq km (1.74 million sq miles), 2.4 million sq km (0.93 million sq miles) are in the island countries of Indonesia and the Philippines, together with eastern Malaysia (Sabah and Sarawak), while the remainder are on the mainland of Asia.

All but the northern extremity of Southeast Asia (northern Myanmar) is within the tropics, the Equator passing through three of the main islands of Indonesia. There are extensive areas of high mountain on the mainland and on most of the islands, contrasting with large areas of lowland, both bordering the coast and in inland basins and valleys on the mainland. In 1990, about 17 per cent (780,000 sq km – 300,000 sq miles) of the total area of Southeast Asia was cultivated, compared with only about 13 per cent in the early 1960s. Much of the growth of cultivated land has been at the expense of forest, almost all of which falls broadly in the category of tropical rainforest, including mountain varieties.

In 1990, 50 per cent (2,270,000 sq km – 876,000 sq miles) of Southeast Asia was still under forest compared with about 61 per cent (2,750,000 sq km – 1,060,000 sq miles) in the early 1960s. The rate of clearance during that period was fastest in Viet Nam, the Philippines and especially Thailand. Since the population of Southeast Asia more than doubled between 1962, when it was about 213 million, and 1993, when it reached 460 million, the loss of forest and the extension of the cultivated area are not surprising. Some of the jobs taken by the increasing population have been in agriculture, which still employs about half of the entire economically active population of the region, numbering almost 100 million in 1991, compared with 84 million only as recently as 1975.

According to my assessment of the natural resource

TABLE 15.1 Calendar for Southeast Asia

1044	Establishment of first Burmese national state at Pagan
1170	Apogee of Srivijaya kingdom in Java under Shailendra Dynasty
c. 1180	Angkor Empire (Cambodia) at greatest extent
c. 1220	Emergence of first Thai kingdom
1349	First Chinese settlement at Singapore; beginning of Chinese expansion in Southeast Asia
c. 1400	Establishment of Malacca as a major commercial port of Southeast Asia
1428	Chinese expelled from Viet Nam
1511	Portuguese take Malacca
1571	Spanish conquer Philippines
1619	Foundation of Batavia (now Jakarta) by Dutch; start of Dutch colonial empire in East Indies
1641	Dutch capture Malacca from Portuguese
1819	British found Singapore as free trade port
1824	British start conquest of Burma
1825	Java war; revolt of Indonesia against Dutch
1863	France establishes protectorate over Cambodia, Cochin China (1865), Annam (1874), Tonkin (1885) and Laos (1893)
1886	British annex Burma
1887	French establish Indo-Chinese Union
1942	Japan overruns most of Southeast Asia
1946–54	Vietnamese struggle against France
1949	Indonesia independent
1954	Geneva conference: Laos, Cambodia and Viet Nam (partitioned) become independent states
1957	Civil war in Viet Nam
1961	Increasing US involvement in Viet Nam
1965	Military take-over in Indonesia; Singapore separates from the Malaysian Federation
1967	Formation of the Association of Southeast Asian Nations (ASEAN), including Thailand, Malaysia, Indonesia, the Philippines, Singapore and Brunei
1973	US forces withdraw from Viet Nam

base per inhabitant of Southeast Asia, the region is poorly endowed, having less than half the average level for the world as a whole. With 8.4 per cent of the total population of the world, it has 5.9 per cent of the bioclimatic resources (with forest well represented), 10.2 per cent of the fresh water, but only 2.8 per cent of the fossil fuels and 4.2 per cent of the non-fuel minerals. Southeast Asia is credited with about 1 per cent of the oil reserves of the world (almost all in Indonesia, Malaysia and Brunei), and with 3 per cent each of the natural gas and coal reserves, the latter restricted to Indonesia. With regard to non-fuel minerals, Southeast Asia is prominent globally only for its tin reserves (nearly 70 per cent of the world total): Malaysia has 35 per cent, Indonesia 20 per cent and Thailand 10 per cent. Nickel ore is the only other non-fuel mineral of which Southeast Asia has more than its 'share', most located in Indonesia and the Philippines. Almost all of the major reserves of minerals in Southeast Asia are in the islands, very few on the mainland.

Until recently, Southeast Asian countries had very little modern industry. One reason, which now no longer applies, was the fact that everywhere but Thailand was part of the colonial empires of Britain (Myanmar, Malaysia), France (Indo-China), the Netherlands (Indonesia) or the USA (Philippines) (see Table 15.1). The prolonged conflict in Viet Nam (from the end of the Second World War until 1973) held back development in that country and in neighbouring Laos and Cambodia. Ever since its independence in 1948, Burma/Myanmar has kept to itself, a striking case of a 'hermit' state. The remaining countries of the region continued to export primary products, the processing of which formed one of the main sectors of their industry.

Even in the 1980s there was still very little heavy industry in Southeast Asia: only 0.4 per cent of the world production of steel and virtually no engineering industry. On the other hand, light manufacturing has expanded rapidly in a limited number of centres in the region since the early 1970s, notably in the larger cities. The largest centres in the region are Jakarta (8 million population) in Indonesia, Bangkok (7 million) in Thailand, Manila (7 million) in the Philippines, Singapore, neighbouring Johor Baharu in Malaysia, and Kuala Lumpur. Such centres provide large sources of cheap, fairly well-educated labour, have good airline connections with Japan, Western Europe and North America, and are run by political leaders who favour industrialisation.

As has been evident from this discussion of Southeast Asia's recent development, the region is one of great physical diversity. Culturally, too, there are big differences and contrasts. Linguistically, Southeast Asia is very complex. Each of the larger countries has an official language. For example, Burmese in Myanmar and Thailandic in Thailand are members of the Chinese-Tibetan language family. The population of the island countries has Malayo-Polynesian languages: Tagalog/Pilipino in the Philippines, Malay and Bahasa Indonesian in Malaysia and Indonesia respectively. In addition, English is widely used in most parts of Southeast Asia and Chinese particularly in Singapore and Malaysia. At the other extreme (Gunnemark and Kenrick, 1985), there are over a thousand different minority languages in Southeast Asia, about a fifth of all the languages in the world. Myanmar has about 100, the Philippines 150, and Indonesia 700, of which over half are in West Irian. This situation reflects the greatly fragmented nature of the region in a political sense, showing the effects of European and US colonisation leaving the more marginal population alone, followed by slow integration in the post-colonial half century since the Second World War. The geographical

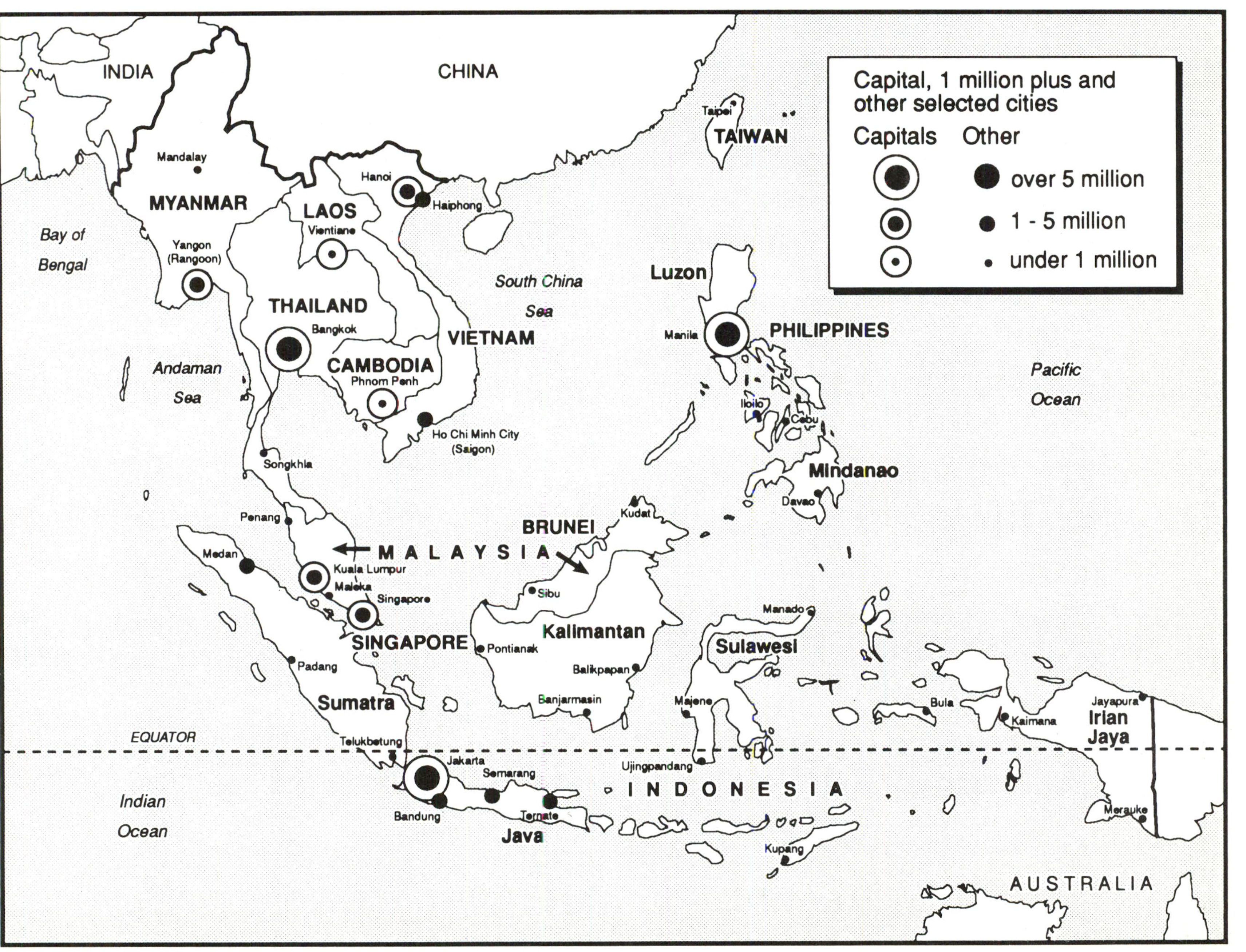

FIGURE 15.1 Countries and cities of Southeast Asia. Population size of cities shown in key

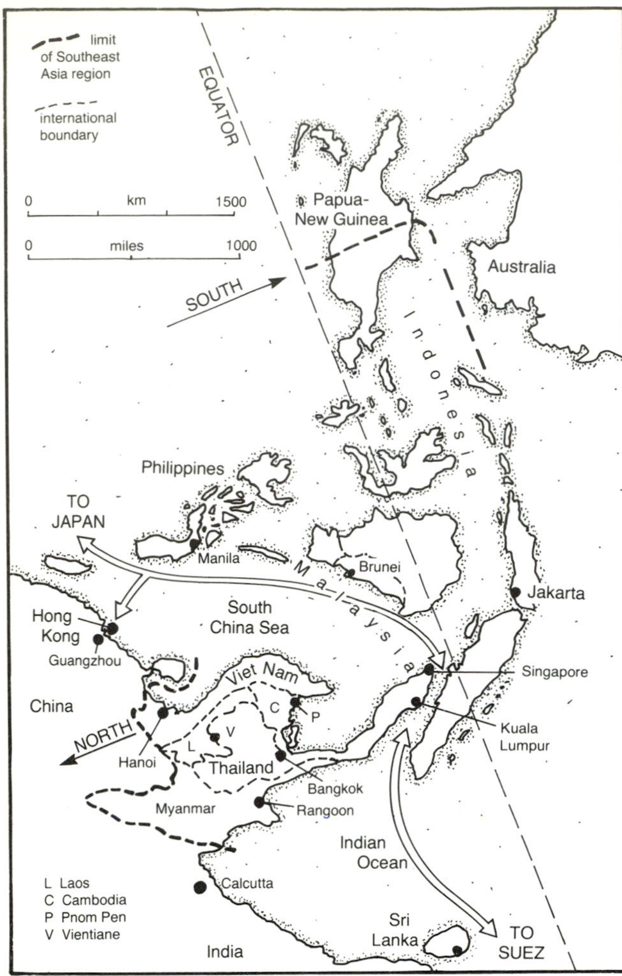

FIGURE 15.2 Southeast Asia and neighbouring regions

fragmentation into many islands and isolated valleys and basins also contributes by making transportation slow and costly. Nevertheless, reference to Table 15.1 shows that some of the major nations of Southeast Asia have been in existence for at least several centuries and in that respect are less artificial than many in Latin America, Africa and Southwest Asia. Many of the latter have come into existence in an arbitrary fashion, after independence movements, or are often quite inappropriately identical in form to previous colonial units.

Southeast Asia also has a variety of religions: Buddhism is very widespread on the mainland (see Box 15.1), Islam strong in most of Malaysia (over half of the population) and Indonesia (85 per cent), and Christianity (85 per cent Roman Catholic) remains strong in the Philippines after four centuries of Spanish rule and almost half a century of US domination. Indian settlers have brought Hindu and Sikh religions into western Malaysia. In spite of their relative proximity, in recent centuries China and Japan have not made much direct impact on Southeast Asia, at least until almost all the area was briefly held by Japan between 1942 and 1945, and since 1949, when Communist China with its Soviet ally was furthering the cause of communism in the region in the 1950s.

Further evidence of differences between and changes in the countries of Southeast Asia will be shown in the commentary that follows on the data in Tables 15.2 and 15.3. In the remaining sections of this chapter, differences within countries will be stressed.

15.2 Discussion of Tables 15.2 and 15.3

Table 15.2

(1)–(4) Since the break-up of the USSR, Indonesia has become the fourth largest country in the world in population. Indeed, it has about 40 per cent of the total population of Southeast Asia on about the same share of the territory. Thailand, the Philippines and Viet Nam have all overtaken France, Italy and the UK in population recently. Singapore, consisting of one main island and some smaller ones, and totalling only 626 sq km (242 sq miles), is only one-fifth the area of Rhode Island (USA), twice the area of the Isle of Wight in England, but it packs more than 3 million people into that space. Even without Singapore, the density of population varies greatly from country to country (Figures 15.3 and 15.4), with less than 20 per sq km in Laos but over 200 in Viet Nam and the Philippines, reflecting to some extent the availability of good farmland and the intensity of forest clearance.

(5)–(6) In Southeast Asia as a whole the annual rate of increase in population was 1.9 per cent in 1993, but with a TFR of 3.4, growth should continue for several decades, certainly for some time after 2025.

(7)–(8) Apart from Singapore, the countries of Southeast Asia still have the age structure characteristic of developing countries, but with the under-15 age group several percentage points below the levels in Africa, Southwest Asia and nearby Bangladesh. Family planning policies in Thailand have clearly been effective in recent years, the TFR there being only 2.4 in 1993.

Column (9) shows that Southeast Asia has a larger proportion of the population working than is found in Southwest Asia or the predominantly Muslim Pakistan and Bangladesh in South Asia. The high level in Viet Nam, Cambodia and Laos reflects the characteristic socialist policy of nearly full employment for all adults, regardless of gender. Thailand has no strong religious barriers to women working outside the home, or indeed to the employment of children at an early age. Singapore's population includes a large proportion of adults of working age.

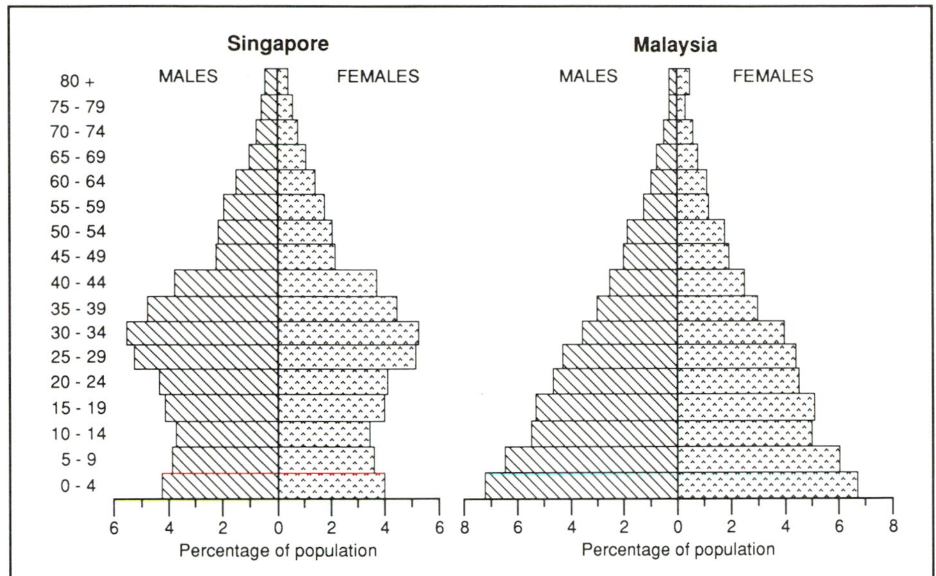

FIGURE 15.3 A comparison of the population structure of Singapore and Malaysia

Table 15.3

(1) In spite of possible inconsistencies in the definition of 'urban', very marked contrasts are evident in the level of urbanisation between the countries of Southeast Asia. Apart from the special case of Singapore, Malaysia and the Philippines stand out as being the most highly urbanised, the mainland countries of Southeast Asia (apart from Malaysia) being less urbanised on account of their large agricultural populations. As noted earlier in this chapter, there is a concentration of urban population in certain large cities, including several of the national capitals, which

BOX 15.1

Notes on Buddhism

Buddhism is regarded as one of the world's great religions, although its followers are fewer than those of Hinduism, Islam and Christianity. Its influence in the world is largely confined to parts of Southeast Asia, where almost all the population is nominally Buddhist in Thailand, Myanmar, Cambodia and Laos, as well as the majority in Sri Lanka. Buddhism also has followers in Korea, Japan and parts of China, including Tibet. Since 1949 Buddhism has, however, been discouraged and at times suppressed by the atheist/secular Communist Party of the People's Chinese Republic.

Like Christianity and Islam, the Buddhist religion originates in the teachings of one man. At the age of 29, Gautama Buddha, whose original name was Prince Siddhartha (563–483 BC), abandoned his life of wealth and comfort as the son of a king ruling in India, near the border of Nepal. The main themes of his teachings fall into 'Four Noble Truths': human life is intrinsically unhappy; unhappiness is caused by human selfishness; individual selfishness and desire can be brought to an end, the escape being through a right approach to life along a given path. The ideas of Buddhism were diffused widely in South, Southeast and East Asia, thanks to the conversion to the religion of the Indian emperor, Asoka, in the third century BC.

Like Christianity and Islam, modern Buddhism has a more fundamentalist and 'pure' strand, Theravada Buddhism, and a more 'liberal' one, Mahayana Buddhism. Perhaps its main difference from Christianity and Islam is the strong pacifist element. Modern countries in which Buddhism is an influence do not, however, have a particularly peaceful record. Internal suppression in Burma and genocide in Cambodia, however, cannot be attributed to the teachings of Buddha any more than Christian teaching can be blamed for the numerous conflicts in which Christians have fought each other. Emphasis in Buddhism is also on the creation of an atmosphere conducive to spiritual development, a feature presumably reconciling the believers to modest levels of material consumption and helping to protect the natural environment, but not an obvious feature of the growing affluent minority in Thailand. Although Buddhism originated in and spread from India, the religion now has very few followers there. Hinduism has to some extent absorbed ideas and principles of Buddhism.

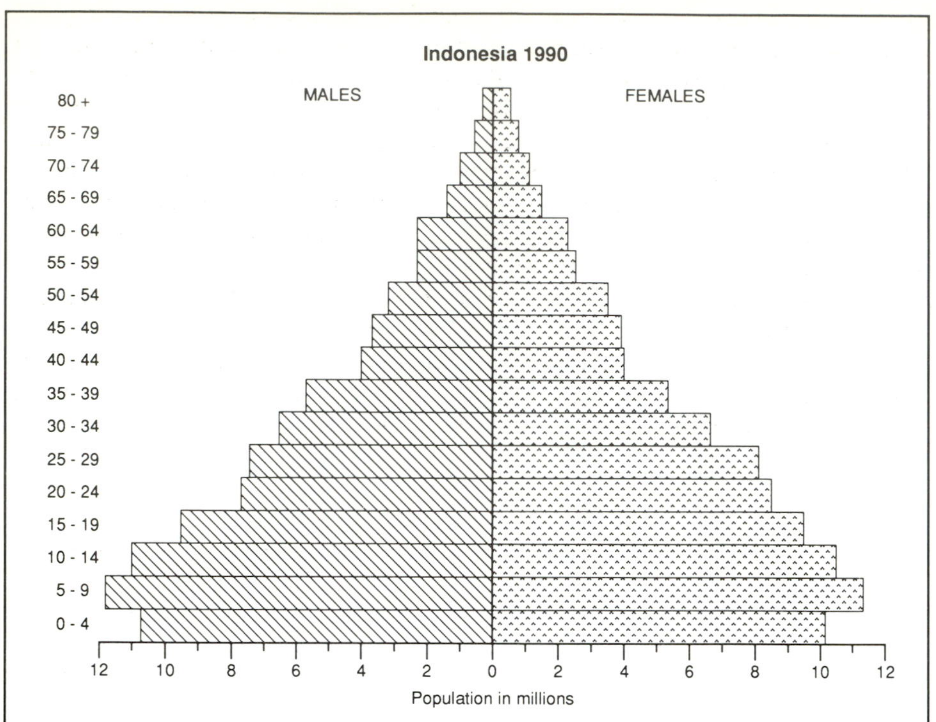

FIGURE 15.4 The population structure of Indonesia

TABLE 15.2 Demographic data for the countries of South, Southeast and East Asia

	(1) Popn rank in world	(2) Popn in millions, 1993	(3) Area in thous. sq km/(miles)		(4) Density persons per sq km/(mile)		(5) Annual % increase in popn	(6) Expected popn 2025 in millions	(7) % under 15	(8) % over 64	(9) % of total population in employment
South Asia											
India	2	897.4	3,288	(1,269)	273	(707)	2.1	1,380	36	4	38
Pakistan	8	122.4	796	(307)	154	(399)	3.1	275	44	4	29
Bangladesh	9	113.9	144	(56)	791	(2,049)	2.4	211	44	3	29
Nepal	44	20.4	141	(54)	145	(376)	2.5	41	42	3	41
Sri Lanka	48	17.8	66	(25)	270	(699)	1.4	24	35	4	37
Southeast Asia											
Myanmar	25	43.5	677	(261)	64	(166)	1.9	70	37	4	43
Thailand	21	57.2	513	(198)	112	(290)	1.4	76	29	5	53
Malaysia	46	18.4	330	(127)	56	(145)	2.3	34	37	4	39
Singapore	129	2.8	0.6	(0.2)	4,667	(12,088)	1.3	3.3	23	6	47
Indonesia	4	187.6	1,905	(736)	98	(254)	1.7	278	37	4	40
Philippines	14	64.6	300	(116)	215	(557)	2.5	101	39	4	36
Cambodia	77	9.0	181	(70)	50	(130)	2.5	17	35	3	45
Laos	106	4.6	237	(92)	19	(49)	2.9	10	45	4	46
Viet Nam	13	71.8	332	(128)	216	(559)	2.2	107	39	5	48
East Asia											
China	1	1,178.5	9,597	(3,705)	123	(319)	1.2	1,546	28	6	60
Hong Kong	93	5.8	1	(0.4)	5,800	(15,022)	0.7	6.7	21	9	53
Taiwan	42	20.9	36	(14)	581	(1,505)	1.1	25	26	7	na
Mongolian PR	135	2.3	1,567	(605)	1.5	(4)	2.7	4.6	44	4	47
North Korea	40	22.6	121	(47)	187	(484)	1.9	32	29	4	52
South Korea	24	44.6	99	(38)	451	(1,168)	1.0	55	26	5	44
Japan	7	124.8	378	(146)	330	(855)	0.3	126	18	13	51

Note: na = not available
Sources: for data in columns (2)–(8): *WPDS 1993*; (9) *FAOPY 1993*

draw in migrants from rural areas and smaller centres. By sealing itself off from Malaysia, Chinese-dominated Singapore has been able to limit migration and prevent the swelling of the cohorts of young adults that contributes to rapid growth in many other cities of Southeast Asia (see Figure 15.3). As in other regions of the developing world, the large cities of Southeast Asia have become at once outposts of the developed world, in which some citizens enjoy living standards similar to those generally found in developed countries, and interfaces between the outside world and their own hinterlands.

(2) Striking contrasts occur between levels of per capita energy consumption among the countries. Apart from Singapore, where processing, manufacturing and a sophisticated urban complex require large amounts of energy, Malaysia also stands out with a high level of consumption. On the other hand, minute amounts of energy are consumed in Myanmar, Cambodia and Laos, reflecting the virtual absence of heavy industry, a minimal amount of motorisation and mechanisation, and very little domestic consumption of commercial energy sources.

(3)–(4) In all the countries of Southeast Asia, the share of agricultural employment in the total economically active population has diminished, although in most countries the absolute number has increased somewhat. Since there is very little industry in

TABLE 15.3 Economic and social data for the countries of South, Southeast and East Asia

	(1)	*(2)*	*(3)*	*(4)*	*(5)*	*(6)*	*(7)*	*(8)*	*(9)*	*(10)*
			% employed in agriculture		*Quality of life*				*Non-primary exports as percentage of all exports*	
	% urban	*Energy consumption (kce) per head*	*1965*	*1991*	*life expectancy in years*	*% adult literacy*	*real GDP dollars per head*	*HDI*[3]	*1970*	*1991*
South Asia										
India	26	320	70	66	59	48	910	297	52	73
Pakistan	28	290	64[1]	49	58	35	1,790	305	57	73
Bangladesh	14	80	84[1]	68	52	35	820	185	64	70
Nepal	8	20	92	92	52	26	900	168	35	89
Sri Lanka	22	130	54	52	71	88	2,250	651	1	65
Southeast Asia										
Myanmar	24	60	62	46	61	81	600	385	5[2]	10[2]
Thailand	19	750	78	64	66	93	3,570	685	8	66
Malaysia	51	1,530	55	31	70	78	5,650	789	7	61
Singapore	100	5,610	7	1	74	88	15,110	848	30	74
Indonesia	31	320	66	48	62	77	2,030	491	2	41
Philippines	43	310	59	46	64	90	2,270	600	7	71
Cambodia	13	30	80	70	50	35	1,000	178	na	na
Laos	19	40	81	71	50	54	1,030	240	na	na
Viet Nam	20	140	83	60	63	88	1,000	464	na	na
East Asia										
China	26	840	73[2]	67	70	73	2,660	612	70	76
Hong Kong	100	1,790	6	1	77	90	15,180	913	96	95
Taiwan	71	na	47	na	74	93+	7,520[2]	900[2]	76	92
Mongolian PR	57	1,840	59	30	63	93	2,000	574	na	na
North Korea	60	2,810	70	30	70	96	2,170	654	na	na
South Korea	74	2,500	54	24	70	96	6,120	871	76	91
Japan	77	4,160	24	6	79	99	14,310	981	93	98

Notes:
na = not available
kce = kilograms of coal equivalent
[1] Data for before the break-up of Pakistan
[2] Author's estimate
[3] Human Development Index, maximum possible score 1,000
Sources: (1) *WPDS 1993*; (2) UN 1991, *Energy Statistics Yearbook*; (3), (4) *FAOPY 1993*; (5)–(8) *UNHDR 1992*, Table 1; (9), (10) *WDR 1993*, Table 16

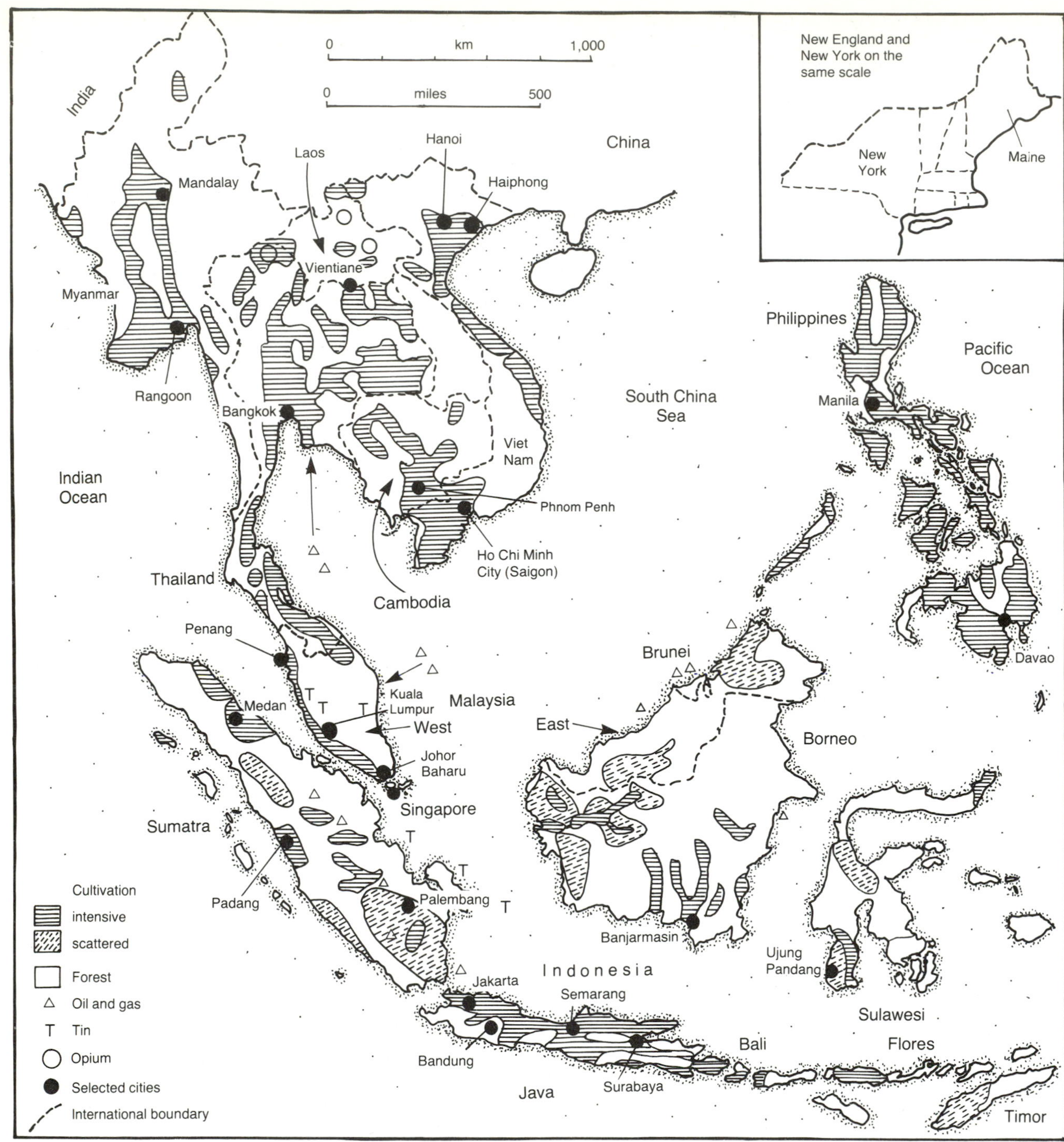

FIGURE 15.5 Economic map of Southeast Asia

Myanmar it is difficult to understand what the 54 per cent not in agriculture find to do. Cambodia and Laos were still heavily dependent on agriculture in the early 1990s.

(5)–(8) The UN Human Development Index brings out strong contrasts between the countries of Southeast Asia. Malaysia and Singapore have life expectancy levels comparable to those in the poorer countries of Western Europe, while Thailand, Singapore, the Philippines and Viet Nam have levels of literacy well above those prevailing in the developing world. Myanmar, Viet Nam, Cambodia and Laos all have levels of real GDP per inhabitant comparable with levels widely found in Africa. The extent to which the conflicts in Viet Nam and Cambodia, and the indirect involvement also of Laos, have held back development in these

PLATE 15.1

a Part of Buddhist temple complex, near Chang Mai, northern Thailand

countries is a matter of speculation, but Viet Nam seems to be making a better recovery than Laos or Cambodia.

Columns (9)–(10) show clearly the rapid rate of industrialisation in five of the countries of Southeast Asia. A comparison with the accompanying data for South and East Asia indicates that the region is 'catching up' industrially. With population growth and changing economic structure, all three regions are making greater use of their own primary products and have less to export than three decades ago. Thus, for example, between 1982 and 1992 Indonesian oil production grew from 69 to 74 million tonnes per year, whereas its consumption grew from 22 to 35 million tonnes.

Although population is not growing as quickly in Southeast Asia as it is in Africa and Southwest Asia, an average of 7–8 million people a year are expected to be added to the total population between 1993 and 2025, increasing the total from 460 million to about 700 million.

15.3 Myanmar and Thailand

Myanmar and Thailand are similar in certain respects, but differ enormously in their roles in and approaches to the rest of the world. Myanmar is rather larger in area, Thailand in population. Each consists of extensive lowland areas, mostly devoted to the cultivation of rice, and rugged, largely forested mountain ranges. The main concentration of population, consisting respectively of the Burmese and Thai majority ethnic groups, is in each case almost encircled by a large number of ethnic minorities. In addition, each country has a long

b Junior school classroom, northern Thailand. A reduction in the fertility rate is one change attributed to the widespread availability of schools and attention to the educational needs of girls as well as boys

TABLE 15.4 Land use and agricultural data for selected countries of South, Southeast and East Asia

	(1)	(2)	(3)	(4)	(5)	(6)	(7)	(8)	(9)	(10)	(11)
	Total popn in thous.		*Arable + permanent crops thous. sq km/(miles)*		*Forest, thous. sq km/(miles)*		*Rice yields, kg per hectare/(acre)*			*Roundwood million cu m*	*Fuelwood million cu m*
	1965	*1991*	c. *1965*	c. *1990*	c. *1965*	c. *1990*	*1961–5*	*1979–81*	*1989–91*	*1990*	*1989*
South Asia											
India	486.7	859.2	1,638 (632)	1,691 (653)	623 (241)	667 (258)	1,480 (600)	1,860 (750)	2,620 (1,060)	275	245
Pakistan	52.1	117.5	197 (76)	208 (80)	28 (11)	36 (14)	1,650 (670)	2,470 (1,000)	2,330 (940)	27	23
Bangladesh	61.2	116.6	85 (33)	91 (35)	14 (5)	19 (7)	1,650 (670)	1,950 (790)	2,680 (1,090)	31	6
Sri Lanka	11.2	17.4	20 (8)	19 (7)	29 (11)	21 (8)	1,910 (770)	2,560 (1,040)	2,950 (1,190)	9	8
Southeast Asia											
Indonesia	105.7	181.4	127 (49)	220 (85)	1,522 (588)	1,134 (438)	1,810 (730)	3,260 (1,320)	4,300 (1,740)	172	131
Viet Nam	34.5	67.6	49 (19)	66 (25)	135 (52)	99 (38)	1,960 (790)	2,120 (860)	3,170 (1,280)	29	24
Philippines	32.3	62.3	85 (33)	80 (31)	118 (46)	104 (40)	1,260 (510)	2,250 (910)	2,780 (1,130)	39	33
Thailand	30.7	58.8	114 (44)	221 (85)	274 (106)	141 (54)	1,590 (640)	1,890 (760)	1,940 (790)	38	26
Myanmar	24.7	42.1	90 (35)	101 (39)	453 (175)	324 (125)	1,640 (660)	2,690 (1,090)	2,860 (1,160)	23	17
Malaysia	9.4	18.3	59 (23)	49 (19)	250 (97)	193 (75)	2,500 (1,010)	2,840 (1,150)	2,580 (1,040)	51	8
East Asia											
China	725.4	1,151.3	1,000 (386)	966 (373)	1,185 (458)	1,265 (488)	na	4,240 (1,720)	5,620 (2,280)	277	178
Japan	97.8	123.8	56 (22)	46 (18)	253 (98)	251 (97)	5,020 (2,030)	5,580 (2,260)	6,120 (2,480)	30	–

Notes:
na = not available
– negligible
Sources: *FAOPY 1970, 1991*; (10) *UNSYB 90/91*; (11) UN 1991, *Energy Statistics Yearbook*

'panhandle' of territory, projecting south.

There are striking differences between the two countries in their political and international backgrounds and attitudes since the Second World War. Between 1986 and 1990 the value of Myanmar's imports averaged 254 million US dollars a year, while the value of its exports was 240 million, each less than 6 dollars per head of the total population, giving an import (or export) coefficient of only 3 per cent of its very small GDP. Timber and rice are the main exports of Myanmar. The corresponding totals for Thailand were over 20 billion US dollars' worth of imports and almost 16 billion of exports, giving imports worth about 350 dollars per head, and equivalent to 22 per cent of Thailand's much larger GDP.

Before 1948 Myanmar (then Burma) was part of the British Empire in South Asia. It is different culturally from Bangladesh and Bengal in being predominantly Buddhist, not Muslim or Hindu. Communications by land between Myanmar and its neighbours are poor, contributing to its isolation. This feature was highlighted by the use of the difficult Burma Road from Mandalay to Chongqing (then Chungking) to supply China during the Japanese occupation of most of China's coastal provinces, and the infamous Burma railway, built by Allied prisoners of war and the local population for the Japanese, to provide a short cut from Thailand (then Siam) into Burma, avoiding the long sea route via Singapore. The Rivers Irrawaddy and Salween, with accompanying roads and railways, are the main north–south lines of movement in the country.

The data in Table 15.4 show that the cultivated area in Myanmar has not increased as fast as population since the 1960s, but great progress has been made in increasing rice yields. Less satisfactory has been a reduction of almost 30 per cent in the extent of forest. Twice as much roundwood is produced in Thailand as in Myanmar, but restrictions in the former country have prevented expansion of the industry there whereas in Myanmar production has risen gradually but steadily during that time. In contrast, rice production in Myanmar declined in the 1980s while it increased in Thailand. Both countries are heavy users of fuelwood, but Thailand is a much more prolific user of fossil fuel energy. Myanmar produces and consumes about 3 million tonnes of coal equivalent of oil and natural gas per year whereas Thailand produces about 16 million tonnes, while importing another 27 million tonnes, exporting 2 million, and using 41 million. The amounts used per inhabitant are 60 kg in Myanmar, 750 in Thailand. Thailand produces almost 25 per cent of the world's natural rubber (in 1991 1.4 million tonnes), compared with one-hundredth of that in Myanmar.

The contrast between the two countries is further highlighted by the growth in Thailand of the light manufacturing sector, exemplified by a comparison of the industrial consumption of cotton (thousands of tonnes): Thailand 1982/3 114, 1991/2 390, Myanmar 17 and 26 respectively. Western consumer tastes contrast too, with 20,000 tonnes of beer produced in Myanmar in the late 1980s compared with 1.8 million in Thailand. Contrasts in transport and communications are illustrated by the 80,000 cars in use in Myanmar compared with 1.2 million in Thailand (see Box 15.2). Bangkok is teeming with motor vehicles. It is also a popular destination for tourists, with over 5 million visitors in 1990, most from the rest of Asia, but over a million from Europe. In 1989 Myanmar attracted a mere 5,000 and thereafter the tiny industry ceased to exist because of internal unrest and conflict.

The contrasts between Myanmar and Thailand have been noted above at some length in order to illustrate the danger of assuming that all developing countries are broadly similar. Myanmar and Thailand are indeed similar in their physical and environmental background, cultural history and general location in the world but there was no American presence in Myanmar after the Second World War, whereas Thailand was an important base for US activities in Viet Nam.

15.4 Malaysia, Singapore and Indonesia

These three countries may be considered together because they have much in common: Singapore provides services for Malaysia and western Indonesia, and Malaysia and Indonesia share the large island of Kalimantan (formerly Borneo). Indonesia dwarfs the other two, stretching 5,300 km (3,300 miles) from northwest Sumatra to the southeast of West Irian, the same distance as from Juneau, Alaska, to Miami, Florida.

Malaysia consists of three distinct territories, all formerly British colonies. In 1993 West Malaysia, a peninsula of mainland Asia, had about 15 million of Malaysia's total population of 18 million, while Sarawak and Sabah (formerly North Borneo) each had 1.5 million. About 15 per cent of the total area of Malaysia is under arable (10,000 sq km – 3,860 sq miles) and permanent crops (40,000 sq km – 15,400 sq miles). Malaysia, Indonesia and Thailand each produce about one-quarter of the world's natural rubber. Tropical rainforest has been cleared since early in the twentieth century for the planting of the Brazilian *Hevea brasiliensis* tree. Although the forest is thereby replaced by a monoculture, at least a tree cover is preserved, in contrast to places where field crops are grown or livestock raised. A large proportion of the permanent crops of Malaysia is under oil-palm trees which, like rubber, provide a tree cover in place of the native forest.

Off the western coast of the Malay Peninsula is the

PLATE 15.2
Thailand

a Fish farms in the coastlands near Bangkok, one of the places visited by tourists on trips

Molucca Strait, along which shipping routes from the Gulf of Arabia (oil), the Red Sea (from Suez) and Southern Africa pass to round Singapore and enter the South China Sea. Penang (Malaysia) and Batan Island (Indonesia) are miniature versions of Singapore, with good locations and a growing infrastructure to encourage industrial development. Kuala Lumpur, the capital of Malaysia, is a fast-growing industrial and financial centre.

Singapore gained independence from the UK in 1959 and until 1965 was part of Malaysia, after which it became a sovereign state within the British Commonwealth. The island of Singapore is only 640 sq km (247 sq miles) in area but it has almost 3 million inhabitants. The smallest state of the USA, Rhode Island, is five times larger in area but has only 1 million inhabitants. Singapore is in effect a completely urbanised community with virtually no agricultural land or mineral resources and inadequate water for its needs. It depends therefore on the export of manufactures and the provision of services to pay for its imports of food, fuel and raw materials, one reason why until recently the policy was to reduce fertility and to stabilise population size. Singapore has demographic indices similar to those of many developed countries: life expectancy 74 years, infant mortality 5 per 1,000 and a total fertility rate below replacement at 1.8.

The population of Singapore is predominantly of Chinese origin (78 per cent), while Malays and Indonesians make up 14 per cent and Indians, mainly from the south of India, 7 per cent. The languages used in Singapore are Chinese, Malay, Tamil and English, underlining the cosmopolitan nature of the country, a feature also of its religious life. As with healthcare, standards in education are high, and the level of adult literacy exceeds 90 per cent.

Singapore depends for its high GDP per capita – 15,900 US dollars in 1990 (cf. USA 21,400, UK 15,800) – on its role as a focus of shipping and air routes, its industries and its financial services. With more than 9 million gross registered tonnes of shipping, its merchant fleet is twice the size of that of the UK and half as large as that of the USA. Shipbuilding and repairing, the handling of containers and the refining of oil are modern activities continuing Singapore's role as a key port in the British Empire. The manufacture of textiles, clothing and electronic goods in Singapore is typical of several of the larger cities of Southeast Asia. Singapore is also a financial centre of growing importance and its international reserves (excluding gold) rose from 8.5 billion US dollars in 1982 to 34 billion in 1991 (cf. Malaysia 10, Japan 72). Malaysia, Japan and the USA are now the main trading partners of Singapore.

Singapore has been described at some length because its experience shows how an island that four decades ago was part of the colonial world can be transformed into a place with development indices similar to those of Japan and the developed countries of Western Europe. If Singapore had remained within Malaysia it would undoubtedly have attracted numerous migrants from both rural and urban settlements on the mainland. As it is, Singapore can regulate migration (see Figure 15.3).

With about 40 per cent of the population of Southeast Asia, 42 per cent of the land and forming part of the largest archipelago in the world, Indonesia forms a series of stepping-stones between the Asian mainland and Australia. As the Dutch East Indies, this area of tropical islands was the main overseas focus of interest for the Netherlands, being developed as the source of a wide range of tropical agricultural and forest products

b Dam and hydroelectric station in the mountains of northern Thailand. The larger hydroelectric stations in developing countries are usually in thinly populated areas, which themselves are able to consume only a small proportion of the current generated. Much may be transmitted over large distances to urban centres, with considerable loss of power as distance increases

for the industrial countries: spices, sugar, coffee, tea, rubber and timber. After the defeat of Japan at the end of the Second World War, the Netherlands could retain control of only the eastern extremity of the Dutch East Indies.

For some time after independence, Indonesia languished under a generally isolationist government, which reduced the influence of the powerful Chinese settlements and banned the Communist Party. Each region of Indonesia was largely self-supporting, and there was little migration between the islands. The data in Table 15.5 show, however, that 60 per cent of the population is concentrated on Java, which constitutes only 7 per cent of Indonesia's area. Such concentrations of population are found in other countries, but in Indonesia the concentration on a single island draws greater attention to the situation. The high density of population on Java can be accounted for by the extensive area of very fertile volcanic and alluvial soils. The total population of Indonesia has doubled in the thirty-three years from 1960 to 1993 (from about 95 to 188 million) and is expected to grow to about 280 million by 2025. Various attempts have been made to move population from Java to other islands, Sumatra being the preferred destination. More recently, settlements have been established in West Irian. Given the need to provide homes and employment for those resettled, the transfers have understandably been small in relation to the growth of population in Java.

In 1990, 60 per cent of the area of Indonesia was covered in forest compared with 78 per cent in the mid-1960s. Some of the reduction in the area of forest can be attributed to the expansion of the cultivated area from 6–7 per cent of the total area of the country to over 11 per cent since the mid-1960s. The rate of cutting of the forest for timber and fuelwood has also grown, with 85 million cubic metres (3,005 million cu feet) of roundwood produced annually in the early 1960s compared with 170 million (6,010 million cu feet) in 1990.

Of the cultivated area of Indonesia, 60,000 sq km (23,200 sq miles) is under permanent crops (rubber, cocoa, coconut), while of the remaining 160,000 sq km (62,000 sq miles) about 100,000 (30,000) is used for the cultivation of rice. Thanks to both the expansion of rice cultivation and a spectacular increase in average yield (1961–5, 1,800 kg per hectare, 1989–91 4,300 kg), rice production has grown faster than population. A comparable increase in production in the next thirty years would indeed be an impressive achievement.

As areas of forest are cleared and hillsides are reclaimed by terracing, new land is being brought into cultivation in Indonesia. Fertiliser consumption is also rising. Behind the scenes, however, change is taking place. Lynch (1977) anticipated the effect of the increased efficiency of new agricultural tools and techniques on the poorer, often landless villagers.

> Like field preparation and transplanting, harvesting too is intensively done by hand. Each head of grain is individually severed using a small cutting knife called *ani-ani* in Indonesian. The *ani-ani* necessitates a large labour force and is used allegedly so as not to offend *Sri Dewi*, the Goddess of Fertility. Old varieties of rice grow to uneven heights and heads are likely to shatter if handled roughly. The *ani-ani* ensures maximum employment and minimum loss of valuable grain.

As recently as 1970, almost the whole of the value of the exports of Indonesia was accounted for by fuels and other minerals (44 per cent) and other primary

BOX 15.2

What the tourist sees: the Thailand experience

In the last thirty years, since the beginning of the era of jet aircraft, charter flights and package tours, many developing countries have encouraged the growth of tourism. Each country has a favourable or unfavourable location in relation to the three main sources of tourists from developed countries, North America, Western Europe and Japan. Each has its own range of attractions to offer, whether natural such as beaches and mountains or human such as historical sites and gambling centres. Thailand is one of the developing world countries that has promoted tourism most vigorously in the last two decades and its experience and attractions serve to illustrate some of the features and problems of modern long-distance tourism (Figure 15.6).

Bangkok is only a few hours' flying time from Tokyo, and is also one of the preferred stopping places on the long flights between Europe and Australia. Between 1986 and 1990 alone, the number of tourists visiting Thailand almost doubled, from 2.8 million to 5.3 million. In 1990 well over half came from South and East Asia, including Japan, and from Australia and New Zealand, but about 1.3 million came from Europe and about 400,000 from the more remote North America. Visitors from developed regions travelling to Thailand for non-business reasons fall into two categories. The majority book a package tour, which usually includes travel from home and accommodation and travel in Thailand. The minority, mostly less affluent, are the young backpackers, who, to the surprise if not annoyance of the local population, usually have very little money to spend. For the more affluent, tour companies offer a great variety of mixes of attractions and itineraries in their tours. Travel within Thailand is usually by air or road but, to please rail enthusiasts, some rail journeys are available. Tourists are usually booked into good-quality hotels or chalets. The geographical features of Thailand are such that the places and attractions visited fall broadly into six types, any of which may be the central feature of the visit or may be combined with others. The following types and examples cover most features:

1. The capital city, Bangkok, offers a sophisticated urban environment, with historical buildings, large pleasantly air-conditioned stores with cheap factory-made goods, as well as more sleazy venues. There are also opportunities to visit local features such as the many canals or the famous Bridge over the River Kwai, built by the forced labour of Allied prisoners and local people for the Japanese in 1942 after they occupied Thailand. About half of Thailand's 2 million motor vehicles in use are concentrated in Bangkok: traffic is extremely heavy, and pollution levels are high.

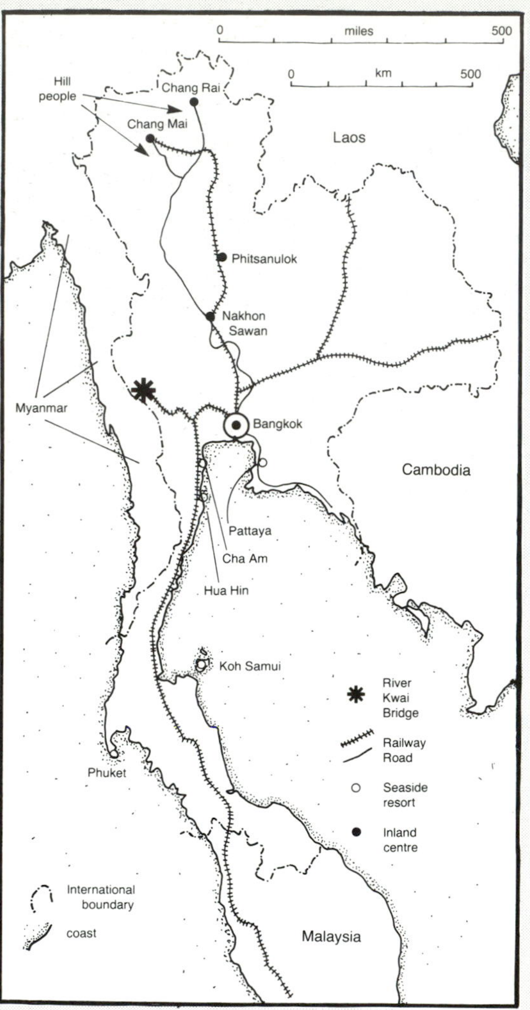

FIGURE 15.6 Centres of tourism in Thailand.

2. Many tourists choose to leave Bangkok for the seaside in such places as Pattaya and Phuket. Here Thailand is selling its tropical climate and attractive beaches.

3. A large number of inland cities, with names that are unpronounceable for the tourists, offer secondary attractions, including innumerable temples, such that when jet-lag has passed, temple-lag can set in.

4. The lowlands of the interior of Thailand offer a picturesque countryside, watery with irrigated rice-fields, unfamiliar to most European and North American visitors.

PLATE 15.3
Villagers of the hill tribes of northern Thailand. Mortality rates are high for all age groups.

5. Chang Mai is the regional centre of the north of Thailand and in its surrounds are numerous craft centres, producing for both the home and tourist markets. Metal-workers produce religious ornaments for local use and various metal items for tourists. Carpenters make elegant furniture from local and Burmese teak. This can be shipped at the drop of a credit card to Osaka, Hamburg or Houston. Lacquer-work, fabrics, leather goods and sundry other products are displayed for sale in large quantities.

6. The more ambitious and discerning tourists are expected to visit the so-called hill tribes. Throughout Southeast Asia, many people lead a life that is semi-independent of the rule of the central government. Unfortunately for many of the hill tribes in northern Thailand, whose austere existence was alleviated somewhat by the commercial cultivation of opium poppies, policy regarding the sale and use of drugs by the Thai government has forbidden this comparatively lucrative enterprise. One way of helping these remote citizens of Thailand is to take tour groups there.

Most tourists want to visit the various attractions of Thailand in comfort and they do not therefore experience the living conditions, discomforts and deprivations experienced by many of the inhabitants of the country. They see elephants, model schools, orchid cultivation, hand weaving, and tropical vegetation. Very occasionally a tourist gets killed or taken hostage, while many suffer stomach and other afflictions, regardless of how careful they are. A more general appraisal of the implications of the tourist industry for the major regions of the world is provided in Box 17.2 (p. 415).

PLATE 15.4
Malaysia

a Typical home of a hill tribe family in the mountainous forested Cameron Highlands of central Malaya, Malaysia, this one located near a fairly busy road

b New mosque in northwest Malaya, Malaysia. In view of the large numbers of people of Chinese and Indian origin and therefore mostly non-Muslims, the Malaysian government preaches tolerance but there is continual construction work to provide more mosques for the Islamic majority

commodities, mainly of bioclimatic origin (54 per cent). The first category was still prominent in 1991, but exports of agricultural products had fallen by then to 14 per cent of the value, while manufactures rose from 1 per cent to 39 per cent, about one-third of them textiles. In both 1970 and 1991, manufactures still accounted for well over three-quarters of the value of imports.

Indonesia's recent emergence as an exporter of manufactured goods reflects the establishment of an industrial base since the 1960s. In 1992, Indonesia was credited with 0.6 per cent of the world's oil reserves, a quantity that would barely last a decade at the rate of output in that year. The prospect of finding further deposits in the vast expanse of territory and in offshore areas seems good. Annual output has remained around 60–75 million tonnes a year but consumption has nearly doubled. Natural gas production has increased almost threefold in the 1980s but most is exported (as LNG), principally to Japan. Indonesia's large coal reserves consist almost entirely of lignite. At all events, the country has a reasonably good energy base for its industrial growth. It also has cheap labour, with concentrations not only in the capital Jakarta (8 million population), but also in several other large cities, including Surabaja, Bandung (each with about 2 million), all three on Java, and Medan (1.5 million) and Semarang (1.2 million) on Sumatra.

c Malaya, Malaysia, a swathe of tropical rainforest cleared for a road improvement scheme

d The risk you run if you leave your shoes outside to visit a temple in Penang, Malaya, Malaysia

The industrial development of Indonesia in the 1980s can be exemplified by the growth of some sectors of heavy industry as well as of light industry. Cement output rose from 5.7 million tonnes in 1981 to 16.0 million in 1990, evidence of the growth in the construction sector, while steel production has grown to 2 million tonnes from its start in the late 1970s. In the light manufacturing sector, the consumption of raw cotton has increased threefold in the 1980s. Apart from the maintenance and repair of equipment, there is virtually no engineering industry in Indonesia. In the early 1990s the level and sophistication of its industrial sector could be compared with those of Brazil half a century ago.

Compared with most countries in South, Southeast and East Asia, Indonesia has a reasonably good natural resource base, but the growth in population of 1.7 per cent per year, although somewhat below the average for Southeast Asia, is leading to increasing pressure on bioclimatic and mineral resources, the excessive growth of Jakarta, and increasing clearance of forest. Various countries in East Asia, above all Japan and increasingly China, can absorb huge quantities of primary products, and Indonesia is conveniently near to supply these.

15.5 The Philippines

In many respects the Philippines is a smaller version of Indonesia. It has one-third of the population on less than one-sixth of the area. In the period of Spanish and Portuguese domination of European colonies in the tropical world, the location of the Philippines was

TABLE 15.5 The islands of Indonesia (1989) and the Philippines (1990)

	Area in thous. sq km/(miles)		*Population*		
			in millions	*per sq km/(mile)*	
Java	130.4	(50)	107.5	824	(2,134)
Sumatra	481.8	(186)	36.9	76	(197)
Kalimantan	549.0	(212)	8.7	16	(41)
Sulawesi	194.4	(75)	12.5	64	(166)
West Irian	419.7	(162)	1.6	4	(10)
Other islands	38.5	(15)	11.9	309	(800)
Indonesia	1,948.7	(752)	179.1	92	(238)
Luzon	108.2	(42)	34.1	315	(816)
Mindanao	99.3	(38)	14.9	150	(389)
Other islands	92.5	(36)	19.7	213	(552)
Philippines	300.0	(116)	68.7	229	(593)

about as marginal as anywhere in terms of distance from Europe. Most of the time for three hundred years, Spain controlled and traded with the Philippines from Panama or from Acapulco in Mexico, in galleons crossing the Pacific. The Philippines remained a Spanish colony until it was acquired by the USA, which defeated Spain in a brief war in 1898. In the later decades of the nineteenth century, the USA was becoming a Pacific power of considerable influence, able to force Japan to open its ports to trade and to put pressure on China to do the same. After the Second World War, however, the Philippines was given independence.

The Philippines still had almost half of its economically active population in agriculture in the early 1990s but, like Thailand, Malaysia and Indonesia, the economy has become much more diversified in the last two decades. There has been little change in the area cultivated, but forest has been steadily cleared by logging so that only one-third of the country is now forested. Agricultural production has increased greatly through a marked increase in rice yields.

The Philippines has a very weak energy base and, in spite of some limited coal and hydroelectric production, has to import about 90 per cent of its energy needs, almost all provided by oil. Although it has a wide range of minerals, none is of outstanding importance. The growth of industry can primarily be accounted for by the presence of cheap labour, with proximity to China and Japan a locational advantage. With 7 million inhabitants, Greater Manila is by far the largest city and industrial centre.

In 1991 about 30 per cent of the imports of the Philippines consisted of food, fuels and other primary commodities, little changed since 1970. On the other hand, its exports of primary commodities dropped from 93 per cent of the total value to 29 per cent in 1991, while exports of manufactures grew to 71 per cent. The expectation is that the population of the Philippines will grow from 65 million in the early 1990s to about 100 million in 2025. If the trends of the last two decades continue, the Philippines will become a net importer of primary products, a characteristic of a growing number of developing countries.

15.6 Viet Nam, Laos and Cambodia

The area now occupied by Viet Nam, Laos and Cambodia was strongly influenced by China for about 1,000 years from 111 BC to AD 939 and intermittently after that until the nineteenth century. In the nineteenth century Spain still held the Philippines, Britain was expanding its empire eastwards from India into Burma and Malaya, while the Dutch were consolidating what is now Indonesia. On the mainland of Southeast Asia, during the period 1859–93, France occupied what was referred to as Indo-China, which extends from Tonkin in the north, adjacent to China, through Laos and Annam, to Cambodia in the south. North from Tonkin, the French established a sphere of influence in China and to the west one in the eastern part of Siam (now Thailand). Compared with some of the other colonies of Southeast Asia, the French-held territories were of limited economic importance.

In the Second World War, after its occupation by Germany, France was not in a position to resist the Japanese invasion of Indo-China in 1942, but in 1945 the colonies returned to French control. Resistance to the French presence grew and after nine years of anticolonial conflict the French withdrew from the region. The revolutionary movement was strongest in North Viet Nam and in the Geneva agreements of 1954 Viet Nam was divided along the parallel of 18° North by a partition reminiscent of the earlier division of Korea.

In the 1950s the USA was concerned over the growing economic and military strength of communist China, while also building up alliances and military strength to face the Soviet Union. The USA did not accept the Geneva agreements and by 1957 it was already building up a counter-revolutionary government in Saigon, capital of South Viet Nam, against the communists of North Viet Nam. Fear that an increasingly militant China could push into Southeast Asia, and influenced by the 'domino theory', according to which the fall of one country to communism could lead to penetration of its neighbour, the USA committed itself to ever-increasing military intervention in the whole region. By the mid-1960s, more than half a million American service troops were committed to defending South Viet Nam. Thus, as a pawn in a global game of chess, Indo-China was the scene of a second period of war, from 1957 to 1973. North Viet Nam was subjected to saturation bombing, while pro-North Vietnamese forces gained control of large mainly thinly populated, mountain areas of South Viet Nam, supplying their forces from the Ho Chi Minh trail through theoretically neutral Laos.

Soon after hostilities ceased following a compromise settlement in 1973, North Viet Nam gained control of South Viet Nam and Viet Nam was unified under a communist government. Although Marxism-Leninism was renounced in 1992 and some private ownership of means of production has been permitted since then, the Communist Party is still in control. Evidence of continuing dissatisfaction with the government and with conditions in Viet Nam was the illegal emigration of about half a million refugees between 1975 and 1984.

Viet Nam, Laos and Cambodia had a combined population in the early 1990s of 85 million, expected to reach 135 million by 2025. These former French colonies have been the scene of one of the longest and most devastating conflicts since the Second World War. Loss of life, and damage to roads, buildings and agricultural land and to the environment in general, have left all three countries with a very low level of real GDP per inhabitant. The UN Human Development Index, however, is made higher for Viet Nam than for Laos and Cambodia on account of its higher levels of life expectancy and adult literacy.

All three countries still have a large labour force in agriculture and, while the share of this sector in total employment has diminished since the 1960s, the absolute number of persons in the agricultural sector has actually increased (e.g. Viet Nam 1975 16.1 million, 1991 19.7 million). The area under cultivation has increased in Viet Nam and Laos, but not in Cambodia. Rice yields have increased substantially in Viet Nam since the end of hostilities in 1973 and also in Laos, but in view of the unstable situation in Cambodia, it is not surprising that this country has the lowest rice yields in Asia.

The energy base of Cambodia and Laos is very limited, as shown by the annual per capita consumption of commercial sources of energy of 30 and 49 kg respectively. Indeed, both consume more fuelwood than fossil fuel energy. Viet Nam, on the other hand, produces several million tonnes of coal a year (5 million in 1990) and since the late 1980s has become an oil producer (2.5 million tonnes in 1990), while also using a comparable amount of fuelwood. The area of forest in all three countries has been greatly reduced since the 1950s. Laos and Cambodia have virtually no modern industry, but Viet Nam is following the examples of Thailand and Malaysia, with the dubious advantage of having even cheaper labour. Hanoi (with its port Haiphong) in the north, and Ho Chi Minh City (formerly Saigon) in the south are the prospective starting places for the establishment of industries.

A down-to-earth view of the changes in Viet Nam, not matched by any of significance yet in Cambodia or Laos, is imparted by the following (*The Times*, 27 February 1995, p. 11):

> Sleepy island means business: A fishing boat puts to sea from Phu Quoc, a sleepy island off the south-western coast of Vietnam with just six cars, unpaved roads and four hours of electricity on a good day. The island is to be transformed into an international financial centre at a cost of £639 million. Mekong Holdings, a Vancouver-based investment firm, is expecting the go-ahead soon from Hanoi for its subsidiary, Phu Quoc Island Development, to draw up a master plan that will include a port, airport and export-processing zone. But the idea of an offshore tax haven has encountered some scepticism. 'Vietnam is still coming to grips with the idea of onshore banking, let alone anything else,' one Western banker said.

Like South Asia, most of Southeast Asia was colonised by European powers and held for varying lengths of time. During the post-colonial decades the new, independent countries have followed divergent paths. On the mainland, Burma has remained largely isolated from the world, Indo-China is gradually recovering from the devastation of a Cold War conflict not of its own making, while Malaysia and Thailand have gone for modernisation and economic growth. In Indonesia and the Philippines, there are marked contrasts in the way different islands have been developed, with resettlement from densely populated areas impinging on the remaining areas still inhabited by the original forest populations. Industrialisation on modern lines is mainly confined to a few large cities and a number of offshore islands, the latter providing infrastructure and tax concessions to firms from outside the region. In 1994 the total population of Southeast Asia was approaching 500 million. The total is expected to exceed 700 million by 2025. Compared with Africa south of the Sahara, which lacks political stability, and with South Asia, which lacks natural resources, Southeast Asia is better placed to face the future. Environmental crises and ethnic conflicts should be of local rather than international extent.

KEY FACTS

1. Southeast Asia is characterised physically by the presence of many ranges of high mountains, with intervening lowland river valleys and basins, and extensive coastal plains. The natural vegetation of most of the area is tropical rainforest.

2. The political geography of the region is characterised by a number of 'middle-sized' sovereign states, mostly formed of colonies established by European powers, but based on earlier cultural entities.

3. Nowhere in Southeast Asia is far from the coast.

4. The economy of most of the region is based on agriculture, but mainly in the last two decades modern industrial enterprises, using cheap labour, have been set up in some of the countries. Most are located in and around the largest cities.

MATTERS OF OPINION

1. The quotation at the head of this chapter illustrates the ambition of many leaders in the countries of Southeast Asia. Is this ambition to be commended?

2. Indonesia and to a lesser extent other countries of Southeast Asia have implemented plans to resettle population from densely populated rural areas and from the large cities in more remote, less 'developed' parts of their countries. Is this policy justified, given that it causes damage both to local communities and to the environment?

SUBJECTS FOR SPECULATION

1. The rate of population growth in Southeast Asia is slowing down. Even so, between 1994 and 2025 it is expected to grow from 484 million to 718 million, thus adding about 230 million people. What changes can be expected in employment structure in three decades' time?

2. The forest cover in Southeast Asia is being cleared both by loggers and to make way for cultivation. Will it be possible to save more than the most remote stands of forest?

3. Most of the countries of Southeast Asia are net importers of fossil fuels and large consumers of fuelwood. How can the lack of large energy sources – except in Indonesia – be overcome?

Chapter sixteen

China and its neighbours

China is a developing country with a huge population but limited cultivated land, inadequate per capita resources and a weak economic foundation. These make up China's basic conditions. A rapidly expanding population will conflict with the country's socio-economic progress and create contradictions between the utilization of resources and environmental protection, thus bringing tremendous pressures on socio-economic development and overall improvement in living standards.

Peng Yu, vice-minister of the State Family Planning Commission, 1994: in Wu Naitao (1994a)

This chapter is primarily about the People's Republic of China, but for convenience four neighbouring or nearby countries are included: Hong Kong, to become part of China in 1997; Taiwan, in the view of its citizens an independent country, but regarded by China as its 31st Province; the Democratic People's Republic of (North) Korea, still solidly communist in 1994; and Russia's former protégé, the Mongolian People's Republic. These four smaller countries are discussed in the last section of the chapter. China's position in the world is shown in Figure 16.1.

16.1 A brief account of the recent history of China (see Table 16.1)

Unlike most of the Americas, India and Africa, China was never occupied by European colonial powers, even though its shores were reached by Portuguese navigators soon after 1500. In 1557 the small territory of Macao was ceded to the Portuguese as a trading post. In 1689 Russian expansion into northern Asia was checked in the direction of China in the Far East by the Treaty of Nerchinsk. Serious attempts to open up China for trade by West European powers, Russia and the USA only began in the nineteenth century. In 1842, by treaty, five ports were opened to trade, and Hong Kong was ceded to Britain. Many other ports were opened later. Various unfortunate encounters with the outsiders, whose overtures were resented by the traditionally minded leaders of China, were followed by the Chinese Revolution in 1911–12, in which the empire came to an end and a republic was established under Sun Yat Sen. Plans for modernisation, including the construction of a very extensive rail network, were put into practice slowly, and when Japan attacked Manchuria in 1931, China was not equipped to repel its old enemy. Between 1937 and 1945 Japan occupied much of eastern China.

When the communists under Mao Zedong gained

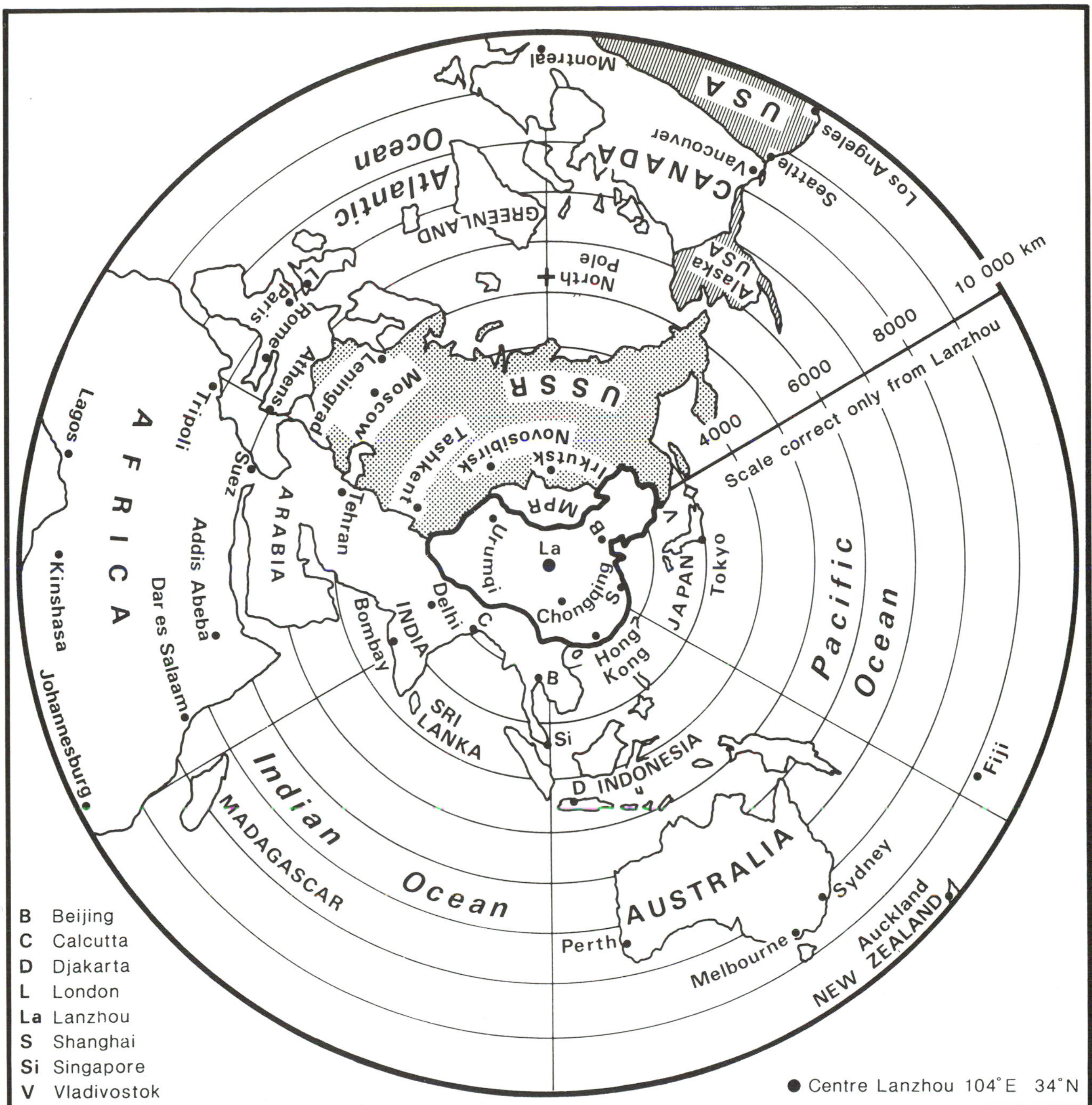

FIGURE 16.1 The position of China in the hemisphere centred on Lanzhou. Note that distances are correct outwards from Lanzhou but not between other pairs of places. Oblique zenithal equidistant projection compiled by the author

control of mainland China in 1949, economic reconstruction and growth began seriously in the new People's Republic. For a decade, Chinese planners were influenced by the Soviet model, and China received Soviet financial and technical assistance. The state took control of virtually all the land, its natural resources, and the means of production. Through national Five-Year Plans, investment was allocated to the various sectors of the economy and to regions. Central planning has continued into the 1990s, although Soviet influence virtually ceased after 1958. Since the death of Mao in 1976, some relaxation of central control has occurred, with greater freedom being given to farms and factories to decide what to produce and how to market the produce. Since the early 1980s, China's leaders have spoken about 'market socialism', a socialist market economy with Chinese characteristics. The term implies that in principle they have not abandoned the communist goal of equality, or at least the removal of gross disparities in income between individuals and

TABLE 16.1 China calendar since 1400

1405–33	Admiral Cheng-ho makes conquests on the shores of the Indian Ocean
1514	The Portuguese reach Macao and in 1557 are given a trading concession
1550	China's population reaches about 130 million
1644	End of Ming Dynasty, start of Ching Dynasty
1689	Treaty of Nerchinsk with Russia
1800	China's population reaches about 300 million
1839–42	Opium War and cession of Hong Kong to Britain and opening of first five treaty ports
1858	Treaties of Tientsin. Britain, France, the USA and Russia put pressure on China to open up for trade. Eleven ports 'opened'
1860	17,000 British and French troops occupy Peking
1894–5	Japan occupies Formosa
1911–12	Chinese Revolution. Emperor abdicates. Sun Yat Sen becomes President of Republic
1921	Communist Party of China founded
1931–2	Japanese occupy and 'protect' Manchuria, establishing industries
1937	Japan begins occupation of all China
1941	Japan brings USA into Second World War
1945	Second World War ends, Japan withdraws leaving Nationalists and Communists to dispute control of China
1949	Communists under Mao gain control of 'Mainland' China. Nationalists (under Chang) are confined to Taiwan (Formosa). Mao Zedong chairman
1950–4	Economic reconstruction
1955–8	Collectivisation, industrial growth
1958	Great Leap Forward. Communes established
Early 1960s	Serious failures in agriculture and in industrial growth, large numbers of deaths
1960	End of close association with USSR
1964	Nuclear fission device exploded ('atom bomb')
Autumn 1966	Cultural Revolution, Red Guards, followed by instability and unrest
1967	Nuclear fusion device exploded ('hydrogen bomb')
1969	Return of 10–15 million people from cities to countryside
1970	Fifth country in world to launch satellite of its own
1971	China admitted to United Nations, replacing China Taiwan
1973	China's Communist Party has 28 million members
1975	Under a new constitution China becomes a 'socialist state of the dictatorship of the proletariat'
1976	Death of Zhou Enlai, prime minister. 'Gang of Four' identified. Death of Mao Zedong, chairman of Communist Party
1978	Reshaping of Chinese economic system with shift from heavy to light industry. Four modernisations proposed. Goal: to increase the production of goods and services about fourfold between 1980 and 2000
1981	China's population passes 1,000 million
1989	Massacre of dissident students in Beijing (Tiananmen Square)
1992	Deng Shao Ping advocates shift from planned to market economy
1997	UK cedes Hong Kong colony to communist China

between regions, but in practice the gap between rich and poor is widening. In 1992 in a speech in southern China, Deng explicitly accepted the market economy, although politically the Communist Party is in control and key aspects of the economy are still managed by it.

16.2 China's natural resources and population

In mid-1994 China had 1,192 million inhabitants, 21.4 per cent of the total population of the world, compared with India's 912 million. Its land area of 9,597,000 sq km (3,705,000 sq miles) makes it similar in extent to Canada, the USA and Europe (including the European part of the former USSR). It occupies just over 7 per cent of the world's land surface. For administrative purposes China is divided into thirty units (see Figure 16.2), of which three are Municipalities, five are Autonomous Regions, and twenty-two are Provinces. In this chapter each of these major administrative divisions starts with a capital letter. In places, for simplicity Province refers to all three types.

Reference is often made in a general way to the vast natural resources of China but, to be meaningful, they have to be considered in relation to the very large population. With 21 per cent of the population of the world, very roughly China has 8 per cent of its bioclimatic resources, 10 per cent of the freshwater resources, 5 per cent of the fossil fuels and less than 4 per cent of the non-fuel minerals. Since the 1960s it has been a net importer of food in some years, and through its rapid industrial programme is increasing its imports of some raw materials. It has a very small surplus of fuel to export.

The area under arable and permanent crops in China has hardly changed since 1949. It occupies just over 10 per cent of the total land area, but about half is double-cropped, making the area sown and harvested much larger. About half is irrigated. Some 13 per cent of China is forested, and over 50 per cent is classed as natural pasture. The remaining 25 per cent is mainly desert and high mountain waste.

China is estimated to have about 114 billion tonnes of coal reserves, over half of them of high quality, giving it

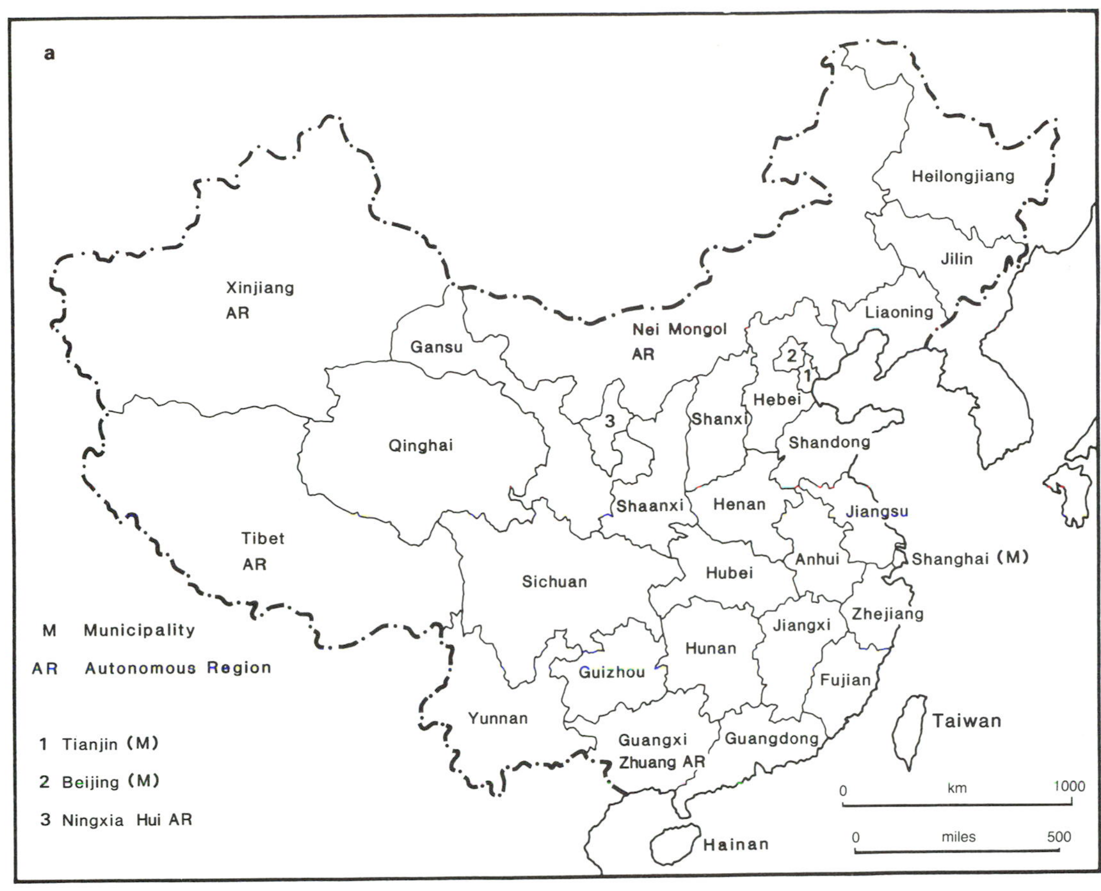

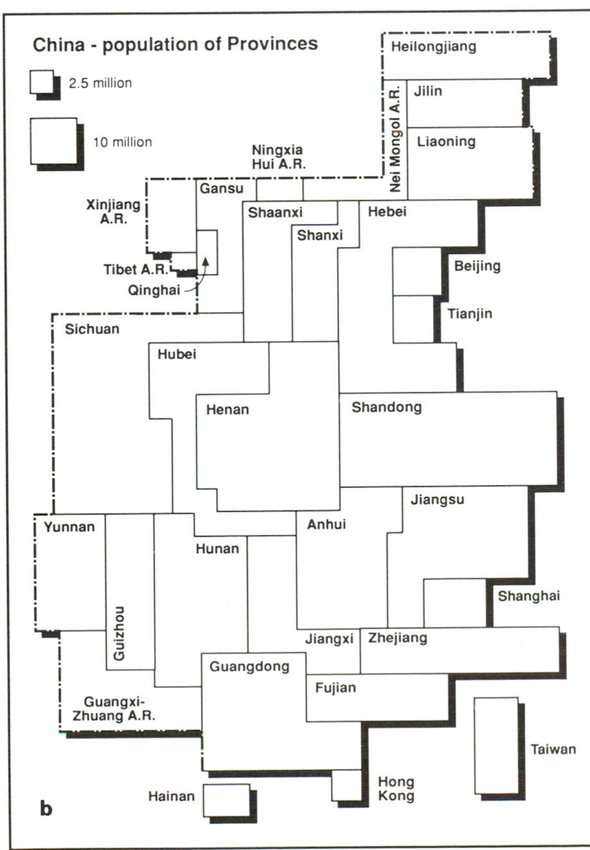

FIGURE 16.2 The Provinces of China represented (*a*) conventionally, (*b*) in the form of a cartogram with Provinces drawn according to the size of population. Note that the thirty major administrative divisions of China, all referred to for convenience as Provinces, consist of twenty-two Provinces, three Municipalities and five Autonomous Regions

11 per cent of the world's total. In 1992 its reserves of oil (3.2 billion tonnes) and natural gas (1.4 trillion cubic metres) are 2.4 and 1 per cent respectively of the world totals. At present rates of production, its coal reserves would last for over a century, but if oil and gas consumption rises sharply, as expected, and in the absence of further major discoveries, these reserves will have a much shorter life than the coal reserves. China also has a large hydroelectric potential, but many sites for dams and power stations are in the remote south-west, and environmental problems could arise from the large construction works needed. As yet, China has only a negligible civil nuclear power programme. Again, although most non-fuel minerals are found in commercial quantities in China, reserves are mostly small, iron ore and tungsten being the only two available in abundance. More thorough exploration, especially in the interior, should reveal further mineral deposits of both fossil fuels and non-fuel minerals.

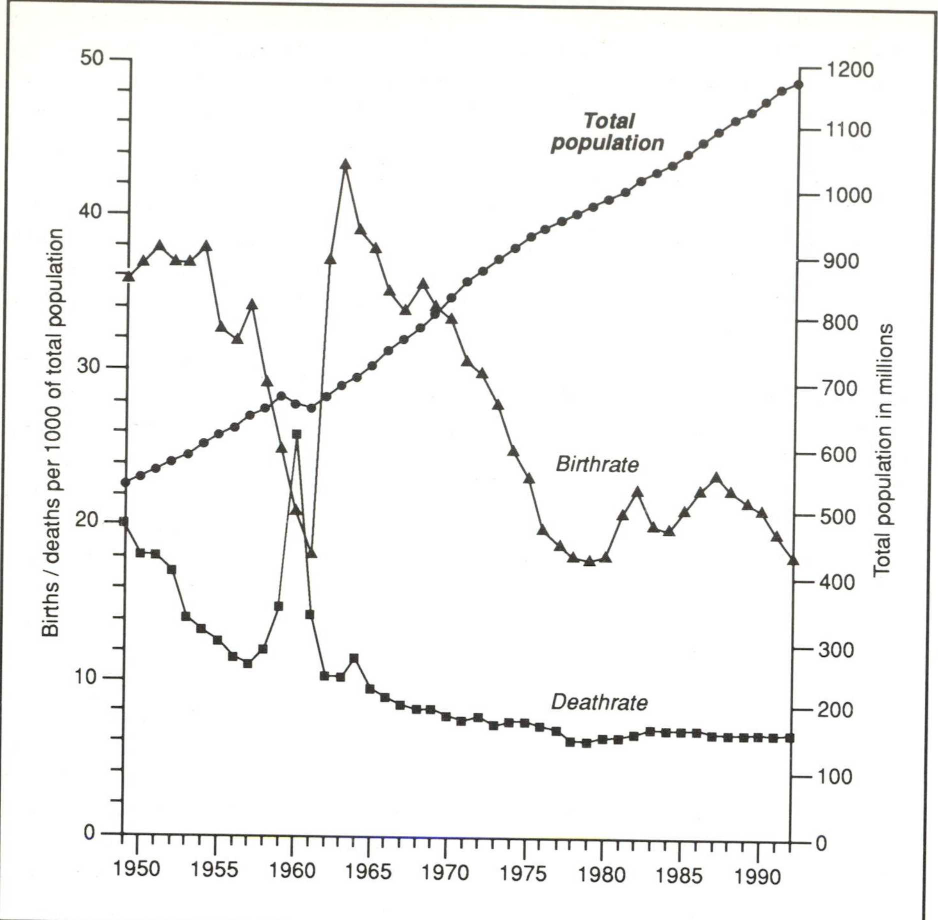

FIGURE 16.3 Population change in China, 1949–92, showing total population, birthrate and deathrate. *Source data: China Statistical Yearbook*, 1992

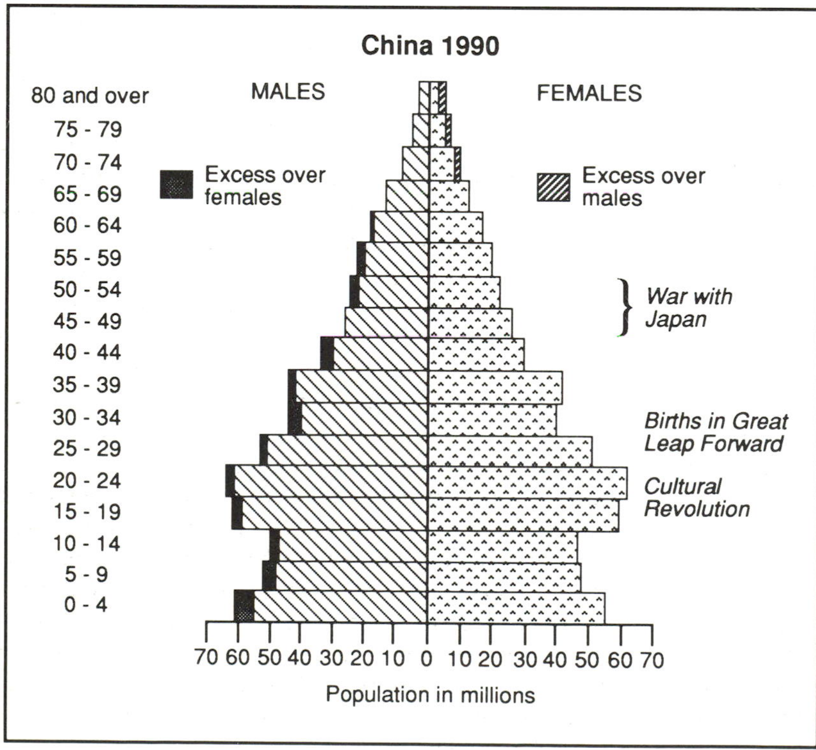

FIGURE 16.4 The population structure of China, 1990

The generally poor natural resource base of China is one reason why there has been concern for two decades now about continuing population growth. During the 1950s, the total fertility rate was high, and large families were encouraged in the belief or assumption that overpopulation was solely a 'capitalist' problem. In military and strategic terms, also, the attitude prevailed that the more people there were, the more able China would be to survive a war, even a nuclear war. However, it was through a combination of unfavourable weather, and the reorganisation and neglect of the agricultural sector, not a war, that in 1959–61 there were about 20 million deaths in excess of those normally experienced. China's pro-natalist policy was revived during the first part of the Cultural Revolution, which started in 1966, and a baby boom ensued. Only in the 1970s did a reversal of policy lead to measures to encourage and in places actively force the one-child family on the generally reluctant population. Figure 16.3 shows the changing gap between birthrate and deathrate in China and the steady and relentless growth of the absolute number of people from the start of the communist period.

The population structure of China is shown in Figure 16.4. The baby booms of the 1950s and the late 1960s show in the differences in the size of 5-year cohorts; the year-by-year structure (see Figure 3.3, p. 57) shows finer features. In 1990, large numbers of females were about to move into the preferred child-bearing age groups. Figure 16.3 shows how the gap between birthrate and deathrate has gradually narrowed, but a decline from an annual natural increase rate of 2 per cent to little over 1 per cent must be interpreted with care because 1 per cent still means an annual gain of 12 million a year from a larger population, whereas an annual increase of 2 per cent in the 1950s also produced a gain of 12 million. In the nineteen years between 1974 and 1993 the population of China increased by more than the *whole* population of the USA (260 million) and in the five years between 1988 and 1993 by more than the whole population of the UK. These massive increases took place in spite of the drastic family-planning programme, with its swingeing sanctions and penalties applicable in theory to any violations except among the non-Han minority populations of China. Most other parts of the developing world have considerably or very much faster rates of population growth than China, but the sheer size of China's population and the publicity given to family planning have given it prominence.

Even when the total population of China passes 1.5 billion around the year 2020, it will still be growing. This is far above a target for the year 2080 of about 700 million, proposed in the early 1980s by Liu Zheng *et al.* (1981) as an optimum population for China, given the natural resource limitations of the country, and in particular the size of its freshwater resources. Without unthinkably harsh and unrealistic demographic engineering, China cannot avoid having a population of 1.5 or 1.6 billion well into the twenty-first century. Box 16.1 shows three possible futures (see Figure 16.5). It will have the added inconvenience of a very uneven age profile, requiring for example far more school places for some periods than for others. Like the developed countries of the world, thanks to the comparatively high life expectancy in China, there will also be a far larger proportion of elderly in a few decades' time than now. A secondary but by no means insignificant demographic problem for China is the continuing excess of males over females, especially among young adults. This is a result partly of the long-standing practice of disposing of female infants at birth or, with modern technology, aborting female foetuses detected by ultrasound examinations, when a son is hoped for in the context of a one- or two-child family.

While the growth and age–sex structure of the population of China causes concern, the uneven distribution of people over the national area also gives rise to problems. About 90 per cent of the population is concentrated in the plains, valleys and low plateaux of the southeastern two-fifths of the country. A very uneven spatial distribution of population is, however, a feature of all the larger countries of the world except India. In the case of China, with a rural population of about 80 per cent of total population, it reflects the very uneven distribution of cultivable land. The cartogram in Figure 16.2b makes the interior Provinces and Autonomous Regions shrink to their true population size. In spite of the resettlement of millions of Han Chinese from the east into the interior, the broad imbalance remains little changed.

Unlike most developing countries, China has had a policy of discouraging or preventing migration from rural to urban areas. As various restrictions on personal freedom are relaxed, it is becoming easier for people from the generally poorer interior regions to move to the cities of eastern China. The proportion of urban to total population is still low in China, even by comparison with many parts of the developing world. The proportion of urban population rose from about 10 per cent in 1949 to almost 20 per cent in 1960, an absolute increase from 58 to 131 million. Thereafter it remained below 20 per cent until 1980, after which an apparent sharp increase has occurred, which was, however, mainly the result of a change of definition, whereby urban status was given to many former rural counties. The greatest control on family size can be exerted in the larger cities, so women in most now have a fertility rate below replacement rate. Figure 3.3e (p. 57) shows the bizarre structure of the population of central Shanghai.

16.3 Agriculture

Even if only the lowest estimate of the economically active population in agriculture in China in the early 1990s is accepted, 350 million workers is a staggering number. It means that for each 100 hectares (247 acres, or 1 sq km) of cultivated land there are on average about 350 workers, whereas in Australia (with 49 million hectares – 12 million acres – and 400,000 workers) each 100 hectares of cultivated land has an average of less than one worker. What is more, between 1978 and 1991, the labour force in farming and associated activities in China rose from 284 million to 350 million. Even when the very intensive nature of cultivation and the lack of mechanisation in China are taken into account, the sector is greatly overmanned, as is shown by a comparison with the Netherlands, where 220,000 farm workers cultivate 930,000 hectares (2,300,000 acres), about 25 workers per 100 hectares. It is not surprising, therefore, that the

BOX 16.1

Demographic engineering in China

In a centrally planned economy such as that of China since 1949 it is in theory possible for the national government to control the size and geographical distribution of population. Although some concern was expressed about the rapid growth of population in China in the 1950s and 1960s, it was not until the 1970s that increasing emphasis was given to reducing and eventually stopping the rate of population growth. In 1982 the demographer Liu Zheng (1981, pp. 26–9) referred to a letter by Frederick Engels, written in 1881, in which Engels stated:

> Of course, there exists the abstract possibility that the number of human beings will increase to a level where it becomes necessary to stipulate a limit to its growth. But if communist society should one day be compelled to regulate the production of human beings, as it regulates the production of goods, then it and it alone will do this without any difficulty.

Liu Zheng and his associates worked out a desirable 'final' population size for China 'based on quantitative studies of economic development, population food requirements, ecological balance and fresh-water resources'. They concluded that such a population in 2080 should number between 650 and 700 million. It will be argued below that to attain such a precise goal is impossible without drastic, very precise, and very strictly controlled demographic engineering. The problem is analogous to that facing Shylock in Shakespeare's *Merchant of Venice*, when he demanded precisely a pound of flesh from the merchant who had borrowed money from him. In the event, family planning is now widely practised in China, with varying degrees of success in the different Provinces. But how can a target of 700 million by reached by the year 2080? Chinese planners may assume, as Engels stated, that they can control population. The sooner population growth is controlled to meet the target, the better.

Three projections based on clear assumptions of future fertility and mortality rates have been made to illustrate how population *could* change in the future. The starting year is 1980 and the population of China is divided into males and females in each of nine age cohorts. The mortality rates are the same for all three projections. For simplicity it is assumed that all births take place to women between the ages of 20 and 29. Only the fertility rate can change. Three alternative projections are shown in Figure 16.5

Projection A shows that, with an average throughout the period 1980–2080 of exactly two children per female, the irregularities in size of age groups of the 1980 population stay throughout the period 1980–2080, while total population rises from about 1 billion in 1980 to reach 1.5 billion in the 2020s, thereafter declining to 1.4 billion by 2080, twice the target population.

Projection B shows that, with an average throughout the period 1980–2080 of exactly 1.5 children per female, the large cohorts of 0–29-year-olds in 1980 will age and will be underpinned by smaller sized cohorts, giving in the year 2020 a very top-heavy pyramid, with 20 per cent over 59 years of age. The total population peaks around 2010 at 1.24 billion and then declines to almost exactly the target of 0.7 billion in 2080, continuing to decline after that unless the fertility rate increases. From 2030 on, over-59-year-olds form about 30 per cent of the population. For the decline in population of some 0.5 billion between the early 2010s and 2080 there has to be an excess of deaths over births each decade of about 70 million! Throughout the 100 years, the shape of the pyramid is inconveniently irregular.

Projection C shows what happens if what might be called demographic surgery rather than demographic engineering is applied. As each ten-year cohort of women enters the age bracket 20–29, births are rationed in such a way that the larger the number of women in the cohort, the smaller the average number of births permitted per female. Thus in the 1980s, an average of 1.5 is allowed, but in the 1990s an average of only one, to allow for the extra large number of births in the 1960s, the post-Great Leap Forward and average value of production of each worker in industry in China is about 12 times that of an agricultural worker, a point elaborated below.

Unlike Japan, which imports much of its food and most of its agricultural raw materials, paying for them mainly by exporting manufactures, China has to be largely self-sufficient in agricultural products for at least two reasons. First, it has little to export except manufactured products, but the world market is becoming saturated with these. Second, there is not a surplus of food world-wide to feed hundreds of millions of Chinese. While the main thrust of economic development in China since 1949 has been to industrialise, great importance has also been given to agricultural production, particularly of grain, cotton and oilseeds. Various forms of land tenure were tried, including collectivisation on Soviet lines, state farms, enormous communes and, most recently, family-run holdings. All developments must take place within the constraints of some 10 per cent of the land and, since

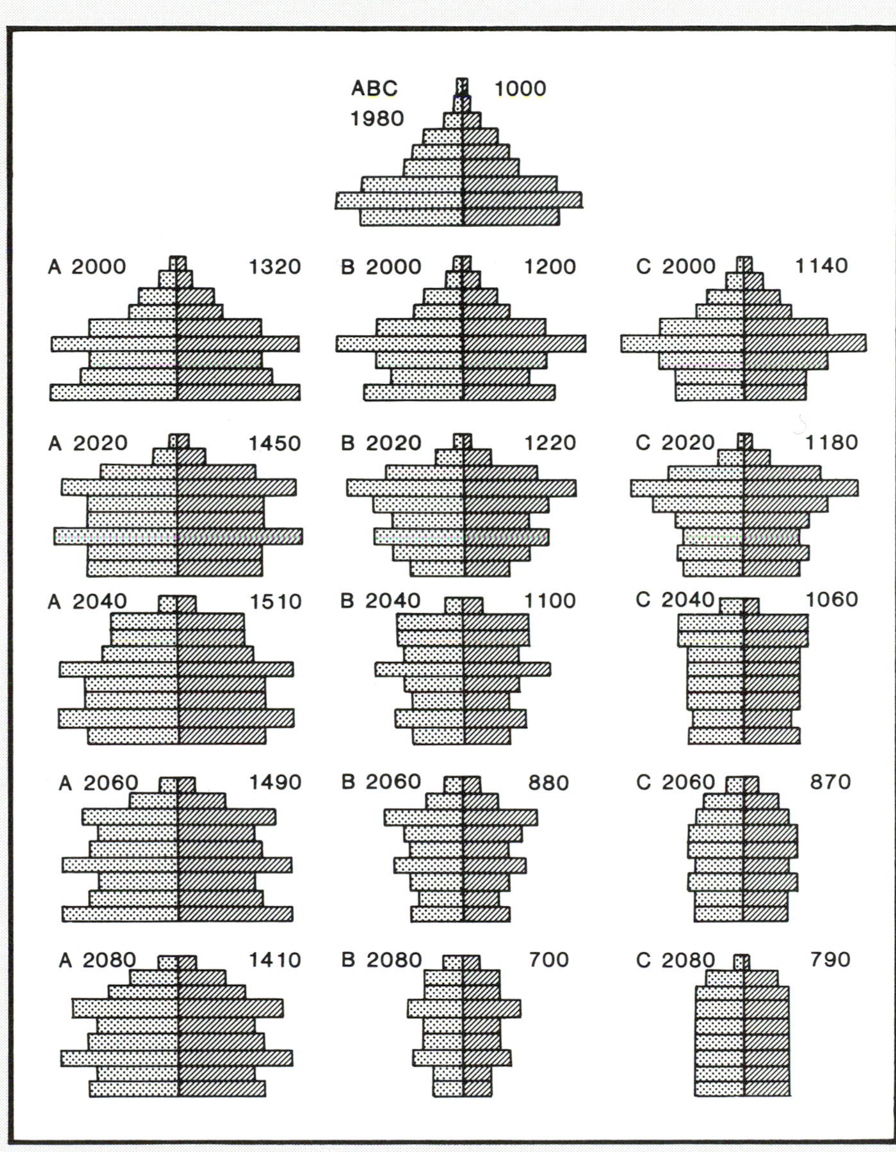

FIGURE 16.5 Three alternative futures for the population of China according to three different sets of assumptions (see text)

the Cultural Revolution decade. After that the level is 1.5 in the 2000s and 1.8 thereafter to 2080. The total population of China peaks around 1.2 billion in 2015 then falls to 0.8 billion in 2080.

The following are some of the problems and issues related to the demographic plans and hopes of China in the next eight to nine decades.

1949, of the state planning guidelines, targets and quotas, of varying influence on production and its disposal.

Most of the land in China is too rugged to cultivate, too dry and/or has very poor soils (see Figure 16.6). Since 1949, a small area has been painstakingly added to the cultivated area through terracing, the levelling of difficult terrain, draining and reclaiming low-lying areas along the coast and in river valleys, and ploughing semi-arid areas. In recent years, however, there has actually been a net loss of cultivated land of about 0.5 per cent a year to construction, forests and pastures. Li Bin (1993) warns that 'China's land resources cannot be misconstrued as boundless. Only 0.087 of a hectare per person is under cultivation.' Much has been done to control and increase water supply to farms and the high summer temperatures allow about half of the cultivated land of China to be double-cropped. In eastern China, the more rugged areas have been extensively planted (or replanted) with trees but there is

BOX 16.1 *continued*

1. Population can most effectively be reduced or limited by influencing fertility rates. To raise mortality levels deliberately would be an unacceptable policy on both humane and political grounds. Moreover, females of childbearing age or under would have to be specifically targeted for extermination for such an unthinkable policy to be effective. Migration into and out of China has a minute effect on population change, a situation unlikely to change.

2. Chinese experience in the last two decades has been that family size can be controlled more easily in and around the larger cities than elsewhere, with many one-child families now to be found especially in the coastal Provinces. In contrast, among the rural Han population and the non-Han population in general, two or more children are common.

Data for the early 1980s show the great differences in family size for different employment categories and levels of education of mothers.

Employment	%	*Number of children*
Peasants (agriculture)	79.5	5.95
Workers (industry)	12.5	4.27
Office workers	2.7	3.10

Educational levels correlate negatively with number of children

Education	%	*Number of children*
Illiterate and semi-literate	37.2	5.86
Primary school	30.4	4.80
Junior middle school	22.3	3.74
Senior middle school	9.6	2.85
University	0.5	2.05

Unfortunately for Chinese planners the citizens with a tendency towards low fertility are only a very small proportion of the total population.

3. If births are controlled and rationed, in a few decades' time China will have a large proportion of elderly, and among the young adults, a large proportion of only children.

4. When the one-child policy is applied, it is not uncommon for female foetuses to be aborted and even for infanticide to be practised on girl babies. Under these circumstances it is inevitable that there will be a marked excess of males over females, a situation fraught with problems.

5. The move towards a market economy in recent years and the resulting greatly increased mobility, especially of the rural poor, complicate the demographic picture. Family-planning measures could be more difficult to apply, but increasing urbanisation could dampen fertility rates.

In practice, if the Population Reference Bureau estimate for the population of China in 1994 and 2025 proves to be correct, there will be an increase from 1.2 to 1.5 billion, at the rate of 10 million a year. Zero growth, if it comes at all, would not therefore be until 2030–40, after which it would be demographically impossible to have an excess of 20 million deaths over births per year between 2040 and 2080 to reach 700 million. Without deaths from famine that would make the 20 million net loss of population around 1960 like a children's party. The population of China could be around 1.3 to 1.4 billion in 2080, twice Liu Zheng's target.

6. The assumption that China could support even 0.7 billion people, let alone 1.4 billion, with material standards comparable to those in Europe, is highly dubious. With about 0.73 billion people during the period 1995–2025, Europe maintains high material standards because it is able to import part of its food needs, about half its energy needs and most of its non-fuel mineral needs. With perhaps half the absolute quantity of natural resources that China has, Australia comfortably supports a mere 20 million people. What, indeed, is the optimum population size for China, given particular criteria and requirements? The 'optimum' 700 million seems excessive.

PLATE 16.1
China: contrasts in the quality of accommodation

a Blocks of apartments in the northwest of Beijing. In the capital and in other cities in coastal areas, lavish apartments have been built to house foreigners temporarily employed in China. In the 1950s, Russian experts working in China stayed in rather austere Friendship Hotels

b A dwelling in Nanjing, typical of the conditions in which much of the urban population of China lives. People accumulate piles of bricks and stones and pieces of timber, gradually improving their homes

evidence, discussed in Box 16.2, that in places the remaining hitherto untouched forests are being cleared, notably on Hainan Island. In the interior of China, the official area of permanent pasture is 200 million hectares (494 million acres), but in general this land has a very low productivity.

When all the constraints and strategies for agriculture are considered, it is evident that since 1949 almost all the increase in agricultural production has come from obtaining higher yields from each unit area of cultivated land, thanks particularly to improved water supply and the use of chemical fertilisers. One even sees vegetables and cereals growing along the side of the road, on railway embankments, in small patches in the heart of cities, and on minute terraces constructed on mountain sides, using soil carried there by hand. In the thirteenth century Marco Polo referred to vast areas of wilderness in eastern China, replete with an abundance of wild animals for hunting. Now every possible patch produces something.

For the last forty years, then, neither the arable area of China, nor its sown area, have changed substantially, even allowing for an increase in double- and triple-cropping. In 1991 the total sown area of China was about 150 million hectares (370 million acres) (50 per cent more than the cultivated area). In 1991, 75 per cent of the total sown area was under grain crops (compared with 85 per cent in 1952), 16 per cent under

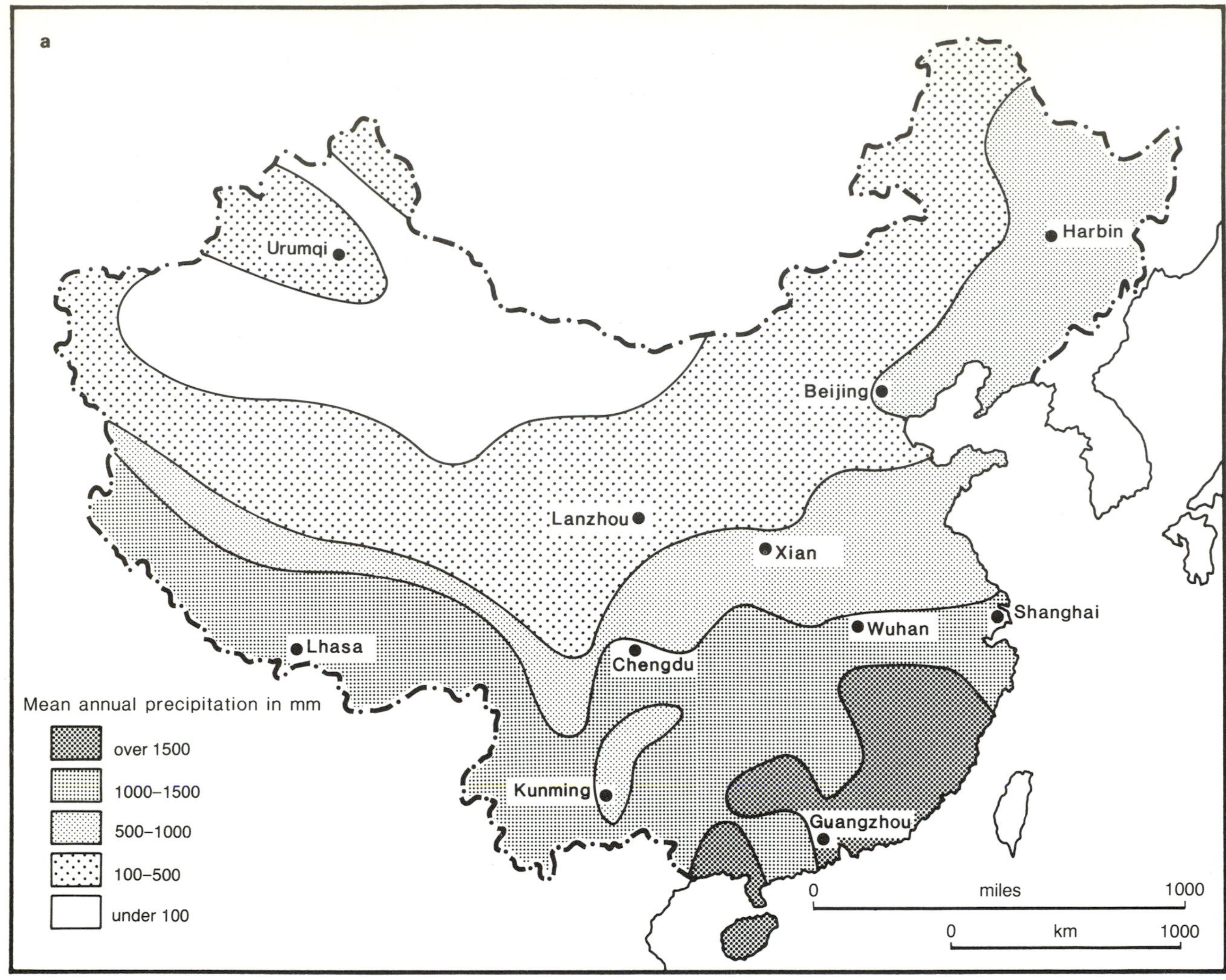

FIGURE 16.6 Aspects of the physical geography of China: (*a*) annual precipitation, (*b*) January temperature and relief features

industrial crops such as cotton and groundnuts (compared with 9 per cent in 1952), while the rest was mainly vegetables. Virtually no forage crops are grown as feed for livestock.

Allowing three years to recover from internal conflicts, 1952 is a reasonable base year for considering the start of China's post-war agricultural progress. Between then and 1991, the area defined as under 'grain', which includes rice, wheat, maize, soy beans and tubers, as well as millet and sorghum, now less widely grown, dropped from 124 million hectares (306 million acres) sown in 1952 to 112 million (277 million) in 1991. The area under rice changed little after the early 1960s, the wheat area increased slightly, but the maize area almost doubled. Between 1952 and 1991, grain output increased from 164 million tonnes to 435 million (an increase of 165 per cent, or 100 to 265). During that period, population rose from 575 million to 1,158 million (an increase of 101 per cent, or 100 to 201). Grain output per inhabitant therefore rose from 285 kg to 375 kg, not a remarkable achievement compared with that in some regions of the world during the same period, but impressive in view of the tight environmental constraints and the lack of incentives to individuals to produce more until the 1980s.

Although industrial crops and vegetables occupy much less sown area in China than grain crops, in value per unit of area they tend to be considerably higher, and in their various ways equally indispensable. In contrast to grain crops, the area under other crops has roughly doubled in the last four decades, while, as with grain, impressive increases in yield are claimed, for cotton (fourfold) and for groundnuts, rapeseed, sugarcane and tobacco (double to treble).

The cultivated land in China is distributed over a wide range of natural conditions. Almost all of it is in the eastern half of the country. Here, precipitation increases from north to south (see Figure 16.6a). In the

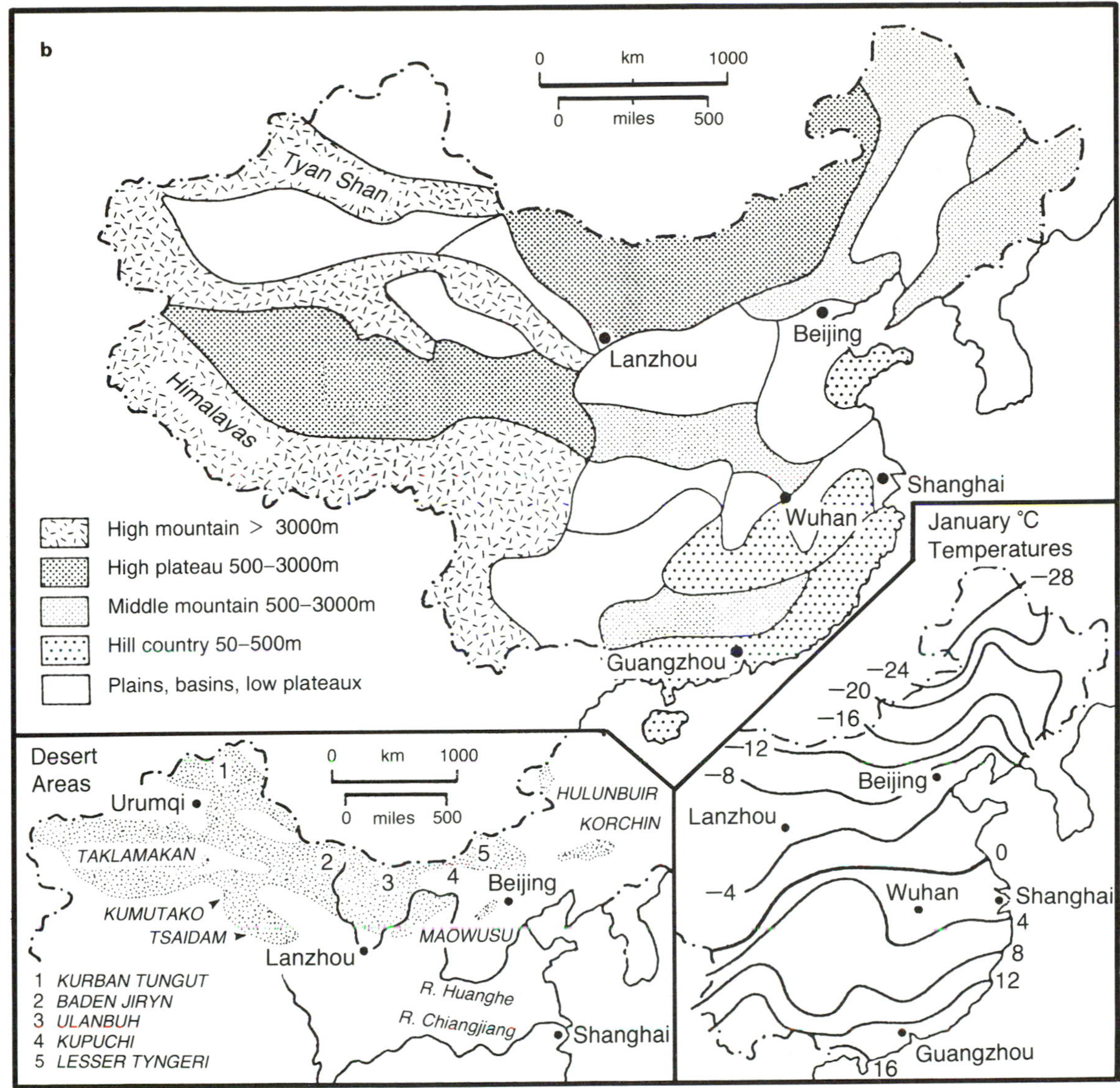

drier north, wheat is the most suitable and widely grown of the cereals, but maize is also grown here. Where water and thermal resources allow, both give way southwards to rice because yields are appreciably higher. Cotton is grown most widely in the north-centre. Sugar-cane and some other subtropical crops are confined to the extreme south.

Central planners have stressed the battle for grain until the 1980s, but with decisions about what is to be grown increasingly left to people on the farms, changes are taking place. As ever, the impression is still one of great effort on the part of an enormous labour force, using limited supplies of electricity and levels of mechanisation. With the help of double-cropping the prodigious yield of 15,000 kg (15 tonnes) of grain per hectare was claimed on one farm southwest of Beijing that I visited in 1992. In the era of the communes, exemplified by the famous model Dachai commune, yields were exaggerated and production inflated to signal the success of farm managers. More recently, production tends to be understated by farmers, to reduce possible taxation.

For two thousand years the wetter southern regions of eastern China have been regarded as highly favoured because yields have been higher and more reliable than further north. In spite of improvements in the last four decades, even at Province level, and more so still at County level, yields vary greatly from region to region. China's average grain yield in kg per hectare was 3,870 in 1991. Three factors largely explain differences in yield. The very high yields (5,800 kg) in Beijing and Shanghai Municipalities reflect the presence of very large cities. Yields well above average at around 5,000 kg per hectare are achieved in well-watered Provinces south of Beijing (e.g. Zhejiang, Hunan, Guangdong). Lower than average yields, around 3,000 kg, are common in drier or more mountainous Provinces in the north and in the southwest. In the dry

PLATE 16.2
China: arable farming

a Material being spread on a field near Nanjing after its extraction from the bottom of a dried-out pond. Many hands make light work?

b New terraces being created on hillsides near Qingdao, Shandong Province, evidence of the pressure of population on the land. Even smaller terraces, as small as 10 sq m in area, are found on some mountain-sides in this part of China

north, Shanxi, Shaanxi, Inner Mongolia and Gansu have yields that are usually below 2,500 kg. If China returned to a national policy and plan for agriculture it could be asked where the necessary efforts to continue increasing agricultural production should ideally be made. The choice is between areas that have the highest yields at present and have the more prosperous members of the farming community, and areas that are less productive and generally poorer.

Whatever the yields obtained, China has far too many people on the land for them to produce enough to make more than a modest living. One estimate cited in 1994 was of an excess of some 200 million workers in the agricultural sector, while Mirsky (1994) cites from China's *Farmers' Daily* an estimated 300 million 'surplus' farmers and a floating rural population of 100 million already. Even the removal of these workers from the land would still leave the average Chinese farmer working only one-hundredth as much cultivated land as an Australian farmer. Whether the surplus workers from the land transfer to other types of employment locally or move to cities, especially the new industrial zones on or near the coast, remains to be seen.

TABLE 16.2 Changes in the production of selected industrial sectors in China, 1949–91

	(1)	(2)	(3)	(4)	(5)	(6)	(7)		(8)	(9)
	Million tonnes			*Million tonnes*						
	coal	*oil*	*Electricity bn kWh*	*steel*	*cement*	*fertiliser*[1]	*Cloth in bn metres/(feet)*		*Bicycles (millions)*	*TV sets (millions)*
1949	32	< 1	4	< 1	1	< 1	–	–	< 1	–
1955	93	< 1	12	3	5	< 1	4	(13)	< 1	–
1960	397	5	59	19	16	< 1	5	(16)	2	–
1965	232	11	68	12	16	2	6	(20)	2	–
1970	354	31	116	18	26	2	9	(30)	4	–
1975	482	77	196	24	46	5	9	(30)	6	0.2
1980	620	106	301	37	80	12	13	(43)	13	2.5
1985	872	125	411	47	146	13	15	(49)	32	16.7
1990	1,080	138	621	66	210	19	19	(62)	31	26.8
1991	1,087	141	678	71	253	20	18	(59)	37	26.9
1992	1,116	142	754	81	308	20	19	(62)	41	28.7

Notes:
– negligible
[1] Fertilisers – active ingredients in millions of tonnes
Sources: *CSY 1992*, pp. 400–6; *CSY 1993*, pp. 402–8

16.4 Industry

Starting from a much more limited base than agriculture, industry in China increased its output between 1952 and 1991 almost 60-fold (see Table 16.2), whereas agriculture achieved just over a 3-fold increase. The rate of growth during that period of construction (20-fold), transport (17-fold) and commerce (7-fold) fell between the two main sectors. The 'gross output value' (*CSY 1992*, p. 50) of agriculture and industry in 1991 was 3,641 billion yuan, of which industry contributed 2,825 billion, or 78 per cent, and agriculture 816 billion or 22 per cent. The social labour force (*CSY 1992*, p. 84) totalled 584 million in 1991, of which 350 million were in agriculture, 99 million in industry. The value of production of the average industrial worker was therefore about 12 times that of the average agricultural worker. Yet China employs about five times as many people in its industrial sector as Japan, while producing less in total. Box 16.3 shows some of the difficulties of interpreting Chinese statistical data.

Throughout the period since 1949, industry has received a much greater share of capital construction investment (*CSY 1992*, p. 45) than agriculture. In 1991, for example, from a total of 212 billion yuan, industry was allocated 54 per cent, agriculture a mere 4 per cent (transport and communications got 16 per cent). During the last forty years, heavy industrial capacity and output expanded faster than those of light industry, the 1952–91 increase claimed in the gross value of output being 107 times against 55.

China has a long and distinguished tradition and record of inventions and technological development in mining, processing and manufacturing. However, the modern technological innovations and applications that have proliferated in Western Europe since the sixteenth century hardly affected China until well into the nineteenth century. As the number of treaty ports increased, so modern industrial processes were gradually applied in China, both by outside investors and by the Chinese themselves. By about 1930, factories using modern equipment had been established, mainly in seaports and along the lower Changjiang River. Little further industrial development was possible for the next two decades except in Manchuria. Here, the Japanese, with their own rather than Chinese interests in mind, produced coal, raw materials, and steel. Thus in 1949 the communists inherited a heavy industrial base established by the Japanese mainly in Liaoning Province (South Manchuria) as well as the predominantly light industries in such coastal centres as Tianjin, Qingdao and Shanghai.

As formerly in the USSR, workers in industry in China have generally been better paid that workers in the agricultural and service sectors, in spite of the ultimate communist goal of equal standards for everyone. Industrial capacity and output in China are far from evenly distributed according to population. The main reason for large disparities in average living standards between the Provinces of China is this disparity. One measure of the distribution of industry is shown in columns (4) and (5) of Table 16.4. In 1991 the value of industrial production was about 2,400 billion yuan, giving an average per inhabitant in the whole of China of over 2,000 yuan. The data in column

(5) show a remarkable difference in industrial output per inhabitant, ranging from over 12,000 in Shanghai Municipality to not much more than one-twentieth of that amount in the Provinces of Hainan and Guizhou (both 657). Like Shanghai, Beijing (6,870) and Tianjin (7,470) are also highly urbanised, and have a relatively small agricultural element.

Regional contrasts in industry in China can be summarised as follows.

- The three Municipalities are highly industrialised, and have a balance between light and heavy industry.
- Shanxi (coal), Inner Mongolia (steel), Liaoning (heavy industry), Jilin (engineering) and Heilongjiang (Datong oilfield) are also highly industrialised, but have a preponderance of heavy industry.
- The coastal Provinces of Shandong, Jiangsu (with Nanjing, Suzhou, Wusi), Zhejiang (with Hangzhou), Fujian (facing Taiwan) and Guangdong (with Hong Kong's influence) are highly industrialised and, between them, have most of the Special Enterprise Zones (SEZs). Light industry is more prominent than heavy industry.
- The remaining Provinces and Autonomous Regions are mostly well below the average level of industrialisation for China, although they contain some large concentrations of industry (Wuhan in Hubei, Chongqing in Sichuan, Lanzhou in Gansu).

A high priority for industry during the first thirty years of communist rule in China was to increase industrial capacity rapidly and, for both social and

BOX 16.2

Environmental concern in China

The need for harmony between humans and the natural environment has been a theme widely expressed throughout Chinese history. During the twentieth century, however, political upheavals, internal conflict, Japanese invaders, population growth, and the need to achieve economic growth have all contributed to conditions in which environmental issues have been ignored. Ever since 1949, when an 'enlightened' Communist Party took over, large projects such as dams and mining sites have had negative impacts. Although a big increase in coal production has been achieved in many Provinces since 1949, China uses about 10 per cent of the fuelwood consumed in the world, in spite of having very limited forest reserves. In this box four examples of environmental issues are briefly referred to.

1. Hainan's dwindling forests Apart from tiny islands in the South China Sea claimed by China, Hainan Island is the most southerly part of the country. Of its 34,000 sq km (13,000 sq miles), roughly the area of Belgium, 20,000 (8,000 sq miles) were tropical rainforest, a type of vegetation found only in the extreme south of China. There were some 500 plant and animal species known to be unique to the island. The indigenous population were hunters and cultivators and they coexisted with the forest ecosystem. Since 1949 and especially since the mid-1960s, large numbers of settlers have come from mainland China, mainly to exploit the timber reserves and to plant rubber trees. By 1987 only 3,800 sq km (1,470 sq miles) of forest were still intact.

Hainan was recently given the status of a Province and was made a Special Economic Zone. In 1992 the population was approaching 7 million, compared with a few hundred thousand before 1949. While professing to live in harmony with nature and to protect wildlife, Chinese leaders and planners have put the strategic location of Hainan Island above considerations of its indigenous population and wildlife. According to Catton (1992, p. 15):

> As in the rest of China, there is no shortage of laws to protect wildlife in Hainan. At the local level, there are 'Regulations for Environmental Protection in Hainan Province', 'Regulations for Nature Reserves', 'Regulations Regarding the Implementation of China's Laws on Protection of Wildlife', 'Rules for Areas of Natural Scenery' ... the list goes on and on. But a law is useless unless it is enforced. In 1991, 24 restaurants in the island's capital city, Haikou, were serving what are legally regarded as endangered species.

2. Prospects for the panda One of the best known and most loved species of rare mammal is the giant panda. Its natural habitat is restricted to the forests of a small part of southern China, the Sichuan Mountains. Its diet is almost exclusively the shoots or leaves of bamboo plants. Only about 700 now live in the wild. Catton (1992, pp. 15–16) cites Zhang Hemin, director of the breeding centre in Wolong reserve, who describes the threat to the species:

> Some lawless business people from abroad smuggle panda furs out of China, and some lawless Chinese people hunt and kill the panda because they cannot resist the lure of the money. There have been many examples where such people have been sentenced to

strategic reasons, to locate much of the new capacity in interior regions. Backward agricultural areas could be modernised, and industrial areas, especially those specialising in military equipment, would be more secure from invasion from the coast or from the USSR than those in the existing concentrations. In the 1950s new iron and steel works were built. The largest were at Baotou in Inner Mongolia and at Wuhan in Hubei. Textile manufacturing and other branches of light industry were also established in many centres away from the coast. Industrial development was particularly difficult and costly in southwest China where, even as recently as the late 1940s, there were no railways. Given the size of the country, it was and still is difficult to move goods and passengers on the very modest rail system, which has a route length roughly equal to that of the French rail system (France has one-twentieth the area and population of China). It was therefore reasonable to plan for a high degree of economic self-sufficiency at provincial and even at more local level. This policy was helped by the fact that coal reserves, although mostly quite small and of low grade, are widely distributed in eastern China, ensuring a local energy supply for many places.

Since about 1980 Chinese trade with the rest of the world has increased enormously. To develop and diversify its own economy, China has to import capital equipment and raw materials. It must therefore increase its exports, but it needs virtually all the food, fuel and raw materials it produces, so it must depend on the export of manufactures. Here, potentially, is a new Japan, very much larger in population, but as yet with a smaller and much less sophisticated industrial base. The trend points to the increasing need to have a

> death. But I feel that poaching is not the main threat to pandas. That is the loss of the panda habitat.
>
> In order to look after the habitat of the panda, China will have to do an enormous amount of work. Panda habitat has one characteristic: where there are pandas, there must also be big tall forests and, underneath, a lot of bamboo. If you cut the forests down, then protecting the panda is a nonsense.

As elsewhere in China, the human population is growing in this part of Sichuan Province. More and more forest is cut for timber and fuelwood and cleared for cultivation.

3. **Milu deer** By the beginning of this century a particular species of deer (the Milu), adapted to a wetland environment, had become extinct in China through over-hunting. A vivid illustration of the lack of natural habitat in China is the fact that a herd of Milu deer, raised near Beijing in a special walled compound, from stock flown in the late 1980s from Woburn in England, and breeding successfully, cannot be released into the wild because China has no suitable natural wetlands for it. Apart from that drawback, it would need to be heavily protected from the local inhabitants of any area in which it was placed.

4. **The Three Gorges dam** (main source Wu Naitao (1994b)) The Chang Jiang (Yangtze Kiang) River flows from southwest China to the coast near Shanghai. As it leaves the Sichuan basin it passes through three gorges, which are navigated with difficulty by vessels proceeding upstream towards the interior. The Gezhouba water conservancy and hydroelectric project was completed at Gezhouba, at the mouth of the Kiling Gorge. The Three Gorges dam project will cost about ten times as much. The hydroelectric potential in the area is about 23 million kW (26×700 kW turbo-generator units), a capacity that, if completely harnessed, could generate as much electricity a year as about 50 million tonnes of coal, equivalent to about half the electricity generated in China in the early 1990s. The feasibility, cost and human and environmental aspects of the project have been extensively debated.

Benefits: a major contribution to the electricity supply of much of China. The need for several major regions of China to be seen to benefit from the project means, however, that much of the electricity generated will be transmitted over large distances to serve eastern, central and northern China as well as Sichuan Province, resulting in heavy loss of current. Secondary benefits include improvements to navigation, and flood control. Yichang and other places in western Hubei Province are immediate benefactors.

Drawbacks: about a million people will have to be resettled from areas to be flooded, adding greatly to the cost of the project. Assuming problems of silting are overcome, the impact on the natural environment should be relatively small.

Cost: as with many large (and presumably smaller) engineering projects, the estimated cost of the project has risen – from 36 billion yuan (about 7 billion US dollars) in 1986 to 57 billion in 1990 and 95 billion in 1993, the latter cost including 50 for water control and 30 for population relocation. This compares with 7 billion yuan for the Gezhouba Project. Although foreign technology is to contribute to the construction, China itself is financing the work, which started in 1992. It involves about 100,000 workers, and should last about seventeen years. The first electricity cannot be expected before 2003, an indication of the doubtful value of allocating large amounts of capital for no immediate returns. See also Yang Zheng (1994).

TABLE 16.3 Demographic and employment data for the Provinces of China around 1990

	(1) Popn (millions)	(2) Area in thous. sq km/(miles)		(3) Minorities millions	(4) Minorities %	(5) Annual increase of total popn/thous.
Beijing	10.9	17	(7)	–	–	2
Tianjin	9.1	11	(4)	–	–	6
Hebei	62.2	188	(73)	1.0	2	10
Shanxi	29.4	156	(60)	–	–	15
Inner Mongolia	21.8	1,200	(463)	4.1	19	10
Liaoning	39.9	146	(56)	2.5	6	5
Jilin	25.1	187	(72)	1.6	6	10
Heilongjiang	35.8	453	(175)	–	–	10
Shanghai	13.4	6	(2)	–	–	1
Jiangsu	68.4	100	(39)	–	–	11
Zhejiang	42.0	100	(39)	–	–	8
Anhui	57.6	140	(54)	–	–	15
Fujian	30.8	120	(46)	–	–	14
Jiangxi	38.7	167	(64)	–	–	14
Shandong	85.7	153	(59)	–	–	9
Henan	87.6	167	(64)	–	–	13
Hubei	55.1	186	(72)	2.0	4	13
Hunan	62.1	210	(81)	2.8	5	13
Guangdong	64.4	178	(69)	0.1	–	15
Guangxi	43.2	230	(89)	16.6	38	15
Hainan	6.7	34	(13)	1.0	15	17
Sichuan	109.0	570	(220)	4.4	4	9
Guizhou	33.2	176	(68)	7.5	23	14
Yunnan	37.8	394	(152)	10.3	27	14
Tibet	2.3	1,200	(463)	2.1	91	16
Shaanxi	33.6	206	(80)	–	–	13
Gansu	22.9	454	(175)	1.5	7	13
Qinghai	4.5	722	(279)	1.6	36	15
Ningxia	4.8	66	(25)	1.6	33	17
Xinjiang	15.6	1,600	(618)	9.5	61	17
China	1,158.2	9,536	(3,682)	70.4	6	13

Note: – negligible or none
Source: *China Statistical Yearbook 1992*, State Statistical Bureau of the People's Republic of China, Beijing 1992, various tables

BOX 16.3

An example of problems of definition in data sets

The contribution of different sectors to the total Chinese economy is calculated in a number of ways, giving conflicting results. These are worth examination as an illustration of the difficulty of obtaining any definitive description of many situations in the world economy. The following sectors are quoted from *China Statistical Yearbook 1992*. In 1991 the GNP of 1,985 billion yuan was produced as follows: 27 per cent from Primary Industry, 46 per cent from Secondary Industry (including Construction) and 27 per cent from Tertiary Industry. Per capita GNP was 1,725 yuan, converting at 5 yuan per US dollar to a mere 345 dollars. The National Income of 1,612 billion yuan was accounted for as follows: 33 per cent from Agriculture, 54 per cent from Industry and only 13 per cent from Services.

(6)	(7)	(8)	(9)	(10)	(11)
		Numbers employed, in millions			
% of employment not agricultural	*total*	*in agriculture*	*in industry*	*% in agriculture*	*% in industry*
59	6.2	1.1	2.3	18	37
55	5.0	1.4	1.9	28	38
14	34.1	26.3	3.9	77	11
21	15.0	9.6	2.8	64	19
30	11.2	7.1	1.9	63	17
41	22.4	10.8	6.2	48	28
38	12.9	7.4	2.7	57	21
41	17.5	9.0	4.1	51	23
65	8.1	0.9	4.1	11	51
19	41.9	24.4	10.7	58	26
15	24.6	13.0	7.4	53	30
14	33.5	26.6	3.4	79	10
16	15.0	9.8	2.8	65	19
17	20.4	15.1	2.8	74	14
13	50.8	39.1	6.0	77	12
12	50.2	41.3	3.9	78	8
19	31.6	22.4	4.3	71	14
14	34.9	27.4	3.7	79	11
22	33.7	20.3	7.4	60	22
13	22.9	19.0	1.6	83	7
19	3.3	2.4	0.4	73	12
14	68.5	56.2	5.8	82	8
12	18.0	15.4	1.1	86	6
12	21.0	17.6	1.5	84	7
13	1.1	0.9	0.1	82	9
18	18.1	13.7	2.1	76	12
16	13.2	10.6	1.2	80	9
28	2.4	1.6	0.4	67	17
23	2.4	1.7	0.3	71	13
28	7.6	4.9	1.3	64	17
19	647.2	456.8	98.1	71	15

A third version of the relationship between the three main sectors of the economy is given by the Total Output Value of Society, 4,380 billion yuan. Agriculture contributed only 19 per cent, Industry and Construction 73 per cent and Services 8 per cent (see Table 16.4; *CSY 1992*, pp. 44, 46).

Whereas agriculture accounts for only about a quarter of total production in China, it has a much larger share of the economically active population. The highest figure in a Chinese source (*CSY 1992*, p. 75) is 457 million in 1991 in Farming, Forestry, Animal Husbandry and Fisheries, which tallies with 460 million in *FAOPY 1992* (p. 27). China has over 40 per cent of all the people employed in agriculture in the world. Confusingly, 350 million is the total given in the same Chinese source for the 'social labour force in agriculture'. The possibility is that many people working part-time are counted in one estimate but not in another.

TABLE 16.4 Economic data for the Provinces of China around 1990

	(1)	(2)	(3)	(4)	(5)	(6)	(7)	(8)	(9)	(10)	(11)
		Value in billions of yuan						Capital construction		Foreign investment	
	Popn in millions	total	agriculture	industry	Industry yuan/total popn	% of heavy industry	1990 national income	bn yuan	yuan per head	million dollars	dollars per head
Beijing	10.9	105.4	7.0	74.9	6,872	56	3,577	8.7	798	300	28
Tianjin	9.1	9.0	5.5	68.0	7,473	51	2,981	6.4	703	261	29
Hebei	62.2	174.9	35.8	112.3	1,805	53	1,148	7.3	117	96	1
Shanxi	29.4	82.9	12.5	53.8	1,830	75	1,124	7.4	252	31	1
Inner Mongolia	21.8	53.5	15.7	26.3	1,206	60	1,080	5.0	229	2	–
Liaoning	39.9	22.7	23.4	160.7	4,028	69	1,990	12.8	321	645	16
Jilin	25.1	87.5	18.9	55.2	2,199	61	1,383	4.0	159	54	2
Heilongjiang	35.8	133.4	24.5	86.4	2,413	68	1,628	7.6	212	30	1
Shanghai	13.4	204.2	6.8	164.3	12,261	50	4,822	10.9	813	330	25
Jiangsu	68.4	379.8	58.1	276.4	4,041	47	1,689	9.1	133	315	5
Zhejiang	42.0	207.1	33.6	143.4	3,414	35	1,717	5.1	121	144	3
Anhui	57.6	121.9	37.1	67.0	1,163	48	933	4.7	82	25	–
Fujian	30.8	92.0	22.9	53.1	1,724	38	1,313	4.8	156	570	19
Jiangxi	38.7	80.9	25.5	42.6	1,101	56	943	3.2	83	51	1
Shandong	85.7	324.9	64.7	220.1	2,568	49	1,372	11.2	131	373	4
Henan	87.6	185.9	50.2	103.7	1,184	55	880	8.0	91	75	1
Hubei	55.1	164.6	40.2	100.8	1,829	53	1,248	6.4	116	141	3
Hunan	62.1	133.2	39.7	71.3	1,148	56	976	5.4	87	103	2
Guangdong	64.4	309.4	60.1	190.2	2,953	34	1,842	21.3	331	2,584	40
Guangxi	43.2	72.3	25.2	35.3	817	45	798	2.9	67	87	2
Hainan	6.7	15.6	6.9	4.4	657	33	1,193	2.5	373	212	32
Sichuan	109.0	227.3	63.7	122.3	1,122	54	903	12.1	111	154	1
Guizhou	33.2	44.0	14.6	21.8	657	56	645	2.6	78	16	–
Yunnan	37.8	66.4	21.2	34.5	913	49	956	3.7	98	21	1
Tibet	2.3	3.4	1.5	0.4	174	71	865	0.9	391	–	–
Shaanxi	33.6	75.7	17.0	43.3	1,318	57	930	4.7	140	36	1
Gansu	22.9	50.5	10.3	27.9	1,218	74	938	3.4	148	5	–
Qinghai	4.5	11.0	2.5	5.5	1,222	70	1,100	1.5	333	–	–
Ningxia	4.8	11.7	2.5	6.5	1,354	73	1,024	1.4	292	3	1
Xinjiang	15.6	45.9	14.5	22.0	1,410	51	1,374	5.9	378	74	5
China	1,158.2	3,803.8	766.2	2,392.4	2,066	51	1,267	190.9	165	6,739	6

Note: – negligible or none
Description of variables:
(2)–(4) Value of service sector included in total. Total output value of society, 3,804 billion yuan was (paradoxically) more than 2½ times higher than total national income, at 1,438 billion
(5) Output in industry per head of total population in yuan
(7) In yuan per head
Source: as for Table 16.3: *CSY 1992*, various tables

reasonably efficient, well-located industrial base, producing goods that other countries will buy.

The advantages of expanding existing industrial capacity and of locating new capacity on or near to the coast are related both to the location and to the resources of many of the coastal regions. On account of the limited transport network and especially the lack of good links between interior regions, a coastal location, although peripheral within the territory of China, can reach and be reached by shipping services along the coast. It is also much more attractive to foreign investors and to Chinese industries when so many raw materials come from overseas and some of the products are to be exported there. In some extreme instances it has actually been difficult or impossible to transport large and/or heavy pieces of equipment from the coast to places inland. The loading gauge of any railway is strictly limited, especially by tunnels, and China does not have a road network on which wide loads can be easily moved. The presence on or near the coast of many cities with an industrial tradition, a population with skills and experience in manufacturing, and research facilities, also makes the coastal Provinces attractive.

Given the capacity of industry to enhance economic performance and material standards in China, at first sight it would seem that the socialist goal of greater equality could be implemented by concentrating investment in the poorest regions of the country. The evidence shows that the opposite is happening. In 1991, capital construction investment of state-owned units was 212 billion yuan, over half of this going to the industrial sector, a negligible quantity to agriculture. The data in column (9) of Table 16.4 show the large variations in investment per inhabitant among the Provinces. Shanghai, Beijing and Tianjin received several times the national average in spite of already being among the most advanced places industrially and the best-off materially. The very Provinces that were already developed in the 1930s, especially Liaoning and Heilongjiang (oil-fields), together with Shanxi (with China's main coal reserves), received above-average investment (165 yuan per inhabitant for the whole of China), as did Guangdong and also the recently created island Province of Hainan, the former comparatively heavily industrialised and prosperous already. The Provinces and Regions receiving above-average investment were the Autonomous Regions of Tibet, Ningxia and Xinjiang, as well as the Province of Quinghai, all with below-average national incomes, but with a combined population of only about 25 million, a mere 2 per cent of China's population, albeit on more than 40 per cent of the land. Well over half of the population of China lives in Provinces which receive a very limited amount of investment yet have a massive number of rural poor.

Equally significant for the longer term development of China is the preference on the part of foreign investors for a few areas. For example, in 1991, over 11.5 billion US dollars were taken up, over 6.7 billion of which were region-specific. The distribution at Province level is shown in column (10) of Table 16.4. More than seven times as much per inhabitant as the Chinese average went to Guangdong Province, most of it to Shenzhen and Guangzhou. With only 10 per cent of the population of China, Guangdong, Shanghai, Beijing, Tianjin and Hainan received 60 per cent of the foreign capital, while Fujian Province, facing Taiwan, is also favoured. Since virtually all the capital comes from countries that are located overseas rather than from land neighbours (such as Russia) it is not surprising that coastal regions are preferred. In 1991, Japan provided 25 per cent of the capital, Hong Kong and Macao 16 per cent, and the European Union 15 per cent.

If this investment policy continues, one potential prospect for China is a coastal zone like a poor Japan and an interior like India. As an increasing share of investment comes from the private sector, in time there is likely to be an even greater concentration of wealth in the coastal Provinces than now, accompanied by the emergence of a new property-owning class. The rural population in the vicinity of the larger cities often shares in this process, producing food for the urban population and taking on piece-work from the larger factories. One of the preferred uses of such increased wealth is to build solid, if not lavish, new homes, but new factories and schools are also springing up around, for example, Beijing, Shanghai and Guangzhou. In one village on the banks of the Changjiang River close to Wusi, every family will soon own a car, although without good roads and facilities for maintaining the vehicles it is difficult to see how the cars will be of much practical use (see Box 16.4). Many families in China do not even own a bicycle and in some interior villages the terrain and roads are such that bicycles would be of no use. The uneven ownership of bicycles was divisive in socialist China. The car could become so in the future. Where all the cars would fit is difficult to see, since at present the parking of bicycles is a major problem in some Chinese cities.

16.5 Healthcare and education

Agriculture still employs about 60 per cent of the economically active population of China while industry accounts for most of the GDP. It is not surprising, therefore, that since 1949 these sectors have received much of the attention of planners and most of the investment in the economy. In the most developed countries of the world the service sector accounts for

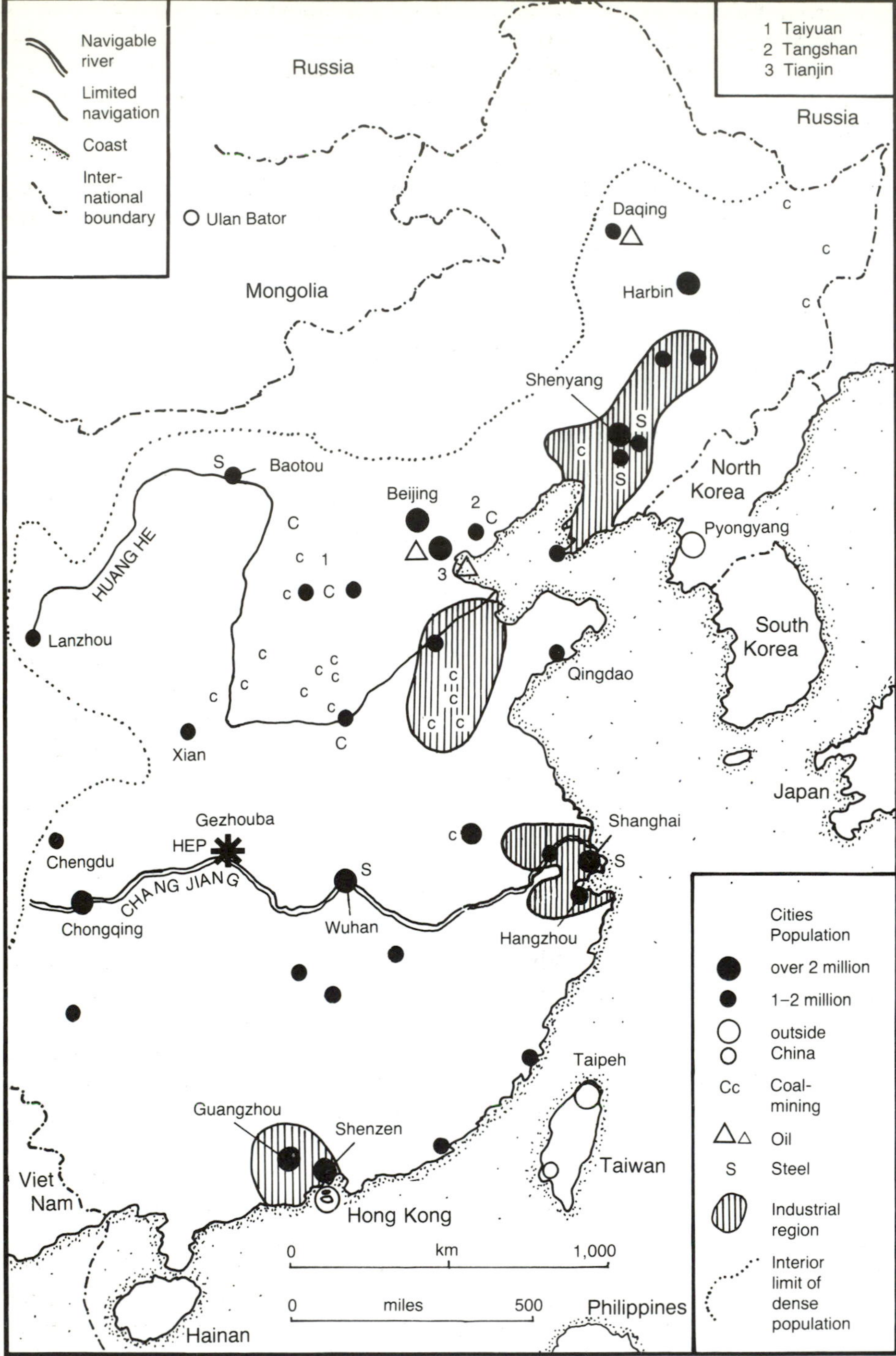

FIGURE 16.7 Main urban centres, industrial concentrations and waterways

around 60–70 per cent of GDP. In China, services have also made key contributions to economic progress, but on limited budgets and in difficult conditions. In this section, healthcare and education will be discussed, but the development of other services has also been crucial, including transport (see Section 16.6).

Healthcare

In 1948, China was extremely poor, services were disrupted, and healthcare facilities were very limited. Evidence of the improvement in standards of health in China includes the increase in life expectancy from around 45 years in 1949 to 70 years in the early 1990s, a decline in the rate of infant mortality from over 150 per 1,000 in 1949 to 35 in the early 1990s, and the increasing proportion of deaths from degenerative causes (e.g. malignant neoplasms, heart diseases) as opposed to infectious and parasitic diseases. According to *CSY 1992* (p. 738), the number of personnel in healthcare institutions has increased from 450,000 in 1949 to over 5 million in 1991. In the 1950s there were more doctors practising traditional Chinese medicine than western medicine, but since the early 1960s the number of traditional practitioners has hardly changed, while the number of practitioners of western medicine has increased more than fourfold. With its average of 650 people per doctor, China is well provided for, compared with almost all other developing countries. The back-up in the form of nursing and other staff and of hospital beds and other facilities is not, however, equally generous.

Medical personnel and hospital facilities are very unevenly distributed according to population in China, as is evident form the availability of hospital beds in different Provinces. The average number of people per hospital bed in 1991 (*CSY 1992*, p. 740) was about 430. In Beijing it was only 185 and in Shanghai 213 whereas in two of the poorest Provinces it was three times as high: Guangxi 608 and Guizhou 615. Everywhere in China healthcare facilities in the cities tend to be superior to those in rural areas in the same Province. This is one reason why the average family size is lower among the urban population than among rural dwellers, a paradox explained by the fact that greater control over family planning can be exerted in urban areas. In the 1980s, the TFR was much higher in predominantly rural Provinces (e.g. 4 in Guizhou) than in cities (e.g. 1.5 in Shanghai, Beijing).

Education

In all post-1945 socialist countries, strong emphasis has been placed on education. With hindsight, however, much of what was taught was of little practical use, while information could be manipulated to favour the state and the Communist Party. In 1949 the level of literacy in China was very low. Even now, about one-third of the adult population cannot read competently. While spoken Chinese is fairly simple grammatically, the written language is difficult to learn because it is not phonetic, making it difficult to relate spoken words to the writing, and making the learning of an adequate number of the written symbols a lengthy process. The written Japanese language borrows heavily from the Chinese symbols, however, and illiteracy there hardly exists.

As with healthcare facilities, educational provision in China is very unevenly distributed in relation to population, while the number completing higher education is still very small. Following the Soviet example, in the 1950s higher education expanded, but in the 1960s the system contracted, as a result of new policy during the Cultural Revolution. Only in the 1980s were levels of the late 1950s passed. For a country aspiring to become an advanced industrial power the need for a sound educational base seems obvious. A less obvious by-product of education in China is the rapid decline in TFR among women with secondary and higher education. Such women usually have only one or two children, while women with only primary education or none at all in the 1970s had four or five children. Unfortunately for Chinese family planning and population goals, the vast majority of women are in the last two categories (see Box 16.1).

16.6 Regional disparities, transport and trade

At present, living standards differ greatly from one part of China to another at regional and even local level, with many consumer goods now widely available in some parts of China but hardly known in others. One reason why such disparities are becoming a problem is that migration from poorer to more prosperous regions is increasing now that restrictions on travel have been reduced. By Chinese standards some interior regions are well endowed with natural resources, as for example the main oil-fields of China in Heilongjiang Province and new ones in Xinjiang, but some of the production goes to coastal Provinces, tending therefore to favour the latter regions. It would have been unthinkable in Mao's time for publicity to be given to regional inequalities. Extracts from the *Beijing Review* and *China Daily* in Box 16.5 show how poor some regions are.

Whether central planning continues to be a major influence in the economic development of China, or the market economy determines the location of economic activity, sound management is necessary since, given its population size, China's capital resources are very limited and its natural resources restricted. Under communism, the economy has grown, although the balance of sectors has not been the most fruitful, much being used to support the military establishment. The location of production has also been unbalanced, and consumption uneven. Under four and a half decades of 'capitalism' instead of communism the ingredients would probably have been rather different, the location of economic activities perhaps more rational, but the shareout of goods and services even less equal. Unless measures are taken to prevent it, much of the expansion of the economy will continue to take place in cities on or near the coast. Transactions between coastal Provinces will increase, while interior Provinces will remain isolated from one another and from the outside world.

Figure 16.8 shows a possible way of managing space that could help to ensure that interior Provinces get some share of development. Indeed *China Daily* (1992) instances the situation in Guangdong where coastal counties are assisting interior ones in the *same* Province. Each of several (seven in this example) coastal regions would be charged with ensuring a good link inland to the next Province or Provinces 'behind' it so that these could have access, initially by rail, later by motorway, to the coast. At the same time, lateral links between interior Provinces should be improved. Already improvements are being made on the important interior rail link Beijing–Wuhan–Guangzhou. Virtually all the foreign trade of China passes through its ports, but some increase in trade may be expected with Russia and Kazakhstan, and the new rail link between Xinjiang and Kazakhstan could carry some container traffic between China and Europe in the future.

Until the twentieth century much of the movement of goods and passengers in China was either on the few stretches of rail or by waterways and tracks. Even in 1949 there were only 80,000 km (50,000 miles) of highway, mostly concentrated in a few of the more modernised parts of the country. By 1992 there were more than 1 million km (620,000 miles), only a small part of which were well surfaced. There is, however, very little long-distance road traffic and until the 1990s there were no private cars.

Li Ningh (1993) refers to a massive plan to improve existing roads and to construct new ones. In 1992 there were only 650 km (404 miles) of expressway and 3,500 km (2,180 miles) of first-class highways. China and the USA occupy roughly similar areas but China has more than four times as many people. The density of the road network is six times as high in the USA and the quality of the roads vastly superior. It could be many decades before even eastern China has a road system comparable to the US interstate system of the 1960s. J. Prynn (*Times*, 3 April 1995, p. 40) reports on the latest Chinese motorway plan, to build a network of toll roads 35,000 km (22,000 miles) in length at a cost of £80 billion (about $130 billion US) by the year 2020. The proposed system would have five north–south and seven east–west dual carriageway arteries and would be roughly as large as the motorway system of the whole of Western Europe. China also still has

BOX 16.4

Material progress in a Chinese village

Traditionally rural Chinese have been regarded among the poorest people in the world. The following extract by Jing Wei (1993) in *Beijing Review*, 10–16 May 1993, describes one village that has achieved material standards far above those associated with rural China under Mao. The experience of Huaxi is common now among villages in certain parts of China, although few can yet have aspired to achieving one car per household. The extract below is followed by a commentary by the author.

> On March 3, the festive mood in Huaxi, a village in the suburb of Wuxi, Jiangsu Province, was heightened by the boom of drums, loud music, and the popping of fireworks. On that day, 50 new Volkswagen Jettas, one-fifth of the suburb's order from the Changchun-based No. 1 Automobile Works in northeast China, drove into the village square. The other 200 cars will be delivered to the village by July.
>
> It is unprecedented for a small Chinese village of only 320 families to purchase so many cars at one time. They have all been ordered by locals at a price of 175,000 yuan each, equivalent to 50 years' salary for a ministry-level official, an astronomical figure for the average workers. 'When all 250 are delivered,' said Zhu Miying, who was in charge of the reception work, 'we will have about one car for every six villagers. This is one of the most exciting events ever in our village.'
>
> Huaxi occupies less than one square km in the middle of the Yangtze River delta in southern Jiangsu Province. The village has been known nationally since the 1970s for its steady high grain yield, the result of the concerted efforts of the residents. In 1989 it was an outstanding agricultural and side-line products producer. At that time it also had three small factories manufacturing steel net, nylon cement bags and pesticide sprayers. Total industrial and agricultural output value for 1982 hit 6 million yuan, and average per capita income was 800 yuan. Per capita living space was 21.7 square metres, all located in two-storey apartment buildings. Children's education, beginning with kindergarten, was provided free. Scholarships were offered to students enrolled in colleges and technical schools. Pensions were also given to retirees.
>
> The decade that has passed has completely changed Huaxi. In the village, a broad square with seating stands has been constructed and an asphalt road leads directly to the village centre, which contains a 500-metre long covered passageway flanked by stores, theatres, restaurants, tea houses and karaoke rooms.
>
> The western part of the village is the industrial district. Among the 38 industrial enterprises in the village, five are Sino-foreign joint ventures. In 1992, total industrial and agricultural output was valued at 516 million yuan, 85 times that of 1982. Industrial output value accounted for 99 per cent of the total.

much to do to improve its rail network.

One final aspect of China's economy will be considered here, its foreign trade. In 1949 China's share of international trade was negligible, and in the 1950s much of its trade was with the USSR and its CMEA partners. In the 1960s and early 1970s, the country was still inward-looking and largely self-sufficient. In the last twenty years China's economy has grown very fast and the ratio of its foreign trade to total GDP has risen. Even so, it is important to keep the scale of its foreign trade in perspective. With about 21 per cent of the total population of the world, in 1981 China accounted for just over 1 per cent of the value of all international trade, in 1990 rather less than 2 per cent. The foreign trade of Hong Kong and Switzerland, with a mere 6 and 7 million inhabitants respectively, each exceeded the Chinese total in both years, although some of Hong Kong's foreign trade is actually carried out on behalf of China. The share of the world's exports originating in the USA was 14 per cent in 1981 and 15 per cent in 1990.

The main features of the composition of China's foreign trade are shown in Table 16.5. Given the size of the country and the great variety of environmental and economic conditions it is not surprising to find that China is both an exporter and an importer of most products. Although in some years it is a net exporter of agricultural products and fuel, only a minute contribution is made to international trade. For most primary products, China is a net importer. Products of the engineering industry make up over two-fifths of all imports, reflecting the limited development of the production of capital goods in China and the current expansion of productive capacity. In contrast, textiles and clothing form the largest group of products exported. The present pattern of trade seems likely to continue well into the next century.

China's main trading partners are shown in Table 16.6. The dominant position of the USSR and CMEA partners no longer characterises Chinese foreign trade. Hong Kong takes almost half of the value of Chinese exports and provides about a quarter of imports. This

The prosperity of Huaxi, where not one person is destitute is widely known. Family incomes vary, but even the poorest is richer than an ordinary urban citizen.

Huang Yonggao is a 43-year-old purchasing agent at the Farmers' Hotel. He moved to Huaxi from a neighbouring village in 1984 and spent 8,000 yuan on a two-storey apartment with four rooms and two-room single-storey house, large enough for the three members of his family. Huang's wife is a factory worker, and his son attends the free kindergarten. They have an annual combined income of over 20,000 yuan, and their apartment includes modern accoutrements like a telephone, air conditioner, television, refrigerator, hot water heater and motorcycles. 'In the last few years I've spent quite a bit on my wedding, the birth of my son, the house and the appliances inside,' said Huang. 'As a result, I now only have about 20,000 yuan in savings.'

Commentary

Although the account of Huaxi sounds improbable, it is unlikely that a situation that goes against the principles of communism would be publicised for the outside world if it were untrue. Even allowing for some licence on the part of the correspondent of *Beijing Review*, how can the situation be explained? The following points may be considered:

- The location of villages is crucial. Those near a large city have been able to concentrate on achieving a surplus from the agricultural land to sell to the urban population. Huaxi's success started with an increase in yields. Its 1 square kilometre of land is of high quality.

- Huaxi is described as a suburb of Wuxi, one of the most highly industrialised cities in China. Villages in the vicinity of such large cities are well placed to establish links with industries in the region and to manufacture whatever parts or whole products are required.

- Initiative is needed on the part of local managers and they need the support of the whole population of the village. In Huaxi, the secretary of the local Party committee, Wu Renbao, was the inspiration. The initiative was presumably encouraged from above or at least permitted.

- Huaxi's progress is unstoppable. At the expense of parts of its minute area of agricultural land, new factories are being built and an enormous increase is expected in the value of industrial output, in billion yuan, from 1 in 1993 to 2 in 1994 and 3.5 in 1995.

- What are the implications for China of the Huaxi syndrome? There are about half a million Huaxi-sized villages in China. Only a very small proportion can follow a similar path. If the whole of China had a level of car ownership equivalent to that in Huaxi, 250 cars for 1,500 people, there would be about 200 million cars in China, more than in the USA.

TABLE 16.5 The composition of China's foreign trade, 1992

	Percentage of value of	
	exports	*imports*
Agricultural products	12.5	5.2
Minerals, chemicals	12.9	15.4
Plastics, hides, wood	7.5	12.8
Textiles	29.0	12.6
Clothing, jewellery	7.2	1.7
Base metals	5.4	9.7
Machines, instruments	18.9	41.6
Other items	6.6	1.0
Total	100.0	100.0

Source: *CSY 1993*, pp. 576–8

trade will be redefined as internal in 1997. Most of China's foreign trade is therefore either with the NICs of East Asia or with the world's leading industrialised regions.

16.7 China's immediate neighbours

For the purposes of this book, four countries adjoining China (or located close to it in the case of Taiwan) – Taiwan, Mongolia, North Korea and Hong Kong – and related culturally or ideologically to it, are included in this section.

1. **Taiwan** is regarded by China as integral to it but it was ceded to Japan in 1895 and remained a Japanese colony until 1945, during which time it was a source of primary products. After the Communist Party under Mao Zedong gained power in China in 1949 and the People's Republic was created, Formosa, reverting later to its name of the Republic of China-Taiwan, has been *de facto* independent of the mainland and strongly supported by the USA. With a population of over 20 million by 1990, a small area, much of it mountainous, and a very high density of population, 560 per sq km compared with China's 120, it has very few natural resources. Only 25 per cent of the total area is cultivated, but half is forested. Taiwan is now highly industrialised, producing both heavy industrial items such as steel (3 million tonnes a year) and cement, and light manufactures such as textiles and clothing, electronic goods and motor vehicles. Like those of Japan and South Korea, the economy of Taiwan depends on the importation of fuel, raw materials, food and some capital goods, and it exports manufactures. About half of Taiwan's foreign trade is with the USA (22 per cent of imports, 36 per cent of exports) and Japan (29 per cent and 13 per cent). Taiwan has maintained a highly favourable balance of trade and has very large reserves, enabling it to invest in projects outside the country, including China. The eventual reconciliation of the two Chinas still seems far off, but in view of the sudden unexpected unification of Germany, who can be sure of anything political in the future? Taiwanese leaders will probably at least wait to see what happens to Hong Kong.

TABLE 16.6 Trading partners of China in 1992

	Percentage of value of trade		
	exports to	*imports from*	*ratio E/I*
Hong Kong	44.1	25.5	173
Singapore	2.4	1.5	160
South Korea	2.9	3.3	88
Japan	13.8	17.0	81
European Union (12)	8.9	12.2	73
USA	7.2	11.0	65
Canada, Australia, NZ	0.8	4.8	17
Taiwan	0.8	7.3	11
Other	19.1	17.4	110
Total	100.0	100.0	100

Source: *SYC 1993*, pp. 579–81

2. **Mongolia,** formerly the Mongolian People's Republic, could hardly be more different from Taiwan. It has no coast, and a population of 2 million living in an area of 1,567,000 sq km (605,000 sq miles), compared with Taiwan's 36,000 (14,000 sq miles). Less than 1 per cent of it is cultivated, less than 10 per cent forested, but almost 80 per cent is defined as natural pasture. Its economy is based on livestock farming: 15 million sheep, together with large herds of cattle and horses. Mongolia is crossed by the most direct rail link from northern China to Russia. The country may still be seen as a buffer state between these two great powers, its affinity with China underlined by the presence to its south of China's Autonomous Region of Inner Mongolia. The former Soviet Buryat-Mongol ASSR dropped the Mongolian part of its name some time ago.

3. **The Democratic People's Republic of Korea,** North Korea, has been strongly influenced since 1945 by both China and the USSR. Together with South Korea it had been a colony of Japan from 1910 to 1945. North Korea has a smaller population and a more favourable natural resource base than South Korea. Following the Soviet example, with state ownership of means of production and central planning, much investment was devoted to building up a heavy industrial base. Initially, industrial growth was faster in North Korea than in South Korea, but the neglect of

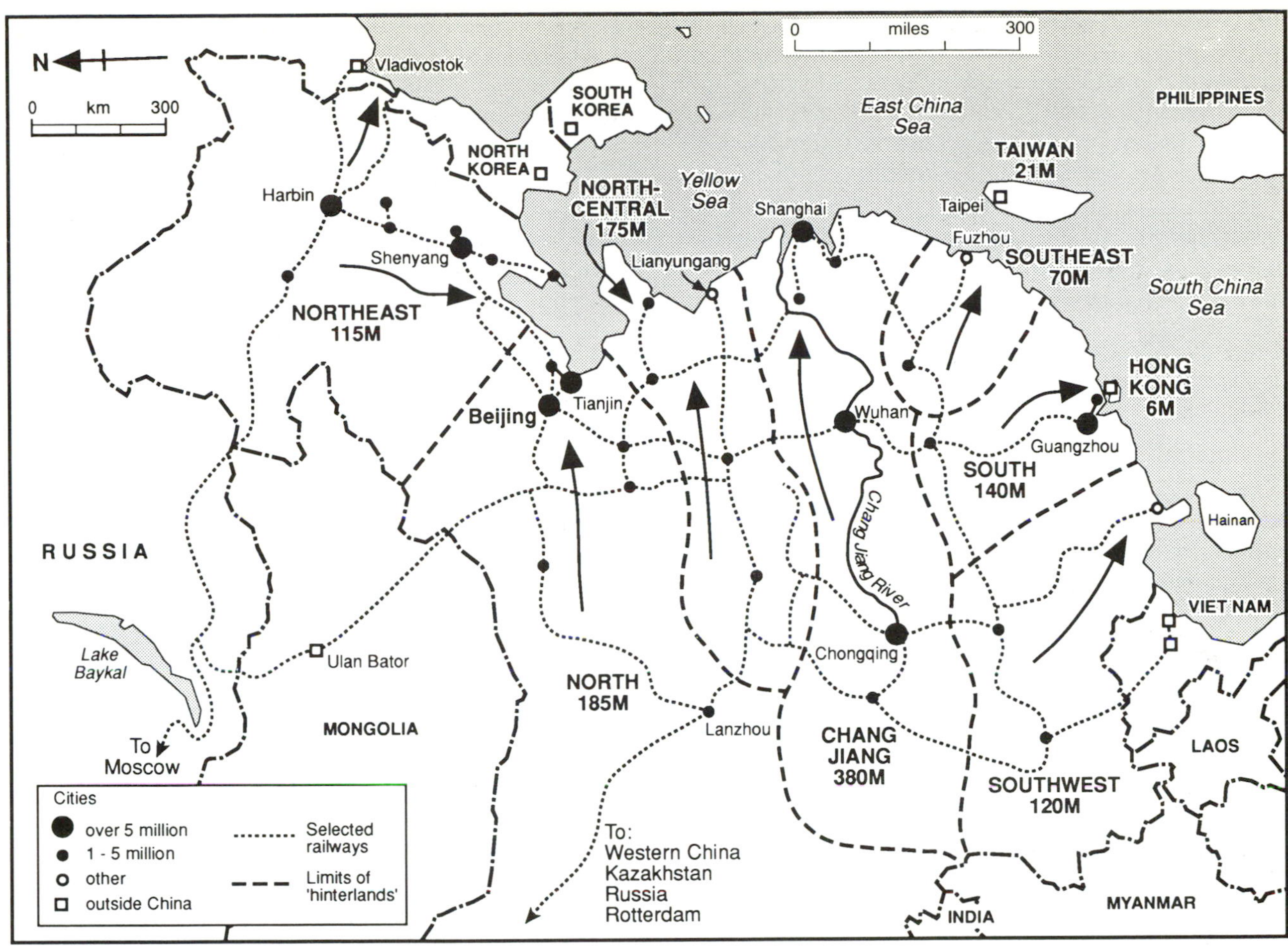

FIGURE 16.8 Seven proposed economic management regions for China based on the principle that each section of the rapidly developing coastal zone should 'adopt' the lagging Provinces behind. The population of each management region is shown in millions

light industry has left the economy very unbalanced. Reliable data are not readily available, but the FAO records prodigiously high rice yields for North Korea, almost 7,500 kg per hectare in 1979–81, compared with the world average at that time of 2,760 and Japan's 5,580, while over 8,000 was claimed for 1989, the highest in the world that year.

The economic gap between the two Koreas is now enormous, with a real GDP per capita in 1989 of 6,120 dollars per head for the South, compared with 2,170 for the North, although the two parts have almost identical levels of life expectancy (70 years) and adult literacy (96 per cent). The reunification of Korea has long been a possibility but, given the cost of the unification of Germany after 1990, South Korea may not entertain the equally daunting burden of paying to make the North Korean economy compatible with its own.

4. **Hong Kong (Xianggang)** resembles Singapore (see Section 15.4) in many respects with regard to its site, its functions, and its history in the British Empire. Hong Kong Island was ceded to Britain by the Treaty of Nanking in 1842 when the industrial powers of the time were beginning to put pressure on China to trade. Further territories on the mainland were added later, making the total area 1,077 sq km (416 sq miles), a third the area of Rhode Island state, USA, but with almost 6 million inhabitants in the mid-1990s. Only a small part of Hong Kong's food supply is provided by its agricultural land and fisheries, while it has no economic minerals and it depends on the adjoining Chinese Province of Guangdong for much of its water supply.

The population of Hong Kong is almost entirely Chinese. Sixty per cent of the population have been born in the colony while 34 per cent are Chinese born. Demographically Hong Kong is similar to developed countries, with high life expectancy, low infant mortality and the lowest total fertility rate in the world, 1.2. The population of Hong Kong has grown from about 600,000 in 1930 to 3 million in 1960 and 6 million in the 1990s, a tenfold increase in sixty years. Since much of the land is mountainous, the pressure on space for commercial, industrial and residential development is acute. About fifty different airlines use Hong Kong airport, the site of which is precariously close to high-rise buildings. A new airport is under construction on an island away from the built-up area. Hong Kong also has an extensive subway system. Like Singapore, Hong Kong is a sophisticated modern city with a high standard of living albeit without the quality of life that a surrounding rural area would give.

Hong Kong's initial growth was related to its function as a trading centre dealing with British interests in the Far East, and since the 1950s, serving as an outlet for much of China's exports. Hong Kong is one of the two top container ports in the world. China, Japan and the USA are its largest trading partners. Since the Second World War industrial growth has been rapid, with specialisation in textiles and clothing, electronic goods, clocks and watches, plastic products

BOX 16.5

Poverty in China

In contrast to the conspicuous consumption in Huaxi village and others in the surroundings of the big coastal cities, the situation in a poor rural area in northeast China is described below in *China Daily* (1 October 1992, p. 1), item A, while Wang Xin (1994) in *Beijing Review* highlights extremes in affluence and poverty in contemporary China in item B.

Item A

Lanzhou (Xinhua) – about 5.5 million farmers in northwest China's Dingxi, Hexi and Xihaigu areas, once the poorest in China, are now able to feed and clothe themselves, following successful implementation in the past decade of a State programme to help the poor.

Commenting on the changes, one of the local people said with gratitude 'This is a meritorious feat by the central authorities for the poor.'

Located on the loess plateau, the region was notorious for poverty, owing to lack of water and adverse natural conditions.

In 1982, about 75 per cent of farm households, with a population of 6 million, lacked enough to eat or wear, and 2.51 million local residents and 2.17 million head of livestock had inadequate drinking-water supplies.

In addition, the State had to supply the areas with nearly 100 million yuan ($18.2 million) worth of relief food and 7 million yuan ($1.27 million) in subsidies for water delivery every year.

The State endowed areas in the Gansu Corridor with good irrigation conditions in that help-the-poor programme. Those who lived in areas taxed by overpopulation and baffled by extremely adverse local conditions would be resettled in the Gansu Corridor.

The programme covered 47 counties with a total population of 10.8 million, although many were already living above the poverty line. The level of concern on the part of the central authorities gave the locals great encouragement, and they immediately plunged into a gigantic battle against the elements.

With State funding, farmers in the poor areas have built 154 water conservation projects, bringing an additional 140,000 hectares of farmland into irrigation, and installed 12,000 kilometres of electric lines, giving 95 per cent of the towns and 80 per cent of the villages in the locality access to electricity supplies, according to Li Zhong, an official in charge of the State programme.

Item B

While the wealthy in Guangzhou and Shenzhen are vying to show off their affluence by hosting banquets costing as much as 200,000 yuan per table, many desperately poor people in the mountainous areas of north Guangdong Province are still living far below the poverty line. In fact, many lack enough grain for three meals a day and use straw for their beds. While students attending 'prominent schools' in Beijing and Chengdu are paying as much as 10,000 yuan in annual tuition fees, many children of their age, like 11-year-old Li Changmei in Schuicheng County, Guizhou Province, have never attended school because their families have been unable to afford a meagre 60 yuan in annual tuition fees. No matter the weather, be it fierce winds or heavy rains, Li has to climb nearby mountains every day to gather firewood for her family.

These, in fact, are the harsh realities confronting China. According to a recent report by the Poverty Relief and Development Leading Group under the State Council, at least 80 million poverty-stricken people in China's rural areas currently have an annual income of less than 300 yuan, and as a result lack adequate food and clothing.

Beginning in 1994, as part of an effort to solve the problem, the State Council approved a seven-year poverty reduction plan designed to solve the basic food and clothing problems for the 80 million stricken by poverty by the end of this century.

and jewellery, products using small amounts of raw materials and energy in relation to value added. With 600,000 workers, manufacturing accounts for about a quarter of total employment. The most recent sector to expand has been the provision of commercial and financial services. After that of Tokyo, its Stock Exchange has the largest turnover of any in the Far East.

Hong Kong reverts to China in 1997. Already a considerable number of people have emigrated (e.g. 66,000 in 1992) from Hong Kong. About 4 per cent of the population have British citizenship so theoretically the UK could expect up to a quarter of a million immigrants in the unlikely event of a mass exodus. For those who remain in Hong Kong the prospect is that they should remain a market economy, part of a 'two-system' China. The potential value of Hong Kong for China is great. It is located next to one of the most economically dynamic Provinces of China, Guangdong. It has excellent transport facilities, enormous

PLATE 16.3
China: children

a Well-heeled children at a commune near Beijing, visited frequently by foreign delegations in the 1980s

b Less well-heeled children in the countryside near Guilin in the Autonomous Region of Guangxi, southern China. One boy carries a cane used to 'herd' the ducks

PLATE 16.4
China: people at work in the transport sector

a Familiar sight: a team of barrow-men at their destination with loads of building materials, near Hangzhou, 1992

b A less familiar sight: one member of a team of men pulling tree-trunks through a back street in Nanjing 1982. There is a picture of a horse doing the same job in England in 1850, in the Ironbridge Gorge Museum

financial resources and increasing experience in sophisticated manufacturing. Cadres of Hong Kong citizens could be persuaded or coerced into settling in the more backward parts of China to assist in economic development. There is no space for a further influx of migrants from Guangdong and elsewhere in China. To some extent Hong Kong is shielded by its understudy, Shenzen, just across the border.

Since the Communist Party came to power in mainland China in 1949, the country has been closely watched by the rest of the world. Its prestigious history as one of the main civilisations of Eurasia, and its very large population, both make it conspicuous in world affairs. Many people with socialist views admired its emphasis on equality and on the importance attached to the rural community; many people in developing countries, especially in Africa and Latin America, saw communist China as a model for their own development. Gradually, two of the basic features of the policy of the 1950s and 1960s have changed. Policy on population has shifted fundamentally from the view that population growth was a problem only under capitalism to one that the whole Han population of the country

PLATE 16.5
China: more modern forms of transport

a A multi-purpose walking tractor, capable of hauling various kinds of instrument and load, near Guilin, south China. Such 'vehicles' are now commonly used in the developing countries of Asia. They might not satisfy required safety standards in most developed countries

b A major engineering feat built by the Chinese in the 1960s: one end of the road and rail bridge that crosses the Chang Jiang River near Nanjing. Note the smoke of the steam locomotive hauling a goods train towards the lower deck of the bridge. The upper deck carries road traffic but there is little sign of motor vehicles although symbolically this bridge links northern and southern China

should if possible consist of one-child families. Policy on the economy has shifted from maintaining public ownership and state control of virtually all economic activities to support for a mixed economy, with the state overseeing development and retaining only the 'commanding heights'. Geographically, current policy is producing two Chinas. The first is situated along the coast and consists of large industrial cities with closely associated surrounding rural areas, in which families are small but population is set to increase through migration from the interior. The second consists of the interior of China, predominantly rural, in which population continues to grow, forming a 'reservoir' from which migrants move with increasing freedom to areas perceived as offering job opportunities. Such a situation is not compatible with the principles of equality and full employment initially set out by the Communist Party of China. To what extent the new tail of Hong Kong will wag the Chinese dragon after 1996 remains to be seen.

PLATE 16.6
Overloaded 'cement boats' carrying liquid sewage along a canal in Suzhou, Jiangsu Province, China. The boats themselves are made of cement but do not carry it, as the name might suggest

PLATE 16.7
Recently completed school in a village about 70 km east of Beijing. When the author visited this school unannounced an orchestra was practising on the steps outside. The school contained a considerable number of pieces of equipment for science teaching, and portraits in the corridors of such western scientists as Copernicus, Galileo and Darwin

KEY FACTS

1. The Chinese civilisation is one of the oldest in the world. Two thousand years ago during its Han Dynasty, it was roughly equal in area and population to the Roman Empire of the time.

2. China has over one-fifth of the total population of the world. In spite of a sharp reduction in the rate of growth, partly at least the result of strong measures to enforce family planning, it continues to grow at about 1.1 per cent a year.

3. China has a wide range of natural resources, but in relation to its population size its bioclimatic and water resources are limited while its proved reserves of oil, natural gas and most non-fuel minerals are very small.

4. The growth of heavy industry in the last four decades has been rapid, and agricultural production in the last three decades has kept up with population growth. Healthcare and educational services are superior to those in most developing countries, but the development of land transport has been characterised by emphasis on rail and neglect of roads.

5. Chinese data show that in general the coastal cities and Provinces are much more highly industrialised than the interior, and material standards are higher, a situation not compatible with socialist aims. Nevertheless, much of the investment in industry, both Chinese and foreign, is being placed in coastal areas.

MATTERS OF OPINION

1. China is large enough and independent enough to establish its own rules on human rights. Such rights and liberties as freedom of speech, religion and movement have not been universally applied in China. Are western human rights any more universally valid than any others? An example of differences in standards is the practice in China of keeping prisoners to be executed alive until a convenient moment. They can then be executed to order so that their kidneys can be used fresh for transplants for influential Chinese and rich foreigners (Lloyd-Roberts 1994).

2. China's ruthless application of its family planning policy has caused concern in other countries. Do the means used – abortion, unofficially even infanticide, and the one-child family – justify the end, a population small enough for the natural resources of the country to be adequate for development?

3. China justifies its occupation of Tibet on historical grounds, yet there is unease in many western countries about its policy aimed to eliminate many of the country's cultural features. Who is ' right'?

SUBJECTS FOR SPECULATION

1. After examining the possible alternative futures for the population of China in Box 16.1, work out what you think is the most probable future scenario.

2. For how long will it be possible for the contradiction to last of having a Communist Party, committed to central planning, running a country increasingly organised economically as a market economy?

3. During considerable periods of its history in the last 2,000 years China has not in practice been a cohesive empire. Since 1949, under communism it has experienced one of the highly centralised periods of its history. Does the break-up of the former USSR have any lessons for Chinese leaders? Which parts of China, if any, are the ones most likely to break away from central control? Note: Russians were only slightly more than half the total population of the USSR whereas the Han Chinese make up 92 per cent of the population of the People's Republic.

Chapter seventeen

Regional prospects

> **The earth-centred perspective assumes that the world population flattens out for a while at 15 billion people, give or take a factor of two (that is a range of 7.5 to 30 billion); the per capita product at 20,000 dollars, give or take a factor of three; and the Gross World Product at about 300 trillion dollars, give or take a factor of five.**
>
> *Herman Kahn et al. (1977, p. 6)*

17.1 Speculating about the future

Some examples were given in Box 1.4 of forecasts that illustrate the pitfalls of predicting future events too precisely. The idea that past events can be useful guides to the future is not a new one, but although it appeals to many people, historians are cautious about the degree to which past experiences and lessons can serve as guidelines for planning the future or be used to forecast future trends and events. A strong argument for speculating about the future, however, is the fact that the past has happened and cannot be changed, although new information about it can be discovered and it can be reinterpreted, whereas at least to some extent the future can be influenced and controlled.

In traditional societies, in which technology and organisation changed slowly, the future could be expected to be broadly 'more of the same'. In the present world situation the growth of the human population, the development of new productive techniques, and the rapidly increasing use of the world's natural resources, many of them non-renewable, have already led to enormous changes in the recent past, and seem set to produce further changes in the future. Concern over current problems and over both short and longer term prospects for future generations makes it particularly relevant to speculate about the future prospects for the world's major regions and for individual countries. How long ahead we should and in practice can look is open to debate. In Box 17.1 methods of speculating about the future are outlined.

In their book *The Year 2000*, Kahn and Wiener (1967) made a major contribution to the establishment of a more 'professional' set of practices and methods for looking at the future. At the same time, the increasing availability of electronic computers since the 1950s has made it possible to experiment with alternative futures for many aspects of the world, from population trends to climatic change. Some good practices in speculation about the future will now be proposed.

- Very little is certain about the future. In general it is more realistic to consider probabilities than certainties and to produce forecasts and projections rather than precise predictions.

- It may be useful to produce alternative futures in order to compare them, rather than to depend on just one projection. For example, the present population of China (about 1,180 million in 1993) can be projected a number of decades ahead, assuming an average of two children per female, just one per female, or, with some demographic engineering and rationing of births, variations through time in the number of children born. The results differ greatly (see Box 16.1).

- After projections are made, they should be monitored as time passes, and if reality turns out to be different they should be abandoned or modified.

- At a more personal level, forecasters should be aware of the danger of making forecasts that produce what they want to happen. For example, economic models setting out 'inevitable' stages of development in the future should be used only with many reservations.

- Plans and intentions already made for the future in, for example, the construction business or the economy should be taken into account in forecasts.

- Although a forecast may be concerned with only one particular future trend, event or prospect, changes in related matters should be taken into account. It could be unrealistic, for example, to anticipate a doubling of motor traffic in some country or region without taking into account future road construction and future supplies of fuel to run the vehicles.

- It is helpful to think about the speed at which change occurs through time. For example, it takes 60–70 years for almost all the population of the world to be replaced, 15–20 years for almost all the 'park' of motor vehicles in use to be replaced. Buildings generally last longer than cars or people, and road layouts, whether in the countryside or in cities, tend to stay in place for centuries. As well as concentrating on the features that change in a given location and period of time it is helpful to focus on what has not changed there. How much change *is* possible over a given span of time?

In systems of government in which the population elects the politicians, most politicians live from day to day or from election to election, formulating and changing policy frequently, and hoping for economic growth to continue or to revive. In both the USSR from the 1920s to the late 1980s and in China from 1949 to

BOX 17.1

Methods of speculating about the future

For the purposes of this chapter, five possible methods of speculating about the future are noted below. They do not cover all possibilities and to some extent they overlap.

1. **The projection of past trends** A danger in this method is that some past trends are very erratic or simply do not make sense when projected. A computer-based projection of the Chilean fish catch made in 1980 continued the decline in the 1970s into the future and ended up in the 1990s with minus half a million tonnes caught each year.

2. **Goals** The feasibility and implication of different goals may be studied. This method is used in Figure 17.2. One must examine the changes needed and the path to be followed through time to produce a given result required or planned for some future date.

3. **Probabilities** Life insurances are based on the average life expectancy of large numbers of individuals. Insurers must also rely on the probable frequency of environmental hazards such as floods and earthquakes to work out risks and premiums. Although it would be pleasing in the economy if every sector were to do well in the future, it is more probable that only some will do well, some will do badly and some will experience near-average performance. Although almost anything is possible in the future, some futures are more probable than others. For example, the demographic scene could be drastically modified if average life expectancy in the world rose over a very short period from around 65 years at present to 130 years, or if suddenly only female or male babies were born, but no one is likely to include such remote possibilities in a projection.

4. **Systems** When it is considered that enough is known about the behaviour and performance of a number of related variables in a region or in the world as a whole, a system of expected changes based on tested or assumed causes and effects can be constructed and projected, preferably with the help of a computer to allow enough calculations. Such a model was used by Forrester (1971), and by Meadows *et al.* (1972), and more recently by Meadows *et al.* (1992), to make a large number of alternative estimates of possible futures for global population, natural resources, production and environmental pollution.

5. **The Delphic method** In some kinds of projection it may be revealing to consult a large number of independent experts about a particular future prospect, obtaining a consensus, a method referred to as the Delphic approach. For example, in the answer to the question 'When will fusion power be developed to generate electricity commercially?' the average estimate could be fifty years, with extremes at twenty years and never.

the present there were one-party systems. The Communist Party professed to be building a future 'perfect' society, but with no final date fixed for it to materialise. Neither approach to the future seriously takes into account long-term trends that could, if they continue unchanged, threaten the existence of human societies, at least in their present form.

Modern professional forecasters tend to fall into two groups. Those in the 'optimistic' camp of Kahn are confident that human ingenuity can overcome all obstacles to development. The pessimistic camp of Meadows and his associates cannot find any projection of bundles of interrelated variables that do not end in disaster for the human population of the world unless drastic changes take place 'immediately' in the way society and consumption are organised. In short, one camp assumes that sustainable development can be maintained indefinitely while the other does not.

In the view of the author, it is simplistic and undiscerning to adhere strictly to either the optimistic or the pessimistic view, neither of which takes into account the enormous regional differences in the world. In 1987, a vehicle-carrying ferry, *The Herald of Free Enterprise*, left Zeebrugge in Belgium for England with its bow doors wide open. In a matter of minutes it sank. If a large meteorite struck the earth or an all-out nuclear war took place, something similar would happen to the population of the world. In contrast, in 1912 the transatlantic liner *Titanic* struck an iceberg. The sea seeped into a number of apertures caused by the collision, leading to the eventual sinking of the ship. In the view of the author, what is happening in the world is analogous to the *Titanic* situation. The pressure of population on the environment, conflicts and economic collapse, are already occurring in various parts of the world.

The aim of the rest of this chapter is to examine the future prospects for the twelve major regions with regard in particular to their demographic and economic prospects. Instead of attempting to fit the sum of human activities into a single global system, selected discrete aspects of human activity and interest are studied in turn, aspects that can readily be compared through time and across space without complications such as the effect of inflation and exchange rates on GDP and on more esoteric economic measures. The following aspects are covered in turn: total population, economically active population in the agricultural sector, arable land, energy reserves and consumption, steel production and car ownership. An attempt is then made to assess the prospects of each of the twelve regions, taking into account various assumptions about the aspirations of their populations and estimating what could actually happen.

17.2 Population futures

In this section the possible future size of the population of the world will be considered. As with other projections, one can only *speculate* about the future size and structure of the population of the world or of any region, and no projection can be claimed to be any more 'correct' than any other one, although some may be more plausible than others. Projections of the population of the world are, however, more straightforward to make than forecasts of the weather or of economic changes such as the price of shares. While both physical and man-made disasters can produce dramatic upsets in population trends locally, as for example the Aberfan tip landslide disaster in Wales in 1966, in which almost all the pupils of one school died, at global level and even at the level of the twelve major regions used in this book, changes are likely to be gradual. The decision to have children is made by a very large number of individuals or couples, planning independently, and the human lifespan is confined within broad limits.

Table 17.1 contains estimates of population made by the author for the twelve major regions. The regions are those used throughout the book. They reflect the state of the world around 1990, with the former GDR included in Western Europe and the Baltic republics transferred to Central Europe from the rest of the USSR. Thus the data for the first three regions will not tally exactly with regional data from other sources unless the same modifications have been made. The table will now be discussed column by column.

(1) The 1993 data from the *World Population Data Sheet 1993* are shown here since they have been used extensively in the book as the most recent and reliable data set available.

(2) The population data for 1750 are very approximate. North America does not include areas now in the USA but then part of the Viceroyalty of New Spain. China includes a number of adjoining and nearby territories. It is likely that the 1750 population figures underestimate rather than overestimate the actual population of the time.

(3) and (4) The population for 1950 of about 2.5 billion is reasonably accurate. Column (14) shows that in that year the six developed regions had about 36 per cent of the total population of the world, compared (column (15)) with just over 13 per cent expected in 2075. Between 1950 and 1975 the population of the world grew in *25 years* by almost 1.6 billion, roughly the amount added in the previous *200 years*.

(5) The population estimated for the year 2000 is based on *WPDS 1989* but more recent trends indicate that the world total then may be 100–150 million less than the 6,323 million indicated.

(6) The population estimate for the year 2025 is

TABLE 17.1 Long-term population change in the twelve major regions of the world

	(1)	(2)	(3)	(4)	(5)	(6)	(7)	(8)	(9)	(10)	(11)	(12)	(13)	(14)	(15)
	Population in millions								*Population change (earlier = 100)*					*% of world total*	
	1993	*1750*	*1950*	*1975*	*2000*	*2025*	*2050*	*2075*	*1950–75*	*1975–2000*	*2000–25*	*2025–50*	*2050–75*	*1950*	*2075*
Western Europe[1]	381	110	300	358	382	384	365	329	119	107	101	95	90	12.5	3.1
Central Europe[2]	131	40	100	122	132	132	125	113	122	108	100	95	90	4.1	1.1
Former USSR[3]	285	45	187	249	304	320	320	304	133	122	105	100	95	7.8	2.9
Japan, S Korea	169	40	102	146	179	181	181	172	143	123	101	100	95	4.2	1.6
North America	287	4	165	237	297	371	427	448	144	125	125	115	105	6.8	4.2
Oceania	28	1	10	17	30	39	43	43	170	176	130	110	100	0.4	0.4
Latin America	460	30	165	324	535	682	784	823	196	165	127	115	105	6.8	7.7
Africa S of Sahara	550	60	155	321	744	1,326	1,989	2,387	207	232	178	150	120	6.4	22.5
N Africa, SW Asia	353	50	121	187	430	748	1,122	1,346	155	230	174	150	120	5.0	12.7
South Asia	1,173	150	450	819	1,388	1,933	2,320	2,436	182	169	139	120	105	18.8	22.9
Southeast Asia	460	60	143	272	549	696	766	766	190	202	127	110	100	5.9	7.2
China plus	1,230	240	512	914	1,353	1,614	1,614	1,453	179	148	119	100	90	21.3	13.7
World	5,506	830	2,410	3,966	6,323	8,425	10,056	10,620	165	159	133	119	106	100.0	100.0

Notes:
1 Includes former GDR
2 Includes Baltic republics, excludes GDR
3 Excludes Baltic republics

Sources: Material in the table has been compiled by the author from a number of sources, particularly McEvedy and Jones (1985) and various years of *WPDS* and *UNSYB*

PLATE 17.1
Demographic futures

a Ethiopia is one of the poorest countries in the world yet it has one of the highest fertility rates in the world. One man and his family

b Demographically, China has 'caught up' with many developed countries to achieve in some Provinces fertility rates below 2. The plaque acknowledges the presence of many one-child families in a village near to Beijing

based on *WPDS 1993*. Thereafter, the estimates have been made by the author. Two basically different approaches can be used to estimate future populations: either a global total can be estimated and then subdivided and allocated appropriately to each region, or each region can be considered separately and a global total then calculated. The latter procedure has been used here. Virtually no allowance has been made for future large-scale interregional migration except into North America and Oceania, principally Australia, which are likely to be net 'importers' of population, willingly or not.

Columns (7), (8) are the populations of the regions for 2050 and 2075 as calculated from the expected increases or decreases shown in (12) and (13) respectively.

Columns (9)–(13) show the rate of population growth observed or expected for each of the twelve regions during the periods 1950–75 (observed), 1975–2000 and 2000–25 (expected). From those trends, estimates of population change are shown in (12) and (13) for 2025–50 and 2050–75 respectively.

Columns (14), (15) show the percentages of the total population expected in each of the major world regions in 2050 and 2075 respectively.

The above projections result in a world population that is moving towards a peak around 2100, after which a state of zero growth at about 11 billion would follow or population would begin to decline. Estimates of a peak between 12 and 14 billion are common in other sources. After the peak is reached it is unlikely that exact zero growth will continue for long. At all events, if demographic trends observed in the twentieth century continue, then the populations shown in Table 17.1 for 2025, 2050 and 2075 are 'reasonable', perhaps conservative. But can the natural resources of the world support such a population, and can the natural environment take the strain? Should this question be restricted to the next hundred years, after which very few people alive now would still be alive, or should it refer, say, to the next 500 years? After all, 1500 was not so long ago. Before natural resources, production and consumption are examined in the remaining sections of this chapter in relation to the population projections in Table 17.1, two aspects of population itself will be discussed: how reliable are the projections of population and how many countries will there be?

1. How confident can one be about the projections? Western and Central Europe, the former USSR apart from its southern republics, and Japan with South Korea, have the prospect of little population change. For example, Japanese demographers (*Nippon 1990/91*, p. 45) have made a projection for the population of their country that rises from about 124 million in 1990 to a peak of about 136 million in 2010 and declines to 124 million again in 2085. In contrast, the demographic future of North America is more problematical, even without the possible addition of Mexico to make an enlarged 'United States'. The populations of Latin America and Southeast Asia seem to be heading towards a state of little change by about 2075. In contrast, the populations of Africa and Southwest Asia are so explosive that, if recent trends continue, they could each increase as much as threefold between 2000 and 2075. During that time the already enormous population of South Asia could add another billion. For Chinese planners hoping for a population of around 600–700 million by 2080 (see Box 16.1 and Liu Zheng *et al.*, 1981), disappointment lies ahead. Only a rigid policy of one child per female for the next seven to eight decades could reduce the expected 1.25 billion in 2000 to that size, short of the unlikely mass extermination of much of the existing population for some reason not imaginable at present.

As indicated in Chapter 3 (see Box 3.5, p. 68), the attainment of a state of zero growth of population in the world, if and when that state is reached, does not automatically mean the end of the world's problems in natural resource depletion, growth in production, overcrowding and pollution. These continue at a given level even with a stable population.

2. How many countries will there be in the world? The future of the population of the world must be considered in relation to the number and size of countries in the future. In this respect, two contradictory trends may be noted. First, the number of countries has grown during certain periods: in the Americas following the end of the War of American Independence (1783), especially with the emergence of new states from former Spanish colonies (1810–20), the proliferation of new states from the former French, British and other colonies for two decades after the Second World War, and, in the early 1990s, new states in Central Europe and the USSR following the break-up of the latter. The opposite trend has been the gradual breaking down of trade and other barriers between groups of countries forming free trade areas or customs unions: the six of the European Economic Community in 1957, the Latin American Free Trade Association in 1961, and others.

3. A world organisation? In spite of the globalisation of many economic activities, a world organisation with substantial powers to influence demographic trends, economic development and interregional transactions, seems a long way off. Cole (1981) showed that *even with* a world government with unlimited powers to plan and manage the population, resources and production of the world it would be physically, logistically and geographically impossible to move enough people and/or capital goods and/or consumer goods between the twelve major regions, or any other set of regions, to produce equal living conditions throughout the world, at least until population growth ceased.

17.3 The economically active population in agriculture

Throughout history until the nineteenth century, in most regions of the world and at most times, the majority of the population has been engaged in agriculture, forestry and fishing, producing food, raw materials and fuel for their everyday needs. In many civilisations, however, the agricultural population has been able to support a minority of people specialising in manufacturing and services. In many countries that minority has now become the majority. In the middle of the nineteenth century, over half of the economically active population was still in the agricultural sector in the USA and France, and over 75 per cent in Japan. In Britain early in the nineteenth century, however, only about 25 per cent were in agriculture.

In virtually every country in the world the per-

TABLE 17.2 Economically active population in agriculture in millions and as a percentage of all economically active population

	All developed		*Latin America*		*Africa*		*India*		*China*	
Year	*millions*	*%*	*millions*	*%*	*millions*	*%*	*millions*	*%*	*millions*	*%*
1965	108	23	34	43	91	74	143	72	–	–
1970	90	19	36	41	97	72	151	69	336	78
1975	80	15	38	36	119	76	172	71	368	76
1980	70	13	39	32	130	74	185	70	406	74
1985	60	10	41	29	141	71	200	68	439	71
1990	50	8	41	26	154	68	215	67	458	68

Note: – = no data
Sources: FAOPY, various years

centage of agricultural employment in the total economically active population has declined in the last few decades. In the view of the author, in the next few decades a major problem in most of the developing countries of the world will be the absolute growth of the labour force in agriculture, in spite of its relative decline in total employment. In the present developed countries, mechanisation and changes in management and ownership of land have made it possible this century for each person engaged in agriculture to work a larger and larger area. *In most parts of the developing world the area worked per person has diminished.*

In developed countries there has been an absolute decline in the agricultural population, but in most developing countries a relative decline has been accompanied by an absolute increase. Table 17.2 shows that the absolute growth of employment in agriculture in Africa, India and China continued between 1965 and 1990, but was accompanied by a percentage decline in all three regions. The contrast between these regions and Latin America is very striking. The four developing countries or regions in Table 17.2 are representative because they contain about 75 per cent of the total population of the developing world, Southwest Asia, Southeast Asia and part of South Asia being excluded. In many countries here a trend similar to that in India and China is also occurring.

The data in Table 17.3 show that in the developing countries as a whole between 1965 and 1990 there was an increase of over 60 per cent in the labour force in agriculture, from about 650 million to 1,050 million, whereas the area of arable and tree crops increased from 6,070,000 sq km (2,344,00 sq miles) to 7,600,000 sq km (2,934,000 sq miles), an increase of only about 25 per cent. In sharp contrast, between 1965 and 1990 the labour force in agriculture in the developed countries declined from about 110 million to 50 million while the area of arable and tree crops grew from 5,450,000 sq km (2,104,000 sq miles) to 6,700,000 sq km (2,600,000 sq miles). Thus, whereas in the developing countries in 1990 on average about 1.5 people worked each hectare of cultivated land, in the developed countries each person worked 13–14 hectares. In both developing and developed regions there were of course great differences between countries.

Projections of the economically active population in agriculture to the year 2010 are shown in Table 17.3. In the developed countries, possibly another 15 million jobs will be lost in agriculture between 1990 and 2010, mainly in Central Europe, the former USSR and the southern countries of the European Union. These people will expect to find jobs in industry or services. The industrial sector itself has lost jobs, however, since the 1970s in many developed countries, while some sectors of services, such as banking and retailing, are now also losing jobs.

In contrast to the developed countries, the total population of the developing countries has grown enormously between 1965 and 1990, increasing by about 1.7 billion, with about that number expected to be added again during the next twenty years. The agricultural, predominantly rural, areas of the developing world are sources of migrants, adding constantly to the population of rapidly growing urban centres. The agricultural employment in developing regions can be expected roughly to double between 1965 and 2010, from 650 to 1,250 million. The number of people employed or seeking employment in the non-agricultural sectors could increase more than fourfold, from less than 300 million to about 1,300 million, a massive number of jobs to create, whether in market economy conditions or with a great deal of state intervention.

The upper graph in Figure 17.1 shows the observed decline in the share of agriculture in all employment during the period 1965–90 and that projected during 1990–2010 in four groups of countries. The lower

TABLE 17.3 Economically active population in agriculture in developed and developing countries, 1965–2010

Year	Total population in millions			All economically active in millions		In agriculture in millions		In agriculture percentage	
	of world	*of all developed*	*of all developing*	*all in developed*	*all in developing*	*developed*	*developing*	*developed*	*developing*
1965	3,276	1,023	2,253	461	931	108	646	23.4	69.4
1970	3,694	1,072	2,622	484	1,100[1]	90	750[1]	18.6	68.2
1975	4,079	1,124	2,955	519	1,244	80	851	15.4	68.4
1980	4,448	1,168	3,280	552	1,404	70	923	12.7	65.7
1985	4,851	1,209	3,642	579	1,584	60	993	10.4	62.7
1990	5,205	1,251	3,954	601	1,765	50	1,051	8.3	59.5
1995	5,665	1,275	4,390	615	1,960	45	1,105	7.3	56.4
2000	6,125	1,295	4,830	630	2,155	40	1,155	6.3	53.6
2005	6,585	1,310	5,275	640	2,350	35	1,200	5.5	51.1
2010	7,040	1,320	5,720	650	2,545	35	1,240	5.4	48.7

Note: [1] Adjustment to total made by author
Source: *FAOPY*, various numbers

graph contrasts the absolute gain in the agricultural sector in developing countries with the absolute loss in developed ones. But what of the decades beyond 2010? Most of the developing world seems destined to continue well into the twenty-first century with such a large labour force in agriculture that each person working in the agricultural sector will be able to provide for the needs of only a few other people, whereas in the USA now each person working in agriculture supplies over 100 others. There is now much speculation in financial circles about the growing affluence and emerging 'middle class' in developing countries and the growth in the demand for consumer goods. The purchasing power of a landless farmer who digs and plants part of a hectare by hand and harvests grain ear by ear with a minute sickle will not suddenly be sufficient to buy sophisticated household goods or a car. In the section that follows, prospects for increasing the area of land under cultivation will be discussed. Will that increase match or exceed the expected increase in the agricultural population outlined above?

17.4 Arable land and agricultural output in the future

In spite of famines in various parts of the developing world since 1945, there has been enough food in the world to supply the growing population (see e.g. Grigg, 1993). Local and regional shortages have occurred, but failure to stop lives being lost through starvation has been blamed less on global shortages than on local and regional problems of distribution, as for example when millions of deaths from starvation occurred in China in the early 1960s, a result of the unwillingness of the government to request foreign charity. Is a time coming when total world food production will not be adequate to support the total population? The actual extent of increasing pressure on agricultural land in recent decades and the further pressure expected in the future in the major regions of the world will now be discussed.

Table 17.4 shows estimates of the arable area per capita in thirteen regions of the world (Barney, 1982), most of them similar to the twelve regions used in this book. As a result of only a modest increase in the arable area of the world during the second half of the twentieth century and more than a doubling of world population, the average arable area available per inhabitant has dropped from half a hectare to a quarter of a hectare. Regional disparities are very marked in all three periods shown. Since the population of the developing regions has grown faster than that of the developed regions, however, the disparity between them has tended to grow with time and will continue to do so if the estimates for 2000 are correct. Thus in all the developed regions except Canada and Australia, the reduction in arable land per capita between the early 1950s and the year 2000 is expected to be by a third. In Latin America, Southeast Asia and China it could be by half, and in Africa, the Middle East and South Asia by well over half. The comparatively favourable prospect for the developed regions is strengthened by the generally higher yields obtained from the same crops compared with those obtained in the developing regions, and by their much greater affluence and therefore ability to import from other regions as and when they need to supplement their own production, provided of course surpluses are available.

Barney's (1982) estimates were made in the late

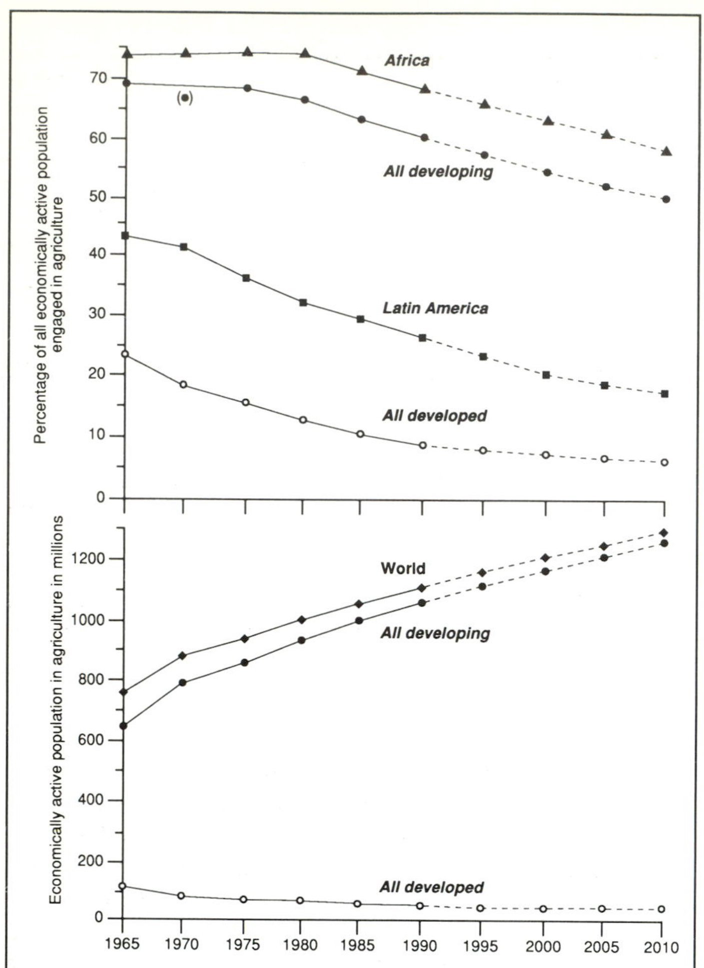

FIGURE 17.1 Economically active population in agriculture in selected regions of the world 1965–2010 (projected)

TABLE 17.4 Arable area per capita, actual and projected in hectares (and in acres)

	1951–5		*1971–5*		*2000*	
Western Europe	0.33	(0.82)	0.26	(0.64)	0.22	(0.54)
Eastern Europe	0.50	(1.24)	0.43	(1.06)	0.36	(0.89)
USSR	1.16	(2.87)	0.93	(2.30)	0.73	(1.80)
Japan	0.06	(0.15)	0.05	(0.12)	0.04	(0.10)
USA	1.17	(2.89)	0.95	(2.35)	0.84	(2.07)
Canada, Australia	1.72	(4.25)	1.58	(3.90)	0.94	(2.32)
Latin America	0.56	(1.38)	0.47	(1.16)	0.28	(0.69)
Africa S of Sahara	0.72	(1.78)	0.62	(1.53)	0.32	(0.79)
N Africa/Middle East	0.68	(1.68)	0.47	(1.16)	0.22	(0.54)
South Asia	0.38	(0.94)	0.26	(0.64)	0.13	(0.32)
Southeast Asia	0.38	(0.94)	0.35	(0.86)	0.20	(0.49)
China	0.19	(0.47)	0.16	(0.40)	0.11	(0.27)
Other East Asia	0.15	(0.37)	0.13	(0.32)	0.08	(0.20)
Developed market	0.61	(1.51)	0.55	(1.36)	0.46	(1.14)
Less developed[1]	0.45	(1.11)	0.35	(0.86)	0.19	(0.47)
World	0.48	(1.19)	0.39	(0.96)	0.25	(0.62)

Note: [1] Excludes China
Source: Barney (1982), modified from p. 99

TABLE 17.5 Arable area, 1962–2025

	Population			*Million hectares/(acres) of arable*			*Hectares/(acres) per inhabitant*		
	1962	*1993*	*2025*	*1961–5*	*early 1990s*	*2025*	*1962*	*1993*	*2025*
Western Europe	349	381	384	102 (252)	91 (225)	85 (210)	0.29 (0.72)	0.24 (0.59)	0.22 (0.54)
Central Europe	106	131	132	55 (136)	58 (143)	55 (136)	0.52 (1.28)	0.44 (1.09)	0.42 (1.04)
USSR	200	285	320	225 (556)	225 (556)	220 (543)	1.13 (2.79)	0.79 (1.95)	0.69 (1.70)
Japan and S Korea	121	169	181	8 (20)	7 (17)	6 (15)	0.07 (0.17)	0.04 (0.10)	0.03 (0.07)
North America	206	287	371	222 (548)	236 (583)	250 (618)	1.08 (2.67)	0.82 (2.03)	0.67 (1.65)
Oceania	17	28	39	36 (89)	50 (124)	50 (124)	2.12 (5.24)	1.79 (4.42)	1.28 (3.16)
Latin America	224	460	682	116 (287)	152 (375)	170 (420)	0.52 (1.28)	0.33 (0.82)	0.25 (0.62)
Africa S of Sahara	214	550	1,326	167 (412)	155 (383)	160 (395)	0.78 (1.93)	0.28 (0.69)	0.12 (0.30)
N Africa and SW Asia	135	353	748	88 (217)	94 (232)	100 (247)	0.65 (1.61)	0.27 (0.67)	0.13 (0.32)
South Asia	586	1,173	1,933	193 (477)	204 (504)	210 (519)	0.33 (0.82)	0.17 (0.42)	0.11 (0.27)
Southeast Asia	230	460	696	60 (148)	78 (193)	90 (222)	0.26 (0.64)	0.17 (0.42)	0.13 (0.32)
China	747	1,230	1,614	94 (232)	100 (247)	100 (247)	0.13 (0.32)	0.08 (0.20)	0.06 (0.15)
World	3,135	5,506	8,425	1,366 (3,375)	1,450 (3,582)	1,496 (3,695)	0.44 (1.09)	0.26 (0.64)	0.18 (0.44)

Source: FAOPY, various numbers, 2025 author's projection

1970s. With data for the early 1990s now available it is possible to extend the projections to 2025 with reasonable confidence (see Table 17.5). With only a small increase assumed in the total arable area of the world, much of it at the expense of tropical rainforest, there would be about 0.18 hectares per inhabitant in 2025 compared with about three times that amount in 1945. In 2025 the area per inhabitant in the developing regions would be cut to half what it was in 1962 in Latin America, Southeast Asia and China but to a third or less in Africa, and in Southwest and South Asia.

The greatest single reason why agricultural production has increased so much since the 1940s, in spite of only a limited increase in the cultivated area, has been through the big increase in resource-augmenting inputs, particularly fertilisers. In *The Global 2000*, Barney (1982) estimates that fertiliser use in the world in tonnes per year was as follows: 18 million between 1951 and 1955, 38 million between 1961 and 1965, 80 million between 1971 and 1975, and would be 220 million for the year 2000. In the period 1951–5 the developed countries used almost 75 per cent of that total, while in the year 2000 they could still be using two-thirds of the much larger total.

Such large expected levels of fertiliser consumption will eventually introduce or aggravate problems of actual availability of reserves of fertiliser minerals as well as causing environmental problems. No less serious a prospect is noted in *The Global 2000* (p. 99):

> Expanding food production through increased use of resource-augmenting inputs, however, is subject to diminishing marginal returns. In highly simplistic terms, the 20 million (metric) ton increase in fertiliser consumption from the early 1950s to the early 1960s was associated with a 200 million ton increase in grain production, suggesting a 10:1 ratio. Growth from the early 1960s through the early 1970s appears to have been at a somewhat lower ratio of 8.5:1.

So the returns continue to diminish as levels of application increase.

In order to plan food production for a world population of about 11 billion towards the end of the twenty-first century, with double the present population, the following prospects and strategies are worth consideration:

- Some of the existing area of land under annual and permanent crops will be lost both to settlements and transport links, and through erosion.

- New areas will be brought into cultivation from existing natural pastures, forest and, with irrigation, even from land at present with no bioclimatic uses. In general, however, the principle of diminishing returns may be expected to apply, since the cost of reclaiming new land will generally be high because of less favourable environmental conditions.

- Yields of crops in existing areas of cultivation can theoretically be increased towards the highest achieved at present somewhere in the world.

• The raising of livestock to produce meat and dairy products could be greatly reduced, allowing the production of more food that can be directly fed to humans, eliminating the amount 'lost' in feeding the livestock.

17.5 Energy

Two aspects of the world energy situation will be discussed in this section: first, the expected life of reserves of fossil fuels and second, contrasting levels of energy consumption between the regions of the world. The scale of the reserves of fossil fuels in the world as a whole and in the major regions was referred to in Chapter 3. In Table 17.6 the reserves/production (R/P) ratios of the three types of fossil fuel are shown. As far as possible the data are for the twelve major regions of the world used in this book.

In 1992, the total amount of primary energy *consumed* in the world in tonnes of oil equivalent was 7,900 million (an average of about 1,460 kg per inhabitant). The three fossil fuels accounted for 7,180 million, almost 91 per cent of the total. The rest was nuclear and hydroelectricity, which does not, however, include fuelwood and other traditional sources of energy or the as yet very small contribution of new sources such as wind power. Of the three fossil fuels (total, 7,180 million), oil accounted for 3,170 million (44 per cent) of consumption, natural gas for 1,839 million (26 per cent) and coal for 2,171 million (30 per cent).

Oil was the most important fossil fuel in the early 1990s in quantity consumed and also on account of its versatility in application. Since the early 1970s concern over both the location and the size of oil reserves has resulted in the price bearing very little relationship to the cost of extraction. The world oil market has been influenced more by political considerations than by economic ones. Oil production was rising fast in almost every region of the world in the 1960s. Until 1972 the discovery of new reserves exceeded the amount of oil extracted, but between 1972 and 1986 production and discoveries were roughly similar, so reserves stood at about 90 billion tonnes during that period. Between 1986 and 1989, however, new reserves took the total to about 140 billion, almost entirely through new discoveries in the Middle East, together with extra reserves claimed by Venezuela on the assumption that its tar sands could be commercially exploited in the future.

Given that many regions of the world remain to be explored, including extensive offshore areas at considerable depth, it seems probable that oil reserves will keep pace with production for some decades to come unless consumption increases greatly world-wide. The fact that oil reserves would last over forty years at present rates of production removes the sense of urgency and concern felt two decades ago about its future, but it does not mean that reserves will last for ever, and the breathing space has been achieved because oil consumption per capita remains very low in most developing countries.

By comparison with oil, natural gas reserves have a considerably longer life at present rates of production (see Table 17.6). The proved coal reserves of the world are so large that the prospect that they might run out is not a matter of concern at all at present, although some of the reserves, especially in Russia, are far from the nearest transport links – rail links in this case.

It seems likely that the consumption of fossil fuel in the developed countries will not change markedly in the next few decades. This is because population is expected to grow only slightly, constant improvements

TABLE 17.6 Fossil fuel reserve 'lives' at 1992 rates of output

	Life in years of					
	oil reserves		*natural gas reserves*		*coal reserves*	
1	Middle East	99.6	Middle East	over 100	Former USSR	over 400
2	OPEC	81.8	Africa	over 100	North America	260
3	Latin America	44.0	OPEC	over 100	OECD	250
4	WORLD	43.1	Latin America	75.8	Latin America	250
5	Africa	24.9	Former USSR	67.8	WORLD	230
6	Central Europe	24.5	WORLD	64.8	Central Europe	230
7	Rest of Asia, Oceania	17.9	Rest of Asia, Oceania	52.5	Africa, Middle East	180
8	Former USSR	17.3	Western Europe	27.6	Rest of Asia, Oceania	180
9	North America	9.8	OECD	16.0	Western Europe	180
10	OECD	9.6	Central Europe	15.7		
11	Western Europe	9.2	North America	12.0		

Source: BP (1993), pp. 2, 18, 26

are being made in the efficiency of the use of fossil fuels, and concern over the effects of burning fossil fuel on both local and global environments is a deterrent to the expansion of their use. Most developing countries have a far lower level of fossil fuel consumption per inhabitant than the developed countries, and it is here that a large increase in fossil fuel consumption may be expected. In China alone in the period 1982 to 1992 the consumption of coal rose from 324 million tonnes of oil equivalent to 527 million, the latter comprising 1,020 million tonnes of hard coal and 90 million tonnes of lignite and brown coal, almost twice as much as that consumed in the whole of Western Europe and slightly above the amount consumed in North America.

The data in Table 17.7 are used to view the world energy situation from a different angle. The consumption of energy in six selected countries or groups of countries is compared with the population in four different years. In Figure 17.2 the six regions are located in the graph with their positions on the horizontal axis determined by total population and their positions on the vertical axis determined by total energy consumed. Their 'paths' through recent time can be plotted and possible future positions marked. It should be noted that *neither* scale is itself a time scale.

The data in Table 17.7 (using coal equivalent, not oil equivalent) and their graphical representation in Figure 17.2 show world energy consumption in a regional perspective between 1960 and 1990. For simplicity only six countries or groups of countries are shown. They accounted for 77 per cent of all world energy

BOX 17.2

International tourism: the material cost of your journey

Travel agencies in western countries are full of brochures giving details of tours to 'remote' parts of the world. Before the 1960s holiday travel to other continents was far too expensive to contemplate except for a small affluent minority. The time constraint made it too long to reach the destination and return in the limits of a normal two- or three-week holiday from work. Since the 1950s, groups of tourists have been accommodated in regular intercontinental jet services, while whole aircraft have also been chartered. It has become possible for considerable numbers of North Americans, West Europeans and Japanese to spend two or three weeks in venues far from their own countries without losing more than two days (and nights) in travelling.

From a geographical point of view long-distance tourism is of interest for two reasons. First, in theory it results in a net transfer of money from rich to poor countries. In practice, much is actually spent on travel with airlines owned by companies in the rich countries, while many hotels in developing countries are owned by citizens of developed countries. Second, it provides the opportunity for generally better-off people of the rich countries to learn something about the developing world. With first-hand experiences they can better appreciate some of the features shown in the proliferation of television documentaries on life in the developing world. In practice, they are usually kept well away from the poorest and most seedy areas.

Do such package tours really benefit the host countries? In *Tourism and Development in the Third World*, Lea (1988) puts arguments for and against. Is it in the long-term interest of the world as a whole that long air journeys for pleasure and leisure are made at all? At the start of my journey to Thailand with my wife, a potentially thought-provoking statement came from the pilot of the Boeing jumbo jet as it set out from London for Thailand: 'We weigh 330 tonnes altogether, of which 150 tonnes are fuel.' The aircraft was carrying about 300 passengers. Most of the fuel would be used up by the time we arrived at Bangkok. Therefore roughly half a tonne (500 kg) of fuel was consumed to carry each passenger to his/her holiday destination, and the same quantity would be needed for the return journey. Put in perspective, and disregarding the fact that the flight would take place anyway, 2 tonnes of fuel (about 1,600 gallons) were used up for two people, roughly the amount used in one year of driving at home in Britain. A holiday in Britain using the car would have taken up only a fraction of the 1,600 gallons. A slogan in materials-poor Britain in the Second World War was 'Is your journey really necessary?'

More sobering still, perhaps, is the fact that the present oil reserves of the world, estimated to be about 140 billion tonnes, average out at 25 tonnes per head of the *present* population of the world, *not counting future generations*. The calculation is admittedly somewhat hypothetical and even spurious, but it does draw attention to what is becoming a growing concern in the world: what are the fuel and raw material consumption rates of various activities? Playing games on a computer is less exacting on primary products than motor racing. My wife and I each used up one twenty-fifth of our 'ration' of the world's oil reserves on that trip.

For key facts, matters of opinion and subjects for speculation related both to this chapter and the whole book, proceed to Chapter 18.

TABLE 17.7 Energy consumption in selected regions and countries of the world, 1960–90

	Energy consumption in millions of tonnes of coal equivalent				*Population in millions*			
	1960	*1970*	*1980*	*1990*	*1960*	*1970*	*1980*	*1990*
USA	1,448	2,279	2,364	2,482	181	203	223	250
EU without GDR	747	1,185	1,400	1,418	279	305	318	327
USSR	613	1,055	1,486	1,931	214	243	266	291
Japan	109	332	433	512	93	104	117	124
China	296	310	545	922	662	830	987	1,143
India	60	98	139	265	430	540	676	827
World	4,238	6,817	9,718	11,550	3,010	3,694	4,414	5,310

Source: United Nations (1993), *Energy Statistics Yearbook 1991*

consumed in 1960 and for 65 per cent in 1990. They are therefore reasonably representative of the world situation. The graph illustrates two distinct features. First, there is a huge gap in consumption per inhabitant between the developed and the developing countries, which can be seen by reference to the lines radiating from the point of origin of the graph. Second, the graph can be used to plot and examine the implication of future goals for energy use. This has been done for China, with possible production levels in the years 2000 and 2010, allowing for an increase in population, giving totals expected to be within the bands indicated. Goals *a* and *b* are for a per capita consumption level of 1,500 kg of energy in the year 2000 or in the year 2010, approximately the level in Japan in the early 1960s, but still less than one-sixth the present US level. Such a level in China would, however, require the use of some 2,000 million tonnes of coal equivalent, almost half the *total world* consumption of energy in 1960. Goals *c* and *d* for China are less ambitious, achieving 1,000 kg (1 tonne) per inhabitant in 2000 or 2010.

The data in Table 17.7 show that energy consumption is levelling out in the developed countries, with the former USSR likely to consume no more in the year 2000 than in 1990, following the experience of the other developed regions. In many developing countries, however, population is growing fast, and energy consumption even faster. A 50 per cent increase in the

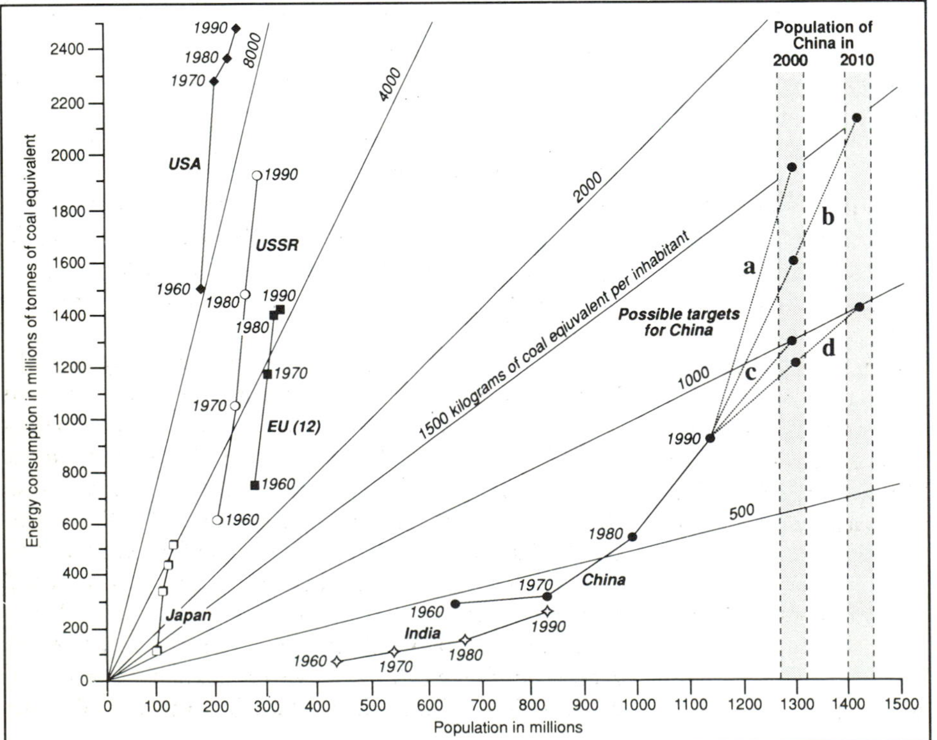

FIGURE 17.2 Energy consumption in selected countries of the world. Note that neither scale shows time but the position of each country at different moments in time can be seen and compared. Note the slowdown in consumption in the USA after 1970 and in the EU after 1980. The diagonal lines radiating from 0 on the axes show levels of consumption per inhabitant for different population and energy consumption sizes

world consumption of energy between 1990 and about 2015 seems a conservative estimate. Nor will growth suddenly cease after that.

17.6 Steel production

As the 'post-industrial' era allegedly supersedes the industrial era in the developed regions of the world, the relative importance of various heavy industrial products declines. Production and consumption per inhabitant of steel, among other products, has gradually declined. The metal is used more efficiently, newer materials replace it, increasing amounts of steel are made from recycled scrap (reducing the amount of pig-iron needed to make steel), and there are fewer construction projects using steel. On the other hand, the developing regions have been needing more steel in recent decades and should need even more in the next few decades, increasing their output greatly and even exporting some of it to the older industrial countries.

The data in Table 17.8 show steel production in the major regions of the world over a considerable time perspective:

Columns (1)–(3) show the output of steel in each of the twelve major regions in 1930, 1958 and 1989. There has been a massive eightfold increase in six decades since 1930.

(4)–(6) show the share of production in each major region in each of the three years. The dominant position of Western Europe and North America in 1930, when together they accounted for over 85 per cent of world steel output, was lost by 1989, although 35 per cent is still a substantial share. Japan and the USSR have taken up most of the difference.

(7)–(9) This index shows the comparative amount per inhabitant in each of the three years, using 100 as the world average. The steel 'gap' in all three years is very marked.

(10)–(14) show a projection of population and of steel output for the year 2025. Only a moderate increase in world output is anticipated because output is expected to decline in Western Europe and in the former USSR, while it is assumed that only Latin America and China will achieve a very large increase.

The gini coefficient of concentration has been calculated for the four distributions and it shows a fairly sharp reduction in concentration between 1930 and 1958, little change between 1958 and 1989, and a further gradual reduction expected between 1989 and 2025, the indices for the four years being:

1930: 0.785; 1958: 0.670; 1989: 0.665; 2025: expected 0.595

The Lorenz curves are shown in Figure 17.3.

17.7 Passenger cars in use

The private car has become one of the leading products of the industrial sector in most of the larger developed countries and in some developing countries. It is the principal means of moving passengers in North America, Western Europe and Oceania and both privately owned cars and public taxis have a prominent place in many of the larger cities of the developing world. It consumes a substantial portion of the world's oil production and contributes generously to local pollution and to the gases purported to be warming the atmosphere. It requires the construction of vast distances of road and the provision of parking space, filling stations and service facilities. Its rise to prominence has happened virtually during the twentieth century.

Table 17.9 shows the distribution of passenger cars in the twelve major regions in 1932, 1958, 1989 and estimated for 2025. The columns in the table show the following:

(1)–(3): cars registered in the three years. The Second World War (1939–45) held back the incipient car boom in Western Europe. Car ownership was still very small in Central Europe, the USSR and Japan at the start of the war.

(4)–(6): the concentration of cars in North America, Western Europe and Oceania, almost 96 per cent of the world total in 1932, is even more marked than the concentration of steel output in those regions. The proportion fell only gradually to 93 per cent in 1958 but more markedly to 72 per cent in 1989, car ownership rising quickly in Japan and Latin America, and also towards the end of the period in Central Europe and the USSR.

(7)–(9): the percentage of population in each of the twelve regions.

(10)–(12) A comparison of cars per population can be made by comparing the score of each country with the world average set at 100, or about 11 persons per car. Throughout the period 1932–89 car ownership in North America has been far higher than elsewhere, but Western Europe and Oceania have reduced the gap considerably.

(13)–(18) The author's estimate is roughly a doubling of car ownership in the world between 1989 and 2025, with a substantial increase in North America from 156 million to 223 million merely to allow for the expected increase in total population and an increase in the level of car ownership to 600 per 1,000 people. North America would still have about 25 per cent of all the cars in use. The level of car ownership would still be thirty times as high in North America as in Africa south of the Sahara, South Asia and China. For every region in the world to have in 2025 a level of car ownership similar to that in North America of 600 per

TABLE 17.8 Steel production in the major regions of the world, 1930–2025

	(1)	(2)	(3)	(4)	(5)	(6)	(7)	(8)	(9)	(10)	(11)	(12)	(13)	(14)
	Production of steel millions of tonnes			*Percentage of production*			*World = 100*			*Situation in 2025*				*World = 100*
	1930	*1958*	*1989*	*1930*	*1958*	*1989*	*1930*	*1958*	*1989*	*popn*	*steel*	*% popn*	*% steel*	
Western Europe	40.7	88.4	161.9	42.5	32.3	21.1	308	283	306	384	140	4.6	15.6	339
Central Europe	3.6	15.0	52.7	3.8	5.5	6.9	76	153	288	132	50	1.6	5.6	350
Former USSR	5.9	54.9	160.1	6.2	20.1	20.8	77	287	400	320	120	3.8	13.3	350
Japan, S Korea	2.3	12.1	131.6	2.4	4.4	17.1	62	110	552	181	120	2.1	13.3	633
North America	42.4	81.3	107.0	44.2	29.7	13.9	680	443	267	371	100	4.4	11.1	252
Oceania	0.3	3.2	6.6	0.3	1.2	0.9	60	240	180	39	10	0.5	1.1	220
Latin America	–	3.0	41.4	0.0	1.1	5.4	0	16	64	682	100	8.1	11.1	137
Africa S of Sahara	–	1.9	9.2	0.0	0.7	1.2	0	11	12	1,326	20	15.7	2.2	14
N Africa, SW Asia	–	0.2	11.9	0.0	0.1	1.5	0	2	29	748	50	8.9	5.6	63
South Asia	0.6	1.8	12.9	0.6	0.7	1.7	3	4	8	1,933	30	22.8	3.3	14
Southeast Asia	–	0.0	0.9	0.0	0.0	0.1	0	0	1	696	10	8.3	1.1	13
China plus	–	11.6	72.2	0.0	4.2	9.4	0	18	42	1,614	150	19.2	16.7	87
World	95.8	273.4	768.4	100.0	100.0	100.0	100	100	100	8,425	900	100.0	100.0	100

Note: – negligible or none
Variables:
In columns (7)–(9) the percentage of the total steel production in each region in columns (4)–(6) is divided by its share of total population (not shown in the table) and then multiplied by 100
Source: *UNSYB*, various years, 2025 author's projection

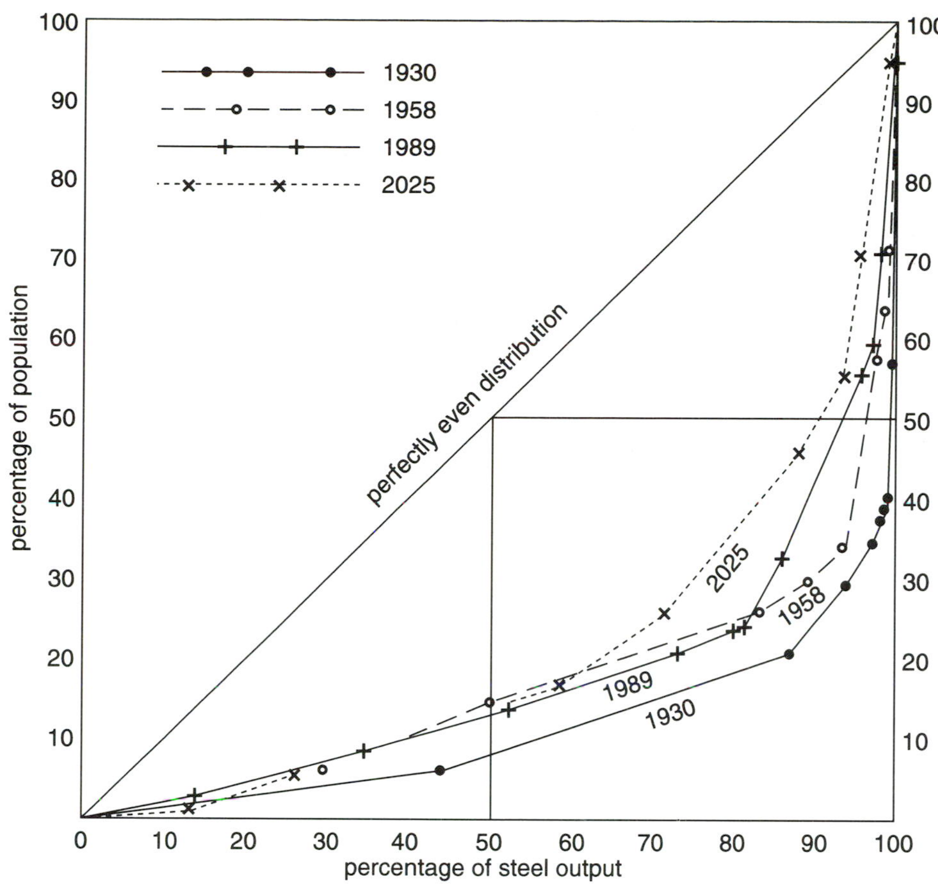

FIGURE 17.3 The distribution of steel production in the twelve regions of the world in 1930, 1958, 1989 and (projected) 2025 shown by Lorenz curves. Gini coefficients are 1930, 0.785; 1958, 0.670; 1989, 0.665; 2025, 0.615. If production per capita is the same in all twelve regions the distribution would follow the diagonal. The greater the concentration of production in one or a few regions, the more the curve sags. A comparison of 1930 and 2025 shows considerable decentralisation during that period

1,000 population, there would have to be over 5 billion cars in use in the world, compared with less than 500 million in 1989.

17.8 Which region?

The main theme of this book has been the diversity of physical conditions and of human activities in the world. Two outstanding questions of our time have been central issues in innumerable debates: how big can the 'cake' of production get before adverse conditions arise and how can the 'cake' be sliced more equably? What follows is no more than an outline of how these questions might be addressed.

Seven criteria have been chosen to represent issues and problems of relevance to all twelve regions (Table 17.10). The choice of criteria is subjective and transparently Eurocentric. For example, Western Europe scores a maximum of 5 points for the parliamentary democracies of its governments, North Africa and Southwest Asia a minimum of 0 (zero) points for the absence of political freedom.

In order to avoid giving a false impression of precision and accuracy, a scale ranging from 0 (unfavourable) to 5 (favourable) is used on all seven attributes, which means also that they all receive equal weight. The maximum combined score of all the attributes is 35. The twelve regions fall into four fairly distinct groups:

Highest scores: North America, Oceania, Western Europe, Japan

Middling scores: Former USSR, Central Europe, Latin America

Low scores: China, Southeast Asia, South Asia, North Africa/Southwest Asia

Very low score: Africa south of the Sahara

The scores on each attribute are based on the following assumptions and judgements:

(1) Population growth is assumed to be unfavourable. Scores range from little or no growth with 5 to fastest growth with 0.

(2) Natural resource scores per inhabitant are based on assessments shown in Table 3.4 (p. 64).

TABLE 17.9 Car ownership in the major regions of the world in 1932, 1958, 1989 and 2025

	(1)	(2)	(3)	(4)	(5)	(6)	(7)	(8)	(9)	(10)	(11)	(12)	(13)	(14)	(15)	(16)	(17)	(18)
													Cars in 2025					
	Cars (millions)			Percentage of cars by region			Percentage of population by region			World = 100			popn 2025 millions	cars per 1,000 population	cars in millions	% of popn	% of cars	World = 100
	1932	1958	1989	1932	1958	1989	1932	1958	1989	1932	1958	1989						
Western Europe	3.8	17.5	143.5	14.0	20.4	33.6	13.8	11.4	6.9	101	179	487	384	500	192	4.6	21.2	461
Central Europe	0.1	0.2	15.5	0.5	0.2	3.6	5.0	3.6	2.4	10	6	150	132	300	40	1.6	4.4	275
Former USSR	–	0.3	16.3	0.2	0.3	3.8	8.1	7.0	5.2	2	4	73	320	200	64	3.8	7.1	187
Japan, S Korea	–	0.3	34.2	0.1	0.3	8.0	3.9	4.0	3.1	3	8	258	181	300	54	2.1	6.0	286
North America	21.8	60.5	155.8	80.0	70.5	36.5	6.5	6.7	5.2	1,230	1,050	702	371	600	223	4.4	24.7	561
Oceania	0.5	2.2	8.8	1.8	2.6	2.1	0.5	0.5	0.5	360	520	420	39	500	20	0.5	2.2	440
Latin America	0.3	2.2	27.9	1.1	2.6	6.5	5.9	6.9	8.4	19	38	77	682	100	68	8.1	7.5	93
Africa S of Sahara	0.2	1.2	6.0	0.7	1.4	1.4	5.6	6.6	10.0	13	21	14	1,326	20	27	15.7	3.0	19
N Africa, SW Asia	0.2	0.6	10.4	0.7	0.7	2.4	3.8	4.8	6.4	18	15	38	748	100	75	8.9	8.3	93
South Asia	0.1	0.4	3.4	0.4	0.5	0.8	17.6	17.9	21.2	2	3	4	1,933	20	39	22.8	4.3	19
Southeast Asia	0.1	0.4	4.8	0.4	0.5	1.1	6.3	7.1	8.4	6	7	13	696	100	70	8.3	7.7	93
China plus	–	–	0.8	0.1	–	0.2	23.0	23.5	22.3	0	0	1	1,614	20	32	19.2	3.6	19
World	27.3	85.8	427.6	100.0	100.0	100.0	100.0	100.0	100.0	100	100	100	8,425	107	904	100.0	100.0	100

Note: – negligible or none
Variables:
In columns (10)–(12) and (18) the share (%) each region has of the world's cars is divided by its share (%) of the world's population and then multiplied by 100
Sources: *UNSYB*, various years; for 1989 *UNSYB 90/91*

TABLE 17.10 Regional scores on a selection of seven attributes, 5 = best, 0 = worst

	(1) Population change	(2) Natural resources per head	(3) Real GDP per head	(4) HDI	(5) Political freedom	(6) Ethnic stability	(7) Regional stability	(8) Total
Western Europe	5	2	4	5	5	4	5	30
Central Europe	5	2	2	5	3	2	2	21
Former USSR	5	4	3	5	3	1	1	22
Japan, S Korea	5	0	4	5	5	5	5	29
North America	3	4	5	5	5	4	5	31
Oceania	3	5	4	5	5	4	5	31
Latin America	2	3	2	4	3	3	4	21
Africa S of Sahara	0	3	0	0	1	0	0	4
N Africa, SW Asia	0	3	2	3	0	1	2	11
South Asia	2	1	0	1	4	2	2	14
Southeast Asia	2	2	1	3	3	2	3	16
China plus	3	1	1	4	0	3	4	16

Description of variables:
(1) Population change: 5 – slowest growth, 0 – fastest growth
(2) Consensus of 5 natural resources (area, water, productive land, fossil fuels, non-fuel minerals)
(3) High to low
(4) High to low
(5) Democracy with universal franchise highest
(6) Ethnic variety, recent tension, high = good
(7) Regional stability, conflicts with neighbouring countries unlikely, high

(3) Real GDP per head: highest 5, lowest 0.

(4) HDI of the United Nations (see Chapter 4) measures the quality of life.

(5) Democratic political system, freedom of speech, free elections and universal vote, 5 to 0.

(6) Ethnic stability is a rough measure of the number of ethnic divisions in a region and the juxtaposition of ethnic groups within a country or across national boundaries, causing conflict at present, or likely to cause conflict in the future.

(7) Likelihood of conflict between sovereign states in the region (e.g. the Iran–Iraq conflict) or between neighbouring states in two different regions (e.g. China–Viet Nam).

The reader may wish to modify the choice of attributes, extend the number of regions by subdividing some of the twelve, or give more weight to some attributes than to others. It could even be justifiable to reverse the scores on some of the variables. For example, it could be assumed that fast population growth is favourable and zero growth unfavourable, or by arguing that GDP per capita is a measure of the intensity of resource depletion and pollution and therefore that high GDP per inhabitant is not desirable or is harmful.

The following points are meant to be thought-provoking rather than final or conclusive.

- North America and Oceania have high scores on natural resource availability, and also on GDP per capita and quality of life (HDI) indicators. So long as natural resources are abundant, they are materially desirable places to be.

- Western Europe and Japan in particular manage at present to achieve high scores on GDP and quality of life (HDI) with comparatively limited natural resources. Their fortunes and future prospects depend on their ability to obtain primary products from other regions of the world. In time the regions supplying Western Europe and Japan may make greater use of their own fuel, raw materials and food than at present. With increasing population and industrialisation, such a prospect seems reasonable. Prices of primary products could then rise relative to prices of manufactured products.

- Central Europe and the former USSR lag behind the four regions mentioned above in material levels and political freedom, but could eventually approach Western Europe and North America respectively in the future. Russia is far more generously endowed with natural resources than Western and Central Europe or Japan.

• Latin America emerges as an intermediate region, a bridge between the developed and developing regions. Its natural resource endowment is reasonably good and the rate of population growth appears to be slowing down in most countries. North Africa and Southwest Asia are fairly similar in their scores to Latin America but their natural resource profile is unbalanced, population is increasing quickly and Muslim fundamentalists are extending their influence, not necessarily a 'bad' trend, but one that could put spiritual considerations above material ones, thereby depressing scores on conventional attributes such as GDP.

• Over 50 per cent of the population of the world lives in the last three regions but they have less than 20 per cent of the natural resources. Once net exporters of primary products, increasingly now they are using these at home. In time they could be competing with Western Europe and Japan for the primary products of other regions to satisfy their growing populations and industrial bases.

• Africa's score in Table 17.10 is lower than expected but it has been achieved quite spontaneously. What might Africa be like now if it had not been carved up and colonised by European powers in the last hundred or so years?

The state of the world in the 1990s, then, is such that enormous contrasts are found in material levels and living standards from country to country and within countries. To divide the world rigidly into developed and developing, First World and Third, or North and South, gives an unacceptably oversimplified and naive dichotomy in the view of the author. Net transfers of population, capital resources and/or products between countries and regions are on a small scale compared with the amounts needed to make substantial reductions in global disparities. Changes in the political situation in various parts of the world, regional conflicts, and economic uncertainty in the developed countries all contribute to a world situation in which it is unlikely that a global strategy will be adopted explicitly to reduce inequalities.

In spite of pious expressions of hope and intention in the last fifty years, very little has been done to reduce inequalities since the Second World War. For the next decade at least, the 'traditional' developing countries are likely to lose out as Central Europe and the former USSR absorb large amounts of assistance from the three remaining major developed regions, North America, Western Europe and Japan. To be sure, some countries currently defined as developing may qualify in due course to be considered developed. If, however, the population projections in Table 17.1 turn out to be correct, in 2050 the three rich regions referred to above would have less than 10 per cent of the population of the world. They will be shrinking demographically as the image on an old-fashioned television screen receded and vanished when the set was turned off.

Chapter eighteen

Twenty-first-century earth

Canadians have used an abundance of natural and human resources to build a prosperous nation. However, an understanding has been growing that the true basis of wealth is not ownership or dominion over the land, but wise stewardship.

Canada Yearbook 1992 (1991, p. 9)

18.1 Preamble

Before looking at this chapter the reader may like to note down the twenty issues of potentially world-wide significance that she or he considers will have the greatest prominence and urgency in the next century. The purpose of the chapter is to set out a number of issues that the author expects will be among those raised and debated in the twenty-first century. For convenience, twenty issues will be discussed in turn in this chapter, following this brief preamble.

The human species has made an enormous impact on the natural world in three stages:

- Initially in a modest way through the use of tools for hunting and other purposes.

- More markedly through agriculture, selecting specific plants and animals to use.

- Explosively through the large-scale use of inanimate sources of energy and of machines and other devices that enhance the natural power of the muscles, the senses and the brain. Humans have rarely been short of self-congratulation about their achievements. After all, who else is there to tell them what a wonderful phenomenon they are?

Apart from some spectacular landings on the moon, human activity has been confined almost exclusively to a few kilometres above and below the earth's finite surface. The enormous cost of transport of people and objects between the earth and the moon has shown how unrealistic it would be to expect anything in the twenty-first century even from the planets of our solar system, let alone from planets elsewhere in our galaxy.

In Chapter 17 the future prospects for each of the twelve major regions of the world were discussed. Africa south of the Sahara appeared to have the worst prospects according to the criteria used. South Asia, Southeast Asia and China have a long way to go to ensure a reasonable standard of living for everyone. Western and Central Europe, together with Japan, could face problems in the future with regard to natural resources.

A similar assessment is made in this chapter of the issues with 'favourable' and 'unfavourable' prospects proposed. Attention focuses on the seven elements first set out in Chapter 3: population, natural resources, means of production, production of goods and services, links, organisation, and international transactions. Issues are numbered 1–20 for convenience.

18.2 Population

Issue 1 Total population It is a fact that the population of the world has increased between threefold and fourfold between 1900 and 1995, from about 1.6 billion to 5.7 billion. It is a matter of speculation

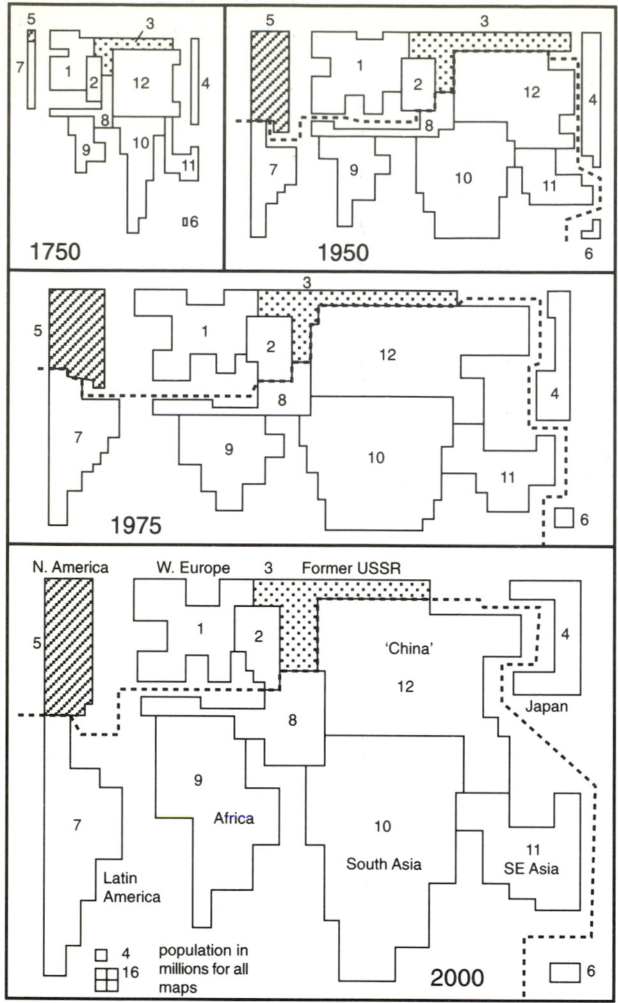

FIGURE 18.1 The world population explosion in perspective: each of the twelve regions used is drawn in the diagrams proportional in area to its estimated population, 1750–2000

whether or not it will grow at the same rate in the next century or so. If it does, the *absolute* gain will be far greater. Figure 18.1 shows graphically the increase in the world's population since 1750 in the major regions of the world (not precisely those used in this book). In the view of the author, the 64,000-dollar (or 64,000-calorie) question that must be raised at some time in the future is: should the production of food and other necessities be made to meet population size or should population size be made to meet the production of food and other necessities?

Issue 2 Distribution of population Until the eighteenth century almost all the world's population was rural and was engaged in farming. Its distribution was closely related to local food supply since, with some notable exceptions such as by the Grand Canal in China, it was only economic to transport food over short distances. Population is still very unevenly distributed over the earth's land surface, but with a growing proportion concentrated in urban centres usually detached from their sources of food, fuel and materials. Population is now very unevenly distributed also in relation to natural resources (e.g. Australia is well endowed, Japan poorly endowed) and to production of goods and services (e.g. USA rich, most of Africa very poor).

In Figure 18.2, an analogy is made with livestock farming. Two points are made in the picture. First, the distribution of livestock does not relate closely to the distribution of fodder. Second, the total number of livestock may need adjusting on some farms unless fodder can be transferred from others, or livestock can be moved from overstocked farms to those carrying fewer animals in relation to the pasture available.

Issue 3 Redeployment of population So can enough people be moved in a short enough time to make the relationship of population to natural resources and/or to production more balanced than it is now? What organisational constraints would have to be removed to achieve this possible aim? Even without such constraints would it be physically possible to effect the massive migrations needed? Inter-continental migration was a conspicuous feature of the world demographic scene between about 1840 and 1930. It has been on the increase again since the 1960s. Most net migration both between countries and within larger ones is of three kinds: from overpopulated rural areas, either to 'empty' regions or to urban centres, and from urban centres to 'empty' regions.

18.3 Natural resources, production and consumption

Issue 4 Natural resource availability Concern over this issue was fashionable in the 1970s, less so since. Water, agricultural land, fossil fuels and non-fuel minerals cover most natural resources. The distribution and availability of water and of productive land are known, but it is likely that many new mineral reserves will be discovered. Water and food cannot be replaced by other products whereas there is considerable flexibility in the use of sources of energy and the prospect that in the twenty-first century fossil fuels can be supplemented if not replaced by renewable sources (solar power, fusion power). Certain non-fuel minerals are limited in availability but there is scope for the substitution of one material with another. Availability of water and agricultural land seems, therefore, to be the constraint to resource use in the next century.

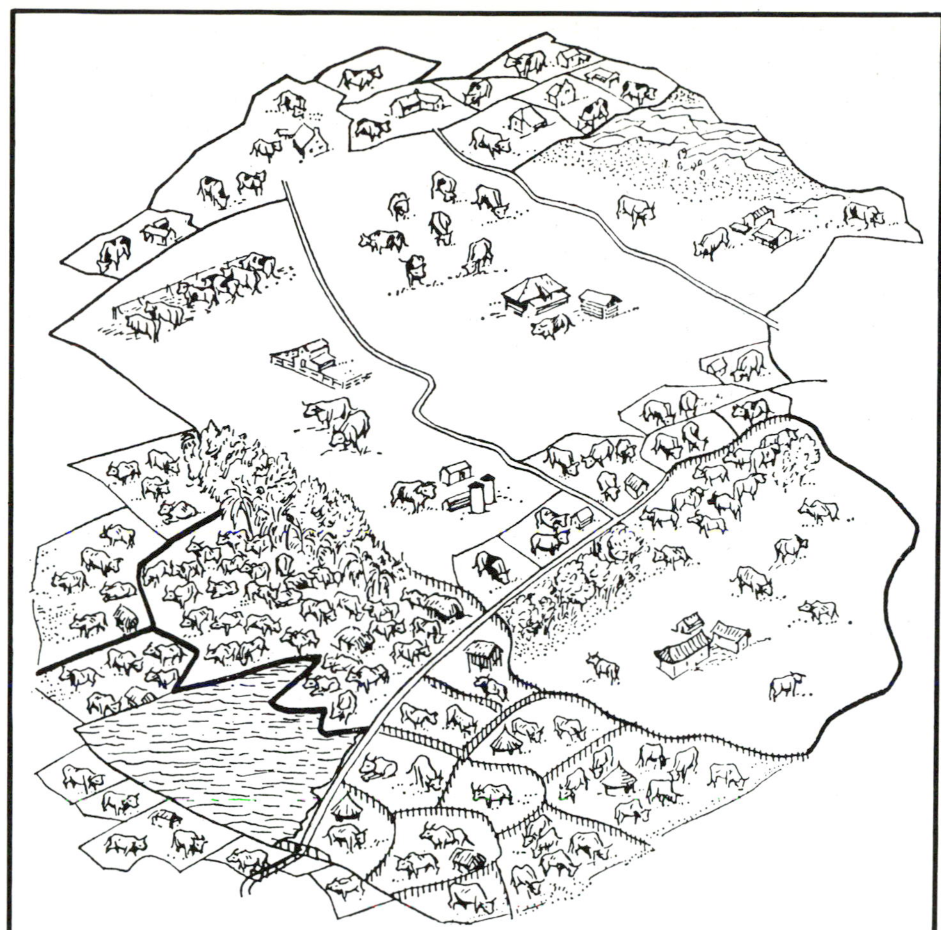

FIGURE 18.2 The farms pictured here are like the world. Population is very badly allocated according to the natural resources and fodder available

Issue 5 Availability of primary products In the nineteenth century the number of countries in the world that were net exporters of manufactured goods was small and these countries were almost all in Europe. Throughout the twentieth century industrialisation has proceeded fast in at least some countries in most of the world's major regions: the USA, Japan, Russia/USSR, some Latin American countries, South Africa, India, China and recently in many other Newly Industrialising Countries. With time, such countries have become net exporters of manufactured goods. In many developing countries, population has grown faster than the production of agricultural and mineral primary products. An increasing proportion is used at home. The amount available for export has therefore diminished.

One example must suffice here: coffee production in Brazil. Table 18.1 shows that between 1950 and 1990 the population of Brazil increased approximately threefold. Coffee production in Brazil fluctuates from year to year, partly due to weather conditions, partly to prices. The data in column (2) are averages for three-year periods. Around 1950, coffee production was about 20 kg per head, but around 1990 it was only about 10 kg. Much of the coffee produced in Brazil is actually consumed internally. The proportion of exports accounted for by coffee has fallen drastically, dropping to less than 10 per cent of the value around 1990. Whether or not coffee is a luxury is a matter of opinion, but the fact is that there is a similar trend in many developing countries for many primary products.

TABLE 18.1 Brazilian coffee production

	(1) Population	*(2)* Coffee total in tonnes	*(3)* Production in kg per head
1950 (48–51)	52	1,077	21
1955 (54–56)	61	1,129	19
1960 (59–61)	70	2,170	31
1965 (64–66)	81	1,363	17
1970 (69–71)	93	1,278	14
1975 (74–76)	106	1,088	10
1980 (79–81)	122	1,465	12
1985 (84–86)	137	1,457	11
1990 (89–91)	150	1,497	10

Source: UNSYB, various numbers

PLATE 18.1 Arguably the availability of fresh water is the ultimate determinant of the number of people a region – and indeed the whole world – could support. The establishment of a new settlement in Gambia, West Africa, depends on the sinking of a deep well in an area where the water level has actually been dropping

The issue: will there be a long-term trend in the twenty-first century whereby more and more countries are competing for the exports of primary products from a few countries?

Issue 6 Economic growth Few politicians, whether elected by their citizens or self-elected, would care or dare in the 1990s to declare a moratorium on economic growth. Increasing the production of goods and services at a faster rate than the growth of population is seen as the only means of increasing living standards and a means of reducing unemployment or of keeping it from increasing. Until the 1970s, economists have regarded the 'development gap' between the rich and poor countries as a temporary phase. Whether on the assumptions of Soviet economists inspired by Marx or of American economists happy with Rostow's stages of economic growth (see Rostow, 1960), many expected the less developed countries to 'catch up' the more developed ones. The expectation as expressed in 1963 by the Economic Commission for Africa (1963, pp. 4–5) is summed up in the following rather sanguine fashion:

> The main object of economic growth in any country has always been to increase the supply of goods and services available for present consumption and of investible resources to assure future development. In this respect, the experience of the industrial countries has considerable relevance for Africa.... To raise the low *per caput* output in Africa and other industrially less developed areas to the output of the industrial countries is now generally accepted as a long term objective of economic development.... The economic transition in Africa would involve ... a twenty-five fold increase in industrial output *per caput* for the whole population.

As far as Africa is concerned, thirty years later this prospect is dead. A few small NICs and a few regions in large developing countries have achieved fast rates of industrial growth. Their impact has significantly affected about 10 per cent of the total population of the developing world.

Issue 7 More rich but also more poor Following from Issue 6, the result of the failure even to narrow, let alone close, the gap in levels of consumption between the rich and poor countries and regions of the world is leading to a bizarre situation and trend. Figure 18.2, admittedly very approximate, and depending on definitions of poverty, nevertheless shows a paradox. For simplicity, two classes of people will be considered, poor and not poor. A small proportion of the population of the developed countries is poor, at least by standards in those countries, while a substantial proportion of the population of the developing countries is poor. Table 18.2 shows the percentage and total number of poor and not poor to total population in developed and developing countries. The author's estimate for 2025 is based on the assumption that the proportion of poor declines relatively between 1995 and 2025 roughly as it did between 1965 and 1995. The inescapable conclusion, if the estimates are reasonable, is that between 1965 and 1995 the proportion of poor in the developing countries declined from 75 to 69 per cent but the number rose from 1.8 to 3.0 billion. With a similar trend between 1995 and 2025 the number would rise from 3.0 to 4.4 billion. Figure 18.3 shows the situation diagrammatically.

b The implement he uses

PLATE 18.2 The ultimate sustainable development situation?

a A Machiguenga Indian in the interior of Peru prepares to fell a giant tree in the tropical rainforest

18.4 Organisational issues

Issue 8 Which are the emerging markets? The end of the Cold War has changed the world scene drastically. One result is the emergence in Central Europe and the former USSR of new 'semi-developed' countries. Another is the removal of a dampener on the globalisation of investment and the activities of transnational companies. Before the 1990s many developing countries, some influenced by the Soviet experience, had strong state sectors and protected industries. Russian influence is now minimal outside the former Soviet Union. Suddenly many developing countries have come to be seen as both safe and attractive places in which to invest.

Table 18.3 shows the assessment of one investment company, Guinness Flight, for its Global Emerging Markets Fund. It is claimed that a 'disciplined approach to asset location' has been used. Three types of country are identified: Approaching maturity, Industrialising and Embryonic. A little patronising, perhaps, but scores for the countries in the first two groups are calculated on the basis of fourteen equally weighted criteria, each ranging between 0 and 10. Half the attributes are related to medium-term influences on the stockmarket, half to fundamental strength of the economy. Attributes include, for example, currency risk, economic strength and infrastructure. For clarity and simplicity, scores out of a possible maximum of 140 (e.g. Malaysia scores 107) have been converted to percentages. India's low score is partly accounted for by its 0 on both state involvement (too much) and infrastructure (too little). Brazil and South Korea both score 0 on political uncertainty.

Other reports and commentators give different appraisals of the emerging countries, an indication of the sudden manner in which investment funds have had to get something together quickly to be ahead in the new economic landscape of the world. The implications for the developed countries of the 'emergence' of developing countries include loss of investment and jobs at home as companies desert their bases in search mainly of low labour costs. That is not a consideration likely to concern transnational companies, of which (see Chapter 4) the USA, Western Europe and Japan each have one-third of the fifty largest. Good news for the emerging countries, in which conditions are not suitable for the accumulation of large amounts of capital.

TABLE 18.2 Poor and not poor 1965, 1995, 2025

		Population					
		percentage			*total in billions*		
		1965	*1995*	*2025*	*1965*	*1995*	*2025*
Developed	not poor	80	83	86	0.8	1.0	1.2
	poor	20	17	14	0.2	0.2	0.2
Developing	not poor	25	31	37	0.6	1.4	2.6
	poor	75	69	63	1.8	3.0	4.4
World total		–	–	–	3.4	5.6	8.4

Issue 9 How many countries in the future? In spite of globalisation, with increasing freedom of movement of capital and the relaxing of tariff barriers through GATT negotiations, the sovereign states still exert great influence on the movement of both goods and people. Will the number of countries increase or decrease? How quickly will the boundaries of existing countries cease to be barriers in the emerging supranational units, notably the EU, which could be extended early in the next century to contain twenty or more? Will other countries break up as the USSR, Yugoslavia and Ethiopia have done in the 1990s?

Issue 10 What follows the end of the Cold War? The superpowers of the 1980s have largely given up opposing each other in the United Nations, using their positions on the Security Council to veto each other's proposals. Is the world heading for a stability engineered by the UN and maintained mainly by the efforts of the one remaining superpower, the USA? In the nineteenth century, much of the time Britain and France 'policed' the world. What incentives does the USA have to carry out such a role in the twenty-first century? There is no way that, even with UN backing, US resources would allow it to intervene in every conflict, broker peace between warring neighbours or impose democratic government everywhere. On the evidence of its performance in the late 1980s and early 1990s, the USA has been careful in selecting where to make its presence felt: Panama for the Canal, Haiti as a source of unwanted immigrants, Iraq on account of the potential threat to half of the world's oil reserves, are all of direct interest. Why Somalia but not Rwanda? What about human rights violations in Indonesia (East Timor) and of course China? Should Georgia (former USSR) remain a Russian preserve? Why should the USA support Muslims and Croats in former Yugoslavia against Bosnian Serbs? Why protect

TABLE 18.3 Guinness Flight assessment of the investment potential of emerging economies

Approaching maturity	*Score (%)*	*Industrialising*	*Score (%)*
Malaysia	76	Chile	71
Taiwan	72	Thailand	57
South Korea	61	Argentina	53
Mexico	60	Portugal	48
South Africa	55	Philippines	42
Israel	51	India	38
		Brazil	35
		Turkey	34
		Greece	31
		Indonesia	31

Embryonic
China (Shenzen), China (Shanghai), China (Hong Kong), Viet Nam, Bangladesh, Pakistan, Sri Lanka, Bolivia, Colombia, Ecuador, Peru, Venezuela, Czech Republic, Hungary, Poland, Egypt, Jordan, Morocco, Gabon, Ghana, Zambia, Zimbabwe
Source: Guinness Flight (1994)

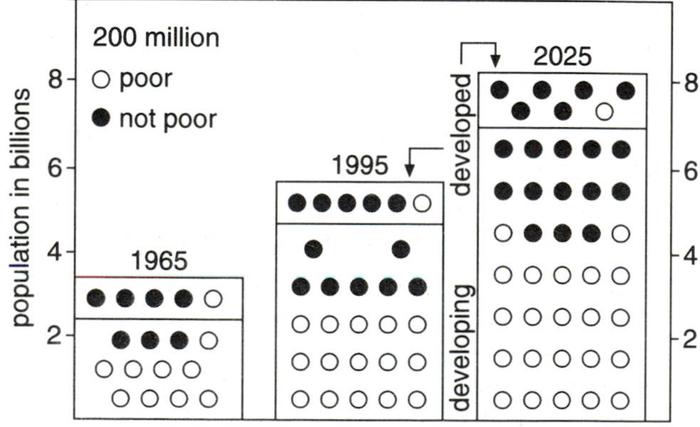

FIGURE 18.3 Absolute numbers and proportions of poor and not poor in developing and developed countries, 1965–2025

the Kurds in Iraq but not in Turkey? Could the concern over peace in the Near East and in Northern Ireland be aimed at parts of the US electorate?

18.5 The dark side of things

Issue 11 Post-colonial hangover The effects of colonial rule and domination have not yet been eliminated. It has been argued that when political control of colonies was relinquished, economic influence remained. France maintains an uneasy relationship with Algeria, partly at least on account of its interest in the oil and gas reserves. Portugal, first in and last out of Africa, left Angola and Mozambique open to become battlegrounds between Soviet-backed interests and South Africa. Russia itself transformed agriculture in its Central Asian republics to grow cotton. Shortly before its colonies in Africa were given independence, the UK was still organising migration schemes for British emigrants and producing primary products such as groundnuts for its own industries. The USA is not averse to using cheap labour in Mexico and the Caribbean.

Issue 12 Animal and (even) plant rights In general, neither religious institutions nor secular governments have cared about the place in the world of the other species of the biosphere. Concern over cruelty to animals is largely confined to private societies, with limited resources. Mostly the activities of animal protection groups consist of damage limitation, and are

BOX 18.1

Trading blocs

The early 1990s have been characterised by two opposing trends, the emergence of many new sovereign states, and the strengthening or creation of large trading blocs. GATT, the temporary structure established in 1948 to oversee world trade and tariffs, has been replaced in 1995 by the World Trade Organisation (WTO). In spite of the success of the nine-year GATT Uruguay Round of negotiations, which resulted in considerable reductions in tariffs and subsidies world-wide, several trading blocs remain, with policies aimed to protect in various ways the economies of member countries. The following trading blocs existed at the beginning of 1995:

- NAFTA, with the USA, Canada and Mexico, and Chile to be considered, in spite of its location in the southern part of *South* America. At a conference in Miami in 1994, attended by all the countries of Latin America except Cuba, as well as the USA and Canada, an all-American trading association was proposed.

- Meanwhile almost all the countries of Latin America belong either to LAIA (Latin American Integration Association), CACM (Central American Common Market) or CARICOM (Caribbean Community).

- EU and EFTA are discussed in detail in Chapter 5. From 1995 there will be fifteen EU countries, with a further twelve under consideration for eventual membership according to the agreement in the European Council reached in Essen at the end of the German Presidency of the EU in December 1994: the four Visegrad countries – Poland, the Czech Republic, the Slovak Republic and Hungary – and Romania, Bulgaria, Slovenia, Malta, Cyprus and the three Baltic republics.

- The CIS is a loose bloc consisting of eleven of the fifteen former Soviet Socialist Republics.

- Arab League: Arab countries of northern and eastern Africa, and Southwest Asia.

- Three groups in the rest of Africa together containing most of the countries: ECOWAS (the Economic Community of West Africa), IDEAC (the Central African Customs and Economic Union), and the South African Development Coordination Conference. The member countries of these African groupings mostly have very small markets and, since they tend to produce broadly similar primary products, do not actually trade much with one another.

- ASEAN (the Association of Southeast Asian Nations).

- Yet another international trading forum was taking shape in 1994. The APEC (Asia-Pacific Economic Co-operation) forum in November was attended by heads of state of eighteen countries, including President Clinton of the USA. This loose bloc contains developed countries (USA, Japan, Australia, New Zealand), newly industrialising ones (e.g. South Korea, Taiwan), and developing ones (e.g. China, Indonesia). The main points of accord include the achievement of free and open trade and investment in the Asia-Pacific region by 2020 or earlier.

Further reading

Evans, R. (1995), 'Brave new world order', *Geographical*, January, vol. 67, no. 1, pp. 39–42.

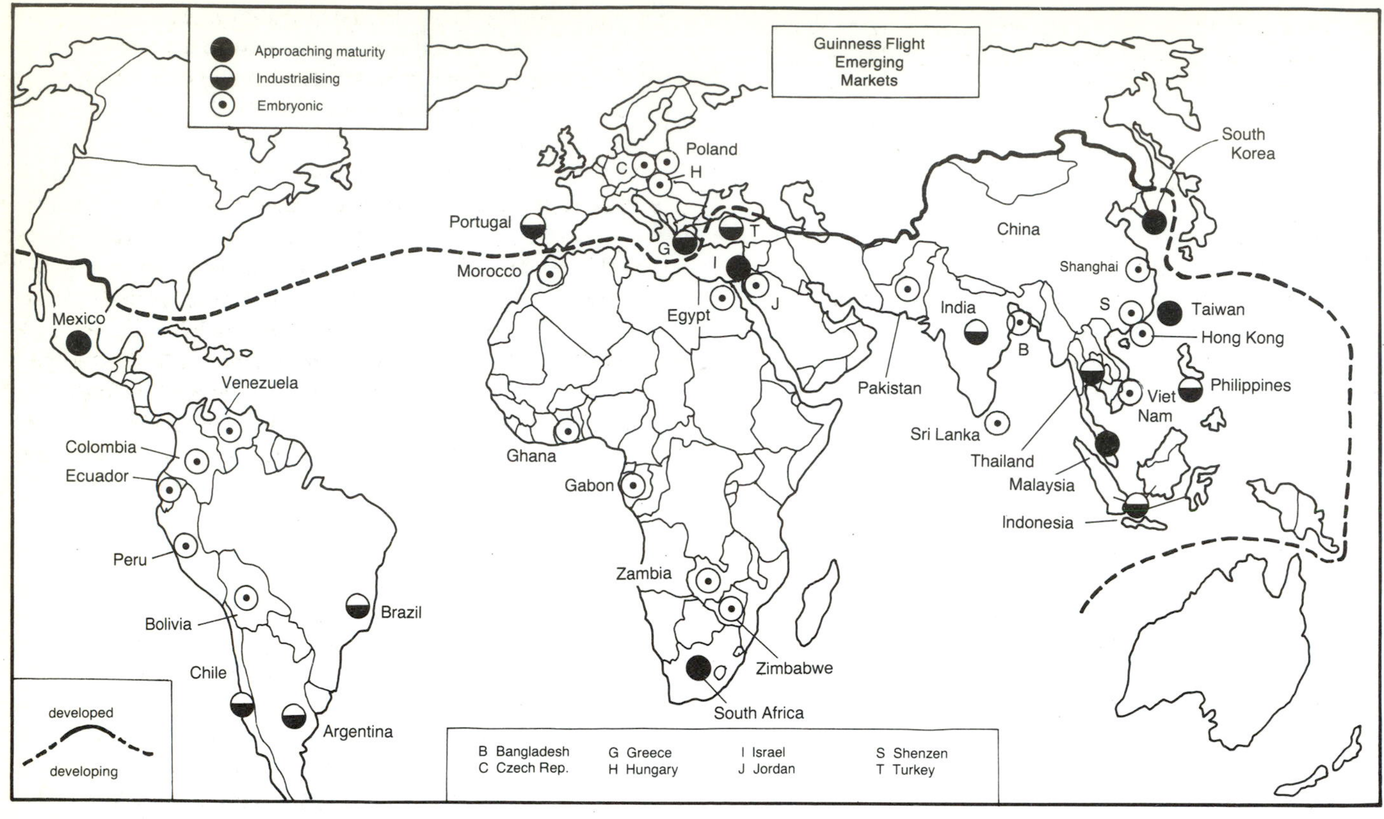

FIGURE 18.4 Guinness Flight emerging markets

extended only to a few of the more attractive species with anthropomorphic characteristics. To the question 'Do anthropoid apes have rights?' most people in the world would be indifferent or hostile. The treatment of domesticated and domestic animals does concern some people, but fish removed from the water are left to pass their dying moments without a thought. Presumably they conveniently do not have feelings.

Plants and animals are increasingly a matter of concern from another quarter, the minute portion of the world's population aware that human interference with the biosphere is causing the irretrievable loss of many species. The concern of biologists for the biosphere may save parts of the remaining areas of natural vegetation and animal habitat, even if these are just fragments of former forests, grasslands and mountain areas.

Issue 13 Environmental pollution This issue has gained prominence in the 1970s and is not likely to go away in the next century. Catastrophic events such as nuclear accidents and oil spills are very newsworthy and news of their occurrence is quickly diffused round the world. The burning of Kuwait's oil wells in 1991, with spectacular plumes of smoke, and a leaking oil pipeline in the northern part of European Russia, have been in and out of the news for long periods.

In the next century the distinction between the effects of pollution confined to a local area (e.g. the Bhopal chemical plant in India) or a regional context (e.g. the Chernobyl nuclear accident), as against pollution on a global scale, should be better understood. Emissions that accumulate in the atmosphere, groundwater and oceans are the concern of everyone. The prospect of climatic change has not gone away, even if some of the most drastic scenarios have been discredited. Private motoring could be the target for restrictions in the next century: higher fuel and road taxes, car sharing, battery-powered cars, restrictions on speed, on movement in city centres and rural roads.

Issue 14 Underclasses In a community of any size many subsets of people can be distinguished. Among others, three seem likely to continue to complicate life and cause distress and conflict in the next century: the issues of gender, race/ethnicity and unemployment.

In the twentieth century the 'gender gap' has narrowed in the developed countries as women have received the vote, increased their participation in higher education and moved into professional jobs. It is disturbing, therefore, to know that, according to Holloway (1994, p. 67), in developed as well as developing countries:

> Women around the world share a common health problem whether they are missing from US clinical drug trials or are nowhere to be seen in a malaria clinic in Thailand. Social customs, health policies and a largely male medical community have tended to treat women as wives or wombs and little else.

Funding and attention has been minimal for such aspects of women's health as nutrition, ageing, response to illnesses such as AIDS, and to diseases that only plague women. The well-being of women is threatened by domestic violence, the damaging effects of illegal abortion, sexually transmitted diseases, and genital mutilation.

> In Bombay one in four deaths of women between the ages of 15 and 24 is caused by 'accidental' burning (murdering a wife in order to get a second dowry from another marriage). In the USA between 22 and 35 per cent of visits to emergency rooms are for injuries caused by domestic violence.

Issue 15 Organised crime Geographers have studied the geography of many features and situations from river networks to central business districts of cities. Attempts to get to grips with the geography of crime have been sporadic, constrained understandably by lack of information and secrecy. The subject has not been covered in this book, but reference is made here to the less salubrious side of the geography of the world's major regions because transnational crime appears to have grown in recent decades and is likely to 'benefit' even more from the globalisation of the world economy and the ever-increasing network of airline services in particular. Endeau (1994) provides a thought-provoking checklist of fifteen major areas of transnational crime and the countries in which various types are most prominent, as well as major flows of various criminal activities.

- Drugs: Western Europe and the USA. Flow into these regions from Latin America, Africa and Asia
- Arms: Central and Southern Europe
- Immigrants: Latin America to the USA; Latin America, Africa and Asia to Western Europe
- Money laundering: venues used include Switzerland, Austria, Luxembourg, UK
- Cars: to Central Europe and the former USSR
- Nuclear material: former USSR, Poland, Germany
- Body parts: to Germany, Italy, Switzerland
- Prostitution: Poland, Russia, developing countries
- Baby trafficking: Latin America to USA and Western Europe, Africa and Asia to Western Europe
- Terrorism: UK, Italy, Turkey, North Africa
- *Objets d'art*: from former USSR
- Rare animals: tropical regions to Western Europe
- Kidnapping: Italy, Latin America
- Piracy: Southeast Asia, Malacca Straits
- Cattle rustling: Africa

Of the larger countries of the world, only the tightly sealed Japan largely escapes transnational crime. The gulags of the Soviet Union are being disbanded, but lack of information plus self-preservation ensured that western geographers specialising in the USSR avoided reference to the subject.

18.6 The bright side of things

Issue 16 The prospect of a nuclear war has receded Although the USA and Russia still have sufficient nuclear weapons to wipe out the human race several times over, they seem less prone now to threaten their use. Several other responsible countries have enough nuclear weapons to make a big dent in the human geography of the world. The problem now is to ensure that nuclear proliferation ceases and that there is no chance that nuclear weapons would even be used as a last resort in a regional or local conflict.

It would be simplistic, however, to expect that in the twenty-first century local conflicts both international and internal using conventional weapons will cease. Meanwhile the trade in arms flourishes as countries such as Saudi Arabia and Indonesia arm themselves to the teeth, requiring their military establishment to protect territory and natural resources.

Issue 17 Human rights Very few countries in the world can claim to have their hands entirely clean. Optimists see a gradual improvement in the situation in some countries. In particular, the ominous intensive surveillance of the activities of the ordinary citizens in the former USSR and Central Europe has abated if not ceased.

Issue 18 Progress in science and technology It is to be hoped that some research effort will be diverted from military to civilian areas. The argument that the spin-off from weapons research has benefited non-military endeavours and that the manufacture of arms creates jobs seems weak, given that arms are either not used at all or, if they are used, destroy existing constructions.

Issue 19 The search for Utopia The fifty years between 1865 and 1915 were characterised by considerable optimism in Europe and North America.

TABLE 18.4 The human development situation in 1990: the extremes and the middle according to the ranking of 173 countries

Rank	Country	(1) Life expectancy	(2) Educational attainment	(3) Real GDP per capita	(4) HDI
1	Japan	78.6	2.87	17,620	983
2	Canada	77.0	2.98	19,230	982
3	Norway	77.1	2.95	16,030	979
4	Switzerland	77.4	2.90	20,870	978
5	Sweden	77.4	2.90	17,010	977
6	USA	75.9	3.00	21,450	976
7	Australia	76.5	2.94	16,050	972
8	France	76.4	2.94	17,410	971
9	Netherlands	77.2	2.86	15,700	970
10	UK	75.7	2.94	15,800	964
83	Kyrgyzstan	68.8	2.25	3,110	689
84	Saudi Arabia	64.5	1.39	10,990	688
85	South Africa	61.7	1.59	4,870	673
86	Sri Lanka	70.9	2.29	2,410	663
87	Libya	61.8	1.40	7,000	658
88	Tadjikistan	69.6	2.25	2,560	657
89	Ecuador	66.0	2.12	3,070	646
90	Paraguay	67.1	2.17	2,790	641
91	North Korea	70.4	2.38	2,000	640
164	Guinea-Bissau	42.5	0.47	840	90
165	Chad	46.5	0.30	560	88
166	Somalia	46.1	0.16	840	87
167	Gambia	44.0	0.26	910	86
168	Mali	45.0	0.36	570	82
169	Niger	45.5	0.25	650	80
170	Burkina Faso	48.2	0.00	620	74
171	Afghanistan	42.5	0.33	710	66
172	Sierra Leone	42.0	0.13	1,070	65
173	Guinea	43.5	0.20	500	45

Description of variables:
(1) Life expectancy at birth in years, 1990: lowest Sierra Leone, highest Japan
(2) Educational attainment: lowest Burkina Faso, highest USA
(3) Real GDP in US dollars per capita (ppp dollars), 1990: lowest Zaire (370), highest USA
(4) Combination of (1)–(3), maximum possible 1,000, minimum possible 0.0
Source: UN, *Human Development Report 1993*

Science and technology were leaping ahead, the world economy was growing. If the influence of Christianity was diminishing in its European heartland, it was expanding over Africa and was strong in the Americas. Darwinian ideas of evolution were erroneously applied to human society. Humans were rapidly evolving into more rational and caring creatures. Karl Marx's version of the road to an ideal society led ultimately to a world of uniformly satisfied, orderly people in a state of communism. The First World War shattered many dreams. Is 1995 a return to 1865? Of the major world powers, only the USA and China seem to be concerned about what happens in the future beyond their own boundaries. The USA hopes for a free world with universal democracy and a market economy. China continues to profess adherence to socialism. The great disparities in human development displayed in Table 18.4 show that for many countries the road to Utopia – always assuming Utopia is to be found in Japan or Canada or Scandinavia – is a very long one.

Issue 20 Planning for the future In 1945 the victors of the Second World War looked ahead to peace and prosperity for all. The last fifty years have been disappointing, to say the least. The developed countries (including the USSR) have experienced very few conflicts, but the poorer countries have been the scene of devastating conflicts (e.g. Viet Nam, the Iran–Iraq War, Angola, Afghanistan), often serving among other things as battlegrounds for the superpowers to test out their latest weaponry. Can the experiences of the last fifty years throw light on the prospects for the next fifty? Many of the readers of this book will still be alive in 2045. Why not compile a map of the world for that year showing the main features you might expect to find and the major changes from 1995 to 2045? It could be illuminating to consult an elderly relative or friend who was old enough in 1945 to remember what the world was like then and to find out how their hopes and expectations for the future differed from the state of the world now. Shall we drift on from crisis to crisis or will it be possible to determine a more positive and constructive route through the twenty-first century?

Glossary

Bioclimatic resources: the water, soil and climatic elements that support plant and animal life, as opposed to mineral resources, which include fossil fuels and metallic and other raw materials.

Biodiversity: the variety of organisms considered at all levels, especially with regard to the communities that exist in given habitats, the physical conditions under which they live, and their interdependence. The biodiversity of a given habitat may be threatened by human interference.

Birthrate: the number of births occurring in a given region (usually administrative) expressed in per thousand of the total population of the region in a given period (usually a year). More specifically, the number of births may be related to the number of members (usually only female) in a given age group.

Central Europe: a geographical term loosely applied to the area between Germany, the Baltic coast, Russia and the southern Balkans. Between 1945 and the 1990s the tendency was to subdivide Europe only into Western and Eastern blocs. Central is back in fashion and roughly includes countries not in the EU or the former USSR.

Cold War: the ideological, cultural, political and economic conflict between western (capitalist) and eastern (communist) countries, the main protagonists being the USA and USSR. It was at its most intense in Europe but was waged globally, in spite of the existence of a Third World of unaligned countries. Officially ended by Gorbachev in 1986.

Country: a general term popularly used to refer to a sovereign (independent) state, a colony, even a subdivision within a sovereign state (e.g. Scotland in the UK, Tibet in China). Not synonymous with nation since many countries in the world have more than one nationally conscious group (e.g. former Yugoslavia).

Deathrate: the number of deaths occurring in a given region (usually administrative), expressed in per thousand of the total population of the region in a given period (usually a year).

Demographic transition model (see Box 3.1, p. 52): an empirically based model in which, during an unspecified period of time, the relationship between birthrate and deathrate changes, usually producing a period in which birthrate exceeds deathrate and population therefore grows.

Development gap: a term sometimes used to refer to the difference, usually in economic or material terms, between the richest and poorest regions (especially countries) of the world.

Discovery: finding something hitherto unknown. It is therefore patronising and inappropriate for the journeys of exploration of Europeans since the fifteenth century to lands already inhabited and therefore known to be referred to as voyages of discovery. A distinction should also be made between discoveries (e.g. how the solar system 'works') and inventions (e.g. the steam engine).

Ecosystem: the organisms living in a particular environment and their interrelationships, and the physical part of the environment that impinges on them.

Emerging nations: a term little used before the 1990s, now a fashionable way of referring to developing countries.

Epidemiological transition model (see also demographic transition model): the ETM focuses on the causes of death, with contagious and parasitic afflictions keeping deathrate high in the early ages but (with improved healthcare) degenerative afflictions gaining in prominence as causes of death, as life expectancy increases.

Fossil fuels: coal, oil (petroleum), natural gas, including different types of coal such as anthracite and lignite, oil in shales and tar sands.

Globalisation: a term fashionable especially in the 1980s and 1990s, referring particularly to the tendency for each transnational company to operate in many countries and even world-wide. Communications have facilitated and accompanied globalisation. Arguably the earliest global enterprise was the activity in the tropical world of the crowns of Spain and Portugal during 1580 and 1640 when they were united.

'Greenhouse' effect: a process perceived by many scientists, resulting from the increasing amount of carbon dioxide and other substances in the atmosphere, caused particularly by the burning of fossil fuels and forests. The sun's rays are 'trapped' in the atmosphere, causing warming. Expectations of what could happen from the process vary among those who assume it is actually happening.

Gross Domestic Product (GDP): the total value of goods and services produced in a given period (usually a year). The total amount earned domestically. Real GDP refers to the GDP of a given country after adjustments have been made to allow for distortions in exchange rates (usually in relation to the US dollar).

Gross National Product (GNP): has a slightly different meaning since it includes income from abroad but excludes imports and property income paid abroad. For the purposes of this book, the difference is not marked enough for it to matter which particular index is being used. Some publications (e.g. the World Bank) use GNP, some (e.g. UN *Human Development Report*) use GDP.

Iberia(n): Spain and Portugal. Also Ibero-America, Latin America except for some small countries (Haiti, Jamaica, Guyanas).

Ice cap/ice sheet: the two terms are largely interchangeable but the former term is generally used for Greenland, the latter for Antarctica. Ice sheets may cover sea areas as well as land. Ice cover on individual mountain ranges is generally referred to as glacier(s).

Industry: loosely used term, at one extreme covering virtually all economic activities, at the other referring specifically to processing and manufacturing. Generally the 'middle' of three broad categories of economic activity: agriculture (with forestry and fishing), industry (including mining) and services ('non-goods'). 'Post-industrial' and 'de-industrialise' are recent concepts.

Least Developed Countries: a United Nations term, determined by the General Assembly (resolution 46/206). See Figure 4.2.

Maghreb: Morocco, Algeria and Tunisia.

Merchandise trade: trade in goods as opposed to 'invisible' transactions in the service sector.

Nation: see 'Country'.

Natural increase of population: the amount by which a population increases (or increases negatively, i.e. decreases) in a given period through a difference between the number of births and deaths. The gain or loss is often expressed as a percentage of the initial total – e.g. the population of a given region is 100,000 in mid-1993, 102,000 in mid-1994, a natural increase of 2.0 per cent, there having been no immigration or emigration.

Natural resources: see Box 3.4 (pp. 62–3).

Needs and wants: assessing the needs of the poor in various regions and countries of the world is a major industry. Even absolutely basic needs differ from one region to another (e.g. clothing in the tropics and the Arctic). Most people want more than they need, some far more. In modern industrial societies, things that were at one time luxuries or non-existent, such as the car, have become necessities in new life styles.

Nordic countries: Denmark, Norway, Sweden, Finland and Iceland.

Oil: a mineral fossil fuel, commonly referred to, especially in the USA, as petroleum, or a vegetable product derived from plants bearing oil-seeds.

Overpopulation: see Section 3.4, (p. 55).

Per capita, per caput, per head, per inhabitant, per person: all having virtually identical meaning. Per caput is strictly more correct than per capita, the Latin singular and plural (nominative) of head respectively.

Political correctness: a term widely applied to the use of words or phrases that may be offensive to particular individuals or groups of people. See Box 1.2 (p. 16).

Pre-Columbian: the (history of the) world before the voyages of Columbus and other European explorers led to a global view of the world rather than an appreciation by different populations of particular parts of it.

Primary products: particularly products of cultivation, grazing, forestry and mining, often forming the ingredients for subsequent processing and/or manufacturing.

Procrustean: physical or mental actions or arguments aiming or tending to produce uniformity by violent and arbitrary methods, after a legendary robber of Attica who made his victims fit the size of his bed by stretching or mutilation. (Who can say he or she has never used procrustean methods!)

Purchasing power parity (ppp): one of the terms used by economists to refer to estimates of the real purchasing power of national currencies after the possible distorting influence of exchange rates has been removed. Equivalents are usually in US dollars.

Reductionism: the tendency to reduce a complex situation to one that focuses on a particular aspect, given undue prominence.

Sahel: loosely used to include various parts of northern Africa, recently applying particularly to the countries on the southern fringes of the Sahara Desert.

Scandinavia: Norway, Sweden and Denmark (but not Finland).

Slavery: see Box 2.1 (pp. 36–7).

Sovereign state: a state (or country) that is independent. Almost all the sovereign states are represented in the United Nations, but each is entirely responsible for its own internal affairs.

State: see 'Country'. The term 'state' is used to describe the major civil or administrative divisions of many sovereign countries, especially federal countries (e.g. USA, Brazil).

Sustainable development: a situation in which some kind of economic change is occurring, particularly with growth in material production and consumption, but without immediate or long-term damage to the environment and loss of natural resources. Commonly used with reference to the possible improvement in living conditions of the poor in resource-poor regions of the developing world. Common sense alone suggests that extractive activities that use up non-renewable natural resources are not sustainable in the long term.

System: a widely used term referring to a situation in which a number of distinct elements influence each other (e.g. a central heating system in a building). Implicit is the concept that the whole (system) is greater than the sum of its parts. To be used and interpreted with caution.

Total fertility rate: the average number of children each woman in a given region bears during her (childbearing) lifetime. This statistic can only be calculated for women who have passed their childbearing period and its application to the estimation of the number of future births must be based on the assumption that there is no change in the observed rate.

Tribe: see Box 12.1 (pp. 284–5).

List of illustrations

Plates

Figures

Tables

Further reading

1 Introduction

De Blij, H. J. and Muller, P. O. (1992), *Geography: regions and concepts*, New York: John Wiley. Broadly similar to the present book, with excellent descriptive material of the world's regions, fine illustrations and useful back-up sections.

Garver, J. B. (1988), 'New perspective on the world', *National Geographic*, vol. 174, no. 6, pp. 910–13, a simple account of some of the problems of representing the whole world on a flat map.

Goudie, A. (1993), *The Nature of the Environment*, Oxford: Basil Blackwell. Various types of physical environment are examined and the processes at work are described with the help of regional examples.

Groombridge, B. (ed.) (1992), *Global Diversity, Status of the Earth's Living Resources*, London: Chapman & Hall, World Conservation Monitoring Unit. Biogeographical background with insights into ecological problems and the need for conservation.

Haggett, P. (various editions), *Geography: a modern synthesis.* A global approach to the subject, emphasising the contribution of models, theories and techniques to an understanding of the geography of the world.

Harvey, D. (1984), 'On the history and present condition of geography: an historical materialist manifesto', *The Professional Geographer*, February, vol. 36, no. 1, pp. 1–11. The Marxist view of the world influenced Soviet geographers, and became fashionable among western geographers in the 1970s and 1980s. The collapse of communism in the USSR has dulled the attraction of this narrow approach.

Kates, R. W. (1994), 'Sustaining life on the earth', *Scientific American*, October, vol. 271, no. 4, pp. 92–9. A useful overview of issues affecting human activities and the prospects for a sustainable future.

Mather, P. M. (ed.) (1993), *Geographical Information Handling – Research and Applications*, Chichester: John Wiley. For reference on this important new development in geography, but not easy reading.

Newson, M. (1994), *Hydrology and the River Environment*, Oxford: Clarendon Press. An examination of the physical processes determining the behaviour of drainage systems and the impact on human activities.

Pattison, W. D. (1964), 'The four traditions of geography', *Journal of Geography*, vol. 63. One of many attempts to present a simple version of the nature of geography.

Union of Concerned Scientists (1993), *World Scientists' Warning Briefing Book*, Cambridge, Mass.: Union of Concerned Scientists. A forthright declaration warning of the impending disasters ahead for humanity if pressure on natural resources and the environment continues to grow.

Wheeler, J. H., Kostblade, J. T. and Thoman, R. S. (various editions), *Regional Geography of the World*, New York: Holt, Rinehart & Winston. Global coverage of the world regions, with useful detail on regional and local situations.

2 A brief history of the world since 1500

Barraclough, G. (ed.) (1979), *The Times Atlas of World History*, London: Times Books, especially Sections 5, The world of the emerging West; 6, The age of European dominance; and 7, The age of global civilisation, covering the voyages of exploration and the colonial conquests of the European powers and the emergence of the present global political system.

Ferguson, R. B. (1992), 'Tribal warfare', *Scientific American*, January, vol. 266, no. 1, pp. 90–5. Summary of the influence of colonial powers on tribal communities.

Hart, M. (1993), *The 100, a Ranking of the Most Influential Persons in History*, London: Simon & Schuster. Thought-provoking if subjective presentation of the evidence that individuals may have shaped the course of history as much as underlying structural situations, contradictions and conflicts.

Rostow, W. W. (1975), *How It All Began*, London: Methuen. A thoughtful account of the crucial technological and economic developments, mainly in Europe and the USA, that led up to the present world situation, in which the central economic task will be the need to achieve a dynamically stable balance between man and his physical environment.

3 Population and resources

United Nations, *Statistical Yearbook, Demographic Yearbook.*

Food and Agriculture Organization, *Production Yearbook.*

British Petroleum (BP), *Statistical Review of World Energy* (annual report).

US Bureau of Mines, *Mineral Facts and Problems.*

McEvedy, C. and Jones, R. (1985 reprint), *Atlas of World Population History*, Harmondsworth: Penguin Books.

Auty, R. (1994), *Patterns of Development: resource endowment, development policy and economic growth*, London: Edward Arnold. A generous natural resource endowment does not necessarily ensure favourable economic performance. The author examines rural neglect, income inequality, hyperurbanisation, unequal terms of trade and the role of government.

Bloom, D. E. and Brender, A. (1993), 'Labour and the emerging world economy', *Population Bulletin*, October, vol. 48, no. 2 (Population Reference Bureau). Focuses on the subject of the changing employment structure and job creation in the world.

Dasgupta, P. S. (1995) 'Population, poverty and the local environment', *Scientific American*, February, vol. 272, no. 2, pp. 26–31, on the differing roles of men and women in the decisions affecting economic activities at local level.

Grigg, D. (1993), *The World Food Problem*, Oxford: Basil Blackwell. The problem of hunger is discussed and the possibilities of raising food production considered. The author's theme in this and other publications is that there is no problem of production, only a problem of distribution.

Hubbard, H. M. (1991) 'The real cost of energy', *Scientific American*, April, vol. 264, no. 4, pp. 18–23.

Johnson, S. P. (1994), *World Population – Turning the Tide. Three decades of progress*, London: Graham & Trotman. The author is explicit in his view that reducing the rate of population growth is a positive achievement. An exhaustive and exhausting inventory of policies and practices throughout the world.

Knox, P. and Agnew, J. (various editions or reprints), *The Geography of the World Economy*, London: Edward Arnold. Focuses mainly on economic geography and has a global approach without a regional breakdown.

Larson, E. E., Ross, M. S. and Williams, R. H. (1986), 'Beyond the era of materials', *Scientific*

American, June, vol. 254, no. 6, pp. 24–31. On the increasingly efficient way in which raw materials and sources of energy have been used in developed countries since the early decades of the Industrial Revolution and the increasing possibilities of substitution of one material by another. Refer also to Clark, J. P and Flemings, M. C. (1986), 'Advanced materials and the economy', *Scientific American*, October, vol. 255, no. 4, pp. 42–9.

Lutz, W. (1994), 'The future of world population', *Population Bulletin*, June, vol. 49, no. 1 (Population Reference Bureau).

Parry, M. (1990), *Climatic Change and World Agriculture*, London: Earthscan Publications. Speculates about the possible impact resulting from global warming and other changes on agriculture in major regions of the world.

Rockett, I. R. (1994) 'Population and health. An introduction to epidemiology', *Population Bulletin*, November, vol. 49, no. 3 (Population Reference Bureau).

Wallace, I. (1992), *The Global Economic System*, London: Routledge. The world economy is treated as a whole, with a historical background and the problems facing both capitalist and state-socialist economies.

4 Global contrasts

Beaumont, P. (1993), *Drylands, Environmental Management and Development*, London: Routledge.

Collins, M. (ed.) (1990), *The Last Rain Forests*, Oxford: Mitchell Beazley.

Cooke, R. U. and Doornkamp, J. C. (1990 2nd edition), *Geomorphology in Environmental Management*, Oxford: Clarendon Press. A thorough examination of the impact of changes in the physical landscape on human activities and the impact of these activities on the geomorphology of different types of natural environment.

Dasgupta, P. A. (1995), 'Population, poverty and the local environment', *Scientific American*, February, vol. 272, no. 2, pp. 26–31. The importance of understanding local economic activities and the decisions affecting them. The differing roles of men and women need to be taken into account.

Goudie, A. (1983), *Environmental Change*, Oxford: Clarendon Press. An examination of environmental changes in the last 3 million years, with emphasis on how human activities have been influenced during that period.

——— (1986), *The Human Impact on the Natural Environment*, Oxford: Basil Blackwell. A study of the human impact on the biosphere, soil, hydrosphere and atmosphere.

Grainger, A. (1993), 'Rates of deforestation in the humid tropics: estimates and measurements', *Geographical Journal*, March, vol. 159, no. 1, pp. 33–44. The difficulties of obtaining accurate data for the deforestation rates of the tropical moist forest. Remote sensing measurements need to be more extensive and regular.

Grove, R. H. (1992), 'Origins of western environmentalism', *Scientific American*, July, vol. 267, no. 1, pp. 22–7. Shows that the European colonial powers of the eighteenth century were already concerned about the threatened ecology of tropical islands.

Harris, N. (1986), *The End of the Third World*, London: I. B. Taurus. The great contrasts in the developing world or Third World make it oversimplified now to talk of a set of similar countries.

Johnson, D. L. and Lewis, L. A. (1995), *Land Degradation: creation and destruction*, Oxford: Basil Blackwell.

National Geographic (1988), a volume devoted to the theme 'Can man save this fragile earth?', December, vol. 174, no. 6. Contains regional examples of conservation problems and an overview of population, plenty and poverty.

Park, C. (1992), *Tropical Rainforests*, London: Routledge.

Repetto, R. (1992), 'Accounting for environmental assets', *Scientific American*, June, vol. 266, no. 6, pp. 64–71. 'When governments calculating their economic performance fail to account for the depreciation of forests, fisheries, minerals or water caused by development, the balance sheets often show growth and prosperity. In reality, the result is usually impoverishment.'

Roberts, N. (1993), *The Changing Global Environment*, Oxford: Blackwell. Covers the main elements of the natural environment, with numerous case studies at local, regional and world scales.

Thrift, N. (1986), 'The geography of international economic disorder', in *A World in Crisis?*, Oxford: Basil Blackwell. A critique and criticism of the disorderly and unjust way in which modern capitalism affects the world. No clear alternative is offered by Thrift.

Todd, E. (1987), *The Causes of Progress*, Oxford: Basil Blackwell, esp. pp. 160–2. A refreshing anthropological approach to development with a tendency for aspects of family structure and intrafamily relationships to be given excessive prominence in influencing the progress.

5 The European Union and EFTA

Blacksell, M. and Williams, A. M. (eds) (1994), *The European Challenge*, Oxford: Oxford University Press. Geography and development in the EC (the former GDR not fully included). The first 87 pages are 'lost', consisting of summaries in five out of the eight non-English EU languages.

Clout, H. *et al.* (1994), *Western Europe: geographical perspectives*, Harlow: Longman. Covers all of Western Europe, not exclusively the EU.

Cole, J. P. and Cole, F. J. (1993), *The Geography of the European Community*, London: Routledge. Deals with the subject topic by topic, emphasising different aspects of EU policy. Former GDR included and EFTA referred to.

Eurostat (1992), *Europe in Figures*, Luxembourg: Office for Official Publications of the European Communities. Glossy publication with a wealth of statistical data for the twelve EU Member States.

Masser, I., Sviden, O. and Wegener, M. (1992), *The Geography of Europe's Futures*, London: Belhaven Press. Thought-provoking but repetitive, concentrating heavily on transportation and communications.

Minshull, G. N. (1990 edn), *The New Europe into the 1990s*, London: Hodder & Stoughton. Good studies of local and regional problems in Western Europe. The former GDR not integrated into the EU.

Williams, A. M. (1994, 2nd edn), *The European Community – The Contradictions of Integration*, Oxford: Basil Blackwell. Focuses on problems and negative aspects of the EU. The former GDR not integrated.

6 Countries of Western and Central Europe

In the absence of a definitive book on the geography of Central Europe for the 'post-Soviet' years of the early 1990s, the *National Geographic* is referred to here because its editors have kept track of the major changes in the region.

Abercrombie, T. J. (1993), 'Czechoslovakia: the velvet divorce', *National Geographic*, September, vol. 184, no. 3, pp. 2–37. Emphasis on the split into the Czech and Slovak Republics.

Doder, D. (1992), 'Albania opens the door', *National Geographic*, July, vol. 182, no. 1, pp. 66–94. 'Europe's poorest country has emerged from nearly five decades of forced isolation under a repressive dictatorship. Now it must struggle to catch up.'

Jones, A. (1994), *The New Germany*, Chichester: Wiley. Treats Germany as a whole, using post-unification material with a spatial breakdown of data below the level of the Länder.

Mountjoy, A. B. (1973), *The Mezzogiorno*, London: Oxford University Press. A detailed account of the first two decades of the work of the Cassa per il Mezzogiorno and other related endeavours in the development of southern Italy.

Szulc, T. (1991), 'Dispatches from Eastern Europe', *National Geographic*, March, vol. 179, no. 3, pp. 2–33. Conveys the feeling of the momentous experience of removing Soviet domination.

Thompson, J. (1991), 'East Europe's dark dawn', *National Geographic*, June, vol. 179, no. 6, pp. 36–69. Dramatic, well-illustrated account of the effects of neglect of the problem of environmental pollution.

Vesilind, P. J. (1990), 'The Baltic nations', *National Geographic*, November, vol. 178, no. 5, pp. 2–37. Charts the effects of fifty years of Soviet domination and the prospects, written when independence was still uncertain.

7 The former USSR

Several books have been published by US and British geographers in the 1980s and should be widely available. The reading material listed here focuses on the 'post-Soviet' generation of material, mainly journal papers, published since 1991, in which the new situation is taken into account. See also sources at the end of Box 7.6 on the Aral Sea.

Bater, J. H., Amelin, V. C. and Degtyarev, A. A. (1994), 'Moscow in the 1990s: market reform and the central city', *Post-Soviet Geography*, May, vol. 35, no. 5, pp. 247–66. Shows how westernisation is proceeding fast in Moscow and gives insights into people's lives.

Belt, D. (1992), 'The world's Great Lake' (Baykal), *National Geographic*, June, vol. 181, no. 6, pp. 2–39. Ecologically and environmentally a key element in the acceptance by the Soviet government that pollution and conservation are matters to be addressed, not swept under the carpet.

Bradshaw, M. J. (ed.) (1991), *The Soviet Union, a New Regional Geography?* London: Belhaven Press. Published too soon to do more than describe the great changes in the Soviet Union *before* its break-up, but the most up-to-date of the 'old' generation of books on the geography of the USSR.

Cole, J. P. and Filatotchev, I. V. (1992), 'Some observations on migration within and from the former USSR in the 1990s', *Post-Soviet Geography*, September, vol. 33, no. 7, pp. 432–53.

Cybriwsky, R. (1992), 'The Ukrainian springtime', *Focus*, vol. 42, no. 2, pp. 31–3. A brief study of the economic and social problems facing the new nation of the Ukraine.

Dunlop, J. B. (1994), 'Will the Russians return from the Near Abroad?' *Post-Soviet Geography*, April, vol. 35, no. 4, pp. 204–15. Russian elites in the Russian Federation are keen to reassemble the former Soviet Union. Expatriate Russians may be encouraged to stay in the other republics rather than participate in an 'ingathering' of Russians and Slavs.

Edwards, M. (1993), 'A broken empire – Russia, Kazakhstan, Ukraine', *National Geographic*, March, vol. 183, no. 3, pp. 2–53. Post-Soviet issues and problems in arguably the three most prominent successor states of the former USSR.

—— (1994), 'Soviet pollution', and 'Chernobyl', *National Geographic*, August, vol. 186, no. 2, pp. 70–99 and 99–115. Excellent illustrations and up-to-date text.

Frazer, G. and Lancelle, G. (1994), *Zhirinovsky – the Little Black Book: making sense of the senseless*, Harmondsworth: Penguin Books. Speeches and off-the-cuff comments of Vladimir Zhirinovsky classified and analysed.

Harris, C. D. (1993), 'The new Russian minorities', *Post-Soviet Geography*, January, vol. 34, no. 1, pp. 1–28. Thorough coverage of the topic.

—— (1994), 'Ethnic tensions in the successor republics in 1993 and early 1994', *Post-Soviet Geography*, April, vol. 35, no. 4, pp. 185–203. Focuses on the current and likely future problems of Russians outside the Russian Federation.

Haub, C. (1994), 'Population change in the former Soviet republics', *Population Bulletin*, December, vol. 49, no. 4 (Population Reference Bureau).

Jensen, R. G., Shabad, T. and Wright, A. W. (1983), *Soviet Natural Resources in the World Economy*, Chicago: University of Chicago Press. Monumental work consisting of an inventory of resources and background material on their size, distribution and exploitation. Not liable to get out of date quickly.

Pryde, P. R. (1994), 'Observations on the mapping of critical environmental zones in the former Soviet Union', *Post-Soviet Geography*, January, vol. 35, no. 1, pp. 38–49. Back to basic geography of the country.

Sagers, M. (1993), 'The energy industries of the former USSR', *Post-Soviet Geography*, June, vol. 34, no. 6. Whole volume devoted to oil, natural gas, coal, electric power. Great detail.

8 Japan and the Korean Republic

Dahlby, T. (1994), 'Kyushu', *National Geographic*, January, vol. 185, no. 1, pp. 88–117. The southernmost island of Japan, with a backwater image, has recently been gaining a share of Japan's research laboratories and high-tech factories.

Dobbs-Higginson, M. S. (1994), *Asia Pacific: its role in the new world disorder*, London: Heinemann. On the complexity of East Asia, including Japan, and its prospects in the world economy.

Smith, P. (1994), 'Inner Japan', *National Geographic*, September, vol. 186, no. 3, pp. 65–95.

A description of conditions in a part of Japan remote in its life style if not in distance from the big cities.

Witherick, M. and Carr, M. (1993), *The Changing Face of Japan*, Sevenoaks: Hodder & Stoughton.

9 The USA

Archer, J. C. *et al.* (1988), 'The geography of U.S. presidential elections', *Scientific American*, July, vol. 259, no. 1, pp. 18–25. Evidence that for a century voting has been related to lasting geographic divisions. Voting results mapped at county level for various elections.

Birdsall, S. S. and Florin, J. W. (1992), *Regional Landscapes of the United States and Canada*, Chichester: Wiley.

Brown, L. J. (1987), 'Hunger in the U.S.', *Scientific American*, February, vol. 256, no. 2, pp. 21–5. Hunger was virtually eliminated in the 1970s but returned in the 1980s because of federal cutbacks. About 8 per cent of the US population may be affected, deprivation in the richest country in the world.

Holloway, M. (1994), 'Trends in biological restoration: nurturing nature', *Scientific American*, April, vol. 270, no. 4, pp. 76–84. Describes an attempt to restore an environment damaged by human activity. Experience for future attempts elsewhere.

Knox, P. L. *et al.* (1988, 1990), *The United States: a contemporary human geography*, New York: Wiley; Harlow: Longman. Team effort covering population, culture and environment, political and economic organisation, rural and metropolitan features, and social problems. Tends to take a negative view of the capitalist system without offering clear alternatives. Prolific number of references – about 600.

Landau, R. (1988), 'U.S. economic growth', *Scientific American*, June, vol. 258, no. 6, pp. 26–35. It is argued that to revive economic growth and compete in the world the USA must adopt the aggressive exploitation of technology and ensure more intense capital investment in the manufacturing sector.

Mairson, A. (1994), 'Great Flood of '93', *National Geographic*, January, vol. 185, no. 1, pp. 42–81. Account of the 'in excess of a 100-year flood' afflicting all or parts of nine US states in 1993, killing about fifty people, causing more than 10 billion dollars' worth of damage and raising questions of flood control.

Martin, P. and Midgley, E. (1994), 'Immigration to the United States: journey to an unknown destination', *Population Bulletin*, September, vol. 49, no. 2. Thorough review of migration into the USA since 1820: identifying four 'waves'. A prognostication of only 52 per cent White non-Hispanic by 2050 if 900,000 immigrants come per year (Population Reference Bureau).

National Geographic Special Edition (1993), *Water: the power, promise and turmoil of North America's fresh water*, vol. 184, no. 5A.

Rubinstein, J. (1988), 'The changing distribution of US automobile assembly plant', *Focus*, vol. 38, no. 3, pp. 12–17. The author discusses the changing site factors of land, capital, and labour costs to explain new plant location.

Szekely, J. (1987), 'Can advanced technology save the U.S. steel industry?' *Scientific American*, July, vol. 257, no. 1, pp. 24–31. In the face of stiff foreign competition, entirely new processing techniques are needed for the long term.

10 Canada and Oceania

Arden, H. (1991), 'Journey into Dreamtime', *National Geographic*, January, vol. 179, no. 1, pp. 2–41. Lavish coverage of Australia's most empty region, the northwestern area of the state of Western Australia.

Chalkley, B. and Winchester, H. (1991), 'Australia in transition', *Geography*, April, vol. 76, no. 2, pp. 97–108.

Hamley, W. (1993), 'Problems and challenges in Canada's Northwest Territories', *Geography*, July, vol. 78, no. 3, pp. 267–80.

Heathcote, R. L. (1994), *Australia*, Harlow: Longman Scientific, Halsted Press. Two centuries of impact on the environment, covering recent developments in conservation, multiculturalism and industrial expansion as well as the future sustainability of agriculture.

Manners, G. (1992), 'Unresolved conflicts in Australian mineral and energy resource policies', *Geographical Journal*, July, vol. 158, no. 2, pp. 129–44.

Pearce, H. (1991), 'The flooding of a nation', *Geographical*, November, vol. 63, no. 11, pp. 18–21. The implications of the James Bay hydroelectric power project for the local population and the environment.

Pelly, D. (1994), 'Birth of an Inuit nation', *Geographical*, April, vol. 66, no. 4, pp. 23–5.

11 Latin America

Colinvaux, P. A. (1989), 'The past and future Amazon', *Scientific American*, vol. 260, no. 5, pp. 68–75. The species diversity of the Amazon basin's rainforest has been favoured by frequent changes in climate and physical structure. Can its resistance to human exploitation be ensured?

Ellis, W. S. (1988), 'Rondonia's settlers invade Brazil's imperiled rain forest', *National Geographic*, December, vol. 174, no. 6, pp. 772–99, the fragmentation of the rainforest of western Brazil by immigrants from other states, and in the same issue L. McIntyre, 'Last days of Eden', pp. 800–17, on the changing life style of the Urueu-Wau-Wau Indians in the same area.

Furley, P. A. (1993), *The Forest Frontier*, London: Routledge. Settlement and change in the state of Roraima, on the northern side of Brazil's Amazonia.

Gilbert, A. (1990), *Latin America*, London: Routledge. Readable, focuses on economic and social aspects.

Green, D. (1991), 'Columbus, commodities and cocaine', *Geographical*, December, vol. 63, no. 12, pp. 14–17 and in the same issue, M. Murray, 'Natural forgiveness', pp. 18–22. Latin America suffers the insecurities of commodity trading. The economic and environmental potential for swapping debt for nature.

Gwynne, R. N. (1985), *Industrialization in Latin America*, London: Croom Helm. Concentrates on the case of Chile.

Idyll, C. P. (1973), 'The anchovy crisis', *Scientific American*, June, vol. 228, no. 6, pp. 22–9. In 1970–2 the excessively large Peruvian fishing catch, mainly of anchovies, fell catastrophically to a third. Natural conditions and human excesses both contributed. The question of sustainability clearly illustrated.

MacEwen Scott, A. (1994), *Divisions and Solidarities: gender, class and employment in Latin America*, London: Routledge. An examination of class analysis and the interrelationship of gender and class which creates a shared interest between men and women in some contexts and a divergence of interest in others.

Repetto, R. (1990), 'Deforestation in the tropics', *Scientific American*, April, vol. 262, no. 4, pp. 18–24. The experience of Costa Rica is a case in point: economic growth achieved without accounting for the depreciation of natural resources.

12 Africa south of the Sahara

Agnew, C. (1990) 'Green belt around Sahara', *Geographical*, April, vol. 62, no. 4, pp. 26–30. Trees to improve soil, water and botanical resources rather than to attempt to hold back the sand.

Anderson, R. M. and May, R. M. (1992), 'Understanding the AIDS pandemic', *Scientific American*, May, vol. 266, no. 5, pp. 20–7. Includes maps of Africa showing the distribution of HIV occurrence.

Bell, M. (1986, 1991), *Contemporary Africa*, Harlow: Longman. Good coverage of economic and social problems. Thin on maps.

Binns, T. (1993), *Tropical Africa*, London: Routledge. Introduction to development issues in Africa.

Caldwell, J. C. and Caldwell, P. (1990), 'High fertility in sub-Saharan Africa', *Scientific American*, May, vol. 262, no. 5, pp. 82–9. Religious and social beliefs promote large families. Improved healthcare could mitigate the fear of dying without descendants.

Caputo, R. (1991), 'Zaire River', *National Geographic*, November, vol. 180, no. 5, pp. 5–35. Vivid description of a journey along one of Africa's largest rivers.

Chapman, G. P. and Baker, K. M. (1992), *The Changing Geography of Africa and the Middle East*, London: Routledge.

Fardon, R. and Furniss, G. (eds) (1993), *African Languages. Development and the state*, London: Routledge. The authors argue the importance of multilinguism in African society as a positive aspect of development.

Glantz, M. H. (1987), 'Drought in Africa', *Scientific American*, June, vol. 256, no. 6, pp. 34–40. Attention has focused on physical aspects of drought. The author argues that better insights will come from greater attention to social, economic and cultural factors as well.

Griffiths, I. L. L. (1990), *An Atlas of African Affairs*, London: Routledge.

Grove, A. T. (1989, 1993), *The Changing Geography of Africa*, Oxford: Oxford University Press.

Luling, V. (1989), 'Wiping out a way of life', *Geographical Magazine*, July, vol. 61, no. 7, pp. 34–7. Vivid account of the attempt to collectivise and resettle the agricultural population of Ethiopia under the ruthless President Mengistu, now no longer in power.

Mazrui, A. A. (1986), *The Africans – a Triple Heritage*, London: BBC Publications. Described as 'Africa's most independent thinker', Mazrui was born in Kenya in 1933 but has studied and carried out research in the USA and the UK as well as Africa. A critique of events in the continent since European involvement started.

Onimode, B. (1992), *A Future for Africa – beyond the Politics of Adjustment*, London: Earthscan Publications.

Palmer, C. (1992), 'The cruelest commerce', *National Geographic*, September, vol. 182, no. 3, pp. 63–91. Frank account of the basic features of the slave trade that flourished for three and a half centuries between the Atlantic coast of Africa and the Americas.

13 North Africa and Southwest Asia

Allen, T. B. (1994) 'Turkey struggles for balance', *National Geographic*, May, vol. 185, no. 5, pp. 2–36. Tradition and attempts to modernise result in uncertainty.

Beaumont, P., Blake, G. H. and Wagstaff, J. M. (1988), *The Middle East: a geographical study*, Halsted Press (Wiley).

Canby, T. Y. (1991), 'After the storm', *National Geographic*, August, vol. 180, no. 2, pp. 2–35. Dramatic pictures of the fires started by Iraqi forces before they left Kuwait, losing a huge quantity of oil and polluting the western side and shores of the Persian Gulf. Update, S. A. Earle (1992), *National Geographic*, February, vol. 181, no. 2, pp. 122–34: 'Assessing the damage one year later'.

Findlay, A. M. (1994), *The Arab World*, London: Routledge. Development issues in the Arab world illustrated by case studies from Saudi Arabia, Yemen, Morocco and Jordan.

Hitchens, C. (1992), 'Struggle of the Kurds', *National Geographic*, August, vol. 182, no. 2, pp. 32–62. In 1987–8 Iraqis attacked Kurdish villages with cyanide and mustard gas. One of many cases of persecution of the Kurdish people.

Hubbard, H. M. (1991), 'The real cost of energy', *Scientific American*, April, vol. 264, no. 4, pp. 18–23. Gas/petrol prices artificially low. Cost to environment, health, military expenditure to protect Gulf oil-fields not taken into account.

Mackenzie, R. (1993), 'Afghanistan's uneasy peace', *National Geographic*, October, vol. 184, no. 4, pp. 58–9. Depressing picture of Afghanistan, showing efforts to rebuild the country after a decade of war, in which 1 million Afghan lives were lost.

Omran, A. R. and Rondi, F. (1993), 'The Middle East population puzzle', *Population Bulletin*, July, vol. 48, no. 1 (Population Reference Bureau).

Robinson, F. (1982), *Atlas of the Islamic World since 1500*, Oxford: Phaidon. Definitive, beautifully illustrated account of the history of Islam, the background to the religion, and associated cultural features.

Vesilink, P. J. (1993), 'Middle East water – critical resource', *National Geographic*, May, vol. 183, no. 5, pp. 38–72. Ataturk Reservoir on the Euphrates. Irrigation. Salinisation and desalinisation.

14 South Asia

Bray, F. (1994), 'Agriculture for developing nations', *Scientific American*, July, vol. 271, no. 1, pp. 18–25. The case against a blanket application of modern western agricultural methods in developing countries, especially those in South, Southeast and East Asia. US and Japanese agriculture compared.

Chapman, G. P. and Baker, K. M. (eds) (1992), *The Changing Geography of Asia*, London: Routledge. Covers India, China, Southeast Asia, Japan and Siberia but not Southwest Asia. A democratic system does not necessarily bring development and progress.

Cobb, C. E. (1993) 'Bangladesh: when the water comes', *National Geographic*, June, vol. 183, no. 6, pp. 118–34. Vivid account of the impact on rural life when paddy fields are submerged by monsoon floods.

Evans, R. (1991), 'India after the Gandhis', *Geographical*, September, vol. 63, no. 9, pp. 28–32. Political and religious problems and conflicts.

Farmer, B. H. (1994), *An Introduction to South Asia*, London: Routledge. The geographical and historical background to the diversities of South Asia, examining how environments, sociology, politics and economics explain development patterns and problems which characterise the region today.

Hobson, C. (1993), 'The sorry side of Shangri-la', *Geographical Magazine*, January, vol. 65, no. 1, pp. 10–14. No Buddhist utopia. The largest ethnic group has been persecuted by the government.

Johnson, B. L. C. (1979), *India, Resources and Development*, London: Heinemann. Very readable account of India in the 1970s, well illustrated, by an author clearly steeped in the ways of Indian life.

——— (1981), *South Asia*, London: Heinemann.

Raju, S. and Bagchi, D. (eds) (1994), *Women and Work in South Asia. Regional patterns and perspectives*, London: Routledge. The contribution of women to the economy in South Asia is underestimated in this region.

Shrivastava, P. (1992), *Bhopal – Anatomy of a Crisis*, London: Paul Chapman. Account of the worst industrial accident in history.

Ward, G. C. (1992), 'India's wildlife dilemma', *National Geographic*, May, vol. 181, no. 5, pp. 2–29. Increasing numbers of poor, land-hungry farmers erode the limited remaining wildlife areas.

15 Southeast Asia

Dixon, C. (1991), *South East Asia in the World Economy, a Regional Geography*, Cambridge: Cambridge University Press. One cannot study Southeast Asia or its members in isolation. They must be seen as interacting parts of the world economy.

Drakakis-Smith, D. (1992), *Pacific Asia*, London: Routledge. Limited material on Japan and China.

Dwyer, D. J. (ed.) (1990, 1992 reprint), *Southeast Asian Development*, Harlow: Longman.

Jomo, K. S. (1993), *Industrializing Malaysia, Policy, Performance, Prospects*, London: Routledge.

Lynch, B. (1977), *Indonesia, Problems and Prospects*, Malvern Victoria: Sorrett Publishing. Remarkable insights into life in the rural areas of the country, showing the impact of 'too many' people living off too little land. Memorable book, brilliantly illustrated with diagrams and pictures. Many questions raised.

Mahathir Bin Mahamad (1994), 'Malaysia 2020', *Scientific American*, April, vol. 270, no. 4, Supplement: Malaysia Advertising Section. Lavish presentation of developments in Malaysia, preparing for the goal of becoming a 'fully developed nation' by 2020.

Rigg, J. (1990), *South-East Asia, a Region in Transition*, London: Routledge.

16 China and its neighbours

Beijing Review, weekly: subscribe to China International Book Trading Corporation, PO Box 399, Beijing, China.

China Statistical Yearbook, annual: subscribe to 342 Hennessy Road, 10–16 Fl., Hong Kong.

Chowdhury, A. and Islam, I. (1993), *The Newly Industrializing Economies of East Asia*, London: Routledge. An examination of market economy and state influences on development.

Court, S. (1993), 'Where opposites attract', *Geographical Magazine*, January, vol. 65, no. 1, pp. 24–7. International free trade area to cost 20 billion US dollars under study near Tumen River mouth, involving China, the two Koreas, Russia and Japan.

Dwyer, D. J. (ed.) (1994), *China: the next decades*, Harlow: Longman. Includes a coverage of the economic organisation, rural and urban features and problems, and environmental problems.

Ellis, W. S. (1994), 'Shanghai', *National Geographic*, March, vol. 185, no. 3, pp. 2–35. Changing conditions in China's largest concentration of industry, likely to become a major financial centre of East Asia.

Leeming, F. (1993), *The Changing Geography of China*, Oxford: Basil Blackwell, IBG Studies in Geography.

Money, D. C. (1990 revised edn), *China: the land and the people*, London: Evans Brothers. Pleasingly illustrated, simple account of the geography of China, emphasising the relationship between people and the natural environment.

Qu Geping and Li Jinchang (1994), *Population and the Environment in China*, London: Paul Chapman. Ways are proposed to improve the relationship between human population and the environment.

Smil, V. (1985), 'China's food', *Scientific American*, December, vol. 253, no. 6, pp. 104–12. Can China continue to feed its 1 'billion' people adequately? Illuminating insights into what the Chinese eat. By 1995 there were 150 million more mouths to feed!

Tien, H. Yuan *et al.* (1992), 'China's demographic dilemmas', *Population Bulletin* of the Population Reference Bureau, June, vol. 47, no. 1. China's fertility realities are at variance with the country's population planning goals. Low fertility would give 1.5 billion around 2050, constant fertility 2.1 billion.

Walsh, J. (1993), 'China – the world's next superpower', *Time*, 10 May, vol. 141, no. 19, pp. 36–77. Welcome overview of a range of features and issues covering personalities, the economy and culture. Parts likely to get out of date quickly.

Zich, A. (1993), 'Taiwan', *National Geographic*, November, vol. 184, no. 5, pp. 2–33. Chinese nationalists made an impoverished island off the China coast a bastion against mainland communism. Taiwan has since become an economic powerhouse – and now a democracy.

17 Regional prospects

Kennedy, P. (1993), *Preparing for the Twenty-First Century*, London: Harper Collins.

Mitchison, A. (1993), 'Will we survive?' *Scientific American*, September, vol. 269, no. 3, pp. 102–8.

Olshansky, S. J. *et al.* (1993), 'The aging of the human species', *Scientific American*, April, vol. 268, no. 4, pp. 18–24.

Rhoades, R. E. (1991), 'The world's food supply at risk', *National Geographic*, April, vol. 179, no. 4, pp. 74–105.

Robey, B. *et al.* (1993), 'The fertility decline in developing countries', *Scientific American*, December, vol. 269, no. 6, pp. 30–7. Economic improvement is not a necessary precondition of falling birth rates. Access to contraception and changes in cultural values and education have caused fertility to decrease.

Scientific American (1989), Special issue: 'Managing planet earth', September, vol. 261, no. 3. Emphasis on atmospheric and biological aspects, population growth, agriculture, energy and manufacturing and the strategies for sustainable economic development.

Thrift, N. (1986), 'The geography of international economic disorder', in *A World in Crisis?*, Oxford: Blackwell. A critique and criticism of the disorderly and unjust way in which modern capitalism affects the world. No clear alternative is offered by Thrift.

Todd, E. (1987), *The Causes of Progress*, Oxford: Basil Blackwell, esp. pp. 160–2. A refreshing anthropological approach to development with a tendency for aspects of family structure and intra-family relationships to be given excessive prominence in influencing progress.

18 Twenty-first-century earth

Brown, L. A. (ed.) (1994), *State of the World 1994*, London: Earthscan. A World Watch Institute report on progress towards a sustainable society.

Lea, J. (1988), *Tourism and Development in the Third World*, London: Routledge.

References

Adams, W. M. (1993), 'Indigenous use of wetlands and sustainable development in West Africa', *Geographical Journal*, July, vol. 159, no. 2, pp. 209–18.

Allcock, J. (1990), 'Death by misadventure', *Times Higher Educational Supplement*, 14 December, p. 15.

Amalrik, A. (1970), *Will the Soviet Union Survive until 1984?* New York: Harper & Row.

Anderson, I. (1992), 'Can Nauru clean up after the colonialists?' *New Scientist*, 18 July, no. 1830, pp. 12–13.

Anderson, R. M. and May, R. M. (1992), 'Understanding the AIDS pandemic', *Scientific American*, May, vol. 266, no. 5, pp. 20–7.

Atlas narodov mira (Atlas of the peoples of the world) (1964), Moscow: Glavnoye Upravleniye Geodezii i Kartografii.

Bailey, H. M. and Nasatir, A. P. (1960), *Latin America*, London: Constable.

Barney, G. O. (ed.) (1982), *The Global 2000 Report to the President*, Harmondsworth: Penguin Books.

Beard, H. and Cerf, C. (1992), *The Official Politically Correct Dictionary and Handbook*, London: Grafton.

Beeston, R. (1995), 'Zhirinovsky the author predicts return to gulag', *The Times*, 3 March, p. 14.

Bennett, N. (1994), 'London beats New York in funds battle', *The Times*, 15 August.

Bigland, J. (1810), *A Geographical and Historical View of the World, Vol. III*, London: Longman.

Bingham, A. (1992), 'Muddy waters', *Geographical Magazine*, August, vol. 64, no. 8, pp. 21–5.

Binyon, M. (1993), 'Territories in the pink cling to rewards of dependence', *The Times*, 27 November, p. 12.

Boeing (1994), *Company Profile*, Seattle: The Boeing Company.

Bongaarts, J. (1994), 'Can the growing human population feed itself?', *Scientific American*, March, vol. 270, no. 3, pp. 18–24.

Boyes, R. (1994). 'Vienna stumbles in waltz towards EU membership', *The Times*, 15 February, p. 11.

BP (1993), *BP Statistical Review of World Energy June 1993*, annual publication of reserves, production, consumption and trade in main fossil fuels.

Branson, C. (1994), 'The importance of being well connected', *The European*, 21–7 October, p. 25.

Brodie, I. (1994), 'Foreign firms find nothing finer than to be in Carolina', *The Times*, 14 January, p. 25.

Bryson, B. (1992), 'Main–Danube canal links Europe's waterways,' *National Geographic*, August, vol. 182, no. 2, pp. 3–31.

Bureau of Mines (1985), *Mineral Facts and Problems*, Bulletin 675, United States Department of the Interior, Washington: US Government Printing Office.

Burton, T. (1994), 'The great white hope of the Red Indians', *Daily Mail*, 3 September, p. 3.

Butler, C. (1994), 'Ferries chart new course for survival', *The European*, 1–7 July, p. 19.

Caldwell, J. C. and Caldwell, P. (1990), 'High fertility in sub-Saharan Africa', *Scientific American*, May, vol. 262, no. 5, pp. 82–9.

Calendario Atlante de Agostini, annual yearbook, Novara: Istituto Geografico De Agostini.

Catton, C. (1992), *Tears of the Dragon: China's environmental crisis*, London: Channel 4 Television.

Cavalli-Sforza, L. L. (1991), 'Genes, peoples and languages', *Scientific American*, November, vol. 265, no. 5, pp. 72–9.

Chalkley, B. and Winchester, H. (1991), 'Australia in transition', *Geography*, April, vol. 76, no. 2, pp. 97–108.

Chapman, G. (1992), 'News of the world', *Geographical Magazine*, October, vol. 64, no. 10, pp. 15–19.

China Statistical Yearbook 1991, State Statistical Bureau of the PRC, Beijing: China Statistical Information and Consultancy Center, available in English since the early 1980s.

Clark, D. (1984), *Post-industrial America*, New York: Methuen.

Clark, J. P. and Flemings, M. C. (1986) 'Advanced materials and the economy', *Scientific American*, October, vol. 255, no. 4, pp. 42–9.

Clayton, G. (1994), 'Electronics guard the Amazon', *Sunday Times*, 31 July, p. 3.10.

Cliff, A. D. and Smallman-Raynor, M. R. (1992), 'The AIDS pandemic: global geographical patterns and local spatial processes', *Geographical Journal*, July, vol. 158, no. 2, pp. 182–98.

Cole, J. P. (1959), *Geography of World Affairs*, Harmondsworth: Penguin Books.

——(1979), *Geography of World Affairs* (5th edn), Harmondsworth: Pelican.

——(1981), *The Development Gap*, Chichester: Wiley.

——(1983), *Geography of World Affairs* (6th edn), London: Butterworths.

——(1991), 'The USSR in the 1990s: which republic?', Working Paper 9, Department of Geography, University of Nottingham.

Cole, J. P. and Cole, F. J. (1993), *The Geography of the European Community*, London: Routledge.

Cole, J. P. and Filatotchev, I. V. (1992), 'Some observations on migration within and from the former USSR in the 1990s', *Post-Soviet Geography*, September, vol. 33, no. 7, pp. 432–53.

Commission of the European Communities (1993), *Community Initiatives: objectives and regions covered by structural assistance (1989–93)*, Luxembourg: Office for Official Publications of the European Communities.

Cornish, E. (1986), 'Colossal cities of the future', *The Futurist*, September/October, pp. 2, 59.

Corves, C. and Bax, S. (1992), 'Amazon skywatch', *Geographical Magazine*, November, vol. 64, no. 11, pp. 33–8.

Couch, I. (1993), 'An urbane solution', *Geographical*, April, vol. 65, no. 4, pp. 20–3.

Coulmas, F. (ed.) (1991), *A Language Policy for the European Community, Prospects and Quandaries*, Berlin–New York: Mouton de Gruyter.

De Blij, H. J. and Muller, P. O. (1992), *Geography: regions and concepts*, New York: John Wiley.

De Bres, K. and Buizlo, M. (1992), 'A daring proposal for dealing with an inevitable diaster? A review of the Buffalo Commons proposal', *Great Plains Research*, vol. 2, no. 2, pp. 165–78.

Degg, M. (1992), 'Natural disasters: recent trends and future prospects', *Geography*, July, vol. 77, no. 3, pp. 198–209.

Desmond, E. W. (1994), 'Beyond the trade game', *Time*, 21 February, pp. 28–31.

Diamond, J. (1995), 'No Incas in Hastings', *The Times Higher Educational Supplement*, 24 February, pp. 15–16.

Dowden, R. (1994), 'The new Africa', *The Independent*, 13 March, pp. 40–2.

Dunlop, J. B. (1994), 'Will the Russians return from the near abroad?' *Post-Soviet Geography*, April, vol. 35, no. 4, pp. 204–15.

Economic Commission for Africa (1963), *Industrial Growth in Africa*, New York: United Nations.

'Energy for Planet Earth' (1960), *Scientific American*, Special Issue, September, vol. 263, no. 3, whole issue.

Energy Statistics Yearbook of the United Nations, New York: UN.

Erickson, J. and Dilks, D. (1994), *Barbarossa, the Axis and the Allies*, Edinburgh: Edinburgh University Press.

European Union (1993), 'Regional profiles', EU duplicated document XVI, A.3.

Evans, M. (1994), 'Barbarossa toll rises to 49 million lives', *The Times*, 16 June, p. 11.

Evans, R. (1993), 'Reforming the Union', *Geographical*, February, vol. 65, no. 2, pp. 24–7.

Extel Financial (1993), *The Times 1000 1994*, London: Times Books. Annual data on the capitalisation and sales of major companies throughout the world, with greater detail for Europe and the UK.

FAOPY – Food and Agricultural Organisation Production Yearbook, Rome: FAO, with data on land use, employment, production and inputs, including data for forestry and fishing.

FAOTY – Food and Agriculture Organisation Trade Yearbook, Rome: FAO.

Fearnside, P. M. (1993), 'Deforestation in Brazilian Amazonia: the effect of population and land tenure', *Ambio*, December, vol. 22, no. 8, pp. 537–45.

Ferguson, R. B. (1992), 'Tribal warfare', *Scientific American*, January, vol. 266, no. 1, pp. 90–5.

Fisher, J. C. (1966), *Yugoslavia – A Multinational State*, San Francisco: Chandler Publishing Co.

Fleming, I. (1957), *The Diamond Smugglers*, London: Jonathan Cape.

Flower, A. R. (1978), 'World oil production', *Scientific American*, March, vol. 238, no. 3, p. 42.

Forrester, J. W. (1971), *World Dynamics*, Cambridge, MA: Wright-Allen Press, Inc.

Frazer, G. and Lancelle, G. (1994), *Zhirinovsky – the Little Black Book: making sense of the senseless*, Harmondsworth: Penguin Books.

Freedman, P. (1993), 'What makes countries count?', *Geographical Magazine*, March, vol. 65, no. 3, pp. 10–13.

Geograficheskiy atlas, dlya uchiteley sredney shkoly (Geographical atlas for middle school teachers) (1982), Moscow: Glavnoye Upravleniye Geodezii i Kartografii pri Sovete Ministrov SSR.

Glantz, M. H. (1987), 'Drought in Africa', *Scientific American*, June, vol. 256, no. 6, pp. 34–40.

Goode's World Atlas (1987), ed. E. B. Espenshade and J. L. Morrison, Chicago: Rand McNally & Co.

Goskomstat Rossii (1992a), *Narodnoye khozyaystvo Rossiyiskoy Federatsii 1992*, Moscow: Russian State Statistical Centre.

———(1992b), *Promyshlennost Rossiyiskoy Federatsii 1992*, pp. 35–8.

Goskomstat SSSR (1990), *Demograficheskiy yezhegodnik SSSR 1990*, Moscow: 'Finansy i statistika'.

———(1991), *Narodnoye khozyaystvo SSSR v 1990 godu*, Moscow: State Statistical Centre for the USSR.

Gottmann, J. (1951). 'Geography and international relations', *World Politics*, vol. 3, no. 2, pp. 153–73.

Grainger, A. (1993), 'Rates of deforestation in the human tropics: estimates and measurements', *Geographical Journal*, March, vol. 159, part 1, pp. 33–44.

Greig, G. (1994), 'From Motown to no town', *Sunday Times*, 20 March, p. 10.2.

Grigg, D. (1992), 'World agriculture: production and productivity in the late 1980s', *Geography*, April, vol. 77, no. 2, pp. 97–108.

———(1993), 'International variations in food consumption in the 1980s', *Geography*, July, vol. 78, no. 3, pp. 251–66.

Groombridge, B. (ed.) (1992), *Global Biodiversity*, London: Chapman & Hall, World Conservation Monitoring Unit.

Grove, R. H. (1992), 'Origins of Western environmentalism', *Scientific American*, July, vol. 267, no. 1, pp. 22–7.

Guinness Flight (1994), *Global Emerging Markets Fund*, Guernsey: Guinness Flight House.

Gunnemark, E. and Kenrick, D. (1985), *A Geolinguistic Handbook*, Goterna: Kungalv.

Ham, P. (1995), 'Swim or sink on the dollar wave', *The Sunday Times*, 12 February, p. 5.3.

Hamley, W. (1993), 'Problems and challenges in Canada's Northwest Territories', *Geography*, July, vol. 78, no. 3, pp. 267–80.

Hanks, J. (1994), 'Just for risk-takers', *The Times*, 14 May, p. 35.

Hanson 1994 (1995), London: Hanson plc.

Harris, C. D. (1993), 'The new Russian minorities: a statistical overview', *Post-Soviet Geography*, January, vol. 34, no. 1, pp. 1–28.

Hart, M. H. (1993), *The 100, a Ranking of the Most Influential Persons in History*, London: Simon & Schuster.

Hauser, P. M. (1971), 'The census of 1970', *Scientific American*, July, vol. 225, no. 1, pp. 17–25.

Hawkes, N. (1993a), 'The grain of a good idea', *Geographical*, August, vol. 65, no. 8, pp. 28–31.

——(1993b), 'London ranked as a world leader in science', *The Times*, 13 February, p. 6.

HDR – Human Development Report, published annually since 1990 for the United Nations Development Programme by Oxford University Press.

Helgadottir, B. (1994), 'Diminutive giant of Japan', *The European*, 4–10 February, p. 10.

Hemming, J. and Johnson, S. (1992), 'Reactions to Rio (Earth Summit)', *Geographical Magazine*, September, vol. 64, no. 9, pp. 23–9.

Hicks, N. (1993), 'A tale of two Chinas', *Geographical Magazine*, January, vol. 65, no. 1, pp. 28–32.

Hitchens, C. (1992), 'Struggle of the Kurds', *National Geographic*, August, vol. 182, no. 2, pp. 32–61.

Hoggart, K. (1991), 'The changing world of corporate control centres', *Geography*, April, vol. 76, no. 2, pp. 109–20.

Holloway, M. (1993), 'Sustaining the Amazon', *Scientific American*, July, vol. 269, no. 1, pp. 76–84.

——(1994), 'Trends in women's health: a global view', *Scientific American*, August, vol. 271, no. 2, pp. 66–73.

Homer Dixon, T. F. *et al.* (1993), 'Environmental change and violent conflict', *Scientific American*, February, vol. 268, no. 2, pp. 16–23.

Howard, K. (1995), 'City in the fast lane' in 'South Korea', a 12-page supplement in *The Times*, 8 March.

Hussey, A. (1991), 'Regional development and cooperation through ASEAN', *Geographical Review*, January, vol. 81, no. 1, pp. 87–98.

——(1993), 'Rapid industrialization in Thailand 1986–1991', *Geographical Review*, January, vol. 83, no. 1, pp. 14–28.

IBGE – Instituto Brasileiro de Geografia e Estatística (1992), *Anuario estatístico do Brasil 1992*, Rio de Janeiro: IBGE.

Idyll, C. P. (1973), 'The anchovy crisis', *Scientific American*, June, vol. 228, no. 6, pp. 22–9.

International Monetary Fund (IMF) (1992), *Direction of Trade Statistics 1992 Yearbook*, Washington, DC.

James, P. E. (1950), *Latin America*, London: Cassell, citation p. 4.

Jing Wei (1993), 'Revisiting Huaxi village', *Beijing Review*, 10–16 May, pp. 13–18.

Johnson, B. L. C. (1979), *India, Resources and Development*, London: Heinemann.

Jones, P. D. and Wigley, T. M. L. (1990), 'Global warming trends', *Scientific American*, August, vol. 263, no. 2, pp. 66–73.

Kahn, H., Brown, W. and Martel, L. (1977), *The Next 200 Years*, London: Associated Business Programmes.

Kahn, H. and Wiener, A. J. (1967), *The Year 2000: a framework for speculation on the next thirty-three years*, New York: Macmillan.

Kalish, S. (1994), 'International migration: new findings on magnitude, importance', *Population Today*, March, vol. 22, no. 3, pp. 1–2 (Population Reference Bureau).

Keating, M. (1993), 'Exploding the trade in mines', *Geographical*, August, vol. 55, no. 8, pp. 24–7.

Keir, M. (1992), 'The strange survival of nationalism (in Europe)', *Geographical Magazine*, July, vol. 64, no. 7, pp. 25–9.

Kiley, S. (1993), 'Eritrea's litany of African failures stuns leaders', *The Times*, 29 June, p. 15.

King-Hall, S. and Dewar, A. (eds) (1952), *History in Hansard*, London: Constable, citation p. 68.

Krushelnycky, A. (1995), 'Two worlds collide in Copenhagen', *The European*, 10–16 March, p. 4.

Langer, W. L. (1972), 'Checks on population growth: 1750–1850', *Scientific American*, vol. 226, no. 2, pp. 92–9.

Larson, E. E., Ross, M. S. and Williams, R. H. (1986), 'Beyond the era of materials', *Scientific American*, June, vol. 254, no. 6, pp. 24–31.

League of Nations (1933), *Statistical Yearbook of the League of Nations 1932/33*, Geneva: Economic Intelligence Service.

Lewis, D. (1992), 'Beating a path to wildlife survival', *Geographical Magazine*, November, vol. 64, no. 11, pp. 10–14.

Li Bin (1993), 'Land, an ignored limited resource', *Beijing Review*, 15–21 March, pp. 17–18.

Li Ning (1993), 'China's expanding highway network', *Beijing Review*, 4–10 October, pp. 11–17.

Lin, H. (1985), 'The development of software for ballistic-missile defense', *Scientific American*, December, vol. 253, no. 6, pp. 32–9.

Liu Zheng, Song Jian *et al.* (1981), *China's Population: problems and prospects*, Beijing: New World Press.

Lloyd-Roberts, S. (1994), 'Killed for their kidneys', *The Times*, 26 October, p. 16.

Lorenz, A. and Randall, J. (1994), 'Ford prepares for global revolution', *Sunday Times*, 27 March, p. 11.

Lynch, B. (1977), *Indonesia, Problems and Prospects*, Malvern, Victoria: Sorrett Publishing.

McEvedy, C. and Jones, R. (1985), *Atlas of World Population History*, Harmondsworth: Penguin Books.

Mackinder, H. J. (1904), 'The geographical pivot of history', *Geographical Journal*, April, vol. 23, no. 4, pp. 421–2.

Mahathir Bin Mahamad (1994), 'Malaysia 2020', *Scientific American*, April, vol. 270, no. 4, Supplement.

Malthus, T. R. (1970), *An Essay on the Principle of Population*, Harmondsworth: Penguin Books. The essay was first published in 1798.

'Managing Planet Earth' (1989), *Scientific American*, Special Issue, September, vol. 261, no. 3, whole issue.

Manners, G. (1992), 'Unresolved conflicts in Australian mineral and energy resource policies', *Geographical Journal*, July, vol. 158, part 2, pp. 129–44.

Mather, M. (1992), 'Polishing the pearl (Uganda)', *Geographical Magazine*, October, vol. 64, no. 10, pp. 20–5.

May, R. M. (1992), 'How many species inhabit the earth?', *Scientific American*, October, vol. 267, no. 4, pp. 18–25.

Mayes, G. M. *et al.* (1992), *Child Sexual Abuse*, Edinburgh: Scottish Academic Press.

Mazrui, A. A. (1986), *The Africans – a Triple Heritage*, London: BBC Publications.

Meadows, D. H. *et al.* (1972), *The Limits to Growth*, London: Earth Island Ltd.

Meadows, D. H., Meadows, D. L. and Randers, J. (1992), *Beyond the Limits*, London: Earthscan Publications.

Mesarovic, M. and Pestel, E. (1975), *Mankind at the Turning Point*, London: Hutchinson.

Metzger, B. M. (1965), *The Oxford Annotated Apocrypha*, New York: Oxford University Press.

Micklin, P. P. (1992), 'The Aral crisis: introduction to the Special Issue', *Post-Soviet Geography*, May, vol. 33, no. 5, pp. 269–82 (citation p. 274).

Miller, R. (1994), 'Take the shock out of stocks', *The Times*, 14 May, p. 35.

Miller, S. K. (1992), 'Mines with a sting in their tailings', *New Scientist*, 18 July, no. 1830.

Mirsky, I. (1994), 'China takes a swing at golf courses', *The Times*, 14 January.

Mitchell, B. R. (ed.) (1980), *European Historical Statistics 1750–1975*, London: Macmillan Reference Books. Also International Historical Statistics:

——(1982), *Africa and Asia* (to 1975).

——(1983), *The Americas and Australasia.*

——(1993), *The Americas 1750–1988.*

Mitchell, G. D. (1968), *A Dictionary of Sociology*, London: Routledge & Kegan Paul.

Mitchison, A. (1993), 'Will we survive?', *Scientific American*, September, vol. 269, no. 3, pp. 102–9.

Morgan, W. B. (1992), 'Economic reform, the free market and agriculture in Poland', *Geographical Journal*, July, vol. 158, part 2, pp. 145–56.

Mounfield, P. R. (1991), *World Nuclear Power*, London: Routledge.

Munchau, W. (1991), 'Economics of a civil war', *The Times*, 7 August.

Murdock, G. P. (1959), *Africa, its Peoples and their Culture History*, New York: McGraw Hill.

Narborough, C. (1994), 'Eastern promise of 6,000 British jobs', *The Times*, 18 October, p. 25.

Naudin, T. (1994), 'Czech enterprise leads the way in former Soviet bloc', *The European*, 21–7 October, p. 18.

Nelan, B. W. (1992), 'How the world will look in 50 years', *Time*, Fall 1992, p. 37 (Special Issue – 'Beyond the year 2000').

Nippon, a Charted Survey of Japan 1993/94 (1993), Tokyo: Kokusei-sha Corporation, annual publication.

Nuttall, N. (1994), 'World's fishing fleets face ruin as catches disappear', *The Times*, 27 July, p. 7.

OECD (1991), *The State of the Environment*, Paris: OECD.

Orwell, G. (1949), *Nineteen Eighty-four*, Harmondsworth: Penguin Books (1954).

Parlement Européen (1993), *Le Comité des régions*, Luxembourg: Revue du Marché commun et de l'Union européenne, September–October, no. 371.

Pattison, W. D. (1964). 'The four traditions of geography', *Journal of Geography*, May, vol. 63, pp. 211–16.

Pearson, L. (chairman) (1969), *Partners in Development* (Report of the Commission of International Development), London: Pall Mall Press.

Pelzer, K. J. (1951), 'Geography and the tropics', in G. Taylor (ed.), *Geography in the Twentieth Century*, London: Methuen.

Pleydell-Bouverie, J. (1993), 'Convenience shipping', *Geographical*, June, vol. 65, no. 6, pp. 16–20.

Popper, D. E. and Popper, F. J. (1987), 'The Great Plains: from dust to dust, a daring proposal for dealing with an inevitable disaster', *Planning*, vol. 53, pp. 12–18.

Potter, R. B. (1993), 'Urbanization in the Caribbean and trends of global convergence–divergence', *Geographical Journal*, March, vol. 159, part 1, pp. 1–21.

Prentice, E. (1995), 'Third World goes unheard at costly UN social summit', *The Times*, 13 March, p. 13.

Raj Krishna (1980), 'The economic development of India', *Scientific American*, September, vol. 243, no. 3, p. 134.

Redclift, M. (1991), 'The multiple dimensions of sustainable development', *Geography*, January, vol. 76, no. 1, pp. 36–42.

Repetto, R. (1992), 'Accounting for environmental assets', *Scientific American*, June, vol. 266, no. 6, pp. 64–71.

Richards, P. W. (1973), 'The tropical rain forest', *Scientific American*, December, vol. 229, no. 6, p. 58.

Rifkin, J. (1992), 'Bovine burden', *Geographical Magazine*, July, vol. 64, no. 7, pp. 10–15.

Rigg, J. (1991), *Southeast Asia, a Region in Transition*, London: Unwin Hyman.

Riley, B. (1993), Survey, *Financial Times*, 24 September, p. II.

Robinson, F. (1982), *Atlas of the Islamic World since 1500*, Oxford: Phaidon.

Robinson, G. (1994), '£9.7 bn airport breaks price barrier', *The Times*, 5 September, p. 11.

Rosenberg, N. and Birdzell, L. E. (1990), 'Science, technology and the western miracle', *Scientific American*, November, vol. 263, no. 5, pp. 18–25.

Rostow, W. W. (1960), *The Stages of Economic Growth*, Cambridge: Cambridge University Press.

Rousseau, Jean-Jacques (1762), *The Social Contract* (1968 edn), Harmondsworth: Penguin Books.

Ryan, D. (1993), 'Environmental lobbying at work', *Essoview*, no. 16, December.

Ryan, S. (1994), 'Scientists dismiss global warming leading to floods', *Sunday Times*, 27 March, p. 1.5.

Schweizer Weltatlas (1993), ed. E. Spiese, Zurich: Konferenz der Kantonalen Erziehungs-direktoren (EDK).

Scrimshaw, N. S. (1991), 'Iron deficiency', *Scientific American*, October, vol. 265, no. 4, pp. 24–31.

Sen, A. (1993), 'The economics of life and death', *Scientific American*, May, vol. 268, no. 5, pp. 18–25.

Serril, M. S. (1994), 'Zapatismo spreads', *Time International*, February, vol. 143, no. 8, p. 33.

Shell (1993), *The 'Shell' Transport and Trading Company plc, Annual Report 1993.*

Short, M. and McDermott, A. (1975), *The Kurds*, Minority Rights Group, Report no. 23.

Showalter, W. C. (1914), 'How the world is fed', *National Geographic*, January, vol. 29, no. 1, pp. 24–5.

Smith, Adam (1776), *The Wealth of Nations*, Harmondsworth: Penguin Classics, 1970.

Specter, M. (1994), 'Death rate rises as Russians worry', *New York Herald*, 7 March, p. 7.

Stamp, L. Dudley (1960), *Our Developing World*, London: Faber & Faber.

Staple, G. (1993), 'World on the line', *Geographical*, September, vol. 65, no. 9, pp. 24–7.

Statesman's Yearbook, annual publication with data for all the countries of the world, London: Macmillan.

Statistical Abstract of the United States, Washington, DC: US Department of Commerce, Bureau of the Census, annual publication.

Statistical Handbook of Japan 1987 (1987), Tokyo: Japan Statistical Association.

Taylor, A. (1993), 'Tunnel handover smooths troubled waters', *Financial Times*, 10 December, p. 7.

Taylor, G. (ed.) (1951), *Geography in the Twentieth Century*, London: Methuen.

Taylor, P. J. (1992), 'Understanding global inequalities: a world-systems approach', *Geography*, January, vol. 77, no. 1, pp. 10–21.

Taylor, T. (1992), 'Diminishing dividend', *Geographical Magazine*, June, vol. 64, no. 6, pp. 20–4.

Thrift, N. (1986), 'The geography of international economic disorder', in *A World in Crisis?*, Oxford: Basil Blackwell.

Tickell, C. (1993), 'The human species: a suicidal success?' *Geographical Journal*, July, vol. 159, no. 2, pp. 219–26.

Tickell, O. (1992), 'Going to seed', *Geographical Magazine*, December, vol. 64, no. 12, pp. 29–32.

The Times Atlas of World History, ed. G. Barraclough, London: Times Books.

Todd, E. (1976), *La Chute finale*, Paris: Robert Laffont.

———(1987), *The Causes of Progress*. Oxford: Basil Blackwell.

Toth, N., Clark, D. and Ligabue, G. (1992), 'The last stone ax makers', *Scientific American*, July, vol. 267, no. 1, pp. 66–71.

Treaties (1987): *Treaties Establishing the European Communities*, Luxembourg: Office for Official Publications of the European Communities.

Tyler, C. (1993), 'Flying in the future', *Geographical*, March, vol. 65, no. 3, pp. 28–32.

Union of Concerned Scientists (eds) (1993) *World Scientists' Warning Briefing Book*, Cambridge, MA: Union of Concerned Scientists.

UNDYB – Demographic Yearbook of the United Nations, latest available demographic data for all the countries of the world plus special topics in some years; indispensable for tracing demographic trends since the Second World War.

United Nations Human Development Report 1992 (1992), United Nations Development Programme (UNDP), Oxford: Oxford University Press.

UNSYB – Statistical Yearbook of the United Nations, published since the late 1940s, normally yearly; normally two or three years out of date when it appears but indispensable for tracing trends since the Second World War.

UNYITS – United Nations Yearbook of International Trade Statistics 1953 (1954), New York: UN.

Vesilind, P. J. (1993), 'The Middle East's water – critical resource', *National Geographic*, May, vol. 183, no. 5, pp. 38–71.

Wagley, C. W. (ed.) (1952), *Race and Class in Rural Brazil*, Paris: UNESCO.

Wallerstein, I. (1983), *Historical Capitalism*, London: Verso.

Wang Xin (1994), 'China strives to eradicate poverty', *Beijing Review*, 16–22 May, p. 7.

Ward, G. C. (1992), 'India's wildlife dilemma', *National Geographic*, May, vol. 181, no. 5, pp. 2–29.

Watts, D. (1988), 'Longest tunnel in world heads towards failure', *The Times*, 26 February, p. 11.

WDR – World Development Report of the World Bank, Oxford University Press, annual publication.

Webb, S. and Webb, B. (1937), *Soviet Communism – a New Civilisation*, London: Victor Gollancz Ltd (Left Book Club), vol. I, 2nd edn, pp. 1214–15.

Westlake, M. (1991), 'The Third World (1950–1990) RIP', *Marxism Today*, August, pp. 14–16.

White, R. M. (1990), 'The great climate debate', *Scientific American*, July, vol. 263, no. 1, pp. 18–25.

Wild, T. (1992), 'From division to unification: regional dimensions of economic change in Germany', *Geography*, July, vol. 77, no. 3, pp. 244–60.

Wilson, A. C. and Cann, R. L. (1992), 'The recent African genesis of humans', *Scientific American*, April, vol. 266, no. 4, pp. 20–7.

Wood, R. C. (1964), *1400 Governments*, New York: Anchor Books.

WPDS – World Population Data Sheet of the Population Reference Bureau, Washington, DC, annual publication.

Wright, D. (1994), 'Uncle Sam leads the way to world growth', *Sunday Times*, 7 August, p. 5.3.

Wu Naitao (1994a), 'How China handles population and family planning', *Beijing Review*, 1–8 August, pp. 8–12.

——(1994b), 'Three Gorges Project proceeds smoothly', *Beijing Review*, 13–19 June, pp. 14–20.

Yanagishita, M. and MacKellar, F. L. (1995), 'Homicide in the U.S.: who's at risk?', *Population Today*, February, vol. 23, no. 2, pp. 1–2 (Population Reference Bureau).

Yang Zheng (1994), 'Yangtze Pearl [Yichang, a town on the Chang Jiang River] foresees new century', *Beijing Review*, 12–18 September, pp. 7–11.

Zelinsky, W., Kosinski, L. A. and Prothero, R. M. (eds) (1970), *Geography and a Crowding World*, New York, London, Toronto: Oxford University Press.

Index

Note: Italicized numbers relate to tables or illustrations.